Springer Series in
OPTICAL SCIENCES
14

founded by H.K.V. Lotsch

Springer Series in
OPTICAL SCIENCES

The Springer Series in Optical Sciences, under the leadership of Editor-in-Chief *William T. Rhodes*, Georgia Institute of Technology, USA, provides an expanding selection of research monographs in all major areas of optics: laser and quantum optics, ultrafast phenomena, optical spectroscopy techniques, optoelectronics, quantum information, information optics, applied laser technology, industrial applications, and other topics of contemporary interest.
With this broad coverage of topics, the series is of use to all research scientists and engineers who need up-to-date reference books.

The editors encourage prospective authors to correspond with them in advance of submitting a manuscript. Submission of manuscripts should be made to the Editor-in-Chief or one of the Editors. See also http://springeronline.com/series/624

Peter Günter   Jean-Pierre Huignard

(Editors)

Photorefractive Materials and Their Applications 2

Materials

With 370 Illustrations

 Springer

Peter Günter
Institute of Quantum Electronics
Nonlinear Optics Laboratory
Swiss Federal Institute of Technology
ETH Hoenggerberg HPF E 8
CH-8093 Zurich
Switzerland
Email: gunter@phys.ethz.ch

Jean-Pierre Huignard
Thales Research and Technology France
RD 128
91767 Palaiseau Cedex
France
Email: jean-pierre.huignard@thalesgroup.com

With 370 figures.

ISBN 978-1-4419-2224-3
e-ISBN 978-0-387-34081-5

Printed on acid-free paper.

9 8 7 6 5 4 3 2 1

springer.com

Preface

In this second volume of the book series devoted to photorefractive effects we examine the most recent developments in the field of photorefractive materials and highlight the factors that govern photoinduced nonlinearity. The availability of materials having the required properties is of major importance for further development of this field, and there are many parameters that have to be considered in the figure of merit of a photorefractive material. As an example the recording slope of the dynamic hologram and the saturation value of the index modulation are specific characteristics of a given material. However, other features, such as spectral sensitivity range, dark storage time, material stability, and power handling capabilities, are also critical parameters to consider in using crystals for advanced applications in laser photonics. There is a large diversity of potential materials that exhibit interesting photorefractive properties, such as ferroelectric and nonferroelectric electrooptical crystals, semi-insulating semiconductors, and electrooptical polymers. If the basic mechanisms for space charge recording are well established, it is now required to have a very precise and extended knowledge of the physics of the charge transfer and related mechanisms that arise in doped materials. Also, we must know the material response for different conditions of hologram recording wavelength, laser intensity, and continuous or pulsed regimes. These research achievements on the physics of photorefractive materials is of great importance in optimizing and tailoring material properties. The main purpose of this second volume is to highlight the advances in materials research including crystal-growing conditions and material preparations and their impact on photorefractive performance. Following this objective, the reader will find in this book very detailed analysis of materials physics: investigations of defects in crystals, growing of stoichiometric $LiNbO_3$ or $LiTaO_3$, a new crystal $Sn_2P_2S_6$ for the near infrared, quantum-well semiconductor structures and sillenites. Besides conventional electrooptical crystals, this volume also deals with organic photorefractive materials. Great progress has been made recently in materials sensitivity and efficiency under an applied electric field. It is undoubtly a class of materials of growing interest. We are confident that new advances will be made in the chemistry and synthesis of polymers for better control and optimization of photorefractive properties. A closely related field is the photorefractive effect in liquid crystals, which exhibit attractive features due to their large photoinduced index modulation. We also outline in this volume two other contributions that have an important impact on applications: the mechanisms of permanent photoinduced gratings in silica-glass fibers used as wavelength-selective Bragg filters and the growing of materials like $LiNbO_3$, which have to be highly resistant to photorefractive damage for electrooptical and nonlinear optical applications.

This volume gives an in-depth review of the present understanding of the fundamental origins of the effect in a variety of materials. All the materials considered in this volume will play a significant role in the development of applications such as presented in the third volume of this series. For further progress in the field of photorefractive nonlinear optics it is therefore most important to stimulate significant research efforts in the basic physical phenomena of different materials. These research achievements may contribute to the discovery of a new class of photorefractive materials or allow the optimization of the performance of existing materials.

We would like to thank all the authors of chapters for the excellent overview of their contributions. We hope that this collection of articles will be a helpful source of scientific information for students, scientists, and engineers working in the field or looking for an extended analysis of photoinduced phenomena and parameters that control the nonlinear behavior of materials.

P. Günter, Zurich
J.P. Huignard, Paris
June 2005

Contents

Contributors

Ivan Biaggio Department of Physics and Center for Optical Technologies, Lehigh University, Bethlehem, PA 18015, U.S.A.

B. Briat, Laboratoire d'Optique Physique, ESPCI, 75231 Paris Cedex 05, France

V.V. Bryksin Physico-Technical Institute, Russian Academy of Sciences, St. Petersburg 194021, Russia

Karsten Buse University of Bonn, Institute of Physics, Wegelerstr. 8, 53115 Bonn, Germany

M. Douay Laboratoire de Physique, Atomes et Molécules (PHLAM), Unité Mixte de Recherche du Centre National de la Recherche Scientifique (CNRS, UMR 8523), Centre d'Etudes et de Recherches Lasers et Applications, Université des Sciences et Technologies de Lille, Bâtiment P5, 59655 Villeneuve d'Ascq, France

Michael Ewart present address: wzw-optic AG, Balgach (Switzerland)

Alexander A. Grabar Institute of Solid State Physics and Chemistry, Uzhgorod National University, 88000 Uzhgorod (Ukraine)

V.G. Grachev, Fachbereich Physik, Universität Osnabrück, D-49069 Osnabrück, Germany
and
Department of Physics, Montana State University, Bozeman, MT 59717, USA

Peter Günter Nonlinear Optics Laboratory, Swiss Federal Institute of Technology, ETH—Hönggerberg, 8093—Zürich (Switzerland)

Hideki Hatano National Institute for Materials Science, 1-1 Namiki, Tsukuba, Ibaraki 305-0044, Japan

Jörg Imbrock Westfälische Wilhelms-Universität, Institute of Applied Physics, Corrensstr. 2, 48149 Münster, Germany

Mojca Jazbinšek Nonlinear Optics Laboratory, Swiss Federal Institute of Technology, 8093 Zurich (Switzerland)

Eunkyoung Kim Department of Chemical Engineering, Yonsei Univ., 134, Sinchon-dong, Seodaemun-gu, Seoul, 120-749, South Korea

Bernard Kippelen School of Electrical and Computer Engineering, Georgia Institute of Technology, Atlanta, GA 30332 404 385-5163.

Marvin B. Klein Hughes Research Laboratories, 30011 Malibu Canyon Road, Malibu, CA 90265, USA. Present address: Lasson Technologies, Inc. 6059 Bristol Parkway, Culver City, CA 90230, USA

Kenji Kitamura National Institute for Materials Science, 1-1 Namiki, Tsukuba, Ibaraki 305-0044, Japan

Eckhard Krätzig University of Osnabrück, Physics Department, Barbarastr. 7, 49069 Osnabrück, Germany

M. Lancry Laboratoire de Physique, Atomes et Molécules (PHLAM), Unité Mixte de Recherche du Centre National de la Recherche Scientifique (CNRS, UMR 8523), Centre d'Etudes et de Recherches Lasers et Applications, Université des Sciences et Technologies de Lille, Bâtiment P5, 59655 Villeneuve d'Ascq, France

Youwen Liu National Institute for Materials Science, 1-1 Namiki, Tsukuba, Ibaraki 305-0044, Japan

L. Lucchetti Dipartimento di Fisica e Ingegneria dei materiali e del Territorio and CNISM, Università Politecnica delle Marche, Ancona, Italy

G.I. Malovichko, Department of Physics, Montana State University, Bozeman, MT 59717, USA

Carolina Medrano Nonlinear Optics Laboratory, Swiss Federal Institute of Technology, ETH—Hönggerberg, 8093—Zürich (Switzerland)

Klaus Meerholz Institut für Physikalische Chemie, Universität zu Köln, Luxemburgerstr. 116, D-50939 Köln, Germany

Germano Montemezzani Nonlinear Optics Laboratory, Swiss Federal Institute of Technology, 8093 Zurich (Switzerland)
and
Present Address: Laboratoire Matériaux Optiques, Photonique et Systèmes (LMOPS, CNRS UMR 7132), University of Metz and Supélec, 57070 Metz (France)

P. Niay Laboratoire de Physique, Atomes et Molécules (PHLAM), Unité Mixte de Recherche du Centre National de la Recherche Scientifique (CNRS, UMR 8523), Centre d'Etudes et de Recherches Lasers et Applications, Université des Sciences et Technologies de Lille, Bâtiment P5, 59655 Villeneuve d'Ascq, France

Konrad Peithmann University of Bonn, Helmboltz Institute for Radiation and Nuclear Physics, Nußallee 14-16, 53115 Bonn, Germany

M.P. Petrov Physico-Technical Institute, Russian Academy of Sciences, St. Petersburg 194021, Russia

B. Poumellec Laboratoire de Physico-Chimie de l'Etat Solide, Unité Mixte de Recherche du Centre National de la Recherche Scientifique (CNRS, UMR 8182),

ICMMO (Institut de Chimie Moléculaire et des Matériaux d'Orsay), Université de Paris Sud-Orsay, Bâtiment 414, 91405 Orsay Cedex, France

N. Rubinina Moscow State University, 117234 Moscow, Russia

O.F. Schirmer, Fachbereich Physik, Universität Osnabrück, D-49069 Osnabrück, Germany

Konstantin Shcherbin Institute of Physics, National Academy of Sciences, Prospect Nauki 46, 03650 Kiev (Ukraine)

Alexander N. Shumelyuk Institute of Physics, National Academy of Sciences, 03650 Kiev (Ukraine)

F. Simoni Dipartimento di Fisica e Ingegneria dei materiali e del Territorio and CNISM, Università Politecnica delle Marche, Ancona, Italy

T. Volk Institute of Crystallography of Russian Academy of Sciences, 117333 Moscow Leninski Prospect 59, Russia

Yulian M. Vysochanskii Institute of Solid State Physics and Chemistry, Uzhgorod National University, 88000 Uzhgorod (Ukraine)

M. Wöhlecke Fachbereich Physik, Universität Osnabrück, D-49069 Osnabrück, Germany

Marko Zgonik present address: Faculty of Mathematics and Physics, University of Ljubljana and Jozef Stefan Institute, Ljubljana (Slovenia)

List of Symbols

κ_0	Debye wave vector
κ_0^{-1}	Debye screening length
κ_d^{-1}	diffusion length
κ	grating wave vector
τ_0	carrier recombination time for continuous-wave recording, free-carrier lifetime for pulsed recording.
τ_{cw}	response time of the photorefractive effect for cw recording
τ_{die}	dielectric relaxation time
τ_{ex}	excitation time (short grating spacing limit of the photorefractive response time)
τ_d	diffusion time
τ_{pulsed}	response time of the photorefractive effect for pulsed, single-shot recording
μ	charge-carrier mobility
σ	conductivity
ϵ_{eff}	effective dielectric constant
N_{eff}	effective trap dinsity
n_c	average density of free-charge carriers
I_0	average light intensity
m	modulation index of the interference pattern
s	probability rate for photoexcitation
α_p	absorption constant caused by the carrier photoexcitation process
N_d	donor density
E_{sc}	amplitude of the space-charge field
E_q	trap-limited field
E_d	diffusion field
E_{ph}	photon-limited field

1

Introduction

P. Günter and J.P. Huignard

Research and development in laser physics and nonlinear optics have been rapidly expanding over the last twenty years, and many applications of photonics are now relevant to the industrial and consumer markets. Also, it is expected that photonics will disseminate in the near future in new, important areas such as medicine, biology, and nanotechnologies. A major factor behind such impressive growth is the advent of high-powered and efficient solid-state lasers in combination with the use of materials that exhibit large second- or third-order nonlinearities. In such conditions, nonlinear optics is becoming an important technology in the design of new laser sources emitting in the visible or near IR. Nonlinear optics also enables the attainment of new functionalities in laser systems and optoelectronic signal transmission and processing. The class of nonlinear phenomena based on the photorefractive effects in electrooptical crystals will undoubtedly play a major role for these different applications of laser photonics.

To briefly introduce this particular field of nonlinear optics, let us recall in the following the basic physical mechanisms and main characteristics: when coherent laser beams interfere in the volume of a material, a photoinduced space charge field is generated that modulates the crystal or polymer refractive index through linear electrooptical effects. In other words, photorefractive materials are very well suited to recording dynamic holograms using two-wave or four-wave mixing interactions involving continuous-wave or pulsed lasers. Since the magnitude of the nonlinear index modulation is proportional to the absorbed incident energy, there is a tradeoff between response time and material sensitivity.

The photorefractive effect has a particularly interesting history. Almost 40 years ago, soon after the mystery of "optical damage" was discovered in $LiNbO_3$, this effect was proposed as a holographic optical memory with huge storage capability. At that time, no basic understanding of the origin of the effect had been discovered. The interest in memory declined in the mid 1970s, mainly because the stored information degraded both during readout and (over longer time scales) in the dark. For a while, only the challenge to understand the basic mechanism giving rise to this effect, with a view to eliminating the problem of wave-front distortion in useful electrooptical materials, drove in research in this area. More and more materials were found to be photorefractive. New electrooptical interactions were discovered. The fields of nonlinear optics, optical spectroscopy, electrooptics, ferroelectrics, electronic transport, and Fourier optics were brought together to develop a reasonably complete understanding of the complex microscopic mechanism involved.

With available the knowledge base, this field has progressed rapidly, and more applications are being demonstrated, including the "renaissance" of optical memories in the 1990s. More precise theory has been developed, which reveals new and interesting aspects of photorefractive interactions not previously recognized. The need for improved materials has become urgent, and it is apparent that different applications require different materials properties: some require fast materials response times; some require high diffraction efficiency. Since the magnitude of the nonlinear index change is proportional to absorbed energy, the usual tradeoff of speed versus sensitivity applies in photorefractive processing.

Nonlinear photorefractive optics is now well established, and it has reached scientific maturity. It stimulates basic research in solid-state physics into the detailed mechanisms of charge transport in different types of ferroelectric and semiconductor crystals. A result has been great interest in growing new doped crystals whose photorefractive properties can be tailored to applications, encompassing materials that have no photorefraction to materials that can exhibit large photoinduced index modulations.

When we published the first edition of this series on photorefractive materials and applications in 1987, it was the result of very intense research activity in the field, as well as in other closely related subjects in the areas of optical phase conjugation and information processing [1]. These volumes remain a very good source of comprehensive coverage of the field. Starting from the discovery of the effects in lithium niobate at Bell Labs in the mid 1960s, the book chapters developed a first extensive analysis of the nonlinear mechanisms and applications in materials such as $LiNbO_3$, $BaTiO_3$, $KNbO_3$, $Bi_{12}SiO_{20}$, and GaAs. The fundamentals of early work on charge-transport models and beam-coupling phenomena in these materials were investigated in detail. Several chapters also highlighted early demonstrations of high-reflectivity phase-conjugate mirrors using photorefractive crystals as well as image amplification through dynamic holography. The capability of achieving high-beam refractive-index changes and two-beam coupling gain at low power levels in most materials was already well developed by analyzing and demonstrating several types of self-induced optical cavities. It was often mentioned in these articles that a critical issue is to identify and select the material having the required characteristics. So it is now important to review the research that has significantly contributed to new advances during the last decade. An important objective is thus to show that the discovery of new effects and extended analysis of the physical phenomena has led to better control of material properties for further development of attractive applications.

One of the important newer research areas since the appearance of the first edition of our volumes on photorefractive materials is the development of new organic crystals [2] and polymers [3, 4] showing pronounced photorefractive effects. On the other hand, new (and also previously known) materials have been developed for applications in new wavelength ranges, mainly in the infrared [5] and ultraviolet [6] or deep ultraviolet [7]. Materials showing photoinduced

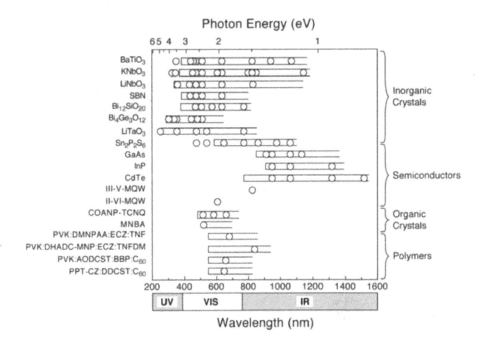

FIGURE 1.1. Wavelengths at which photorefractive gratings have been observed in a se-lection of photorefractive materials (abbreviations of polymer composites: see Chapters 13 and 14). The horizontal bars indicate the transmission range of a material, with the band-edge wavelength at their left end. Data for inorganic crystals are limited to bulk materials.

refractive index changes via space-charge-driven electrooptical effects presently cover a wavelength range from 257 nm [7] to 1.55 μm [8]. Figure 1.1 shows the possible wavelength ranges of the most important ones and the wavelengths for which photorefractive effects have been investigated in these materials. The short-wavelength region from the near-ultraviolet to the blue-green spectral range is presently dominated by inorganic crystals. Organic crystals and photorefractive polymers occupy mainly the red to near-infrared border region, while semicon-ductors and $Sn_2P_2S_6$ respond at infrared wavelengths larger than 800 nm. Multiple quantum-well photorefractive devices (MQW) differ from the other materials by the fact that holographic gratings can be recorded and read out for a spectral band only a few nanometers wide corresponding to the excitonic resonances.

The detailed physical properties of most of these materials are treated in this volume.

For many applications the most important materials parameters are the pho-toinduced refractive index changes (dynamic range), the two-wave mixing gain, and the recording time of the photoinduced refractive index changes or the pho-tosensitivity of the materials.

The refractive index changes mainly depend on the electrooptical figure of merit n^3r (n: refractive index; r: electrooptical coefficient) and the effective density of

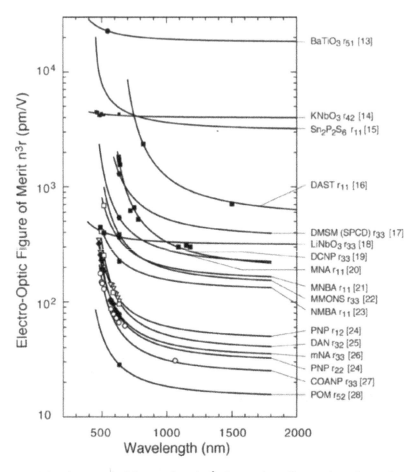

FIGURE 1.2. Electrooptical figure of merit n^3r for a series of inorganic and organic crystals. For the full name of organic crystals we refer to the references given next to the acronym. The solid curves are according to the polarization-potential model of Wemple and DiDomenico [29].

trapped photomigrated charges. Figure 1.2 shows a comparison of this figure of merit for a series of photorefractive materials. The recording time depends mainly on the migration time (drift and diffusion) of photoexcited charge carriers, which also depends on the trap density and on other parameters.

A brief graphical overview of the performance of the most important materials (inorganic and organic crystals and polymers) is shown in Figure 1.3 [9]. Here the maximum diffraction efficiency η of photoinduced gratings and the recording time of these gratings.

The higher the diffraction efficiency and the shorter the response time, the better the material. The diagram of Figure 1.3 shows the steady-state diffraction efficiency and the inverse response time obtained in a selection of inorganic and

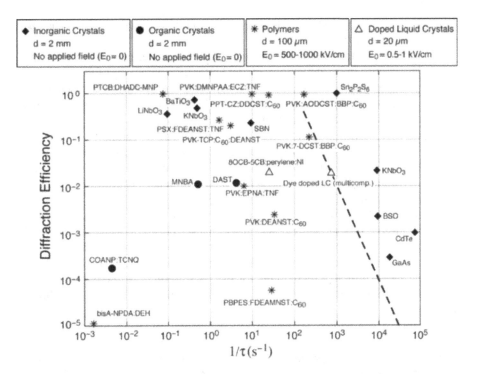

FIGURE 1.3. Diffraction efficiency versus growth rate τ^{-1} for a selection of photorefractive crystals, polymers, and doped liquid crystals (abbreviations of organic crystals: see Figure 2; polymer composites: see Chapters 13 and 14). The values are scaled to 1 W/cm^2 light intensity and to a typical thickness of 2 mm for crystals, 100 μm for polymer films, and 20 μm for liquid crystal cells. The dashed line connects points of equal sensitivity [9].

organic crystals, photorefractive polymers, and dye-doped liquid crystals. The diffraction efficiencies are scaled to a typical thickness d of actual samples, i.e. $d = 100$ μm for polymers, $d = 20$ μm for liquid crystalline cells, and $d = 2$ mm for inorganic and organic crystals. The response time is scaled to a light intensity of 1 W/cm^2. The data for polymers are taken under an applied electric field of typically 500–1000 kV/cm and those for liquid crystals under a field of 0.5–1 kV/cm, while those for inorganic and organic crystals are under zero-field conditions. The dashed line in the diagram connects points of equal sensitivity. However, the same sensitivity for the different categories of materials in general is not on the same line because of the different values assumed for the sample thickness. The region of the best materials (the upper right corner in the diagram) is still occupied mainly by inorganic crystals; however, the best polymers and liquid crystals are approaching this region. The best organic crystals have made it to the midfield region. We also include a new inorganic crystal, Sn$_2$P$_2$S$_6$ (tin hypothiodisulphate), a wide-bandgap semiconducting ferroelectric with a high sensitivity, combining fast recording times and large diffraction efficiencies as reported recently [10].

This second volume, dealing with photorefractive materials, shows the large worldwide activity in the development and better understanding of this group of materials. There are 17 chapters included in this volume dealing with discussions of basic materials parameters, photorefractive defects in electrooptical materials, and the techniques of their investigation.

Chapter 2 reviews inorganic oxidic crystals and the investigation of defects in these crystals. Investigation of these defects has been performed since the discovery of the photorefractive effect in $LiNbO_3$, and we now have reached a good understanding of the defects structure and its influence on the photorefractive properties in several materials.

In Chapter 3, I. Biaggio discusses the grating recording speed under cw and pulsed excitation. It is shown in this chapter, that the photorefractive effect under ideal situations does not necessarily have to be slow, but that with pulsed excitation, recording times down to nanoseconds can be observed [11]. In this chapter all optical photorefractive techniques for the measurement of the photoconductivity, the diffusion length, the free-carrier lifetime, and the free-carrier mobility are presented.

In Chapter 4, the photorefractive effects in $LiNbO_3$ and $LiTaO_3$ are reviewed. Due to space limitations the chapter can present only the main results from the enormous amount of work performed in the group of Prof. E. Krätzig at the University of Osnabrück and later by Prof. K. Buse at the University of Bonn.

The growth of stoichiometric $LiNbO_3$ and $LiTaO_3$ and their photorefractive properties are treated in Chapter 5 by H. Hatomo and Y. Liu of K. Kitamura's group focusing on the difference between such properties and those of classical congruent materials.

In Chapter 6, T. Volk et al. discuss damage-resistant $LiNbO_3$. They show that the photorefractive effect (optical damage) in this crystal can be drastically reduced by doping with suitable ions, which leads to larger photoconductivity in this material and to lower photoinduced space-charge fields and smaller optical damage at high laser intensities.

In Chapter 7, M. Zgonik et al. discuss the photorefractive effects in $KNbO_3$, a material with both relatively large electrooptical coefficients and relatively large photoconductivity and short recording times in reduced samples.

Chapter 8, on $BaTiO_3$, has been reproduced from our former edition. The chapter gives a detailed discussion of the properties of reduced and oxidized $BaTiO_3$. Most recent experiments done with Rh-doped $BaTiO_3$, which show a very large sensitivity in the infrared up to 1.06-μm wavelength, are of course not covered in this contribution to the former volume. However, the editors believe that this chapter contains sufficient basic information for the photorefractive community to merit republication in this new book series. Moreover, the defect structure and the properties of Rh-doped $BaTiO_3$ are covered in Chapter 2 of this volume and in several chapters of Volume 3 of this book series, which deals with applications of the photorefractive effect.

Petrov et al. discuss photoinduced space-charge waves and sillenites in Chapter 9.

In chapter 10, the photorefractive properties of $Sn_2P_2S_6$, a semiconducting ferroelectric with enhanced photorefractive properties (large electrooptical coefficients and large photoconductivity) in the infrared, are reviewed, and Chapter 11 gives an overview of the photorefractive properties of semiconductors and multiple quantum wells. In addition, K. Shcherbin describes progress in semiconductor photorefractive crystals in Chapter 12.

Two groups who contributed substantially to the development of novel polymeric photorefractive polymers describe the results of their work in Chapters 13 and 14.

Since the discovery of the photorefractive effect by Ashkin et al. [11], the basic understanding of photorefraction has also led to the application of these basic effects to other groups of materials. Space-charge-induced effects and photoinduced structural changes leading to photoinduced refractive index changes having some similarities with the photorefractive effect have been described not only in organic crystals and polymers but also in inorganic glasses and glass fibers, in liquid crystals, and in photochromic materials. A brief status report of research into photoinduced effects in these materials is given finally in Chapters 15 to 17.

In conclusion, this collection of chapters provides a broad survey of the most advanced materials developments and our general understanding of the photorefractive effects in these materials. The different contributions to this volume show that a rather complete understanding of the properties of these materials has been reached, at least from a phenomenological point of view. A series of materials with attractive properties at visible, infrared, and ultraviolet wavelengths has been proposed. These materials could be useful for future applications in photonics. Some of the proposed applications will be described in Volume 3 of this book series.

The authors of this volume are major scientists in the field, and their contributions bring a full complement of research efforts in materials science, dynamic holography, nonlinear optics, and their applications. The volume should serve the needs of the scientific and engineering communities interested in multidisciplinary aspects of photorefractive nonlinear optics.

The editors of this volume express their warm regards to all the authors for their outstanding contributions and very fruitful cooperation for the preparation of this volume, which should help to stimulate further developments of the field. We also thank Mrs. Lotti Nötzli for her valuable secretarial support.

References

1. P. Günter and J.P. Huignard (eds.). "Photorefractive Materials and Their Applications I and II," in *Topics in Applied Physics*, vols. 61 and 62, Springer-Verlag, Berlin, 1988.
2. K. Sutter, J. Hulliger, and P. Günter. *Solid State Commun.* **74**, 867 (1990).

3. S. Ducharme, J.C. Scott, R.J. Twieg, and W.E. Moerner. *Phys. Rev. Lett.* **66**, 1846 (1991).
4. See Chapters 13 and 14 of this volume.
5. See Chapters 10, 11, and 12.
6. G. Montemezzani, S. Pfändler, and P. Günter. *J. Opt. Soc. Am. B* **9**, 1110 (1992).
7. Ph. Dittrich, B. Koziarska-Glinka, G. Montemezzani, P. Günter, S. Takekawa, K. Kitamura, and Y. Furukawa. *J. Opt. Soc. Am. B* **21**, 632 (2004).
8. Ph. Delaye, L.A. de Montmorillon, I. Biaggio, J.C. Launay, and G. Roosen. *Opt. Commun.* **134**, 580 (1997).
9. G. Montemezzani, C. Medrano, M. Zgonik, and P. Günter. "The photorefractive effect in inorganic and organic materials," in *Nonlinear Optical Effects and Materials*; ed. P. Günter, Springer Series in Optical Sciences, vol. 72, 2000.
10. M. Jazbinsek, G. Montemezzani, P. Günter, A.A. Grabar, I.M. Stoika, and Yu.M. Vysochanskii. *J. Opt. Soc. Am. B* **20**, 1241 (2003).
11. M. Ewart, I. Biaggio, M. Zgonik, and P. Günter. *Phys. Rev. B* **49**, 5263 (1999).
12. A. Ashkin, G.D. Boyd, J.M. Dziedzic, R.G. Smith, A.A. Ballmann, H.J. Levinstein, and K. Nassau. *Appl. Phys. Lett.* **9**, 72 (1966).
13. M. Zgonik, P. Bernasconi, M. Duelli, R. Schlesser, P. Günter, M. H. Garett, D. Rytz, Y. Zhu, and X. Wu. *Phys. Rev. B* **50**, 5941 (1994).
14. M. Zgonik, R. Schlesser, I. Biaggio, E. Voit, J. Tscherry, and P. Günter *J. Appl. Phys.* **74**, 1287 (1993).
15. D. Haertle, G. Caimi, G. Montemezzani, P. Günter, A.A. Grabar, I. M. Stoika, and Yu. M. Vysochanskii. *Opt. Commun.* **215**, 333 (2003).
16. R. Spreiter, Ch. Bosshard, F. Pan, and P. Günter. *Opt. Lett.* **22**, 564 (1997).
17. T. Yoshimura. *J. Appl. Phys.* **62**, 2028 (1987).
18. M. Jazbinsek and M. Zgonik. *Appl. Phys. B* **74**, 407 (2002).
19. S. Allen, T.D. McLean, P.F. Gordon, B.D. Bothwell, M.B. Hursthouse, and S.A. Karaulov. *J. Appl. Phys.* **64**, 2583 (1988).
20. G.F. Lipscomb, A.F. Garito, and R.S. Narang. *J. Chem. Phys.* **75**, 1509 (1981).
21. G. Knöpfle, Ch. Bosshard, R. Schlesser, and P. Günter. *IEEE J. Quantum Electron.* **30**, 1303 (1994).
22. J.D. Bierlein, L.K. Cheng, Y. Wang, and W. Tam. *Appl. Phys. Lett.* **56**, 423 (1990).
23. R.T. Bailey, G.H. Bourhill, F.R. Cruickshank, D. Pugh, J.N. Sherwood, G.S. Simpson, and K.B.R. Varma. *J. Appl. Phys.* **71**, 2012 (1992).
24. Ch. Bosshard, K. Sutter, R. Schlesser, and P. Günter. *J. Opt. Soc. Am. B* **10**, 867 (1993).
25. P. Kerkoc, M. Zgonik, Ch. Bosshard, K. Sutter, and P. Günter. *Appl. Phys. Lett.* **54**, 2062 (1989).
26. J.L. Stevenson. *J. Phys. D: Appl. Phys.* **6**, L13 (1973).
27. Ch. Bosshard, K. Sutter, and P. Günter *J. Opt. Soc. Am. B* **6**, 721 (1989).
28. M. Sigelle and R. Hierle *J. Appl. Phys.* **52**, 4199 (1981).
29. S.H. Wemple and M. DiDomenico, Jr. "Electrooptical and nonlinear optical properties of crystals," in *Appl. Solid State Science*, vol. 3, ed. R. Wolfe, p. 264, Academic Press, New York, 1972.

2

Defects in Inorganic Photorefractive Materials and Their Investigations

B. Briat,[1] V.G. Grachev,[2,3] G.I. Malovichko,[3] O.F. Schirmer,[2] and M. Wöhlecke[2]

[1] Laboratoire d'Optique Physique, ESPCI, 75231 Paris Cedex 05, France
Briat.Bernard@wanadoo.fr
[2] Fachbereich Physik, Universität Osnabrück, D-49069 Osnabrück, Germany
Schirmer@uos.de
[3] Department of Physics, Montana State University, Bozeman, MT 59717, USA
Malovichko@physics.montana.edu

2.1 Introduction

The role of defects in the photorefractive effect is known in principle: their photoionization in the brighter regions of a photorefractive crystal, illuminated by an inhomogeneous light pattern, causes quasifree charge carriers. These are separated from their home sites by diffusion, by the bulk photovoltaic effect, or by drift in an external electric field. Eventually they are trapped at empty levels of defects in the darker regions, and a space charge field is created. Crystals with a linear electrooptic effect transform this into a refractive index pattern. Theoretically, also the rates of the ionization process, the transport, and the trapping can be formulated quantitatively. This chapter gives information on how this general picture may be filled by real defects in existing photorefractive crystals.

A few classes of inorganic materials are available—lithium niobate and related compounds, the ferroelectric oxide perovskites, the sillenites, some acentric semiconductors, and various other compounds—that show properties favorable for photorefractive operation, as far as the host lattice is concerned. There is a large number of possibilities to optimize their performance by introducing suitable defects. In this way the following features of a photorefractive material can be influenced: magnitude and spectral dependence of the photorefractive sensitivity, speed of the charge transport, lifetime of carrier trapping, etc.

Optical absorption of a photon by a defect is the primary step in triggering the photorefractive effect. Most often, however, the absorption bands by themselves do not indicate which lattice perturbation causes them; they have to be assigned first to their origin by a method sensitive to the microscopic structure of defects.

Studies of electron paramagnetic resonance (EPR), together with its extensions, are unsurpassed in furnishing such information, and methods have been developed to transfer this knowledge to the optical absorption phenomena. They will be outlined below. Detailed knowledge of defects in photorefractive materials and the reliable interpretation of their absorption bands thus rests almost exclusively on EPR and EPR-related investigations. The Mössbauer effect may be cited as an exception; unfortunately, it is essentially restricted to iron-containing defects. Therefore, the chapter mainly describes investigations by EPR and related methods.

In ideal situations, studies of defects furnish knowledge of their structure, including the chemical identity, geometry, and charge state, of the incorporation site in the lattice, the nature of the electronic ground and excited states and their energies, the optical and thermal excitation mechanisms, the light-induced transfer of charges to the valence and conduction bands, and, on this basis, the prediction of the photorefractive performance of a material. In some cases these aims could be achieved rather closely; generally, however, many questions remain unanswered and offer opportunities for further research.

Several chapters in the two volumes on photorefractive materials edited by Günter and Huignard in 1988 [1] contain information on defects in such compounds. Since then, new methods for defect investigation have been developed and the range of results obtained in the field has vastly expanded. This chapter will give a survey of the present status of the studies. It starts with a brief general overview on the properties of defects and their classification; then an introduction to the experimental methods employed will follow. The later sections will deal with the defect-related results obtained for the various classes of photorefractive inorganic materials. Also a short section covering the properties of hydrogen in oxide materials is included. The brevity necessary for covering a large field of research in a short chapter will be compensated by an extended list of references. For a recent review on defects in inorganic photorefractive materials with an emphasis on applications, an article by Buse [2] can be consulted.

2.2 Classification and General Properties of Defects

A defect is anything that perturbs the translational symmetry of a crystal. In this chapter the term "defect" will be used in a narrower sense: only pointlike perturbations will be treated, i.e., cases in which an ion of the lattice is missing or lattice sites are replaced by nonregular ions. Also, small clusters of such point defects may be included. If only ions are involved that belong to the ideal crystal, the defects are *intrinsic*; otherwise, they are *extrinsic*. Examples for intrinsic defects are vacancies or *antisite* defects. A Bi ion replacing Si in the sillenite $Bi_{12}SiO_{20}$, labeled Bi_{Si}, is an example of the latter. In this article we mostly use this type of labeling; the chemical symbol for the ion present is appended by a subscript, marking the site of replacement. The letter V is used in this context as a symbol for *vacancy*, e.g., V_O stands for oxygen vacancy. If the charge of the defect is to be indicated, a corresponding superscript is added. Sometimes the charges refer to that of the replaced ion, then Bi_{Si}^x is used for an antisite defect if the replacing and the replaced ion have the same charges. If the replacing one is negative (positive), the notation is Bi_{Si}' ($Bi_{Si}^\bullet$). Also the label Bi_{Si}^{4+}, indicating the electronic configuration of the defect ion, may be used for the neutral case and Bi_{Si}^{3+} (Bi_{Si}^{5+}) for the negative (positive) one. For an overview on defect notation see [3].

As a consequence of the broken translational symmetry, defects can introduce levels in the gap between valence and conduction band, which represent the eigenenergies of an ideal crystal. Depending on the position of the Fermi level in the crystal, such defect levels may be occupied by electrons or be empty. For oxide materials, reduction and oxidation are convenient means to shift the Fermi level. The photorefractive effect is based on the fact that the level population can also be changed by illumination, especially in oxide materials often in a metastable manner. A level introduced by a defect X, where the charge state 0 is assumed to coexist with the charge state $-$, is labeled $X^{0/-}$ (Figure 2.1), in analogy to the notation for redox pairs in electrochemistry, see, e.g., [5]. This means that if the

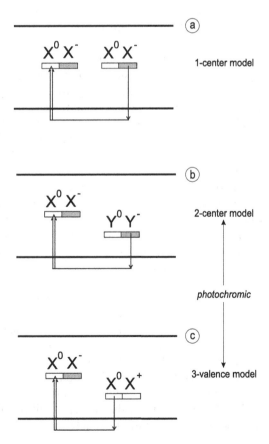

FIGURE 2.1. Definition of defect levels, illustrated with the three basic models for charge transfer between defects in photorefractive crystals [4]. The special cases are shown in which electrons are transferred from the valence band to the defect levels. Electron transfer from defect levels to the conduction band would lead to complementary schemes. Double arrows indicate light-induced transfer, single arrows recombination of a defect electron with a valence band hole. (a): one-center model: under illumination the total concentrations X^0 and X^- are not changed. The model therefore does not lead to photochromicity. (b), (c): Here the shallow levels can be metastably populated after optical excitation. The concentrations of the defect charge states change, leading to photochromicity.

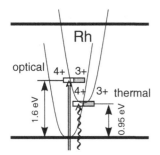

FIGURE 2.2. Discrimination between optical and thermal levels, exemplified for $Rh^{4+/3+}$ in $BaTiO_3$ (see Section 2.5). The elastic lattice energies, corresponding to the electronic groundstate (upper edge of valence band) and to the electron excited to the defect, are shown as depending on a configuration coordinate Q [6]. Double arrow: transition to optical level, vertical in Q-space. Wavy arrow: vibrational transition to thermal level.

level lies at an energy E_X above the valence band edge, E_X must be expended in order to excite a valence band electron to the defect X^0, transforming it into X^-.

Especially in oxide crystals, the charge carriers tend to couple strongly to the lattice. Since optical excitations take place under Franck–Condon conditions, i.e., with the "lattice kept fixed" [5, 6], in the case of strong coupling, *thermal* levels must be distinguished from *optical* ones (for an example see Figure 2.2): The final state reached by the optical transition ends in the *optical* level, lying higher than the vibrational ground state, the *thermal* level; for an example, see Section 2.5. In oxide crystals energy differences between both types of levels up to about 2.3 eV [7] have been found!

Such charge transfer transitions (e.g., from a valence band oxygen ion to a defect) or intervalence transitions (from a defect to a conduction band ion) usually are rather strong, because the electron moves through a considerable distance, corresponding to a large transition moment [5]. The range of the transfer is limited by the covalent mixture between the states of the initial and final ions; this is strongest between those nearest to each other. On the other hand, internal transitions of crystal field type occur, e.g., among the d-states of one transition metal ion. For ions at crystal sites having inversion symmetry, such excitations thus are parity forbidden [5]. They become easily observable in situations without inversion symmetry, e.g., at the tetrahedral sites of sillenite crystals. Such transitions can also be stronger if the excited state is resonant with the conduction band. As a general rule it can still be stated that the photorefractive effect is triggered most decisively by charge transfer or intervalence transitions, both because they are strong and because they lead to defect photoionization.

A further important consequence of lattice coupling is the formation of *polarons*. This term is related to the equivalence of corresponding lattice sites in crystals (Figure 2.3). The tunneling of a charge carrier between these sites competes with the lattice distortion, tending to break the equivalence by spontaneously localizing the carrier (Figure 2.3). For more details see [8]. The interplay between tunneling and the lattice distortion [8] decides whether a polaron is of large size (e.g., for electrons in the conduction band of the sillenites [9]) or of small or intermediate size (e.g., conduction electrons in $BaTiO_3$ [10]). The features of such polarons determine the carrier mobility, influencing the speed of the photorefractive effect. Under favorable conditions, two polarons can combine into

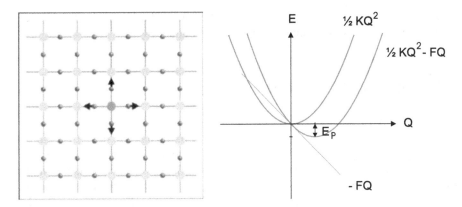

FIGURE 2.3. Left: Sketch of a crystal lattice with translational symmetry with one added electron. If the tunneling from site to site is not too strong, an electron can be self-localized at one lattice site by repelling its neighbors, spontaneously breaking the equivalence of sites. Right: The total energy of a polaron is the sum of the lattice elastic energy $\frac{1}{2}KQ^2$, and the lowering of the electronic energy by repelling the neighbors $-FQ$. For two electrons in a bipolaron, the lowering of the electronic energy is doubled, $-2FQ$, and the total energy lowering quadrupled, $4E_P$. If the energy excess for two paired electrons, $4E_P$, as compared to two separated electrons, $2E_P$, overcompensates the Coulomb repulsion, a bipolaron is stable.

bipolarons [8]. This occurs when the Coulomb repulsion between the carriers is overcompensated by the surplus stabilization energy caused by their joint lattice distortion.

Also, polarons usually lead to strong optical absorptions. For small polarons, where the carrier is self-trapped at essentially one lattice ion, the corresponding transitions occur from the initial trapping site of a carrier to a final equivalent one. This represents a special case of a charge transfer or intervalence transition. Therefore, also, large transition moments are involved, leading to high absorption intensities. The absorption energies are finite, in spite of the equivalence of initial and final sites, because the transitions occur under Franck–Condon conditions.

Charge carriers bound to a defect can also exhibit polaron effects [7, 11]. Consider, e.g., a Bi^{3+} ion replacing a Si^{4+} ion in the sillenite $Bi_{12}SiO_{20}$. This is the negatively charged antisite defect Bi'_{Si}, see Figure 2.12 in Section 2.6. Photoionization takes an electron not from Bi^{3+} but from the four tetrahedrally arranged equivalent O^{2-} ions surrounding Bi. The created hole is self-trapped at one of these oxygen ions, forming a *bound small polaron* by spontaneously breaking their equivalence. Also such bound polarons lead to strong optical absorptions [7, 11], due to the long transition dipole between the initial and a neighboring oxygen ion in the tetrahedron. For more details, see Section 2.6.

Some of the photorefractive crystals, such as congruently melting $LiNbO_3$ (LN), $Sr_{1-x}Ba_xNb_2O_6$ (SBN), or $Ba_{1-x}Ca_xTiO_3$ (BCT), are strongly disordered. This leads to considerable spatial fluctuations of the defect levels, causing wide lines in all spectroscopic studies [12]. Furthermore, the mobility of quasifree charge carriers tends to be reduced in such materials.

2.3 Methods of Defect Investigation

We start by giving a qualitative overview of the electron paramagnetic resonance (EPR) method. The electronic ground state of a paramagnetic defect is characterized by its spatial distribution and by its spin number. Within the spatial range of their wave function, the unpaired electrons collect all interactions by which they can couple with their surroundings, among these spin-spin and spin-orbit coupling, also in combination with crystal fields, and hyperfine interaction of the electrons with the "visited" nuclei, representing local probes in the crystal. If a static external magnetic field is applied, Zeeman interaction is also active. EPR methods probe the energy splittings between the lowest states caused by these couplings. The external magnetic field also provides a reference direction and thus allows the symmetry of the interactions to be identified. They are of tensorial character if the defect as a whole or the positions of the interacting nuclei are noncubic. This analysis leads to the most essential information supplied by the method: the symmetry of the coupling tensors and the orientation of their principal axes with respect to the crystal lattice, strongly narrowing down possible defect models. For a recent overview on the application of EPR and related techniques, especially optical ones, to the elucidation of defect properties, see, e.g., [13].

It is an advantage of the EPR method that it allows one to determine the concentration of paramagnetic defects [14], opening the way to the quantitative analysis of the performance of a photorefractive material on the basis of EPR studies alone. For further details see Section 2.5. EPR can "count" defect densities down to a few ppm. Only neutron activation analysis, see for example [15], is more sensitive. But there it is not possible to detect intrinsic defects or to differentiate between the various charge states that the same defect element can assume, such as, e.g., Fe^{3+} and Fe^{5+}, a task that can be solved by EPR.

All interactions probed by the electron(s) are usually summarized by a spin-Hamiltonian, for example

$$H = \mu_B \mathbf{BgS} + \mathbf{SDS} + \sum \mathbf{SA_iI_i} + \cdots, \tag{2.1}$$

where only three representative terms are given for illustration. The first one describes the Zeeman interaction, the second a crystal field interaction, and the last one the hyperfine interaction. Here μ_B is the Bohr magneton, $\mathbf{B}$ the external magnetic field vector, $\mathbf{g}$ the tensor of the Zeeman interaction, $\mathbf{S}$ the spin operator ($2S + 1$ is the multiplicity of the lowest level considered), $\mathbf{D}$ a crystal field tensor, $\mathbf{I_i}$ the spin operator of nucleus i, and $\mathbf{A_i}$ the tensor of the hyperfine interaction. The Hamiltonian usually operates only on the $(2S + 1)$ lowest states and is thus called a spin-Hamiltonian.

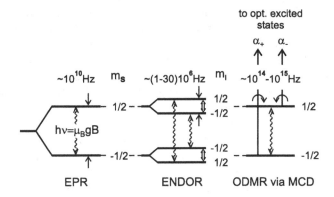

FIGURE 2.4. Basic schemes for paramagnetic resonance and related methods, demonstrated for one electron spin, $S = \frac{1}{2}$.

EPR: One transition occurs between the $m_s = \pm \frac{1}{2}$ states, Zeeman-split by a static magnetic field B. Obtained information: magnitude and angular dependence (symmetry) of splitting factor g and value of crystal field (for $S > \frac{1}{2}$ only). In general: spin value S, related to defect charge state.

ENDOR: If the electron interacts with one nucleus, assumed to have nuclear spin $I = \frac{1}{2}$, the indicated additional nuclear splittings occur. The double arrow nuclear transitions are detected by changes of the EPR signals (wavy arrows). Information: nuclear spin(s), nuclear splitting factor g_n, and the magnitude and angular dependence of the hyperfine interaction.

ODMR via MCD: Optical excitations with left- and right-circular polarized light originate from different Zeeman-EPR sublevels, as shown. The EPR transition (wavy arrow) decreases the population difference of these sublevels. The EPR is detected by the change of the MCD signal, $\Delta \alpha = \alpha_+ - \alpha_-$. Information: EPR parameters and the optical absorption bands originating from the ground-state Zeeman levels.

The energy splittings described by a Hamiltonian of this type are generally monitored by unbalancing a microwave bridge circuit when the supplied microwave energy matches the energy splittings, i.e., when the resonance condition is fulfilled (Figure 2.4). Because the population difference of the levels, Δn, behaves as

$$\Delta n = n_0 tanh(\mu_B g B / 2kT), \tag{2.2}$$

low temperatures and high magnetic fields increase the sensitivity. Also the heating of crystal specimens by the resonant absorption of microwaves can be used, leading to *thermally detected EPR* [16]. Double resonance methods, to be introduced in the following, constitute further ways to detect the EPR transitions.

Important information on the defect wave function is furnished by hyperfine interaction. If a hyperfine splitting is resolved, the resulting $(2I_i + 1)$ lines allow the spin I_i of the corresponding nucleus to be identified. This gives a strong hint of the chemical identity of this nucleus. The tensor $\mathbf{A}_i$ in (2.1) partly depends on the density of the wave function at the local probe represented by nucleus i. If hyperfine interaction originates from parts of the wave function with low probability density, the corresponding small splittings are usually not resolved. Then the *electron nuclear double resonance (ENDOR)* [13] technique (Figure 2.4) may help: Here, using a special experimental scheme, the highly resolved nuclear magnetic resonances, lying at radio frequencies, are detected by changes of the intensities of the corresponding EPR signals. If applicable, this technique leads to the most detailed information about a defect wave function, e.g., its spatial distribution and the nuclei it encompasses.

We consider now magnetic circular dichroism (MCD), i.e., the differential absorbance, $\Delta\alpha = \alpha_+ - \alpha_-$, presented by a cubic or uniaxial sample for left (σ_+) and right (σ_-) polarized light propagating along the direction of an applied magnetic field. In general [17, 18, 19], the MCD signal associated with an isolated electronic transition contains two main contributions. The diamagnetic term (S-shaped and temperature-independent) results from the difference in energy of the circularly polarized components. Although always present down to relatively low temperatures in the case of very sharp lines (e.g., lanthanide ions [18]), it can be safely ignored in the case of the broad bands at low temperature. The paramagnetic term (absorption-like shape, temperature dependent) monitors the magnetization in the ground state. In the case of spin $S = \frac{1}{2}$ (Figure 2.4) it is proportional to the difference in relative populations at equilibrium as in (2.2) between its two Zeeman sublevels. Experiments at very low temperatures (pumped helium) thus furnish the largest MCD signals. The technique is very sensitive, since the smallest detectable absorbance is about 10^{-5}, i.e., roughly two orders of magnitude smaller than with a classical spectrometer. In the case of the sillenites, Fe and Cr impurities could be monitored down to the ppm level.

The term ODMR has often been used in connection with the detection of EPR by various features of photoluminescence transitions [13]. Since a study of the photorefractive effect requires the assignment of the optical absorption bands, we concentrate rather on the optical detection of magnetic resonance (ODMR) via the magnetic circular dichroism (MCD). The signal $\Delta\alpha$ is measured as a function of B/T and a dip is observed ($|\Delta n|$ in (2.2) is reduced) in the saturation curve, whenever the resonance conditions are fulfilled.

The great advantage of the MCD-ODMR method is its ability to connect optical absorption features to their microscopic origins in the logically most stringent way. It is thus ideally suited to the analysis of the optical absorption properties of defects as related to the photorefractive effect. When several broad absorption bands are overlapping, as is most often the case for defects, especially in oxides, MCD-ODMR allows such a superposition to be deconvoluted by identifying exactly that one among the bands that is linked to a definite EPR signal. An example is given at the end of Section 2.5.

A necessary, but unfortunately not sufficient, precondition for the application of EPR and related methods is the paramagnetism of the defects. Such procedures therefore can be applied to only about one-half of all defects, the EPR-active ones. Since the photorefractive effect involves all types of defects, independent of whether they are EPR-active or EPR-silent, additional information is necessary to circumvent this problem. Actually, in the case of cubic crystals, the very absence of an MCD signal associated with a given absorption band is proof that the responsible defect is diamagnetic. In favorable cases (see Section 2.6), transition metal ions show internal transitions in the near-infrared, which are characteristic of the site symmetry and charge state of the defect. If a material containing the investigated defect is gyrotropic, such as the sillenites, then it is possible to study EPR-silent defects by their natural circular dichroism (CD) (see Section 2.6). The CD signals, however, do not provide any knowledge about the structure of the defects, usually derived from their magnetic properties. Altogether, a combination of techniques proves necessary for a reliable labeling of defects.

Light-induced absorption changes (LIAC) and their correlation with EPR or MCD changes have been largely exploited to label defects, both EPR-silent and EPR-active ones [20, 21, 22], and to identify between which defects charge carriers are transferred by light. This approach was introduced recently as the basis for the quantitative prediction of the performance of photorefractive materials. An example will be given in Section 2.5.

Measurements of optical absorptions induced in photorefractive crystals by specific dopings have sometimes been used to draw conclusions about the nature of the resulting defects. If used critically and cautiously, such results can give hints about the nature of the responsible defects. The absorption signals usually do not carry information on the charge state and the incorporation site of the defect, whether it is isolated or associated with some partner defect. Such caveats are also necessary when one is interpreting measurements of PIXE, channeling [23], neutron activation analysis (see for example [15]), etc., induced by specific dopings.

2.4 Defects in LiNbO$_3$ (LN)

Since the discovery of the photorefractive effect, LN has played a major role in the development of this field. Correspondingly, great efforts have gone into the elucidation of the function of defects in this material. A considerable number among them, intrinsic and extrinsic, could be identified by EPR or related methods. In the following we start by describing intrinsic defects in the usually employed Li-deficient material, then we give introductory information on the structure of stoichiometric LN crystals. This is followed by an overview on extrinsic defects in both types of crystals. In previous publications we have given reviews on defects in Li-deficient LN [24, 25]; see also the relevant sections in [26]. For brevity we shall often cite these papers and shall concentrate here on the newer results of EPR-based studies of defects in LiNbO$_3$.

2.4.1 Intrinsic Defects

LiNbO$_3$ tends to crystallize with a Li content below that of its stoichiometric composition, where the Li fraction $x = [Li]/([Li] + [Nb])$ is expected to equal

0.5. In the most often employed congruently melting composition of LN, the Li fraction in the crystal, x_c, is equal to that in the melt, x_m; both are 0.484 [27]. As a consequence there are many lithium vacancies, V'_{Li}, and, compensating them, $Nb_{Li}^{4\bullet}$ antisite defects, in the simplest model. The composition of such a congruently melting crystal is therefore expressed by $[Li_{1-5y}Nb_y]_{Li}Nb_{Nb}$, and $x_c = 0.484$ thus corresponds to an antisite content $y \approx 1\%$; i.e., about each fiftieth unit cell contains a Nb_{Li} antisite defect. In the present context the study of the intrinsic defects is necessary in order to assess their role in the photorefractive effect, but also because they facilitate the doping with aliovalent extrinsic defects, possibly improving the photorefractive performance: the charge misfits of such dopings are easily compensated by the available reservoir of intrinsic defects [12, 28]. The high density of intrinsic defects in congruent LN represents strong perturbations of the crystal lattice. This causes rather wide signals in EPR studies of such samples, see, e.g., Figure 2.8, tending to conceal the wanted information on the structure of the defects. Consequently, the obtainable spectral resolution has increased tremendously, Figure 2.7, when stoichiometric crystals became available essentially free of intrinsic defects.

In transmission electron micrographs of congruent LN, clusters of intrinsic defects could be observed that were consistently interpreted as consisting of Nb_{Li}, V_{Li}, and V_{Nb} as well as possibly Nb at the structural vacancy of $LiNbO_3$, Nb_V [30]. This may indicate that the scenario of intrinsic defects could be more involved than previously modeled [31, 32]; for a discussion see [24]. Among such defects only Nb_{Li} has been identified definitely, using EPR and related studies. In the ground state of a congruent crystal, Nb_{Li} is present in the diamagnetic, EPR-silent charge state Nb_{Li}^{5+} ($4d^0$). After two-photon or X-irradiation [33] or reduction and subsequent illumination of the crystal [34]—details will be given below—the paramagnetic configuration, Nb_{Li}^{4+} ($4d^1$), can be studied [35]. A model of the electronic ground state and its orientation with respect to the crystal axes is shown in Figure 2.6a.

The optical absorption of the Nb_{Li}^{4+} defect is characterized by a wide band peaking at 1.6 eV (Figure 2.5b). This assignment has been proved in the most compelling way by MCD-ODMR studies [36]. The absorption is attributed to an intervalence transition from Nb_{Li}^{4+} to Nb_{Nb}^{5+}, i.e., from a localized level to the conduction band. This transfer leads to comparatively high photovoltaic currents [37, 38].

As has been stated, the EPR of the Nb_{Li} defect and its optical absorption can be observed with a reduced crystal if it is illuminated. The reduced state of congruent LN is characterized by an absorption band peaking near 2.5 eV (Figure 2.5c) and a diamagnetic ground state. Optically pumping with light energies in the range of this band creates the paramagnetic state Nb_{Li}^{4+} and the corresponding absorption (Figure 2.5b). Because of its diamagnetism, the ground state of the reduced crystal cannot furnish direct EPR information on the defect causing its optical absorption. On the basis of various circumstantial evidences [24] we have assigned the absorption (Figure 2.5c) to a bipolaron, proposing as a model system two electrons with antiparallel spins at two neighboring Nb ions, $Nb_{Li}^{4+}-Nb_{Nb}^{4+}$

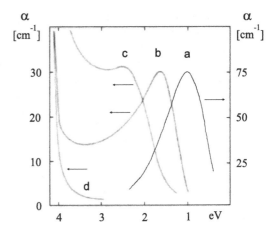

FIGURE 2.5. The various optical absorption bands of reduced LN in a schematic overview. a: Nb_{Nb}^{4+} free polarons [29]. b: polarons localized as Nb_{Li}^{4+} c: bipolarons bound as Nb_{Li}^{4+}–Nb_{Nb}^{4+}. Band c has been observed with a congruent crystal after reduction for 1 h at 900 °C in the dark; band b after illuminating this crystal with the unfiltered light of a xenon arc at 80 K. Band a results from reducing a congruent crystal, doped with 6% Mg, for 9 h at 500 °C.

(Figure 2.6c); here one Nb ion replaces Li, and the other one is part of the regular lattice. On account of the high density of Nb_{Li} in Li-deficient LN, one in each fiftieth unit cell, there are many such preformed pairs of Nb_{Li} and Nb_{Nb}; both Nb positions are distinguished only by the slightly different Madelung potentials active at the respective sites. The model has to explain that two electrons jointly occupying both Nb sites are more stable than if they were distributed over two separated and isolated Nb_{Li} ions, because light energy has to be fed into the system to create Nb_{Li}^{4+} from the diamagnetic precursor. Since this diamagnetic state is present in the ground state of a reduced crystal in spite of the Coulomb interaction between the two Nb^{4+} electrons (Figure 2.6c), this repulsion must be

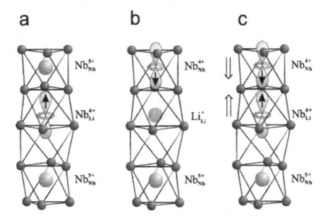

FIGURE 2.6. Schematic models of the three Nb^{4+}-containing defects in LN. a: The ground-state orbital of Nb_{Li}^{4+}. b: Ground-state orbital of Nb_{Nb}^{4+}. c: Two electrons with antiparallel spins trapped at a preformed Nb_{Li}–Nb_{Nb} pair. By relaxing toward each other, indicated by the double arrows, the covalent bond between both orbitals is strengthened. In this way the electronic energy driving the bipolaron formation (see Section 2.2) is lowered.

overcompensated by their joint lattice distortion. Such a situation is typical for a bipolaron. It is most likely that both partners relax toward each other (double arrows in Figure 2.6c), thereby lowering the electronic part of the total energy by strengthening the covalent bond in the pair; this is quite similar to the dynamics of a H_2 molecule. The axial orientation of the bipolaron (Fig. 2.6c) is expected to be stabilized by the overlap of the Nb^{4+} orbitals, each being aligned along the c-axis by the local crystal fields. The creation of isolated Nb_{Li}^{4+} by illumination into the 2.5 eV band has the effect that the bipolaron is optically dissociated; the electron ionized from Nb_{Nb}^{4+} will be trapped rather rapidly at a further empty Nb_{Li} defect. This dissociated state is metastable at low temperatures. Starting near 200 K, thermal dissociation of the bipolarons begins, and the 2.5 eV band decreases while the 1.6 eV absorption rises (Figure 2.5b). For the enthalpies involved in these processes see [25]. At room temperature a sizable portion of the 2.5 eV band still is present [24] in thermal equilibrium. At this temperature the described optical switching process—between bands c and b in Figure 2.5— could thus be utilized for optically gated holographic recording [39]: Illumination with energies in the range of the 2.5 eV band sensitizes a reduced crystal for photorefractive operation at the lower energies corresponding to the 1.6 eV band.

It has sometimes been postulated [34, 40] that the defect responsible for the 2.5 eV band rather is an oxygen vacancy, filled with two diamagnetically paired electrons. Among the arguments against this model [25], a strong one is the observation [29] that the reduction of congruent LN, in which the presence of Nb_{Li} is prevented by strong Mg doping (see below), does not lead to the 2.5 eV band. Instead, a band peaking near 1 eV (Figure 2.5a) is found, having all features characteristic of free Nb_{Nb}^{4+} polarons [29]. If reduction would lead to V_O as a stable defect in LN, this would be created also in the absence of Nb_{Li}. In assessing this and related arguments [25], it should be kept in mind that the defect chemistry of LN is different from that of its relatives, the oxide perovskites [32]. In such materials acceptor defects are compensated by V_O [32]. In LN, however, it was found from density measurements rather early [31] that the acceptors V_{Li}' are not compensated by $V_O^{\bullet\bullet}$, but by $Nb_{Li}^{4\bullet}$. Here, evidently, the formation of Nb_{Li} costs less energy than that of V_O. Thus the natural intrinsic donor in LN is Nb_{Li}. As a further difference between LN and the ABO_3 oxide perovskites, such as $BaTiO_3$, it should be remarked that the perovskites do not support the formation of B_B–B_A pairs, corresponding to Nb_{Li}–Nb_{Nb}. Accordingly, bipolaron-type absorption bands are not observed. Reduction in these cases leads only to almost free single polarons with absorption bands peaking near 0.7 eV.

Of course, under reducing conditions oxygen atoms evaporate from a congruent LN crystal according to the reaction

$$1\, LiNbO_3 + 2\, V_{Li}' \rightarrow \frac{3}{2}O_2 + Li_{Li} + Nb_{Li}^{4\bullet} + 6\, e'$$

(on the basis of [31]). This means that three oxygen atoms per formula unit leave back six free electrons, which can be captured at the numerous preexisting Nb–Nb pairs. The cations Li^+ and Nb^{5+} left over after the departure of the three oxygen ions recombine with two V_{Li} in the crystal, creating one additional Nb_{Li} defect. The kinetics by which this reaction proceeds to the crystal ground state has not yet been studied.

Also, the Li vacancy V_{Li} is an intrinsic defect; it is expected to be four times more abundant than Nb_{Li}, since formally four monovalent V'_{Li} will compensate one $Nb_{Li}^{4\bullet}$. The attempt to transform the diamagnetic V'_{Li} to the neighboring paramagnetic charge state V_{Li}^x, likely to be detectable by EPR, was not successful under illumination with light energies near and above the fundamental absorption edge at about 3.7 eV. Holes possibly created under such illumination would be expected to be trapped at V'_{Li}. Only two-photon [33], X-ray [33], or high-energy electron irradiation [41] have led to EPR signals typical for holes trapped near acceptor defects [33, 41, 42, 43, 44]. The available information, however, is not yet sufficient to decide whether a hole is situated near V_{Li} or near a conceivable other intrinsic acceptor defect, such as V_{Nb}.

A Li-deficiency of LN crystals, $x_c < 0.5$, is found also if the melt composition x_m is higher than the congruent composition, 0.484, even for $x_m > 0.5$ [12, 45], if conventional growth methods are used. Often specimens grown from high x_m have erroneously been called stoichiometric in the past. It came as a great surprise [46] when a procedure was identified that could produce exactly stoichiometric specimens with $x_c = 0.5000 \pm 0.0015$ [47]. Chapter 4 will deal with the respective growth methods. However, even $x_c = 0.5$ does not exclude the presence of stoichiometry-preserving intrinsic defects, such as interchanges of Li and Nb, $Nb_{Li} + Li_{Nb}$. By probing the crystallographic order of stoichiometric specimens by EPR and ENDOR, using a low concentration of Cr^{3+} ions as paramagnetic probes, it was demonstrated [12, 48] that such crystals are really "regularly" ordered.

2.4.2 Extrinsic Defects

First the dopings are treated that prevent or reduce the formation of Nb_{Li} during crystal growth from Li-deficient melts, such as Mg or Zn with concentrations in the percent range. Further information on such "optical damage resistant" dopings is given in Chapter 6. As has been mentioned above, the reduction of crystals of this type leads to the formation of almost free electrons, transformed to Nb_{Nb}^{4+} polarons by coupling to the lattice. This is an indication of the absence of Nb_{Li} as trapping centers. Several additional dopings in such crystals were investigated by EPR. The corresponding signals are characterized by their large width, resulting from the addition of the doping-induced disorder to the intrinsic one. Table 2.1 also contains information on the extrinsic defects investigated in such crystals. It was found that additional extrinsic ions tend to be incorporated at the Nb sites of the lattice; apparently it is more favorable that the Mg ions, rather abundant in the melt, replace Li ions.

Here it has to be noted that it is rather difficult to determine from basic arguments at which site, Li or Nb, cation dopings will enter LN. Both sites have trigonal symmetry and offer similar chemical backgrounds, oxygen octahedra, for incorporation (see Figure 2.6). It is only by identifying the next cation neighbors

TABLE 2.1. Extrinsic defects in LiNbO$_3$

Extrinsic defects in congruent LN:

Defect	References		Defect	References
Ti^{3+}_{Li}	[16, 49, 50]			
Cr^{3+}_{Li}	[51, 52, 53, 54]		$2Cr^{3+}_{Li}$	[55]
Mn^{2+}_{Li}	[56, 57, 58, 59, 60]			
Fe^{3+}_{Li}*	[57, 61, 60, 62, 63, 64, 65]		Fe^{2+}_{Li}**	[66]
Co^{2+}_{Li}	[67, 68, 69]			
Ni^{+}_{Li}	[70, 71]		Ni^{2+}_{Li}	[72]
Cu^{2+}_{Li}	[71]		Nd^{3+}_{Li}	[73, 52]
$Gd^{3+}_{?}$	[74, 75]		$Tb^{4+}_{?}$	[76]
$Dy^{3+}_{?}$	[52]		$Er^{3+}_{?}$	[77, 52, 78]
$Yb^{3+}_{?}$	[73, 79]			

* For Mössbauer studies of Fe see Chapter 5 of this volume.
** investigated by thermally detected EPR.

Extrinsic defects in congruent LN, codoped with Mg or Zn:

Defect	References
Ti^{3+}_{Nb}	[80]
Cr^{3+}_{Nb}	[81, 82, 83]
$Fe^{3+}_{?}$	[41, 84, 28]

Extrinsic defects in stoichiometric LN:

Defect	References
Cr^{3+}_{Nb}	[85]
$Fe^{3+}_{?}$	[64]
Mn^{2+}_{Li}	[86]
$Tb^{3+}_{?}$	[76]
$Nd^{3+}_{?}$	[12]
$Yb^{3+}_{?}$	[12, 87]

of a paramagnetic extrinsic ion by careful ENDOR measurements that conclusive evidence can be obtained. With this method, e.g., the behavior of Cr^{3+} has been studied in great detail [48, 85]. In Mg-doped and stoichiometric crystals it replaces Nb, and enters on Li sites in Li-deficient crystals. Figure 2.7 shows as examples that the EPR spectra of the extrinsic defects Cr^{3+} and Fe^{3+} are quite different for congruent and stoichiometric crystals, indicating different sites of the ions.

The interrelation of aliovalent extrinsic defects with their intrinsic compensators is demonstrated with Figure 2.8. It shows one EPR line of isolated Cr^{3+}_{Li} as depending on the Li fraction x_c of the host crystal [88]. It is seen that the signal becomes considerably sharper with increasing Li content. Due to these smaller line widths many satellite lines could be resolved with a crystal grown from a

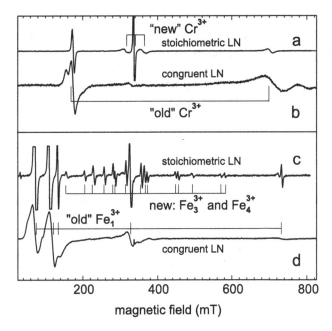

FIGURE 2.7. Comparison of EPR signals of LN:Cr and LN:Fe in congruent and stoichiometric LN. Top: "New" Cr_{Nb}^{3+} signals (a), characterized by a weak axial crystal field typical for Nb replacement, as compared to those (b) of "old" Cr_{Li}^{3+} with typical strong axial crystal field [88]. Bottom: In congruent LN the "old" Fe_{Li}^{3+} (d) is observed. In stoichiometric material two additional types of Fe^{3+} spectra arise (c) and all lines become narrower [64].

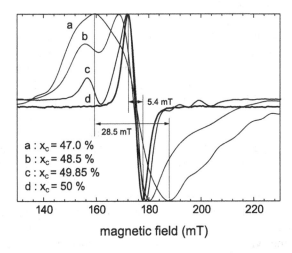

FIGURE 2.8. Strong decrease of EPR line width of the main Cr_{Li}^{3+} transition with increasing Li content [88]. Small satellite lines can be seen, caused by complexes consisting of Cr_{Li}^{3+} and compensating intrinsic defects.

melt with $x_m = 0.4985$ (Figure 2.8). For this composition the lines resulting from Cr_{Li}^{3+} associated with intrinsic defects at various discrete closer lattice distances were identified in addition to axial spectra related to Cr_{Li}^{3+} compensated by distant defects. It was possible to identify intrinsic compensator positions up to the ninth cation shell around the central Cr_{Li}^{3+} [88]. By comparison of the EPR and optical spectra it could furthermore be shown that the low-symmetry short-distance associations have higher oscillator strengths than the isolated Cr_{Li}^{3+}.

This study of Cr^{3+} represents a sample case for other extrinsic cation dopants, usually also leading to low-symmetry satellite lines in the EPR spectra such as Mn^{2+}, Fe^{3+}, Gd^{3+}, Yb^{3+}, Nd^{3+} [89], and Er^{3+} [77]. ENDOR investigations of Cr_{Nb}^{3+} in stoichiometric crystals revealed that in this case protons in the nearest neighborhood are present to compensate the effective negative charge of Cr_{Nb}^{3+}. They are located asymmetrically between two O^{2-} ions of the plane perpendicular to the c-axis of the oxygen octahedron next to Cr_{Nb}^{3+}. This finding coincides with results obtained by IR optical spectroscopy (see Section 2.8).

The advantages of MCD-ODMR studies have been used for LN not only to link the EPR of the intrinsic defect Nb_{Li}^{4+} to its optical absorption [36] at 1.6 eV (Figure 2.5) but also to study the optical properties of the extrinsic defects Ti_{Li}^{3+} [36], Cr^{3+} [90], Cu^{2+} [91], Fe^{3+} [91], and Mn^{2+} [91]. Also Cr^{3+} in congruent LN:Mg has been investigated with this technique [90].

2.5 Defects in Oxide Perovskites

2.5.1 BaTiO₃ (BT)

The properties of $BaTiO_3$ (BT) as a photorefractive host material are well known [92]. It may suffice in the present context to remind the reader that it is acentric below about 120°C, where a transition from the cubic to the ferroelectric tetragonal phase takes place, stable at room temperature and down to about 8°C. Its large electrooptic coefficients [93] allow comparatively few optically transposed charge carriers to create measurable index changes. The features of many defects in $BaTiO_3$ have been identified by EPR, usually at low temperatures in the rhombohedral phase (T ≤ 185 K), and for some of them their role in the photorefractive behavior of the material has been elucidated in detail.

Especially for defects in BT it is often observed that they can be recharged metastably to neighboring valencies under illumination; this allows the access to numerous EPR-active charge states with little preparatory effort. Since defects with changed charges have altered optical absorption characteristics, BT crystals thus usually are photochromic. On this basis a further EPR/optical method was developed, which is able to assign optical absorption bands to their microscopic origin. An outline will be given below. Here we state already that it has the following useful consequences: (1) The EPR information, usually obtained at low temperatures, can be transferred to room temperature, where photorefractive devices are supposed to operate. (2) Also EPR-silent defects can be identified.

(3) The question can be answered between which defects charge carriers are transferred under illumination. (4) The photorefractive performance of a material can be predicted quantitatively relying only on EPR-based defect studies. We introduce this EPR/optical method at the opening of this section because several results presented later will depend on it.

The development of the method was started in order to unravel why Rh doping of BT sensitizes the material for operation in the infrared. Since then the procedure has been applied to several further problems connected with the role of defects in the photorefractive effect [94, 95, 96, 97]. The useful influence of Rh on the photorefractive properties of BT was discovered in 1993 by Ross et al. [98] and was intensely studied in the following years [99, 100]. EPR/optical investigations on this system [21, 101] showed that a main part among the occurring photoinduced charge transfers involves the three defects Rh^{3+}, Rh^{4+}, and Rh^{5+}, fulfilling the "3-valence model" (Figure 2.1) [102]. On this basis and relying on experimentally determined values of the effective trap density N_{eff}, Huot et al. [103] and Corner et al. [104] analyzed the charge transfer properties of the system quantitatively. However, because the available experimental information was not sufficient to determine all relevant parameters, a simplified theoretical basis was employed. Later, using the EPR/optical method, a complete solution of the problem was possible [105, 106].

The procedure starts with investigating the wavelength dependence of the photochromic coloration of a BT:Rh crystal [98], induced by a series of rising pump light energies. The result is plotted over the field of the pump light, E_{pump}, and probe light, E_{probe}, energies (Figure 2.9c). Here and in the following only a rather brief sketch of the method is given. For further details see [105, 106]. The main features in Figure 2.9c are a strong light-induced transparency at $E_{probe} = 1.9$ eV and pronounced absorption increases at 1.6 eV and 3.0 eV. Simultaneous measurements of the EPR of Rh^{4+}, the only EPR-active Rh charge state in "as grown" BT:Rh, observable at $T \leq 20$ K, show that the intensity of this EPR signal has an identical dependence on E_{pump} as that of the transparency along $E_{probe} = 1.9$ eV. This assigns the band at 1.9 eV to Rh^{4+}. Further EPR studies [107] indicate that the Rh^{4+} intensity is decreased by the transfer of a valence-band electron to Rh^{4+}. In this way the EPR-silent charge state Rh^{3+} is created. The band at 3.0 eV is attributed to Rh^{3+}, lying higher than Rh^{4+} because of its lower charge. The hole created in the valence-band by the electron transfer to Rh^{4+} is expected to be captured by another Rh^{4+}, causing Rh^{5+}, which is also EPR-silent. Because of its higher charge, less energy is needed to excite a valence-band electron to Rh^{5+}. Therefore the other strong feature in Figure 2.9c, at 1.6 eV, is assigned to Rh^{5+}. In a similar way the further structures in Figure 2.9c are attributed to various charge states of Fe; this element is usually present in BT as an unintended background impurity. These assignments fulfill systematic topological constraints typical for plots of the type of Figure 2.9c [108].

Summarizing this part: By combined EPR/optical absorption studies, based on the photochromic behavior of BT, the optical absorption bands indicated by vertical dashed lines in Figure 2.9c, have been identified. Among these only the

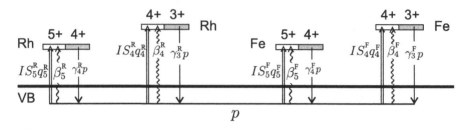

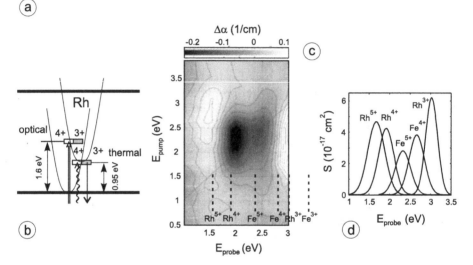

FIGURE 2.9. a: Schematic representation of the four levels in BT:Rh, defined by the three charge states of Rh and by three states of the background contamination Fe. The chosen energies are not to scale. The levels exchange holes with the valence band by optical excitation (double arrows), thermal excitation (wavy arrows) and recombination (single arrows). The corresponding rates are indicated beside the arrows. Here the parameters have the following meanings: S, absorption cross-sections; q, quantum efficiencies for ionization after absorption; β, thermal transition rates; γ, recombination parameters.

b: Discrimination between optical level (end of optical excitation in sketched configuration diagram) and thermal level, both for the special case of $Rh^{3+/4+}$ level. For simplicity this distinction is not included in Figure 2.9a.

c: Dependence of the absorption changes (grey scale) induced by pump light (vertical energy scale). The probe light energy is given as the abscissa. Monitoring pump light-induced EPR changes allows one to assign the absorption features to specific defects, indicated by vertical dashed lines.

d: The deconvolution of the absorption changes in c, compared with the defect densities derived from EPR, leads to the shown absorption cross-sections.

FIGURE 2.10. MCD spectra of various Fe-containing defects in $KaTO_3$:Fe [143] assigned to the indicated species by ODMR.

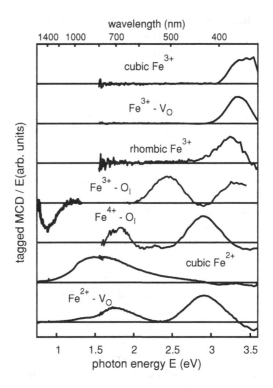

defects Rh^{4+}, Fe^{5+}, and Fe^{3+} are EPR-active. The position of the Fe^{3+} band, covered by the fundamental absorption, has been inferred from its peak energy in $KTaO_3$, Figure 2.10. The other optical bands, belonging to the EPR-silent defects Rh^{3+}, Rh^{5+}, and Fe^{4+}, are assigned by consistency arguments of the type as forwarded for Rh^{3+} and Rh^{5+}. This identification of the optical absorption bands allows them to be used as "fingerprints" of the corresponding defects. They can be employed likewise at room temperature; EPR measurements, on the other hand, usually require low temperatures. As shown, also absorption bands corresponding to EPR-silent defects could be identified. The shapes of the absorption bands are derived from a deconvolution of the photochromic absorption in the plot of Figure 2.9c into five Gaussian bands [106] (Figure 2.9d). This is possible for all values of E_{pump} with a maximal deviation between experiment and model of 10^{-3} cm^{-1}.

In addition, the charge transfer processes causing the photochromic changes have been clarified. They are summarized by the scheme in Figure 2.9a: two "3-valence systems," containing three charge states of Rh and of Fe, respective, are coupled to each other via the valence band, with which both systems can exchange holes. This figure also symbolizes the optical and thermal processes by which valence band electrons are excited to the levels, and their recombination with holes in the valence band. The parameters attached to the arrows, expressing the rates by which these processes occur, are explained in the figure caption. The indicated scheme can be cast into rate equations [106] for the populations of the various occurring charge states, similarly to those given by Kukhtarev et al. [109] for the one-center model.

After the qualitative step, indicating the defects involved and the charge transfer ways connecting them, as summarized before, it is desirable to obtain also quantitative information on the photorefractive performance of such a system. This means first: What are the values of the parameters in Figure 2.9a? And what is the size of the space charge fields E_{SC} that can be reached with such a system? If the photoionized charges are transported only by diffusion, their dependence on the grating wave vector k is given by $E_{SC} \propto k/(1 + k^2/k_0^2)$ with $k_0^2 \propto N_{eff}$ [109], the effective trap density, with known parameters of proportionality. E_{SC} thus is fixed, once N_{eff} has been determined. For a "3-valence" system based on Rh, as an example, N_{eff} depends on the defect densities in the following way [103]: $N_{eff}^{Rh} = Rh_{tot} - Rh^{4+}(I) - (1/Rh_{tot})(Rh^{3+} - Rh^{5+})^2$ [105]. Here $Rh^{4+}(I)$ is the concentration of Rh^{4+} under illumination with intensity I; the other specified densities are those in the equilibrated dark state. An analogous relation holds for the Fe "3-valence" subsystem, both leading to $N_{eff} = N_{eff}^{Rh} + N_{eff}^{Fe}$. In order to determine E_{SC}, the densities of the identified defects must be known quantitatively (see [105, 106]). For the EPR-active charge states the concentrations are easily determined from the intensity of the corresponding EPR signals. Indirectly, this is also possible for the EPR-silent charge states [105, 106]. This information can be transformed into the absorption cross-sections $S(E)$ (Figure 2.9d) of the defects taking part in the transfers. These quantities allow one to obtain the density N of a defect at room temperature when only its absorption band can be detected, since the absorption function $\alpha(E)$ is given by $\alpha(E) = N S(E)$. On this basis the defect densities in BT crystals containing various Rh dopings have been determined (Table 2.2); in all cases the effective trap density, N_{eff}, a measure of the space charge fields, could be well predicted (see Table 2.2). By investigating the time-dependence of the absorption changes under varying illumination, we were furthermore able to obtain all the parameters occurring in Figure 2.9a necessary to describe the kinetic behavior of the system.

The advantage of the approach essentially consists in the fact that the densities of the six defects appearing in Figure 2.9a can be determined; previously [103, 104], the evaluations had to rely on only one number representing a concentration, the experimentally determined trap density, N_{eff}. In the present method this is not an input parameter but can be predicted from the analysis (Table 2.2). The earlier studies [103] furthermore took only one Fe level ($Fe^{3+/4+}$) into account and did not include the quantum efficiency q as a free parameter.

Results obtained with the described method, beyond those discussed so far, have been included in an overview, Table 2.3, of those defects in BaTiO₃ for which EPR and related studies led to definite microscopic models. Those EPR-silent defect charge states identified only by the EPR/optical method are underlined.

Discussing the entries in Table 2.3, we start by making some remarks on the intrinsic defects. By EPR the oxygen vacancy, V_O, is identified only if it is associated with an extrinsic defect, as in the case $Fe^{3+}-V_O$ and further examples in Table 2.3.

TABLE 2.2. Numerical values of the parameters of the BT:Rh system

(A) Parameters valid for all BT:Rh systems
(see also the absorption cross-sections shown in Figure 2.9d)

	Rh	Fe
q_4	0.02 ± 0.01	0.15 ± 0.10
q_5	1	1
γ_3 (10^{-13} cm^3 s^{-1})	1.7 ± 0.8	1.1 ± 0.7
γ_4 (10^{-11} cm^3 s^{-1})	2.7 ± 1.2	3.3 ± 2.5
β_4 (10^{-3} s^{-1})	14.5 ± 5.5	2.8 ± 2.1
β_5 (10^{-1} s^{-1})	15.0 ± 7.0	1.2 ± 0.9

(B) Parameters valid for a specific crystal
(example: BT:Rh grown from 1000 ppm Rh in the melt)

Defect densities (ppm):

Rh^{3+}	Rh^{4+}	Rh^{5+}	Fe^{3+}	Fe^{4+}	Fe^{5+}
17.5	4.5	1.5	10	<1.0	<1.0

Comparison of predicted and experimentally determined N_{eff}:

N_{eff} (predicted)	$(1.7 \pm 1.2) \cdot 10^{17}$ cm^{-3}
N_{eff} (measured)	$(1.7 \pm 0.3) \cdot 10^{17}$ cm^{-3}

TABLE 2.3. Defects identified in $BaTiO_3$. The underlined charge states are EPR-silent. Their presence was proved indirectly by combined EPR/optical absorption studies. Especially in these cases the related optical absorption bands and the charge transfer processes in which these defects are involved have been determined

Extrinsic defects:

$\underline{Cr^{2+}}$ [96], Cr^{3+} [110], Cr^{5+} [111, 112]
Mn^{2+} [113], Mn^{4+} [114]
$\underline{Fe^{2+}}$ [115], Fe^{3+} [116, 112], $\underline{Fe^{4+}}$ [108], Fe^{5+} [117, 108], Fe^{3+} - V_O [117], Fe^{4+} -V_O [112]
Co^{2+} [118], Co^{3+}–V_O [112], Co^{4+}-V_O [112]
Ni^+_{Ba} [119]
Mo^{5+} [120]
Rh^{2+} [112], $\underline{Rh^{3+}}$ [108], Rh^{4+}, $\underline{Rh^{5+}}$ [108]
Ir^{4+} [112]
Nd^{3+} [121]
Gd^{3+} [122]
Er^{3+} [25]
Ce^{3+} [123]
Na^+_{Ba}–O^- [124], K^+_{Ba}–O^- [124], Al^{3+}–O^- [112]

Intrinsic defects:

Ti^{3+} (bound to various unidentified defects [125] or free as a conduction band polaron [10]), Ti^{3+}–Nb^{5+} [112]

Isolated V_O has so far not been detected. For a discussion of this unsolved problem, see [126]. After strong chemical reduction of $BaTiO_3$, assumed to create V_O, only almost-free Ti^{3+} conduction band polarons, slightly bound to various unidentified lattice perturbations [127], are found. Such Ti^{3+} carriers, also of various association types, likewise can result from optical excitation from deep defects or by bandgap illumination [127]. The appearance of signals of this kind is the experimental proof for light-induced electronic processes. Similarly, optical hole transfers are monitored by the occurrence of the O^- defects listed in Table 2.3, observable at low temperatures. By doping with shallow alkali acceptors such as Na_{Ba} and additional oxidation [107], the Fermi level can be lowered.

Table 2.3 represents a large toolbox of defects whose possible influence on the photorefractive performance can be assessed on a microscopic basis. In this context the dopings Co and Ce, in addition to Rh, have received considerable attention. With Co-doped $BaTiO_3$, having a wide absorption band at 2.25 eV, larger two-beam coupling gains were observed, using 515 nm light, than with Fe, Cr, or Mn doping [128]. The presence of $Co^{3+}-V_O$ and of Co^{2+} was identified in the crystals used in this investigation. The $Ce^{3+/4+}$ level lies near midgap, and holes trapped there thus have a rather long lifetime [123, 129]. Together with codoped Rh, introducing a $Rh^{3+/4+}$ level more shallow than $Ce^{3+/4+}$, the material offers potential for nonvolatile holographic storage using gated two-wave illumination [123].

2.5.2 $Ba_{1-x}Ca_xTiO_3$ (BCT)

The major drawback of $BaTiO_3$ for its application as a photorefractive material is the fact that its phase transition from the tetragonal to the orthorhombic phase near 8°C tends to deteriorate the optical quality of the crystals. The replacement of part of Ba by Ca forms the isostructural $Ba_{1-x}Ca_xTiO_3$ (BCT), with the congruently melting composition at $x = 0.23$ [130]. BCT is tetragonal between 100°C and at least 50 K [131] and thus avoids the phase transition problem impeding the use of BT. The photorefractive properties of BCT as a host material are similarly favorable as those of BT [132]. The EPR detection of defects in BCT meets with difficulties: Because of the irregular lattice positions caused by the statistical replacement of Ba^{2+} (161 pm) by the much smaller Ca^{2+} (134 pm), the local crystal fields at the defect sites are rather nonuniform, leading to superpositions of rather diverse signals. The resulting wide and odd-shaped EPR patterns are hard to evaluate with respect to the underlying defect models. Furthermore, for so far unknown reasons, the presence of Ca causes strong microwave losses. In spite of these difficulties, it has been possible to identify several defects in BCT:Rh by EPR [133]. Such crystals usually have sizable background concentrations of iron. In all of the investigated BCT:Rh samples, Fe^{3+}, isolated as well as in various low-symmetry associations, was identified.

Besides BCT:Fe [134, 135], the system BCT:Rh has so far received the most attention in photorefractive studies, because also in BCT doping with Rh increases the infrared sensitivity. Initial investigations of BCT:Rh are described by Veenhuis et al. [136] and by Bernhardt et al. [137, 138]. Because of the similarity of BCT:Rh to BT:Rh, the optical absorption bands in both cases are also quite similar. Based on the EPR information on defects in BT:Rh [139] it thus was possible to elucidate all photorefractively relevant light-induced charge-transfer processes in BCT:Rh qualitatively [138] as well as quantitatively [106] in the same way as described above for BT:Rh. On the basis of the determined absorption cross-sections, the defect densities, and the charge transfer parameters, again the photorefractive performance of the system is predicted. Bernhardt et al. [138] discovered that rather high pump light intensities—about hundred times higher than in BT:Rh— are necessary for the photoconductivity of the material to outweigh the dark conductivity. Under this condition the space charge fields are saturated at their maxima. The authors attributed this fact to the higher relative background Fe content in BCT:Rh. Meyer et al. [106], however, could show that the Fe content in the samples is not as high as assumed by Bernhardt et al. [138]. It is proposed that the high dark conductivity of BCT might instead be caused by shallow hole traps induced by disorder [106]. With reduced BCT:Fe it was shown [135], using combined EPR/optical studies, that Fe^{2+} has a wide absorption band peaking at 1.9 eV.

2.5.3 KNbO₃ (KN)

Chapter 7 deals with the favorable photorefractive properties of this material in detail. In $KNbO_3$, orthorhombic below $\approx 200°C$, the defects Mn^{2+}, Fe^{3+}, $Fe^{3+}-V_O$, Co^{2+}, $Co^{2+}-V_O$, Ir^{4+}, $Ti^{4+}-O^-$ could be identified with EPR [25]. An extended EPR and combined EPR/optical absorption study of $KNbO_3$:Fe is included in [86]. This doping is known to improve the photorefractive properties of $KNbO_3$ [140]. Besides isolated Fe^{3+}_{Nb} and $Fe^{3+}_{Nb}-V_O$, several low-symmetry associations of Fe^{3+}_{Nb} with unknown partners are reported. As grown, oxidized and reduced samples were investigated. Light-induced changes of EPR and optical absorption were observed in some cases, but only in one case with the same defect. Definite relations between Fe defects and the photorefractive behavior of the crystals could not be established. Nor was it possible to decide experimentally whether Fe^{2+} in $KNbO_3$ absorbs at 2.55 eV [140]. With EPR/optical absorption it was established that the optical absorption of Fe^{2+} has its peak at 2.1 eV in BT [115], at 1.9 eV in BCT [141], and at 1.5 eV in $KTaO_3$ [142] (Figure 2.10).

In view of the obstacles to obtaining detailed EPR information about defects in $KNbO_3$, it was decided to investigate the similar $KTaO_3$:Fe as a model system. Such crystals are cubic at all temperatures and thus facilitate the investigation by ODMR-MCD. At least eleven different Fe-containing defects, most of them low-symmetry associations, were found [143]. Figure 2.10 shows a selection among them, characterized by their MCD spectra. Usually such bands are overlapping in the same crystal; with the ODMR technique it is possible to select those bands originating from a definite defect. Since the frequency ranges of the MCD structures are identical to those of the related absorption bands, this plot shows at what energies to expect absorption bands of the given defect configurations and charge states, also in the similar $KNbO_3$. Analogous MCD-ODMR studies have been performed with $KTaO_3$:Ni [142]. Three low-symmetry defects in various charge states were found.

2.6 Defects in the Sillenites $Bi_{12}MO_{20}$ (BMO, M = Si, Ge, Ti)

In their review article, Arizmendi et al. [144] drew attention to the lack of knowledge concerning the nature and role of defects in these materials. The situation has been clarified to a large extent, new spectroscopic evidence being the main object of this section.

2.6.1 Intrinsic Defects in Undoped Crystals

The most precise information has been obtained by combined optical absorption, MCD and ODMR studies, performed with thermally bleached (e.g., 1/2 hour at 500°C) as well as optically colored crystals in Paris [145, 146, 147, 148, 149, 150] and in Osnabrück [149, 151]. Several important conclusions are summarized in [152] and in a forthcoming article [153]. As illustrated in the central part of Figure 2.11 for BGO as an example, an undoped and thermally bleached BMO sample does not exhibit any MCD signal in the near-IR to UV spectral range. This means that the "shoulder" observed near 3 eV in the corresponding absorption

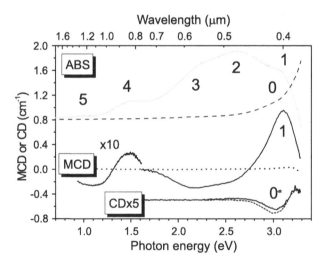

FIGURE 2.11. Upper part (arbitrary linear ordinate scale): Bleached state absorption of BGO at 80 K (dashed line) and additional absorption (10.7 cm^{-1} at maximum) induced by an illumination with blue light (solid grey line). Lower parts: CD and MCD spectra (1.4 K) in the bleached state (dotted lines) and in the colored state (solid lines).

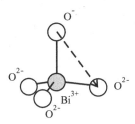

FIGURE 2.12. Model of the Bi_M^{4+} antisite defect. ODMR measurements show that the hole is delocalized to an O^{2-} neighbor of Bi^{3+}. There it is stabilized by lattice distortion, caused by the decreased attraction between O^- and Bi^{3+}. Because of the initial equivalence of the tetrahedrally arranged oxygen ions, the system can be called a bound small-hole polaron. The process of optical absorption is indicated by the dashed line: the hole is transferred under Franck–Condon conditions from its initial O^- site to a neighboring O^{2-} ion. Because of the equivalence of the three final O^{2-} ions, a tunneling of the excited hole among them is expected, leading to two optical absorption bands of nearly equal intensities and widths [11]. The dashed hole transition can alternatively be viewed as a transition of an electron from the O^{2-} ions, being part of the valence band, to O^-. This explains the transitions 2 and 3 in Figure 2.13.

spectrum (top of Figure 2.11) is necessarily correlated with a diamagnetic defect. The "shoulder" has obviously a counterpart in the natural circular dichroism spectrum (CD, 0 in Figure 2.11) with a minimum at the same energy. Illuminating the bleached BMO crystal with light energies lying within the range of this CD band causes a slight decrease of its amplitude. Simultaneously intense additional absorption (bands 1–3) in the visible region and at least two components (4, 5) in the near IR are created. Qualitatively similar results concerning this photochromic behavior in BGO and BSO were reported by another group [154, 155]. Strong MCD signals are associated with bands 1–5, demonstrating the paramagnetic character of the related defect(s), in contrast to the situation observed in the bleached state.

The ODMR [151, 149], performed with the MCD signals 1–3 could be attributed to one unpaired spin ($S = 1/2$) with a strong hyperfine coupling to the nuclear spin (9/2) of the 100% abundant ^{209}Bi. All evidence led to the assignment of the MCD signals to an antisite defect [156] Bi_M^{4+} or, in more detail, Bi_M^{3+} "dressed" by a hole, delocalized on one of the four tetrahedral oxygen neighbors. Although the observed symmetry of the defect was essentially isotropic, trigonal distortion was not excluded [157]. As concluded from the ODMR-MCD studies, all bands "1–3" in Figure 2.6 have the same Bi_M^{4+} initial state. Bands "4 and 5", in contrast, have to be assigned to different paramagnetic species [146] showing a very broad ODMR signal for microwaves at 35 GHz [153]. The magnetic and optical properties of Al_{Si}^{4+} [150] and Ga_{Si}^{4+} [153] were also established for the first time by ODMR.

The transitions leading to the observed optical properties of all these defects are shown in Figure 2.13. In the bleached state only the charge state Bi_M^{3+} is present.

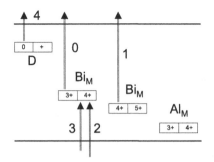

FIGURE 2.13. Spectroscopic model for sillenite crystals. The notation 0–4 for the various absorption processes is the same as that used in Figure 2.11.

The CD band (0) associated with the absorption shoulder reflects the spectral dependence of the photoionization of an electron from Bi_M^{3+} to the conduction band; its onset around 2.4 eV at 77 K or lower and around 2.2 eV at room temperature indicates the thermal energy of the $Bi_M^{3+/4+}$ level with respect to the conduction band. The ionized electron is then trapped at one of the donor states D^+, which are all empty previously, and the paramagnetic D^0 is formed. Photoabsorption and photo-MCD experiments have shown that the bands in the visible region can be partly bleached at low temperature by a further illumination either at 1.5 eV (800 nm) or, more effectively, at 2.2 eV (560 nm). In the first instance, the electron ionized from the $D^{0/+}$ level is trapped at the $Bi_M^{3+/4+}$ level. In the second process, electrons are excited from the valence band to the $Bi_M^{3+/4+}$ level.

Band 1 in the MCD and absorption spectra (colored state) is assigned [153] to the ionization of an electron from the $Bi_M^{4+/5+}$ level to the conduction band, i.e., to the second ionization of the antisite defect. This is supported by the following observations: (1) as shown by ODMR, the initial state of the transition is the Bi_M^{4+} state; (2) the band does not occur under strong acceptor (e.g., Al) doping. This removes all Bi_M^{4+} charge states. Since the $Al_{Si}^{3+/4+}$ level was shown earlier, using thermal bleaching studies [150], to lie at most 0.5 eV above the valence band, one has to conclude that the $Al_{Si}^{3+/4+}$ level lies below the $Bi_M^{4+/5+}$ level, as shown in Figure 2.13. The position of the former at about $E_C - 2.7$ eV at 4.2 K is taken from the onset of band 1. Finally, the broad MCD and optical absorption features 4 and 5 were tentatively assigned [146] to oxygen vacancies.

Figure 2.11 shows that bands 2 and 3 dominate the optical absorption in the colored state. As shown above, they are caused by electron excitation from the valence band to Bi_M^{4+} or, alternatively, to the excitation of the Bi_M^{4+} hole to the valence band. A hole at one of the tetrahedral oxygen ions next to Bi_M^{3+} represents a small-hole polaron bound in tetrahedral symmetry [11]. By the self-localization at one oxygen site the symmetry is lowered to trigonal. It is a natural consequence [11, 158] of such a system that its optical absorption consists of two strong, equally wide bands of nearly identical intensities, separated by some tenths of an eV. One of the two excited states is orbitally degenerate, spin-orbit coupling causing the occurrence of two oppositely signed MCD contributions for band 2. Bands 2 and 3 are thus likely to be attributed to optical excitations of a small-hole polaron bound to the antisite defect Bi_M^{3+}. Also the Al_M^{3+} defect, similarly having the structure

Al_M^{4+}–O^-, has optical features supporting the assignment to optical excitation of a hole polaron next to Al_M^{3+} [150].

Newer neutron [159] and X-ray diffraction [160] studies of the crystal structure of the sillenites indicate that the large Bi_M^{3+} ion, consisting of a xenon core and a $(6s^2)$ lone electron pair, can hardly fit into the space of a tetrahedrally coordinated M^{4+} ion. There is evidence that Bi_M^{3+} is in fact neighbored by an oxygen vacancy, V_O, accommodating the $6s^2$ lone pair. Also in this situation optical absorption and MCD spectra of the type indicated above are expected. The trigonal symmetry is now induced by the vacancy, the hole captured near the antisite defect replacing one of the $6s^2$ electrons since these are the least tightly bound.

2.6.2 Additional Observations

In the case of BTO, it was found experimentally that a complete dark erasure of light-induced absorption requires months at room temperature [161]. Quite consistently, the MCD spectra of as-received BMO samples invariably demonstrate that Bi_M^{4+} is present [147, 148] to an amount varying from sample to sample. So usually one operates with specimens far from equilibrium, unless they are carefully thermally bleached. There are several additional observations that are based on this fact.

From thermal bleaching studies [162] it was concluded that three relatively shallow electron traps (0.65–0.97 eV) play a major role in the case of BGO. These are certainly correlated with the absorption and MCD bands observed in the low-energy region of Figure 2.11. Studies of the absorption kinetics with changes of temperature [163] or illumination [161] indicate that two-step processes among the donor levels occur. Also, the time dependence of the relaxation of photorefractive gratings depends on the charge density stored in the shallow levels [164, 165].

The wavelength dependence of the photoionization processes can be estimated grossly from a deconvolution of the CD, absorption, and MCD spectra. Quite clearly, however, the strength of the bands (the concentration of defects) depends on the thermal and optical history of the sample. Only electron excitations are possible for bleached crystals, starting from Bi_M^{3+} Figure 2.12. After an illumination with a wavelength near or below 514.5 nm, Bi_M^{4+} and D^0 defects are created. Electron and hole transitions can then occur simultaneously at certain wavelengths. Furthermore, electron excitations from occupied shallow levels become possible in the near IR. Complementary gratings were observed in BSO at 785 nm [166], a spectral region where bands 3 and 4 (Figure 2.11) overlap at room temperature.

Photorefractivity in the sillenites was usually studied with green or blue light, but there is a need to improve their properties in the red and near-IR in order to benefit from laser diodes in this spectral region. The spectroscopic results reported above suggest that the shallow defects leading to bands 4 and 5 in Figure 2.6 can be used for that purpose. Their electron population can be raised substantially via a preliminary blue homogeneous illumination. For example, an enhancement of the beam coupling at 1.06 μm was observed [167] in undoped BTO by preexposure to visible light. Consistently with the possible assignment of bands 4 and 5 to oxygen vacancies, crystals grown under argon show an unusual photorefractive sensitivity and a remarkable operation speed in this spectral region [168]; the photoconductivity in red light is two orders of magnitude higher than in crystals grown in the presence of oxygen [169].

2.6.3 Extrinsic Defects

In principle, the introduction of extrinsic defects offers more choices for tailoring the photorefractive properties of the sillenites. A few encouraging results were obtained by doping with transition metal ions. Cr improves the diffraction efficiency and increases the response speed of BSO at 633 nm [170]. Holographic gratings were recently recorded in the near IR in ruthenium-doped BTO [171] and BSO [172] samples. In the case of $Bi_{12}Ti_{0.76}V_{0.24}O_{20}$, homogeneous illumination and the application of an external electric field after writing enhance the diffraction efficiency by a factor of almost 40 [173] at 514 nm. The mechanisms underlying these experimental observations are far from being understood.

In spite of intensive research with absorption and CD spectroscopies [174], most progress with respect to the definite assignment of absorption bands has been reached by MCD and ODMR studies [175]. Many defects have been identified in this way and their optical excitations have been greatly clarified: Cr_M^{4+} and Cr_M^{5+} [146], Mn_M^{5+} and Mn_M^{4+} [146, 176, 177], Cu_M^{2+} [146] and Cu_{Bi}^{2+} [175, 178], Fe_M^{3+} [145, 146], V_M^{4+} [177], Co_M^{2+} and Ni_M^{2+} [146], and Ru_{Bi}^{n+} ($n = 3 - 5$) [172, 179]. Until now, V_M was thought to be always in the 5+ diamagnetic state since no EPR signal could be found by conventional means [180].

As has been demonstrated above with the intrinsic defects, most detailed information is obtained when experiments are carried out in the bleached and colored states, but also in the near IR (likely to be connected with internal transitions) as well as in the visible (charge transfer or intervalence transitions). Figure 2.14 illustrates such information related to Cr doping [146, 175]. In the initial state, MCD-active Bi_M^{4+} is absent and the broad features observed in the visible are thus associated with chromium. Very characteristic MCD (and CD) sharp features are detected for Cr_{Ge}^{4+} ($^3A_2 \rightarrow {}^1E$ transition) and Cr_{Ge}^{5+} ($^2E \rightarrow {}^2T_2$) in the near IR. The coexistence of both charge states indicates that the Fermi level is pinned at the $Cr_{Ge}^{4+/5+}$ level in thermal equilibrium. Illumination with $E < 2.4$ eV (process 1) leads to the "optical" reduction of Cr_{Ge}^{5+}. The photoionized hole is trapped at Bi_{Ge}^{3+}, causing Bi_{Ge}^{4+}. Blue light with $E \geq 2.4$ eV leads to the alternative process 2. The onset of the first broad MCD band observed in the bleached state therefore provides an upper limit for the energy of the $Cr_M^{4+/5+}$ level at approximately 2.0–2.3 eV above the valence band edge. The $Mn_M^{4+/5+}$ level is located in the same energy range [146, 175]. The case of Fe doping is of special interest because Fe (as well as Cr) is always present as a background impurity. The MCD spectrum of Fe_M^{3+} peaks at 3.22 eV [25, 145] with a threshold around 2.7 eV. It is not yet established firmly whether this ion acts as a donor, with Fe^{4+} as the final state, or

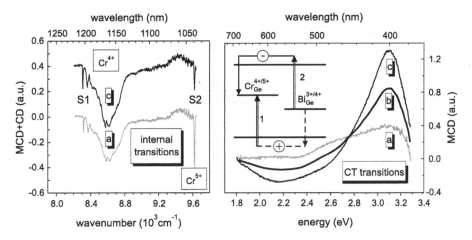

FIGURE 2.14. Light-induced MCD (or MCD+CD) changes for BGO:Cr in the visible (right) and near IR (left). The specimen had the following history:
(a) after bleaching;
(b) after illumination at 2.2 eV, corresponding to process 1;
(c) after illumination at 2.7 eV, corresponding to process 2. The sharp line S1 in the IR-part monitors the concentration of Cr^{4+}, the line S2 that of Cr^{5+}.

as an acceptor (Fe^{2+} final state). As usual, the Fermi-level position can be raised [174] by annealing under reducing conditions or by codoping with donors (e.g., phosphorous). Finally, high resolution FTIR spectroscopy has proved very useful in monitoring the amount of various elements (P^{5+}, V^{5+}, Mn^{5+}, S^{6+}, OH^-) in BMO samples [181].

2.7 Defects in Other Photorefractive Materials

2.7.1 $Sn_2P_2S_6$ (SPS)

Chapter 9 deals with the various aspects related to the development of the material for efficient photorefractive operation in the infrared. Here we confine ourselves to the information gained so far with respect to the light-induced charge transfer phenomena as studied by EPR and combined EPR/optical absorption studies.

Optical excitation across the bandgap (≈ 2.3 eV at room temperature, ≈ 2.5 eV at 4.2 K) creates several paramagnetic defects in all specimens, independent of their doping. It is known [182] that such illumination leads to persistent conductivity; i.e., the electrons and holes created in this way are connected with such large-lattice relaxations that the consequent small vibrational overlaps between excited and ground state are strongly impeding recombination. The extreme metastability of the optically excited state allows one to study the hole and electron systems independently. All the identified paramagnetic defects are of hole type. The complicated monoclinic crystal structure of SPS, where all ionic sites do not have any point symmetry except the identity, renders the definite assignment of the EPR signals rather difficult. The following defect types have been identified at 10 K [95, 183]: (1) Two species with properties compatible with two slightly different Sn^{3+} ($5s^1$) ions, Sn_1^{3+} and Sn_2^{3+}; they are likely to be caused by hole self-trapping at the two crystallographically inequivalent Sn^{2+} ions in the lattice. (2) S^- ions probably resulting from hole self-trapping at the many different S^{2-} sites in the crystal. (3) Fe^{3+}, attributed to hole trapping at Fe^{2+} background impurities at a low-symmetry crystal site characteristic of SPS.

So far, one can only speculate why the excited electrons in or near the conduction band are EPR-silent. Since the conduction band has mainly Sn^+ character ($5p^1$, spin $S = 1/2$), concluded from a rigid lattice-band calculation [184], an electron in such a state should lead to an EPR signal. Since the Sn ions are rather loosely bound in wide cages formed by the surrounding P_2S_6 units, a conduction band charge carrier can lower its energy by large-lattice relaxation, i.e., it should form a polaron. Possibly each two of them pair diamagnetically to EPR-silent bipolarons. Since the mobility of bipolarons is much lower than that of single polarons, this would add to the metastability of the excited state and the slow component [185] of the photorefractive effect.

The excitation across the bandgap causes wide optical absorption bands ranging from ≈ 1 eV (or lower) up to the fundamental absorption with a maximum for most samples near 1.9 eV. These features are at least partly related to the above hole-trapping defects. This is demonstrated in [185] with combined EPR/optical absorption investigations of the light-induced hole states.

2.7.2 $Pb_5Ge_3O_{11}$

This material has been known since 1990 [186] as promising for photorefractive applications. Lead germanate crystals are ferroelectric with $T_c = 178°C$ and have trigonal symmetry. Two competing photorefractive gratings have been found [187]; a fast one with a buildup time of about 1 s and a slow one responding within minutes to hours. There are indications that the responsible defects are of intrinsic type.

Under bandgap (≈ 3.2 eV) illumination, electrons and holes are formed [188]. As established from the analysis of the EPR data, the latter are trapped at one of the six inequivalent intrinsic Pb^{2+} sites of the lattice, forming Pb^{3+}, or rather $Pb^{2+} + O^-$, i.e., the trapped hole has highest density at one of the O^{2-} neighbors. Furthermore, a Cu^{2+} background defect is identified in all investigated crystals, which, however, cannot be recharged by illumination. The hole trapping at Pb^{2+} is metastable below ≈ 120 K and leads to a wide and strong optical absorption, at 2 K starting at 2 eV and monotonically increasing toward the fundamental absorption. As shown by combined optical absorption and ODMR-MCD measurements, this coloration is caused by a transition connected with $Pb^{2+}–O^-$. The thermal decay time of this metastable defect is of the same order as that of the fast photorefractive grating.

The concentration of the Pb trapped holes has been measured to be of the order of several 10^{16} cm^{-3}, much less than the available Pb^{2+} lattice sites. Therefore the obtainable hole concentration is limited by the concentration of the compensating donors. All electrons not trapped at these donors recombine immediately with the holes, putting an upper limit to those surviving. All attempts have failed so far to identify these donors. If the material could be doped with a higher concentration of them, once the donors are known, a sensitive material with a high density of photorefractively active hole centers would be available.

2.7.3 $Sr_{1-x}Ba_xNb_2O_6$ (SBN)

The material crystallizes in the tetragonal tungsten bronze structure, melting congruently for $x = 0.61$. Only five of the six lattice sites favorable for incorporation of Sr and Ba are occupied in ideal crystals; the statistical distribution of the corresponding empty cation sites causes strongly fluctuating crystal fields, e.g., at the incorporation sites of EPR-active defect ions. Thus only Ce^{3+} ($4f^1$), where the paramagnetic f-electrons are shielded by the outer electrons from the influences of the crystal surroundings, was identified by EPR [189]. This doping is rather efficient in sensitizing the photorefractive effect for light in the energy range 2 to 3 eV [190, 191]. The EPR data indicate that Ce^{3+} enters the crystal at only one of the two crystallographically different alkaline earth sites: one is 12-fold, the other one 15-fold coordinated by oxygen ions. Whatever its lattice site, EPR shows that Ce^{3+} moves off-center, destroying all symmetry. This is more likely at the spacious 15-fold coordinated site than at the other one. Under optical illumination, e.g., with 488-nm light, the Ce^{3+} EPR signal decreases. This is consistent with the lowering of the Ce^{3+} 4f–5d optical absorption band under such irradiation, peaking near 2.5 eV [190]. Simultaneously, a wide absorption band rises [192], with maximum near 0.8 eV, which is typical for optical absorption by Nb^{4+} ($4d^1$) conduction band polarons. They are created by electrons photoionized from Ce^{3+}. EPR of the paramagnetic Nb^{4+} could not be identified. Probably the EPR signals are too wide to be detected; this is expected to result from the spatially fluctuating crystal fields at the Nb sites, where the Nb 4d electrons are not shielded from their surrounding distorted octahedra of oxygen ions. Also, crystals doped with Cr were studied with EPR. Only rather faint and wide signals typical for Cr^{3+} were observed. The resonances could serve only to identify the presence of Cr^{3+}, probably incorporated at Nb sites. A further analysis was not possible.

2.7.4 Terbium Gallium Garnet Doped with Cerium ($Tb_3Ga_5O_{12}$:Ce)

In this centric material a novel photorefractive effect was identified [193], which results from photochromic absorption, leading to Kramers–Kronig conjugated index changes. To detect EPR signals from paramagnetic defects embedded in a bulk paramagnetic sample is an unusual situation for plain EPR measurements.

Here the strong paramagnetism of the host lattice tends to drive the microwave bridge circuit employed in such an experiment out of balance and prevents the registration of the defect resonances. However, using MCD-ODMR, the resonances of the dopant could be detected on their optical bands, lying at optical frequencies different from those of the host lattice [194]. In this way the electronic structure and the optical absorption properties of Ce^{3+} and its interactions with the host lattice were clarified in spite of the paramagnetic background, and it was proved that the photorefractive activity of the material arises from this defect.

2.7.5 $Bi_4Ge_3O_{12}$ (Eulytite Structure)

Bismuth germanate is a well-established scintillator used in high-energy photon and particle detectors. It is also one of the few crystals in which photorefractive gratings have been recorded in the near-UV (330–350 nm) [195]. Photorefractive and photochromic effects have been investigated in crystals doped with Cr [196, 197], Mn and Fe [198, 199], and Co [200].

The cubic structure of $Bi_4Ge_3O_{12}$ is very close to that of $Bi_{12}GeO_{20}$, with distorted tetrahedra (S_4 site symmetry at Ge^{4+}) and octahedra (C_3 at Bi^{3+}) in the former crystal [201]. From extensive EPR work in Madrid [201], the dominant defects in crystals doped with 3d ions were found to be Fe_{Ge}^{3+}, Cr_{Ge}^{4+}, Co_{Bi}^{2+}, and Mn_{Bi}^{2+}.

MCD was used in Paris to monitor the photochromic behavior of the internal transitions of Cr_{Ge}^{4+} [197], Co_{Bi}^{2+}, and Mn_{Bi}^{2+} (unpublished). Near-band-gap (≈ 4.1 eV at room temperature, ≈ 4.5 eV at 4.2 K) illumination at room temperature reduces their intensity (only 5% in the case of Cr) and creates new paramagnetic defects (Co and Mn). At this time, the electron or hole trapping character of these dopants is not clearly established. Comparison of absorption and MCD spectra of Fe- and Cr-doped crystals [202] led to the conclusion that diamagnetic defects are also present. ODMR experiments were carried out only for V-doped $Bi_4Ge_3O_{12}$ [22]. It was shown that UV illumination creates paramagnetic V_{Ge}^{4+} (absent in the bleached state) at the expense of diamagnetic V_{Ge}^{5+}, and an energy level scheme was proposed. Cr and Fe impurities could be detected at the ppm level.

2.7.6 Bismuth Tellurite Bi_2TeO_5 (CaF_2-based Lattice)

At high-write beam intensities, this relatively new material has been announced to be competitive for data storage with the known best materials ($LiNbO_3$:Fe or $BaTiO_3$), with the advantage of the self-fixing of volume holograms [203, 204, 205].

Bi_2TeO_5 crystallizes in an orthorhombic structure with a large number ($\approx 17\%$) of structural oxygen vacancies with a high concentration (about 17%) of structural oxygen vacancies. It is an interesting host for dopants since it has three different cation sites with different coordination spheres, Bi(1) with 8, Bi(2,3) with 7, and Te with 5 oxygens; a good incorporation was observed for Cr, V, Mo, substituting for Te, and rare earth ions substituting for Bi [206]. The exposure of Bi_2TeO_5:Cr crystals to white light induces photochromism. The partial transformation of Cr^{6+} into Cr^{5+} was inferred from photoabsorption experiments [207] but no EPR result has yet been reported.

2.8 Hydrogen

Just as the defects treated before, hydrogen is a tool to tailor the photorefractive properties of a material as well as to investigate its properties. This element is present in all compounds containing oxygen as a constituent, with concentrations ranging from a few ppm up to several percent, after special treatments depending on the crystal. The presence of hydrogen is not always an advantage, for example in fibers. Therefore procedures have been invented to reduce the hydrogen content, in particular in $LiNbO_3$ [208, 209]. Usually hydrogen is bonded to an oxygen ion and performs a vibration, referred to as OH^- stretch mode, with an energy of about 3200–3700 cm^{-1}. Detailed spectroscopic data have been compiled and reported recently [210]. Besides basic studies of the spectroscopic properties of the OH^- stretch mode, it has served very often as a tool to investigate properties of the host. Because of its easy observation by absorption spectroscopy, a qualitative and in favorable cases even a quantitative measure of the hydrogen content is possible. Again $LiNbO_3$ served as an example. It was found that the OH^- absorption is polarized perpendicularly to the c-axis of LN in accordance with the microscopic model of proton incorporation established by ENDOR, see Section 2.4. An absorption strength per ion of about 9×10^{-18} cm was found (for details of the definition see [211, 212]). By monitoring the change of the hydrogen isotope concentrations in thermal processes the diffusivity of these isotopes can be determined. In the last two decades strong efforts have been devoted to studying the diffusion of hydrogen isotopes in insulating materials, especially in oxides; see [213] and references therein.

Protons can be involved in thermal fixing of holograms (LN, KTN, BSO). However, Buse et al. concluded from spatially resolved optical data especially for $LiNbO_3$ that their presence is not mandatory for this purpose [214]. Strongly related to hologram fixing is the production of very-narrow-bandwidth interference filters by hydrogen doping. This subject has been reviewed by Cabrera et al. [215].

Some changes of physical properties are reflected in the position or shape of the OH^- absorption spectra. An example for such strong influence is given in Chapter 6 of this book, were the stretch mode was used to monitor changes induced in $LiNbO_3$ crystals doped with damage-resistant impurities such as Mg, Zn, In, and Sc.

2.9 Summary

We have spread out the rich field of defects in most of the inorganic photorefractive materials investigated at present. The aim is to contribute to the improvement of their performance in their various applications. Emphasis was therefore first on the elucidation of the identity of the defects and their microscopic defect structures. Studies by EPR and related methods, such as ODMR and ENDOR, are unsurpassed for this goal. Second, in order to connect these data to the photorefractive effect, close links to the optical defect phenomena have been established where possible. Here a strong tool is ODMR via the MCD of a defect. Furthermore, it is seen that the analysis of the changes of optical absorption, EPR, and MCD in photochromic materials under varying pumplight illuminations can furnish a systematic approach to the identification of defects and the paths of light-induced charge transfers between them. Also, a quantitative assessment of the photorefractive performance of a material can be derived from this method. It applies to all defects, independent of whether they are identified by EPR or not. While the results presented in this chapter are more of a basic nature, it is anticipated that they will establish a foundation on which the further chapters in the book, focusing on the details of the photorefractive effect in most of the materials treated here, can be linked to the microscopic origins of the phenomena.

References

1. Topics in applied physics, vols. 61 and 62. In P. Günter and J.P. Huignard, editors, *Photorefractive Materials I and II*. Springer, 1988.
2. K. Buse. *Appl. Phys.* B, 64:391, 1997.
3. W. Hayes and A.M. Stoneham. *Defects and Defect Processes in Nonmetallic Solids*. Wiley, 1985.
4. K. Buse. *Appl. Phys. B*, 64:273, 1997.
5. P.W. Atkins. *Physical Chemistry*. Oxford University Press, 1986, p. 275.
6. W.B. Fowler. Electronic states and optical transitions of color centers. In W.B. Fowler, editor, *Physics of Color Centers*, p. 53. Academic Press, 1968.
7. O.F. Schirmer. *Z. Physik* B, 24:235, 1976.
8. D. Emin. *Phys. Rev.* B, 48:13691, 1993.
9. I. Biaggio, R.W. Hellwarth, and J.P. Partanen. *Phys. Rev. Lett.*, 78:891, 1997.
10. S. Lenjer, O.F. Schirmer, H. Hesse, and T.W. Kool. *Phys. Rev.* B, 66:165106, 2002.
11. O.F. Schirmer. *J. Phys. Condens.* Matter (submitted)
12. G.I. Malovichko, V.G. Grachev, and O.F. Schirmer. *Appl. Phys.*, B, 68:785, 1999.
13. J.M. Spaeth and V. Overhof. *Point Defects in Semiconductors and Insulators*. Springer, 2003.
14. J.A. Weil, J.R. Bolton, and J.E. Wertz. *Electron Paramagnetic Resonance*. Wiley, 1994.
15. Th. Woike, G. Weckwerth, H. Palme, and R. Pankrath. *Solid State Commun.*, 102:743, 1997.
16. S. Juppe and O.F. Schirmer. *Phys. Lett.* A, 117:150, 1986.
17. P.J. Stephens. *Adv. Chem. Phys.*, 25:197, 1976.
18. J. Badoz, M. Billardon, A.C. Boccara, and B. Briat. *Symposium of the Faraday Society*, 3:27, 1970.
19. S.B. Piepho and P.N. Schatz. *Group Theory in Spectroscopy, with Application to Magnetic Circular Dichroism*. Wiley, New York, 1983.

20. O.F. Schirmer, W. Berlinger, and K.A. Müller. *Sol. State Commun.*, 16:1289, 1975.
21. H. Kröse, E. Possenriede, R. Scharfschwerdt, T. Varnhorst, O.F. Schirmer, H. Hesse, and C. Kuper. *Opt. Materials*, 4:153, 1995.
22. B. Briat, A. Watterich, F. Ramaz, L. Kovács, B. Forget, and N. Romanov. *Opt. Materials*, 20:253, 2002.
23. A. Kling, J.C. Soares, and M.F. da Silva. In F. Agulló-López, editor, *Insulating Materials for Optoelectronics*, p. 175. Word Scientific, 1995.
24. O.F. Schirmer, O. Thiemann, and M. Wöhlecke. *J. Phys. Chem. Solids*, 52:185, 1991.
25. O.F. Schirmer, H.J. Reyher, and M. Wöhlecke. In F. Agulló-López, editor, *in Insulating Materials for Optoelectronics*, p. 63. World Scientific, 1995.
26. A. Räuber. In E. Kaldis, editor, *Current Topics in Material Sciences*, p. 481. 1978.
27. P.F. Bordui, R.G. Norwood, C.D. Bird, and G.D. Calvert. *J. Crystal Growth*, 113:61, 1991.
28. V. Grachev, G. Malovichko, and E. Kokanyan. *Ferroelectrics*, 258:423, 2001.
29. B. Faust, H. Müller, and O.F. Schirmer. *Ferroelectrics*, 153:297, 1994.
30. C. Leroux, G. Nihoul, G. Malovichko, V. Grachev, and C. Boulesteix. *J. Phys. Chem. Solids*, 59:311, 1998.
31. P. Lerner, C. Legras, and J. P. Dumas. *J. Cryst. Growth*, 3:2331, 1968.
32. D.M. Smyth. *Prog. Solid St. Chem.*, 15:145, 1984.
33. O.F. Schirmer and D. von der Linde. *Appl. Phys. Lett.*, 33:335, 1978.
34. K.L. Sweeney and L.E. Halliburton. *Appl. Phys. Lett.*, 43:336, 1983.
35. H. Müller and O.F. Schirmer. *Ferroelectrics*, 125:319, 1992.
36. H.J. Reyher, R. Schulz, and O. Thiemann. *Phys. Rev. B*, 50:3609, 1994.
37. O.F. Schirmer, S. Juppe, and J. Koppitz. *Cryst. Latt. Def. and Amorph. Mat.*, 16:353, 1987.
38. F. Jermann and J. Otten. *J. Opt. Soc. Am. B*, 10:2085, 1993.
39. L. Hesselink, S.S. Orlov, A. Liu, A. Akella, D. Lande, and R.R. Neurgaonkar. *Science*, 282:1089, 1998.
40. L. Arizmendi and F. Agulló-López. *MRS Bulletin*, March 1994, p. 33.
41. K.L. Sweeney, L.E. Halliburton, D.A. Bryan, R.R. Rice, R. Gerson, and H.E. Tomaschke. *J. Appl. Phys.*, 57:1036, 1985.
42. G. Corradi, K. Polgár, I. M. Zaritskii, L. G. Rakitina, and N. I. Deryugina. *Sov. Phys. Stat. Sol.*, 31:1540, 1989.
43. I.M. Zaritskii, L.G. Rakitina, G. Corradi, K. Polgár, and A.A. Bugai. *J. Phys.: Condens. Matter*, 3:8457, 1991.
44. I.M. Zaritskii, L.G. Rakitina, and K. Polgár. *Sov. Phys. Sol. State*, 37:1037, 1995.
45. U. Schlarb, M. Wöhlecke, B. Gather, A. Reichert, K. Betzler, T. Volk, and N. Rubinina. *Optical Materials*, 4:791, 1995.
46. G.I. Malovichko, V.G. Grachev, L.P. Yurchenko, V.Y. Proshko, E.P. Kokanyan, and V.T. Gabrielyan. *Phys. Stat. Sol.* (a), 133:K29, 1992.
47. G.I. Malovichko, V.G. Grachev, E.P. Kokanyan, O.F. Schirmer, K. Betzler, B. Gather, F. Jermann, S. Klauer, U. Schlarb, and M. Wöhlecke. *Appl. Phys. A*, 56:103, 1993.
48. G.I. Malovichko, V.G. Grachev, A. Hofstaetter, E.P. Kokanyan, A. Scharmann, and O.F. Schirmer. *Phys. Rev. B*, 65:224116, 2002.
49. K.K. Ziling, V.A. Nadolnii, and V.V. Shashkin. *Sov. Reports of the AS of USSR, Inorganic Materials*, 16:701, 1980.
50. L.G. Rakitina, I.M. Zaritskii, G. Corradi, and K. Polgár. *Sov. Phys. Sol. State*, 32:654, 1990.
51. G. Burns, D.F. O'Kane, and K.S. Title. *Phys. Rev.*, 167:314, 1968.

52. N.F. Evlanova, L.S. Kornienko, L.N. Rashkovich, and A.O. Rybaltovskii. *Sov. JETP*, 53:1920, 1967.
53. D.J. Rexford, Y.M. Kim, and H.S. Story. *J. Chem. Phys.*, 52:860, 1970.
54. G.I. Malovichko, V.G. Grachev, and S.N. Lukin. *Sov. Phys. Solid State*, 28:553, 1986.
55. V.G. Grachev, G.I. Malovichko, and V.V. Troitskii. *Sov. Phys. Solid State*, 29:349, 1987.
56. J.C. Danner, U. Ranon, and D.N. Stamires. *Chem. Phys. Letters*, 2:605, 1968.
57. J.B. Herrington, B. Dischler, and J. Schneider. *Sol. State Commun.*, 10:509, 1972.
58. D.G. Rexford and Y.M. Kim. *J. Chem. Phys.*, 57:3094, 1972.
59. G. Corradi, H. Söthe, J.M. Spaeth, and K. Polgár. *J. Phys.: Condens. Matter*, 2:6603, 1990.
60. G.I. Malovichko and V.G. Grachev. *Sov. Phys. Sol. State*, 27:1678, 1985.
61. H.H. Towner, Y.M. Kim, and H.S. Story. *J. Chem. Phys.*, 56:3676, 1972.
62. W. Keune, S.K. Date, U. Gonser, and H. Bunzel. *Ferroelectrics*, 13:443, 1976.
63. V.G. Grachev and G.I. Malovichko. *Sov. Phys. Sol. State*, 27:443, 1985.
64. G.I. Malovichko, V.G. Grachev, O.F. Schirmer, and B. Faust. *J. Phys.: Condens. Matter*, 5:3971, 1993.
65. H. Söthe and J.-M. Spaeth. *J. Phys.: Condens. Matter*, 4:9901, 1992.
66. S. Juppe and O.F. Schirmer. *Sol. State Commun.*, 76:299, 1990.
67. H.J. Donnerberg and O.F. Schirmer. *Sol. State Commun.*, 63:299, 1987.
68. A.A. Mirsakhanian and A.K. Petrosyan. *Sov. Phys. Sol. State*, 28:1593, 1986.
69. Y.N. Choi, I.W. Park, S.S. Kim, S.S. Park, and S.H. Choh. *J. Phys.: Condens. Matter*, 11:4723, 1999.
70. J. Rosa, K. Polak, and J. Kubatova. *Phys. Stat. Sol.(b)*, 111:K85, 1982.
71. G. Corradi, K. Polgár, A.A. Bugai, I.M. Zaritskii, V.G. Grachev, and N.J. Deryugina. *Sov. Phys. Sol. State*, 28:412, 1986.
72. A.A. Mirsakhanyan. *Sov. Phys. Solid State*, 23:2452, 1981.
73. G. Burns, D.F. O'Kane, and R.S. Title. *Phys. Rev.*, 147:314, 1967.
74. B. Dischler, J.R. Herrington, and A. Räuber. *Solid State Commun.*, 12:737, 1973.
75. P.F. McDonald, C.P. Tam, and Y.W. Mok. *J. Chem. Phys.*, 56:1007, 1972.
76. M. Lee, I. Gyoo Kim, S. Takegawa, Y. Furukawa, Y. Uchida, K. Kitamura, and H. Hatano. *J. Appl. Phys.*, 89:5311, 2001.
77. T. Nolte, T. Pawlik, and J.-M. Spaeth. *Sol. State Commun.*, 104:535, 1997.
78. D. Bravo, A. Martin, and F.J. López. *Sol. State Commun.*, 112:541, 1999.
79. C. Bonardi, R.A. Carvalho, H.C. Basso, M.C. Terrile, G.K. Cruz, L.E. Bausa, and J.G. Sole. *J. Chem. Phys.*, 111:6042, 1999.
80. G. Corradi, I.M. Zaritskii, A. Hofstaetter, K. Polgár, and L.G. Rakitina. *Phys. Rev. B*, 58:8329, 1998.
81. J. Diazcaro, J. García-Solé, D. Bravo, T.P.J. Han, F. Jaque, and B. Henderson. *Ferroelectrics Letter Section*, 23:27, 1997.
82. G. Corradi, H. Söthe, J.-M. Spaeth, and K. Polgár *J. Phys.:Condens. Matter*, 3:1901, 1991.
83. G. Corradi, H. Söthe, J.-M. Spaeth, and K. Polgár. *Ferroelectrics*, 125:295, 1992.
84. T.H. Yeom, S.H. Lee, S.H. Choh, and D. Choi. *J. Korean Phys. Soc.*, 32:S647, 1998.
85. V.G. Grachev and G.I. Malovichko. *Phys. Rev. B*, 62:7779, 2000.
86. G. Malovichko. *OSA TOPS*, 27:59, 1999.
87. G.I. Malovichko, V.G. Grachev, E. Kokanyan, and O.F. Schirmer. *Ferroelectrics*, 239:357, 2000.

88. G.I. Malovichko, V.G. Grachev, E.P. Kokanyan, and O.F. Schirmer. *Phys. Rev.* B, 59:9113, 1999.
89. G.I. Malovichko, V.G. Grachev, E.P. Kokanyan, and O.F. Schirmer. *Ferroelectrics*, 239:357, 2000.
90. O. Thiemann. PhD thesis, University of Osnabrück, 1993.
91. M. Pape, H:-J. Reyher, and O.F. Schirmer. *J. Phys.: Condens.* Matter 17, 6835 (2005).
92. M.B. Klein. Topics in applied physics, vol. 61. In P. Günter and J. P. Huignard, editor, *Photorefractive Materials I*, p. 195. Springer, 1988.
93. M. Zgonik, K. Nakagawa, and P. Günter. *J. Opt. Soc. Am.* B, 12:1416, 1995.
94. O.F. Schirmer, M. Meyer, A. Rüdiger, and C. Veber. *Opt. Materials*, 18:1, 2001.
95. S. Rüdiger, O.F. Schirmer, S. Odoulov, A. Shumelyuk, and A. Grabar. *Opt. Materials*, 18:123, 2001.
96. O.F. Schirmer. *Rad. Eff. Def. in Solids*, 149:1, 1999.
97. H. Kröse, R. Scharfschwerdt, A. Mazur, and O.F. Schirmer. *Appl. Phys. B*, 67:79, 1998.
98. G.W. Ross, P. Hribek, R.W. Eason, M.H. Garrett, and D. Rytz. *Opt. Commun.*, 101:60, 1993.
99. B.A. Wechsler, M.B. Klein, C.C. Nelson, and R.N. Schwartz. *Opt. Lett.*, 20:1850, 1994.
100. M. Kaczmarek and R.W. Eason. *Opt. Letters*, 20:1850, 1995.
101. H. Kröse, R. Scharfschwerdt, O.F. Schirmer, and H. Hesse. *Appl. Phys. B*, 61:1, 1995.
102. K. Buse and E. Krätzig. *Appl. Phys. B*, 61:2, 1995.
103. N. Huot, J.M.C. Jonathan, and G. Roosen. *Appl. Phys. B*, 65:489, 1997.
104. L. Corner, R. Ramos-Garcia, A. Petris, and M.J. Damzen. *Opt. Commun.*, 143:165, 1997.
105. C. Veber, M. Meyer, O.F. Schirmer, and M. Kaczmarek. *J. Phys.: Condens. Matter*, 15:415, 2003.
106. M. Meyer, O.F. Schirmer, and R. Pankrath. *Appl. Phys. B, in print*, 79:395, 2004.
107. R. Scharfschwerdt, O.F. Schirmer, H. Hesse, and D. Rytz. *Appl. Phys. B*, 68:807, 1999.
108. O.F. Schirmer, M. Meyer, A. Rüdiger, and C. Veber. *Opt. Materials*, 18:1, 2001.
109. N.V. Kukhtarev, V.B. Markov, S.G. Odoulov, M.S. Soskin, and V.l. Vinetskii. *Ferroelectrics*, 22:949, 1979.
110. K.A. Müller, W. Berlinger, and J. Albers. *Phys. Rev.* B, 32:5837, 1987.
111. E. Possenriede, O.F. Schirmer, J. Albers, and G. Godefroy. *Ferroelectrics*, 107:313, 1990.
112. E. Possenriede, P. Jacobs, and O.F. Schirmer. *J. Phys.: Condens. Matter,*, 4:4719, 1992.
113. H. Ikushima and S. Hayakawa. *J. Phys. Soc. Japan*, 19:1986, 1964.
114. K.A. Müller, W. Berlinger, K.W. Blazey, and J. Albers. *Solid State Commun.*, 61:21, 1987.
115. A. Mazur, O.F. Schirmer, and S. Mendricks. *Appl. Phys. Lett.*, 70:2395, 1997.
116. T. Sakudo and H. Unoki. *J. Phys. Soc. Japan*, 19:2109, 1964.
117. E. Possenriede, O.F. Schirmer, H.J. Donnerberg, G. Godefroy, and A. Maillard. *Ferroelectrics*, 92:245, 1989.
118. K. Zdansky, H. Arend, and F. Kubec. *phys. stat. sol.*, 20:653, 1967.
119. S. Lenjer, R. Scharfschwerdt, T.W. Kool, and O.F. Schirmer. *Sol. State Commun.*, 116:133, 2000.

120. R.N. Schwartz, B.A. Wechsler, and L. West. *Appl. Phys. Lett.*, 67:1352, 1995.
121. E. Possenriede, O.F. Schirmer, and G. Godefroy. *Phys. Stat. Sol.*, (b) 161:K55, 1990.
122. L. Rimai and G.A. de Mars. *Phys. Rev.*, 130:145, 1963.
123. R.N. Schwartz and B.A. Wechsler. *Phys. Rev.* B, 61:8141, 2000.
124. T. Varnhorst, O.F. Schirmer, H. Kröse, R. Scharfschwerdt, and Th. W. Kool. *Phys. Rev.* B, 53:116, 1996.
125. R. Scharfschwerdt, A. Mazur, O.F. Schirmer, H. Hesse, and S. Mendricks. *Phys. Rev.* B, 54:15284, 1996.
126. S. Lenjer, O.F. Schirmer, H. Hesse, and T.W. Kool. *Phys. Rev. B*, 70: in print, 2004.
127. R. Scharfschwerdt, A. Mazur, O.F. Schirmer, H. Hesse, and S. Mendricks. *Phys. Rev. B*, 54239:15284, 1996.
128. D. Rytz, B.A. Wechsler, M.H. Garrett, C.C. Nelson, and R.N. Schwartz. *J. Opt. Soc. Am. B*, 7:2245, 1990.
129. H. Song, S.X. Dou, M. Chi, Y. Zhu, and P. Ye. *Appl. Phys. B*, 70:543, 2000.
130. C. Kuper, R. Pankrath, and H. Hesse. *Appl. Phys.* A, 65:301, 1997.
131. C. Veber. unpublished, 1998.
132. C. Kuper, K. Buse, U. van Stevendaal, M. Weber, T. Leidlo, H. Hesse, and E. Krätzig. *Ferroelectrics*, 208-209:213, 1998.
133. G. Malovichko, V. Grachev, R. Pankrath, and O. Schirmer. *Ferroelectrics*, 258:169, 2002.
134. H. Veenhuis, T. Börger, K. Buse, C. Kuper, H. Hesse, and E. Krätzig. *J. Appl. Phys.*, 88:1042, 2000.
135. A. Mazur, C. Veber, O.F. Schirmer, C. Kuper, and H. Hesse. *J. Appl. Phys.*, 85:6751, 1999.
136. H. Veenhuis, T. Börger, K. Peithmann, M. Flaspöhler, K. Buse, R. Pankrath, H. Hesse, and E. Krätzig. *Appl. Phys. B*, 70:797, 2000.
137. S. Bernhardt, P. Delaye, H. Veenhuis, D. Ritz, and G. Roosen. *Appl. Phys. B*, 70:789, 2000.
138. S. Bernhardt, H. Veenhuis, P. Delaye, R. Pankrath, and G. Roosen. *Appl. Phys. B*, 74:287, 2002.
139. C. Veber, M. Meyer, O.F. Schirmer, and M. Kaczmarek. *J. Phys.: Condens. Matter*, 15:415, 2003.
140. P. Günter and F. Micheron. *Ferroelectrics*, 18:27, 1978.
141. A. Mazur, C. Veber, O.F. Schirmer, C. Kuper, and H. Hesse. *J. Appl. Phys.*, 85:6751, 1999.
142. N. Hausfeld. PhD thesis, University of Osnabrück, 1999.
143. H.J. Reyher, N. Hausfeld, and M. Pape. *J. Phys.: Condens. Matter*, 12:10599, 2000.
144. L. Arizmendi, J.M. Cabrera, and F. Agulló-López. *Intern. Jour. Optoelectronics*, 7:149, 1992.
145. B. Briat, J.C. Fabre, and V. Topa. In O. Kanert and J.M. Spaeth, editors, *Defects in Insulating Materials*, p. 1160. World Scientific, 1993.
146. B. Briat, A. Hamri, F. Ramaz, and H. Bou Rjeily. *SPIE*, 3178:160, 1997.
147. B. Briat, C. Laulan-Boudy, and J.C. Launay. *Ferroelectrics*, 125:467, 1992.
148. A. Hamri, M. Secu, V. Topa, and B. Briat. *Optical Materials*, 4:197, 1995.
149. B. Briat, A. Hamri, N.V. Romanov, F. Ramaz, J.C. Launay, O. Thiemann, and H.J. Reyher. *J. Phys.: Condensed Matter*, 7:6951, 1995.
150. B. Briat, T.V. Panchenko, H. Bou Rjeily, and A. Hamri. *J. Opt. Soc. Am.* B, 15:2147, 1998.

151. H.J. Reyher, U. Hellwig, and O. Thiemann. *Phys. Rev.*, B47:5638, 1993.
152. B. Briat, A. Hamri, F. Ramaz, and H. Bou Rjeily. *SPIE*, 3178:160, 1997.
153. B. Briat et al. unpublished.
154. J.J. Martin, I. Földvári, and C.A. Hunt. *J. Appl. Phys.*, 77:7554, 1991.
155. D.W. Hart, C.A. Hunt, D.D. Hunt, J.J. Martin, M.T. Harris, and J.J. Larkin. *J. Appl. Phys.*, 73:1443, 1993.
156. R. Oberschmid. *phys. stat. sol. (a)*, 89:657, 1985.
157. H.-J. Reyher, S. Ruschke, and F. Mersch. *Rad. Eff. Def. Solids*, 136:1039, 1995.
158. H.P. Jeon, H.P. Gislason, and G.D. Watkins. *Phys. Rev. B*, 48:7872, 1993.
159. S.F. Radaev, L.A. Muradyan, and V.I. Simonov. *Acta Cryst. B*, 47:1, 1991.
160. M. Valant and D. Suvorov. *Chem. Mat.*, 14:3471, 2002.
161. O.V. Kobozev, S.M. Shandarov, A.A. Kamshilin, and V.V. Prokofiev. *J. Opt. A: Pure Appl. Opt.*, 1:442, 1999.
162. H. Marquet, J.-G. Gies, and J.C. Merle. *Europhys. Lett*, 46:389, 1999.
163. P. Mersch, K. Buse, W. Sauf, H. Hesse, and E. Krätzig. *phys. stat. sol. (a)*, 140:273, 1993.
164. P. Tayebati and D. Mahgerefteh. *J. Opt. Soc. Am. B*, 8:1053, 1991.
165. A. Attard. *J. Appl. Phys.*, 69:44, 1991.
166. M.C. Bashaw, T.-P. Ma, R.C. Barker, S. Mroczkowski, and R.R. Dube. *Phys. Rev. B*, 42:5641, 1990.
167. S. Odoulov, K.V. Shcherbin, and A.N. Shumeljuk. *J. Opt. Soc. Am. B*, 11:1780, 1994.
168. E. Raita, O. Kobozev, A.A. Kamshilin, and V.V. Prokofiev. *Optics Letters*, 25:1261, 2000.
169. E.V. Mokrushina, M.A. Bryushinin, V.V. Kulikov, A.A. Petrov, and I.A. Sokolov. *J. Opt. Soc. Am.*, 16:57, 1999.
170. T.S. Yeh, W.J. Lin, I.N. Lin, L.J. Hu, S.P. Lin, S.L. Tu, C.H. Lin, and S.E. Hsu. *Appl. Phys. Lett.*, 65:1213, 1994.
171. V. Marinova, S.H. Lin, V. Sainov, M. Gospodinov, and K.Y. Hsu. *J. Optics A: Pure Appl. Opt.*, 5:S500, 2003.
172. F. Ramaz, L. Rakitina, M. Gospodinov, and B. Briat. *Optical Materials*, 27:1547, 2005.
173. E. Rickermann, S. Riehemann, K. Buse, D. Dirksen, and G. von Bally. *J. Opt. Soc. Am. B*, 13:2299, 1996.
174. V.I. Burkov, A.V. Egorysheva, and Yu.F. Kargin. *Crystallography Reports*, 46:312, 2001.
175. A. Hamri. PhD thesis, University of Paris XI, 1996.
176. F. Ramaz, A. Hamri, B. Briat, V. Topa, and G. Mitroaica. *Rad. Eff. Def. Solids*, 136:1009, 1995.
177. B. Briat, M.T. Borowiec, H. Bou Rjeily, F. Ramaz, A. Hamri, and H. Szymczak. *Rad. Eff. Def. Solids*, 157:989, 2002.
178. V. Chevrier, J.M. Dance, J.C. Launay, and R. Berger. *J. Materials Science Letters*, 15:363, 1996.
179. H. Bou Rjeily, F. Ramaz, D. Petrova, M. Gospodinov, and B. Briat. *SPIE*, 3178:169, 1997.
180. J.F. Carvalho, R.W.A. Franco, C.J. Magon, L.A.O. Nunes, and A.C. Hernandes. *Optical Materials*, 13:333, 1999.
181. R. Capelletti, P. Beneventi, L. Kovács, and A. Ruffini. *Ber. Bunsenges. Phys. Chem.*, 101:1282, 1997.
182. A.A. Grabar. *Ferroelectrics*, 192:155, 1997.

183. A. Rüdiger. PhD thesis, University of Osnabrück, 2001.
184. K. Küpper, B. Schneider, V. Caciuc, M. Neumann, A.V. Postnikov, and A. Rüdiger. *Phys. Rev.*, B 67:115101, 2003.
185. S. Odoulov, A.N. Shumelyuk, U. Hellwig, R.A. Rupp, A.A. Grabar, and I.M. Stoika. *J. Opt. Soc. Am.* B, 13:2352, 1996.
186. W. Krolikowski, M. Cronin-Golomb, and B.S. Chen. *Appl. Phys. Lett.*, 57:7, 1990.
187. X. Yue, S. Mendricks, Y. Hu, H. Hesse, and D. Kip. *J. Appl. Phys.*, 83:3473, 1998.
188. M. Pape, H.J. Reyher, and N. Hausfeld. *J. Phys.: Condens. Matter*, 13:3767, 2001.
189. J. Wingbermühle, M. Meyer, O.F. Schirmer, R. Pankrath, and R.K. Kremer. *J. Phys.: Condens. Matter*, 12:4277, 2000.
190. K. Megumi, H. Kozuga, M. Kobayashi, and Y. Furukawa. *Appl. Phys. Lett.*, 30:631, 1977.
191. R. Neurgaonkar, W. Cory, J. Oliver, W. Hall, and M. Ewbank. *Opt. Engineering*, 26:3392, 1987.
192. G. Greten, S. Hunsche, U. Knüpfer, S. Pankrath, U. Siefker, N. Wittler, and S. Kapphan. *Ferroelectrics*, 185:289, 1996.
193. B. Sugg, N. Nürge, B. Faust, E. Ruza, R. Niehüser, H.J. Reyher, R.A. Rupp and. L. Ackermann. *Opt. Materials*, 4:343, 1995.
194. H.J. Reyher, B. Faust, B. Sugg, R. Rupp, and L. Ackermann. *J. Phys.: Condens. Matter*, 9:9065, 1997.
195. G. Montemezzani, St. Pfändler, and P. Günter. *J. Opt. Soc. Am.* B, 9:110, 1992.
196. E. Moya, L. Contreras, and C. Zaldo. *J. Opt. Soc. Am.* B, 5:1737, 1988.
197. E. Moya, C. Zaldo, F.J. López, B. Briat, and V. Topa. *J. Phys. Chem. Solids*, 54:809, 1993.
198. C. Zaldo, E. Moya, L.F. Magana, L. Kovács, and K. Polgár. *J. Appl. Phys.*, 73:2114, 1993.
199. L. Kovács, E. Moya, K. Polgár, F.J. López, and C. Zaldo. *Appl. Phys.* A, 52:307, 1991.
200. C. Zaldo and E. Diéguez. *Opt. Materials*, 1:171, 1992.
201. D. Bravo and F.J. López. *Opt. Materials*, 13:141, 1992.
202. B. Briat, V. Topa, and C. Zaldo. In O. Kanert and J.M. Spaeth, editors, *Defects in Insulating Materials*, p. 1157. World Scientific, 1993.
203. I. Földvári, H. Liu, R.C. Powell, and Á. Péter. *J. Appl. Phys.*, 71:5465, 1992.
204. I. Földvári, C. Denz, Á. Péter, J. Petter, and F. Visinka. *Optics Commun.*, 177:105, 2000.
205. G. Berger, C. Denz, I. Földvári, and Á. Péter. *J. Opt. A: Pure and Appl. Opt.*, 5:S444, 2003.
206. Á. Péter, O. Szakács, I. Földvári, L. Benes, and A. Munoz. *Materials Research Bulletin*, 31:1067, 1996.
207. I.I. Földvári, L.A. Kappers, R.H. Bartram, and Á. Péter. *Opt. Materials*, 10:47, 1998.
208. H. Nagata, T. Sakamoto, H. Honda, J. Ichikawa, E.M. Haga, K. Shima, and N. Haga. *J. Mater. Res.*, 11:2085–2091, 1996.
209. R. González, E.R. Hodgson, C. Ballesteros, and Y. Chen. *Phys. Rev. Letters*, 67:2057–2059, 1991.
210. M. Wöhlecke and L. Kovács. *Critical Reviews Solid State Materials*, 26:1–86, 2001.
211. S. Klauer, M. Wöhlecke, and S. Kapphan. *Phys. Rev. B*, 45:2786–2799, 1992.

212. S. Kapphan and A. Breitkopf. *phys. stat. sol. (a)*, 133:159–166, 1992.
213. R. Gonzalez. Hydrogen-related effects in oxides. In G. Borstel, A. Krumins, and D. Millers, editors, *Defects and Surface-Induced Effects in Advanced Provskites*, pp. 305–316, Dordrecht, 2000. NATO Science series, 3. High Technology; Vol. 77, Kluwer Acad. Publ.
214. K. Buse, S. Breer, K. Peithmann, S. Kapphan, M. Gao, and E. Krätzig. *Phys. Rev. B*, 56:1225, 1997.
215. J.M. Cabrera, J. Olivares, M. Carrascosa, J. Rams, R. Müller, and E. Diéguez. *Advances in Physics*, 45:349, 1996.

3

Recording Speed and Determination of Basic Materials Properties

Ivan Biaggio

Department of Physics and Center for Optical Technologies, Lehigh University, Bethlehem, PA 18015, U.S.A. `biaggio@lehigh.edu`

This chapter presents a discussion of the recording speed in photorefractive materials, the parameters that influence it under various circumstances of continuous wave and pulsed recording, and the experimental conditions under which the recording speed is determined in the simplest way by readily identifiable material parameters. This leads to all-optical techniques for the measurement of the photoconductivity, the diffusion length, the free-carrier lifetime, and the free carrier mobility. Two useful examples that will be discussed at greater length are photoinduced space-charge relaxation (PSCR) using continuous-wave excitation at long grating spacing to determine the conductivity, and holographic time of flight (HTOF) using short-pulse excitation at short grating spacings to determine the free carrier mobility.

3.1 Introduction

The observation of the speed of photorefractive recording and erasure under carefully controlled experimental conditions is an important all-optical technique to determine selected material properties. Although other characteristics of the photorefractive response can be used to determine material properties (the reader may be familiar with the measurement of the photorefractive gain in two-beam coupling to study the electrooptical properties and the effective trap density), this chapter focuses on the response time exclusively. The advantage of studying the response time is that it can open more diverse windows on material parameters, depending on experimental conditions, and that it is often easier to measure than, e.g., absolute diffraction efficiencies.

The manner in which the photorefractive response time depends on material parameters is fundamentally related to the origins of the photorefractive effect. In contrast to, e.g., off-resonant third-order nonlinearities, where three photons combine to produce a fourth one in an interaction that is basically a perturbation of the ground state of matter, the photorefractive effect intrinsically relies on the excitation of electrons to higher excited states, typically from an impurity level into the conduction or valence band of an insulator or semiconductor. As such, the speed at which a certain refractive index change is produced depends both on how fast the required number of charge carriers is excited and on how fast they are able to move to establish a space-charge modulation.

This chapter builds on the material presented in Volume I of this book series, but makes a special effort to develop the fundamental understanding that is so important for an effective application of these techniques to fundamental research. Because of this, it adopts a didactic style that should make it also accessible to those wishing to enter the field or to scientists in other fields interested in the photorefractive effect as a material characterization tool. The basic band transport model with one donor/trap level will be used to highlight the experimental conditions that allow limiting cases to emerge, in which few specific material parameters determine the photorefractive response. Under these circumstances the photorefractive response becomes very simple, and therefore much of this chapter will concentrate on a detailed and careful discussion of the simplest expressions for the photorefractive response time. This is due not only to the choice of making this chapter accessible to newcomers to the field, but also to the fact that it is through these simple expressions that the best ways to determine material parameters are realized.

It is an aim of this chapter to provide both a review of the fundamental properties of the photorefractive response time under different experimental conditions and a quick-start guide to using the response time as an investigative tool.

It is not the aim of this chapter to be a review of all the experimental or theoretical work on photorefractive time dynamics, especially when it relates to experimental conditions that are intermediate between the limiting cases discussed here. Even though some references to past contributions to the field will appear, they will do so only when they give a good example of the particular behavior that is being discussed, and the list is far from comprehensive. The reader is referred to Volume I of this book series for detailed derivations of some of the expressions presented in this chapter.

3.1.1 Experimental Parameters and Material Parameters

For the purpose of the determination of basic material properties, one needs to identify regions in the space of *experimental parameters*—laser beam intensities and wavelength, exposure time, grating wave vector—that make some characteristics of the photorefractive response dependent on only a subsection of *material parameters*—charge-carrier mobilities, dielectric properties, trap and donor structure.

A discussion and careful selection of these limiting cases in the choice of experimental parameters will be useful both for highlighting and understanding the mechanisms that determine the photorefractive response time and to identify the experiments that can be used to study some material parameters without interference from the others. A summary of the experimental regimes that will

TABLE 3.1. Regions in experimental parameter space and the corresponding material parameters that can be investigated

Excitation type	Measurement conditions	Material parameter
continuous wave	long grating spacings	photoconductivity
	varying grating spacings	diffusion length
		effective trap density
	varying grating orientation	mobility anisotropy
	at long grating spacings	elastic contributions
		to dielectric constant
short pulse	short grating spacings	carrier mobility
	long grating spacings	carrier lifetime

be discussed in this chapter and of the related material parameters that can be determined is given in Table 3.1 As seen from the table, major subdivisions in experimental parameter space are characterized by the dichotomies *short grating spacings/long grating spacings*, and *short-pulse recording/continuous-wave recording*.

It is especially useful to contrast the two limiting cases of photorefractive recording in which photoexcitation is going on all the time—continuous-wave (cw) illumination—and in which the photoexcitation happens once during a very short time and is immediately switched off—short-pulse illumination. "Short-pulse" for the purposes of this chapter means that no significant charge transport can take place during the pulse length. Under cw illumination, excitation and charge displacement happen concurrently, and the response time is given by the time needed to reach equilibrium between excitation, charge carrier diffusion and drift, and recombination. In the limit of short-pulse illumination, photoexcitation is immediate, and the response time depends on the efficiency of charge transport. The interest of contrasting cw recording and short-pulse recording is highlighted by a typical duality that can be observed in large regions of experimental parameter space. For cw recording, the response time depends on the intensity, while the steady-state grating amplitude does not. For short-pulse recording it is the other way around: the response time does *not* depend on intensity, but the photoinduced transient grating amplitude does. From these fundamental qualitative differences between the two cases one can already suspect that they can provide interesting complementary insights into different material properties.

For each different recording condition, the limits of short and long grating spacings allow one to distinguish between the different material parameters that influence the photorefractive response. As an example, the limit of long grating spacings highlights the influence of the (photo-) conductivity for cw recording, and of the free-carrier lifetime for short-pulse recording. Conversely, one can choose to work at the very short grating spacings obtained with almost counterpropagating beams to highlight the influence of the effective trap density and photoexcitation probability for cw recording, or the influence of the free-carrier mobility for short-pulse recording.

3.1.2 Overview

The next section presents a review of the basic theoretical models used to describe the photorefractive response time under both cw and pulsed conditions. Pulsed and cw response will be presented in two separate subsections with closely matched structures, which should make it easy to compare these completely different mechanisms of photorefractive grating buildup. Section 3.2.1 discusses the photorefractive response time under continuous-wave recording and erasure conditions, including the short and long grating spacing limits, and a discussion of possible fundamental limits to the speed of the photorefractive response. A very short overview of the effects of an applied electric field on the time dynamics is also included. Section 3.2.2 goes on to describe what happens when photoexcitation is confined to an ultrashort time interval and the buildup of a photorefractive grating is a one-shot event that happens in the dark. A short discussion of speed limits and the effects of an electric field is included in this section, too.

The second part of this chapter is dedicated to the actual investigative techniques that can be used to determine important material parameters and that are based on studying the photorefractive time dynamics under carefully selected experimental conditions. Mirroring the theoretical part, the investigative techniques will be presented according to the type of excitation. Section 3.3.1 presents cw photoexcitation techniques, while Section 3.3.2 presents techniques relying on pulsed excitation. The discussion will concentrate on two particular examples in which cw and pulsed recording are exploited most efficiently: photoinduced space-charge relaxation (PSCR) to determine photoconductivities and their anisotropies and holographic time of flight (HTOF) to measure free-carrier mobilities and lifetimes separately.

3.2 Theoretical Review

This section provides a discussion of the phenomena that determine the photorefractive response time under various experimental conditions. This will be done by describing the basic mechanisms and by highlighting the experimental conditions that allow limiting cases to emerge. To do this, it is sufficient to consider the simplest band-transport model for the photorefractive effect. The following discussions will assume one type of charge carrier (electrons) and one impurity level that has N_d occupied levels and N_A empty levels in the dark at zero temperature, corresponding to an effective trap density $N_{eff} = N_d N_A/(N_d + N_A)$. The illumination pattern that creates the photorefractive grating is described by

$$I(z) = I_0[1 + m \cos(kz)], \tag{3.1}$$

where z is the coordinate along the direction of the grating wave vector $\mathbf{k}$, m is a modulation index, and I_0 is the average intensity in the interference pattern. The following discussion will emphasize the experimental limits where any change of I_0 or m that occurs at a time $t = 0$ will result in a modification of the electric space-charge field amplitude that occurs with an exponential time dynamics:

$$E_{sc}(t) - E_{sc}(0) = E_{sc}\left[1 - e^{-t/\tau}\right],\tag{3.2}$$

where E_{sc} is the photoinduced change in space-charge field amplitude and τ is a photorefractive response time that will be discussed for various experimental situations in the following. Note that in general, the space-charge field amplitude can be described by a complex-valued function and that the response time τ can become complex when an electric field is applied to the photorefractive crystal, leading to an oscillatory behavior in the buildup dynamics. In the following I will mostly concentrate on the cases in which there is no applied field and τ is real.

3.2.1 Continuous-Wave Recording

When a photorefractive material is exposed to the interference pattern between two laser beams with an intensity constant in time, the phenomena of excitation, transport, and trapping start to occur at the same time. A spatially inhomogeneous charge distribution matching the interference pattern starts to grow, and will keep growing until an equilibrium is reached that matches the exposure conditions (given by the intensities of the two writing beams and the grating spacing). The spatial dependence of the light intensity and the corresponding photorefractive grating can be approximated by a plane wave with a wave vector k, as used in (3.1). In case of a subsequent change in the illumination level, the system will move from the old equilibrium with a space-charge field amplitude $E_{sc}(0)$ to a new equilibrium condition with a space-charge field amplitude $E_{sc}(\infty)$. In the limit where both spatial and temporal changes in the exposure pattern are a small perturbation of an established equilibrium the approach to equilibrium is described by (3.2), with $E_{sc} = E_{sc}(\infty) - E_{sc}(0)$, both when the change in exposure causes a decay of the grating amplitude and when it causes a buildup. The rise time and decay time of a photorefractive grating under cw illumination are identical, provided that the average light intensity during writing and erasing is the same. As shown in Volume I of this book series, the photorefractive response time for cw recording and in the absence of any applied electric fields is given by

$$\tau_{cw} = \tau_{die}\frac{1 + k^2/k_d^2}{1 + k^2/k_0^2},\tag{3.3}$$

where

$$\tau_{die} = \frac{\epsilon_0\epsilon_{eff}}{\sigma},\tag{3.4}$$

$$k_d^2 = \frac{e}{\mu\tau_0 k_B T},\tag{3.5}$$

$$k_0^2 = \frac{e^2 N_{eff}}{\epsilon_{eff}\epsilon_0 k_B T}.\tag{3.6}$$

Here, k is the modulus of the grating wave vector k, ϵ_0 is the permittivity of vacuum, $\epsilon_{eff}(k)$ is the effective dielectric constant [1, 2], $\sigma(I_0)$ is the total average conductivity for an average illumination intensity I_0, e is the unit charge, μ is the mobility of the charge carriers, τ_0 is their recombination time, k_B is Boltzmann's constant, and T is the absolute temperature.

The grating spacing dependence of the response time (3.3) is best expressed as a function of the grating wave vector k, which must be compared to the inverse of two well-known lengths, the diffusion length $k_d^{-1} = \sqrt{D\tau_0}$ ($D = \mu k_B T/e$ is the diffusion constant) and the Debye screening length $k_0^{-1} = \sqrt{D\epsilon_0 \epsilon_{eff}/(e\mu N_{eff})}$. The inverse diffusion length k_d depends only on the product between effective carrier lifetime τ_0 and mobility μ. The inverse Debye screening length k_0, which will often be called Debye wave vector in the following, depends on the effective trap density and the effective dielectric constant.

In the above expressions τ_{die} is the dielectric relaxation time. It depends on the dielectric properties of the material and on the average conductivity σ, which in general contains contributions from both the dark conductivity and the photoinduced conductivity. The dielectric properties determine the amplitude of the electric field induced by a given space-charge modulation. Since the space-charge is modulated like a plane wave in space, there are constraints on the stress and strain pattern that is established in a piezoelectric crystal by the space-charge field. Because of these constraints, the appropriate value of the dielectric constant that one must use in this case is neither the strain-free one nor the stress-free one, but is something in between, an *effective dielectric constant* ϵ_{eff} [1, 2].

It is important to note that, although the above expression for the photorefractive response time is derived from the simplest band transport model with only one level responsible for charge carrier generation and trapping, its form can be applied in more general cases to describe small changes around an equilibrium state. The only requirement to be able to analyze the response time in more complex systems using (3.3) is to establish a well-defined initial condition and measure the response time when moving to a slightly different illumination condition. Depending on the situation, this can be done in several different ways, of which the following are just a few:

- Illuminate a crystal with a strong beam, switch on a second, weaker beam, to write a photorefractive grating, and observe the dynamics of the diffraction efficiency of a probe beam.
- Write a photorefractive grating in the presence of a strong background illumination. Observe the dynamics of beam coupling or diffraction efficiency when abruptly changing the background illumination by a small amount.
- After having written a photorefractive grating with small modulation, substitute the two writing beams with a homogeneous erasing beam away from the Bragg angle, and observe the grating decay by Bragg diffraction of a weak probe beam.
- Write a photorefractive grating with small modulation, and use an electro-optic modulator or a mirror mounted on a piezoelectric crystal to vibrate the phase of one beam faster than any photorefractive response time. Stop and restart the vibration to observe how the photorefractive grating builds up or decays (in this experiment the average intensity is always the same).

The main aim of such experimental methods is always to enforce small intensity modulations both in space and in time. To understand this issue, consider the opposite case of a photorefractive grating written by illuminating a crystal that was previously in the dark. The amplitude of the photorefractive grating will grow exponentially with a time constant (3.3) only if a local equilibrium between photoexcitation and recombination, leading to a constant average conductivity, can be established in a time much shorter than the photorefractive response time. For the simplest model with only one trap level, the time needed to establish a constant photoconductivity is essentially given by the recombination time τ_0, and it can be very short. However, in the presence of multiple levels, equilibrium between photoexcitation and recombination (plus thermal excitation from shallow traps and any other excitation/trapping dynamics) may need a longer time to be established. This can lead to an additional time dependence of the photoconductivity, complicated nonexponential buildups, a sublinear relationship between photoconductivity and light intensity, and an intensity dependence of both the diffusion length and the Debye wave vector. But even for a situation like this the relaxation time measured for small changes around a given illumination intensity can often be described by (3.3). In such a case quantities such as the Debye wave vector and the diffusion length will correspond to some appropriate intensity-dependent averaging of the various excitation and recombination processes, but they will still be able to broadly describe the grating spacing dependence of the photorefractive response time as described above. This applies in particular to the conductivity, which can always be determined by measuring the photorefractive response time in the long-grating spacing limit.

Further insight into the different forces that determine the speed of the photorefractive response can be gained by considering the limiting cases of short and long grating spacings. This will be done in the next two subsections before we move on to a discussion of the full grating-spacing dependence of the response time.

Short Grating Spacings

At grating wave vectors much longer than both the Debye wave vector k_0 and the inverse diffusion length k_d, the photorefractive response time becomes

$$\tau_{cw}(k \gg k_0, k_d) = \tau_{ex} = \tau_{die}\frac{k_0^2}{k_d^2} = \frac{N_{eff}\tau_0}{n_c}, \qquad (3.7)$$

where n_c is the total density of charge carriers. In this limit $\tau_{ex}/\tau_{die} = eN_{eff}\mu\tau_0/(\epsilon_{eff}\epsilon_0)$ and depending on the material parameters, τ_{ex} can be either larger or smaller than the dielectric relaxation time.

The density n_c appearing in (3.7) consists in general of those charge carriers that are already present without illumination (corresponding to the dark conductivity) and of those that have been photoexcited (photoconductivity). But assuming that the dark conductivity is negligible, n_c is given by the equilibrium of photoexcitation and recombination. Writing the electron photoexcitation rate as sI_0N_d, with s a photoexcitation probability rate, I_0 the light intensity, and N_d the density of bound electronic states that can be photoexcited to the conduction band, and setting it equal to the recombination rate n_c/τ_0, one gets $n_c = \tau_0 s I_0 N_d$ and the short grating spacing limit of the photorefractive response time can be written as

$$\tau_{ex} = \frac{1}{s I_0 N_d} N_{eff} = \frac{\hbar\omega}{I_0 \alpha_p} N_{eff}, \qquad (3.8)$$

where $\alpha_p = \hbar\omega s N_d$ is the absorption coefficient caused by the charge carrier photoexcitation process, and $\hbar\omega$ is the energy of a photon.

The time τ_{ex} can be recognized as an excitation time. In this limit the photorefractive response time corresponds to the time it takes to photoexcite a number of charges equal to the effective trap density. Note that this does not mean that it is really necessary to excite and displace this number of charges to reach the new equilibrium state. In fact, the above expression is valid in the limit where the modulation in the interference pattern is arbitrarily small and the space-charge modulation corresponds to an arbitrarily small number of displaced charges. The meaning of (3.8) is only that the time needed to reestablish equilibrium between excitation, diffusion, drift, and trapping depends on the effective trap density compared to the excitation rate $s I_0 N_d$ (or the recombination rate n_c/τ_0).

Long Grating Spacings

At grating wave vectors much shorter than both the Debye wave vector k_0 and the inverse diffusion length k_d, the photorefractive response time becomes equal to the dielectric relaxation time. There is a well-known analogy between this long grating spacing limit and a system consisting of a series of capacitors connected by wires with constant resistivity where the dielectric relaxation time is recognized as an "RC time constant."

As for the excitation time, by neglecting dark conductivity and assuming a charge carrier excitation rate proportional to the intensity, one finds that the dielectric relaxation time is inversely proportional to the light intensity,

$$\tau_{die} = \frac{\epsilon_0 \epsilon_{eff}}{e\mu n_c} = \frac{\epsilon_0 \epsilon_{eff}}{e\mu \tau_0 s I_0 N_d}, \qquad (3.9)$$

but in contrast to the excitation time it is also inversely proportional to the mobility-lifetime product and depends on the charge transport efficiency. However, for purposes of material characterization it is better to keep writing the dielectric relaxation time as in (3.4), which highlights the fact that its determination can be used to accurately measure the conductivity σ. As mentioned above, this is always true in the long grating spacing limit, even when the photorefractive effect is determined by multiple trap levels and when the photoconductivity is not linearly proportional to the intensity. A measurement of the photorefractive response time as a function of average light intensity is the simplest way to determine the intensity dependence of the total conductivity.

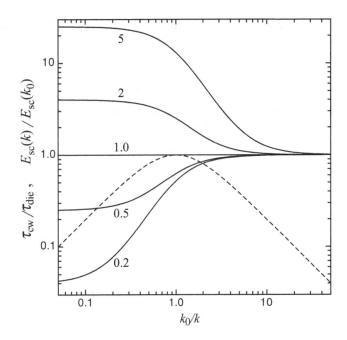

FIGURE 3.1. Solid curves: Normalized photorefractive response time, in units of the dielectric relaxation time, vs. the normalized grating spacing. The curve parameter is the ratio k_0/k_d between diffusion length and Debye screening length. The response time at long grating spacings is in all cases given by the dielectric relaxation time. Dashed curve: Normalized space-charge field amplitude vs. the normalized grating spacing. The normalized grating spacing is expressed as k_0/k and is in units of 2π times the Debye screening length.

Grating Spacing Dependence

The grating spacing dependence of the photorefractive response time (3.3) is shown in Figure 3.1 for different values of the ratio between diffusion length and Debye screening length. For completeness, Figure 3.1 also plots the space-charge field amplitude for the simplest band-transport model, which can be written as

$$E_{sc} = m E_q \frac{1}{1 + k_0^2/k^2}, \tag{3.10}$$

where m is the modulation index in the interference pattern and $E_q = eN_{eff}/(k\epsilon_0\epsilon_{eff})$ is the well-known "trap-limited field." The grating amplitude reaches a maximum when the grating wave vector k equals the Debye wave vector k_0, at which point the trap-limited field becomes equal to the "diffusion field" $E_d = kk_B T/e$.

In order to understand the grating spacing dependence of the response time it is useful to consider some alternative ways of writing (3.3). First, using $\tau_{ex}/\tau_{die} = k_0^2/k_d^2$ it is possible to write (3.3) in a manner that highlights how the photorefractive response time changes with respect to the excitation time [instead of the dielectric relaxation time as in (3.3)]:

$$\tau_{cw} = \tau_{ex} \frac{1 + k_d^2/k^2}{1 + k_0^2/k^2}. \tag{3.11}$$

In addition, it is also possible to rewrite (3.3) and (3.11) in a way that highlights the regions where the photorefractive response time is more influenced by the grating wave vector. The result is

$$\frac{\tau_{cw}}{\tau_{die}} = 1 + \frac{k^2/k_0^2}{1 + k^2/k_0^2} \left[\frac{k_0^2}{k_d^2} - 1 \right], \tag{3.12}$$

$$\frac{\tau_{cw}}{\tau_{ex}} = 1 + \frac{k_0^2/k^2}{1 + k_0^2/k^2} \left[\frac{k_d^2}{k_0^2} - 1 \right]. \tag{3.13}$$

These expressions show that the Debye wave vector, besides determining the grating spacing at which the space-charge field amplitude is a maximum, also determines the grating spacing at which the photorefractive response time is midway between excitation time and dielectric relaxation time. This fact is not immediately seen from the log-log plots of Figure 3.1, but is evident from (3.12) and (3.13): The factor multiplying the expressions in square brackets on the right-hand sides of (3.12) and (3.13) varies between 0 and 1 and is equal to $\frac{1}{2}$ at $k = k_0$.

Thus, the plots in Figure 3.1 and the above expressions show that a measurement of the photorefractive response time vs. grating spacing can deliver the ratio k_d/k_0 when measurements at very short and very long grating spacings are compared, but it can also provide a way to determine k_0—and thus the effective trap density— if the measurements can be performed at grating wave vectors around k_0.

Speed Limits

Identifying a fundamental limit to the speed of the photorefractive effect is attractive because it would make it possible to compare the potential for improvement in different photorefractive materials. From the above treatment it is clear that for a given intensity, the photorefractive response time becomes shorter for high photoconductivities, small effective trap densities, and large photoexcitation probabilities. However, these are all material-dependent quantities. Is there a fundamental limit for the shortest response time that can ever be achieved, in any material?

One could argue that the buildup time of a given space-charge modulation is limited by the time needed to photoexcite enough charges. This argument was used by Yeh to introduce his definition of a fundamental limit of the speed of photorefractive effects [3]. The minimum time needed to excite a number of carriers N_p is similar in form to the excitation time (3.8) discussed above:

$$\tau_{Yeh} = \frac{\hbar\omega}{I_0 \alpha_p} N_p. \tag{3.14}$$

Yeh defined N_p as the number of carriers corresponding to the grating amplitude that is required for a practical application. The resulting time is inversely proportional to the light intensity, and there is no question that it represents a true fundamental limit. However, this limit may be *too short*, and may therefore not be very useful for judging the efficiency of a material. As an example, if the given space-charge modulation is achieved very early during the buildup of a space-charge grating, this limit can become much smaller than the photorefractive time introduced above, which is the time needed to reach steady state. Note that using the steady-state charge modulation to calculate N_p in (3.14) delivers a time that is also a fundamental limit to the time needed to reach steady state, but that can still become arbitrarily small for small modulations, while the photorefractive response time does not depend on the modulation.

Glass et al. rejected the fundamental limit proposed by Yeh on the grounds that it does not depend on charge *transport* properties [4], and they concentrated their discussion on the time needed by a charge carrier to travel half a grating spacing. This time would correspond to the response time if all charge carriers were excited instantaneously and traveled independently afterwards, and it can be associated with the diffusion time discussed below in the section on short-pulse excitation [see (3.16)]. In contrast to the fundamental limit proposed by Yeh, this limit does not depend on the photoexcitation process and is based only on charge-transport efficiency. But like Yeh's fundamental limit, it can be much smaller than the real photorefractive response time. It can range from picoseconds to nanoseconds in most common photorefractive materials from semiconductors to oxides, depending on charge-carrier mobility.

Thus, both the time introduced by Yeh [3] (time needed to photoexcite a given charge density) and that introduced by Glass et al. [4] (time needed to move a charge by half a grating spacing) are true lower limits to the photorefractive response time, but they can be quite far from the real response times observed in cw experiments. In addition, the physical models with which they were derived are far from what really goes on under continuous-wave illumination, where the response time is determined by an approach to an equilibrium. The discussion between Yeh and Glass et al. [3, 4] is a good example of the pitfalls of arguments based on attractive, but oversimplified, approaches. The real issues with the photorefractive response time are much better described by the simple expression (3.3) and by its two limiting cases (3.7) and (3.4): The lower limit to the photorefractive response time is the dielectric relaxation time for materials with a long diffusion length, where it is observed at long grating spacings, and it is the excitation time for materials with a short diffusion length, where it is observed at short grating spacings.

The excitation time (3.8) is similar in concept to the time introduced by Yeh [3], because it is also equal to the time needed to excite a specific number of charges. But in (3.8) this number of charges is given by the effective trap density, not by a more or less arbitrary choice of grating amplitude. It follows that the excitation time (3.8) does not depend on the modulation index in the interference pattern, and depends only on the average light intensity and on material parameters related to the impurity levels that enable the photorefractive effect in the first place. It is valid at short grating spacings and it corresponds to the lower limit of the photorefractive response time for all materials where $k_0 < k_d$ (that is, where the diffusion length is smaller than the Debye screening length). These materials are typically represented by as-grown $BaTiO_3$. But notice that in this case, (3.8) is the lower limit only because the mobility-lifetime product in these materials is so small that it makes charge transport inefficient. To improve on the photorefractive response time it is in general necessary to improve the mobility-lifetime product of the photoexcited carriers. An ideal material from the point of view of photorefractive speed will have a large diffusion length, and in such a case the minimum photorefractive response time is the dielectric relaxation time (3.4). This is the case for semiconductors like CdTe and GaAs, which have large charge-carrier mobilities, and also for the sillenite $Bi_{12}SiO_{20}$, which has smaller charge-carrier mobilities but long (effective) charge-carrier recombination times.

One concludes that the speed limit for optimized photorefractive materials is the dielectric relaxation time (3.4), and it is determined by the photoconductivity. Obviously, in addition to this generalization one also has to consider the particular application in which the speed needs to be optimized. As an example, for high-density holographic recording one might be more interested in the excitation time (3.8) observed at short grating spacings, and consider the tradeoff between a high effective trap density (needed to obtain a large space-charge field amplitude at small grating spacings) and a longer response time (the excitation time is proportional to the effective trap density).

The Influence of an Applied Electric Field

The photorefractive response time introduced above is valid in the absence of applied electric fields. The effect of the electric field complicates the situation by changing the simple exponential response that is observed without an applied field, with the real-valued time constant (3.3), into an oscillatory buildup dynamics where the time constant becomes complex. An expression for this complex time constant and its dependence on the applied field can be found elsewhere. A discussion of some limiting cases can be found in [5]. Here is a summary of the basic characteristics of the photorefractive response in the presence of an applied electric field:

– The response time for long grating spacings is still given by the dielectric relaxation time, also in the presence of an applied electric field.
– At shorter grating spacings, an applied electric field of magnitude E starts having an influence on the response speed when the drift length $\ell_E = \mu\tau_0 E$ becomes larger than the diffusion length k_d^{-1}.
– The photorefractive speed can be increased or decreased by an applied field, depending on the relative size of diffusion length k_d^{-1} and Debye screening length k_0^{-1}. For $k_0 > k_d$ (large diffusion length) the response becomes slower.

An acceleration of the response is seen only for small mobility-lifetime products $(k_0 < k_d)$.

- In order to observe an oscillatory behavior for the buildup of a photorefractive grating, the drift length must also be significantly larger than the grating spacing, but this is not the only condition. For too-large drift lengths the oscillatory behavior disappears again. The same happens for too-large diffusion lengths or when the Debye screening length is too close to the diffusion length.
- The limiting value for the response time when the applied field goes to infinity is the excitation time (3.7), the same as the short grating spacing limit with no applied field.

As can be seen by the behavior described in this short list, there is no region in experimental parameter space that makes the (continuous-wave) photorefractive response time in the presence of an applied field particularly easy to interpret in relation to only one material parameter, or even a few of them. Effective trap density and mobility-lifetime product interact in complex ways to determine the kind of response observed in the presence of an applied field. Even though experiments with an applied field can give complementary information as to, e.g., the relative size of diffusion and drift length, there is no simple way to extract a material parameter from the observation of the photorefractive response time alone. For this reason, the case of an applied electric field under cw illumination is not discussed further in this chapter.

3.2.2 Short-Pulse Recording

The preceding section showed that the cw response depends on the equilibrium between several processes that are going on concurrently, and as such does not allow the separate determination of fundamental material parameters such as the carrier lifetime and carrier mobility.

The interest of short-pulse recording for material characterization is the possibility to temporally separate the photoexcitation process from the transport process and simplify both modeling and interpretation of the effect and its dynamics. This temporal separation occurs whenever photoexcitation is caused by a laser pulse that is so short that the photoexcited carriers do not have the time to move during the pulse duration.

This section will focus on this aspect, discussing the limit of laser pulses that can be considered delta functions in time from the point of view of the other processes that lead to a photorefractive grating. The reader is alerted to the fact that the case of longer pulses that are intermediate between the short-pulse limit considered in this section and cw recording can be much more difficult to interpret. The same can be said for the case in which the photorefractive response to a train of pulses is to be studied. Here, I prefer not to consider these cases at all. The main reason for this is the usual one: experiments performed under such conditions cannot deliver a limit for which the dynamics can be simply related to only a few material parameters.

In contrast to the previous section on continuous-wave recording, which discussed a general photorefractive response time that can be observed both in writing a photorefractive grating and in erasing it, this section is completely devoted to the *writing* of a photorefractive grating by short pulses. The decay of the photorefractive grating that happens later is governed by the same mechanisms that determine the decay of gratings written under cw illumination, and will not be discussed further here.

In the limit of short pulse excitation discussed here, the interference pattern flashes on and off to set up a modulated distribution of charge carriers that acts as the initial condition for a subsequent development of a space-charge field. The typical signature of this effect, which consists in the buildup of a photorefractive grating taking place in the dark, has been observed in several materials [6, 7, 8, 9, 10, 11].

When an otherwise undisturbed crystal is exposed to the interference pattern between two short pulses, diffusion of the photoexcited carriers takes place after the writing pulses left the crystal and leads to an exponential growth of the space-charge modulation [12, 13, 8, 11, 14] of the form (3.2), with an exponential time constant

$$\tau_{pulsed} = \left(\tau_d^{-1} + \tau_0^{-1}\right)^{-1}, \tag{3.15}$$

where

$$\tau_d = \frac{e}{k^2 \mu k_B T} = \frac{1}{k\mu E_d} \tag{3.16}$$

is the diffusion time, which can be expressed in terms of the well-known diffusion field $E_d = k_B T k/e$, and τ_0 is the free-carrier lifetime.

Using $\tau_0/\tau_d = k^2/k_d^2$ it is possible to write (3.15) as

$$\tau_{pulsed} = \tau_0 \frac{1}{1 + k^2/k_d^2}, \tag{3.17}$$

which has the same form as the limit of (3.3) for $k_d \gg k$, but depends on the diffusion length instead of the Debye screening length. The buildup described by (3.15) or (3.17) is a one-shot event that happens after illumination with one pulse, and depends only on how fast the photoexcited carriers diffuse, and how long their lifetime is. The space-charge grating starts growing immediately after the writing pulses have left the crystal, and will go on growing until there are free charges, or until free-carrier diffusion becomes compensated by the drift caused by the growing space-charge electric field.

Expression (3.15) for the buildup time is valid provided that [11, 14]:

1. The photoexcitation pulses are so short that no charge transport takes place during the pulse duration.
2. Only one type of mobile carrier is excited by photoionization of impurity centers with energies in the gap between valence and conduction band. Photoexcitation takes place in the bulk of the material.

3. The density of photoexcited carriers is so small that there are no saturation effects during photoexcitation and that the space-charge field that is built up by subsequent charge transport has a negligible influence on the transport dynamics itself.

4. The average lifetime τ_0 of the photoexcited carriers is a constant (i.e., the distribution of any trapping centers can be assumed to be homogeneous).

The above conditions can be easily fulfilled by appropriately selecting the experimental parameters. For a more detailed discussion of the range of validity of these assumptions and how to enforce them, see [11] and [14]. Here it is useful to discuss requirement 3 and the meaning of the free-carrier lifetime appearing in requirement 4, since they relate to the differences between short-pulse recording and cw recording.

Requirement 3 means that a laser pulse of low enough energy must be used to write the grating. After the writing pulses have left the crystal, the free-carrier distribution that is left behind will have a given average number of carriers n_c and a given modulation index m. The maximum space-charge field that can be built up by carrier diffusion in this situation corresponds to the case in which the free-carrier distribution becomes homogeneous and is given by mE_{ph}, where

$$E_{ph} = \frac{en_c}{\epsilon_{eff}\epsilon_0 k} \tag{3.18}$$

is a "photon-limited field" that plays the role of the trap-limited field for pulsed experiments. But in contrast to the cw case it is not a material parameter. It depends on the average number of photoexcited carriers (instead of the effective trap density) and it can be tuned by the fluence in the writing pulse. Requirement 3 is fulfilled by choosing writing pulse fluences so low that $E_{ph} \ll E_d$. It is interesting to note that by rearranging terms this condition can also be written as $\tau_d \ll \tau_{die}$. This way of writing might appeal to somebody accustomed to cw photorefractive effects, but the dielectric relaxation time does not play any direct role for short-pulse recording (and the diffusion time does not play any direct role for cw recording). Comparing diffusion and dielectric time in this way (the mobility drops out of the comparison) does not relate as directly to short-pulse recording as comparing the field limited by the amount of photoexcited charges—the photon-limited field E_{ph}—to that limited by the equilibrium of diffusion and drift—the diffusion field E_d.

Requirement 4 makes it clear that short-pulse recording does not depend on the origin of the carrier lifetime. For a crystal that contains one single trap level, τ_0 is both the free-carrier lifetime and the recombination time that was introduced in the previous section about cw recording. However, because of the very common presence of shallow traps and multiple levels, this is very often not the case in practice. The free-carrier lifetime can be caused by trapping at impurities that are not related to those responsible for the electron excitation and recombination leading to a photorefractive grating in the cw case. It is not uncommon to observe in a short-pulse experiment a free-carrier lifetime that is shorter by orders of magnitude than the recombination time that would appear to be appropriate to describe the cw photorefractive response. This can be easily understood in the simple case of a crystal with a single deep trap level and a shallow trap level in which electrons can be captured and thermally released. In short-pulse experiments at long grating spacings, diffusion will lead to a decreasing modulation of the free-carrier distribution and to the growth of a space-charge field until all photoexcited carriers are captured by shallow traps. At this point the buildup described by (3.15) is finished. The thermal reexcitation time from shallow traps will be in general much longer than (3.15). In cw experiments the average photoexcited carrier would be trapped and thermally reemitted several times before it finally recombines into a deep trap, leading to a much longer deep trap recombination time and to the much smaller trap-limited mobility of an electron that is continually captured and thermally reemitted from shallow traps while it moves from one deep trap to another [15, 16, 17]. Interestingly, in this simple case of one shallow and one deep level it can be shown that the mobility-lifetime product calculated from trapping time into shallow traps and band mobility is the same as that calculated from deep trap recombination time and trap-limited mobility [18, 17]. This example shows that it is in general not advisable to try to relate the material parameters that determine the photorefractive grating buildup after a single-shot short-pulse excitation to those used to describe the cw photorefractive effect. One can argue that the "lifetime" concept used in cw experiments has to be seen from the point of view of the trapping centers where the space-charge modulation is stored, and can be described by a deep trap recombination time. On the other hand, the lifetime concept in the short-pulse experiments described here is to be seen from the point of view of a free carrier in the conduction or valence band, whose movement is stopped by any mechanism that transforms a mobile carrier into a practically immobile one.

Short Grating Spacings

At grating wave vectors much longer than the inverse diffusion length k_d the buildup time for pulsed illumination becomes equal to the diffusion time τ_d. In this limit the speed of the buildup is completely determined by the charge-carrier mobility and is not influenced by any other material parameter.

This is a significant simplification compared to the cw case, and it allows the determination of a fundamental parameter describing charge transport that cannot be accessed in cw photorefractive effects and can also be difficult to determine with any other technique.

Long Grating Spacings

At grating wave vectors much shorter than the inverse diffusion length k_d the buildup time is directly given by the carrier lifetime. This is again a spectacular simplification in comparison to the cw case. The lifetime determined in such an experiment is a real free-carrier lifetime that can be difficult to observe with any other experiment.

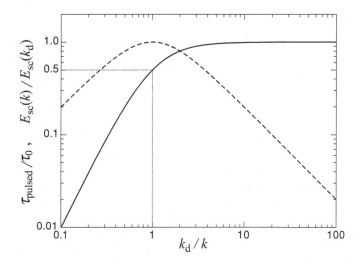

FIGURE 3.2. Solid curve: normalized buildup time, in units of the free-carrier lifetime, of a photorefractive grating after illumination with one short pulse. Dashed curve: normalized space-charge field amplitude vs. the grating spacing. The grating spacing is expressed as k_d/k. The space-charge field reaches a maximum when the grating spacing is equal to $2\pi k_d = 2\pi\sqrt{\mu\tau_0 k_B T/e}$. At the same grating spacing the buildup time is one-half of the free-carrier lifetime, observed at long grating spacings.

Grating Spacing Dependence

As a function of the grating spacing, the buildup time moves from the diffusion time observed at short grating spacings to the free-carrier lifetime observed at long grating spacings. Since the diffusion time is grating spacing dependent, there is no limit to the decrease of the buildup time when the grating spacing is decreased. Contrast this to the cw case, in which the response time tends toward the constant excitation time. For short-pulse experiments the excitation time does not have any meaning because all the relevant excitation takes place during the short pulse. Figure 3.2 shows the grating spacing dependence of the buildup time. An experimental confirmation of this behavior can be found in [9].

For completeness, Figure 3.2 also gives the grating spacing dependence of the space-charge field amplitude obtained after short-pulse exposure, which can be written as

$$E_{sc} = \frac{mE_{ph}}{1+\tau_d/\tau_0} = \frac{mE_{ph}}{1+k_d^2/k^2} = m\frac{en_c}{\epsilon_0\epsilon_{\mathit{eff}}}\frac{k}{1+k_d^2/k^2}, \tag{3.19}$$

where m is the modulation index in the interference pattern and E_{ph} is the photon-limited field introduced above. The grating amplitude reaches a maximum of $mE_{ph}/2$ when the grating wave vector equals the inverse diffusion length, $k = k_d$.

Speed Limits

Writing with a short pulse eliminates the influence of charge-carrier excitation from the issue of the photorefractive response time. The excitation time, or the fundamental limit introduced by Yeh [3], does not have any importance for pulsed experiments.

However, the buildup of a photorefractive grating after illumination with short pulses can be stopped prematurely by a short free-carrier lifetime. This leads to short buildup times, but it also leads to inefficient grating writing, and is clearly something to be avoided. It is more interesting to look at the case in which $\tau_d \ll \tau_0$ and the buildup time is determined by the diffusion time. In this case the charge-transport properties determine the speed of the photorefractive effect, in a manner that is closely related to the limiting time introduced by Glass et al. [4].

The diffusion time can be much shorter than any photorefractive response time observed in cw experiments. In GaAs the diffusion time is on the order of 25 ps at a grating spacing of one micrometer for holes, which have a mobility on the order of $400 \ \mathrm{cm^2 V^{-1} s^{-1}}$, and it is on the order of 2 ps or less for electrons. In $KNbO_3$ and $Bi_{12}SiO_{20}$, with mobilities of $0.5 \ \mathrm{cm^2 V^{-1} s^{-1}}$ and $3.4 \ \mathrm{cm^2 V^{-1} s^{-1}}$, respectively [9, 10], one observes short-pulse-induced photorefractive grating rise times in the nanosecond range for micrometer-sized grating spacings.

The Influence of an Applied Electric Field

The effect of an applied electric field on short-pulse recording is much more straightforward to understand and analyze than for cw recording. This is again due to the fact that photoexcitation does not play any role during the buildup, which is driven solely by diffusion and drift of free carriers.

A discussion of the buildup dynamics in the presence of an electric field and under the same assumptions used above for the buildup time without applied field can be found in [13, 15, 16, 14]. Since the essential features of the photorefractive response speed in short-pulse recording are already well described by the case without applied fields, only the basic facts are given here.

In the case of pulsed excitation, and in contrast to the cw case, the application of an electric field in the direction of the grating wave vector will always tend to make the buildup faster. An oscillatory time dependence will be superimposed on top of the exponential buildup observed without electric field in such a way that the minima of the oscillations lie on the buildup curve without field. The frequency of the oscillations is proportional to the drift velocity of the photoexcited carriers in the applied electric field [13, 14]. This is most easily observed with a grating wave vector equal to the inverse diffusion length. As in the no-applied-field case, the buildup dynamics depend only on free-carrier mobility and lifetime.

Another effect of the electric field is to strongly enhance the final photorefractive grating amplitude at long grating spacings [14]. It is interesting to note that this enhancement depends only on the electric field, grating spacing, and temperature, and is independent of any material parameter [14]. It could therefore be used to determine the strength of the applied electric field inside the sample, which may be difficult to estimate from the applied voltage because of the influence of shadow regions and space-charge trapped near the electrodes.

3.3 Determination of Material Properties

The following sections are dedicated to a description of the experimental methods that allow for the determination of material quantities from a measurement and analysis of the photorefractive response time under various circumstances. The experimental methods discussed here are confined to those that rely on the limits in experimental parameter space that were discussed above and that allow one to highlight the influence of a given material parameter on the response speed.

The experimental methods based on cw excitation will be discussed first, while those relying on short-pulse excitation will be discussed later, following the same sectioning scheme that was used to introduce the theoretical expressions describing the response time.

3.3.1 Continuous-Wave Excitation

The most interesting direct measurement that can be performed with cw excitation is that of the photoconductivity. Since the photorefractive response time tends to the dielectric relaxation time for increasing grating spacings, it is in principle straightforward to determine the conductivity and photoconductivity of a sample by the photorefractive response time at long grating spacings for various average intensities. To do this, one only has to make sure that the chosen grating spacing is large enough by estimating the diffusion length and the Debye screening length. This can be done by measuring the photorefractive response time as a function of the grating spacing.

Once the long grating spacing limit has been established, it is normally best to measure the response time by writing the photorefractive grating with two weak beams while illuminating the crystal with a third beam that establishes a strong, controlled background illumination. Instead of measuring beam coupling, it is also better to use a probe beam at a different wavelength to determine the grating amplitude by Bragg diffraction. This avoids possible instabilities in the buildup caused by the sensitivity of the beam-coupling gain to vibrations and allows flexibility in blocking both writing beams to measure the decay in the presence of a good homogeneous illumination.

I like to call this measurement technique *photoinduced space-charge relaxation* (PSCR) because it is a name that does not rely on terminology characteristic of the field of photorefractivity, and is therefore more readily accessible to the solid-state physicists who might be interested in it. The main interest of PSCR is that it is an all-optical technique that does not require contacting the sample to measure a current and does not require a particular sample shape.

One difficulty with PSCR is that in order to determine the absolute value of the photoconductivity one needs to know the effective dielectric constant $\epsilon_{eff}(\mathbf{k})$ for the particular direction of the grating wave vector one is using. This cannot be done in general without also knowing all the elastic and piezoelectric properties of a crystal [2]. Since this information is not necessarily available for all materials, it may be possible only at some particular orientations of the photorefractive grating to decide from symmetry considerations whether ϵ_{eff} is close to the strain-free or the stress-free value. Otherwise, the uncertainty in ϵ_{eff} is bound to affect the precision with which the photoconductivity can be determined in the form of a systematic error. There are two reasons why this difficulty is often not so important: (1) the errors in a conventional measurement of photoconductivity could be higher than the errors in PSCR caused by an uncertainty in ϵ_{eff}; (2) The uncertainty in ϵ_{eff} matters only when the absolute value of the photoconductivity needs to be determined for a given light intensity, but will not affect relative measurements such as the intensity dependence of the photoconductivity.

Photoinduced Space-Charge Relaxation

Conventional electrical measurements bring with them the necessity of well-defined sample shapes, of ohmic contacts, of well-controlled homogeneous illumination, and a possible sensitivity to surface states and surface conductivity that makes it difficult to access the real, bulk conductivity of a sample. In PSCR the drift direction is defined by the wave vector of the interference pattern, and the photoconductivity can be measured in the middle of the bulk of an irregularly shaped sample, simply by choosing the place where the two writing beams cross.

In general, the photoconductivity depends both on the polarization of the light and the drift direction selected in an experiment. The ability to measure all components of the photoconductivity tensor simply by choosing the orientation of the photorefractive grating is an interesting advantage of PSCR. Contrast this with the necessity to cut samples and apply contacts in selected directions to measure conductivity in different directions electrically.

To give an example of the issues connected with SPCR I will discuss in the following some recent measurements of the anisotropy of the hole-mobility in BaTiO$_3$. This problem was first attacked in [19, 20] by measuring the photoinduced decay of photorefractive gratings with wave vector both perpendicular and parallel to the c-axis of the crystal but with a relatively short grating spacing that required taking into account the value of the Debye wave vector, which was measured separately. Similar measurements were then repeated in [21] in the long grating spacing limit where the photorefractive response time is the dielectric relaxation time, and for many different directions of the grating wave vector. In this context it is important to note that the Debye wave vector depends on the direction of the grating wave vector via the dielectric constant, and that to do this kind of measurement one needs to make sure that the grating spacing is long enough for all grating directions. This is particularly true for a material like BaTiO$_3$, where for a grating wave vector perpendicular to the c-axis one observes an effective dielectric constant larger than 4000 [22].

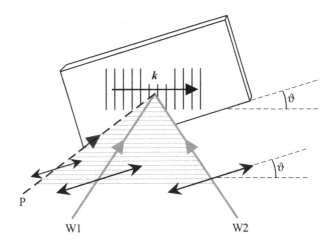

FIGURE 3.3. Scheme of an experiment for the all-optical determination of the drift-direction dependence of the photoconductivity by continuous-wave four-wave mixing. Two interfering beams W1 and W2 induce a space-charge modulation, and a probe beam P is used to detect it by diffraction at the corresponding refractive index grating. The hatched surface represents the incidence plane of the beams. The arrows at an angle ϑ from the incidence plane indicate the polarization of the beams. The sample and the polarizations are rotated synchronously by the angle ϑ, thus changing the direction of the space-charge wave vector in the sample (corresponding to the drift direction that matters for space-charge relaxation) while at the same time keeping the magnitude and direction of the optical electric field in the sample reference-frame constant.

Once the above conditions are satisfied, it is possible to determine the photoconductivity by PSCR for different orientations of the grating. This is done using the arrangement shown in Figure 3.3. The photoconductivity is obtained from (3.4):

$$\sigma(\vartheta, I_0) = e\mu(\vartheta)n_c(I_0) = \frac{\epsilon_0 \epsilon_{\mathit{eff}}(\vartheta)}{\tau_{die}(\vartheta)}, \tag{3.20}$$

where I_0 is the average light intensity and ϑ is the angle between the grating wave vector and the c-axis. An important aspect of this experiment is that the polarization of the light is held constant while the direction of the grating is varied. In this example the polarization remains along c. This ensures that the photoexcitation cross section is a constant independent of ϑ, and implies a constant $n_c(I_0)$ that does not depend on the orientation of the grating. The dependence of the observed response time from the orientation of the wave vector k is then described by

$$\frac{\tau_{die}(\vartheta)}{\tau(0)} = \frac{\epsilon_{\mathit{eff}}(\vartheta)}{\epsilon_{\mathit{eff}}(0)} \frac{\mu(0)}{\mu(\vartheta)}, \tag{3.21}$$

where $\mu(\vartheta)/\mu(0)$ gives the angular dependence of the charge carrier mobility.

In general, the mobility is described by a second-rank tensor. Although the static space-charge field E_{sc} is always parallel to the grating wave vector k, the drift current is generally not. From the component of the current density parallel to the grating wave vector, which is responsible for the decay of the space-charge modulation, one can calculate the effective scalar mobility $\mu(\vartheta)$ relevant for the decay of the space-charge modulation induced by an interference pattern with k in the direction ϑ. The projection of the drift velocity vector v_D $(v_{Di} = \mu_{ij} k_j E_{sc}/k)$ on k is $k \cdot v_D/k$ and therefore

$$\mu(\vartheta) = \frac{k_i \mu_{ij} k_j}{k^2}, \tag{3.22}$$

where the Einstein summation convention over repeated indices is used. It follows that the ϑ-dependence of the mobility ratio $\mu(\vartheta)/\mu(0)$ appearing in (3.21) is

$$\frac{\mu(\vartheta)}{\mu_c} = \cos^2(\vartheta) + \frac{\mu_a}{\mu_c} \sin^2(\vartheta), \tag{3.23}$$

where $\mu_a = \mu(\vartheta = \pi/2)$ and $\mu_c = \mu(\vartheta = 0)$ are the two components of the diagonal mobility tensor in the plane in which k is rotated.

For the experiments of [21], the grating spacing was $\approx 45\ \mu m$ ($k = 0.14\ \mu m^{-1}$) to ensure the validity of the long grating spacing limit for all orientations of the grating. Once the space-charge modulation photoinduced by the two writing beams reached steady state, it was erased in a controlled way by switching off the two writing beams and switching on a homogeneous erasing beam. The latter was kept polarized in the same direction in the sample reference frame while the sample was rotated. Since the space-charge grating that is erased by the homogeneous beam amounts to a small disturbance, the decay of the space-charge amplitude caused by the increased photoconductivity was a perfect exponential over at least 10 decay time constants. From all the space-charge decay curves measured for different ϑ the drift-direction dependence of the dielectric relaxation time (3.4) could be determined. Figure 3.4 shows the observed response time vs. the angle ϑ between the grating wave vector k and the c-axis in the BaTiO$_3$ crystal. In BaTiO$_3$ the effective electrooptical coefficient that describes diffraction of the probe beam vanishes when the grating wave vector is at $\vartheta = \pi/2$, but this did not impede the measurement of the full ϑ-dependence because it was possible to obtain data near enough to the $\vartheta = \pi/2$ position. The larger error bars around this value of ϑ in Figure 3.4 are due to the smaller diffraction efficiency and correspondingly lower signal-to-noise ratio.

As mentioned above, the dependence of the effective dielectric constant on the direction of k is nontrivial in a noncentrosymmetric material, and it must be calculated from the elastic and piezoelectric properties of BaTiO$_3$ [22]. While the angular dependence of the effective mobility is simply given by the projection of the mobility tensor components in the main-axes system and is described by (3.23), the angular dependence of the effective dielectric constant $\epsilon(\vartheta)$ is more complicated. A simple projection of the tensor elements of the standard dielectric constant along the direction of k would lead to a functional dependence that corresponds to the dashed curve in Figure 3.4. The bad correspondence with the experiment is due to the fact that such a curve does not take into account that

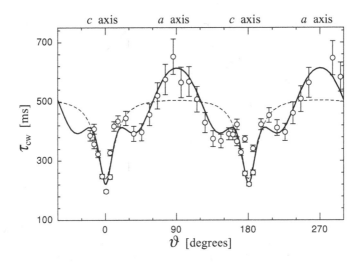

FIGURE 3.4. Relaxation time of the photoinduced space-charge grating in a nominally pure BaTiO$_3$ crystal as a function of the angle between $\boldsymbol{k}$ and the c-axis of the crystal (from [21]). The solid line represents the fit with (3.21) using the $\epsilon_{eff}(\vartheta)$ derived from the elastic and piezoelectric properties of BaTiO$_3$. The dashed curve is the fit obtained when using the uncorrected dielectric constants. The photoexcited charge carriers in this case are holes.

elastic deformations in the crystal contribute to the value of $\epsilon(\vartheta)$, and that the kind of deformations that are allowed is ϑ-dependent [2]. The following is a short discussion of the angular dependence of $\epsilon(\vartheta)$.

The effective dielectric constant is given by the projection of an effective, $\boldsymbol{k}$-dependent dielectric tensor on the grating wave vector: $\epsilon(\vartheta) = \epsilon_{ij}(\boldsymbol{k})k_i k_j / k^2$. The full expression for the effective dielectric tensor is [2, 23]

$$\epsilon_{ij}(\boldsymbol{k}) = \epsilon_{ij}^S + \frac{1}{\epsilon_0} \frac{k_n k_k}{k^2} e_{imk} e_{jln}(A^{-1})_{ml}, \qquad (3.24)$$

where $A_{ik} = C_{ijkl}^E k_i k_j / k^2$, C_{ijkl}^E is the elastic stiffness tensor at constant electric field, e_{ijk} is the piezoelectric tensor, and ϵ_{ij}^S is the strain-free dielectric tensor. All these material parameters are known in BaTiO$_3$ [22], so that $\epsilon(\vartheta)$ can be calculated. Figure 3.5 shows the difference between the correct value of the dielectric constant $\epsilon^{eff}(\vartheta)$ and the one that would be obtained by simply projecting the strain-free dielectric tensor along the direction of $\boldsymbol{k}$ (as in $\epsilon_{ij}^S k_i k_j / k^2$). This difference causes the small ripples in the $\tau_{die}(\vartheta)$ data plotted in Figure 3.4.

The angular dependence of the experimental $\tau_{die}(\vartheta)$ is governed by the single parameter μ_a/μ_c. The scaling parameter $\tau_{die}(0)$ does not influence the shape of the curves, and $\epsilon_{eff}(\vartheta)/\epsilon_{eff}(0)$ is known [22]. Therefore, a least-squares fit of (3.23) to the experimental $\tau_{die}(\vartheta)$ allows a precise determination of the mobility anisotropy μ_a/μ_c. The solid curve in Figure 3.4 is the result of such a fit, which delivered $\mu_a/\mu_c = 19.6 \pm 0.6$. The agreement with the data is excellent. The

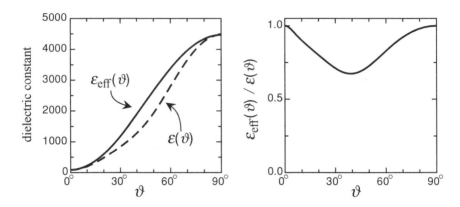

FIGURE 3.5. **Left:** Effective dielectric constant calculated from (3.24) and the data in [22] (solid curve), and dielectric constant calculated from the anisotropy of the standard dielectric tensor (dashed curve). **Right:** Corresponding relative change in the magnitude of the dielectric constant.

reason why the hole mobility anisotropy is so high in $BaTiO_3$ is still an open question [20, 21].

In the above results, the combination of the measurement of the dielectric relaxation time and of its analysis in terms of effective mobility and effective dielectric constant is also an indirect measurement of the ϑ-dependence of the effective dielectric constant that must be used to describe photorefractive effects and that is displayed in Figure 3.5. This was the first direct experimental confirmation of the validity of the expressions for the effective dielectric constant given in [1, 2].

Determination of Effective Trap Density and Diffusion Length

By measuring the full grating-spacing dependence of the response time, one can get the ratio between diffusion length and Debye screening length from the ratio between the response times at short and long grating spacings. The grating spacing that corresponds to halfway between these two times then determines the Debye wave vector. A fit of the experimental dependence of the photorefractive response time from the grating spacing is a relatively easy way to determine the diffusion length and the effective trap density at the same time, delivering complementary information to what can be obtained by a measurement of the two-beam coupling gain vs. the grating spacing. In order to obtain good results from such an experiment, one must be able to easily access the grating wave vectors around the Debye wave vector, and it is often advisable to perform a measurement at the very small grating spacings obtained using counterpropagating beams.

3.3.2 Pulsed Excitation

As mentioned above, when photoexcitation is confined to a short enough pulse, it is demoted from a continuous driving force of the charge-transport process to an initial condition, so that the only material parameters that influence the one-shot buildup of the photorefractive grating after short-pulse excitations are the mobility of the charge carriers and their lifetime.

This obviously opens the door to a very attractive opportunity for the actual determination of these material parameters, which are otherwise not accessible by cw photorefractive studies and can in general be quite difficult to measure also with other techniques, especially in photorefractive crystals and other insulators characterized by small mobilities and lifetimes.

The mobility of photoexcited carriers can be obtained from the short-pulse photorefractive response in two basically different ways: (1) by doing an experiment without applied electric fields and at short grating spacings, so that the diffusion time is much smaller than the lifetime, or (2) by doing an experiment with an applied field and at a grating wave vector near the inverse diffusion length.

In both of these methods one relies on the refractive index change induced by charge migration to detect the time dynamics of the photorefractive grating and from it the charge-carrier mobility.

These techniques have been called *holographic time of flight* (HTOF) [12, 10] in analogy to conventional time-of-flight techniques—where the drift of a thin sheet of photoexcited charges from one side of a sample to the other is measured by observing the corresponding electric current—and to highlight the underlying holographic principles.

Holographic Time of Flight

The strong interest of the holographic time of flight (HTOF) method is that it allows the contactless observation of charge displacement with picosecond time resolution (from nanoseconds down to the duration of the optical pulses used), and the measurement of the charge-carrier mobility during the first few pico- or nanoseconds after photoexcitation of a low density of carriers, over transport lengths from hundreds of micrometers to hundreds of nanometers. It is important to note that the length scale determining the time dynamics of the transport is controlled by the period of the sinusoidal excitation pattern (instead of sample thickness as in conventional time of flight) and can be varied very easily.

When the charge-separation is induced by drift in an external electric field one has the case of *drift-mode* HTOF [12, 13]. This is the technique that most closely parallels conventional time-of-flight. In drift-mode HTOF the space-charge field reaches a maximum when the mobile carriers have drifted by half a grating spacing, to a position of anticoincidence with the immobile distribution of the donors from where they were photoexcited. Further drift causes a decrease of the space-charge field until coincidence is reached again and so on. From the oscillations in the signal amplitude one derives the drift velocity, and from it the carrier mobility. HTOF in drift mode was first demonstrated with relatively large transport lengths of several 100 μm by Partanen et al. [12, 13], when it was used to determine the average electron mobility caused by shallow traps in $Bi_{12}SiO_{20}$ and its temperature dependence [15].

When the charge separation is caused only by diffusion, one has the case of *diffusion-mode* HTOF [8, 10]. Diffusion causes the space-charge field to grow exponentially after the excitation. From the time constant of the exponential buildup one derives the diffusion time, and from it the charge-carrier mobility. HTOF in diffusion mode was used to determine the electron mobility in $KNbO_3$ [8, 9] and the temperature dependence of the intrinsic electron mobility and its origin in $Bi_{12}SiO_{20}$ [10, 24, 11].

When the transport length (i.e., the grating spacing) is long, the mobility small, and the lifetime large, charge transport can be observed using a continuous-wave probe beam [13, 15]. For short transport lengths (which can be as small as a fraction of a μm) and larger mobilities, charge migration must be followed using a delayed probe pulse in a pump & probe, degenerate four-wave mixing setup [8, 9, 10, 11].

The latter possibility is particularly interesting because it covers those cases in which the free-carrier lifetime is so small that conventional techniques for the determination of charge-carrier mobilities, such as the time-of-flight method, cannot be applied. In contrast, an optical pump & probe measurement has a time resolution that is limited only by the pulse length, so that HTOF in this regime can determine the mobility of charge carriers that live less than a nanosecond after they are photoexcited. The only requirement is that their mobility be high enough, so that the region where $\tau_d < \tau_0$ can be accessed. As an example, using green light for photoexcitation (532 nm) and an angle of $90°$ between the writing beams, the diffusion time is 1.4 ns for a mobility of $1\ cm^2 V^{-1} s^{-1}$. Using counterpropagating beams at the same wavelength in a material with a refractive index of 2 reduces the diffusion time for the $1\ cm^2 V^{-1} s^{-1}$ mobility to ≈ 150 ps.

These abilities of HTOF make it particularly well adapted for the study of materials whose concentrations of impurities and defects are relatively high (the short and tunable transport lengths used in HTOF allow one to observe the charge carriers in the conduction band before they meet their first impurity) and that are difficult to produce with the geometry and dimensions required for conventional charge-transport measurements (HTOF in diffusion mode measures the mobility at the point of intersection of the optical beams, independently of the shape of the sample).

A review of HTOF can be found in [14]. Here I limit the discussion to the example of the HTOF measurement of the electron mobility in $Bi_{12}SiO_{20}$. This example highlights the high time resolution that can be gained with a pump & probe setup and the advantages of accessing the mobility all-optically.

The basic experimental four-wave-mixing configuration for HTOF is shown in Figure 3.6. A beam splitter sends part of a pulse into a delay line, to act as a probe beam, and the other part is split again into two write pulses that arrive simultaneously in the sample. The phase grating produced by diffusion of the photoexcited charges is detected by diffraction of the time-delayed probe pulse 2. The sample is homogeneously illuminated all the time by a cw laser beam. In the time interval between two measurements, this illumination erases the space-charge grating created by the write pulses. During the measurement time (a few ns)

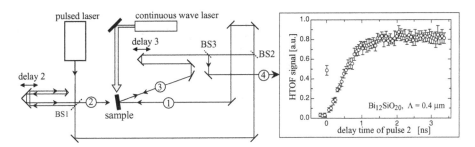

FIGURE 3.6. Degenerate four-wave-mixing setup for HTOF and experimental result in $Bi_{12}SiO_{20}$ at a grating spacing of 0.4 μm and no applied electric field. Beams 1 and 3 write the photorefractive grating with a single shot. Pulse 2, which can be delayed thanks to a computer-controlled delay line, diffracts off the refractive index grating that grows while the photoexcited carriers diffuse, and leads to diffracted pulse 4, whose energy is represented in the plot as a function of its delay time. A continuous-wave laser illuminates the sample all the time to erase the photorefractive grating during the interval between laser pulses (adapted from [10] and [11]).

this erase beam deposits six orders of magnitude less energy than the write pulses, and it does not affect the space-charge dynamics in the nanosecond time scale.

The result of a measurement performed in $Bi_{12}SiO_{20}$ using 30 ps, 532 nm pulses at a repetition rate of 5 Hz from a frequency-doubled Nd:YAG laser, and a cw illumination with an argon laser beam delivering 0.1 W/cm^2 at 514 nm, is also shown in Figure 3.6 [10]. When the probe pulse precedes the write pulse, there is only a small signal. A relatively strong signal caused by third-order nonlinear optical effects is observed when the three pulses are present in the crystal at the same time. When the probe pulse is delayed (positive times in the figure), the diffracted signal increases as the photoexcited electrons diffuse away from the positively ionized donors from which they were photoexcited, thus creating a space-charge field and consequently a refractive index grating.

In order to determine the mobility accurately, two requirements must be fulfilled. First, one must establish the validity of the low-fluence limit on which the simple model used to interpret the data is based. This is necessary because otherwise saturation effects occur, which will deform the single-exponential buildup and lead to wrong estimations of the buildup time [9]. Second, the transport length (i.e., the spatial period of the interference pattern, given by the angle between the beams) must be such that charge-carrier diffusion can have its effect on the charge-carrier distribution during the free-carrier lifetime ($\tau_d \ll \tau_0$). To satisfy these requirements, one must perform measurements for different crossing angles between the beams, and for different pulse energies [10]. This allows the selection of the optimal spatial modulation period in the interference pattern, and of the optimal pulse energies, which still give a good signal-to-noise ratio, but which are well below the limits imposed by the necessity to avoid saturation effects [9]. For the $Bi_{12}SiO_{20}$ samples used in [10], it was found that the fluence in the interference pattern that photoexcites the electrons must be of the order of

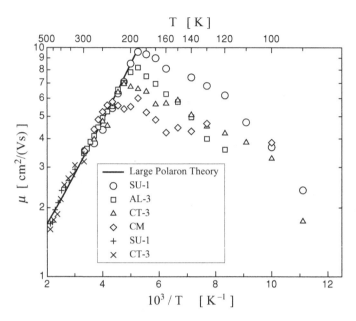

FIGURE 3.7. Temperature dependence of the electron-band mobility in various $Bi_{12}SiO_{20}$ samples as determined by HTOF in diffusion mode. Adapted from [11]. The data points at temperatures higher than room temperature represented by diagonal and vertical crosses come from [10]. The remaining data points are from [11]. The solid line represents the prediction for large polarons, calculated using the model given in [24] from the optical phonon spectrum of $Bi_{12}SiO_{20}$ and using an effective electron mass of 2 electron masses.

1 mJ/cm^2 or less, and that the diffusion time dominates the buildup at a spatial period $2\pi/k$ around 0.4 μm. This means that the electron avoids shallow and deep traps for so long that the electron distribution can become almost homogeneous because of diffusion, so that $\tau_d \ll \tau_0$. An analysis of the rise times as a function of the grating period confirmed the expected quadratic dependence and allowed the determination of the diffusion time τ_d, and from this a mobility value of 3.4 ± 0.5 cm^2/V^{-1}s^{-1} was derived at room temperature [10].

This HTOF determination of the electron mobility in $Bi_{12}SiO_{20}$ was subsequently extended to a larger temperature range between 100 K and 500 K [11], with the results shown in Figure 3.7. Various $Bi_{12}SiO_{20}$ samples, despite their different origins, show an identical temperature dependence between 250 K and 500 K that appears to be well described by an almost straight line in the log μ vs. 1/T plot of Figure 3.7. The order of magnitude of the mobility in $Bi_{12}SiO_{20}$ and its temperature dependence in this region could be explained by showing that the electrons in the conduction band of $Bi_{12}SiO_{20}$ form large polarons, while the decrease of the mobility at temperatures lower than 200 K is probably due to extrinsic effects [10, 24, 11].

It is instructive to note that the buildup of a space-charge grating in $Bi_{12}SiO_{20}$ after pulsed photoexcitation was first observed in [6], and that the authors also tried to determine the mobility from the rise time of the signal. However, they did not consider the experimental boundary conditions under which diffusion-mode HTOF can be performed, and their buildup was stopped by a too-small free-carrier lifetime. In fact, their estimated mobility value of 50 $cm^2/(Vs)$ was much too large.

To conclude this section on HTOF, it is also important to note that photoexcitation and detection of sinusoidal distributions of free carriers have a long history. The amplitude of a sinusoidally modulated free carrier distribution can in general be detected optically when the presence of the free carriers modifies the optical properties of the material and creates a refractive index or absorption grating [25]. So at first sight there does not seem to be a fundamental difference between the use of short-pulse photorefractive effects to determine free-carrier mobilities in HTOF techniques, and what has been done before in the detection of photoinduced carrier dynamics. However, there are at least two important advantages of HTOF that are connected to its basic feature, the fact that the linear electrooptical effect is used to detect space-charge *displacement*, not just the existence of a spatially modulated carrier concentration: (1) In HTOF it is generally possible to detect the movement of a much smaller concentration of free carriers than what is required to significantly modulate refractive index or absorption. In contrast, the necessity to excite larger numbers of carriers in order to detect them in a centrosymmetric material might make it impossible to use carrier photoexcitation from impurity levels, and can also create possible ambiguities in the interpretation of experimental results, e.g., because of saturation effects, or the strong space-charge field induced by the diffusion of the carriers, heating, heat diffusion, etc. (2) In HTOF the signal caused by the movement of the photoexcited carriers grows from zero, and it is a unique signature of charge transport. In contrast, Bragg diffraction from a free-carrier grating is sensitive to only its amplitude. The signal would decay exponentially with both diffusion and recombination of the free carriers, but would not change if the free-carrier grating were drifting in an electric field, unless special techniques related to homodyne detection were used. Moreover, besides charge diffusion, other effects such as the decay of a temperature grating could lead to the same signal dynamics.

3.4 Conclusions

This chapter has described the regimes in which the photorefractive dynamics is particularly simple and in which it can be used to determine interesting material parameters. Table 3.2 concludes the discussion in this chapter by presenting the material parameters that have been successfully determined in some of the most important photorefractive materials using techniques based on the detection of the photorefractive response time under appropriate experimental conditions. (Please

TABLE 3.2. Intrinsic charge carrier mobility data obtained by the PSCR and HTOF techniques presented in this chapter

Material	Carrier type	Band mobility data [$cm^2V^{-1}s^{-1}$]	Ref.
KNbO$_3$	holes	$\mu_a/\mu_c = 1.05 \pm 0.06$	[21]
		$\mu_b/\mu_c = 2.9 \pm 0.3$	[21]
	electrons	$\mu_a/\mu_c = 1.15 \pm 0.9$	[21]
		$\mu_b/\mu_c = 3.0 \pm 0.3$	[21]
		$\mu_a = 0.6 \pm 0.2$	[9, 21]
		$\mu_b = 1.0 \pm 0.3$	[9, 21]
		$\mu_c = 0.5 \pm 0.1$	[9]
BaTiO$_3$	holes	$\mu_a/\mu_c = 19.6 \pm 0.6$	[21]
Bi$_{12}$SiO$_{20}$	electrons	$\mu = 9 \pm 1$ ($T = 190$ K)	[11]
		$\mu = 3.4 \pm 0.5$ ($T = 300$ K)	[10]
		$\mu = 1.7 \pm 0.3$ ($T = 480$ K)	[10]

note that there is a typographical error in [21]: the values of the anisotropies for electrons and holes in KNbO$_3$ given in the abstract are exchanged.)

Table 3.2 focuses only on mobility data that can be argued to be an *intrinsic* property of the charge carriers in the conduction or valence band of these materials, and that are not influenced by impurities. Some of the data, such as the large mobility anisotropy observed in BaTiO$_3$, are still unexplained [20, 21], while the smaller anisotropy observed in KNbO$_3$ could in principle be modeled on the basis of simple small-polaron models [21]. Table 3.2 does not contain the results of careful HTOF measurements in drift mode in Bi$_{12}$SiO$_{20}$, where the value and temperature dependence of the trap-limited mobility in this material was determined [15, 16], as well as several newer mobility measurements that have been performed with HTOF in drift mode in photorefractive polymers and glasses [26, 27, 28, 29, 30].

Historically, fundamental parameters such as the charge-carrier mobility and mobility anisotropy have always been little known in photorefractive materials, despite the fact that a controlled investigation of the photorefractive response itself, when done under the appropriate experimental conditions, can be one of the principal methods for determining exactly these parameters. It is to be hoped that investigations similar to those that provided the data in Table 3.2 will close this knowledge gap in the future.

References

1. P. Günter and M. Zgonik. *Opt. Lett.* **16**, 1826 (1991).
2. M. Zgonik et al. *J. Appl. Phys.* **74**, 1287 (1993).
3. P. Yeh. *Appl. Opt.* **26**, 602 (1987).
4. A. Glass, M. Klein, and G. Valley. *Appl. Opt.* **26**, 3189 (1987).
5. M. Ewart. Ph.D. thesis, Swiss Federal Institute of Technology, Zurich, 1997, diss. ETH N. 12484.

6. J. Jonathan, G. Roosen, and P. Roussignol. *Opt. Lett.* **13**, 224 (1988).
7. M. Zgonik, I. Biaggio, U. Bertele, and P. Günter. *Opt. Lett.* **16**, 977 (1991).
8. I. Biaggio, M. Zgonik, and P. Günter. *J. Opt. Soc. Am. B* **9**, 1480 (1992).
9. M. Ewart, I. Biaggio, M. Zgonik, and P. Günter. *Phys. Rev. B* **49**, 5263 (1994).
10. I. Biaggio, R.W. Hellwarth, and J.P. Partanen. *Phys. Rev. Lett.* **78**, 891 (1997).
11. M. Wintermantel and I. Biaggio. *Phys. Rev. B* **67**, 165108 (2003).
12. J. Partanen, J. Jonathan, and R. Hellwarth. *Appl. Phys. Lett.* **57**, 2404 (1990).
13. J. Partanen, P. Nouchi, J. Jonathan, and R. Hellwarth. *Phys. Rev. B* **44**, 1487 (1991).
14. I. Biaggio. In *Photo-Excited Processes, Diagnostics and Applications*, edited by A. Peled (Kluwer Academic Publishing, Dordrecht/Boston/London, 2003).
15. P. Nouchi, J. Partanen, and R. Hellwarth. *J. Opt. Soc. Am. B* **9**, 1428 (1992).
16. P. Nouchi, J.P. Partanen, and R.W. Hellwarth. *Phys. Rev. B* **47**, 15581 (1992).
17. I. Biaggio and G. Roosen. *J. Opt. Soc. Am. B* **13**, 2306 (1996).
18. G. Pauliat and G. Roosen. *J. Opt. Soc. Am. B* **7**, 2259 (1990).
19. C. Tzou, T. Chang, and R. Hellwarth. *Proc. SPIE Int. Soc. Eng.* **613**, 58 (1986).
20. D. Mahgerefteh et al. *Phys. Rev. B* **53**, 7094 (1996).
21. P. Bernasconi, I. Biaggio, M. Zgonik, and P. Günter. *Phys. Rev. Lett.* **78**, 106 (1997).
22. M. Zgonik et al. *Phys. Rev. B* **50**, 5941 (1994).
23. I. Biaggio. *Phys. Rev. Lett.* **82**, 193 (1999).
24. R.W. Hellwarth and I. Biaggio. *Phys. Rev. B* **60**, 299 (1999).
25. R.I. Devlen and E.A. Schiff. *J. of Non-cryst. Solids* **141**, 106 (1992).
26. G.G. Malliaras, V.V. Krasnikov, H.J. Bolink, and G. Hadziioannou. *Phys. Rev. B* **52**, R14324 (1995).
27. G.G. Malliaras, V.V. Krasnikov, H.J. Bolink, and G. Hadziioannou. *Appl. Phys. Lett.* **67**, 455 (1995).
28. J. Wolff et al. *J. Opt. Soc. Am. B* **16**, 1080 (1999).
29. S.J. Zilker et al. *Chem. Phys. Lett.* **306**, 285 (1999).
30. A. Leopold et al. *Appl. Phys. Lett.* **76**, 1644 (2000).

4

Photorefractive Effects in LiNbO$_3$ and LiTaO$_3$

Karsten Buse,[1] Jörg Imbrock,[2] Eckhard Krätzig,[3] and Konrad Peithmann[4]

[1] University of Bonn, Institute of Physics, Wegelerstr. 8, 53115 Bonn, Germany, **kbuse@uni-bonn.de**
[2] Westfälische Wilhelms-Universität, Institute of Applied Physics, Corrensstr. 2, 48149 Münster, Germany, **imbrock@umi-muenster.de**
[3] University of Osnabrück, Physics Department, Barbarastr. 7, 49069 Osnabrück, Germany, **eckhard.kraetzig@uos.de**
[4] University of Bonn, Helmboltz Institute for Radiation and Nuclear Physics, Nußallee 14-16, 53115 Bonn, Germany, **peithmann@physik.uni-bonn.de**

4.1 Introduction

Lithium niobate (LiNbO$_3$) and lithium tantalate (LiTaO$_3$) crystals have gained increasing interest during recent years because of their fascinating physical properties, such as piezoelectricity, ferroelectricity, pyroelectricity, birefringence, electrooptical, and nonlinear-optic effects. These properties have been successfully utilized for applications in various fields. Basic experiments on acoustic surface wave propagation, electrooptical modulation, optical wave guiding, second harmonic generation, and pyroelectric detection have been performed with these crystals.

LiNbO$_3$ and its isomorphous compound LiTaO$_3$ are not found in nature. The first successful growth of large single crystals by the Czochralski technique was reported by Ballman [1]. A peculiarity results from the fact that the congruently melting composition (about 48.4 mol % Li$_2$O) does not coincide with the stoichiometric one (50 mol % Li$_2$O). The crystals have a perovskite-like structure with oxygen octahedra and are trigonal with only one structural phase transition (paraelectric–ferroelectric) at the Curie temperature T_c of about 1150 °C for LiNbO$_3$ and about 610 °C for LiTaO$_3$. The nonpolar high-temperature phase belongs to the space group R$\overline{3}$c (point group $\overline{3}$m) and at T_c the transition to the ferroelectric low-temperature phase of the space group R3c (point group 3m) occurs. At room temperature LiNbO$_3$ is negatively birefringent. The birefringence of LiTaO$_3$ is very small at room temperature and positive for the congruently melting composition. More information about growth and fundamental properties of LiNbO$_3$ can be found in [2, 3].

In 1966, Ashkin et al. [4] discovered light-induced refractive-index changes in $LiNbO_3$ and $LiTaO_3$ ("optical damage"). Two years later, Chen et al. [5, 6] successfully utilized the refractive-index changes for the storage of volume phase holograms leading to the term "photorefractive effect." In 1975, Staebler et al. [7] superimposed 500 holograms at the same site under different angles. Since this time, the crystals have been intensively studied and a vast amount of pioneering work has been performed. Among the large variety of photorefractive crystals, $LiNbO_3$ and $LiTaO_3$ are still the first choice for holographic applications requiring large storage times and robust samples of good optical quality. Many details and characteristic features of the underlying physical processes have been elucidated during the study of these crystals. New applications have been developed, too, as can be seen in part III of the book series.

In this chapter we want to summarize the present knowledge on photorefractive effects in $LiNbO_3$ and $LiTaO_3$. We start with a short introduction of the fundamentals necessary for the description of photorefractive effects. Then we discuss in detail the light-induced charge transport: Only with this knowledge it is possible to optimize the photorefractive properties and to find out the performance limits as described in the following section. We treat effects at high light intensities where intrinsic centers play an important role. Finally, we briefly discuss thermal fixing and holographic scattering effects in these materials and end with some conclusions.

4.2 Fundamentals of Photorefractive Effects

The physical content of the word "photorefractive" is evident: Any effect by which illumination changes the refractive index of a material is called a "photorefractive effect." In this sense all materials are photorefractive, and observation of photorefractive effects depends on the optical intensity used and the transparency of the material. Light-induced polymerization [8], light-induced reorientation of polymers [9–12], and the Franz–Keldysh effect [13–16] are a few examples of photorefractive processes. However, the "photorefractive effect in electrooptical crystals" names a special process that is illustrated in Figure 4.1.

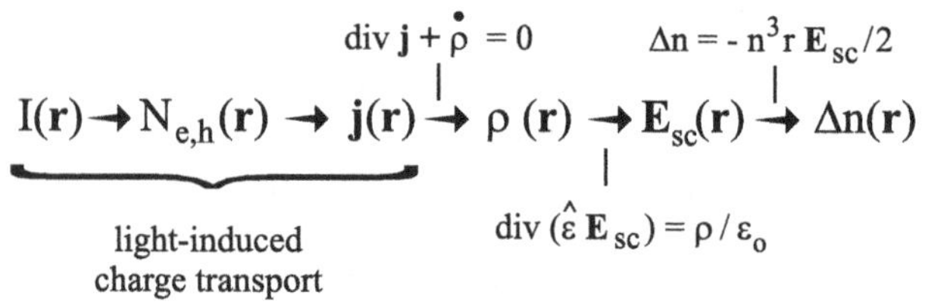

$$\text{div } \mathbf{j} + \dot{\rho} = 0 \qquad \Delta n = - n^3 r\, \mathbf{E}_{sc}/2$$

$$I(\mathbf{r}) \to N_{e,h}(\mathbf{r}) \to \mathbf{j}(\mathbf{r}) \to \rho(\mathbf{r}) \to \mathbf{E}_{sc}(\mathbf{r}) \to \Delta n(\mathbf{r})$$

light-induced
charge transport

$$\text{div}\,(\hat{\varepsilon}\,\mathbf{E}_{sc}) = \rho/\varepsilon_0$$

FIGURE 4.1. Photorefractive effect in electrooptical crystals (I, light intensity; $N_{e,h}$, concentrations of electrons in the conduction band/holes in the valence band; $\mathbf{j}$, electric current density; ρ, space-charge density; $\mathbf{E}_{sc}$, electric space-charge field; Δn, change of the refractive index; $\hat{\varepsilon}$, permittivity tensor; n, refractive index; r, electrooptical coefficient; $\mathbf{r}$, spatial coordinate).

TABLE 4.1. Refractive indices for ordinarily and extraordinarily polarized light, n_o and n_e, as well as electrooptical coefficients r_{13} and r_{33} (contracted indices) for LiNbO₃ and LiTaO₃ crystals at the light wavelength 633 nm.

Material	n_o	n_e	r_{13} (pm/V)	r_{33} (pm/V)
LiNbO₃	2.29 [21]	2.20 [21]	11 [23]	34 [23]
LiTaO₃	2.18 [24]	2.19 [24]	8 [23]	31 [23]

Before going into detail, we will review here the principal steps of photorefractive recording. Inhomogeneous illumination (intensity pattern $I(r)$, where r is the spatial coordinate) excites electrons into the conduction band and/or holes into the valence band (densities of electrons/holes in conduction/valence band are denoted by N_e and N_h). These free-charge carriers migrate, e.g., because of drift, bulk photovoltaic effects [17], and diffusion. A spatially modulated electric current (density $j(r)$) arises. After a while, an appreciable charge density modulation $\rho(r)$ is built up (continuity equation div $j + \dot{\rho} = 0$). The resulting electric space-charge field $E_{sc}(r)$ (Poisson equation div$(\hat{\epsilon} E_{sc}) = \rho/\epsilon_0$, where $\hat{\epsilon}$ is the dielectric tensor) acts against further charge transport, and finally a steady-state situation is achieved. The space-charge field modulates the refractive index because of the linear electrooptical effect. For a space-charge field that is aligned along the optical axis of the material we get [18]

$$\Delta n_{o,e} = -\frac{1}{2}n_{o,e}^3 r_{13,33} E_{sc} . \tag{4.1}$$

Higher-order terms of the electrooptical effect can be neglected because in LiNbO₃ no quadratic electrooptical effect can be observed for applied electric fields of up to 65 kV/mm [19]. The maximum space-charge field that can be achieved in iron-doped LiNbO₃ is about 10 kV/mm [20]. Some numbers for parameters such as refractive indices n and electrooptical coefficients r can be found in Table 4.1. Especially for LiNbO₃ these coefficients are very well known for various wavelengths [21, 22].

4.3 Light-Induced Charge Transport

Detailed knowledge of light-induced charge transport is required for the optimization of LiNbO₃ and LiTaO₃ crystals for holographic applications. We present information on extrinsic and intrinsic centers and discuss bulk photovoltaic effects, photoconductivity, and electron–hole competition.

4.3.1 Extrinsic Centers

The crucial influence of extrinsic centers on the photorefractive properties of LiNbO₃ and LiTaO₃ was discovered very early [25,26]. Especially transition metal dopants may occur in different valence states, making possible the generation of space-charge fields. Usually the dopants are added to the melt, for example as oxides. Distribution coefficients between 0.1 and 1.9 and concentrations of dopants up to about 1 mol % have been reported. In addition, diffusion doping is possible: Metal films are vacuum deposited onto crystal wafers, which are heated to temperatures above 1000 °C for several days.

The valence states of transition metal dopants in $LiNbO_3$ and $LiTaO_3$ can be greatly influenced by suitable thermal annealing treatments [25, 27]. Heating in an oxygen atmosphere to temperatures of about 1000 °C for several hours tends to oxidize the impurities, for example to Fe^{3+} or Cu^{2+}. This process is reversible, and heating in an argon atmosphere tends to reduce the impurities, for example to Fe^{2+} or Cu^+. In the case of $LiTaO_3$ the crystals have to be poled again after these annealing treatments because the Curie temperature is exceeded. In the case of $LiNbO_3$ this is usually not necessary.

Certain valence states have been identified by various methods. Electron spin resonance (ESR, EPR) experiments have been used to determine the spectra of the transition metal ions Fe^{3+} [25, 28], Mn^{2+} [28, 29], Cu^{2+} [30], and Cr^{3+} [31]. An especially powerful method for the determination of valence states of Fe is Mößbauer spectroscopy. Both Fe^{3+} and Fe^{2+} spectra are observed, thus yielding the Fe^{2+}/Fe^{3+} ratio [32, 33]. By this means the changes of the valence states in $LiNbO_3$:Fe and $LiTaO_3$:Fe have been carefully studied, showing that only Fe^{2+} and Fe^{3+} centers are present in the crystals, and the decrease of Fe^{3+} on heating the crystals in a reducing atmosphere corresponds to the increase of Fe^{2+}. Experimental results are illustrated in Figure 4.2. ENDOR [34] and X-ray experiments [35] clearly indicate that the Fe centers occupy Li sites.

Of special importance for photorefractive effects are optical absorption processes. Pure $LiNbO_3$ and $LiTaO_3$ crystals are transparent in the near-IR, the visible, and the near-UV regions up to about 3.8 eV ($LiNbO_3$) and 4.6 eV ($LiTaO_3$), where the fundamental absorption begins. Impurities cause characteristic bands. Figure 4.3a shows the spectrum of a $LiNbO_3$:Fe crystal. The band at 1.1 eV is assigned to crystal field transitions $^5A-^5E$ of Fe^{2+} ions, the broad band centered at about 2.6 eV to intervalence transfers $Fe^{2+}-Nb^{5+}$ (leading to the creation of electrons in the conduction band formed by Nb^{5+} ions), the band beginning at about 3.1 eV and extending to higher photon energies to charge transfers from oxygen π-orbitals to Fe^{3+} ions (leading to the creation of holes in the valence band formed by O^{2-} ions), and finally, the two small lines at 2.55 eV and 2.95 eV to spin-forbidden d–d transitions of Fe^{3+} ions [37]. The absorption spectrum is characteristically changed by reducing treatments (Figure 4.3b): Bands associated with Fe^{2+} increase, while bands associated with Fe^{3+} decrease.

The absorption spectrum of $LiTaO_3$:Fe is very similar to that of $LiNbO_3$:Fe. However, the charge transfer band $O^{2-}-Fe^{3+}$ is quite well separated from the fundamental absorption. For this reason the determination of the Fe^{2+}/Fe^{3+} ratio from absorption measurements is possible [38]. In strongly oxidized samples more than 99 % of the Fe ions are present as Fe^{3+}. The decrease of the Fe^{3+}

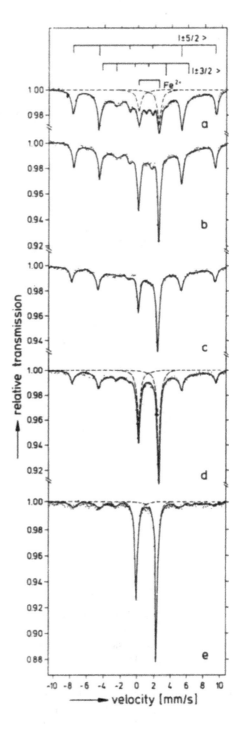

FIGURE 4.2. Mößbauer spectra of LiNbO₃:^{57}Fe at 80 K, (a) in the as-grown state, (b) after annealing at 1000 °C in 1 bar Ar atmosphere for 50 h, (c) after additional annealing at 1000 °C in 1 bar Ar atmosphere for 50 h and rapid cooling, (d) after additional annealing at 1000 °C in 67 mbar Ar atmosphere for 10 h, (e) after additional annealing at 1000 °C in 0.3 mbar Ar atmosphere for 20 h. Source 50 mCi ^{57}Co in Rh [33].

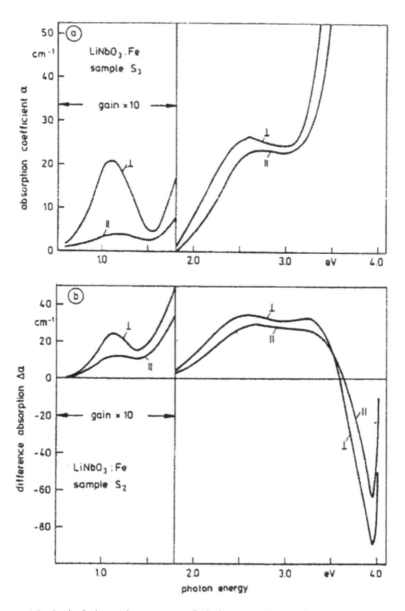

FIGURE 4.3. Optical absorption spectra of LiNbO$_3$:Fe, (a) sample S$_3$ (342 wt. ppm Fe) reduced by annealing in Ar atmosphere, 1 bar, (b) difference absorption $\Delta\alpha$ between reduced and oxidized sample S$_2$ (142 wt. ppm Fe) [33] (light polarization perpendicular ($\perp$) and parallel ($\parallel$) to the crystallographic axis).

charge transfer band caused by reducing treatments reflects the fraction of Fe^{3+} ions transferred into the divalent state. From these measurements the oscillator strengths of the absorption bands have been derived. Results are summarized in Table 4.2.

TABLE 4.2. Oscillator strengths of Fe and Cu bands in LiNbO$_3$
and LiTaO$_3$ [33, 36, 38]

Crystal	Peak [eV]	Transition	Oscillator Strength
LiNbO$_3$:Fe	1.1	$^5A - {}^5E$ (Fe^{2+})	4×10^{-4}
	2.6	Fe^{2+} – Nb^{5+}	1×10^{-2}
	≈ 4	O^{2-} – Fe^{3+}	5×10^{-2}
	2.55 and 2.95	$d - d$ (Fe^{3+})	$<10^{-5}$
LiTaO$_3$:Fe	1.1	$^5A - {}^5E$ (Fe^{2+})	5×10^{-4}
	3	Fe^{2+} – Ta^{5+}	1×10^{-2}
	4	O^{2-} – Fe^{3+}	5×10^{-2}
	2.6 and 2.9	$d - d$ (Fe^{3+})	$<10^{-5}$
LiNbO$_3$:Cu	1.2	$^2E - {}^2T_2$ (Cu^{2+})	2×10^{-4}
	3.3	Cu$^+$ – Nb^{5+}	4×10^{-2}

The absorption spectra of Cu-doped LiNbO$_3$ crystals (Figure 4.4) have also been utilized for the determination of the Cu$^+$/Cu^{2+} ratio and the oscillator strengths of the corresponding bands [36]. The band at about 1.2 eV results from ^2E–$^2T_2$ transitions of Cu^{2+} ions. The band at about 3.3 eV, on the other hand, is caused by Cu$^+$–Nb^{5+} intervalence transfers. As can be seen from the lower part of Figure 4.4, the Cu^{2+} band decreases and the Cu$^+$ band increases with reducing treatments, the decrease of Cu^{2+} again being proportional to the increase of Cu$^+$. These results clearly indicate that only Cu$^+$ and Cu^{2+} centers are involved.

The experiments discussed up to now relate to relatively moderate reductions that influence only the extrinsic impurities and leave the host crystals essentially unchanged. By this means, more than about 80 % of the Fe and Cu ions can be transformed into Fe^{2+} and Cu$^+$, respectively. Stronger reductions, such as heating in vacuum to temperatures near the melting point, produce additional intrinsic centers and absorption bands, which will be treated in the next section.

Further confirmation of the assignment of absorption processes is provided by photochromic effects in doubly doped LiNbO$_3$ crystals containing Fe/Mn or Cu/Mn [39, 40]. Figure 4.5 shows the photocurrent and absorption spectra of LiNbO$_3$:Fe/Mn after illumination with UV or visible light. The change of the spectra is described by the processes

$$\text{Mn}^{2+} + \text{Fe}^{3+} \underset{\text{vis}}{\overset{\text{uv}}{\rightleftarrows}} \text{Mn}^{3+} + \text{Fe}^{2+} . \tag{4.2}$$

For LiNbO$_3$:Fe/Cu the analogous relation reads

$$\text{Cu}^+ + \text{Fe}^{3+} \underset{\text{vis}}{\overset{\text{uv}}{\rightleftarrows}} \text{Cu}^{2+} + \text{Fe}^{2+} . \tag{4.3}$$

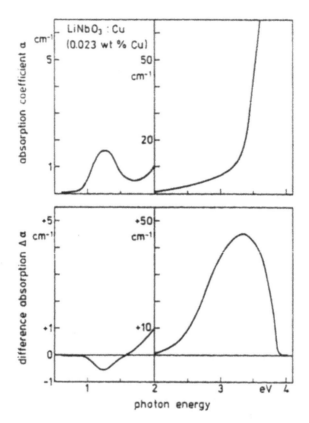

FIGURE 4.4. Optical absorption spectra of $LiNbO_3$:Cu for light polarized perpendicular to the c-axis [36]. Top: Absorption coefficient α of an oxidized sample ($c_{Cu^+}/c_{Cu^{2+}} < 0.05$). Bottom: Difference absorption $\Delta\alpha$ between a partially reduced ($c_{Cu^+}/c_{Cu^{2+}} = 0.6$) and an oxidized ($c_{Cu^+}/c_{Cu^{2+}} < 0.05$) sample.

The increase and decrease of the corresponding absorption bands can be clearly seen in the spectra.

4.3.2 Intrinsic Centers and Stoichiometric Materials

Since congruently melting $LiNbO_3$ and $LiTaO_3$ crystals have a Li concentration of about 48.4 mol % Li_2O, they contain many intrinsic defects (see Chapter 2 of this book). In principle, the following defects may exist in $LiNbO_3$ crystals: Nb^{5+} at Li site ($Nb_{Li^+}^{5+}$), Li vacancy (V_{Li^+}), Nb vacancy ($V_{Nb^{5+}}$), Li on vacancy (Li_V^+), Nb on vacancy (Nb_V^{5+}), and oxygen vacancy ($V_{O^{2-}}$). One possible charge compensation can be $Nb_{Li^+}^{5+} + 4\,V_{Li^+}$. Considering this model, congruently melting $LiNbO_3$ and $LiTaO_3$ crystals contain about $10^{26}\,m^{-3}$ $Nb_{Li^+}^{5+}$ and Ta_{Li}^{5+} antisites, respectively. This concentration is much larger than usual extrinsic defect concentrations. Under illumination, electrons that are excited can be trapped at Nb_{Li}^{5+} or Ta_{Li}^{5+} defects forming small polarons (Nb_{Li}^{4+} or Ta_{Li}^{4+}) [41, 42]. Since the electron

FIGURE 4.5. Photocurrent and absorption spectra of a LiNbO₃: Fe/Mn after illumination with UV light (activated) and visible light (inactivated) [40].

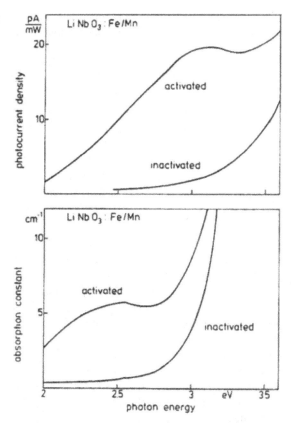

is located at the antisite it is called a bound polaron. The lifetime of a Nb_{Li}^{4+} polaron in the dark depends on the concentration of deep extrinsic traps, on the Li concentration, and on temperature. At room temperature the lifetime of small polarons in iron-doped LiNbO₃ and LiTaO₃ is of the order of microseconds [43,44]. Small polarons induce a broad absorption band in the red and near-infrared spectral regions, respectively. The absorption peak of the Nb_{Li}^{4+} polaron is centered at 1.63 eV [41] and the peak of the Ta_{Li}^{4+} polarons at 2.18 eV [45] (Table 4.3). Besides bound polarons, also free polarons can occur. In reduced LiNbO₃ doped with 6 mol % MgO, an absorption peak at 0.96 eV appears that is attributed to an electron trapped by the self-induced lattice distortion (Nb_{Nb}^{4+}) [46]. If nominally pure LiNbO₃ or LiTaO₃ is reduced, e.g., in vacuum, bipolarons can build up. A bipolaron in LiNbO₃ consists of a combination of a small free polaron (Nb_{Nb}^{4+}) and a small bound polaron (Nb_{Li}^{4+}) [42,47]. The absorption peak of the bipolarons in LiNbO₃ is centered at 2.5 eV [41] and the absorption peak of the bipolarons in LiTaO₃ at 2.7 eV [45]. Illumination of such a crystal with visible light can dissociate a bipolaron into two polarons, which recombine in the dark to a bipolaron.

TABLE 4.3. Intrinsic defects in $LiNbO_3$ and $LiTaO_3$ and their absorption peaks

Crystal	Defect	Peak [eV]	Type
$LiNbO_3$	bound polaron	1.63 [41]	Nb_{Li}^{4+}
	free polaron	0.96 [46]	Nb_{Nb}^{4+}
	bipolaron	2.5 [41]	$Nb_{Li}^{4+} Nb_{Nb}^{4+}$
$LiTaO_3$	bound polaron	2.18 [45]	Ta_{Li}^{4+}
	bipolaron	2.7 [45]	$Ta_{Li}^{4+} Ta_{Ta}^{4+}$

To reduce the concentration of intrinsic defects several methods have been developed:

- *Growing crystals from a Li-rich melt.* Near-stoichiometric $LiNbO_3$ crystals that are grown from a Li-rich melt by the conventional Czochralski method have poor optical quality and compositional homogeneity [48]. A better crystal quality is achieved with the double-crucible Czochralski method, which allows one to grow near-stoichiometric $LiNbO_3$ and $LiTaO_3$ crystals [48–52] (see Chapter 4 of this book).
- *Doping with elements that are incorporated at Li site.* Elements like Mg, Zn, Sc, and In are incorporated at Li site [53–58]. They occur just in one valance state and are not directly involved in light-induced charge transport. Large concentrations of, e.g., Mg can decrease in $LiNbO_3$ the concentration of Nb_{Li} antisites and in $LiTaO_3$ the concentration of Ta_{Li} defects [59] (see Chapter 6 of this book).
- *Growing crystals from a melt containing potassium.* $LiNbO_3$ crystals with a larger Li concentration than 48.8 mol % can be grown from a melt containing potassium, while no potassium is incorporated in the crystal [60, 61]. $LiTaO_3$ has yet not been grown from a potassium-rich melt.
- *Vapor transport equilibration (VTE).* Congruently melting crystals can be treated in a Li-rich gas atmosphere until the Li concentration of the gas and of the crystal are in equilibration. The Li concentration of $LiNbO_3$ [62–64] and $LiTaO_3$ [65, 66] can be tuned in a wide range using VTE treatments.

Many physical properties of $LiNbO_3$ and $LiTaO_3$ depend on the Li concentrations. Measuring for example, the fundamental absorption edge [67, 68], the refractive indices and the birefringence [69–71], the Curie temperature [65,72,73], or the coercive field [51, 74] allows one to determine the Li concentration. In the case of $LiTaO_3$ it is also possible to determine the temperature with vanishing birefringence [66].

Among the intrinsic defects, especially the antisites can influence the photorefractive properties of $LiNbO_3$ and $LiTaO_3$. Their influence is of minor importance in congruently melting crystals if small continuous wave laser intensities are used, because then the concentration of polarons is small. The concentration of polarons is larger at higher light intensities if laser pulses or focused cw laser light is used. Typical effects that indicate the contribution of intrinsic defects are light-induced absorption changes, a photoconductivity that increases super-linearly with growing intensity, and light-induced refractive-index changes that increase with increasing intensity. Furthermore, gated holographic recording via a two-step process is possible only if polarons are present (see Part I, Chapter 8). In Section 4.5 of this chapter, effects at high light intensities are discussed in the framework of a charge-transport model with two centers.

4.3.3 Bulk Photovoltaic Effects

The discovery of a new bulk photovoltaic effect characteristic for pyroelectrics additionally stimulated the investigation of light-induced charge transport in these crystals. Already rather early photoinduced currents and voltages were observed in the absence of external fields in noncentrosymmetric crystals [75]. Glass et al. [17] pointed out in 1974 that a new bulk photovoltaic effect is involved that cannot be explained by conventional photovoltaic processes in crystals containing macroscopic inhomogeneities (pn junctions) or by Dember voltages.

Homogeneous illumination of single-domain doped LiNbO$_3$ and LiTaO$_3$ crystals in thermal equilibrium yields steady-state currents in a short-circuit configuration and steady-state voltages and fields up to 10^7 Vm^{-1} in an open circuit [40]. Bulk photovoltaic effects are induced not only by visible and uv light but also by X-rays [76]. The effects do not depend on electrode material or on illumination of the electrodes.

In first investigations a bulk photovoltaic current has been observed along the polar c-axis ($+c$-end to $-c$-end through the crystal) [17]. The current density $j_{\mathrm{phv},z}$ along this axis is proportional to the absorbed power density,

$$j_{\mathrm{phv},z} = \alpha \kappa I, \tag{4.4}$$

where I is the light intensity, α the absorption coefficient, and κ a constant describing the anisotropic charge transport, called the Glass constant.

The bulk photovoltaic current density j_{phv} depends linearly on light intensity and hence sesquilinearly on the light field e as pointed out by Belinicher et al. [77, 78],

$$j_{\mathrm{phv},i} = \beta_{ikl} e_k e_l^*; \quad \beta_{ikl} = \beta_{ilk}^*; \quad i, k, l = 1, 2, 3. \tag{4.5}$$

The complex quantities β_{ikl} represent the bulk photovoltaic tensor. The real part is symmetric and the imaginary part antisymmetric in k and l; for a linearly polarized light wave the real part of β_{ikl} is involved, for circular polarization also the imaginary part of β_{ikl}.

In the case of LiNbO$_3$ and LiTaO$_3$ (3m) the photovoltaic tensor has four nonvanishing independent components: β_{333}, $\beta_{311} = \beta_{322}$, $\beta_{222} = -\beta_{112} = -\beta_{121} = -\beta_{211}$, and $\beta_{113} = \beta_{131}^* = \beta_{232} = \beta_{223}^*$. The tensor description of photovoltaic effects has been completely confirmed experimentally [79–81]. Though

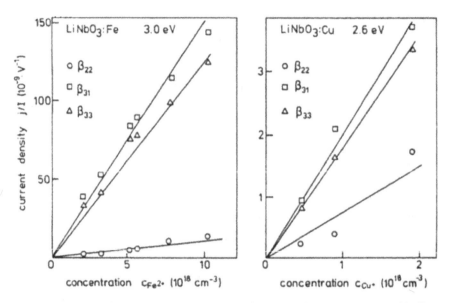

FIGURE 4.6. Dependence of the bulk photovoltaic tensor elements (contracted indices) β_{22} (○), β_{31} (□), and β_{33} (△) on the concentration of Fe^{2+} ions for 3.0 eV and Cu^+ ions for 2.6 eV [80].

the largest current densities are obtained along the c-axis, the currents perpendicular to the c-axis, which are about an order of magnitude smaller, have also been identified unambiguously by the dependence on the polarization angle.

Bulk photovoltaic currents are completely determined by the number of filled impurity traps, e.g., Fe^{2+} or Cu^+ ions. Experimental results are summarized in Figure 4.6 for $LiNbO_3$:Fe and $LiNbO_3$:Cu. Crystals with concentrations of filled traps up to about 10^{19} cm^{-3} have been investigated [80]. In all cases a linear dependence of the current density on $c_{Fe^{2+}}$ or c_{Cu^+} is observed. The elements β_{311} and β_{333} are of the same order of magnitude, and both elements are considerably larger than β_{222}.

The linear dependence of the current density on the concentration of filled traps supports the phenomenological model proposed by Glass et al. [17,82]. The main point is the asymmetric charge transfer from the absorbing center to the neighboring ions during optical excitation. The charge carriers are excited with preferred momentum. When they are scattered, i.e., thermalized, they lose their directional properties and contribute afterward to drift and diffusion currents but not to the bulk photovoltaic current. Recombination may also be asymmetric, but in general the recombination current will not cancel the excitation current, since different states are involved after relaxation. The current density along the c-axis may be written as [17, 82]

$$j_{\mathrm{phv},z}/I = \alpha\kappa = qr_{\mathrm{phv}}\alpha e/h\nu = \mu\tau' E_{\mathrm{phv}}\alpha/h\nu, \qquad (4.6)$$

where q denotes the quantum efficiency, r_{phv} the photovoltaic mean path length, e the electron charge, $h\nu$ the photon energy, μ the mobility, τ' the time in which the excited carriers contribute to the anisotropic charge transport, and E_{phv} a phenomenologically introduced local field acting on the charge carriers. For LiNbO$_3$:Fe and LiNbO$_3$:Cu [80], κ values up to 3×10^{-9} cmV^{-1} are obtained, yielding for the product qr_{phv} of quantum efficiency and photovoltaic mean path length values up to 0.1 nm.

According to this interpretation it seems reasonable to assume that the field E_{phv} depends only on the local asymmetry of filled traps and on the polarization of incident light. The mobility μ is determined by the band structure. Then the experimental results indicate that the time τ' is also independent of the concentration of filled and empty traps. This behavior of photovoltaic effects is completely different from that of photoconductivity described in the next section.

Photovoltaic effects have been observed only for intervalence transfers from the filled traps to the conduction band. No photovoltaic currents have been detected in LiNbO$_3$ and LiTaO$_3$ that result from charge transfers and subsequent hole migration. An estimate of the measuring accuracy indicates that photovoltaic effects due to hole migration—if they exist at all—are at least three orders of magnitude smaller than those of electron migration [83]. Detailed approaches for the theoretical description of the bulk photovoltaic effect and of the underlying physical charge transport mechanism have been given by several authors [84–89].

The photovoltaic tensor element $\beta_{113} = \beta_{131}^*$ has been utilized by Odoulov [90] to demonstrate a new kind of holographic recording, the so-called anisotropic recording. Orthogonally polarized waves, an ordinary and an extraordinary wave, have been superimposed in LiNbO$_3$:Fe. By this means spatially oscillating photovoltaic currents are generated. The symmetric real part β_{113}^s of β_{113} records a refractive index grating in phase with the polarization pattern, and the antisymmetric imaginary part β_{113}^a a grating phase-shifted by $\pi/2$. The gratings are tilted with respect to the crystal surface even for symmetric incidence of the recording beams and can be read isotropically under the corresponding Bragg angle. This recording process is of importance for the explanation of anisotropic holographic scattering patterns described in Section 4.7.

4.3.4 Photoconductivity

The one-center charge-transport model, as shown in Figure 4.7, was developed in the early days of research on photorefractive effects, and it still holds for many situations. Photons (energy $h\nu$) excite electrons from filled traps, C^0, generating free conduction-band electrons and empty traps, C$^+$. The electrons migrate and are trapped elsewhere by empty traps, C$^+$, converting them to filled traps, C^0. For LiNbO$_3$:Fe, here C^0 and C$^+$ represent Fe^{2+} and Fe^{3+}, respectively. Equations with expressions for generation and recombination of free-charge carriers were used already by Amodei in 1971 [91]. The first complete rate equation considering concentrations for filled and empty traps of charge carriers was introduced in 1975 by Vinetskii and Kukhtarev [92].

Because iron is by far the most important photorefractive dopant of LiNbO$_3$ and LiTaO$_3$, we will use in the following Fe^{2+} as a symbol for filled centers and

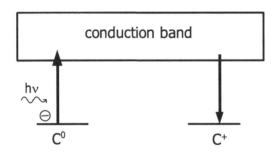

FIGURE 4.7. One-center charge-transport model. C^0 and C^+, filled and empty electron traps; $h\nu$, photon energy.

Fe^{3+} as a symbol for empty centers. However, the same equations hold for other systems such as $Cu^{+/2+}$, $Mn^{2+/3+}$, and $Co^{2+/3+}$.

Excitation of electrons from Fe^{2+} and recombination of conduction-band electrons with Fe^{3+} are described by the rate equation

$$\partial N_{Fe^{2+}}/\partial t = -qsI/(h\nu)N_{Fe^{2+}} + \gamma(N_{Fe} - N_{Fe^{2+}})N_e , \qquad (4.7)$$

where N_e is the concentration of conduction band electrons, q is the quantum efficiency of excitation of an electron upon absorption of a photon, s is the photon absorption cross section, I is the light intensity, $h\nu$ is the photon energy, $N_{Fe^{2+}}$ is the concentration of Fe^{2+} ions, γ is the recombination coefficient, N_{Fe} is the entire Fe concentration, and $N_{Fe} - N_{Fe^{2+}}$ is the concentration of empty iron centers, i.e., of Fe^{3+} ions. The parameter s depends, in general, on the light polarization and wavelength.

Electrons migrate and are trapped elsewhere. Thus a spatially modulated space-charge density ρ builds up. Charge conservation yields the equation

$$\rho = e(N_{Fe^{2+}} + N_e - N_c), \qquad (4.8)$$

where e is the elementary charge, and N_c is a constant concentration of immobile compensation charge that ensures overall charge neutrality.

For illumination with continuous-wave (cw) light, the conductivity σ of photorefractive crystals is typically of the order of 10^{-9} $(\Omega m)^{-1}$ [93]. With a charge carrier mobility of $\mu = 10^{-4}$ $m^2 V^{-1} s^{-1}$ [94], a density of free-charge carriers $N_e = \sigma(e\mu)^{-1} \approx 10^{14}$ m^{-3} follows. This value is much smaller than typical concentrations of photorefractive centers, e.g., 50 mol ppm $\approx 10^{24}$ m^{-3}. Although the mobility value was determined for highly reduced and highly conducting samples and hence the "real" mobility might be less by some orders of magnitude, the free-charge-carrier density is still negligible compared with the concentrations of photorefractive centers. Illumination with high-intensity light pulses can excite many more charge carriers. Depending on the technique of pulse generation

(Q-switching, mode-locking, pulse compression), pulses with typical durations of 10 ns, 30 ps, and 50 fs are used, which are referred to as ns, ps, and fs light pulses. With ps and fs pulses, N_e values comparable with the concentrations of photorefractive centers might be achieved [95–97], while for ns light pulses, N_e is still negligible. Thus in (4.8) the concentration N_e can be neglected in most cases.

However, in the continuity equation

$$\partial j/\partial x + \partial \rho/\partial t = 0, \tag{4.9}$$

the time derivative $\partial \rho/\partial t$ occurs, and because of $\partial N_c/\partial t = 0$ it is necessary to compare $\partial N_e/\partial t$ with $\partial N_{Fe^{2+}}/\partial t$. Switching on the light yields a quick rise of N_e. But then N_e and $N_{Fe^{2+}}$ change on the same time scale, because for constant light intensity further changes of N_e result only from changing concentrations of the photorefractive charge-carrier sources and traps. Then, because of $N_e \ll N_{Fe^{2+}}$, the component $\partial N_e/\partial t$ can also be neglected in $\partial \rho/\partial t$, which is the so-called quasi-steady-state approximation or adiabatic approximation [98]; the concentration N_e reaches nearly instantaneously the equilibrium value that is determined by the actual concentrations of filled and empty traps. The initial rise time of N_e is supposed to be in the ps region for LiNbO₃ crystals. During the initial rise, the adiabatic approximation does not hold. However, for usual light intensities no appreciable fraction of the space-charge field builds up during this short time. For illumination with short high-intensity light pulses the situation changes; the adiabatic approximation cannot be applied if the pulse duration is less than or equal to the rise time of N_e [99].

In terms of the adiabatic approximation, (4.8) yields $\partial N_{Fe^{2+}}/\partial t = 0$. This gives together with (4.7) the solution

$$N_e = \frac{qsI}{\gamma h\nu} \frac{N_{Fe^{2+}}}{(N_{Fe} - N_{Fe^{2+}})}. \tag{4.10}$$

From this equation we get the actual concentration $N_e(r)$ of free electrons in the conduction band inside the crystal at any time and at any place. We just need to know the local light intensity $I(r)$ and the actual concentration of $N_{Fe^{2+}}(r)$. Please note that here $\partial N_{Fe^{2+}}/\partial t = 0$ does not imply that the steady-state situation has been reached. It is just a way to get within the adiabatic approximation a good estimate for the concentration of the electrons in the conduction band. There is a small term $\partial N_{Fe^{2+}}/\partial t \neq 0$ that slowly builds up the spatial modulations of the Fe²⁺ and Fe³⁺ concentrations because of the charge transport.

For the photoconductivity we can write

$$\sigma = e\mu \frac{qsI}{\gamma h\nu} \frac{N_{Fe^{2+}}}{(N_{Fe} - N_{Fe^{2+}})} = e\mu \frac{qsI}{\gamma h\nu} \frac{N_{Fe^{2+}}}{N_{Fe^{3+}}}, \tag{4.11}$$

i.e., the photoconductivity is proportional to the light intensity I as well as to the Fe²⁺/Fe³⁺ concentration ratio. This is excellently confirmed by experiments, as Figure 4.8 shows. For high doping levels (more that 20×10^{24} m⁻³ Fe in LiNbO₃)

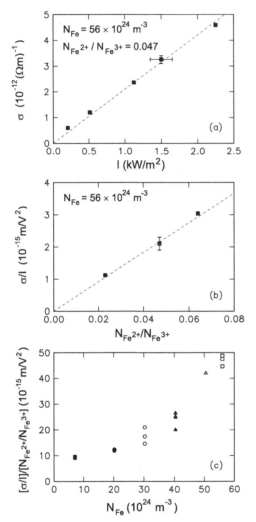

FIGURE 4.8. LiNbO$_3$: (a) Photoconductivity σ vs. light intensity I, (b) specific photoconductivity σ/I vs. Fe^{2+}/Fe^{3+} concentration ratio $N_{Fe^{2+}}/N_{Fe^{3+}}$, and (c) specific photoconductivity that is normalized to the Fe^{2+}/Fe^{3+} concentration ratio vs. overall iron concentration N_{Fe} [20].

an increase of the photoconductivity is observed that cannot be explained by the model described above. Most probably light-assisted tunneling of electrons between iron centers is responsible for this dependence [20].

It is known that near-stoichiometric crystals, i.e., crystals with an increased Li content, and that crystals doped with several percent Mg, Zn, or In show an increased photoconductivity [53, 57, 100–103]. Two possible explanations are that the trapping rate decreases, i.e., that the coefficient γ is reduced, and that the excited charge carriers stay longer in the band. An alternative explanation is that the charge-carrier mobility increases because less-intrinsic crystal defects (Li$^+$

vacancies and Nb^{5+} ions on Li sites) disturb the charge transport, and hence the charge carrier mobility μ is increased.

Besides the photoconductivity, the dark conductivity is another important property. For standard doping levels of transition metals like Fe, Cu, and Mn, the dark conductivity is not influenced by these centers. Instead, an ionic conductivity dominates the processes in the dark [104, 105], e.g., protons (H$^+$ ions) are present in most of the crystals and migrate in electric fields. This influences and limits fixation of holograms (see Section 4.6.3 and Chapter 12, Volume I). Usually, the photoconductivity at low intensities (100 mW/cm^2) exceeds the dark conductivity by many orders of magnitude. Thus for recording and buildup of space-charge fields and of refractive-index changes the dark conductivity does not play a role in LiNbO$_3$ and LiTaO$_3$. An exception is the situation in heavily reduced crystals. In such samples many polarons (electron trapped by Nb^{5+} giving Nb^{4+}) or even bipolarons (Nb$_{Li}^{4+}$ and Nb$_{Nb}^{4+}$) exist, which increases the dark conductivity to a great extent [106–108].

4.3.5 Electron–Hole Competition

So far, we have considered electron transport only. However, excitation of holes is possible as well. Photons excite electrons from the valence band into, e.g., Fe^{3+} and therefore generate Fe^{2+} plus a hole in the valence band. This hole can migrate and trap an electron from another Fe^{2+} center elsewhere.

Conventional Hall measurements fail most of the time because of insufficient conductivity. However, holography allows also to determine the relative contributions of electrons and holes to the charge transport [83]: Electrons and holes both diffuse from the bright areas, where they are present in a high concentration, to the darker areas, where their concentration is smaller. Because of the different charge-carrier signs the electric diffusion currents have different directions, and hence the generated diffusion field depends on whether electrons or holes diffuse. Interference of two laser beams generates a light pattern inside the crystal that is transposed to a refractive-index pattern. However, the response is nonlocal, i.e., the refractive-index pattern can be shifted with respect to the light pattern. Because the recording light fulfills the Bragg condition, even during recording, part of the light is diffracted from one recording beam into the other recording beam. Thus behind the crystal each beam is the superposition of two waves: a transmitted and a diffracted one. These waves interfere, and the phase shift between light and refractive-index pattern determines whether, e.g., constructive or destructive interference takes place. Since this phase shift is influenced by the diffusion field and thus by the sign of the charge carriers, the intensity coupling between the two recording beams gives information about the sign of the charge carriers that are responsible for the charge transport [83].

Figure 4.9 shows results obtained with LiNbO$_3$. For oxidized crystals (larger Fe^{3+} concentration) the excitation of holes dominates, while for more reduced

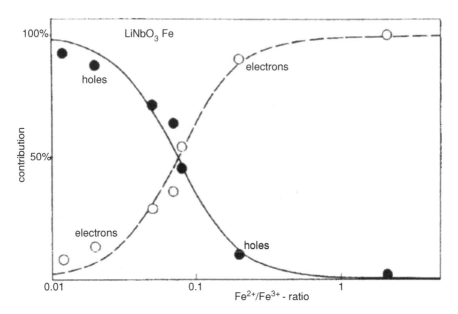

FIGURE 4.9. Relative contribution of electrons and holes to the charge transport in LiNbO₃:Fe for various oxidization/reduction states, i.e., Fe^{2+}/Fe^{3+} concentration ratios. These experiments were performed with ultraviolet light at the wavelength 350.7 nm [83].

ones (larger Fe^{2+} concentration) the electrons dominate. These measurements were done with ultraviolet light. For visible light electrons dominate the charge transport in LiNbO₃:Fe over the entire oxidization/reduction range.

Detailed descriptions of the charge transport, considering simultaneous electron and hole currents, were presented in 1986 by Valley [109] and by Strohkendl, Jonathan, and Hellwarth [110]. Such effects can be observed in many photorefractive materials. However, in LiNbO₃ and in LiTaO₃, in most of the cases electron transport clearly dominates.

4.4 Photorefractive Properties and Performance Limits

This section focuses on the key properties of LiNbO₃ and LiTaO₃ for applications. The light-induced charge transport yields the buildup of an electrical space-charge field E_{sc}, which happens in a characteristic time τ_{sc}. We will show the dependences of these parameters on the charge-transport properties that have been introduced in the last two sections and derive two interesting figures of merit: the saturation values of the refractive-index changes Δn_s that can be obtained in the material and the sensitivity S as a metric for the storage speed that can be reached.

4.4.1 Steady-State Space-Charge Field

To describe the buildup of the electric space-charge field and to obtain information about dynamics and magnitude, the rate equation (4.7), continuity equation (4.9), and Poisson equation (Figure 4.1) must be solved considering the space-charge density ρ (4.8) and current density

$$j = \sigma E + \kappa' I N_{Fe^{2+}} + k_B T \mu_e \nabla N_e . \qquad (4.12)$$

The three terms in (4.12) result from drift, bulk photovoltaic effect, and diffusion. The equations hold for pure electronic transport and negligible hole conductivity, which is valid for LiNbO$_3$ and LiTaO$_3$ in most of the cases. In general, solution of this set of equations can be performed only by computer simulations. However, for the case of a sinusoidal interference pattern, since it is created by two plane waves, an analytic result can be derived. The light intensity of two superimposed plane waves can be written as

$$I(z) = I_0[1 + m \sin(Kz)] , \qquad (4.13)$$

where z is the spatial coordinate, I_0 is the averaged light intensity, m is the modulation degree of the light pattern, and $K = (2\pi)/\Lambda$ is the spatial frequency, with Λ the period length of the grating determined by the light wavelength and the experimental geometry. The calculation is limited to the z direction and thus to one dimension in space.

The solution was first obtained by Kukhtarev et al. in 1976 and 1979 [98, 111, 112]. The idea is to perform a Fourier development of all quantities that depend on the spatial coordinate and to use in the calculations the coefficients of the Fourier developments instead of the full spatially dependent quantities.

The light intensity distribution can be written as

$$I(z) = I_0 + (1/2)[I_1 \exp(iKz) + c.c.] \qquad (4.14)$$

with $I_1 = m I_0$ and $c.c.$ as the abbreviation for complex conjugate. Zero and first Fourier components are denoted by the indices 0 and 1. The same Fourier development is performed, e.g., for the concentrations $N_{Fe^{2+}}$ and $N_{Fe^{3+}}$, for the current density j, for the charge density ρ, and for the space-charge field E_{sc}.

Calculations are done with zero- and first-order Fourier components. Higher Fourier orders are neglected, which is appropriate in case of small modulation degrees $m \ll 1$. The reason is simple: The kth Fourier order decreases with m^k. Thus the second-order terms are a factor of m smaller than the first-order terms. However, it does not make sense to neglect the first-order terms compared to the zero-order ones: We are interested in the modulated part of the space-charge field, i.e., in E_1. This information would be lost if we just dealt with the zero-order components. In the following we will present the results of such an analysis.

For the nonconstant part of electric space-charge field E_{sc}, we get

$$E_{sc} \approx E_1 = -\frac{E_0 + E_{phv} + iE_D}{1 + E_D/E_q - iE_0/E_q - iE_{phv}/E_q'}, \tag{4.15}$$

$$E_{phv} = \frac{j_{phv}}{\sigma_{ph}} = \frac{\kappa'\gamma h\nu}{e\mu qs}N_{Fe^{3+}}, \tag{4.16}$$

$$E_D = \frac{k_B T}{e}K, \tag{4.17}$$

$$E_q = \frac{e}{\epsilon\epsilon_0 K}\left(\frac{1}{N_{Fe^{2+}}} + \frac{1}{N_{Fe^{3+}}}\right)^{-1}, \tag{4.18}$$

$$E_q' = \frac{e}{\epsilon\epsilon_0 K}N_{Fe^{2+}}, \tag{4.19}$$

where E_0 is an externally applied electric field, E_{phv} is the bulk photovoltaic field, E_D is the diffusion field, and E_q and E_q' are space-charge limiting fields. For short-circuited conditions, the field E_0 has to be set to zero. The imaginary unit i in front of E_D in (4.15) indicates that there is a 90° phase shift between the light pattern (4.14) and the diffusion-driven modulated component of the space-charge field, while the modulated component of E_{sc} that is caused by drift in an external field and by the bulk photovoltaic effect is in phase with respect to the light pattern. The absolute amplitude of the entire modulated space-charge field can easily be derived from from (4.15):

$$E_{sc} = \left[\frac{(E_0 + E_{phv})^2 + E_D^2}{(1 + E_D/E_q)^2 + (E_0/E_q + E_{phv}/E_q')^2}\right]^{1/2}. \tag{4.20}$$

During recording, the modulation degree m of the light pattern is larger than zero. Setting later m equal to zero yields the condition for erasure, i.e., illumination with homogeneous light intensity. Solving the above-mentioned equations yields an important outcome: Buildup and decay follow monoexponential laws in time with the time constant

$$\tau_{sc} = \frac{1 + (E_D - iE_0)/E_M}{1 + E_D/E_q - iE_0/E_q - iE_{phv}/E_q'}\tau_M, \tag{4.21}$$

with

$$E_M = \frac{\gamma N_{Fe^{3+}}}{\mu}\frac{1}{K}, \tag{4.22}$$

$$\tau_M = \frac{\epsilon\epsilon_0}{\sigma_0}, \tag{4.23}$$

$$\sigma_0 = e\mu\frac{qsI_0}{\gamma h\nu}\frac{N_{Fe^{2+}}}{N_{Fe^{3+}}}. \tag{4.24}$$

Hence a τ_{sc} with a significant imaginary part causes an oscillatory component of the evolution of the space-charge field [98].

4.4.2 Simplified Solutions

In most practical situations, some assumptions can be applied that simplify the formulas (4.15) to (4.24):

- The sample is short-circuited, i.e., no external field is applied ($E_0 = 0$).
- The bulk photovoltaic field is much larger than the diffusion field ($E_{phv} \gg E_D$). This is the case for sufficient doping ($N_{Fe^{3+}}$ in (4.16) must be large enough) and for gratings that do not have a very small period length Λ ($K = 2\pi/\Lambda$ in (4.17) should be small).
- The concentrations of filled traps $N_{Fe^{2+}}$ and of empty traps $N_{Fe^{3+}}$ are large enough to ensure $E_q \gg E_D$ and $E'_q \gg E_{phv}$. This implies that no space-charge-field-limiting effects are present: enough filled and empty traps are available to support the buildup of the space-charge field.
- The mentioned assumptions immediately imply $E_M \gg E_D$.

For this situation, the following relations are valid for recording and erasure of the space-charge field and thus for the refractive-index changes:

$$E_{sc} = -E_{phv}[1 - \exp(-t/\tau_M)] \qquad \text{(recording)}, \qquad (4.25)$$

$$E_{sc} = E_{sc,0}\exp(-t/\tau_M) \qquad \text{(erasure)}. \qquad (4.26)$$

Here $E_{sc,0}$ denotes the initial space-charge field at the beginning of the erasure process. The recording and erasure dynamics are determined by the constant $\tau_{sc} \approx \tau_M$ and hence by the conductivity σ. The saturation value of the space-charge field is given by the bulk photovoltaic field in (4.16).

From the presented analysis, some characteristic dependences follow that can be checked experimentally. On the one hand, such a check establishes whether the model is correct; on the other hand, it yields interesting material parameters.

Important dependences provided by (4.1), (4.4), (4.11), and (4.16) are

$$j_{phv} \propto I N_{Fe^{2+}}, \qquad (4.27)$$

$$\tau_M^{-1} \propto \sigma_0 \propto I_0(N_{Fe^{2+}}/N_{Fe^{3+}}), \qquad (4.28)$$

$$\Delta n_s \propto E_{sc} \propto N_{Fe^{3+}}. \qquad (4.29)$$

Here the index "s" indicates the saturation value in the steady-state situation. These predictions have been experimentally verified for moderate doping levels $N_{Fe} < 20 \times 10^{24}\,\text{m}^{-3}$: (1) The bulk photovoltaic current density j_{phv} is proportional to the light intensity I and to the Fe^{2+} concentration $N_{Fe^{2+}}$ [17,36]. (2) The conductivity σ is proportional to the light intensity I and to the concentration ratio of filled and empty traps $N_{Fe^{2+}}/N_{Fe^{3+}}$ [36,113–115]. We also refer to Sections 4.3.3 and 4.3.4, where more detailed results are shown. (3) The saturation value of the refractive-index amplitude Δn_s is proportional to the concentration of empty traps $N_{Fe^{3+}}$ [115,116] (Fig. 4.10).

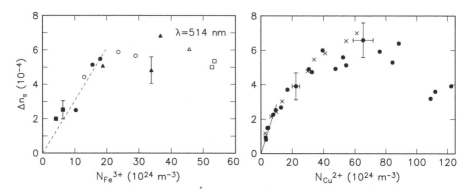

FIGURE 4.10. Saturation values of the refractive-index changes Δn_s for ordinarily polarized light vs. concentration of empty traps. Left: iron-doped $LiNbO_3$ with $N_{Fe^{3+}}$ as trap concentration. Right: copper-doped $LiNbO_3$ with $N_{Cu^{2+}}$ as trap concentration.

4.4.3 Refractive-Index Changes

For high diffraction efficiencies η, large values for the refractive-index changes Δn are necessary [117]:

$$\eta = \sin^2 \left(\frac{\Delta n \pi d}{\lambda \cos \theta} \right) , \qquad (4.30)$$

where λ is the vacuum wavelength and θ is half the angle between the two interfering writing beams in the material. Following the predictions of (4.29), high doping levels yield large values of Δn_s. However, experiments reveal that (4.29) is valid up to moderate doping levels only. Figure 4.10 shows Δn_s for iron-doped $LiNbO_3$ (left part) and copper-doped $LiNbO_3$ (right part) vs. the concentration of empty traps [20,118]. The data were obtained using ordinarily polarized light. The linear dependence derived from the one-center model is valid up to doping levels of $N_{Fe} \approx 20 \times 10^{24}$ m^{-3} and $N_{Cu} \approx 10 \times 10^{24}$ m^{-3}. The largest value obtainable in singly doped $LiNbO_3$ is $\Delta n_s = 7 \times 10^{-4}$. Increasing the doping level does not boost the refractive-index changes. In $LiTaO_3$ doped with iron the situation is similar: A linear dependence $\Delta n_s \propto N_{Fe^{3+}}$ is valid up to doping levels about 30×10^{24} m^{-3}, and a maximum value of Δn_s of 3.5×10^{-4} can be obtained.

The reason for this unexpected behavior is a growth in the specific conductivity $f_{Fe} = (\sigma/I)/(N_{Fe^{2+}}/N_{Fe^{3+}})$ and $f_{Cu} = (\sigma/I)/(N_{Cu^+}/N_{Cu^{2+}})$. The specific conductivity increases with larger doping levels (see Section 4.3.4 and Figure 4.8), whereas the specific bulk photovoltaic effect remains almost unaffected. Thus the values for Δn_s cannot be improved by increasing the total doping level. The physical reason for the increase of f is direct tunneling of electrons between the traps [104].

Applying very high external electric fields, however, helps increasing Δn_s in iron-doped $LiNbO_3$. Luennemann et al. applied 65 kVcm^{-1} and demonstrated $\Delta n_s = 11 \times 10^{-4}$ for ordinarily polarized light [119].

4.4.4 Sensitivity

The sensitivity S is a measure for the buildup speed of holograms normalized to the light intensity I_0 and the crystal thickness d:

$$S = \frac{1}{Id} \left. \frac{\partial \sqrt{\eta}}{\partial t} \right|_{t \ll \tau_M} . \tag{4.31}$$

Equation (4.31) can be rewritten using the (4.1), (4.25), (4.30) as

$$S = \frac{\pi n_{o,e}^3 r_{13,33}}{2I\lambda \cos \theta} \left| \frac{E_{sc}}{\tau_{sc}} \right| . \tag{4.32}$$

For the conditions treated in Section 4.4.2, the sensitivity S is proportional to the concentration of filled traps $N_{Fe^{2+}}$:

$$S \approx \frac{\pi n_{o,e}^3 r_{13,33} \kappa' N_{Fe^{2+}}}{2\lambda \cos \theta \epsilon \epsilon_0} . \tag{4.33}$$

No other parameters allow one to change S for the specific crystal. Typical values that can be obtained are up to 10 cmJ^{-1} [20]. The sensitivity of LiTaO$_3$, however, is smaller: For blue light ($\lambda = 488$ nm), typical values of $S < 1$ cmJ^{-1} are state of the art, but in LiTaO$_3$ the specific bulk photovoltaic coefficient κ' has its maximum at $\lambda = 400$ nm. For both materials, the dependence $S \propto N_{Fe^{2+}}$ has been experimentally confirmed.

In the case of an externally applied, very large electric field E_0 the situation is different: Here (4.32) together with (4.20) yields $S \propto E_0$. Luennemann et al. confirmed this dependence [119]. Using ordinarily polarized light, they increased the best values for S up to 40 cmJ^{-1} with an applied field of 65 kVcm^{-1} in LiNbO$_3$:Fe.

4.4.5 Problems

The photorefractive properties of singly doped LiNbO$_3$ that have been experimentally found are in excellent agreement with the predictions of the one-center model for typical continuous-wave light intensities and not too high doping levels (20×10^{24} m^{-3} for iron doping). Therefore, the one-center model has become widely accepted.

For practical applications, however, singly doped LiNbO$_3$ has drawbacks. If holograms are stored, the readout has to be done using light with the same wavelength as for recording. This consequently erases the stored information. Furthermore, since absorption is needed to create photoconductivity, the same absorption is present when the readout light reconstructs the hologram, which depletes the transmitted light intensity and thus limits the diffraction efficiencies that can be reached. Furthermore, the saturation values of the refractive-index changes are limited, such that an upper limit of the storage capacity cannot be exceeded [20]. Performing the hologram readout with a different light wavelength that is not absorbed works well with elementary holographic gratings, i.e., sinusoidally modulated refractive-index changes, but complete reconstruction of more complex holograms fails due to the Bragg condition. To overcome this problem in $LiNbO_3$ and $LiTaO_3$ two methods have been developed: thermal fixing (see Section 4.6 and Chapter 12 of Volume I) and two-step recording (see Section 4.5.5 and Chapter 8 of Volume I).

4.5 Effects at High Light Intensities

The light-induced charge transport in doped $LiNbO_3$ and $LiTaO_3$ is often described in the frame of a band model with one photorefractive center, e.g., $Fe^{2+/3+}$, $Cu^{+/2+}$, $Mn^{2+/3+}$ (Figure 4.7). This is mostly sufficient if the intensities used of the laser light are not too high ($I < 1\,W/cm^2$). At high light intensities, e.g., if laser pulses or focused cw laser light is used, effects can occur that cannot be explained with only one photorefractive center. Light-induced absorption changes, a photoconductivity that increases superlinearly with increasing intensity, and refractive-index changes that depend on light intensity can be understood with a two-center charge transport model. In this case, the intrinsic defect, $Nb^{4+/5+}$ at Li site ($Ta^{4+/5+}$ at Li site in $LiTaO_3$), is also involved in the light-induced charge transport. Two-step recording in $LiNbO_3$ and $LiTaO_3$ is a further example in which intrinsic defects have to be considered (Chapter 8, Volume I).

The first model with two photorefractive centers was proposed by Chen et al. [120], and Valley could explain a nonexponential decay of a refractive index grating during erasure with a second center [121]. Brost et al. presented a theory to describe light-induced absorption changes in $BaTiO_3$ [122], and Holtmann et al. could explain with this model a photoconductivity that increases sublinearly with increasing intensity in $BaTiO_3$ [123, 124]. However, holographic experiments with iron-doped $LiNbO_3$ using laser pulses could be described only if direct electronic transitions between Fe^{2+} and Nb_{Li}^{5+} ions were taken into account [125]. This is justified because congruently melting $LiNbO_3$ contains many intrinsic Nb_{Li} defects (Section 4.3.2), i.e., the distance between Fe and Nb_{Li} ions is small. This might be different in nearly stoichiometric crystals or in crystals doped with, e.g., Mg or Zn.

4.5.1 Two-Center Charge Transport Model

The charge-transport model with two photorefractive centers proposed by Jermann and Otten is illustrated in Figure 4.11 [126]. In the following, we assume $Fe^{2+/3+}$ as a deep center and $Nb_{Li}^{4+/5+}$ as a shallow center, but the description is also valid for other dopants like Cu and Mn and for $LiTaO_3$ with $Ta_{Li}^{4+/5+}$ as shallow center. In the dark at room temperature the shallow levels are thermally depopulated. The different electronic transitions are described with the following

FIGURE 4.11. Two-center charge-transport model. C^0 and C$^+$, filled and empty deep electron traps; X^0 and X$^+$ filled and empty shallow electron traps; $h\nu$, photon energy.

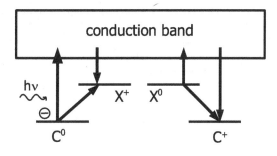

set of rate equations:

$$\frac{\partial N_{Fe^{2+}}}{\partial t} = -[s_{Fe}q_{Fe}I/(h\nu) + s_{FeX}q_{FeX}(N_X - N_{X^0})I/(h\nu)]N_{Fe^{2+}}$$
$$+ (\gamma_{Fe}N_e + \gamma_{XFe}N_{X^0})(N_{Fe} - N_{Fe^{2+}}), \qquad (4.34)$$

$$\frac{\partial N_{X^0}}{\partial t} = -[\beta_X + s_Xq_XI/(h\nu) + \gamma_{XFe}(N_{Fe} - N_{Fe^{2+}})]N_{X^0}$$
$$+ (\gamma_X N_e + s_{Fe}X q_{FeX}I/(h\nu)N_{Fe^{2+}})(N_X - N_{X^0}). \qquad (4.35)$$

The symbols and their meanings can be found in Table 4.4. Excitations of electrons from the valence band are not considered. Nevertheless, in oxidized samples it is also possible to generate holes with ultraviolet light (Section 4.3.5).

4.5.2 Light-Induced Absorption Changes

Under illumination, electrons can be excited from Fe^{2+} either directly or through the conduction band into empty shallow centers (Nb$_{Li}^{5+}$), forming small bound polarons (Nb$_{Li}^{4+}$). If the concentration of polarons N_{X^0} is large enough, the absorption coefficient around 760 nm increases significantly. The light-induced absorption α_{li}^λ at a wavelength λ is $\alpha_{li}^\lambda(I) = \alpha^\lambda(I) - \alpha_0^\lambda$, with the absorption coefficient in the presence of illumination $\alpha^\lambda(I)$ and the absorption coefficient without strong illumination α_0^λ. The light-induced absorption depends mainly on the polaron concentration:

$$\alpha_{li}^\lambda(I) \approx (s_X - s_{Fe})N_{X^0}. \qquad (4.36)$$

The absorption cross section of the Fe^{2+} ions s_{Fe} at 760 nm is nearly zero. The absorption cross section of the polarons at 760 nm was measured by Sweeney et al. at a temperature of $T = 77$ K [41]: $s_X \approx 7 \times 10^{-22}$ m^2. Jermann et al. determined an absorption cross section of $s_X \approx 4 \times 10^{-22}$ m^2 at room temperature [126]. Since the photoconductivity $\sigma_{ph} = e\mu N_e$ is always proportional to the concentration ratio $N_{Fe^{2+}}/N_{Fe^{3+}}$, we assume that the population of shallow centers is mainly

TABLE 4.4. Symbols and their meanings used in the rate
equations (4.34) and (4.35) describing the two-center model

Symbol	Meaning
$N_{Fe^{2+}}$	concentration of Fe^{2+} ions
N_{Fe}	concentration of Fe ions
$s_{Fe}q_{Fe}$	absorption cross section and quantum efficiency for absorption and excitation of an electron from Fe^{2+} to the conduction band
γ_{Fe}	coefficient for recombination of an electron from the conduction band with Fe^{3+}
N_{X^0}	concentration of filled shallow levels (Nb_{Li}^{4+}; Ta_{Li}^{4+})
N_X	concentration of all shallow levels (Nb_{Li}; Ta_{Li})
β_X	thermal excitation rate of an electron from a filled shallow level (Nb_{Li}^{4+}; Ta_{Li}^{4+}) to the conduction band
$s_X q_X$	absorption cross section and quantum efficiency for absorption and excitation of an electron from a filled shallow level (Nb_{Li}^{4+}; Ta_{Li}^{4+}) to the conduction band
$s_{FeX}q_{FeX}$	absorption cross section and quantum efficiency for absorption and excitation of an electron from Fe^{2+} to an empty shallow level (Nb_{Li}^{5+}; Ta_{Li}^{5+})
γ_X	coefficient for recombination of an electron from the conduction band with an empty shallow level (Nb_{Li}^{5+}; Ta_{Li}^{5+})
γ_{XFe}	coefficient for recombination of an electron from a filled shallow level (Nb_{Li}^{4+}; Ta_{Li}^{4+}) with Fe^{3+}
N_e	concentration of electrons in the conduction band
I	light intensity

achieved by direct transitions of electrons between Fe^{2+} and Nb^{5+} ions ($\gamma_X \approx 0$).
If the variation of the concentrations $N_{Fe^{2+}}$, $N_{Fe^{3+}}$, and N_{X^+} under illumination is
small, (4.35) directly yields the time dependence of the polaron concentration:

$$N_{X^0} = N_{X^0}^{sat}[1 - \exp(-t/\tau)], \qquad (4.37)$$

with the saturation value

$$N_{X^0}^{sat} = s_{FeX}q_{FeX} I/(h\nu) N_{Fe^{2+}} N_X \tau \qquad (4.38)$$

and the time constant

$$\tau = [\beta_X + \gamma_{XFe} N_{Fe^{3+}} + s_X q_X I/(h\nu)(N_{Fe^{2+}} + N_X)]^{-1}. \qquad (4.39)$$

Relaxation of Light-Induced Absorption Changes:

According to the two-center model, the relaxation should be exponential with a typical time constant $\tau_X = 1/(\beta_X + \gamma_{XFe} N_{Fe^{3+}})$. However, the best fit to the relaxation process is achieved with a stretched exponential function

$$\alpha_{li}(t) = \alpha_{li,0} \exp\left[-(t/\tau_X)^\beta\right], \tag{4.40}$$

with a stretch factor β varying between 0 and 1 [43, 127]. Numerical simulations have shown that the relaxation process can be described with a stretched exponential function if the probability of the recombination of an electron from a filled shallow center with an empty deep center depends on the distance between these two centers [43]. A typical value of the stretch factor β is 0.3. The consequence of a distance-dependent relaxation process is that the time constant τ_X strongly depends on the concentration of deep electron traps [127]:

$$\tau_X \propto (N_{Fe^{3+}})^{-1/\beta}, \tag{4.41}$$

with the stretch factor β of (4.40). The lifetime τ_X of the polarons in the dark also depends on the total concentration N_X of antisite defects and on temperature T. The temperature dependence of the relaxation process can be described with an Arrhenius fit $\tau_X \propto \exp[E_A/(k_B T)]$ with the activation energy E_A and the Boltzmann constant k_B. In congruently melting iron-doped LiNbO₃ and LiTaO₃ crystals the time constant τ_X is of the order of microseconds and the activation energy is about 0.17 eV in LiNbO₃:Fe and about 0.26 eV in LiTaO₃:Fe [43, 127]. On the other hand, the activation energy in nearly stoichiometric crystals (49.5 mol % < c_{Li} < 50.0 mol %) is much larger ($E_A \approx 0.6$ eV) [43]. At room temperature the lifetime of the polarons in nearly stoichiometric crystals is on the order of milliseconds or higher and therefore larger than τ_X in congruently melting crystals. A large lifetime can lead to a high concentration of polarons under illumination with cw laser light.

Illumination with CW Laser Light:

Illuminating a LiNbO₃:Fe crystal with green continuous wave laser light leads to the generation of Nb_{Li}^{4+} polarons and therefore to a light-induced absorption change α_{li}. The polaron concentration reaches the saturation value $N_{X^0}^{sat}$ (4.38). If the laser intensity I is not too high, N_{X^0} increases linearly with increasing lifetime τ_X and intensity. Equations (4.36) and (4.38) yield the saturation value of light-induced absorption changes:

$$\alpha_{li}^{sat} \approx (s_X - s_{Fe}) s_{FeX} q_{FeX} I/(h\nu) N_{Fe^{2+}} N_X \tau_X. \tag{4.42}$$

Since the polaron lifetime in congruently melting crystals is small, focused laser light with an intensity larger than 100 kW/m² is needed to induce absorption changes on the order of 1 m⁻¹. In nearly stoichiometric doped crystals, light-induced absorption changes occur even at small intensities. Such crystals can be used for two-step recording with cw laser light [108, 128, 129]. It is obvious that "real" stoichiometric crystals ($c_{Li} = 50\,\text{mol}\,\%$) with no intrinsic antisite defects ($N_X = 0$) show no absorption changes at all.

Illumination with Laser Pulses:

If the pulse duration t_P is much smaller than the lifetime of the polarons τ_X—this is the case for nanosecond pulses—no considerable recombination of polarons with empty deep centers takes place during illumination. Then, the polaron concentration does not depend on the lifetime τ_X, and the saturation value of light-induced absorption changes for moderate pulse laser intensities is ((4.36) and (4.38))

$$\alpha_{li}^{sat} \approx (s_X - s_{Fe})N_{X^0} \quad \text{with} \quad N_{X^0} = s_{Fe}q_{FeX}\,I/(h\nu)\,N_{Fe^{2+}}N_Xt_P. \quad (4.43)$$

This dependence has been experimentally verified for iron-doped LiNbO$_3$ and LiTaO$_3$ [44, 126, 130]. In a LiNbO$_3$:Fe crystal, containing about $10^{24}\,\text{m}^{-3}$ Fe^{2+} ions, absorption changes of about $100\,\text{m}^{-1}$ at $\lambda = 633\,\text{nm}$ are achieved with pulse intensities of the order of $100\,\text{GW/m}^2$. The polaron concentration under pulse illumination is smaller in LiNbO$_3$:Fe if the crystals are doped additionally with Mg or Zn [130] because Mg and Zn are incorporated at Li site and reduce the concentration of intrinsic antisite defects (smaller N_X).

4.5.3 Photoconductivity

The photoconductivity $\sigma_{ph} = e\mu N_e$ is always proportional to the concentration ratio $N_{Fe^{2+}}/N_{Fe^{3+}}$ even at high laser intensities [125]. Considering charge conservation and the adiabatic approximation ($\partial N_e/\partial t = 0$), the rate equations (4.34) and (4.35) yield the steady-state concentration of electrons in the conduction band:

$$N_e = \frac{s_{Fe}q_{Fe}I/(h\nu) + (\beta_X + s_Xq_XI/(h\nu))N_{X^0}}{\gamma_{Fe}N_{Fe^{3+}}}. \quad (4.44)$$

For small intensities the polaron concentration can be neglected ($N_{X^0} = 0$) and the photoconductivity is the same as if only one center were considered ((4.11)). But at high intensities the photoconductivity can be better described with an additional quadratic intensity dependence, because the polaron concentration N_{X^0} increases with increasing intensity ((4.43)) [125, 126], and hence two-step excitations become possible, i.e., excitation of electrons from polarons to the conduction band generates an additional contribution to the conductivity.

4.5.4 Refractive-Index Changes

In iron-doped LiNbO$_3$ and LiTaO$_3$ the saturation values of refractive-index changes Δn_s are mainly determined by the bulk photovoltaic space-charge field $E_{phv,Fe}$. If under illumination the concentration of filled shallow centers is very large, the polarons can increase the space-charge field because they contribute to the bulk photovoltaic charge transport. The bulk photovoltaic coefficient of the Nb$_{Li}^{4+}$ polarons at $532\,\text{nm}$ is about one order of magnitude larger than the bulk photovoltaic coefficient of Fe^{2+} in LiNbO$_3$ [126]. The space-charge field is a sum of the bulk photovoltaic field $E_{phv,Fe}$ and the bulk photovoltaic field $E_{phv,X}$ due to the polarons:

$$E_{sc} = E_{phv,Fe} + E_{phv,X} = j_{phv,Fe}/\sigma + j_{phv,X}/\sigma . \qquad (4.45)$$

For small light intensities the bulk photovoltaic current density of the Fe ions dominates ($j_{phv,Fe} \propto I N_{Fe^{2+}}$), while at higher intensities ($I > 10\,\text{MW/m}^2$) a contribution to the current density appears that increases quadratically with light intensity ($j_{phv,X} \propto I N_{X^0} \propto I^2$) [131]. The quadratic component is lowered if the LiNbO$_3$:Fe crystals are additionally doped with Mg or Zn, or if their Li concentration is increased via vapor transport equilibration. Considering the intensity dependence of the photoconductivity σ ((4.44)), the space-charge field and hence the refractive-index change depend on light intensity. For small light intensities Δn_s is constant, increases with increasing intensity, and saturates for very high intensities. This is also the case if focused cw laser light is used. For light intensities higher than $10\,\text{MW/m}^2$, refractive-index changes for extraordinarily polarized light as high as 3×10^{-3} can be achieved [132].

An additional effect that influences the refractive-index changes at high light intensities should be mentioned here. Illumination heats a crystal by light absorption, and thermal gratings can build up [133]. Temperature modulations yield transient refractive-index changes due to the pyroelectric and thermooptic effect. During hologram recording with many laser pulses it is also possible that an additional contribution to the space change field builds up, if simultaneously free-charge carriers are generated that can compensate for the pyroelectric field. However, the pyroelectric space-charge field can usually be neglected in LiNbO$_3$ and LiTaO$_3$ because its value is typically one order of magnitude smaller than the bulk photovoltaic space-charge field.

4.5.5 Two-Step Recording with Infrared Light

Small bound polarons can be used to record holograms with infrared pulses via a two-step process (see also Chapter 8 of Volume I). This has been demonstrated in iron-doped [134] and copper-doped [135] LiNbO$_3$ as well as in iron-doped LiTaO$_3$ [44, 136]. LiNbO$_3$:Fe is homogeneously illuminated with a green pulse ($\lambda = 532\,\text{nm}$) to sensitize the crystal for hologram recording with infrared light ($\lambda = 1064\,\text{nm}$). The green pulse excites electrons from Fe^{2+} to Nb$_{Li}^{5+}$ forming Nb$_{Li}^{4+}$ polarons. Subsequent illumination with two superimposed infrared pulses excites electrons from the filled shallow centers to the conduction band. The electrons migrate and are finally trapped in deep empty centers. The stored hologram can be read nondestructively with infrared pulses because the photon energy of the infrared light is not sufficient to excite electrons from Fe^{2+} and all shallow centers are empty during readout.

In this two-step process, the photoconductivity is mainly determined by the intensity of the green sensitizing light ($\sigma_{ph} \propto I_{532}$), while the saturation value of refractive-index changes increases with increasing intensity of the recording light ($\Delta n_s \propto I_{1064}$). The current density depends mainly on the bulk photovoltaic current due to the polarons $j \propto I_{1064} N_{X^0}$. If the intensity of the green pulses is not too high, the polaron concentration is $N_{X^0} \propto I_{532}$ ((4.43)). This yields the dependence of Δn_s on the intensity of the recording light: $\Delta n_s \propto E_{sc} = j/\sigma \propto I_{1064}$.

4.6 Thermal Fixing

Using lithium niobate as a holographic storage material for either elementary holographic gratings or more sophisticated holograms, a crucial problem appears: The readout of a stored hologram by illuminating the crystal with the reference light erases the previously stored information; the recorded refractive-index changes decay. Thus a method for nondestructive readout is highly desirable. The most important solution for this problem was discovered by Amodei and Staebler in 1971: They heated the crystals after or during the recording of a hologram. After cooling down the crystal, no or just very small refractive-index changes can be detected [26, 137]. Homogeneous illumination, however, now yields the buildup ("development") of strong refractive-index changes that are stable against the readout light. In this section we will describe the details of the processes taking part in thermal fixing during the recording, development and readout of thermally fixed holograms. In our description we will focus on LiNbO$_3$, but the processes in LiTaO$_3$ are similar [38].

4.6.1 Recording Process

Heating a LiNbO$_3$ crystal during the recording of holograms to a temperature of $T \approx 180\,^{\circ}\text{C}$ yields just small diffraction efficiencies right after the recording and cooling down process. The two possible schedules are illustrated in Figure 4.12. One method is shown in the *left part* (this method is sometimes called "low-high-low"): First a holographic grating is recorded at room temperature. Heating of the crystal follows; the refractive-index changes vanish. After cooling down, homogeneous illumination yields a permanent grating. During the heating of the crystal, the electric space-charge field decays, but the reason cannot be an erasure of the electronic grating, because efficient thermal excitation of electrons takes place at temperatures above 200 °C [138]. Thus a different mechanism compensates the space-charge field. The *right part* of Figure 4.12 illustrates true high-temperature recording (also known as the "high-low" method): The crystal is heated during the recording process. In this phase, almost no refractive-index changes build up. However, all electron transport forces, in particular the bulk photovoltaic effect, are still at play, so that a redistribution of electrons takes place even at the elevated temperature. Thus the space-charge field of the electrons must be compensated by a suitable charge-compensation mechanism that is effective

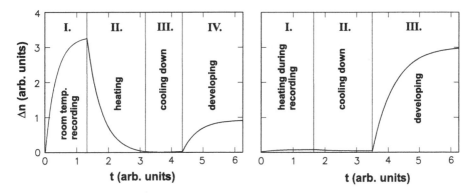

FIGURE 4.12. Schedules for thermal fixing of holograms. **Left:** First, a grating at room temperature is recorded, and a space-charge field builds up (I). Then the crystal is heated (II) and the field decays. After cooling down (III), the hologram can be developed (IV) by homogeneous illumination. **Right:** High-temperature recording yields almost no refractive-index changes during the writing process (I); the space-charge field is instantaneously compensated. After cooling down (II), developing (III) yields the stored hologram.

during the heating. After cooling down, again, developing with homogeneous illumination reveals the stored hologram.

Different suggestions have been made to explain this compensation mechanism: partial reversal of domains [139], mobile silicon [140], moving intrinsic defects or mobile protons [141–143]. Vormann et al. found the solution: They recorded a low-spatial frequency pattern into an iron-doped LiNbO$_3$ crystal and checked spatially resolved the proton concentration N_{H^+} in the material [144]. They reported a slight change in the proton concentration within the illuminated area, but it took another 15 years until Buse et al. investigated this process in more detail [145]. They recorded macroscopic light patterns in copper- and iron-doped LiNbO$_3$ and checked spatially resolved the concentration of filled traps (N_{Cu^+} or $N_{Fe^{2+}}$) and of protons N_{H^+}. The crystals were illuminated with a small light stripe with an $1/e^2$ size of 1 mm and homogeneous intensity perpendicular to the c-axis. The peak intensity of the ordinarily polarized light (wavelength $\lambda = 514$ nm) was 250 kW/m^2. Figure 4.13, left part, shows a photograph of a copper-doped sample (doping level $N_{Cu} = 12 \times 10^{24}$ m^{-3}, size 5×6 mm^2, thickness 1 mm) after illumination for 47 h at 163 °C. A bright stripe where the crystal bleaches out can be observed at the left edge of the formerly illuminated region, whereas at the right edge a dark stripe appears. The crystal color is related to the concentration of electrons in filled traps Cu$^+$, which can be calculated by measuring the absorption at 477 nm [36, 145]. Obviously, a large number of electrons can be redistributed; a modulation of filled traps close to 1 can be established. No electric space-charge field stops the redistribution process. This is completely different from the case of room-temperature recording, in which only a small fraction of the electrons available have to be redistributed to create a large space-charge field. Repeating

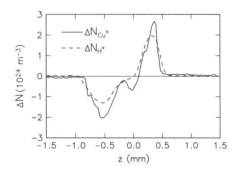

FIGURE 4.13. **Left:** Photograph of a LiNbO$_3$:Cu crystal that has been illuminated with a light stripe (horizontal $1/e^2$-size is 1 mm with vertically homogeneous intensity) for 47 h with a light peak intensity of 250 kW/m^2. A bleached stripe at the left edge of the previously illuminated region and a dark stripe at the right edge can be observed. The c-axis of the crystal points from the left to the right [145]. **Right:** Changes in the concentration of filled copper traps ΔN_{Cu^+} and protons ΔN_{H^+} vs. spatial coordinate z (along the c-axis) in the copper-doped LiNbO$_3$ crystal [145].

the experiment with a similar sample at room temperature yields no changes in the crystal color.

Checking the absorption at 2870 nm caused by protons yields the proton concentration N_{H^+} [146]. In Figure 4.13, right part, the concentration changes of filled traps ΔN_{Cu^+} and protons ΔN_{H^+} are shown versus the spatial coordinate along the crystal c-axis. It can be clearly seen that the compensation of the electronic field is caused by protons.

In other samples, however, the total number of protons is much too small to explain the compensation of the space-charge field in these crystals. Thus a different compensation mechanism appears if no protons are available. Nee et al. chose neutron diffraction experiments to reveal the nature of the contributing ions, using the different coherent neutron scattering lengths of different elements [147]. They found that most probably the movement of lithium ions takes place if no protons are present.

4.6.2 Developing of Thermally Fixed Holograms

After the recording process, an electronic grating and an ionic replica of this grating are present in the crystal, clearly observable in the experiments described above. In contrast to room-temperature recording, the density of electrons in filled traps is strongly modulated with modulation degrees up to 1. Illuminating the crystal with homogeneous light creates spatially modulated bulk photovoltaic currents due to the modulation of filled traps. Thus the development of thermally fixed gratings can be described, taking the standard one-center model and its equations when the spatial dependence of electrons in the filled and empty traps is taken into account [138, 145, 148–150]. In presence of homogeneous illumination,

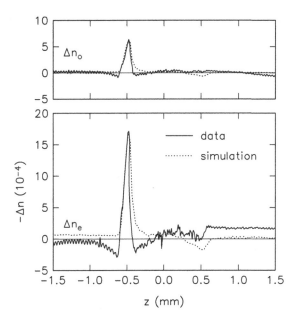

FIGURE 4.14. Refractive-index changes Δn_o and Δn_e for ordinarily (top plot) and extraordinarily polarized light (bottom plot). The solid line shows measured data after developing the pattern with homogeneous illumination; the dotted curves represent the results of a computer simulation using the standard one-center model, taking into account modulated concentrations of filled and empty traps [145].

electrons are redistributed again by the bulk photovoltaic effect, diffusion, and drift, with modulated current densities. Because the ions are immobile now, no compensation of the space-charge field takes place, and the field can be buildup until saturation is reached. To check whether this explanation is valid, the data $\Delta N_{Fe^{2+}}(z)$ is fed into a computer simulation modeling the space-charge field buildup and thus the refractive-index distribution $\Delta n(z)$ that has to be expected. This is compared to the experimentally measured Δn after development of the pattern. The result is shown in Figure 4.14 for two different light polarizations. The resulting refractive-index changes are well described using the one-center model, taking into account the modulated concentration of filled and empty traps.

4.6.3 Long-Term Stability of Thermally Fixed Holographic Gratings

During the recording, a huge number of ions that compensate the electronic space-charge field are redistributed. Because the modulation of the ionic pattern is large, the storage time at room temperature is long. The decay of stored holograms is caused by drift of the protons or lithium ions. In proton-rich crystals, estimations yield storage times. In the order of 10 years [145, 151, 152] for a fully developed

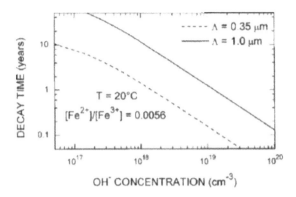

FIGURE 4.15. Calculated decay time for fully developed holograms with two different fringe spacings, depending on the proton concentration [151].

grating. Figure 4.15 shows a calculation performed by de Miguel-Sanz et al. [151] for the lifetime depending on the grating period and the proton concentration for a fully developed, i.e., permanently illuminated hologram. Note that the smaller the grating period, the shorter the lifetime; the hologram lifetime τ is proportional to the square of the grating period Λ ($\tau \propto \Lambda^2$).

4.6.4 Conclusions

Thermal fixing in $LiNbO_3$ is a very efficient tool for creating permanent holograms that are stable against the readout light. As a result of the long storage times, attractive applications using thermally fixed gratings in $LiNbO_3$ such as mirrors for atom-lithography experiments [153] or as Bragg filters for optical telecommunication networks (WDM-components) became possible [154–157].

4.7 Holographic Scattering

Holographic or nonlinear scattering [158] is based on light-induced refractive-index changes. Noisy volume phase gratings are recorded in electrooptical crystals by incident pump waves and waves scattered from volume or surface imperfections of the crystal. Subsequently the scattered light may be amplified because of direct coupling of two waves [159] by a shifted grating or as a result of parametric mixing of more than two waves [160]. Sometimes phase-matching conditions lead to characteristic scattering patterns [161, 162]. Here anisotropic diffraction—the polarization of the diffracted beam differs from that of the incident beam—has also to be taken into account [161–163].

If a photorefractive crystal is illuminated by a single laser beam, two kinds of anisotropic scattering cones have been observed that yield scattering rings on a screen behind the crystal: The first cone was discovered by S. Odoulov et al. [164, 165] and attributed to the wave vector diagram shown in Figure 4.16

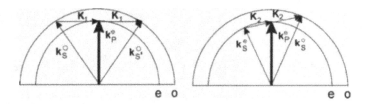

FIGURE 4.16. Wave vector diagram for anisotropic scattering cones generated by a single pump wave in a negatively birefringent crystal.
Left: Outer cone of the scattering pattern [164, 165]; k_p^e wave vector of the extraordinary pump wave; k_s^o and $k_{s'}^o$ wave vectors of the ordinary scattered waves; K_1 grating vector.
Right: Inner cone of the scattering pattern [165–168]; k_p^e wave vector of the extraordinary pump wave; k_s^e and k_s^o wave vectors of the scattered extraordinary and ordinary waves; K_2 grating vector.

(left). The cone (half) angle Θ_1^a (in air) is then given for a negatively birefringent crystal (e.g., LiNbO$_3$) by the relation

$$\sin^2 \Theta_1^a = n_e^2 - n_o^2 \approx 2(n_o - n_e)n_o. \tag{4.46}$$

Here n_o and n_e denote the refractive indices of ordinary and extraordinary waves. The second cone was simultaneously described by several groups [165–168] for LiTaO$_3$, LiNbO$_3$, and BaTiO$_3$ and attributed to the wave vector diagram shown in Figure 4.16 (right). The cone (half) angle Θ_2^a (in air) is then given for a positively birefringent crystal by the relation

$$\sin^2 \Theta_2^a = \left(10n_e^2 n_o^2 - 9n_o^4 - n_e^4\right)/16n_o^2 \approx (n_o - n_e)n_o. \tag{4.47}$$

The scattering cones lead to rings on a screen behind the crystal. Figure 4.17 illustrates the situation in LiNbO$_3$:Cu.

The recording mechanism in the case of the scattering cone of [164, 165], called the outer scattering cone, is based on the parametric coupling of the waves with k_p^e, k_s^o, and $k_{s'}^o$ (Figure 4.16 (left)). The pump wave with k_p^e and the scattered wave with k_s^o record a grating with K_1 via spatially oscillating bulk photovoltaic currents [12]; the pump wave k_p^e is diffracted from this grating and amplifies the wave $k_{s'}^o$; this wave $k_{s'}^o$ and the pump wave k_p^e record again the grating with K_1 via spatially oscillating bulk photovoltaic currents, k_p^e is diffracted from this grating, and so on. In this way the waves k_p^e, k_s^o, and $k_{s'}^o$ are coupled parametrically [164, 169], and the scattered waves k_s^o and $k_{s'}^o$ are amplified.

For the scattering cone of [165–168], called the inner scattering cone, the recording mechanism has also been investigated [170]. Many gratings are isotropically recorded by the incident pump wave and waves scattered from inhomogeneities in the sample and on the surface. The gratings are amplified by two-beam coupling [159] in one half-space of the scattering cone and attenuated in the other. Among these gratings, the grating fulfilling the phase-matching condition for anisotropic diffraction is responsible for the inner scattering cone. For the investigated crystals with moderate bulk photovoltaic fields (relatively low

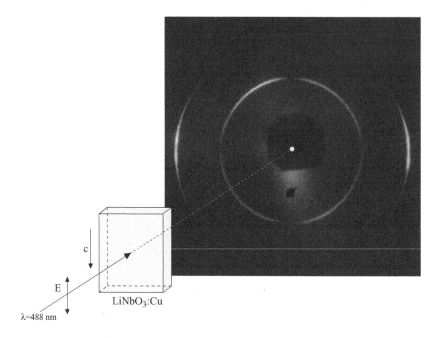

FIGURE 4.17. Outer and inner anisotropic scattering rings of ordinary polarization on a screen behind a LiNbO$_3$:Cu crystal (negative birefringence) illuminated by an extraordinarily polarized pump wave.

impurity concentrations) no experimental indications could be found that parametric coupling of the recording waves and the anisotropically diffracted wave occurs as in the case of the outer scattering cone.

Anisotropic holographic scattering experiments have been successfully utilized to explore material properties. Relatively simple measurements of the angle of a scattering cone yield the Li/Nb ratio of LiNbO$_3$ crystals [171] and the Li/Ta ratio of LiTaO$_3$ crystals [172].

Isotropic wide-angle scattering (equal light polarization of diffracted and incident beam) under illumination with one pump beam has also been investigated in detail for doped LiNbO$_3$ and LiTaO$_3$ crystals with dominating local photovoltaic charge transport [173, 174]. Steady-state amplification of the scattering patterns results from a nonzero shift in the temporal frequency between the coherent optical noise and the pump beam [175]. Competition of photovoltaic and diffusion contributions in the photorefractive effect leads to a spatial and temporal asymmetry of the scattering patterns.

A wealth of characteristic scattering patterns is observed on a screen behind a photorefractive crystal illuminated by two pump beams. A detailed review of the parametric four-wave processes involved has been given in [162]. These processes result from two different phase-matching conditions. According to the classification scheme of Sturman, Odoulov, and Goulkov [162, 176] we speak of

A and B processes for two pump waves:

$$A : k_{p1}^t + k_{p2}^u = k_{s1}^v + k_{s2}^w, \qquad B : k_{p1}^t - k_{p2}^u = k_{s1}^v - k_{s2}^w, \qquad (4.48)$$

where k_{p1}, k_{p2} denote the wave vectors of the pump waves, k_{s1}, k_{s2} those of the scattered waves, and t, u, v, w denote o (ordinary) or e (extraordinary) polarization. The scheme

$$T : (tu \to vw) \qquad (4.49)$$

is used for the description of a T process ($T = A, B$), A processes yield rings on a screen behind the crystal (axis nearly perpendicular to the screen), B processes unclosed lines (axis nearly parallel to the screen). There exist nine elementary A processes and ten B processes.

LiNbO₃ and LiTaO₃ crystals are especially well suited for the investigation of these processes, as can be seen from the above review [162] and the references therein. In the present contribution we will restrict attention to the influence of impurities on light-induced scattering patterns [177]. This influence is of particular importance if refractive index gratings are recorded by orthogonally polarized waves via the bulk photovoltaic tensor element $\beta_{113} = \beta_{131}^s + i\beta_{131}^a$. The antisymmetric imaginary part β_{131}^a is responsible for a phase shift of $\pi/2$ between the polarization grating and the refractive index grating. For LiNbO₃:Fe, the relation $\beta_{131}^a < 0$ is valid and extraordinary waves are amplified by ordinary waves; for LiNbO₃:Cu, however, ordinary waves are amplified by extraordinary waves because of $\beta_{131}^a > 0$ [178].

Figure 4.18 shows scattering patterns for LiNbO₃:Fe in a configuration with the c-axis perpendicular to the plane of incidence of two ordinarily polarized pump beams that form an angle $2\Theta_p = 110°$ (in air). Two extraordinarily polarized rings appear (marked as 1 and 2 in Figure 4.18) inside the area of the pump beams. The phase-matching conditions of the involved processes A: (oo → oe) and A: (oo → ee) and the relations between cone and pump angles have been deduced [177]. For LiNbO₃:Cu these anisotropic rings do not appear. But now two ordinarily polarized rings (in Figure 4.19 again marked as 1 and 2) are observed outside the area of the pump beams when the crystal is illuminated with two extraordinary pump beams. These rings have been attributed to the two processes A: (ee → oo) and A: (ee → eo). Phase conditions and relations between cone and pump angles have been deduced [177]. The further scattering patterns shown in Figure 4.19 have been identified, too. In addition, bright dots are observed at the intersections of scattering rings and lines in Figure 4.19. It has been demonstrated that these dots are caused by constructive interference of four parametric scattering processes [179].

Finally, we want to touch briefly the phenomena of mirrorless oscillation and transverse instabilities. In special geometrical configurations oscillation of light waves may occur without feedback by external mirrors. This effect is called mirrorless oscillation or self-oscillation. An internal feedback is generated by counterpropagating waves [160]; besides forward propagating waves, backward propagating waves have to be involved that meet the same phase-matching

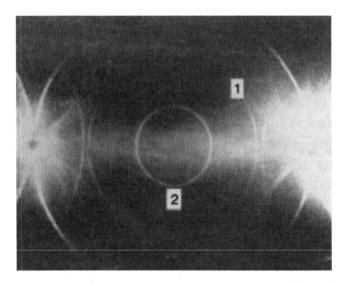

FIGURE 4.18. Light-induced scattering pattern for two ordinarily polarized pump beams ($\lambda = 440$ nm) observed on a screen behind an x-cut $LiNbO_3$:Fe crystal with the c-axis perpendicular to the plane of incidence. The extraordinarily polarized rings are marked as 1 and 2 [177].

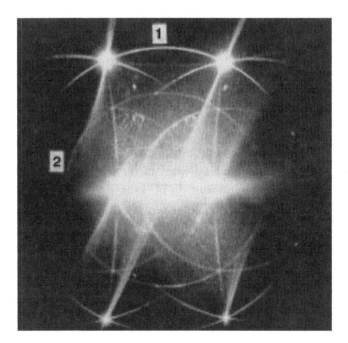

FIGURE 4.19. Light-induced scattering pattern for two extraordinarily polarized pump beams ($\lambda = 514$ nm) observed on a screen behind an x-cut $LiNbO_3$:Cu crystal with the c-axis in the plane of incidence. The ordinarily polarized rings are marked as 1 and 2 [177].

conditions. The appearance of the oscillation may be illustrated as follows: A strong pump and a weak scattered wave, both propagating in the forward direction, record a grating from which the pump wave is diffracted. This leads to the amplification of the scattered wave and to an increase of modulation. At the entrance face of the crystal the grating is weak, at the exit face stronger. At this face now the backward propagating waves are incident, which interact with this grating, leading to a further amplification. The forward-propagating waves meet this amplified grating and amplification is continued until the process is limited by the finite intensity of the pump waves. Self-oscillation has been demonstrated experimentally in LiNbO$_3$ for a configuration in which forward-wave and backward-wave parametric mixing processes are combined [180, 181].

Transverse instabilities lead to the formation of small-angle structures around counterpropagating light beams in photorefractive crystals. The effect was discovered in LiNbO$_3$ as diffuse and labile ring structures [182]. Further research in this field was stimulated by the discovery of circular as well as hexagonal patterns in KNbO$_3$ [183]. Later, hexagonal formation was discovered in LiNbO$_3$, too, and the influence of seed beams was elucidated [184].

4.8 Conclusions

The investigation of photorefractive effects in LiNbO$_3$ and LiTaO$_3$ during the last decades has revealed many unusual phenomena such as thermal fixing, two-beam coupling, parametric amplification, mirrorless oscillation, and hexagon formation. Though many details of these processes are well understood, important questions are still open. This is especially true for the microscopic processes involved.

Of particular importance for the understanding of photorefractive effects is the light-induced charge transport. It can be described in the framework of a band-transport model with one or two photorefractive centers depending on the experimental parameters.

The one-center model cannot explain some effects at high light intensities, e.g., light-induced absorption changes and two-step recording, and effects that appear in highly doped crystals, e.g., an enlarged photoconductivity and dark conductivity. Here secondary centers, in LiNbO$_3$ especially Nb$_{Li}$ and in LiTaO$_3$ the Ta$_{Li}$ antisites, must be considered.

With knowledge about the photorefractive properties of LiNbO$_3$ and LiTaO$_3$ it is possible to tailor the crystals for the desired applications. This can be achieved, e.g., by choosing the right dopants and additional thermal treatments. Due to large dark storage times, LiNbO$_3$ and LiTaO$_3$ are favorable candidates for long-term holographic storage, though the response of these materials is relatively slow. Furthermore, thermal and optical fixing of holograms allows one to read out information nondestructively for a long time.

References

1. A.A. Ballman: *J. Amer. Chem. Soc.* **48**, 1886 (1946).
2. A.M. Prokhorov, YuS. Kuzminov: *Physics and Chemistry of Crystalline Lithium Niobate*. The Adam Hilger Series on Optics and Optoelectronics, 1990.
3. YuS. Kuzminov: *Lithium Niobate Crystals*. Cambridge International Science Publishing, 1999.
4. A. Ashkin, G.D. Boyd, J.M. Dziedzic, R.G. Smith, A.A. Ballman, J.J. Levinstein, K. Nassau: *Appl. Phys. Lett.* **9**, 72 (1966).
5. F.S. Chen, J.T. LaMacchia, D.B. Fraser: *Appl. Phys. Lett.* **13**, 223 (1968).
6. F.S. Chen: *J. Appl. Phys.* **40**, 3389 (1969).
7. D.L. Staebler, W.J. Burke, W. Phillips, J.J. Amodei: *Appl. Phys. Lett.* **26**, 182 (1975).
8. H. Franke, H.G. Festl, E. Krätzig: *Colloid and Polymer Science* **262**, 213 (1984).
9. M. Eich, J.H. Wendorff, B. Reck, H. Ringsdorf: *Makromolekulare Chemie Rapid Communications* **8**, 59 (1987).
10. M. Eich, J.H. Wendorff: *Makromolekulare Chemie Rapid Communications* **8**, 467 (1987).
11. M. Chen, L.P. Yu, L.R. Dalton, Y.Q. Shi, W.H. Steier: *Macromolecules* **24**, 5421 (1991).
12. S.J. Zilker, T. Bieringer, D. Haarer, R.S. Stein, J.W. van Egmond, S.G. Kostromine: *Adv. Mater.* **10**, 855 (1998).
13. D.A.B. Miller, D.S. Chemla, D.J. Eilenberger, P.W. Smith, A.C. Gossard, W.T. Tsang: *Appl. Phys. Lett.* **41**, 679 (1982).
14. D.A.B. Miller, D.S. Chemla, T.C. Damen, T.H. Wood, C.A Burrus, A.C. Gossard, W.Wiegmann: *IEEE Journal of Quantum Electronics* **21**, 1462 (1985).
15. D.A.B. Miller, D.S. Chemla, T.C. Damen, A.C. Gossard, W.Wiegmann, T.H. Wood, C.A. Burrus: *Phys. Rev. B* **32**, 1043 (1985).
16. D. Nolte, M. Melloch: *MRS Bulletin* **19**, 44 (1994).
17. A.M. Glass, D. von der Linde, T.J. Negran: *Appl. Phys. Lett.* **25**, 233 (1974).
18. J.F. Nye: *Physical Properties of Crystals*. Oxford University Press (London), 1979.
19. M. Luennemann, U. Hartwig, G. Panotopoulos, K. Buse: *Appl. Phys. B* **76**, 403 (2003).
20. K. Peithmann, A. Wiebrock, K. Buse: *Appl. Phys. B* **68**, 777 (1999).
21. S. Fries, S. Bauschulte, E. Krätzig, K. Ringhofer, Y. Yacoby: *Opt. Commun.* **84**, 251 (1991).
22. M. Jazbinšek, M. Zgonik: *Appl. Phys. B* **74**, 407 (2002).
23. K. Onuki, N. Uchida, T. Saku: *J. Opt. Soc. Am.* **62**, 1030 (1972).
24. W.L. Bond: *J. Appl. Phys.* **36**, 1674 (1965).
25. G.E. Peterson, A.M. Glass, T.J. Negran: *Appl. Phys. Lett.* **19**, 130 (1971).
26. J.J. Amodei, W. Phillips, D.L. Staebler: *IEEE J. Quant. Electron.* **7**, 321 (1971).
27. W. Phillips, J.J. Amodei, D.L. Staebler: *RCA Rev.* **33**, 94 (1972).
28. J.B. Herrington, J. Schneider, B. Dischler: *Solid State Commun.* **10**, 509 (1972).
29. T. Takeda, A. Watanabe, K. Sugihara: *Phys. Lett.* **27A**, 114 (1968).
30. B. Dischler, A. Räuber: *Solid State Commun.* **17**, 953 (1975).
31. N.F. Evlanova, L.S. Kornienk, L.N. Rashkovich, A.O. Rybaltovski: *Sov. Phys. JETP* **26**, 1090 (1968).
32. W. Keune, S.K. Date, I. Deszi, U. Gonser: *J. Appl. Phys.* **46**, 3914 (1975).

33. H. Kurz, E. Krätzig, W. Keune, H. Engelmann, U. Gonser, B. Dischler, A. Räuber: *Appl. Phys.* **12**, 355 (1977).
34. H. Söthe, J.-M. Spaeth: *J. Phys.: Condens. Matter* **4**, 9901 (1992).
35. T. Gog, P. Schotters, J. Falta, G. Materlik, M. Grodzicki: *J. Phys: Condensed Matter* **7**, 6971 (1995).
36. E. Krätzig, R. Orlowski: *Ferroelectrics* **27**, 241 (1980).
37. M.G. Clark, F.J. DiSalvo, A.M. Glass, G.E. Peterson: *J. Chem. Phys.* **59**, 6209 (1973).
38. E. Krätzig, R. Orlowski: *Appl. Phys.* **15**, 133 (1978).
39. D.L. Staebler, W. Phillips: *Appl. Phys. Lett.* **24**, 268 (1974).
40. E. Krätzig, H. Kurz: *J. Electrochem. Soc.* **124**, 131 (1977).
41. K.L. Sweeney, L.E. Halliburton: *Appl. Phys. Lett.* **43**, 336 (1983).
42. O.F. Schirmer, O. Thiemann, M. Wöhlecke: *J. Phys. Chem. Solids* **52**, 185 (1991).
43. D. Berben, K. Buse, S. Wevering, P. Herth, M. Imlau, T. Woike: *J. Appl. Phys.* **87**, 1034 (2000).
44. J. Imbrock, S. Wevering, K. Buse, E. Krätzig: *J. Opt. Soc. Am. B* **16**, 1392 (1999).
45. L.A. Kappers, K.L. Sweeney, L.E. Halliburton, J.H.W. Liaw: *Phys. Rev. B* **31**, 6792 (1985).
46. B. Faust, H. Müller, O.F. Schirmer: *Ferroelectrics* **153**, 297 (1994).
47. O.F. Schirmer, S. Juppe, J. Koppitz: *Cryst. Latt. Def. and Amorph. Mat.* **16**, 353 (1987).
48. Y. Furukawa, M. Sato, K. Kitamura, Y. Yajima: *J. Appl. Phys.* **72**, 3250 (1992).
49. K. Kitamura, J.K. Yamamoto, N. Iyi, S. Kimura, T. Hayashi: *J. Cryst. Growth* **116**, 327 (1992).
50. S. Kimura, K. Kitamura: *J. Ceramic Soc. Jap.* **101**, 22 (1993).
51. K. Kitamura, Y. Furukawa, K. Niwa, V. Gopalan, T.E. Mitchell: *Appl. Phys. Lett.* **73**, 3073 (1998).
52. Y. Furukawa, K. Kitamura, E. Suzuki, K. Niwa: *J. Cryst. Growth* **197**, 889 (1999).
53. D.A. Bryan, R. Gerson, H.E. Tomaschke: *Appl. Phys. Lett.* **44**, 847 (1984).
54. B.C. Grabmaier, F. Otto: *J. Cryst. Growth* **79**, 682 (1986).
55. B.C. Grabmaier, W. Wersing, W. Koestler: *J. Cryst. Growth* **110**, 339 (1991).
56. T. Volk, N. Rubinina, M. Wöhlecke: *J. Opt. Soc. Am. B* **11**, 1681 (1994).
57. T.R. Volk, N.M. Rubinina: *Ferroelectrics Lett.* **14**, 37 (1992).
58. J.K. Yamamoto, K. Kitamura, N. Iyi, S. Kimura, Y. Furukawa, M. Sato: *Appl. Phys. Lett.* **61**, 2156 (1992).
59. F. Nitanda, Y. Furukawa, S. Makio, M. Sato, K. Ito: *Jpn. J. Appl. Phys.* **34**, 1546 (1995).
60. G.I. Malovichko, V.G. Grachev, L.P. Yurchenko, V.Ya. Proshko, É. P. Kokanyan, V.T. Gabrielyan: *Phys. Stat. Sol. (a)* **133**, K29K32 (1992).
61. G.I. Malovichko, V.G. Grachev, E.P. Kokanyan, O.F. Schirmer, K. Betzler, B. Gather, F. Jermann, S. Klauer, U. Schlarb, M. Wöhlecke: *Appl. Phys. A* **56**, 103 (1993).
62. D.H. Jundt, M.M. Fejer, R.L. Byer: *IEEE J. Quantum Electron.* **26**, 135 (1990).
63. D.H. Jundt, M.M. Fejer, R.G. Norwood, P.F. Bordui: *J. Appl. Phys.* **72**, 3468 (1992).
64. P.F. Bordui, R.G. Norwood, D.H. Jundt, M.M. Fejer: *J. Appl. Phys.* **71**, 875 (1992).
65. P.F. Bordui, R.G. Norwood, C.D. Bird, J.T. Carella: *J. Appl. Phys.* **78**, 4647 (1995).
66. Ch. Bäumer, D. Berben, K. Buse, H. Hesse, J. Imbrock: *Appl. Phys. Lett.* **82**, 2248 (2003).
67. L. Kovacs, G. Ruschhaupt, K. Polgar, G. Corradi, M. Wöhlecke: *Appl. Phys. Lett.* **70**, 2801 (1997).

68. Ch. Bäumer, C. David, A. Tunyagi, K. Betzler, H. Hesse, E. Krätzig, M. Wöhlecke: *J. Appl. Phys.* **93**, 3102 (2003).
69. U. Schlarb, S. Klauer, M. Wesselmann, K. Betzler, M. Wöhlecke: *Appl. Phys. A* **56**, 311 (1993).
70. V.V. Atuchin: *Opt. Spectrosk.* **67**, 1309 (1989).
71. M. Nakamura, S. Higuchi, S. Takekawa, K. Terabe, Y. Furukawa, K. Kitamura: *Jpn. J. Appl. Phys.* **41**, L465L467 (2002).
72. H.M. O'Bryan, P.K. Gallagher, C.D. Brandle: *J. Amer. Ceram. Soc.* **68**, 493 (1985).
73. Y. Furukawa, K. Kitamura, K. Niwa, H. Hatano, P. Bernasconi, G. Montemezzani, P. Günter: *Jpn. J. Appl. Phys.* **38**, 1816 (1999).
74. V. Gopalan, T.E. Mitchell, Y. Furukawa, K. Kitamura: *Appl. Phys. Lett.* **72**, 1981 (1998).
75. A.G. Chynoweth: *Phys. Rev.* **102**, 705 (1956).
76. O.F. Schirmer: *J. Appl. Phys.* **50**, 3404 (1979).
77. V.I. Belinicher, V.K. Malinovskiĭ, B.I. Sturman: *Sov. Phys. JETP* **46**, 362 (1977).
78. B.I. Sturman, V.M. Fridkin: *The Photovoltaic and Photorefractive Effect in Noncentrosymmetric Materials.* Gordon and Breach Science Publishers, 1992.
79. V.M. Fridkin, R.M. Magomadov: *JETP Lett.* **30**, 686 (1977).
80. H.G. Festl, P. Hertel, E. Krätzig, R. von Baltz: *Phys. Stat. Sol. (b)* **113**, 157 (1982).
81. S.I. Karabekian, S.G. Odoulov: *Phys. Stat. Sol. (b)* **169**, 529 (1992).
82. A.M. Glass, D. von der Linde, D.H. Austin, T.J. Negran: *J. Electron. Mater.* **4**, 915 (1975).
83. R. Orlowski, E. Krätzig: *Solid State Commun.* **27**, 1351 (1978).
84. H. Heyszenau: *Phys. Rev. B* **18**, 1586 (1978).
85. R. von Baltz: *Phys. Stat. Sol. (b)* **89**, 419 (1978).
86. W. Kraut, R. von Baltz: *Phys. Rev. B* **19**, 1548 (1979).
87. R. von Baltz: *Ferroelectrics* **35**, 131 (1981).
88. H. Presting, R. von Baltz: *Phys. Stat. Sol. (b)* **112**, 559 (1982).
89. V.I. Belinicher, B.I. Sturman: *Sov. Phys. Uspekhy* **23**, 199 (1980).
90. S.G. Odoulov: *JETP Lett.* **35**, 10 (1980).
91. J.J. Amodei: *RCA Rev.* **32**, 185 (1971).
92. V.L. Vinetskii, N.V. Kukhtarev: *Sov. Phys. Solid State* **16**, 2414 (1975).
93. R. Sommerfeldt, E. Krätzig: *SPIE* **1126**, 25 (1989).
94. Y. Ohmori, M. Yamaguchi, K. Yoshino, Y. Inuishi: *Jpn. J. Appl. Phys.* **15**, 2263 (1976).
95. A.L. Smirl, G.C. Valley, R.A. Mullen, K. Bohnert, C.D. Mire, T.F. Boggess: *Opt. Lett.* **12**, 501 (1987).
96. A.L. Smirl, K. Bohnert, G.C. Valley, R.A. Mullen, T.F. Boggess: *J. Opt. Soc. Am. B* **6**, 606 (1989).
97. I. Biaggio, M. Zgonik, P. Günter: *J. Opt. Soc. Am. B* **9**, 1480 (1992).
98. N.V. Kukhtarev: *Sov. Tech. Phys. Lett.* **2**, 438 (1976).
99. G.C. Valley: *IEEE J. Quant. Electron.* **QE**, 1637 (1983).
100. Y. Furukawa, K. Kitamura, S. Matsumura, P. Bernasconi, G. Montemezzani, P. Günter: *Topical Meeting on Photorefractive Materials, Effects and Devices* **27**, 153 (1997).
101. R. Sommerfeldt, L. Holtmann, E. Krätzig, B.C. Grabmeier: *Phys. Stat. Sol. (a)* **106**, 89 (1988).
102. R. Sommerfeldt, L. Holtmann, E. Krätzig, B.C. Grabmaier: *Ferroelectrics* **92**, 219 (1989).

103. T.R. Volk, V.I. Pryalkin, N.M. Rubinina: *Opt. Lett.* **15**, 996 (1990).
104. I. Nee, M. Müller, K. Buse, E. Krätzig: *J. Appl. Phys.* **88**, 4282 (2000).
105. Y.P. Yang, I. Nee, K. Buse, D. Psaltis: *Appl. Phys. Lett.* **78**, 4076 (2001).
106. J. Koppitz, O.F. Schirmer, A.I. Kuznetsov: *Europhys. Lett.* **4**, 1055 (1987).
107. F. Jermann, M. Simon, E. Krätzig: *J. Opt. Soc. Am. B* **12**, 2066 (1995).
108. L. Hesselink, S.S. Orlov, A. Liu, A. Akella, D. Lande, R. Neurgaonkar: *Science* **282**, 1089 (1998).
109. G.C. Valley: *J. Appl. Phys.* **59**, 3363 (1986).
110. F.P. Strohkendl, J.M.C. Jonathan, R.W. Hellwarth: *Opt. Lett.* **11**, 312 (1986).
111. N.V. Kukhtarev, V.B. Markov, S.G. Odulov, M.S. Soskin, V.L. Vinetskii: *Ferroelectrics* **22**, 949 (1979).
112. N.V. Kukhtarev, V.B. Markov, S.G. Odulov, M.S. Soskin, V.L. Vinetskii: *Ferroelectrics* **22**, 961 (1979).
113. E. Krätzig, R. Orlowski, V. Doormann, M. Rosenkranz: *SPIE* **164**, 33 (1978).
114. E. Krätzig: *Ferroelectrics* **21**, 635 (1978).
115. E. Krätzig, R. Sommerfeldt: *SPIE* **1273**, 2 (1990).
116. E. Krätzig, H. Kurz: *Optica Acta* **24**, 475 (1977).
117. H. Kogelnik: *Bell Syst. Tech. J.* **48**, 2909 (1969).
118. K. Peithmann, J. Hukriede, K. Buse, E. Krätzig: *Phys. Rev. B* **61**, 4615 (2000).
119. M. Luennemann, U. Hartwig, K. Buse: *J. Opt. Soc. B* **20**, 1643 (2003).
120. C.-T. Chen, D.M. Kim, D. von der Linde: *Appl. Phys. Lett.* **34**, 321 (1979).
121. G.C. Valley: *Appl. Opt.* **22**, 3160 (1983).
122. G.A. Brost, R.A. Motes, J.R. Rotgé: *J. Opt. Soc. Am. B* **5**, 1879 (1988).
123. L. Holtmann: *Phys. Stat. Sol. (a)* **113**, K89K93 (1989).
124. L. Holtmann, M. Unland, E. Krätzig, G. Godefroy: *Appl. Phys. A* **51**, 13 (1990).
125. F. Jermann, E. Krätzig: *Appl. Phys. A* **55**, 114 (1992).
126. F. Jermann, J. Otten: *J. Opt. Soc. Am. B* **10**, 2085 (1993).
127. S. Wevering, J. Imbrock, E. Krätzig: *J. Opt. Soc. Am. B* **18**, 472 (2001).
128. H. Guenther, R. Macfarlane, Y. Furukawa, K. Kitamura, R. Neurgaonkar: *Appl. Opt.* **37**, 7611 (1998).
129. J. Imbrock, D. Kip, E. Krätzig: *Opt. Lett.* **24**, 1302 (1999).
130. M. Simon, F. Jermann, E. Krätzig: *Opt. Mater.* **3**, 243 (1994).
131. M. Simon, S. Wevering, K. Buse, E. Krätzig: *J. Phys. D* **30**, 144 (1997).
132. O. Althoff, A. Erdmann, L. Wiskott, P. Hertel: *Phys. Stat. Sol. (a)* **128**, K41 (1991).
133. F. Jermann, K. Buse: *Appl. Phys. B* **59**, 437 (1994).
134. K. Buse, F. Jermann, E. Krätzig: *Ferroelectrics* **141**, 197 (1993).
135. K. Buse, F. Jermann, E. Krätzig: *Appl. Phys. A* **58**, 191 (1994).
136. H. Vormann, E. Krätzig: *Sol. State Commun.* **49**, 843 (1984).
137. J.J. Amodei, W. Phillips, D.L. Staebler: *Appl. Opt.* **11**, 390 (1972).
138. L. Arizmendi, P.D. Townsend, M. Carrascosa, J. Baquedano, J.M. Cabrera: *J. Phys.: Condens. Matter* **3**, 5399 (1991).
139. V.I. Kovalevich, L.A. Shuvalov, T.R. Volk: *Phys. Stat. Sol. (a)* **45**, 249 (1978).
140. B.F. Williams, W.J. Burke, D.L. Staebler: *Appl. Phys. Lett.* **28**, 224 (1976).
141. G. Bergmann: *Solid State Commun.* **6**, 77 (1967).
142. R.G. Smith, D.B. Fraser, R.T. Denton, T.C. Rich: *J. Appl. Phys.* **39**, 4600 (1968).
143. W. Bollmann, H.-J. Stöhr: *Phys. Stat. Sol. (a)* **39**, 477 (1977).
144. H. Vormann, G. Weber, S. Kapphan, E. Krätzig: *Solid State Commun.* **40**, 543 (1981).
145. K. Buse, S. Breer, K. Peithmann, S. Kapphan, M. Gao, E. Krätzig: *Phys. Rev. B* **56**, 1225 (1997).

146. S. Kapphan, A. Breitkopf: *Phys. Stat. Sol. (a)* **133**, 159 (1992).
147. I. Nee, K. Buse, F. Havermeyer, R.A. Rupp, M. Fally, R.P. May: *Phys. Rev. B* **60**, R9896R9899 (1999).
148. W. Meyer, P. Würfel, R. Munser, G. Müller-Vogt: *Phys. Stat. Sol. (a)* **53**, 171 (1979).
149. V.V. Kulikov, S.I. Stepanov: *Sov. Phys. Solid State* **21**, 1849 (1979).
150. P. Hertel, K.H. Ringhofer, R. Sommerfeldt: *Phys. Stat. Sol. (a)* **104**, 855 (1987).
151. E.M. de Miguel-Sanz, M. Carrascosa, L. Arizmendi: *Phys. Rev. B* **65**, 165101 (2002).
152. L. Arizmendi, E.M. de Miguel-Sanz, M. Carrascosa: *Opt. Lett.* **23**, 960 (1998).
153. M.Mützel, S. Tandler, D. Aubrich, D. Meschede, K. Peithmann, M. Flaspöhler, K. Buse: *Phys. Rev. Lett.* **88**, 083601 (2002).
154. R. Müller, M.T. Santos, L. Arizmendi, J.M. Cabrera: *J. Phys. D: Appl. Phys.* **27**, 241 (1994).
155. V. Leyva, G.A. Rakuljic, B. O'Conner: *Appl. Phys. Lett.* **65**, 1079 (1994).
156. S. Breer, K. Buse: *Appl. Phys. B* **66**, 339 (1998).
157. S. Breer, H. Vogt, I. Nee, K. Buse: *Electron. Lett.* **34**, 2419 (1998).
158. R. Magnusson, T.K. Gaylord: *Appl. Opt.* **13**, 1545 (1974).
159. D.L. Staebler, J.J. Amodei: *J. Appl. Phys.* **43**, 1042 (1972).
160. A. Yariv, D. Pepper: *Opt. Lett.* **1**, 16 (1977).
161. S.I. Stepanov, M.P. Petrov, A.A. Kamshilin: *Sov. Tech. Phys. Lett.* **3**, 345 (1977).
162. B.I. Sturman, S.G. Odoulov, M.Y. Goulkov: *Physics Reports* **275**, 198 (1997).
163. M. Röwe, J. Neumann, E. Krätzig: *Opt. Commun.* **170**, 121 (1999).
164. S. Odoulov, K. Belabaev, I. Kiseleva: *Opt. Lett.* **10**, 31 (1985).
165. K.G. Belabaev, I.N. Kiseleva, V.V. Obukhovskii, S.G. Odulov, R.A. Taratuta: *Sov. Phys. Solid State* **28**, 321 (1986).
166. D.A. Temple, C. Warde: *J. Opt. Soc. Am. B* **3**, 337 (1986).
167. R.A. Rupp, F.W. Drees: *Appl. Phys. B* **39**, 223 (1986).
168. M. Ewbank, P. Yeh, J. Feinberg: *Opt. Commun.* **59**, 423 (1986).
169. S.G. Odoulov: *J. Opt. Soc. Am. B* **4**, 1333 (1987).
170. S. Schwalenberg, F. Rahe, E. Krätzig: *Opt. Commun.* **209**, 467 (2002).
171. U. van Olfen, R.A. Rupp, E. Krätzig, B.C. Grabmaier: *Ferroelectrics Lett.* **10**, 133 (1989).
172. K. Bastwöste, S. Schwalenberg, Ch. Bäumer, E. Krätzig: *Phys. Stat. Sol. (a)* **199**, R1 (2003).
173. M.A. Ellabban, R.A. Rupp, M. Fally: *Appl. Phys. B* **72**, 635 (2001).
174. M. Goulkov, S. Odoulov, Th. Woike, J. Imbrock, M. Imlau, E. Krätzig, C. Bäumer, H. Hesse: *Phys. Rev. B* **65**, art. no. 195111 (2002).
175. B.I. Sturman: *Sov. Phys.: JETP* **73**, 593 (1991).
176. B.I. Sturman, M.Y. Goulkov, S.G. Odoulov: *J. Opt. Soc. Am. B* **13**, 577 (1997).
177. M. Goulkov, G. Jäkel, E. Krätzig, S. Odoulov, R. Schulz: *Opt. Mater.* **4**, 314 (1995).
178. S.G. Odoulov: *Ferroelectrics* **91**, 213 (1989).
179. M.Y. Goulkov, S.G. Odoulov, B.I. Sturman, A.I. Chernykh, E. Krätzig, G. Jäkel: *J. Opt. Soc. Am. B* **13**, 2602 (1996).
180. A.D. Novikov, V.V. Obukhovskii, S.G. Odoulov, B.I. Sturman: *JEPT Lett.* **44**, 538 (1986).
181. A.D. Novikov, V.V. Obukhovskii, S.G. Odoulov, B.I. Sturman: *Opt. Lett.* **13**, 1017 (1988).
182. V.V. Lemeshko, V.V. Obukhovskii: *Sov. Tech. Phys. Lett.* **11**, 573 (1985).
183. T. Honda: *Opt. Lett.* **18**, 598 (1993).
184. S. Odoulov, B. Sturman, E. Krätzig: *Appl. Phys. B* **70**, 645 (2000).

5

Growth and Photorefractive Properties of Stoichiometric LiNbO$_3$ and LiTaO$_3$

Hideki Hatano, Youwen Liu, and Kenji Kitamura

National Institute for Materials Science, 1-1 Namiki, Tsukuba, Ibaraki 305-0044, Japan
Hatano.Hideki@nims.go.jp, LIU.Youwen@nims.go.jp,
KITAMURA.Kenji@nims.go.jp

Since the photorefractive effect was discovered in LiNbO$_3$ and LiTaO$_3$, a great number of reports have appeared concerning the optical absorption bands originating from intrinsic defects or such extrinsic impurities as transition metals and rare-earth elements [1–2]. The continuing interest in this topic primarily lies in revealing the basic mechanisms underlying the photorefractive and other optically induced properties of this material, as well as in incorporating them into more efficient and stable holographic recording.

The photorefractive effect is expected to depend substantially on the structures and densities of intrinsic defects in materials. Although commercially available congruent LiNbO$_3$ and LiTaO$_3$ crystals contain a large number of nonstoichiometric defects, the dependence of the photorefractive effect on nonstoichiometry has not been clarified. Recently, novel technologies to grow near-stoichiometric LiNbO$_3$ and LiTaO$_3$ crystals have been developed. Such near-stoichiometric crystals have improved properties compared with congruent ones, and they are now available for characterization [3]. In the strict sense, these crystals are not completely stoichiometric but are very close to stoichiometric, with Li/Nb or Li/Ta ratios over 49.7/50.3. They are hereinafter referred to as "stoichiometric LiNbO$_3$ (SLN)" or "stoichiometric LiTaO$_3$ (SLT)" to distinguish them from congruent crystals (CLN or CLT).

One of the major applications of stoichiometric LiNbO$_3$'s photorefractivity is its use in gated two-color holographic recording [4–5]. The idea of photon-gated holography or two-photon holography was first proposed in the 1970s. However, it was not until the late 1990s that a practical gated two-color holography system was developed with a continuous oscillation laser of relatively low power and the use of SLN [6]. In photon-gated holography, tailoring of multiple energy bands in the forbidden gap is essential. Crystals with fewer nonstoichiometric defects and increased structure sensitivity have made it possible to control multiple energy levels, which make them suitable for multiphoton holography [4–5]

In this chapter, the photorefractive properties of stoichiometric LiNbO$_3$ and LiTaO$_3$ are described, focusing on their differences from the conventional congruent crystal and their application to holographic recording.

5.1 Growth and Basic Properties of Stoichiometric LiNbO$_3$ and LiTaO$_3$

5.1.1 Stoichiometry Control by Crystal Growth

History of LN and LT Crystal Growth

Lithium niobate (LiNbO$_3$ or abbreviated LN) and lithium tantalite (LiTaO$_3$:LT) are typical ferroelectric materials used to make single crystals. Both have a pseudoilmenite structure possessing noncentrosymmetric C3v point group symmetry at room temperature. Their ferroelectricity was first reported by Matthias and Remeika in 1949 [7]. Great attention has been given to their properties since Ballman et al. successfully grew large single crystals using the Czochralski method [8]; in particular, their electrooptical, nonlinear-optic, piezoelectric, and pyroelectric properties have been investigated for a broad spectrum of applications. Their first commercial uses were as substrate for surface acoustic wave (SAW) devices for tuners in TVs, video recorders, and mobile phones.

Oxygen-octahedral ferroelectrics including the perovskite structure group, LN, LT, etc., are interesting materials exhibiting many useful properties. However, it is generally difficult to grow large high-quality single crystals of these materials because they have complicated phase transitions at temperatures close to room temperature. Among them, only LN and LT have a simple 180-degree polarization along the c-axis, with their paraelectric to ferroelectric transition being at comparatively high temperatures (1200 and 690°C, respectively). The ability to commercially produce large single crystals has been seen as a big advantage of using LN and LT in various devices. To improve their crystal quality and homogeneity using the conventional CZ method, the phase relations between composition and temperature have been investigated in detail [9–11]. These studies have revealed that LN and LT phases have considerably wide variability (non-stoichiometry) at high temperature with respect to their [Li]/[Nb] and [Li]/[LT] ratios.

The region of nonstoichiometric solid solubility of LN and LT at high temperature mainly extends toward excesses in their Nb and Ta components, respectively, as schematically shown in Figure 5.1. Therefore, in both cases, congruently melting compositions (congruent compositions) where solid and liquid phases coexist in equilibrium are shifted from stoichiometry to the Li deficient side. The conventional congruent compositions of LN and LT are approximately 48.5 Li$_2$O mol% (i.e., 51.5 Nb$_2$O$_5$ mol%).

FIGURE 5.1. Schematic phase diagram of Li_2O-Nb_2O_5 pseudobinary system in the vicinity of LN. Nonstoichiometric solid solubility region of LN expands mainly toward the Nb component excess side.

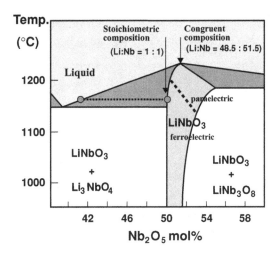

Using the conventional CZ method, single crystals should be grown from a melt of the congruent composition. Since Ballman et al. demonstrated successful crystal growth with this method, "stoichiometry control" has meant growth in which the congruent composition is determined precisely. Due to the limitations of the crystal growth method, most commercial LN and LT crystals have had congruent compositions, and consequently, until recently, studies on single crystals have used only them.

Stoichiometry Control During Crystal Growth

The congruent compositions of LN and LT contain excess Nb and Ta components, respectively, forming nonstoichiometric defects. Such excess cations occupy the Li-ion sites (antisite defect) forming vacancies at these sites to maintain electric charge neutrality [12]. Congruent LN crystal consequently contains huge numbers of such defects; approximately 1% of the Li ion sites are occupied by the excess Nb ions and 4% of these sites are vacant. Although such high densities of defects don't cause serious degradation in SAW device applications, they strongly influence their optical and electrical properties. From this point of view stoichiometric LN and LT crystals grown under nonstoichiometry control have been studied, and various comparisons of congruent and near-stoichiometric crystal properties have been reported [13–16].

It is thermodynamically impossible to obtain perfectly stoichiometric crystals by growth at a high temperature, but so far, three methods to obtain near-stoichiometric LN and LT crystals have been reported. One is the vapor transport equilibration (VTE) method [17, 18], in which a 0.5-mm-thick congruent wafer crystal is put in a container with a Li-rich powder. The container is heated at a temperature higher than 1000°C for longer than 100 hours. The Li component diffuses into the crystal during this time, and the crystal becomes nearly stoichiometric.

FIGURE 5.2. Near-stoichiometric LT single crystal of 3 inches diameter grown by the double crucible CZ method.

The second method uses a melt containing K_2O [19, 20]. The solid solubility of LN in the Li_2O-Nb_2O_5-K_2O pseudoternary system shrinks significantly by increasing the K_2O component. Although the crystals are grown from a solution containing K_2O of more than 10 mol%, K_2O concentration is negligibly small in the resulting crystal. However, it is very difficult to grow large single crystals without inclusions and scattering centers with this method, even at a considerably reduced growth rate.

The third method, the double crucible CZ method (DCCZ), is based on the fact that near-stoichiometric LN coexists in equilibrium with a Li-rich melt (58–60 Li_2O mol%), which is close to a eutectic composition, as shown in Figure 5.1. The conventional CZ method cannot be used to grow large single crystals with high homogeneity from a melt of an almost eutectic composition or a composition much different from a congruent one. An ordinary CZ furnace equipped with a radio frequency generator is used in the double crucible CZ method of Kitamura et al. [21, 22]. The platinum or iridium crucible has a double chamber structure, which divides the melt into two parts. The inner platinum crucible is on the bottom of the outer crucible. The outer melt can enter the inner crucible through a gap between the bottoms of the inner and outer crucibles. To grow near-stoichiometric LN crystals, the melt composition should be kept Li-rich (about 58.5 Li_2O mol%), and to keep it constant, an automatic supply system supplies stoichiometric LN powder to the outer melt. A load cell or electric balance is included in the apparatus to monitor the weight change of the growing crystal, and this information is used for automatic control of the diameter and the powder supply rate.

Recently, stoichiometric LN and LT crystals have become commercially available. Figure 5.2 shows an example of a 3-inch diameter near-stoichiometric LT crystal grown by the DCCZ method.

5.1.2 Composition Determination from the Curie Temperature

It is almost impossible to determine the [Li]/[Nb] or [Li]/[Ta] ratio in crystal by doing a conventional wet chemical analysis. Therefore, a parameter that is sensitive to nonstoichiometry is used for estimating the composition of grown crystals.

FIGURE 5.3. Relationship between composition of LN sintered samples and Curie temperature.

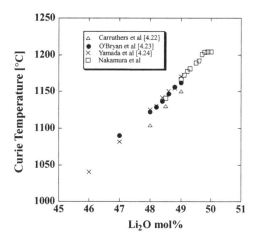

The Curie temperature, the phase transition temperature from the paraelectric to the ferroelectric phase, is a convenient parameter to estimate the composition.

The relationship between LN Curie temperature and composition has been measured by differential thermal analysis (DTA) [23–25]. However, the purpose of these reports had been how to find accurate congruent compositions. The relationship in the vicinity of a stoichiometric composition was not given attention. Recently, instead of DTA, differential scanning calorimetry (DSC), which can easily measure high Curie temperatures, has been used to determine the composition of crystals. Figure 5.3 shows the relationship between sintered samples of Li_2CO_3 and Nb_2O_5 mixtures and the Curie temperature measured by DSC. The Curie temperature of congruent LN is about 1140°C, while that of the stoichiometric composition is 1206°C. Near-stoichiometric LN grown with the double crucible CZ method using a melt of 58 Li_2O mol% has a Curie temperature of 1198–1200°C, from which the grown LN crystal was calculated to have a [Li]:[Nb] ratio of approximately 49.9:50.1. This means that the defect density in the crystal is reduced by about one order of magnitude compared with the congruent LN.

In the case of LT, the Curie temperature changes depending on the melt composition only in the vicinity of congruent composition [10, 26]. Generally, to determine the relationship between composition and Curie temperature, sintered samples of known composition are prepared. However, the Li component is easily evaporated during sintering at 1200°C. Nakamura et al. prepared sintered samples of Li_2CO_3 and Ta_2O_5 mixtures by sealing the powders in a Pt ampoule and determined the relationship between the Curie temperature and composition as shown in Figure 5.4 [27]. The commercially produced congruent LT exhibits a Curie temperature of about 600°C, while that of stoichiometric LT is 695°C. Near-stoichiometric LT grown from a melt of 60 Li_2O mol% exhibits a Curie temperature of about 690°C. This Li_2O mol% of grown LT is thus estimated to be larger than 49.9%.

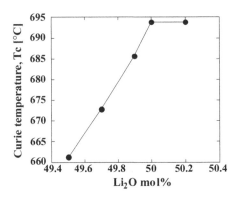

FIGURE 5.4. Relationship between composition of sintered samples and Curie temperature in the vicinity of stoichiometric LT composition [27].

In this chapter, "stoichiometric" LN (SLN) means LN exhibiting a Curie temperature of from 1190 to 1200°C, while "stoichiometric" LT (SLT) is LT having a Curie temperature of from 685 to 690°C. Although the nonstoichiometric defect densities are thus about 1 to 1.5 orders of magnitude smaller than in congruent crystals, there are still more than 10^{18} defects in 1 cm^3 of these "stoichiometric" crystals.

5.1.3 Optical Absorption Spectrum and Schematic Band Diagram

The photorefractive effect [28–29] is a phenomenon consisting of elementary processes such as optically induced charge generation, charge transport, charge trapping, and the electrooptical effect. Energy level design is essential to optimizing the photorefractive properties. The photorefractive material must have multiple energy levels in the forbidden gap, and the properties of the charges at these levels should be controlled according to the purpose of the application. Introducing intrinsic or extrinsic defects by changing crystal composition, doping impurities, and/or thermal annealing is usually carried out for this purpose.

Near-stoichiometric crystal has a structure-sensitive nature because of its decreased nonstoichiometric defects compared with congruent crystal. Consequently, these crystals are sensitive to the introduction of small amounts of impurities or defects. This feature is especially important in holographic data storage because too much doping or reduction treatment has the adverse effect of increasing the dark conductivity of the material and thus of limiting the dark storage time of the stored information [30].

This section describes the important dependencies of the optical absorption spectra of LiNbO$_3$ and LiTaO$_3$ on nonstoichiometric composition, impurities, and heat treatment. A band diagram is illustrated for the purpose of energy band tailoring to achieve holographic applications of the photorefractive effect.

FIGURE 5.5. Absorption spectra of (a) nondoped LiNbO₃, and (b) nondoped LiTaO₃ depending on crystal composition. Nonpolarized light was used for the measurement.

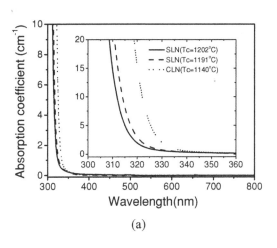

(a)

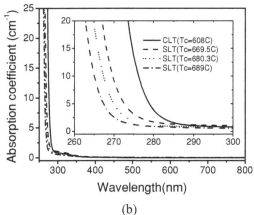

(b)

Undoped LiNbO₃ and LiTaO₃

Figure 5.5 shows the optical absorption spectra of nominally undoped LiNbO₃ and LiTaO₃ crystal using the Curie temperature as a parameter. The crystal composition, defined by the Li/(Li+Nb) or Li/(Li+Ta) atomic ratio, was estimated from the Curie temperature see [5.1.2]. As the composition changes from congruent to stoichiometric, the absorption edge shifts to shorter wavelengths. This effect of widening the band gap in stoichiometric crystals is favorable for many optoelectronic applications using ultraviolet (UV) light. Regarding the composition dependence of the absorption edge, see also [31] (LN) and [32] (LT).

Reduced Nominally Pure LiNbO₃

Figure 5.6 shows the absorption spectrum change in nominally pure stoichiometric LiNbO₃ caused by reduction heat treatment. After reduction, a broad absorption band appears. The origin of this absorption is attributed to bipolaron defects

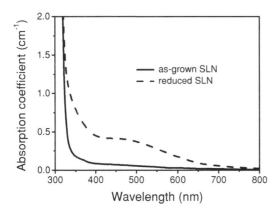

FIGURE 5.6. Change in absorption spectrum by reducing the nondoped stoichiometric $LiNbO_3$. Nonpolarized light was used for the measurement.

$(Nb_{Li}^{4+} Nb_{Nb}^{4+})$ [2]. When the bipolaron absorption band is excited by the illumination of visible light, bipolaron defects dissociate into small polarons (Nb_{Li}^{4+}) [33]. Small polarons are intrinsic defects, which cause a metastable absorption band to form centered at 1.6 eV and decay with a time constant of 10^{-9} to seconds depending on the nonstoichiometric defect density, type of impurity (e.g., Fe), and oxidation/reduction state [4–5, 34]. The reason is that high defect densities may result in a substantial lattice distortion and fast, phonon-assisted decay of excited Nb_{Li}^{4+} small polaron levels [2]. As the crystal composition becomes closer to stoichiometric, the lifetime of excited small polaron states will increase. After reduction, because more filled bipolaron sites are produced and contaminant acceptor ions are converted into donors, the lifetime increases.

Fe-Doping in $LiNbO_3$

Fe is the most common impurity used as a photorefractive (PR) center in $LiNbO_3$ and takes the form of either Fe^{2+} or Fe^{3+} in $LiNbO_3$. The behavior of Fe in CLN has already been investigated [35] and is basically the same as in SLN [36]. However, in near-stoichiometric crystals, in which the optical bandgap is larger, the effect of even a slight amount of Fe doping can be significant. Figure 5.7 shows the absorption spectrum change caused by doping with Fe and thermal annealing in Tb: SLN. Because Tb doping itself doesn't create an observable absorption band in the forbidden gap at room temperature [37], the spectrum change shown in the figure is responsible for the Fe doping and oxidation state change.

The existence of several absorption bands in $LiNbO_3$ is described in [38]. Here we discuss the two typical absorption bands using the notation introduced by Dishler et al. [39].

C-band: This absorption band begins at ≈ 3.1 eV and extends to higher photon energies. The C-band behaves as very deep center very close to the valence band, and its absorption increases dramatically by increasing the Fe content. This causes an apparent shift of absorption edge toward longer wavelengths when the

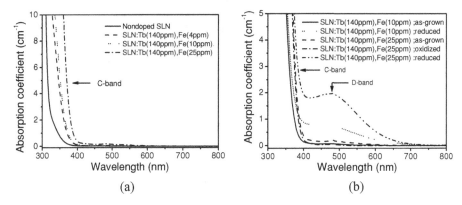

FIGURE 5.7. Comparison of absorption spectra in stoichiometric LiNbO₃ codoped with Tb and Fe. (a) Fe content dependence in as-grown crystals, (b) reduction or oxidation heat treatment dependence. Nonpolarized light was used for the measurement.

Fe content exceeds $\approx$20 ppm. Absorption in the C-band is due to charge transfer from the oxygen π-orbital to Fe^{3+} ions (leading to the creation of holes in the valence band formed by O^{2-} ions) [40].

D-band: This absorption band is centered at about 2.6 eV and originates from the transition Fe^{2+}-Nb^{5+}. D-band absorptions create free electrons and are responsible for the PR effect. They increase with Fe^{2+} concentration, therefore they are sensitive to reductive heat treatment and/or optical reduction by UV (see [4]).

Mn Doping in LiNbO₃

Mn doping creates an absorption band at deeper energies than the D-band [41]. Figure 5.8 is the absorption spectrum change caused by Mn doping in SLN. Mn doping can be used to exploit the PR effect in the near-UV region, because the Mn concentration can control the absorption according to the wavelength and the required absorption value.

Schematic Band Diagram

A schematic energy band model was estimated using a computer simulation of optical transitions based on the facts mentioned above and is illustrated in Figure 5.9. It will be used as an aid for understanding the PR effects that are described in the subsequent sections. The bipolaron level, Fe level, and Mn level are used as donors/acceptors of electrons in the PR process. The small polaron level acts as a metastable intermediate center in two-color gated holograms.

The reader should note that these energy levels are not line spectra, but compose an energy band with a finite width. As a result, two different levels may overlap. This may cause a tradeoff between holographic sensitivity and readout volatility,

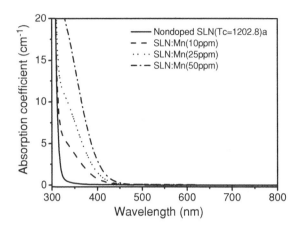

FIGURE 5.8. Change in absorption spectra by Mn-doping in stoichiometric LiNbO₃. Nonpolarized light was used for the measurement.

which comes from the charge separation between the charge source level and charge destination level.

To overcome this problem, it might be possible to form energy levels that consist of narrow bands and possess large absorption cross section. This would ensure effective charge generation and charge separation. However, to the author's knowledge, there have been no reports on development of such a material.

5.2 Photorefractive Properties of Stoichiometric LiNbO₃

5.2.1 Light-Induced Birefringence Change

The light-induced changes of the birefringence $\delta(n_e - n_o)$ (abbreviated $\delta \Delta n$ in the following) of congruent and stoichiometric crystals have been measured [20] at light intensities between 10 and 10^4 W/cm² using the optical phase-compensation technique [42]. The index change $\delta \Delta n$ of congruent LiNbO₃ increases rapidly with rising intensity and does not show a tendency to saturation. For the stoichiometric crystal, on the other hand, $\delta \Delta n$ is higher than that in the congruent crystal at the weak intensities (<200 W/cm²) and becomes independent at the strong intensities. Most importantly, it is found that the maximum index change of the stoichiometric sample is about 5 times lower at high intensities ($\approx 3 \times 10^3$ W/cm²) than in congruent LiNbO₃ indicating that the optical damage is suppressed in stoichiometric material at high intensities. Such an anomalous behavior in photorefractive properties at high intensities was explained by taking into account the contribution of antisite Nb_{Li} to light-induced charge transport [43].

FIGURE 5.9. Schematic band diagram of stoichiometric LiNbO₃.

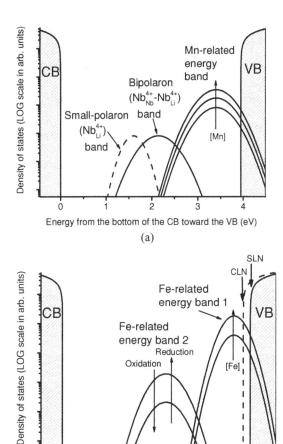

(a)

(b)

5.2.2 Exponential Gain Properties

Kitamura et al. [44] measured the exponential gain coefficient in order to characterize the basic photorefractive properties of nondoped and Fe-doped congruent and stoichiometric LN crystals. A conventional two-beam coupling setup using an LD-pumped frequency-doubled Nd:YAG laser (wavelength is 532 nm) was used. The two plane waves of reference and signal beams were symmetrically incident on the 2-mm-thick Y-plate LN sample. The initial intensity ratio (reference beam to signal beam) was 100. The total intensity of the two beams was about 1 W cm⁻². The diameter of the signal beam was 1 mm, and the spacing of the generated interference fringe was 1.6 μm.

Table 4.1 summarizes the experimental results obtained from the two-wave mixing experiments. Nondoped congruent samples exhibited only a negligible

TABLE 5.1. Results of optical absorption and two-beam coupling experiments

Sample	Curie Temp. (°C)	Fe conc. in crystal (wt. ppm)	Absorption at 532 nm, α (cm^{-1})	Exponential gain coeff., Γ (cm^{-1})	Net amplification, Γ-α (cm^{-1})	Buildup time (s)
Nondoped SLN	1200	<5[a]	0.04	25	25	60
Nondoped CLN	1138	5[a]	0.06	>1	0	800–900
Fe-doped SLN	1198	270[b]	5.8	27	21	0.6
Fe-doped CLN	1137	540[b]	9.3	15	6	1.5

[a] Flameless atomic absorption spectroscopy analysis. (b) ICP-AES analysis.

amplification, which agrees with results obtained previously [20]. The Fe-doped congruent LN had an exponential gain coefficient (Γ) of 15 cm^{-1}. This value compares well with data previously reported [29, 45]. In nondoped stoichiometric LiNbO$_3$, a gain coefficient of 25 cm^{-1} was measured. In Fe-doped stoichiometric LiNbO$_3$, an exponential gain coefficient of 27 cm^{-1} was measured.

As for the response time, the experimental results show that the buildup in the stoichiometric samples is several times shorter than in the congruent ones. For example, nondoped stoichiometric LiNbO$_3$ had a buildup time of 60 s, whereas congruent LiNbO$_3$ had a buildup time of 15 minutes. The situation is even better in Fe-doped samples. Congruent grown crystals had buildup times of 1.5 s, whereas stoichiometric LiNbO$_3$ had buildup times of 0.6 s.

These results suggest that the photorefractive effect in LiNbO$_3$ is enhanced by reducing the nonstoichiometric defects. The same improvement in diffraction gain and buildup time was observed at low pump beam intensities, 50 mW/cm^2, and at a short grating period of 0.4 μm.

5.2.3 Electrooptic Constant Depending on Nonstoichiometry

The photorefractive effect in LiNbO$_3$ crystals is caused by the electrooptic effect, and therefore, quantitative measurement of the electrooptic (EO) coefficients in stoichiometric LiNbO$_3$ can clarify the origin of its large photorefractivity. In addition, a possible increase in the EO coefficients based on LiNbO$_3$ crystals has to be confirmed for photonic devices with higher efficiency EO effect than that to date.

The composition dependencies of the electrooptic coefficients of LiNbO$_3$ crystals were investigated using a Mach-Zehnder interferometer operating at the wavelength of 0.633 μm [46–47]. Anisotropic changes in the EO coefficients, r_{33} and r_{13}, have been observed between stoichiometric and congruent LiNbO$_3$ crystals [46]. The measured values of r_{13} for congruent and stoichiometric crystals were 10.0 ± 0.8 pm/V and 10.4 ± 0.8 pm/V, respectively, and were rather insensitive to nonstoichiometry. On the other hand, the values of r_{33} for congruent and stoichiometric crystals were 31.5 ± 1.4 pm/V and 38.3 ± 1.4 pm/V, respectively. The measured value of r_{33} in the stoichiometric sample was 20% larger than that of the congruent one. Kondo et al. [47] reports that the electrooptic coefficient

r_{33} increases as the Li/Nb ratio increases and that r_{33} in stoichiometric LiNbO$_3$ crystal was 14% larger than in conventional congruent LiNbO$_3$ crystal. The value of r_{33} is also increased by doping with certain ions; an increase in the electrooptic coefficients of more than 30% (compared with nondoped congruent LiNbO$_3$) can be obtained by doping with Ce or Fe ions in SLN.

From the above measurements, we can deduce that the larger photorefractive sensitivity of SLN compared with CLN is correlated to the increased Pockels effect for the extraordinary polarization.

5.2.4 Photoconductive and Photovoltaic Properties

Photovoltaic and photoconductive properties were characterized at the wavelength of 532 nm using a frequency-doubled Nd:YAG laser [36, 48]. The crystal surfaces perpendicular to the z-axis (= c-axis) were contacted with Au electrodes. The samples were homogeneously illuminated by an expanded laser beam with a power density of 1 W/cm^2. The polarization direction of the incident beam was parallel to the z-axis of the Y-plate sample (the extraordinary polarization). A photovoltaic current flowed along the c-axis without applying an electric field. The Glass coefficient κ was calculated from the formula $\kappa = j_{pvc}/(\alpha I)$, where j_{pvc} is photovoltaic current density, α is absorption coefficient, and I is the light power density. The photoconductivity was measured by applying a DC field along the z-axis of the sample. From the relationship between the current and the applied field, the photoconductivity can be calculated by canceling the photovoltaic current.

First, the photovoltaic current density and photoconductivity of nondoped (nominally pure) LN crystals with different [Li]/[Nb] ratios were investigated. The results summarized in Table 4.2 suggest that the photovoltaic effect in the nominally pure LN is enhanced in the stoichiometric sample. The values of photoconductivity are roughly consistent with those reported by Jermann et al. [49], which were calculated from the time dependence of the photoinduced birefringence change. It is worth noting that the photovoltaic effect slightly depends on the polarization direction (it is somewhat larger under the extraordinary polarization) while there is no significant dependence in the photoconductivity.

For various Fe-doped crystals and oxidation levels, the measured Glass coefficient values for the ordinary and extraordinary polarizations were $0.8 \pm 0.1 \times 10^{-9}$ Acm/W and $1.3 \pm 0.1 \times 10^{-9}$ Acm/W, respectively, and are similar to the Glass constant values for Fe-doped LN reported in [50]. There was no

TABLE 5.2. Photovoltaic constants and photoconductivities of CLN and SLN

	Congruent LiNbO$_3$		Stoichiometric LiNbO$_3$	
Polarization	Ordinary	Extraordinary	Ordinary	Extraordinary
Photovoltaic constant [A cm W^{-1}]	1.5×10^{-12}	2.5×10^{-12}	4.1×10^{-11}	4.9×10^{-11}
Photoconductivity [Ω^{-1} cm^{-1}]	2.0×10^{-15}	2.2×10^{-15}	2.6×10^{-14}	2.7×10^{-14}

Conditions: Wavelength: 532 nm, Beam intensity: 10.5 W/cm^2

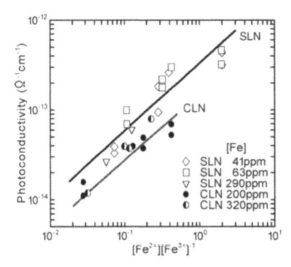

FIGURE 5.10. Photoconductivity as a function of $[Fe^{2+}]/[Fe^{3+}]$ ratio in Fe-doped SLN and CLN.

distinguishable difference between Fe-doped SLN and Fe-doped CLN within the accuracy of the measurement.

Photoconductivity measurements for a sample with a different oxidation state of Fe are summarized in Figure 5.10. We can see that there is a distinct difference between SLN and CLN. Photoconductivity is generally described by the relationship $\sigma_{ph} = e n_{ph} \mu$, where e is the electric charge, n_{ph} is the concentration of photogenerated carriers, and μ is the mobility of the carrier. n_{ph} can be written as $n_{ph} = g\tau$, where g is the carrier generation rate and is proportional to light absorption, and τ is the lifetime of the carrier. If we remember that $[Fe^{2+}]$ and $[Fe^{3+}]$ respectively correspond to the light absorption coefficient and trap density, g should be proportional to $[Fe^{2+}]$, and τ should be proportional to $[Fe^{3+}]^{-1}$. As a result, photoconductivity turns out to be proportional to $[Fe^{2+}]/[Fe^{3+}]$ and mobility. The difference between SLN and CLN crystals in Figure 5.10 probably originates from the difference in mobility between SLN and CLN. It is reasonable that stoichiometric LN has a larger mobility than that of congruent LN because of its lower number of nonstoichiometric defects. This larger photoconductivity can contribute to the larger photorefractive sensitivity of SLN.

5.3 Holographic Properties of Stoichiometric LiNbO$_3$

5.3.1 One-Color Holography in Fe-Doped Stoichiometric LiNbO$_3$

As described in Section 5.2.1, stoichiometric LiNbO$_3$ has a fast-response capability that can be exploited for writing holograms [51]. In fact, Kitamura et al. have reported that Fe-doped stoichiometric LiNbO$_3$ has a fast-response capability

FIGURE 5.11. αL dependence of photorefractive sensitivity $S_{\eta 2}$ in one-color holography for variously Fe-doped SLN and CLN crystals, where α is the optical absorption coefficient and L is the thickness of the sample. The total writing intensity was 260 mW/cm^2 at 532 nm. Extraordinary polarization was used.

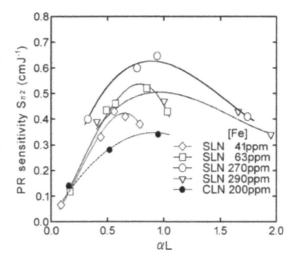

in writing digital holograms at the geometry using the extraordinary polarization [48]. Here, their approach can be extended to a wide range of Fe dopant level [36]. The behavior of Fe in LiNbO$_3$ has been studied for two or three decades, ever since the bulk photovoltaic effect was observed. However, a systematic analysis has not been done on stoichiometric LiNbO$_3$ for the purpose of using it as a medium for digital holographic storage. In this section, we compare the oxidation state of Fe in Fe: SLN with that of Fe: CLN from the viewpoint of writing speed of digital holographic data storage.

Figure 5.11 shows the relationship between photorefractive sensitivity $S_{\eta 2}$ and αL measured in a two-wave mixing experiment using 532-nm plane waves and depending on Fe concentration in the crystals. Here α is the optical absorption coefficient, and L is the thickness of the sample. The accuracy of αL in this measurement was ± 0.05. This figure also shows the characteristic difference between stoichiometric and congruent crystals. The optical polarization is extraordinary. The general tendency is that large sensitivities are obtained in stoichiometric LN (SLN). With increasing αL, the photorefractive sensitivity reaches a maximum value, then it gradually decreases. The maximum sensitivity appears around the αL of 1.0, except for the sample with the small Fe concentration. The SLN had a maximum sensitivity of 0.63 cm/J, nearly two times larger than that of CLN. The αL dependence of the photorefractive sensitivity was measured during digital recording as well. The general trend was similar to that of Figure 5.11. The increased sensitivity in SLN compared with CLN was attributed to the higher photoconductivity and linear electrooptic constant r$_{33}$.

Figure 5.12 shows the relationship between $M/\#$ and αL. In stoichiometric LiNbO$_3$, $M/\#$ decreases with increasing αL and increases with total Fe concentration. Therefore, the maximum $M/\#$ value can be obtained with a large Fe concentration and low αL. A tradeoff exists between sensitivity and $M/\#$. There was no significant difference between stoichiometric and conventional LiNbO$_3$

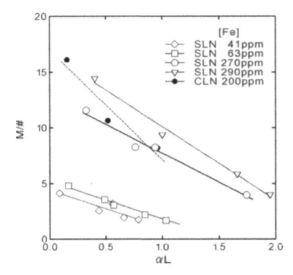

FIGURE 5.12. αL dependence of $M/\#$ in one-color holography for variously Fe-doped SLN and CLN crystals, where α is the optical absorption coefficient and L is the thickness of the sample. The total writing intensity was 260 mW/cm^2 at 532 nm. Extraordinary polarization was used.

with respect to $M/\#$. This comes from the fact that the erasure time constant for SLN is smaller than that of CLN at nearly the same oxidation level of Fe because of the increased photoconductivity. Our measurements were made using the transmission geometry. For the 90-degree geometry, Burr and Psaltis found that $M/\#$ reaches a maximum at an absorption coefficient of about 1.0 [52].

5.3.2 Two-Color Holography Using Stoichiometric LiNbO$_3$

The retrieval of information stored in photorefractive crystals with homogeneous illumination leads to erasure effects, i.e., volatile readout. This is a barrier to practical implementation of a reversible write/read holographic storage system based on photorefractive media. Three schemes have been proposed for nonvolatile readout in photorefractive media. The first is thermal fixing [45], where the stored index grating is copied by thermally activating ion diffusion, which creates an optically stable complementary ion grating. Protons play the major role as mobile ions in the thermal process [53]. Electrical fixing also has been demonstrated [54], where an external electric field smaller than the coactive field is applied to convert the space charge pattern into a ferroelectric domain pattern. Besides the complexity of *in situ* processing, these two methods lead to a loss of recording reversibility, because rapid optical refreshing of the memory is impossible. The third is an attractive all-optical recording scheme using two wavelengths of light, called two-color holography or photon-gated holography [4–6, 55–59], which meets both requirements of nonvolatile readout and reversibility.

Two-color holography relies on the existence of two sets (shallow and deep) of traps with different energy levels in the band gap, as schematically illustrated in Figure 5.13. Holographic recording is performed using the simultaneous presence

FIGURE 5.13. Schematic diagram of charge transition in two-color and one-color holography. In near-stoichiometric LiNbO$_3$, the shallow traps are attributed to small polarons, and the deep traps are attributed to bipolarons and impurities if they exist.

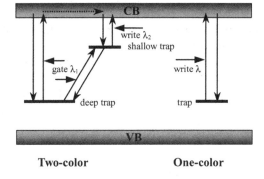

of a gating beam of shorter wavelength (or incoherent light with broad band) and two recording beams of longer wavelength. The gating light can be thought of as sensitizing the material. In the absence of the gating light, the crystal is insensitive to light of the writing wavelength, and the readout is nonvolatile. The processes are optically reversible; i.e., erasure is possible by illumination with the gating light. Figure 5.14 shows the qualitative difference between conventional one-color and two-color holographic recording in near-stoichiometric LiNbO$_3$ crystal.

In early researches, pulsed lasers with intensities on the order of 10^7–10^9 W/cm^2 were needed in order to efficiently perform two-color recording [55–56]. For practical application, low-power cw lasers are needed. In recent years, a variety of materials, including undoped and doped near-stoichiometric LiNbO$_3$ crystals, have been discovered in which two-color recording can be realized at moderate intensities, thus allowing the use of cw laser sources [4–6, 59].

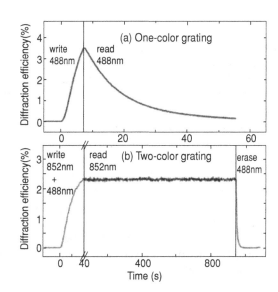

FIGURE 5.14. Typical write-read-erase curves for holographic recording in lithium niobate crystals. (a) conventional one-color holography, (b) nonvolatile two-color holography.

Undoped Stoichiometric Lithium Niobate

Lithium niobate crystals are in general not stoichiometric, since the chemical formula is $[Li_{1-5x}Nb_xV_{4x}][Nb]O_3$ ($x = 0.028$ for congruent composition). In this model, Li vacancies are compensated by antisite defects (Nb_{Li}^{5+}) that reach a concentration of 1% in the congruent composition [2]. Electrons trapped at adjacent Nb_{Li} and Nb_{Nb} sites form $Nb_{Li}^{4+}Nb_{Nb}^{4+}$ bipolarons. Illumination with a photon energy higher than green light dissociates the bipolarons, and these electrons become self-trapped on the Nb site, forming small polarons Nb^{4+}. The 1.6-eV (780 nm) metastable absorption band is used to write the two-color holograms with near-infrared light [4–5]. The depth of the small polaron level below the conduction band was measured to be 0.7 eV from the temperature of the induced absorption at 1.6 eV [5].

The recording process becomes more effective with increasing small polaron lifetime, and the small polaron lifetime increases with increasing Li content and decreasing Li defects. The lifetime in $LiNbO_3$ at room temperature changes from <40 ms for the congruent composition (48.6 Li_2O mol%) to 400 ms for as-grown $LiNbO_3$ with a lower defect density (49.5 Li_2O mol%) [4, 6]. Postgrowth reduction of near-stoichiometric crystals increases the lifetime of small polarons to as long as several seconds and increases the concentration of filled bipolarons. Therefore both sensitivity and dynamic range are enhanced [4–5]. However, excessive reduction leads to very long polaron lifetimes of tens of seconds, where holograms exhibit a substantial dark decay right after writing. The sensitivity has a two-order-of-magnitude increase with increasing Li_2O mol% from 48.6 up to 49.6% in lightly reduced $LiNbO_3$, after which it saturates. It seems that the refractive index change increases when the crystal composition approaches the stoichiometric one [4].

The sensitivity is dependent on the gating wavelength and the gating intensity [5]. A gating beam with a shorter wavelength leads to a higher sensitivity, for example, a factor of 10 increase for wavelengths between 514 nm and 400 nm. The sensitivity increases linearly at low intensity, then saturates at high intensity. The dynamic range expressed by the $M/\#$ (and corresponding to the saturation index change Δn, since they are linearly related to each other) increases with increasing writing intensity at a constant gating intensity, and is proportional to the writing intensity if the reference and object beams are of equal intensity. Increased gating intensity leads to erasure of the grating written with near-infrared light, that is, a decrease in the dynamic range [4].

The gating ratio expresses the resistance to erasure during readout, defined as the ratio of sensitivities in the presence of gating light to the absence thereof [5]. This parameter depends on the wavelength of writing light. Typical values of the gating ratio are in the range of 2000–5000 for writing light of 852 nm in undoped, slightly reduced crystals of 49.7 Li_2O mol%, but the value is only 5 at 670 nm, because the self-gated contribution becomes appreciable at shorter wavelengths. A larger reduction also leads to a substantial increase in self-gated recording and loss of nonvolatility.

Stoichiometric Lithium Niobate Doped with Fe, Mn

Extrinsic dopants, Fe, Mn, may play essentially the same role as bipolarons, provided that their energy levels are sufficiently deep in the band gap of the crystal. Introduction of deep extrinsic dopant provides a means to control the concentrations of deep traps and shortens the lifetime of small polarons; the sensitivity is maintained because direct excitation and recombination between deep and shallow traps seem to be efficient in these cases. A slight improvement in two-color photorefractive properties is possible with Fe-doped stoichiometric crystal (49.9 Li$_2$O mol% and Fe concentration of 0.01 wt%) [4]. Compared with the typical one-color (green) sensitivity (0.06–0.8 cm/J) of Fe-doped LN [36], the sensitivities of reduced stoichiometric LiNbO$_3$ [4–5]) are low (10^{-3}–10^{-2} cm/J). Moreover, postgrowth reduction treatment limits the storage time of holograms by increasing the dark conductivity and decreases nonvolatility.

As-grown stoichiometric LiNbO$_3$ (49.6 Li$_2$O mol%) doped with 8 ppm of Mn has much better two-color photorefractive properties [60]. The absorption coefficient at 350 nm has a moderate value of 3.4 cm^{-1}, which means that 350-nm light can both pass through the crystal and efficiently gate for two-color recording, and no apparent absorption structure of bipolarons around 500 nm is seen in the visible region. Two-color holograms were recorded using a 778-nm IR laser for writing and a 350-nm UV laser for gating. A high two-color sensitivity of 0.21 cm/J with a gating intensity of 1.5 W/cm^2 was obtained (Figure 5.15). This value is much larger than that of reduced SLN, and is comparable to that of one-color holography of Fe-doped LN using a green laser [36]. The sample's light-induced absorption change of 0.3 cm^{-1} in the near IR is much larger than the 0.1 cm^{-1} found in reduced SLN at the pump intensity of 1 W/cm^2 [4–5, 60], and thus

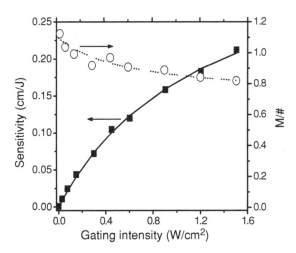

FIGURE 5.15. Dependence of two-color sensitivity and $M/\#$ on gating intensity at 350 nm in Mn-doped stoichiometric LiNbO$_3$. The total writing intensity is 20.5 W/cm^2 at 778 nm.

results in high sensitivity for two-color recording. $M/\# \approx 1$ is expected for a 1-cm thickness at writing intensity of 4 W/cm^2. At this intensity, $M/\#$ is slightly larger than it is for reduced SLN.

The energy levels of Mn lie below those of Fe and the bipolarons in the forbidden band gap [4]; therefore they are further from the shallow traps of small polarons. A smaller photoionization cross-section for near-IR light for Mn traps should be expected compared with Fe and bipolarons, and therefore, Mn doping should reduce the volatility during the readout. On the other hand, it is not necessary to reduce crystals, and dark conductivity is not increased, which leads to a long dark decay time of 0.6 years [60].

Optimization of the crystal composition shows that as-grown stoichiometric LiNbO$_3$ (49.9 Li$_2$O mol%) doped with 10-ppm Mn can further improve the two-color sensitivity and dynamic range [61]. The two-color sensitivity is about 30% higher, and the dynamic range is about 50% higher than that of the stoichiometric LiNbO$_3$ with 49.6 Li$_2$O mol%. The two-color recording performance, especially sensitivity, depends on the population of shallow traps (small polarons), that is, light-induced absorption at the near-IR recording wavelength. The lifetimes of small polarons increased from about 0.3 s to 1.2 s by changing the composition from 49.6 to 49.9 Li$_2$O mol%, which in turn leads to a 100% increase in induced absorption.

5.4 Holography Using Photochromism in Stoichiometric LiNbO$_3$

5.4.1 Enhanced Photochromism in Stoichiometric LiNbO$_3$

Heat treatment has generally been utilized to modify the defect structures in LiNbO$_3$ and the corresponding absorption bands. In particular, thermal reductions, which can increase the optical absorptions in many undoped and doped LiNbO$_3$ systems, have been employed to improve the photorefractive sensitivities to visible or near-infrared recording light for holographic storage using LiNbO$_3$ [4–5, 62]. However, the reduction treatment also has the adverse effect of increasing the dark conductivity of the material and thus of limiting the dark storage time of holographic information. In this section, another method will be presented to modify the absorption band by the light-induced absorption change, which is prominently observed in near-stoichiometric LiNbO$_3$ (SLN). This method is promising in the sense that it is an all-optical means and no thermal process is required.

The SLN crystals were grown using the top-seeded solution growth method, in which the growing crystal is in equilibrium with a Li-rich melt of 59 mol% Li$_2$O and 41 mol% Nb$_2$O$_5$. The crystal composition estimated from the Curie temperature was Li/Nb $\approx 49.7/50.3$. Three different types of crystal were used for the experiment: nominally undoped, Tb codoped (Tb 70 ppm or Tb 140 ppm and Fe ≈ 2 ppm as natural impurity), and Tb, Fe-doped (Tb 140 ppm and Fe

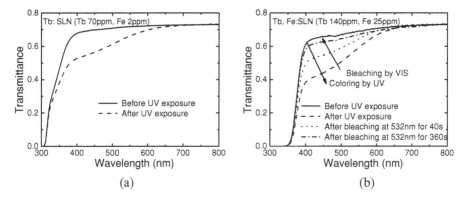

FIGURE 5.16. Typical transmission spectra of as-grown crystals measured before and after irradiation with UV light. The crystal thickness was 3 mm. (a) Tb: SLN (Tb 70 ppm and Fe 2 ppm as natural dopant). $\lambda_{UV} = 313$ nm, $I_{UV} = 50$ mW/cm^2, irradiation time was 1 min. (b) Tb, Fe: SLN (Tb 140 ppm, Fe 25 ppm). The full spectrum of a Hg-Xe lamp with $I_{UV} = 0.7$ W/cm^2 was used as the UV source, and the irradiation time was 10 sec. Bleaching process is shown as well 532-nm light of 0.4 W/cm^2 was used as the bleaching light.

25 ppm, or Fe 10 ppm, or Fe 5 ppm). The doping levels represent the actual concentrations in the crystals, which were estimated from the nominal doping concentrations and effective segregation coefficients.

These crystals exhibited a reversible photochromic effect at room temperature [37, 63]. Irradiation by ultraviolet (UV) light induced a visible absorption band extending from $\lambda \approx 650$ nm to the absorption edge, which could be bleached by subsequent irradiation with visible light. Figure 5.16 shows typical transmission spectra measured before and after irradiation with UV light at room temperature. The filtered 313 nm or whole spectrum from a Hg-Xe lamp was used as the UV source. This photoinduced absorption band was observed in the oxidized samples as well as in the as-grown ones. The response to UV light was very fast, and the induced absorption was almost saturated after several seconds at the UV intensity of ≈ 0.7 W/cm^2. The bleaching process was much slower than the coloration, under similar light intensity. Although the photochromic effect was observed in every SLN crystal, the effect was enhanced and the colored state was stabilized according to the Tb doping and Fe doping. Little photochromic effect was observed in the undoped and Tb-doped congruent LiNbO$_3$ crystals.

An electron paramagnetic resonance (EPR) study was carried out to investigate the charge transfer mechanism [37, 64]. Figure 5.17 shows the typical EPR measurements in Tb, Fe: SLN at room temperature, illustrating a decrease in Fe^{3+} caused by UV excitation. Based on these results, the charge transfer process has been assumed as follows [37]. These materials have three different types of energy level: UV absorption centers just above the valence band, metastable shallow electron traps slightly below the conduction band (presumably attributed to small polaron states), and deep Fe^{3+} traps. The UV light photoexcites electrons from

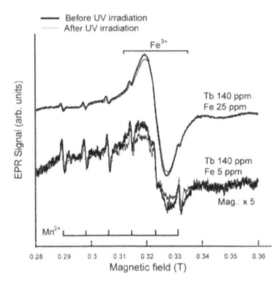

FIGURE 5.17. EPR signals from Tb, Fe: SLN samples of different Fe concentrations at 298 K. The magnetic field was perpendicular to the c axis. Both samples revealed a reduction in the Fe^{3+} intensity after UV irradiation. The relative degree of intensity reduction decreased with increasing Fe doping concentration. No change in the Mn^{2+} intensity was observed.

the UV absorption centers to the conduction band (CB). After iterated trapping and detrapping processes between CB and shallow traps, these electrons are ultimately trapped on the Fe^{3+} levels, giving rise to a visible absorption band. In other words, the photochromic effects observed in the SLN samples are due to the valence change from Fe^{3+} to Fe^{2+} that is caused by the UV-induced charge transfer via the conduction band.

With increasing Fe doping level, the shallow trap lifetime shortened with a simultaneous increase in the dark decay time of the deep-trapped charges, which again indicates that Fe^{3+} ions act as deep traps. The shallow trap can be used as a metastable intermediate level in two-color gated holography, as described in a later section. The absence of any noticeable photochromic effect in congruent $LiNbO_3$ crystals might be because the UV-absorption centers are screened by the shrunken band gap, and the electron–hole pair generated by the direct integrand transition tends to quickly recombine. It was observed only when the wavelength of UV light was shorter than about 330 nm, indicating that the UV absorption centers are located just above the valence band.

The UV-light-induced absorption observed in SLN can be effectively used in holographic storage. Figure 5.18 illustrates a charge transfer model for possible holographic recording schemes.

5.4.2 One-Color Quasinonvolatile Holography

Holograms can be recorded using the charges created at Fe^{2+} centers by UV irradiation. The charge transfer model is illustrated in Figure 5.18(a). The fast coloration by UV enables a fast erasure of the stored information. UV irradiation

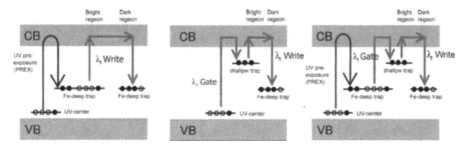

FIGURE 5.18. Schematic energy band diagram of holographic recording based on UV-exposed charge transfer in Tb, Fe: SLN. (a) one-color recording from the UV-preexposed state, (b) UV-gated two-color recording, (c) two-color recording from the UV-preexposed state.

at any time erases the stored gratings, returning the crystal back to the original preexposed state. On the other hand, bleaching by visible light (which corresponds to recording) is much than the UV-induced coloration, suggesting the capability of storing many-multiplexed information.

One-color recording from the UV preexposed state was investigated using a conventional two-wave mixing technique [64–66], and the possibility of quasi-nonvolatile storage was demonstrated [65]. A Tb, Fe: SLN (Tb 140 ppm, Fe 25 ppm) crystal with a thickness of 3.3 mm was used. The crystal was exposed to UV light from a Hg-Xe lamp of about 0.7 W/cm² for about 10 s. After UV preexposure, two-wave mixing was carried out to measure recording parameters such as sensitivity and erasure time constant with a continuous-wave green laser ($\lambda = 532$ nm) at the extraordinary polarization.

Figure 5.19 shows the experimental results comparing sensitivities and erasure time constants of the UV-exposed state and bleached state for different oxidation conditions of Fe. For multiplexed recording from the UV exposed state, the induced absorption gradually became bleached, and as a result, the sensitivity decreased with the number of stored holograms. The figure clearly shows the dynamic change of the parameters arising from the transient nature of the charges created by UV exposure. It also suggests that the fully oxidized crystal is preferable as a holographic recording medium, because it has a wider sensitivity change and longer erasure time constant, leading to a larger dynamic range and quasi-nonvolatility when it is used as a storage medium. This comes from the depletion of charges at Fe deep trap levels by oxidation. The no-absorption feature in the recorded condition has another merit that little optical damage or fanning noise occurs.

$M/\#$, defined as the sum of the square root of diffraction efficiency of the multiplexed holograms, is usually used as a parameter of multiplexing in holographic storage. Because of the transient feature of induced absorption, $M/\#$ cannot be estimated simply by measuring the sensitivity and erasure time constant of a single hologram. It was found, however, that in this particular photochromic material, the decreasing sensitivity was almost completely compensated by the

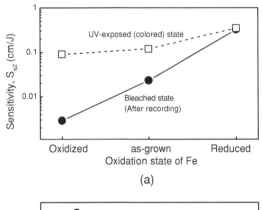

FIGURE 5.19. Comparison of (a) sensitivity and (b) erasure time constant between UV-exposed state and bleached state for different oxidation conditions of Fe for Tb, Fe: SLN (Tb 140 ppm and Fe 25 ppm). Recording intensity was 300 mW/cm^2.

increased erasure time constants of later-written holograms. This made it possible to use a conventional recording scheme of multiplexing. Fifty angle-multiplexed holograms were recorded after UV preexposure using the modified multiplexing schedule. $M/\#$ was estimated to be 1.8 [66].

5.4.3 Multicolor Nonvolatile Holography

Gated Two-Color Recording

One of the most critical issues in holographic recording with photorefractive materials is the gradual erasure of information during readout. As mentioned in Section 5.3.2, two-color gated recording is a promising technique to overcome this problem [55].

We found that the photoinduced charge transfer process via metastable shallow traps in Tb-doped SLN can be used for nonvolatile two-color recording at near-IR wavelengths with UV gating [37, 59]. Figure 5.18(b) shows the relevant charge transfer model. An as-grown SLN crystal with 90 ppm Tb and 2 ppm Fe was chosen for the measurement, and it had a shallow trap lifetime of 3–4 s. Two-color recording was carried out with a CW semiconductor diode laser ($\lambda = 852$ nm) used for the recording beams and a filtered Hg-Xe lamp ($\lambda = 313$ nm) used for the gating beam. The diffraction efficiency was up to several percent, and the two-color sensitivity was 0.01–0.02 cm/J. These values are comparable to those reported for the bipolaron-based two-color recording [4–5]. Nonvolatile digital holographic storage was also demonstrated [59]. A random digital pattern was stored and reconstructed using a TFT-LCD spatial light modulator with 768 × 512 pixels and a CCD detector. No reduction in diffraction efficiency was observed during continuous readout for over 4 h, and the bit error rate of the reconstructed data remained less than 10^{-5} during this period.

For the charge transfer model's configuration of energy levels, although nonvolatile recording can be achieved using UV gating and IR recording light, hologram multiplexing is still challenging from a practical viewpoint. No matter where the gratings are written, they are accessible by UV light, and the response of this material to UV light is very fast. Thus, the UV gating beam not only aids the recording process by sensitizing the material but also will erase the stored holograms rather quickly. In fact, we have observed that early-written holograms almost completely disappear after more than 50 holograms have been stored.

To obtain optimum performance, the recording parameter dependence of two-color recording was investigated. In particular, the intensity dependences of the dynamic range and two-color sensitivity were studied using 780-nm beams from a Ti-sapphire laser for writing, of which the wavelength is close to the absorption peak of small polaron, and 350-nm light from a krypton ion laser for gating [67]. It was found that the saturated diffraction efficiency increased with total writing intensity and reached a saturated value of 56% at 20 W/cm^2. The sensitivity increased linearly with gating intensity at low intensity and saturated at 0.08 cm/J for intensities higher than 1.6 W/cm^2. Fifty plane-wave holograms could be recorded with the angle-multiplexing method, and an $M/\#$ of 1.1 was obtained.

Two-Color Recording from the Colored State

Two-color recording from the colored state may be an alternative way of multiplexing many nonvolatile holograms [37]. Figure 5.18(c) shows the relevant charge transfer model. A visible gating beam can be used in this recording scheme, and consequently, more holograms can be multiplexed because a visible gating light will lead to a longer erasure time. Furthermore, fast erasure of recorded data by UV is a merit as in one-color recording from the colored state.

The recording characteristics were investigated in an angular multiplexed recording with a visible gating light and IR recording lights [68–69]. The Tb: SLN crystal (70 ppm Tb and 2 ppm Fe) was exposed to UV light ($\lambda = 313$ nm, I = 130 mW/cm^2) for 45 seconds prior to recording. After UV preexposure, two-color recording was carried out using a continuous-wave IR laser ($\lambda = 850$ nm) and a gating light from either a filtered Hg-Xe lamp or an SHG-YAG laser (532 nm).

A saturated diffraction efficiency larger than 10% was obtained in single-hologram storage, regardless of the wavelength of the gating light. Figure 5.20 shows the diffraction efficiency profile after the multiplexed recording was

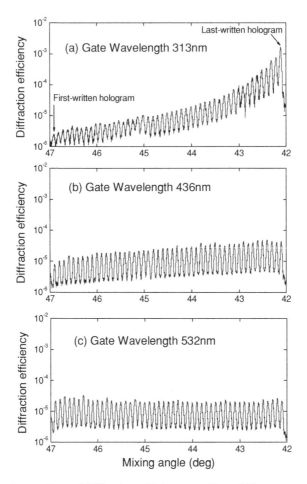

FIGURE 5.20. Comparison of diffraction efficiency profiles of 50 plane-wave holograms stored using the gating light with a wavelength of (a) 313 nm, (b) 436 nm, and (c) 532 nm.

completed, in which 50 angle-multiplexed holograms were sequentially recorded with an equal exposure time of 2.5 s for each hologram. The gating intensity was fixed at 50 mW/cm^2. When the gating light was 313 nm, the recording from the colored state was basically the same process as that from the clear state. As shown in Figure 5.20, the recording has a higher sensitivity, but there is also a quick erasure of early-written holograms. Figure 5.20 also shows the diffraction efficiency profile of 50 holograms recorded with visible gating light, 436 nm or 532 nm. As illustrated, the diffraction efficiency became more equalized, compared with the case of 313-nm-gated recording. This is because the holograms stored with the visible gating light now have a prolonged erasure time, and the total exposure time (125 sec) is shorter than the time taken for complete bleaching at the gating intensity of 50 mW/cm^2. If a small number of holograms are to be recorded and

FIGURE 5.21. Variation in diffracted signal intensity during continuous readout. The readout intensity is 4 W/cm² (half of the total recording intensity) at λ = 850 nm.

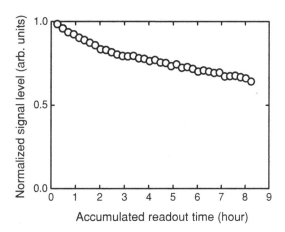

Accumulated readout time (hour)

the total exposure time is not long, the stored holograms will have nearly equal diffraction efficiencies, even when each is recorded with an identical exposure time. However, if many holograms of equal strength need to be stored, we need to take into account the dynamic nature of sensitivity and erasure time.

By considering the dynamic changes in the sensitivity and erasure-time constant as mentioned above, we have been able to record 100 holograms of nearly equivalent strength using a 436-nm gating light [68–69]. An $M/\#$ of 0.33 was obtained for 0.5-cm-thick crystal. Figure 5.21 illustrates the variation in the diffracted signal during continuous reading, in which the readout was carried out with intensity as high as 4 W/cm^y. The 1/e decay time was measured to be 1.1×10^5 s at this reading intensity. Considering that about 4,000 photons per CCD pixel are usually required for signal recognition, the measured decay time indicates that 80 million readouts are possible with a bit rate of 1 Gbit/s before the signal intensity falls to one-half of its initial value.

5.5 Holography Using Undoped Stoichiometric LiTaO₃

Lithium tantalate (LiTaO₃) crystals are isomorphic to LiNbO₃. However, very little attention has been paid to LiTaO₃. In comparison with LiNbO₃, LiTaO₃ has the advantage of long storage time (10 years) because of its wider bandgap [70]. In general, near-ultraviolet light is required for sensitive recording using LiTaO₃, and the holographic sensitivity is one order of magnitude better than that of LiNb₃:Fe. Thus LiTaO₃ is favorable for storage applications.

5.5.1 Photorefraction of Stoichiometric LiTaO₃ [71–72]

The near-ultraviolet photorefractive properties of nominally undoped congruent and stoichiometric lithium tantalate crystals at 364 nm were compared

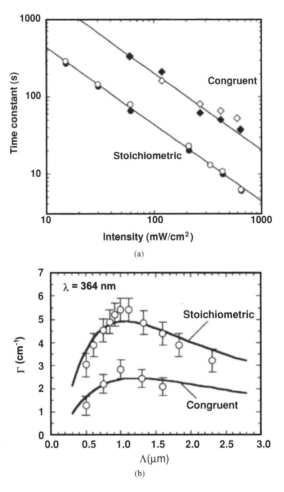

FIGURE 5.22. (a) Building (empty) and erasure (solid) time constants for congruent (diamonds) and stoichiometric (circles) LiTaO$_3$. Gating period: $\Lambda = 1.1$ μm. (b) Two-wave mixing exponential gain in congruent and stoichiometric LiTaO$_3$. The solid lines are theoretical fits.

by performing conventional two-wave-mixing experiments. Near-stoichiometric LiTaO$_3$ was grown with the DCCZ method. These crystals had a Curie temperature of 690°C and a Li ratio of 49.75%. Congruent LiTaO$_3$ was grown with the conventional Czochralski method. Figure 5.22(a) compares the time constants for grating buildup and erasure in the two kinds of crystal. The grating response is about five times faster in the near-stoichiometric crystal than in the congruent one. Figure 5.22(b) shows the steady-state two-wave mixing exponential gain Γ as a function of the grating period Λ. The gain in the near-stoichiometric crystal is about twice as large as in the congruent sample.

The photovoltaic field was estimated by direct measurement of the photovoltaic current under UV illumination and calculation of the photoconductivity, which is inversely proportional to the photorefractive response time. In this way, we could estimate photovoltaic fields of 15 kV/cm and 0.3 kV/cm for the congruent and the near-stoichiometric crystals, respectively, which are consistent with the theoretical fitting results from the above two-wave mixing experiments

5.5.2 Two-Color Holography Using As-Grown Stoichiometric LiTaO$_3$ [73, 74]

Similar to LiNbO$_3$, lithium tantalate has a nonstoichiometric nature (Li-deficient). However, it has a large number of intrinsic antisite defects (Ta$_{Li}^{5+}$). Irradiation with visible or ultraviolet light induces a metastable absorption band (at room temperature) of small polarons (Ta^{4+}) with a maximum absorption occurring at 570 nm [75]. The lifetime of small polarons is about 12 ms for congruent LiTaO$_3$ and hundreds of μs for Fe-doped congruent LiTaO$_3$ [6, 76]. The lifetime is as long as 3–4 s for as-grown undoped stoichiometric LiTaO$_3$ (49.65 Li$_2$O mol%) because of the low defect density and reductive growth atmosphere. The saturated light-induced absorption change at the near infrared wavelength is on the order of 0.1 cm^{-1}.

Nonvolatile two-color holograms were recorded in as-grown near-stoichiometric LiTaO$_3$ with a Li ratio of 49.7%, which was grown along the c-axis using the DCCZ method. Figure 5.23 shows the evolution of the refractive index change during a typical write-read-erase process. The fast decay time

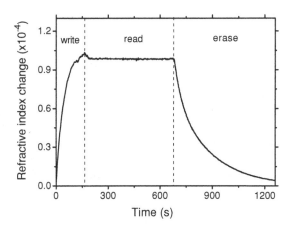

FIGURE 5.23. Evolution of refractive index change during a typical write-read-erase process in a 2-mm-thickness undoped stoichiometric LiTaO$_3$ crystal. The total writing intensity is 11.2 W/cm^2 (equally divided into two beams) at 778 nm, while the gating intensity is 0.031 W/cm^2 at 350 nm.

constant right after writing is nearly the same as the small polaron lifetime, which means the antisite defects ($Ta_{Li}^{4+/5+}$) act as shallow traps. The deep traps are assigned to be unintentional impurities such as Fe [74] or bipolarons ($Ta_{Li}^{4+}\ Ta_{Ta}^{4+}$) produced during crystal growth in a reductive atmosphere by measuring UV-induced absorption spectra at room temperature and low temperature of 77.3 K. It deserves to be noticed that stoichiometric $LiTaO_3$ crystals don't need to be annealed reductively.

Compared with Fe-doped congruent $LiTaO_3$ [77], the obtained refractive index change of 1×10^{-4} (diffraction efficiency of 41% in the 2-mm-thick sample) is one order higher than the obtained maximum value of 1×10^{-5}. The $M/\#$ of 2.0 was calculated from Figure 5.23. $M/\#$ scales linearly with crystal thickness in the absence of absorption, suggesting that $M\# \approx 3.6$ for the 1-cm-thick sample at a total writing intensity of 4 W/cm^2. At this intensity, $M/\#$ is much larger than those of reduced undoped SLN and as-grown Mn-doped SLN [5, 60].

The two-color sensitivity may be increased by increasing the gating intensity, and the sensitivity was 0.086 cm/J with the gating intensity 1.0 W/cm^2 at 350 nm in this sample. This sensitivity is much higher than the 0.01–0.04 cm/J of the reduced undoped SLN, and is also higher than the obtained maximum value of 0.07 cm/J from Fe-doped congruent $LiTaO_3$. Because the absorption edge of this undoped SLT is near 265 nm and the absorption is low in the range of 300–400 nm, we can also use a light of shorter wavelength for gating in order to increase the sensitivity. The sensitivity doubled using a 313-nm gating light from a Hg-Xe lamp in place of the 350 nm one at the same intensity of 22 mW/cm^2. Additionally, because the bandgap of $LiTaO_3$ is wider than that of $LiNbO_3$ and the absorption peak of small polarons (Ta_{Li}^{4+}) is at around 570 nm, we can increase the sensitivity using light of a longer wavelength for writing without sacrificing readout nonvolatility. The sensitivity doubled when the 722-nm writing beam was used in place of the 778-nm one with the gating intensity of 22 mW/cm^2. The sensitivity at 722 nm was 0.18 cm/J with the writing intensity of 10 W/cm^2, and the gating intensity at 313 nm was 0.2 W/cm^2.

If there was no gating light in the writing process, no measurable hologram could be recorded in this crystal; i.e., the gating ratio was very high, and therefore, the resistance to IR erasure was very strong. The nonvolatile IR readout behaviors of the undoped SLT, undoped SLN, and Mn-doped SLN are compared in Figure 5.24, where the reading intensity is 5.6 W/cm^2 at 778 nm. This nonvolatile readout performance of the undoped SLT is the best among the three crystals. The projected lifetime of holograms at room temperature in the undoped SLT crystal is longer than five years. The performances of nonvolatile IR readout and dark decay of stoichiometric $LiTaO_3$ are the best of all materials reported so far [73, 74].

Optimization of crystal composition is the most effective method to improve two-color photorefractive properties. An optimal near-stoichiometric crystal composition of around 49.65 % for both sensitivity and dynamic range together with good readout performance is observed [74].

FIGURE 5.24. Comparison of nonvolatile IR readout behaviors in as-grown undoped SLT, Mn-doped SLN, and undoped SLN. The reading intensity is 5.6 W/cm² at 778 nm.

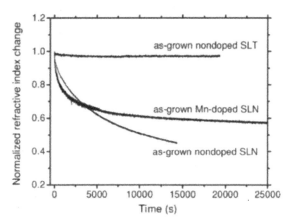

5.5.3 Dark Decay and Proton Concentration [74, 78]

Protons, in the form of OH⁻ ions, are always present as impurities in LiNbO₃ grown in air by the Czochralski method. Protons have been proved to be responsible for thermal fixing in doped LiNbO₃ [53]. The presence of protons in the crystal can be detected by measuring the absorption spectrum of the OH stretching vibration in the IR range near 3480 cm⁻¹ [79]. The OH absorption band consists of several components, and the peaks and the shape of the absorption curve are dependent on the nonstoichiometry of the crystal [80]. In the congruent LiNbO₃ crystal, the absorption peak appears at about 3484 cm⁻¹; however, in near-stoichiometric crystals the absorption peak appears at about 3366 cm⁻¹ without significant change in the vibration frequency of several components [81]. LiTaO₃ has the same crystal structure and a similar OH absorption band as LiNbO₃, but the peak of the OH absorption band shifts from about 3484 cm⁻¹ for the congruent composition to 3461 cm⁻¹ for the near-stoichiometric composition. The OH absorption coefficients of samples were characterized at 3466 and 3461 cm⁻¹ for stoichiometric LiNbO₃ and LiTaO₃, respectively.

For LiNbO₃:Fe crystals, two different dark decay mechanisms have been verified; one is identified as proton compensation and the other is due to electron tunneling between sites of Fe²⁺ and Fe³⁺ [30]. In crystals with doping concentration less than 350 ppm Fe, proton compensation dominates the dark decay and extrapolation of lifetimes using an Arrhenius law to room temperature is valid. For Mn-doped congruent LiNbO₃ crystals, this critical doping density is much higher (0.2 at% Mn) [82]. In our experiments, the lifetimes of holograms written in all samples were based on this extrapolation of high-temperature data. This method is reliable, because all the samples were undoped or slightly doped crystals.

Nonvolatile two-color holograms were written in ten undoped and doped stoichiometric LiNbO₃ and LiTaO₃ crystals using a 778-nm near-IR laser for writing and a 350-nm ultraviolet beam for gating under the same experimental conditions.

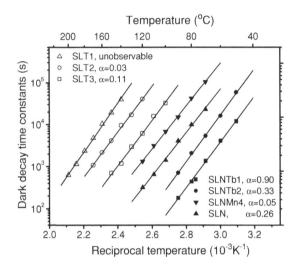

FIGURE 5.25. Arrhenius plot of dark decay-time constants of two-color holograms in stoichiometric $LiNbO_3$ and $LiTaO_3$.

After the holograms were written at room temperature, the crystals were heated to a certain temperature in the dark, and an attenuated beam of 778 nm was used to monitor the dark decay process from time to time while the sample was kept at a steady temperature. The dark decay time constants were fitted excellently using an exponential decay function. After each experiment, the crystal was heated to 245°C and was kept at this temperature under uniform illumination of UV for 40 min to erase the gratings completely.

Figure 5.25 shows the dependence of the measured dark decay time constants on temperature. The dark decay-time constants in all of these samples obey an Arrhenius-type dependence on absolute temperature, $\tau = \tau_0 \exp(E_a/k_B T)$, where τ_0 is a preexponential factor, k_B is the Boltzmann constant, and E_a is the activation energy. The average activation energy is about 1.06 eV. This value is close to the proton activity energies reported in the literature [83], and nearly the same as the activation energy of proton compensation in Fe-doped and Mn-doped congruent $LiNbO_3$ [30, 82]. The data were extrapolated to get the lifetime of the hologram at room temperature. Figure 5.26 plots the dependence of projected room temperature t on OH^- absorption coefficient (i.e., proton concentration) on a double-logarithmic scale. The lifetime is inversely proportional to proton concentration. These results show that the proton compensation mechanism dominates the dark decay of nonvolatile two-color holograms in undoped and slightly doped near-stoichiometric $LiNbO_3$ and $LiTaO_3$. It is seen that the lifetime of holograms in near-stoichiometric $LiTaO_3$ is an order of magnitude longer than that in near-stoichiometric $LiNbO_3$, which is consistent with the proton-exchange waveguide measurement showing that the proton exchange diffusion coefficients in $LiNbO_3$ are generally one order of magnitude higher than the values in $LiTaO_3$ [84].

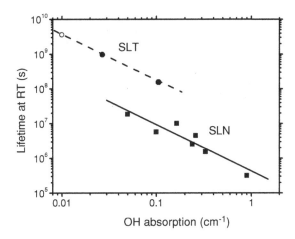

FIGURE 5.26. Dependence of projected hologram lifetime at room temperature on OH⁻ absorption (proton concentration).

5.A Appendix

Sensitivity and M/#

We introduce two figures of merit ($S_{\eta 2}$ and $M/\#$) commonly used to describe holographic storage performance. $S_{\eta 2}$ is photorefractive sensitivity [1], defined as the slope of the square root of diffraction efficiency as a function of time at the very initial stage of hologram formation, divided by the product of light power density and crystal thickness:

$$S_{\eta 2} = \partial \sqrt{\eta}/\partial t \big|_{t=0}/(I_W \cdot L), \qquad (5.A1)$$

where η is the diffraction efficiency, t is the time, I_W is the light power density, and L is the thickness of crystal.

$M/\#$, termed the M number, is a system-material figure of merit for the dynamic range in photorefractive multiplexed recording, which is defined as the coefficient of proportionality between the square root of hologram diffraction efficiency and the number of holograms:

$$M/\# = M\sqrt{\eta_M}, \qquad (5.A2)$$

where M is the multiplexed page number and η_M represents the diffraction efficiency when the Mth page is written. In a photorefractive crystal, a consequence of the read-write ability is that recorded holograms are slowly erased during subsequent recording exposures. To store multiple equal-strength holograms, one follows a schedule of decreasing exposure times. The first hologram is recorded to a strong diffraction efficiency. It is erased exponentially during the remaining exposures, and it finishes at the same diffraction efficiency that the last hologram reached with its short exposure time. $M/\#$ can be found as the product of the

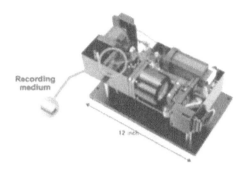

FIGURE 5.27. The holographic digital data storage system.

Recording medium

12 nch

recording slope A_0/τ_r and the erasure time constant τ_e, where A_0 is the saturated value of the square root of diffraction efficiency and τ_r is the recording time constant. Further details on $M/\#$ are given in [85]. These figures of merit and system performance are connected as follows. If η is fixed for the recording system, the writing time required to write one page, τ_M, and number of pages multiplexable, M, can be obtained as a function of the figures of merit:

$$\tau_M = \eta^{1/2}(S_{\eta 2} I_W L)^2, \qquad (5.A3)$$

$$M = M/\# \big/ \eta^{1/2}. \qquad (5.A4)$$

Recording materials require both large $S_{\eta 2}$ values for fast writing and large $M/\#$ values for a large dynamic range.

System Demonstration

A miniaturized holographic data storage system was implemented in order to demonstrate the performance of SLN crystals as a digital holographic recording medium. Figure 5.27 is a photograph of the demonstrator (12 in. × 4.8 in.). The optical system's configuration was based on that of 4-f Fourier transform holography. The one-color holography from the UV-preexposed colored state was used for recording. The recording light from a doubled Nd:YAG laser with continuous

FIGURE 5.28. An example of a digital image as part of the reconstructed digital movie.

wave (70 mW) was split into an object beam and a reference beam. The object beam was expanded to overfill the input aperture of the spatial light modulator (SLM). A liquid crystal panel with 768×512 pixels (26 μm × 26 μm pixel size) was used for the SLM. The page capacity was 24.6 KB. The light transmitted from the SLM was spatially modulated by a signal image pattern encoded with binary differential modulation code and projected into the crystal via a Fourier transform lens and mixed with the reference beam with ≈3-mm diameter at a crossing angle of 30 ± 20°. By angular multiplexing with a separation angle of 0.05 degrees, multiplexing of 800 pages was possible. Figure 5.28 shows an example of reconstructed digital data. The recorded data was a digital motion picture file in Windows AVI format. The bit-error rate was on the order of 10^{-5}, a practically insignificant magnitude.

References

1. P. Günter, J.-P. Huignard (ed.): *Photorefractive Materials and Their Applications I*, (Springer, 1988).
2. O.F. Schirmer, O. Thiemann, M. Woehlecke: *J. Phys. Chem. Solids* **52**, 185 (1991).
3. K. Kitamura, Y. Furukawa, S. Tanaka, T. Yamaji, H. Hatano: *Proc. SPIE* **3291**, 151 (1998).
4. L. Hesselink, S.S. Orlov, A. Liu, A. Akella, D. Linde, R. Neurgaonkar: *Science* **282**, 1089 (1998).
5. H. Guenther, R. Macfarlane, Y. Furukawa, K. Kitamura, R. Neurgaonkar: *Appl. Opt.* **37**, 7611 (1998).
6. Y.S. Bai, R. Kachru: *Phys. Rev. Lett.* **78**, 2944 (1997).
7. B.T. Matthias, J.P. Remeika : *Phys. Rev.* **76**, 1886 (1949).
8. A.A. Ballman: *J. Am. Ceram. Soc.* **48**, 112 (1965).
9. P. Lerner C. Legras, J.P. Duman: *J. Crystal Growth* **3/4**, 231 (1968).
10. L.O. Svaased, M. Eriksrud, G. Nakken, A.P. Grande: *J. Crystal Growth* **22**, 230 (1974).
11. S. Miyazawa, H. Iwasaki: *J. Crystal Growth* **10**, 276 (1971).
12. N. Iyi, K. Kitamura, F. Izumi, S. Kimura, J.K. Yamamoto: *J. Solid State Chem.* **101**, 340 (1992).
13. K. Kitamura, Y. Furukawa, Y. Ji, M. Zgonik, C. Medrano, G. Montemezzani, P. Günter: *J. Appl. Phys.* **82**, 1006 (1997).
14. T. Fujiwara, M. Takahashi, M. Ohama, A.J. Ikushima, Y. Furukawa, K. Kitamura: *Electron. Lett.* **35**, 499 (1999).
15. V. Gopalan, T.E. Mitchell, Y. Furukawa, K. Kitamura: *Appl. Phys. Lett.* **72**, 1981 (1998).
16. K. Kitamura, Y. Furukawa, K. Niwa, V. Gopalan, T.E. Mitchell: *Appl. Phys. Lett.* **73**, 3073 (1998).
17. P.F. Bordui, G. Norwood, D.H. Jundt, M.M. Fejer: *J. Appl. Phys.* **71**, 875 (1992).
18. P.F. Bordui, R.G. Norwood, C.D. Bird, J.T. Carella: *J. Appl. Phys.* **78**, 4647 (1995).
19. G.I. Malovichko, V.G. Grachev, L.P.Yurchenko, Y.Ya. Proshko, E.P. Kokanyan , V.T. Gabrielyan: *Phys. Stat. Sol. (a)* **133**, K29 (1992).
20. G.I. Malovichko, V.G. Grachev, E.P. Kokanyan, O.F. Schirmer, K. Betzler, B. Gather, F. Jermann, S. Klauer, U. Schlarb, M. Woehlecke: *Appl.Phys.* A**56**, 103, (1993).

21. K. Kitamura, J.K. Yamamoto, N. Iyi, S. Kimura, T. Hayashi: *J. Crystal Growth* **116**, 327 (1992).
22. K. Kitamura, Y. Furukawa, N. Iyi: *Ferroelectrics* **202**, 21 (1997).
23. J.R. Carruthers, G.E. Peterson, M. Grasso, P.M. Bridenbaugh: *J. Appl. Phys.*, **42**, 1846 (1971).
24. H.M. O'Bryan, P.K. Gallagher, C.D. Brandle: *J. Am. Ceram. Soc.* **68**, 493 (1985).
25. K. Yamada, H. Takemura, Y. Inoue, T. Omi, S. Matsumura: *Japan. J. Appl. Phys.* **26**, 218 (1987).
26. Y. Fujimoto, H. Tsuya, K. Sugibuchi: *Ferroelectrics*, **2**, 113 (1971).
27. M. Nakamura, S. Takekawa, Y. Furukawa, K. Kitamura: *J. Cryst. Growth* **245**, 267 (2002).
28. A. Ashkin, G.D. Boyd, J.M. Dziedzic, R.G. Smith, A.A. Ballmann, K. Nassau: *Appl. Phys. Lett.* **9**, 72 (1966).
29. N. Kukhtarev, V.B. Markov, S.G. Odulov, M.S. Soskin, V.L. Vinetsky: *Ferroelectrics* **22**, 949, 961 (1979).
30. Y. Yang, I. Nee, K. Buse, D. Psaltis: *Appl. Phys. Lett.* **78**, 4076 (2001).
31. L. Kovacs, G. Ruschhaupt, K. Polgar, G. Corradi, M. Wöhlecke: *Appl. Phys. Lett.* **70**, 2801 (1997).
32. Ch. Bäumer, C. David, A. Tunyagi, K. Betzler, H. Hesse, E. Krätzig, M. Wöhlecke: *J. Appl. Phys.* **93**, 3102 (2003).
33. J. Koppitz, A.I. Kuznetsov, O.F. Schirmer, M. Wölecke, B.C. Grabmaier: *Ferroelectrics* **92**, 233(1989).
34. D. Berben, K. Buse, S. Wevering, P. Herth, M. Imlau, Th. Woike: *J. Appl. Phys.* **87**, 1034 (2000).
35. E. Krätzig, O.F. Schirmer: Chapter 5 in *Photorefractive Materials and Their Applications. I*, edited by P. Günter, J.-P. Huignard (Springer, 1988), p. 131.
36. H. Hatano, T. Yamaji, S. Tanaka, Y. Furukawa, K. Kitamura: *Japan. J. Appl. Phys.* Pt. 1, **38** 1820 (1999).
37. M. Lee, S. Takekawa, Y. Furukawa, Y. Uchida, K. Kitamura, H. Hatano, S. Tanaka: *J. Appl. Phys.* **88**, 4476 (2000).
38. H. Kurz, E. Krätzig, W. Keune, H. Engelmann, U. Gonser, B. Dishler, A. Räuber: *Appl. Phys.* **12**, 355 (1977).
39. B. Dishler, J.R. Herrington, A. Räuber: *Solid State Commun.* **14**, 1233 (1974).
40. E. Krätzig: *Ferroelectrics* **21**, 635 (1978).
41. O. Thiemann, O.F. Schirmer: *Proc. SPIE* **1018**, 18 (1988).
42. F.S. Chen: *J. Appl. Phys.* **40**, 3389 (1969).
43. F. Jermann, J. Otten: *J. Opt. Soc. Am.* B**10**, 2085 (1993).
44. K. Kitamura, Y. Furukawa, Y. Ji, M. Zgonik, C. Medrano, G. Montemezzani, P. Günter: *J. Appl. Phys.* **82**, 1006 (1997).
45. J. Amodei, D.L. Staebler: *Appl. Phys. Lett.* **18**, 540 (1971).
46. T. Fujiwara, M. Takahashi, A.J. Ikushima, Y. Furukawa, K. Kitamura: *Electron. Lett.* **35**, 499 (1999).
47. Y. Kondo, T. Fukuda, Y. Yamashita, K. Yokoyama, K. Arita, M. Watanabe, Y. Furukawa, K. Kitamura, H. Nakajima: *Japan. J. Appl. Phys.* **39**, 1477 (2000).
48. K. Kitamura, Y. Furukawa, S. Tanaka, T. Yamaji, H. Hatano, *Proc. SPIE*, **3291**, 151 (1998).
49. F. Jermann, M. Simon, E. Krätzig: *J. Opt. Soc. Am.* B**12**, 2066 (1995).
50. E. Krätzig, H. Kurz: *Optica Acta* **24**, 475 (1977).

51. Y. Furukawa, K. Kitamura, Y. Ji, G. Montemezzani, M. Zgonik, C. Medrano, P. Günter, *Opt. Lett.* **22**, 501 (1997).
52. G.W. Burr, D. Psaltis, *Opt. Lett.* **21**, 893 (1996).
53. H. Vormann, G. Weber, S. Kapphan, E. Krätzig: *Solid State Commun.* **40**, 543 (1981).
54. F. Micheron, G. Bismuth: *Appl. Phys. Lett.* **20**, 79 (1972).
55. D. von der Linde, A.M. Glass, K.F. Rodgers: *Appl. Phys. Lett.* **25**, 155 (1974).
56. K. Buse, F. Jermann, E. Krätzig: *Opt. Mater.* **4**, 237 (1995).
57. K. Buse, A. Adibi, D. Psaltis: *Nature* **393**, 665 (1998).
58. Y. Liu, L. Liu, C. Zhou, L. Xu, *Opt. Lett.* **25**, 908 (2000).
59. M. Lee, S. Takekawa, Y. Furukawa, K. Kitamura, H. Hatano, S. Tanaka: *Appl. Phys. Lett.* **76**, 1653 (2000).
60. Y. Liu, K. Kitamura, S. Takekawa, G. Ravi, N. Nakamura, H. Hatano, T. Yamaji: *Appl. Phys. Lett.* **81**, 2686 (2002).
61. Y. Liu, K. Kitamura, G. Ravi, S. Takekawa, M. Nakamura, H. Hatano: in *Proceedings of International Symposium on Optical Memory* (Nara Japan, 2003), p. 182 (submitted to *J. Appl. Phys.*).
62. D. Staebler, W. Phillips: *Appl. Opt.* **13**, 788 (1974).
63. M. Lee, S. Takekawa, Y. Furukawa, K. Kitamura, H. Hatano: *J. Appl. Phys.* **87**, 1291 (2000).
64. M. Lee, I.G. Kim, S. Takekawa, Y. Furukawa, Y. Uchida, K. Kitamura, H. Hatano: *J. Appl. Phys.* **89**, 5311 (2001).
65. M. Lee, S. Takekawa, Y. Furukawa, K. Kitamura, H. Hatano: *Phys. Rev. Lett.* **84**, 875 (2000).
66. M. Lee, S. Takekawa, Y. Furukawa, K. Kitamura, H. Hatano, S. Tao: *Opt. Lett.* **25**, 1334 (2000).
67. K. Kitamura, Y. Liu, S. Takekawa, M. Nakamura, H. Hatano, T. Yamaji: *Proc. SPIE* **4930**, 338 (2002).
68. H. Hatano, S. Tanaka, T. Yamaji, M. Lee, S. Takekawa, K. Kitamura: in OSA TOPS Vol. 62, *Photorefractive Effects, Materials, and Devices*, D. Nolte, G. Salamo, A. Siahmakoun, S. Stepanov, eds. (OSA, 2001), p. 171.
69. M. Lee, H. Hatano, S. Tanaka, T. Yamaji, K. Kitamura, S. Takekawa: *Appl. Phys. Lett.* **81**, 4511 (2002).
70. E. Krätzig, R. Orlowski: *Appl. Phys.* **15**, 133 (1978).
71. P. Bernasconi, G. Montemezzani, P. Günter, Y. Furukawa, K. Kitamura: *Ferroelectrics* **223**, 373 (1999).
72. Y. Furukawa, K. Kitamura, K. Niwa, H. Hatano, P. Bernasconi, G. Montemezzani, P. Günter: *Japan. J. Appl. Phys.* **38**, 1816 (1999).
73. Y. Liu, K. Kitamura, S. Takekawa, N. Nakamura, Y. Furukawa, H. Hatano: *Appl. Phys. Lett.* **82**, 4218 (2003).
74. Y. Liu, K. Kitamura, S. Takekawa, M. Nakamura, Y. Furukawa, H. Hatano: *J. Appl. Phys.* **95**, 7637 (2004).
75. L.A. Kappers, K.L. Sweeney, L.E. Halliburton, J.H.W. Liaw: *Phys. Rev. B* **31**, 6792 (1985).
76. S. Wevering, J. Imbrock, E. Krätzig: *J. Opt. Soc. Am. B* **18**, 472 (2001).
77. J. Imbrock, D. Kip, E. Krätzig: *Opt. Lett.* **24**, 1302 (1999).
78. K. Kitamura, Y. Liu, S. Takekawa, H. Hatano, Y. Furukawa, *Proc. SPIE* **5362**, 107 (2004).
79. J. R. Herrington, B. Bischler, A. Räuber, J. Schneider: Solid State Commun. **12**, 351 (1973).

80. L. Kovacs, M. Wöhelecke, J. Jovanovic, K. Polgar, S. Kapphan: *J. Phys. Chem. Solids* **52**, 797 (1991).
81. Y. Watanabe, T. Sota, K. Suzuki, N. Iyi, K. Kitamura, S. Kimura: *J. Phys.: Condens. Matter.* **7**, 3627 (1995).
82. Y. Yang, D. Psaltis, M. Luennemann, D. Berben, U. Hartwig, K. Buse: *J. Opt. Soc. Am. B* **20**, 1491 (2003).
83. S. Klauer, M. Wohlecke, S. Kapphan: *Phys Rev. B* **45**, 2786 (1992).
84. P.J. Matthews, A.R. Mickelson: *J. Appl. Phys.* **72**, 2562 (1992).
85. F.H. Mok, G.W. Burr, D. Psaltis: *Opt. Lett.* **21**, 896 (1996).

6

Optical Damage Resistance in Lithium Niobate

T. Volk,[1] M.Wöhlecke,[2] and N. Rubinina[3]

[1] Institute of Crystallography of Russian Academy of Sciences, 117333 Moscow Leninski Prospect 59, Russia
Volk@ns.crys.ras.ru
[2] Fachbereich Physik, Universität Osnabrück, D-49069 Osnabrück, Germany
manfred.woehlecke@uos.de
[3] Moscow State University, 117234 Moscow, Russia

6.1 Introduction

Lithium niobate is a universal material for optical applications (optical frequency conversion, optical and acoustooptical light modulation, lasing, photorefractive holography, etc.). This enormous versatility of optical properties is due to a pronounced dependence of its properties on composition, namely, on the crystal composition and the doping type. In spite of widespread potentials of $LiNbO_3$, the optical frequency conversion still remains one of the most important. Although $LiNbO_3$ is lower than other materials in rank, for example, potassium titanium oxide phosphate (KTP) with larger values of the nonlinear optical coefficients, nevertheless, the possibility to obtain a noncritical (90°) phase matching and a rather broad angular width of the phase synchronism makes $LiNbO_3$ one of the most attractive materials for nonlinear optics. The use of $LiNbO_3$ for optical frequency conversion on a regular pattern of 180° ferroelectric domains in the quasiphase-matching (QPM) mode of operation is within the bounds of possibility. Detailed descriptions of the properties of $LiNbO_3$ and its optical applications may be found in numerous reviews and monographs [1, 2, 3].

In nonlinear optics the performance of $LiNbO_3$ is limited by three optically induced effects. The first of them, optical (or photorefractive) damage, is a reversible photoinduced change of the refractive indices that appears at relatively low light intensities, for example in $LiNbO_3$:Fe even at power densities as low as a tenth of a mW/cm^2, e.g., [1]). In undoped (congruent) $LiNbO_3$ a saturated photoinduced change δn of the refractive indices is as high as $2 \cdot 10^{-5}$–10^{-4}. The consequences of the optical damage are, for example, a distortion of the wave front of the transmitted light, and a loss of lasing when using the material as a laser medium. The second effect limiting the use of many crystals in optics is the *dark (or gray) trace effect*, which is a light-induced coloration occurring in

LiNbO$_3$ under Nd-YAG laser (1.06 μm) intensities of about 100 MW/cm^2, e.g., [4] and reducing the optical transmission by about 10–15% after several hundreds of light pulses. Finally, the third effect confining the range of used laser intensities is the laser-induced damage, which is an irreversible mechanical destruction of the crystals. This effect, inherent to all solid-state materials, occurs in LiNbO$_3$ in the GW range, e.g., at about ten and one GW/cm^2 for 1053 and 526 nm radiation, respectively, at a pulse duration of 1 ns [5]. Unfortunately, these thresholds are noticeably lower than in KH$_2$PO$_4$ (KDP) or KTP. A principal distinction of the optical damage from the laser damage is that the former obeys the law of interchangeability [6] and has no intensity threshold. With this short background information the importance of combating all these undesirable effects in LiNbO$_3$, particularly the optical damage, is obvious.

As shown below, the most efficient and technologically simple method of suppressing the optical damage in LiNbO$_3$ is provided by doping with *optical-damage-resistant* ions like Mg^{2+}, Zn^{2+}, In^{3+}, and Sc^{3+}. In studies of optical-damage-resistant compositions of LiNbO$_3$ several routes may be distinguished. The detection of the first optical damage-resistant impurity Mg [8] and initial investigations [9, 10] stimulated a wide search for practical potentials of LiNbO$_3$:Mg crystals. Later, other dopants acting in a very similar manner, like Zn [11, 12], In [13, 14], and Sc [15, 16] were detected. Additionally to a suppressed optical damage, one of the most exciting features of these dopants is that they do not induce any absorption bands (Figure 6.1) in the whole spectral range

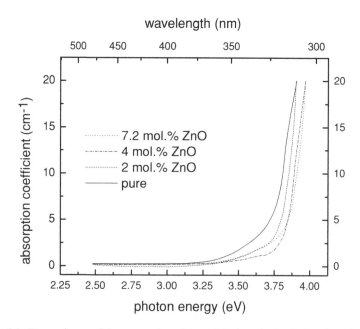

FIGURE 6.1. Dependence of the spectral position of the band-edge in Zn-doped LiNbO$_3$, according to [7].

of LiNbO$_3$ (from 0.3 to 5 μm), a demand for optical frequency conversion. A great similarity of many properties of members of this *family* stimulated several researchers to investigate the influence of the intrinsic defect structure of LiNbO$_3$ on these properties. A possibility to control the optical damage by varying the crystal composition [17, 18, 19] generated an interest in LiNbO$_3$ compositions combining both an increased [Li]/[Nb] ratio and doping by an optical damage-resistant impurity. The aim of this review is to summarize a large number of results available on these topics. The first attempt at such a summarizing was undertaken by us five years ago [20]. Since then a considerable body of new data has been published, the list of compositions under study has been significantly extended, and new steps toward practical applications have been made. In the present review we will discuss the mechanism of optical damage resistance from the viewpoint of fundamental changes in the charge transport produced by doping with these ions. At the end we will briefly dwell on the subjects closely related to practical potentials, namely, on some optical characteristics of optical-damage-resistant LiNbO$_3$ crystals (birefringence, phase-matching conditions of optical frequency doubling, electrooptic effect, etc.).

6.2 Impurity- and Composition-Controlled Optical Damage Resistance in LiNbO$_3$

6.2.1 Introductory Remarks

As is known, optical damage is caused by a space-charge field formed in the illuminated region. In LiNbO$_3$ it is due either to the bulk photovoltaic effect or the diffusion mechanism. Optical damage resistance is required for operating in the nonlinear optical mode regime. If there is no spatial modulation of the light intensity, only the photovoltaic effect controls the optical damage, and it may be expressed in a simplified scalar form proposed by Glass [1]:

$$\delta n = R_{eff} E_{sc} = R_{eff} \frac{j_{ph.v.}}{\sigma_{ph} + \sigma_d}, \tag{6.1}$$

where R_{eff} is the effective electrooptic coefficient, $j_{ph.v.}$ is the bulk photovoltaic current, σ_{ph} and σ_d are photoconductivity and dark conductivity, respectively. As seen from (6.1), there are several ways to suppress the optical damage. Primarily, this may be achieved by growing crystals free of photovoltaically active transition metal impurities, e.g., [21]. However, at present it becomes evident that besides a photovoltaic current caused by an impurity, also an *intrinsic* component exists, obviously being related to intrinsic defects and thus unavoidable. It is of interest that the photovoltaic current in LiNbO$_3$ depends on the sign of the carrier and strongly deceases if the photoconductivity is converted from electron to hole one. This was demonstrated with a strongly oxidized low-doped LiNbO$_3$:Fe [22] sample.

Further, an optical damage resistance may be obtained by increasing the dark conductivity, as observed in strongly reduced [23] or Ag-doped [24] LiNbO$_3$ crystals and in hydrogen-exchanged optical waveguides on the LiNbO$_3$ surface [24, 25]. In LiNbO$_3$ the most efficient suppression of optical damage is achieved by increasing its photoconductivity, which is the dominant reason for the optical damage resistance in LiNbO$_3$ doped with Mg Zn, In, and Sc and also contributes to optical damage resistance in Li-enriched crystals. Note that in LiTaO$_3$ the effects of Mg-doping on the optical damage are similar to those in LiNbO$_3$, e.g., [26].

6.2.2   Optical-Damage-Resistant Impurities (Mg, Zn,In, Sc) in LiNbO$_3$; Their Threshold Concentrations

A reduced optical damage in LiNbO$_3$ doped with 4.6% MgO was noticed by Zhong et al. in 1980 [8] and studied in some detail later by Bryan and coworkers [9, 10]. They were the first to account for the optical damage resistance to an increased photoconductivity and they detected a critical (*threshold*) concentration of about 5 mol % MgO (for the congruent melt), above which the optical damage drastically falls off and becomes negligible. Several optical properties as well reveal anomalies in the same concentration range. Some properties of LiNbO$_3$:Mg lead to the assumption that Mg^{2+} belongs to a group of elements acting in the same manner. Because Zn^{2+} is closely related to Mg^{2+} with respect to crystal chemistry, Zn-doping was supposed to produce similar effects. Optical damage resistance in LiNbO$_3$:Zn was found by Volk et al. [11, 12] for concentrations exceeding 7% ZnO (in the congruent melt). The dependencies of many optical properties on the Zn concentration are very similar to those in LiNbO$_3$:Mg; particularly they revealed anomalies in the concentration range at about this *threshold* of 7 % ZnO. The so-called qualitative crystal-chemistry *principle of the diagonal row* within the periodic table of elements permits one to predict other impurities of this type. These are in the first instance In^{3+} and Sc^{3+} cations, which are analogues of Mg and Zn. Optical damage resistance was actually proved in LiNbO$_3$:In [13, 14, 27] and LiNbO$_3$:Sc [15, 16]. In these cases a drastic decrease of the optical damage occurs already at lower concentrations of about 1–1.5% of the oxides in the congruent melt and can be compared with the effect of 5% MgO. Dependencies of other optical properties on In and Sc concentrations bear a similarity to LiNbO$_3$:Mg and LiNbO$_3$:Zn and show typical anomalies at about 1–1.5% In$_2$O$_3$ or Sc$_2$O$_3$. These lower values of the threshold concentrations for trivalent dopants are simply explained in terms of the charge compensation; see below. Up to now the crystals LiNbO$_3$:In and LiNbO$_3$:Sc have been studied only very slightly. In addition to this group of dopants, a decrease of the optical damage was found in Na-doped LiNbO$_3$ [28], but the results are ambiguous.

Optical damage resistance in LiNbO$_3$:Mg [4] and LiNbO$_3$:Zn [29] is observed in an intensity range up to pulse-laser power not less than 100 MW/cm^2. In LiNbO$_3$ doped with low concentrations of transition metal or rare earth ions, a codoping with concentrations of MgO or ZnO above the threshold value also

FIGURE 6.2. Optical damage $\delta\Delta n$ in LiNbO$_3$ and LiNbO$_3$:Fe vs. Zn and Mg concentrations. The graphs 1 and 2, guides for the eye, correspond to LiNbO$_3$ and LiNbO$_3$:0.02%Fe:Mg, respectively. The two plus markers represent data for LiNbO$_3$:0.02%Fe:Zn. Optical damage was induced by an Ar-ion laser, $\lambda = 488$ nm, $I = 20$ Wcm^{-2}.

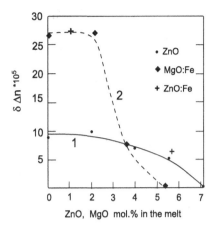

strongly reduces the optical damage, as exemplified by LiNbO$_3$:Fe:Mg [30], LiNbO$_3$:Fe:Zn [31] or LiNbO$_3$:Nd:Mg (e.g. [32]). Figure 6.2 illustrates the effects of Mg and Zn on the optical damage in some LiNbO$_3$ compositions. The concentration range of the optical damage resistance depends on the crystal composition and is shifted to lower impurity concentrations with increasing Li$_2$O content in the crystal. In LiNbO$_3$:Mg grown from a melt with [Li]/[Nb] $= 1.2$ the optical damage resistance occurs at a concentration of about 3% MgO [10, 33]. As repeatedly mentioned, the main reason for a decrease of the optical damage is an enhanced photoconductivity. Figure 6.3 presents a summary of data on the dependencies of the specific photoconductivity σ_{ph}/I in a congruent LiNbO$_3$ on the Mg, Zn, and In concentrations, respectively (in crystals grown from the same

FIGURE 6.3. The effect of Mg, Zn, In on the photoconductivity in LiNbO$_3$.

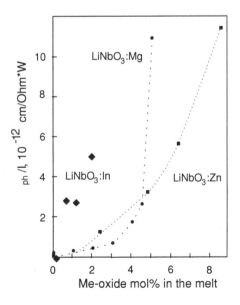

charge [34]). Precisely, the dependencies $\sigma(I)$ in these crystals are slightly sublinear $\sigma_{ph} \approx I^x (x = 0.8 - 0.9)$ [35]. However, in the given presentation this sublinearity has been neglected. The main feature of the concentration dependencies on the photoconductivity in LiNbO$_3$:Mg and LiNbO$_3$:Zn is their nonmonotonic character, namely, steeper slopes of the curves for concentrations exceeding the values close to the threshold at 5% MgO and 7% ZnO, respectively. The dark conductivity in optical-damage-resistant crystals increases, too, as has been shown in LiNbO$_3$:5.5% Mg [36]. The photovoltaic properties of optical-damage-resistant crystals are yet unclear. The direct measurements and estimates of the photovoltaic coefficients in these crystals, as well as in undoped LiNbO$_3$, are hampered by their high transparency, i.e., by a smallness of the photovoltaic currents. Measurements of $j_{ph.v.}$ in LiNbO$_3$:Fe codoped with Mg [30] or Zn [12, 31] revealed no action of these ions on the photovoltaic properties. However, in optical-damage-resistant crystals free of Fe, the photovoltaic properties seem to be modified compared to nominally pure LiNbO$_3$, as follows from the direct measurements of the photovoltaic currents in LiNbO$_3$:Mg [37, 38] (for more details see below). Anyhow, in spite of possible changes in different components of (6.1), the main reason for the optical damage resistance in LiNbO$_3$ crystals under discussion is an essential increase of the photoconductivity.

It seems that there is no pronounced dependence of the optical damage resistance and the photoconductivity close to the *threshold* impurity concentration (Figures 6.2 and 6.3). In Mg-doped crystals free of transition metal admixtures, the optical damage decreases according to a study of Furukawa et al. [21] more smoothly than in reports by Bryan et al. and Sweeney et al. [9, 10]. But concentration dependencies of other properties reveal much more pronounced anomalies in these ranges. To illuminate this with an example, we present in Figure 6.4 a

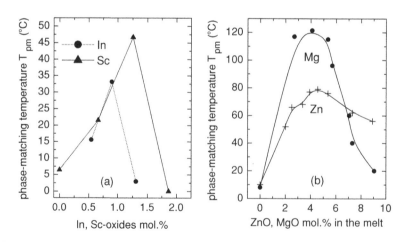

FIGURE 6.4. Phase-matching temperatures for collinear noncritical SHG in LiNbO$_3$ for 1064 nm vs. In, Sc (a) and Mg, Zn (b) concentrations. The data for Mg and Zn are taken from [34], for In from [14], for Sc from [16].

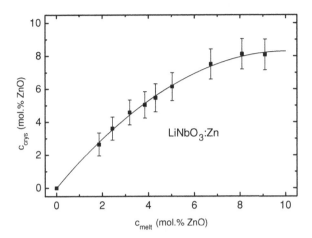

FIGURE 6.5. Variation of the Zn concentration in the crystal as a function of ZnO in the melt. The curve represents a fit of these data using the equation shown [39].

combination of the curves T_{pm} for $\lambda = 1.06\ \mu$m (Nd-YAG laser radiation) versus concentrations of the optical damage-resistant dopants (where T_{pm} is the temperature of the noncritical 90° synchronism for the optical frequency doubling). We emphasize that the concentration ranges for maxima in plots of T_{pm} coincide with those for bends in graphs of the optical damage and photoconductivity. At the end of this section we briefly dwell on the distribution coefficients of optical damage-resistant ions in LiNbO$_3$. The ionic radii of Mg^{2+}, Zn^{2+}, Sc^{3+}, and In^{3+} in the 6-fold coordination are $72, 74, 74.5$, and 80 pm, respectively [40]. The distribution coefficients of these ions in LiNbO$_3$ are practically the same ($K_{eff} = 1.2$–1.3) at low concentrations [2]. With increasing concentrations the values of K_{eff} decrease to about 1 for MgO [41] and ZnO [39] in the range 5–6 mol% and to about 0.9 for In$_2$O$_3$ in the range of about 1.5–2 mol%. The concentration of Zn in the LiNbO$_3$:Zn crystal (c_{crys}) versus the concentration in the melt (c_{melt}) is shown in Figure 6.5 and may be described with good accuracy with the expression $c_{crys} = 1.649\,c_{melt} - 0.082\,c_{melt}^2$.

6.2.3 Optical Damage in LiNbO$_3$ of Different Composition

The dependence of the optical damage on the Li$_2$O content in LiNbO$_3$ was for the first time reported by Anghert et al. [17]. The optical damage was shown to increase with increasing [Li]/[Nb] ratio. The results of subsequent more detailed studies by Furukawa et al. (1992 and 1996) [18, 42], and Furukawa et al. (1998 and 2000) [37, 38] are in agreement with the first observations [17], i.e., a larger optical damage for nearly stoichiometric LiNbO$_3$ (Figure 6.6). A similar conclusion was drawn from a comparison of the two-beam coupling gain Γ in congruent and stoichiometric LiNbO$_3$ codoped with Fe [43, 44]. All these results are valid for relatively low intensities ($I \leq 10 - 20$ W/cm^2). However, observations of

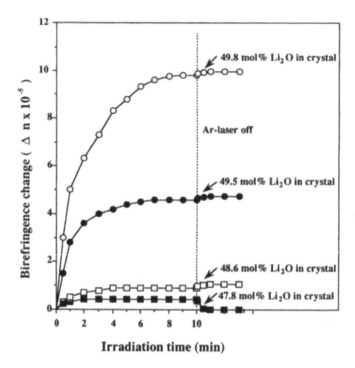

FIGURE 6.6. Optical damage in LiNbO₃ of different Li-content; $\lambda = 488$ nm, $I = 20\,\text{Wcm}^{-2}$ according to [18].

the optical damage at higher intensities [19, 45] differ from these regularities (Figure 6.7). As seen, the value of $\delta\Delta n$ in stoichiometric crystals exceeds that in congruent ones at $I \le 200$ W/cm², whereupon the situation is reversed and stoichiometric crystals become more damage-resistant. This reverse of the dependencies is due to a steep increase of $\delta n(I)$ in the range of high intensities in the congruent crystal, whereas in stoichiometric ones $\delta\Delta n$ grows with the light intensity only very smoothly. Just these results for high intensities generated a wide-spread belief of the optical damage resistance in stoichiometric LiNbO₃. Note that anyhow the influence of compositional changes on the optical damage are incomparably lower than the effects of optical-damage-resistant impurities described above.

To comment on studies of optical damage at high intensities [19, 45] one remark is worthy of noting. As seen in Figure 6.7, at low intensities the value of $\delta\Delta n$ in the congruent crystals studied by Malovichko et al. [19] is rather high, which testifies to the presence of uncontrollable admixtures (particularly, of transition metal ions, e.g., [21]). If one assumes the presence of such random traps, one may suggest that their nonequilibrium repopulation at high intensities would lead to various dependencies of $j_{ph.v.}$, σ_{ph} on the light intensity and to a certain dependence of $\delta n(I)$. These effects are well known in LiNbO₃ [46, 47]. In

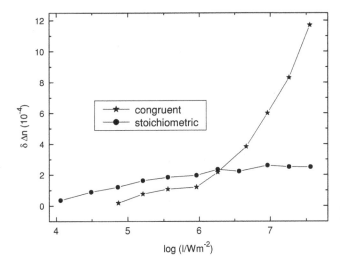

FIGURE 6.7. Intensity dependencies of the optical damage in congruent and stoichiometric LiNbO₃ [19]. The stoichiometric crystal was grown by the Czochralski technique by adding 10.9 mol % K₂O into the melt.

summary, an increase of the optical damage with increasing crystal stoichiometry at moderate light intensities is well established, i.e., stoichiometric crystals are less optical-damage-resistant than congruent (Li-deficient) ones. The regularities and comparisons for high intensities require more detailed investigations.

Now we discuss the reason for the increase of the optical damage with increasing [Li]/[Nb] ratio. Following (6.1), this could be due to a decreased photoconductivity. However, all experiments found an increase of σ_{ph} in stoichiometric crystals. This follows both from direct measurements and from kinetics of hologram recording and erasure, since the speeds of these processes, that is, $\sigma_{ph}(1/\tau_M = \sigma_{ph}/\epsilon\epsilon_0)$, are higher in stoichiometric crystals [19, 38, 43, 44, 45]. Figure 6.8 shows as an example data of the photoconductivities in LiNbO₃ of different compositions taken from [38]. (In Figures 6.8–6.10, the names St-I and St-II denote stoichiometric crystals grown either by means of adding 10.9 mol% K₂O to the charge [48] or from a melt enriched by Li₂O.) One may see an increase of σ_{ph} in stoichiometric crystals compared to congruent ones. This apparent mismatch between the dependencies of the optical damage and the photoconductivity on the Li content was enlightened by measurements of the photovoltaic currents [37, 38].

The authors report a surprising increase of $j_{ph.v.}$ in a stoichiometric crystal by two orders of magnitude compared to a congruent one (Figure 6.9, Table 6.1). Therefore, in spite of an increased photoconductivity, a significantly increased photovoltaic current in stoichiometric crystals makes them less optical-damage-resistant than congruent ones. Fundamentally new results on concentration dependencies of the optical damage were obtained in crystals enriched by Li and doped

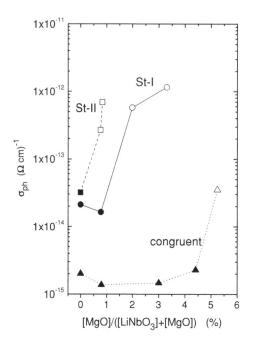

FIGURE 6.8. Photoconductivity vs. MgO concentration in congruent and stoichiometric (StI and StII) LiNbO₃ crystals, redrawn with data from [38].

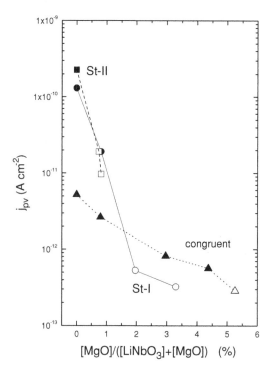

FIGURE 6.9. Photovoltaic current vs. MgO concentration in congruent and stoichiometric (StI and StII) LiNbO₃ crystals [38].

TABLE 6.1. Comparison of the photoconductivity (σ_{ph}), photovoltaic current (j_{pv}), and space-charge fields (E_{sc}) in congruent (CLN), stoichiometric (SLN), and optically damage-resistant CLN:5.1% Mg and SLN:2% Mg ($I = 10\,\mathrm{Wcm}^{-2}$, $\lambda = 488\,\mathrm{nm}$). The data for CLN and CLN:5.1% Mg are taken from [34], the data for SLN and SLN:2% Mg from [38]. The bottom line presents the Glass constant calculated according to $k = E_{sc}\sigma_{ph}/(\alpha I)$ assuming absorption coefficients $\alpha = 0.025\text{--}0.03\ \mathrm{cm}^{-1}$ for $\lambda = 488\,\mathrm{nm}$ (according to experimental estimates)

property	CLN	SLN	CLN:5.1% Mg	SLN:2% Mg
σ_{ph}, 10^{-14} $(\Omega\,\mathrm{cm})^{-1}$	0.2	2–3	3	60
j_{pv} 10^{-12} A/cm^2	6	200	0.3	0.5–0.6
E_{sc} V/cm	2500	7000	10	1
k 10^{11} cm/V	2.4	80	0.1	2

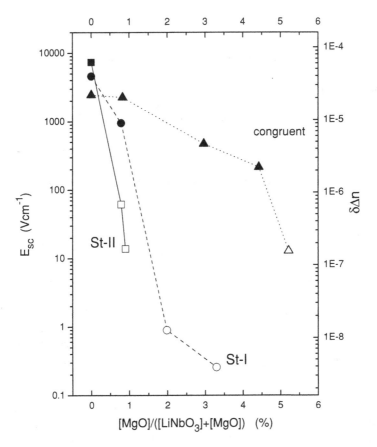

FIGURE 6.10. Saturated space-charge field and optical damage vs MgO concentration in congruent and stoichiometric (St-I and St-II) LiNbO$_3$ crystals [38].

with an optical-damage-resistant ion [37, 38, 49, 50]. As mentioned above, in Li-enriched crystals the threshold concentration of Mg is decreased and the optical damage resistance is observed at Mg concentrations lower than 5% [10, 33].

In detailed studies of Furukawa and coworkers [37, 38, 49] it was found that for the optical damage the combination of an increased Li content and a doping with Mg is not additive. Figure 6.10 presents the results of Furukawa et al. [38] on the dependence of the optical damage on the Mg concentration in Li-enriched (stoichiometric) and congruent LiNbO$_3$. In the starting stoichiometric material (free of Mg) the [Li]/[Nb] ratio is ≈ 0.99. In both stoichiometric crystals a drastic reduction of the optical damage (the threshold value) is observed in the concentration range 1–2% MgO. The effect is enormous, because the optical damage drops to $\delta \Delta n \approx 10^{-8}$, which means that it becomes lower by two orders of magnitude than in congruent LiNbO$_3$ doped with 5.5% MgO. Two reasons may account for this drastic nonadditive effect (Table 6.1). First, as discussed above, for an increasing Li content the photoconductivity is essentially increased. Second, as seen in Figure 6.9, a doping of Li-enriched LiNbO$_3$ with Mg is accompanied by a significant (by more than orders of magnitude) decrease of the photovoltaic current. In Figure 6.9 one may see a decrease of $j_{ph.v.}$ with Mg concentration in congruent crystals, too, but this effect is much less pronounced.

In summary, at present the most efficient suppressing of optical damage in LiNbO$_3$ is provided by a combination of doping with an optical-damage-resistant impurity and a certain increase of the Li content.

6.3 Incorporation of Optical-Damage-Resistant Ions into the LiNbO$_3$ Lattice

6.3.1 Background: The Intrinsic Defect Structure of the Congruent (Li-Deficient) LiNbO$_3$

Prior to the discussion of the microscopic origin of the effects of optical-damage-resistant ions on the charge transport, we should consider their incorporation into the lattice. We now briefly summarize the background for the current model of the LiNbO$_3$ defect structure. Lithium metaniobate belongs to the pseudoilmenite structure formed by distorted niobium–oxygen octahedra; the chains formed by them are aligned along the polar axis z (Figure 6.11). In the nonpolar centrosymmetric phase at $T > T_c \approx 1270\,°C$ the Li ions are on average located in the oxygen planes and Nb ions are placed in the center of the octahedra (between oxygen planes). In the polar ferroelectric phase with point group C$_{3v}$ the Li ions are shifted along the z axis by about 44 pm with regard to oxygen planes, and Nb ions are shifted by about 26 pm. As a result, the octahedral interstitials are one-third filled with Li ions, one-third with Nb and one-third are empty, so that the alternation of cation sites may be schematically depicted as Li-Nb-□ $\cdots$ Li-Nb-□ $\cdots$, where □ denotes the empty octahedron (referred to sometimes as a

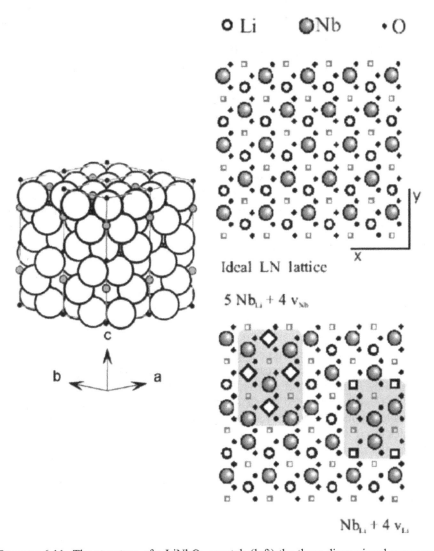

FIGURE 6.11. The structure of a LiNbO₃ crystal: (left) the three-dimensional representation of the structure; the close hexagonal packing (large white balls), 2/3 of octahedral cavities of which are filled with Nb and Li cations (black and gray balls, respectively). (right) The intrinsic defect Nb$_{Li}$ and alternatives of its compensation by Li or Nb vacancies (a projection on the (001) plane) according to [51].

vacancy). The Li octahedron is larger than the Nb one, because the distances between a Li ion and the closest O ions are 206.8 and 211.2 pm, whereas those for Nb are 188.9 and 211.2 pm, respectively [52]. A larger size of the Li octahedron may qualitatively account for a predominant incorporation of impurity ions onto Li sites, as discussed below.

Congruently melting $LiNbO_3$ with a Li_2O concentration of about 48.6 mol % ([Li]/[Nb] = 0.945) is grown by the Czochralski technique, *Li-deficient* crystals ([Li] / [Nb] < 0.94) are grown by a floating-zone technique [53, 54]. Stoichiometric crystals may be prepared by means of the VTE technique (vapor transport equilibration) [55, 56] or grown by adding about 10.9 mol % K_2O to the congruent melt [57, 58], by a double-crucible method [59], or by a modified top-seeded solution method [48]. In the lattice of the congruently melting $LiNbO_3$ crystals about 6% of the Li sites are not occupied. Since the Nb^{5+} ion has a lesser ionic radius than Li^+ (64 and 76 pm, respectively, in the 6-fold coordination [40]), for Nb the probability of entering a Li site is very high. The existence of Nb on Li sites has been repeatedly proved by investigations of the structure [52, 54, 60, 61]. However, estimates of their amount in the lattice give different values. According to [52], the *empty* sites are mostly populated by Nb_{Li}, because Nb occupies 4.9% of the Li sites in the lattice. According to [54], only 1% of the Li sites are filled by Nb, whereas about 4% of the Li sites are empty (Li vacancies) and no Nb vacancies are observed. The existence of a high concentration of Li vacancies in congruent $LiNbO_3$ was proved later by studies of the structure [60, 61] and NMR [62] investigations. All structure determinations [52, 54, 61] are in agreement with model calculations [63, 64] and indicate that by contrast to other oxides, a congruent as-grown $LiNbO_3$ is free of oxygen vacancies. The existence of the *intrinsic* defect $(Nb_{Li})^{4\bullet}$ is the cornerstone of all current models of the defect structure. For the charge compensation two alternative mechanisms have been suggested. The Nb site vacancy model proposed by Peterson et al. [65] and developed by Donnerberg et al. [64, 66] with a compensation achieved via Nb vacancies $(V_{Nb})^{5'}$, supported the structure results of Abrahams and Marsh [52]. The Li site vacancy model [67] assuming the compensation of Nb_{Li} by four Li vacancies $(V_{Li})'$ was the base for a discussion of the structure data of Iyi et al. [54]. For the Li site vacancy model the formula for the congruent crystal looks like $[Li_{1-5x}Nb_x\square 4x]NbO_3$ (where the group within braces corresponds to the population of the Li sites). This model supported by experimental data [54, 61] is now commonly accepted, so the structure results on optical-damage-resistant $LiNbO_3$ crystals are discussed below in this framework.

6.3.2 Defect Structure in LiNbO₃ Doped with Damage-Resistant Impurities Studied by Structure and Spectroscopic Methods. The Structure Origin of the Threshold Concentrations

The effects of optical-damage-resistant impurities on numerous properties of $LiNbO_3$ resemble qualitatively the consequences of the increasing Li content. For example, an increase of the [Li]/[Nb] ratio is followed by a shift of the band edge to shorter wavelengths [68], and by an increase of the crystal density and of T_c [2, 41]. Doping with Mg leads to an increase of the phase transition temperature T_c and of the crystal density [41], and the band edge is shifted to the UV range in all optical-damage-resistant crystals (Figure 6.1). Due to this similarity the effects of optical-damage-resistant ions were a priori explained by a substitution of the intrinsic defects Nb_{Li} by impurity ions on Li sites. Incorporation of optical-damage-resistant ions into the $LiNbO_3$ lattice was investigated by means of chemical microanalysis in $LiNbO_3$:Mg [53, 69, 70] and $LiNbO_3$:Sc [71]. In $LiNbO_3$:Zn and $LiNbO_3$:In it was studied by means of X-ray diffraction [72, 73] and neutron diffraction [74]. The a priori expected incorporation of these ions onto Li sites was proved convincingly, for example, in $LiNbO_3$:Mg [53, 69, 70] and $LiNbO_3$:Zn [72, 73]. However, the microscopic mechanism of the *thresholds* was a subject of long discussion. In the first publications the existence of these specific concentrations was related to a complete removal of Nb_{Li} from the lattice [66, 69, 75]. Model calculations [66] suggested that simultaneously with the disappearance of Nb_{Li}, the impurity ions incorporate partially onto Nb sites. A change of the lattice site of Mg ions was concluded from Raman spectra [76]. In this context a difference in the threshold concentrations of di- and trivalent cations can be easily accounted for by different charge compensation conditions assuming the Li site vacancy model for compensation

$$5\,MgO + 2\,Nb_{Li}^{4\bullet} + 8\,V'_{Li} \quad \leftrightarrow \quad 5\,Mg_{Li}^{\bullet} + 3\,V'_{Li} + Nb_2O_5, \tag{6.2}$$

$$5\,Sc_2O_3 + 6\,Nb_{Li}^{4\bullet} + 24\,V'_{Li} \quad \leftrightarrow \quad 10\,Sc_{Li}^{\bullet\bullet} + 20\,V'_{Li} + 3\,Nb_2O_5. \tag{6.3}$$

The number of trivalent cations required for a removal of Nb_{Li} is less than for divalent ones, which means that the threshold is lower for the former ones. The relation of the thresholds to the disappearance of Nb_{Li} was probed in optical-damage-resistant crystals by means of spectral tests for Nb_{Li}. Let us recall briefly the essence of these tests. A relatively slight reduction of a congruent $LiNbO_3$ results in the appearance of a broad dichroic absorption band with a maximum at about 500 nm; for references see [77]. A bleaching with visible light at $T \leq 80\,K$ is followed by a reversible shift of this band to 760 nm [78, 79] and is accompanied by an electron paramagnetic resonance (EPR) signal.

These phenomena are still under discussion and interpreted either in the framework of the polaron model (for references see [63, 64, 80]) or in terms of oxygen vacancies [78, 79]. According to the polaron model, in reduced crystals the initial band at 500 nm is related to a diamagnetic bipolaron (a stable bonded electron pair located on neighboring lattice sites, i.e., $(Nb_{Nb}\text{-}Nb_{Li})^{2-}$) formed by capturing free electrons produced in a reduction process. This is supported by the fact that in a reduced near-stoichiometric $LiNbO_3$ (with a lowered Nb_{Li} content) the band at 500 nm is much lower than in congruent crystals [81]. Under an illumination $(\hbar\omega = 2.5\,eV)$, at low temperature bipolarons dissociate and, neglecting details of the charge compensation, form isolated small polarons

$$[Nb_{Li} - Nb_{Nb}]^{2-} \propto \left[Nb_{Li}^{4\bullet}\right]^- + Nb_{Nb}\,, \tag{6.4}$$

which are paramagnetic and responsible for the band at 760 nm and the EPR spectrum. The latter is assigned to an intervalence transition from Nb_{Li}^{4+} to

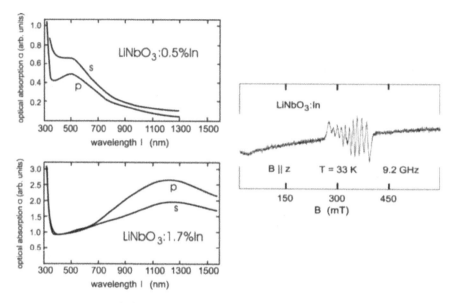

FIGURE 6.12. Spectral characteristics of reduced LiNbO₃:In. Optical absorption spectra for In concentrations below and above the threshold (left). EPR spectrum induced by illumination at low temperature of a crystal with an In concentration below the threshold (right).

one of the surrounding Nb_{Nb} ions [77, 80]. Therefore, in reduced LiNbO₃ the light-induced transformation of the absorption spectrum and a fine structure in the EPR are tests for the presence of Nb_{Li}. Optical-damage-resistant crystals LiNbO₃:Mg [10, 80, 82], LiNbO₃:Zn [83], [34], and LiNbO₃:In [14] with different concentrations of Mg, Zn, and In were reduced and studied with the aid of these tests. The results are exemplified by curves obtained in LiNbO₃:In (Figure 6.12). In the whole concentration range below the threshold concentrations ($< 5.5\%\,MgO$, $< 7.5\%\,ZnO$, $< 1.5\%\,In_2O_3$) the spectral characteristics are similar to those in congruent LiNbO₃, i.e., evidence for the presence of Nb_{Li}. In contrast, for concentrations exceeding these thresholds no phenomena described above were observed. In these compositions a strong reduction produced a stable broad dichroic band with a maximum at about 1300 nm (Figure 6.12, bottom). It was interpreted [84, 85] as a small free polaron Nb_{Nb}^{4+} formed by capturing electrons produced under reduction of a regular Nb_{Nb}^{5+} ion and a coupling to an unknown perturbation in the lattice. This band is specific for all optical-damage-resistant crystals with impurity concentrations above the threshold and is assumed to indicate missing Nb_{Li}. On the basis of these experiments it was concluded that in reduced optical-damage-resistant crystals the threshold concentrations are actually related to the disappearance of Nb-antisites, in conformity with the initial microscopic model. However, later studies by different methods demonstrated a conventionality of these tests. According to a precise chemical analysis [53, 70],

TABLE 6.2. Occupancy coefficients for atoms in $LiNbO_3$:Zn and $LiNbO_3$:In crystals

atom	site symmetry	coordinates	Zn concentrations in the crystal, at.%				
			0	2.87	5.2	7.6	8.2
Li	3	00z	93.8(3)	90.8(6)	89.8(4)	93.9(4)	94.7(4)
Nb_{Li}	3	00z	1.2(1)	0.7(1)			
Zn_{Li}	3	00z		2.9(3)	5.2(4)	6.0(4)	6.1(4)
Nb	3	00z	100.0(3)	100.0(3)	100.0(3)	98.0(9)	97.9(9)
Zn_{Nb}	3	00z				1.5(1)	2.2(2)
O	1	xyz	100.0	100.0	100.0	100.0	100.0

	In concentrations in the crystal, at.%		
	0	1	2.7
Li	93.8(3)	97.0(6)	96.0(6)
Nb_{Li}	1.2(4)		
In_{Li}		1.01(3)	1.15(3)
Nb	100.0(3)	100.0(2)	98.6(3)
In_{Nb}			1.40(4)

a complete removal of Nb antisites occurs at about 3% Mg. For a further increase of the concentration, Mg ions were supposed to substitute for regular Li ions. Similarly, in $LiNbO_3$:Sc the results of Shimamura et al. [71] were interpreted as a complete removal of the Nb antisites at 2% Sc. Therefore, in contrast to a conclusion from spectroscopic results [80], chemical microanalysis indicates a disappearance of Nb_{Li} in $LiNbO_3$:Mg at concentrations much lower than the threshold. A final conclusion of the origin of the threshold concentrations was drawn from precise structure studies of $LiNbO_3$:Zn and $LiNbO_3$:In [72, 73, 74]. The Zn impurity is most convenient for such studies, because due to its largest threshold value (7.5%) among all damage-resistant impurities it provides the highest faithfulness of estimates.

Table 6.2 presents occupancy coefficients of atoms in $LiNbO_3$:Zn and $LiNbO_3$:In calculated from the refinement of the structures obtained by precise X-ray diffraction measurements [72, 73]. Table 6.3 shows corresponding chemical formulae of $LiNbO_3$:Zn based on the Li site vacancy model assuming $(Zn^{\bullet}_{Li})$ to be compensated by Li vacancies. From structure experiments, presented in Tables 6.2 and 6.3, which are supported for Zn-doped $LiNbO_3$ by neutron diffraction studies [74], we conclude that for threshold concentrations and those above them Zn and In ions incorporate partially onto Nb sites. Additionally, under certain assumptions, it was inferred that for $LiNbO_3$:Zn the Nb antisites disappear from the lattice in the concentration range 3% < Zn < 5%, which means far below the threshold. These structure measurements established unambiguously the microscopic origin of the threshold concentrations and related it to a partial incorporation of impurity ions into Nb octahedra, which is valid obviously for all optical-damage-resistant ions. The following comment is worth noting. Microscopically more probable is an incorporation of these relatively large ions into

TABLE 6.3. Chemical formulae for $LiNbO_3$:Zn and $LiNbO_3$:In crystals, the symbol □ denotes vacancies

Zn [at.%]	Formulae
0	$[Li_{0.940}Nb_{0.012}\square_{0.048}][Nb]O_3$
2.87	$[Li_{0.908}Nb_{0.007}Zn_{0.029}\square_{0.056}][Nb]O_3$
5.2	$[Li_{0.898}Zn_{0.052}\square_{0.05}][Nb]O_3$
7.6	$[Li_{0.939}Zn_{0.06}\square_{0.001}][Nb_{0.98}Zn_{0.015}\square_{0.005}]O_3$
8.2	$[Li_{0.95}Zn_{0.06}][Nb_{0.98}Zn_{0.022}]O_3$

In [at.%]	Formulae
1	$[Li_{0.970}In_{0.01}\square_{0.02}][Nb]O_3$
2.7	$[Li_{0.96}In_{0.012}\square_{0.028}][Nb_{0.986}In_{0.014}]O_3$

the Li octahedron in view of its larger dimensions [52]. When an impurity incorporates onto both cation sites, the lattice is brought to a *critical state*. Indeed, as seen in Figure 6.5, in the range above the threshold Zn concentration in the crystal the distribution coefficient practically saturates, which means that this Zn concentration in the crystal is close to a tolerable limit.

A simultaneous incorporation of impurity ions onto both cation sites requires a modification of the charge compensation conditions; see (6.2) and (6.3). For such a case the model calculations of Donnerberg et al. [66] suggested a high probability for a self-compensation mechanism

$$20\, ZnO + 8\left[Nb_{Li}^{4\bullet} - 4V'_{Li}\right] = 15\, Zn_{Li}^{\bullet} + 5\, Zn_{Nb}^{3'} + 4\, Nb_2O_5. \tag{6.5}$$

For a divalent ion it requires a concentration ratio of the self-compensating ions of $[Zn_{Nb}^{3'}]/[Zn_{Li}^{\bullet}] = 1 : 3$. Since this ratio is roughly fulfilled for $LiNbO_3$:Zn with the above-listed threshold concentrations (Tables 6.2 and 6.3), the self-compensation, which requires no more Li vacancies for the charge compensation, may be accepted. According to (6.5) the threshold concentration corresponding to a change of the lattice site of the impurity atom is followed by the removal of V_{Li} from the lattice. A schematic dependence of the defect structure on the Zn concentration, deduced on the basis of structure data and on qualitative considerations outlined above, is presented in Figure 6.13. This scheme illustrates drastic changes in the defect structure occurring at the threshold impurity concentration.

6.3.3 Effects of Optical-Damage-Resistant Ions on the Incorporation of Transition Metal and Other Impurity Ions into the $LiNbO_3$ Lattice

The search for the microscopic origin of optical damage resistance requires investigations of the influence of Mg or Zn on the incorporation of transition metal (TM) impurities, which are responsible for the photovoltaic activity, particularly Fe. Additionally, optically stable laser media based on $LiNbO_3$ activated either by Cr^{3+} or rare earth ions (RE) need optical damage resistance. These considerations have

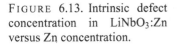

FIGURE 6.13. Intrinsic defect concentration in LiNbO$_3$:Zn versus Zn concentration.

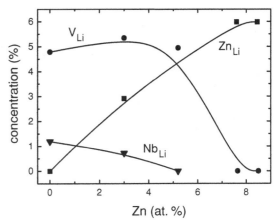

motivated numerous studies of doubly doped crystals (LiNbO$_3$:RE, LiNbO$_3$:Cr, LiNbO$_3$:Fe codoped with Mg or Zn). Prior to a brief description of the current state of this research, we remind the reader of the main concept of the incorporation of TM ions into LiNbO$_3$. Studies of the Fe site in LiNbO$_3$ have been repeatedly performed both by spectroscopic methods (EPR, electron nuclear double resonance (ENDOR); for references see the reviews [80, 86]) and by extended X-ray absorption fine structure (EXAFS) [87]. Iron in LiNbO$_3$ exists in the two charge states Fe^{2+} and Fe^{3+} [88]. Fe^{2+} centers are characterized by absorption bands at 470 and 1100 nm; EPR studies provide a probe for Fe^{3+} sites, and EXAFS points to Fe in general irrespective of the charge state. Tests with these methods gave evidence that in agreement with old conclusions [88] in congruent LiNbO$_3$, both Fe^{2+} and Fe^{3+} are incorporated onto Li sites. The same conclusion is now drawn for Cr^{3+} in congruent LiNbO$_3$ [89, 90, 91]. Note that RE ions are also believed to incorporate into Li octahedra and to be shifted by about 30–50 pm along the z axis with regard to the regular Li$^+$ sites [92, 93].

In further studies the effect of the [Li]/[Nb] ratio on the incorporation of the impurities has been reported. An analysis of EPR spectra in stoichiometric LiNbO$_3$ led to the conclusion that Fe^{3+} and Cr^{3+} in stoichiometric (free of intrinsic defects) crystals occupy partially Nb sites [83, 89, 90, 91]. This result has a direct relationship to the problem of the incorporation of TM ions into optical-damage-resistant LiNbO$_3$ crystals. In LiNbO$_3$ doped with Fe or Cr a codoping with optical-damage-resistant impurities provides effects comparable to those of the crystal composition.

Figure 6.14 demonstrates the influence of Mg or Zn doping on the EPR spectra of Fe^{3+} in congruent LiNbO$_3$:Fe; the spectra coincide for comparable Mg or Zn concentrations. Recall that an asymmetric shape of Fe^{3+} EPR spectra in a congruent LiNbO$_3$ is attributed to a distortion of the crystal field at the Fe^{3+} site by surrounding intrinsic defects reducing its symmetry from C$_3$ (a subgroup of the point group symmetry C$_{3v}$) to C$_1$ [19]. As seen in Figure 6.14, for MgO

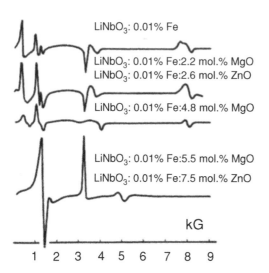

FIGURE 6.14. EPR spectra of LiNbO$_3$:Fe-doped with Mg^{2+} or Zn^{2+} according to [29].

or ZnO concentrations up to the thresholds, the shape of the EPR lines varies smoothly. For MgO and ZnO concentrations close to 5 or 7 mol%, respectively, an abrupt change of the line positions and shapes occurs. This change of the EPR spectra is one more example of a drastic threshold behavior of properties. Below the threshold the EPR spectra correspond to Fe^{3+} on Li sites. The shape of the spectrum at Mg concentrations above the threshold was assigned to a partial incorporation of Fe^{3+} ions onto Nb sites [33, 83, 94] with a possible self-compensation $\left[Fe_{Li}^{3+}\right]^{\bullet\bullet} - \left[Fe_{Nb}^{3+}\right]''$ [66, 94]. These conclusions may doubtless be expanded to LiNbO$_3$:Zn:Fe. Note that spectral characteristics of Fe^{2+} are not affected by Mg [30] or Zn [34], which is evidence of the unalterableness of the Fe^{2+} locations in the Li octahedron independent of the presence of an optical-damage-resistant dopant.

Similar conclusions were drawn about the incorporation of other TM-ions into LiNbO$_3$ codoped with optical-damage-resistant impurities. Codoping of LiNbO$_3$:Cr with Mg or Zn concentrations above the threshold is accompanied by fundamental changes of EPR, emission, and absorption spectra of Cr^{3+} ions (for details see [90] and the bibliography therein [91, 95, 96, 97]), and it is attributed by most authors to a partial incorporation of Cr^{3+} onto Nb sites. Qualitatively, analogous effects of a high Mg doping on the shape and line widths of EPR spectra were found for Mn^{2+} [98].

It is worth mentioning some results of spectroscopic studies in LiNbO$_3$:RE codoped with Mg or Zn, mainly for RE: Nd^{3+}, Yb^{3+}, Er^{3+} in view of an interest of these activators for laser light generation. These results have yet to find an unambiguous explanation, because the interpretation of spectral characteristics of RE^{3+} ions in LiNbO$_3$ is rather obscure. According to the more commonly accepted viewpoint, a RE^{3+} ion in LiNbO$_3$ forms several inequivalent centers in the Li octahedron characterized by different off-center displacements (for Nd^{3+}, e.g., [99, 100, 101, 102], for Yb^{3+}, e.g., [103] and [104]). Codoping with an above-threshold concentration of Mg or Zn modifies the properties of these centers. In $LiNbO_3$:Nd a strong Mg-doping affects the fine structure of the absorption (excitation) band and induces an additional line shifted to higher energies [99, 100, 102], which was attributed to the formation of a new Nd-Mg center. EPR spectra of Er^{3+} [101] [105] and Yb^{3+} [104] in $LiNbO_3$ reveal a qualitative change in the shape and positions of lines when doped with Mg or Zn above the threshold, which is qualitatively similar to effects in $LiNbO_3$:Fe:Mg and $LiNbO_3$:Fe:Zn (Figure 6.14). These changes of the EPR spectra were assigned to the incorporation of Yb^{3+} and Er^{3+} ions [104, 105], respectively, onto Nb sites in analogy to Fe^{3+} or Cr^{3+}. Doping with Mg produces an additional new line in the absorption and photoemission spectra of Yb^{3+} [103], but in contrast to Bonardi et al. [104], these authors attribute it to the formation of a new Yb^{3+} center on a Li site. From rather scattered results on $LiNbO_3$:RE crystals codoped with Mg or Zn one may draw a practically important conclusion that the main effect of codoping is a suppression of the optical damage, whereas the spectroscopic properties and laser characteristics themselves are only weakly affected.

Summarizing the results presented in Section 6.3, one may see that a high (above-threshold) doping with optical-damage-resistant impurities leads not only to a fundamental reconstruction of the intrinsic defect structure, but to a strong perturbation of the properties of other impurity ions, mainly TM ions. Therefore, fundamental changes of many properties in these concentration ranges find a qualitative explanation. This is especially true with respect to an altered location of Fe^{3+} in $LiNbO_3$ highly doped with Mg or Zn, which leads to a fundamental change in the charge transport scheme and thus is responsible for optical damage resistance.

6.4 Microscopic Origin of Optical Damage Resistance

6.4.1 The Current Microscopic Model of Light-Induced Charge Transport in $LiNbO_3$

The formation of the photoinduced space-charge field in $LiNbO_3$ crystals caused by the photovoltaic effect is due to the presence of TM impurities (Fe, Cu, Mn). The model of the photoinduced charge transport in $LiNbO_3$ was developed in detail for the impurities Fe and Cu (for references see [1, 106] and a recent review [107]), and it is based on the approximation of a compensated semiconductor [108]. This model describes the charge-transport process in terms of a one-center scheme in which Fe^{2+} (or Cu^+) are the electron donors and Fe^{3+} (or Cu^{2+}) are the electron traps. Both the photovoltaic current and the photoconductivity are due to the photoelectron transport between a single donor and a single trap:

$$Fe^{2+} + h\nu = Fe^{3+} + e^-. \qquad (6.6)$$

Photoconductivity in such a one-center model is described by the expression

$$\sigma_{ph} = g\tau e\mu, \tag{6.7}$$

where the photogeneration rate is given by

$$g = q(\alpha I / h\nu), \tag{6.8}$$

and the lifetime of the free carriers by

$$\tau = (S \nu N_C)^{-1}. \tag{6.9}$$

In (6.7)–(6.9) the parameter q is the quantum efficiency; e, μ, and ν are carrier charge, mobility, and thermal velocity, respectively; I means the light intensity; $\alpha = s [N_D]$ is the optical absorption coefficient (where s is the probability of the donor photoexcitation); $S = (\gamma_r)/\nu$ is the cross-section of the electron for a trap (where γ_r is the recombination coefficient); N_C and N_D are the trap and donor concentrations, respectively. In this simplified presentation the expression for the photoconductivity looks like

$$\sigma_{ph} = (I/h\nu)qe\mu(s/\gamma_r)[N_D]/[N_C] = A\,I\frac{[N_D]}{[N_C]}. \tag{6.10}$$

For LiNbO$_3$:Fe and LiNbO$_3$:Cu the linearity of σ_{ph} versus the concentration ratio of donors to traps was proved experimentally [109, 110]:

$$\sigma_{ph}/I = ([Fe^{2+}]/[Fe^{3+}])\,10^{-12}\ \mathrm{cm}\,\Omega^{-1}\,\mathrm{W}^{-1},$$

$$\sigma_{ph}/I = ([Cu^{+}]/[Cu^{2+}])\,6.5\,10^{-12}\ \mathrm{cm}\,\Omega^{-1}\,\mathrm{W}^{-1}.$$

The photovoltaic current j_{pv} is linearly proportional to the concentration of donors (Fe^{2+} or Cu^{+}). For the one-center charge transport scheme a linear dependence of the photoconductivity on the light intensity is characteristic. Provided that the crystal is photovoltaic, a saturation of $\delta n \propto E_{pv} = k_G\alpha/A$ (for A see 6.10) occurs at relatively low intensities and the sensitivity $S_1 = (1/I)dn/dt = r_{eff}k_G\alpha/\epsilon\epsilon_0$ does not depend on I. However, the behavior of nominally pure LiNbO$_3$ and of weakly Fe-doped LiNbO$_3$ appeared to diverge under high light intensities from these rules, namely, S_1 depends on the light intensity resulting in abnormally high values for δn at $I \geq 10^3$ Wcm^{-2} [111, 112] and δn depends on the pulse duration, e.g., [46, 113]). These data required a refinement of the charge-transport scheme.

The most probable explanation is given by an assumption that photoinduced charge transport involves a second shallow level, empty in the equilibrium state. Its nonequilibrium population under high light intensity results in a growth of α with I, a sublinear $\sigma_{ph}(I)$, and an increase of δn with I. This simplified two-center model was developed and verified for BaTiO$_3$, in which these anomalies were observed for the first time (e.g., [114, 115, 116, 117]). For LiNbO$_3$ and LiNbO$_3$:Fe a two-center approach was elaborated by various groups [45, 46, 118, 119]. Jermann et al. [45] have shown that in nominally pure (Fe-free) LiNbO$_3$ exposed to high light intensities, the rules predicted by this model are fulfilled. For LiNbO$_3$:Fe the two-center scheme was modified by the assumption of a direct electron transfer from the donor (Fe^{2+}) to a shallow trap (hypothetically Nb_{Li}) [46, 118, 119]. Estimates of secondary trap concentrations, responsible for the observed anomalies, led to very high values 10^{18}–10^{19} cm^{-3}. Since the presence, both in $LiNbO_3$ and $LiNbO_3$:Fe, of uncontrollable admixtures of such concentrations is unlikely, the most reasonable suggestion (assuming n-type photoconductivity) was that of secondary electron traps in $LiNbO_3$ like intrinsic defects Nb_{Li}, whose concentration in a congruent crystal lies approximately in this range (see above). This suggestion is supported by computer simulations [64], according to which Nb_{Li} is the most probable electron trap in $LiNbO_3$. Some experimental results as well favor this suggestion. For example, a chemical reduction of a stoichiometric $LiNbO_3$ [81] or of congruent ones highly doped with Mg or Zn [34, 80], i.e., crystals with a low Nb_{Li} content, is hampered compared to congruent crystals. Additional evidence is given by an enhanced photoconductivity in a stoichiometric $LiNbO_3$ (see above), which is usually interpreted in terms of a decreasing Nb_{Li} content [45]. Starting from these experimental observations and calculations, the suggestion of the role of Nb_{Li} as electron traps is now commonly accepted. The resulting two-center transport scheme in $LiNbO_3$ (electrons as charge photocarriers, Fe^{2+} donors, Fe^{3+} traps, and Nb_{Li} shallow traps with a capture cross-section much less than that of Fe^{3+}) forms the basis for a discussion of the characteristics of optical damage.

6.4.2 A Fundamental Change of the Charge Transport in LiNb O₃ and LiNbO₃:Fe Produced by Optical-Damage-Resistant Dopants

We now discuss the microscopic reasons for an increased photoconductivity in optical-damage-resistant $LiNbO_3$ in terms of the two-center model cited above. Two cases—optical-damage-resistant compositions free of TM admixtures and compositions containing low concentrations (or traces) of TM ions—will be discussed separately.

Following 6.10, an increase of σ_{ph} may be due either to an increase of the charge mobility μ or to a change of the transport scheme. In principle, the value of μ depending on the scattering of the carriers by defects might be changed after Mg or Zn doping, as was proposed by Jackel et al. [30]. However, temperature dependencies of the conductivity in $LiNbO_3$, $LiNbO_3$:5% Mg, and $LiNbO_3$:5%Mg:Fe may be described by an identical expression

$$\sigma = 1620 \exp\left(\frac{-E_a}{kT}\right) \qquad (6.11)$$

with the same activation energy $E_a = 1.2$ eV independent of the composition [120]. From this it was concluded that an increase of σ in no case is related to a change of μ. Direct measurements of μ have not been performed so far. The more probable reason is a variation in the charge transport scheme supported by some results presented above and below. We discuss separately the origin of the optical damage in Fe-free and Fe-codoped optical-damage-resistant compositions. The lux-ampere characteristics in low-doped $LiNbO_3$:Zn (without Fe) are slightly sublinear $\sigma_{ph} \propto I^x$ ($x = 0.8$–0.9), as seen in Table 6.1 in the work of Volk et al. [35]. This is an indication for a two-center scheme. Therefore, in the framework of the foregoing model it may be discussed in terms of the behavior of the assumed Nb_{Li} electron traps. If, for example, we use $LiNbO_3$:Zn to correlate a concentration dependence of σ_{ph} (Figure 6.3) with a schematic modification of the intrinsic defect structure produced by Zn doping (Figure 6.13), then one may ascribe a smooth increase of σ_{ph} in the concentration range Zn $< 5\%$ to a gradually decreasing concentration of Nb_{Li}: $\sigma_{ph} \propto 1/[Nb_{Li}]$. Note that an increase of σ_{ph} cannot be related to an alternative electron trap $[Zn_{Li}]^{\bullet}$, because this would require a lower photoconductivity for a higher Zn concentration.

A steeper slope of σ_{ph} for Zn concentration above the threshold is not related to Nb_{Li}, which is missing in this range. On the other hand, Li vacancies in $LiNbO_3$ are probably hole traps [64]. Since the concentration of V_{Li} drastically falls at about the Zn threshold concentration, one may suggest that an increase of σ_{ph} stems from an increasing hole component at the expense of a decrease of the concentration of hole traps: $\sigma_{ph} \propto 1/[V_{Li}]$. This rather speculative consideration finds an experimental support, because an analysis of the sign of the photocarriers by means of a holographic method [22, 121] in $LiNbO_3$:Mg [122] and $LiNbO_3$:Zn [123] has shown that at low impurity concentrations the photoconductivity is of n-type, whereas above the thresholds it changes its sign and becomes dominantly p-type, the contribution from the hole component increasing with Mg or Zn concentrations.

$LiNbO_3$:Mg and $LiNbO_3$:Zn codoped with Fe also demonstrate the optical damage resistance for Mg or Zn concentrations above the threshold (Figure 6.2). However, the microscopic origin of the damage resistance in this case differs from that proposed above. In contrast to optical-damage-resistant crystals free of Fe, the lux-ampere characteristics in $LiNbO_3$:Mg:Fe and $LiNbO_3$:Zn:Fe [35] are strictly linear, which points to a one-center charge-transport scheme. Since the charge states of iron (Fe^{2+} and Fe^{3+}) in $LiNbO_3$:Mg:Fe are conserved [30], in this case the charge transport may be described by the traditional scheme (see (6.6)), and the role of iron is dominant again. We now discuss the reason for an increasing σ_{ph} in these crystals in terms of expression (6.10) for the one-center scheme. The concentration ratio $[Fe^{2+}]/[Fe^{3+}]$ at high Mg or Zn concentrations decreases by about a factor of two compared to the low-concentration range, as concluded from optical absorption measurements [36] and Mössbauer spectra [124]. Therefore an increase of σ_{ph} above the threshold cannot be accounted for by concentration arguments and is thus related to an influence of the optical-damage-resistant ions on inherent properties of Fe^{2+} or Fe^{3+}. As mentioned above, the shape of the characteristic absorption band of Fe^{2+} is not affected by Mg, Zn, or In doping, which means that the properties of the electron donor Fe^{2+} are preserved and the photoexcitation probability $s_{Fe^{2+}}$ in (6.10) is unchanged. By contrast, as was shown in Section 6.3.3 for Mg or Zn concentrations above the thresholds (Figure 6.14), the electron trap Fe^{3+} changes its lattice site and partially incorporates onto a Nb site, as was concluded from EPR measurements [94].

Studies of X-ray-induced optical absorption spectra [125, 126] proved that Fe^{3+} on the Nb site loses its acceptor properties. In these experiments it was found that in $LiNbO_3$:Mg:Fe the X-ray-induced optical absorption spectra are fundamentally different for Mg concentrations below and above the threshold. In the former, X-irradiation induces a stable absorption band with a maximum at 480 nm, which is due to a formation of extra Fe^{2+} according to the scheme $Fe^{3+} + e^- = Fe^{2+}$. If the Mg concentration exceeds 5.5 %, then this band does not appear and the X-ray-induced spectrum contains only an unstable band at about 360 nm quickly decaying in the darkness. Therefore, Fe^{3+} is no longer an electron trap. Estimates of Gerson et al. [120] also show that the main reason for a drastic growth of the photoconductivity at high Mg concentrations is a decrease of the capture cross-section $S = \gamma_r / v$ of electrons by Fe^{3+} traps by more than two orders of magnitude. A rough estimate for $[Fe_{Li}^{3+}]^{2\bullet}$ gives $S = 1.3 \ 10^{-14} \ cm^2$ [127] and for $[Fe_{Nb}^{3+}]^{2'}$ the value $S < 10^{-15} \ cm^2$ [120]. So, in optical-damage-resistant crystals codoped with iron the main reason for the optical damage resistance is a fundamental change in the charge transport scheme (see (6.6)) because of an alteration of the electron trap. Summarizing these data for optical-damage-resistant $LiNbO_3$ both codoped with Fe and without it, one may conclude that an optical damage resistance for impurity concentrations above the thresholds is caused by fundamental changes of the charge transport scheme.

This consideration may be qualitatively expanded to the case of stoichiometric $LiNbO_3$; its increased photoconductivity is usually related to a decreasing concentration of Nb_{Li} electron traps [45]. However, an increased σ_{ph} in a Li-enriched $LiNbO_3$:Fe in comparison with a congruent $LiNbO_3$:Fe which was detected from measurements of the two-beam coupling gain [43] may be accounted for by the reasons proposed above for $LiNbO_3$:Mg:Fe. Recall that Fe^{3+} in stoichiometric $LiNbO_3$ incorporates partially onto the Nb site [19, 86]. Therefore, one may suggest that in Fe-doped stoichiometric crystals Fe^{3+} loses its acceptor properties and an increased σ_{ph} is due to a fundamental change of the transport scheme.

Finally, we comment very briefly, on the concentration dependencies of the photovoltaic current in optical-damage-resistant and stoichiometric $LiNbO_3$ crystals (Figure 6.9). First of all, a significant increase of j_{pv} in a stoichiometric crystal compared to a congruent one finds no explanation within the current microscopic model. At the same time, a decreasing j_{pv} at high Mg concentrations in congruent and Li-enriched crystals might be explained by the following reasons. As mentioned above, a conversion of σ_{ph} from n-type to p-type in a strongly oxidized $LiNbO_3$:Fe eliminates j_{pv} [22], i.e., no hole photovoltaic current exists. According to [122] and [123] the photoconductivity in $LiNbO_3$:Mg and $LiNbO_3$:Zn at Mg and Zn concentrations above the threshold is predominantly of p-type. So, a decreasing photovoltaic current, especially in Li-enriched $LiNbO_3$ might be ascribed to an altered sign of the main carriers. Remember that this effect additionally to an increased σ_{ph} contributes essentially to the optical damage resistance.

6.5 Optical Properties of LiNbO$_3$ Crystals Doped with Optical-Damage-Resistant Ions

In this section we briefly characterize optical properties that are sensitive to doping with optical-damage-resistant ions and those that are useful for practical applications. We emphasize again that the influence of optical-damage-resistant ions resembles that of an increased Li content in LiNbO$_3$ crystals.

6.5.1 Optical Absorption Spectra in the Near IR

As seen in Figure 6.1, doping of LiNbO$_3$ with optical-damage-resistant ions shifts the UV band edge to shorter wavelengths. Additionally to this effect the shape and the spectral position of the IR absorption band are affected. This band, the OH stretch mode absorption due to an incorporation of water vapor during crystal growth, has been observed in various oxides; for a recent review see [128]. In congruent LiNbO$_3$ the maximum of the IR band has been observed at about 3480–3485 cm^{-1} and is shifted to 3466 cm^{-1} in stoichiometric LiNbO$_3$ (Table 6.4). Similarly, doping with Mg-type impurities shifts this band to shorter wavelengths. The microscopic origin of this band in LiNbO$_3$ and of its shift

TABLE 6.4. Energies of the OH-stretch mode in optical-damage-resistant LiNbO$_3$

Composition	Energies (cm^{-1})	Reference
Congruent	3478, 3491	[129]
Stoichiometric	3466	[130], [131]
LiNbO$_3$:Mg [Li]/[Nb] = 0.945		
<4.5% MgO	3484	[9]
>4.5% MgO	3534	[9]
	3507, 3537	[132]
LiNbO$_3$:Mg; [Li]/[Nb] = 1.1		
<3.5% MgO	3485, 3530, 3540	[133], [134]
>3.5% MgO	3540	[133], [134]
LiNbO$_3$:Mg; [Li]/[Nb] = 1.2		
<2% MgO	3485, 3530, 3540	[133], [134]
>2.5% MgO	3540	[133], [134]
LiNbO$_3$:Zn		
<7% Zn	3484	[29], [135]
>7% Zn	3506, 3535	
LiNbO$_3$:In		
>1.5%In	2506, 3508	[13], [136]
LiNbO$_3$:Sc	3497	[137]

FIGURE   6.15. IR   transmission
spectra of Zn-doped LiNbO$_3$ below
the threshold (a), in the threshold
region (b), and above the threshold
(c).

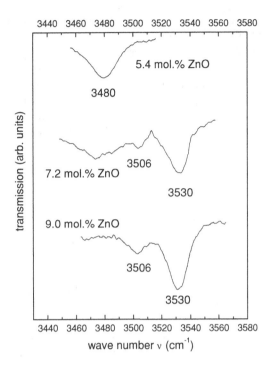

after doping are widely discussed in the literature (for references see [138]),
because of their relation to the intrinsic defect structure. These problems are
beyond our consideration, so we summarize very briefly the data on IR absorption
in optical-damage-resistant crystals (Table 6.4). The position of the IR band is
sensitive to the concentration of optical-damage-resistant ions and demonstrates
a drastic threshold behavior. In the whole concentration range below the threshold
concentration both the shape of the band and its spectral position coincide with
those in congruent LiNbO$_3$. At threshold concentrations the band is shifted in a
step-like manner to shorter wavelengths, the amount of the shift depending on the
type of the optical-damage-resistant ion (di- or trivalent), on the [Li]/[Nb] ratio,
and on the codoping with RE or Cr^{3+} ions.

Figure 6.15 presents an example for LiNbO$_3$:Zn [11]. The OH bands have no
practical importance, but, since its shift correlates with the occurrence of optical
damage, it provides a convenient tool for crystal characterization. Namely, on
the basis of the position of the IR band, one may conclude beforehand whether a
given composition is optical-damage-resistant.

6.5.2 Refractive Indices and Birefringence: A Generalized Sellmeier Equation

Doping with optical-damage-resistant impurities as well as variations of the
[Li]/[Nb] ratio result in variations of the refractive indices and the birefringence.

In the literature several parameter sets for a Sellmeier equation for LiNbO$_3$ of different composition, e.g., [139, 140, 141], and for congruent Mg-doped [5, 82, 142, 143, 144] crystals have been reported. Recently data for Li-enriched (near stoichiometric) LiNbO$_3$:Mg were reported [145]. The results are to some extent scattered. We will dwell on those publications [39, 142, 146, 147] in which a general approach is developed for a description of the dispersion of $n_e(\lambda)$, $n_o(\lambda)$ in LiNbO$_3$ of different composition or doped with optical-damage-resistant ions. The unique feature of this approach is the physical interpretation of each term of the Sellmeier equation with respect to the intrinsic defects. The index of refraction is described as a sum of several oscillators,

$$n^2 = 1 + \sum_j \frac{A_j}{\lambda_j^{-2} - \lambda^{-2}}, \tag{6.12}$$

where λ_j^{-2} describes the resonance frequency of an oscillator and A_j is its strength proportional to the oscillator concentration (assuming their independence). Optical properties of LiNbO$_3$ in the near-UV range depend on the single oscillator represented by a NbO$_6$ octahedron [148]. The approaches reported in [139, 140, 142] are based on estimations of the change of the oscillator strength A_0 resulting from a substitution for a regular Nb$_{Nb}$ atoms by defects Nb$_{Li}$. Their concentrations depend on the Li content or on the concentration of the damage-resistant Mg ions controlling the content of Nb$_{Li}$. Later it was expanded to other optical-damage-resistant dopants [39, 146, 147]. The consideration is based on the Li site vacancy model; a difference in the resonance energies of Nb$_{Nb}$, Nb$_{Li}$, and Mg$_{Li}$ (Zn$_{Li}$, In$_{Li}$) is neglected. This idea led to the following generalized Sellmeier equation for all optical-damage-resistant LiNbO$_3$ compositions:

$$n^2 = \frac{A_0 + A_{Nb_{Li}} c_{Nb_{Li}} + A_{Mg_{Li}} c_{Mg_{Li}} + A_{In_{Li}} c_{In_{Li}}}{(\lambda_0 + \mu_0 [f(T) - f(T_0)])^2 - \lambda^{-2}} + A_{UV} - A_{IR} \lambda^2, \tag{6.13}$$

where A_{UV} is the contribution from UV plasmons, $A_{IR} \lambda^2$ is the contribution from dipole-active phonon oscillations, and $f(T) = (T + 273)^2 + 4.023 \, 10^5 \, (\coth[261.6/(T + 273)] - 1)$ takes into account the temperature dependence of the resonance frequency ($T_0 = 24.5 \,°C$). The concentration of Nb$_{Li}$ as a function of the concentrations of optical-damage-resistant ions is given by

$$c_{Nb_{Li}} = \frac{2}{3(50 - c_{Li})} - \frac{c_{Mg}}{\alpha_{Mg}} - \frac{c_{Zn}}{\alpha_{Zn}} - \frac{c_{In}}{\alpha_{In}}, \tag{6.14}$$

where α_{Mg}, α_{Zn}, α_{In} are fitting parameters corresponding to the experimental values of the threshold concentrations of Mg, Zn, In (5.0, 6.5, 1.5%), respectively. This presentation of the Nb$_{Li}$ concentration is due to the fact that the authors assumed the threshold concentrations to correspond to the removal of Nb$_{Li}$ and the following substitution of regular Li atoms by optical-damage-resistant ions. An incorporation of these ions onto Nb sites found later [72, 73, 74] was of course not considered. According to (6.14) a change of n_e, n_o at threshold concentrations

TABLE 6.5. Parameters for the Sellmeier equation (6.13)

	λ_0	μ_0	A_0	A_{IR}	$A_{Nb_{Li}}$
n_o	223.219	$1.1082\ 10^{-6}$	$4.5312\ 10^{-5}$	$3.6340\ 10^{-8}$	$-7.2320\ 10^{-8}$
n_e	218.203	$6.4047\ 10^{-6}$	$3.9466\ 10^{-5}$	$3.0998\ 10^{-8}$	$11.8635\ 10^{-7}$

	A_{Mg}	A_{Zn}	A_{In}	A_{UV}
n_o	$-7.3648\ 10^{-8}$	$6.7963\ 10^{-8}$	$-2.4\ 10^{-7}$	2.6613
n_e	$7.6243\ 10^{-8}$	$1.9221\ 10^{-7}$	$4.7\ 10^{-7}$	2.6613

means a vanishing oscillator strength for the resonator, and a linear increase of the strengths of Mg_{Li}, Zn_{Li}, and In_{Li} resonators.

Parameters of the Sellmeier equation (6.13) are shown in Table 6.5. In spite of some restrictions, particularly of a postulate on a relation between the disappearance of Nb_{Li} and the threshold concentrations, which lacked support from structure measurements [72, 73, 74], the Sellmeier equation (6.13) may be applied as a good approximation to experimental data on $n_e(\lambda)$, $n_o(\lambda)$ for Mg [142], Zn [39], and In [146] doping for wavelengths up to 1200 nm and even for a double-doping with (Zn + In) [147]. An example is given in Figure 6.16 presenting the concentration dependencies of $n_e(\lambda)$, $n_o(\lambda)$ and of the birefringence for LiNbO$_3$:Zn, where the solid curves are calculated with equation (6.13). Note that both $n_e(\lambda)$ and $n_o(\lambda)$ are growing with Zn concentration in contrast to Mg doping, which results in a decreasing $n_e(\lambda)$ [21, 82, 142]. This specific observation for Zn doping is of importance in practice (see Section 6.6 below). In near-stoichiometric LiNbO$_3$:Mg where a threshold of the optical damage resistance is shifted to 1 % Mg, the extraordinary index is almost independent of Mg [145]. The condition of the noncritical (90°) phase matching of the type I for optical frequency doubling is

$$\Delta n = n_e(\lambda_{\frac{1}{2}}, T_{pm}) - n_o(\lambda_1, T_{pm}) = 0, \tag{6.15}$$

where λ_1, $\lambda_{\frac{1}{2}}$ are fundamental and doubled frequency, respectively, and T_{pm} is the temperature of the phase synchronism. The kinks in the concentration dependencies of the birefringence (Figure 6.16), clearly detected in the range of the threshold concentrations for all optical-damage-resistant crystals [39, 142, 146], predict extrema in concentration dependencies of noncritical phase-matching temperatures T_{pm} and of phase-matching angles Θ_{pm} at a given temperature (maxima and minima, respectively). This indeed can be seen in Figure 6.4. Calculations of T_{pm} and Θ_{pm} versus the concentrations of Mg, Zn, In on the basis of (6.13) are in good agreement with experimental data [39, 142, 146]. At the end of this section we show a dependence of the coefficients of the linear electrooptical effect $r_c = r_{33} - (n_o/n_e)^3 r_{13}$ and r_{22} versus Zn concentration in LiNbO$_3$:Zn (Figure 6.17) [149]. The dependencies of r_c and r_{22} versus Zn (143) [149] are non-monotonic and additionally to maxima in the range of the threshold concentrations reveal anomalies in the low-concentration ranges. Speculatively, in LiNbO$_3$:Zn it correlates with a disappearance of Nb_{Li} (Figure 6.13). From a practical viewpoint,

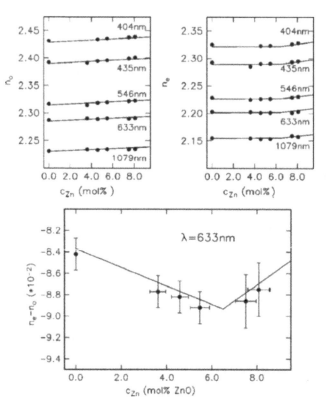

FIGURE 6.16. Ordinary and extraordinary refractive indices (top) and birefringence of Zn-doped congruently melting LiNbO$_3$ versus Zn concentration (bottom). The lines are calculated using (6.13), experimental values are represented by dots.

it is important that in an optical-damage-resistant concentration range the values of r_{ij} are not decreased compared to an undoped congruent crystal.

6.6 An Outline of Practical Potentials of Optical-Damage-Resistant LiNbO$_3$ Crystals

In this section we mention very briefly the main lines in applications of optical-damage-resistant LiNbO$_3$ crystals doped with Mg, Zn, In, or Sc impurities. These crystals with impurity concentrations above the thresholds combine a set of properties appropriate for optical-frequency conversion: a strongly decreased optical damage together with a slightly reduced dark trace effect [150]; sufficient nonlinear-optical coefficients (e.g., $d_{eff} = 5.4$ pm/V in LiNbO$_3$:7% Mg for $\lambda = 1064$ nm [151] in comparison with 7.4 and 1.2 pm/V in KTP [152] and LBO [153], respectively); a possibility of controlling the temperature T_{pm} of the

FIGURE 6.17. Electrooptical coefficients r_c and r_{22} versus Zn concentration in LiNbO$_3$:Zn. Dashed lines are guides for the eye.

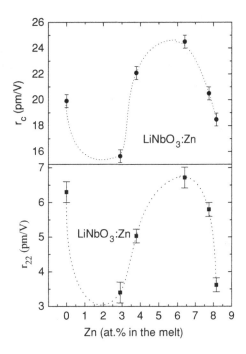

90° phase synchronism by varying the dopant concentration (see Figure 6.4). These properties are relevant as well for lasing structures, because electro- and nonlinear-optical properties of the LiNbO$_3$ matrix itself provide a possibility of a self-conversion of laser oscillations stimulated on RE activators introduced into the bulk. An instability of lasing caused in LiNbO$_3$:RE by optical damage is overcome by codoping with Mg-like impurities, so these doubly doped LiNbO$_3$ crystals may find applications in compact laser systems with self-doubled stimulated laser radiation. Highly doped LiNbO$_3$:Mg [4, 154, 155] and LiNbO$_3$:Zn [11, 50, 156] showed themselves as efficient doublers of the $\lambda = 1064$ nm radiation with an angle of phase synchronism Θ_{pm} close to 90° at room temperature for 7% Mg [151] and 9% Zn [155] (and at lower dopant concentrations in Li-enriched crystals [50]). No degradation caused by the optical damage was detected up to 100 MW/cm^2. In LiNbO$_3$:Zn η is somewhat higher than in LiNbO$_3$:Mg because of a lowered dark-trace effect [12, 150, 156]. An efficient SHG of $\lambda = 1064$ nm was observed as well in waveguiding structures produced by Ti indiffusion into LiNbO$_3$:Mg [157]. There are several publications on OPOs on the basis of LiNbO$_3$:Mg [158, 159, 160, 161]. The examples of bulk lasing systems with self-frequency doubling were demonstrated in LiNbO$_3$:Nd:5% Mg [162, 163]. A. Cordova-Plaza et al. [163] constructed miniature lasers (with dimensions in the range of millimeters), in which the self-doubling of the stimulated laser oscillation ($\lambda_{2\omega} = 547$ nm) was achieved by means of a temperature tuning. A self-frequency doubling was observed as well in LiNbO$_3$:Nd:0.5%Sc$_2$O$_3$ at room temperature

[164]. J. Capmany et al. [165] obtained a cw stable action with a self-frequency doubling in $LiNbO_3$:Nd:8% Zn. Laser oscillation was demonstrated in waveguiding structures on $LiNbO_3$:Nd:Mg substrates obtained by either proton exchange [166] (and bibliography therein) or Ti-indiffusion [167].

A new era in the search for applications of optical-damage-resistant $LiNbO_3$ started after the demonstration of the possibility of an optical frequency conversion in the quasi-phasematching (QPM) mode of operation in periodically poled $LiNbO_3$ crystals (PPLN) (for references see, e.g., the reviews [168, 169]). As is well known, PPLN structures permit one to expand the spectral range of the converted radiation, eliminate the problem of the walk-off angle, and require no temperature or angle tuning of the phase synchronism, but as well as bulk $LiNbO_3$ crystals suffer from optical damage. The structures produced on the basis of $LiNbO_3$:Mg (PPLN:Mg) provide advantages over undoped PPLN, because apart from a suppressed optical damage they exhibit a noticeably reduced coercive field E_c. In a congruent $LiNbO_3$ a field of $E_c \approx 220$ kV/cm [170, 171] is needed, whereas in $LiNbO_3$:5% Mg switching fields of 44 [172] or 60 [173] kV/cm were reported. Since PPLN patterns are prepared more often by means of varied field actions (a field applying to patterned electrodes, electron beam, or corona discharge methods, etc), than by means of growth techniques (e.g., [174] for PPLN:Mg), so a decreased E_c facilitates the preparation of such structures. Examples of the formation of PPLN:Mg structures and their use for QPM optical frequency doubling may be found in reports concerning either bulk [173, 175, 176] or planar (waveguiding) [177, 178, 179] PPLN:Mg structures. In all publications an increased stability of PPLN:Mg against the optical damage in comparison to PPLN is emphasized. The practical potentials mentioned above are common to all optical-damage-resistant $LiNbO_3$ crystals. However, there are some applications that are specific to certain optical-damage-resistant ions, particularly for Zn. Zn-indiffused waveguides on $LiNbO_3$ or $LiNbO_3$:Mg substrates withstand the optical damage and guide both extraordinary and ordinary light polarization [180, 181]. The latter unique feature is due to the fact that Zn-doping of $LiNbO_3$, unlike other optical-damage-resistant ions, leads to an increase of both n_e and n_o (Figure 6.16). Zn-indiffused waveguides provide advantages over H-exchanged waveguides supporting n_e polarization only and Ti-indiffused ones suffer from optical damage. Several preparation techniques of Zn-indiffused waveguides are presented in [181, 182, 183, 184]. Laser oscillations stimulated by Er^{3+} and Nd^{3+} ions in Zn-indiffused channel waveguides are reported in [185] and [186].

At the end of this section we dwell on alternative practical potentials of $LiNbO_3$ compositions under discussion, namely, on using their photorefractive rather than optical-damage-resistant properties. Particularly, a shift of the band edge to shorter wavelengths (Figure 6.1) forms a basis for a UV-recording. In $LiNbO_3$:9% Mg the holographic recording was realized at $\lambda = 351$ nm [187] with a high speed and a stationary two-beam coupling gain $\Gamma = 15$ cm^{-1} indicating the diffusion mechanism of recording.

6.7 Conclusion

The results summarized in the present review show that $LiNbO_3$ crystals doped with optical-damage-resistant impurities are of interest both for practical applications and fundamental studies. Doping with these impurities contributes to an improvement of the optical performance of $LiNbO_3$ for its traditional applications, particularly for optical frequency conversion. A serious limitation is imposed by a deterioration of the uniformity of the crystals for high doping levels and, as a consequence, a decrease of the threshold of the optical breakdown (laser damage). However, this shortcoming is obviously not due to any fundamental reason and may be overcome by means of modified or new growth technology. This was exemplified by the growth of $LiNbO_3$:Mg crystals with an enhanced Li content, which permits one to lower the required concentration of the dopant. Plausible practical applications of optical-damage-resistant $LiNbO_3$ may be based on the fact that Mg-like impurities serve as controllers of the Nb_{Li} content and therefore permit the realization in these crystals after a chemical reduction of an IR-recording in a gating mode of operation by analogy with a reduced stoichiometric $LiNbO_3$. From the fundamental viewpoint, microscopic reasons for connections of properties, especially charge transport to the intrinsic defects, are of great importance. The question of the charge-carrier sign in relation to the crystal composition, particularly of the optical-damage-resistant dopants, is still open. A mysterious dependence of the photovoltaic current on the composition recently found when studying Li-enriched $LiNbO_3$:Mg and the existence of the intrinsic photovoltaic effect are of special interest.

Acknowledgments. We acknowledge the support by the Russian Foundation for Basic Research (RFBR) projects N 96 02-18851 and 00-02-16624, as well as that of the INTAS Foundation, projects N 94 1080 and 96 0599.

References

1. M.E. Lines and A.M. Glass. *Principles and Applications of Ferroelectrics and Related Materials*. Clarendon Press, Oxford, 1977.
2. A. Räuber. Chemistry and physics of lithium niobate. In E. Kaldis, editor, *Current Topics in Materials Science*, page 481. North Holland Publishing Company, 1978.
3. Yu. S. Kuz'minov. *Lithium Niobate Crystals*. Cambridge International Science Publishing, Cambridge CB1 6AZ, UK, 1999.
4. J.L. Nightingale, W.J. Silva, G.E. Reade, A. Rybicki, W.J. Kozlovsky, and R.L. Byer. *SPIE, Laser and Nonlinear Optical Materials*, 681:20, 1986.
5. Y. Furukawa, A. Yokotani, T. Sasaki, H. Yoshida, K. Yoshida, F. Nitanda, and M. Sato. *J. Appl. Phys.*, 69:3372, 1991.
6. F.S. Chen. *J. Appl. Phys.*, 40:3389, 1969.
7. T. Volk, M. Ivanov, V. Pryalkin, and N. Rubinina. *Ferroelectrics*, 126:57, 1992.

8. G.G. Zhong, J. Jian, and Z.K. Wu. In *Proceedings of the Eleventh International Quantum Electronics Conference, IEEE*, page 631, New York, 1980.

9. D.A. Bryan, R. Gerson, and H.E. Tomaschke. *Appl. Phys. Lett.*, 44:847, 1984.

10. K.L. Sweeney, L.E. Halliburton, D.A. Bryan, R. Rice, R. Gerson, and H.E. Tomaschke. *J. Appl. Phys.*, 57:1036, 1985.

11. T.R. Volk, N.M. Rubinina, V.I. Pryalkin, V.V. Krasnikov, and V.V. Volkov. *Ferroelectrics*, 109:345, 1990.

12. T.R. Volk, V.I. Pryalkin, and N.M. Rubinina. *Optics Letters*, 15:996, 1990.

13. T.R. Volk and N.M. Rubinina. *Ferroelectrics Letters*, 14:37, 1992.

14. T.R. Volk, M. Wöhlecke, N. Rubinina, N.V. Razumovskiĭ, F. Jermann, C. Fischer, and R. Böwer. *Appl. Phys. A*, 60:217, 1995.

15. J.K. Yamamoto, K. Kitamura, N. Iyi, S. Kimura, Y. Furukawa, and M. Sato. *Appl. Phys. Lett.*, 61:2156, 1992.

16. J.K. Yamamoto, T. Yamazaki, and K. Yamagishi. *Appl. Phys. Lett.*, 64:3228, 1994.

17. N.V. Anghert, V.A. Pashkov, and N.M. Solov'yeva. *Zh. Exp. Teor. Fiz.*, 26:1666, 1972.

18. Y. Furukawa, M. Sato, K. Kitamura, Y. Yajima, and M. Minakata. *J. Appl. Phys.*, 72:3250, 1992.

19. G.I. Malovichko, V.G. Grachev, E.P. Kokanyan, O.F. Schirmer, K. Betzler, B. Gather, F. Jermann, S. Klauer, U. Schlarb, and M. Wöhlecke. *Appl. Phys. A*, 56:103, 1993.

20. T. R. Volk and M. Wöhlecke. Optical damage resistance in lithium niobate crystals. *Ferroelectrics Review*, 1:195, 1998.

21. Y. Furukawa, M. Sato, F. Nitanda, and K. Ito. *J. Cryst. Growth*, 99:832, 1990.

22. R. Orlowski and E. Krätzig. *Sol. St. Commun.*, 27:1351, 1978.

23. I.B. Barkan, S.I. Marennikov, and M.V. Entin. *Phys. Stat. Solidi (a)*, 45:K17, 1978.

24. J.L. Jackel, D.H. Olson, and A.M. Glass. *J. Appl. Phys.*, 52:4855, 1981.

25. I. Savatinova, S. Tonchev, and P. Kircheva. *Ferroelectrics*, 249:257, 2001.

26. F. Nitanda, Y. Furukawa, S. Makio, M. Sato, and K. Ito. *Jpn. J. Appl. Phys.*, 34:1546, 1995.

27. Y. Kong, J. Wen, and H. Wang. *Appl. Phys. Lett.*, 66:280, 1995.

28. Y.F. Kong, J.K. Wen, and Z.F. Yang. *Chin. Phys. Letters*, 14:629, 1997.

29. T.R. Volk and N.M. Rubinina. *Sov. Phys. Solid State*, 33:674, 1991.

30. R. Sommerfeldt, L. Holtmann, E. Krätzig, and B. Grabmaier. *Phys. Stat. Solidi (a)*, 106:89, 1988.

31. M. Simon, F. Jermann, T. Volk, and E. Krätzig. *Phys. Stat. Solidi (a)*, 149:723, 1995.

32. L.D. Sshearer, M. Leduc, and Z. Zachorowski. *IEEE J. Quantum Electronic*, QE-23:1996, 1987.

33. H.X. Feng, J.K. Wen, H.F. Wang, S.Y. Hand, and Y.X. Xu. *J. Phys. Chem. Solids*, 51:397, 1990.

34. T. Volk, N. Rubinina, and M. Wöhlecke. *J. Opt. Soc. Am. B*, 11:1681, 1994.

35. T.R. Volk, N.V. Razumovskiĭ, A.V. Mamaev, and N.M. Rubinina. *J. Opt. Soc. Am. B*, 13:1457, 1996.

36. J. Koppitz, O.F. Schirmer, M. Wöhlecke, A.I. Kusnetsov, and B.C. Grabmaier. *Ferroelectrics*, 92:233, 1989.

37. Y. Furukawa, K. Kitamura, S. Takekawa, K. Niwa, and H. Hatano. *Opt. Lett.*, 23:1892, 1998.

38. Y. Furukawa, K. Kitamura, S. Takekawa, A. Miyamoto, M. Terao, and N. Suda. *Appl. Phys. Lett.*, 77:2494, 2000.

39. U. Schlarb, M. Wöhlecke, B. Gather, A. Reichert, K. Betzler, T. Volk, and N. Rubinina. *Optical Materials*, 4:791, 1995.

40. R.D. Shannon and C.T. Prewitt. *Acta Cryst. B*, 25:925, 1969.
41. B.C. Grabmaier and F. Otto. *J. Cryst. Growth*, 79:682, 1986.
42. Y. Furukawa, M. Sato, M.C. Bashaw, M.M. Fejer, N. Iyi, and K. Kitamura. *Jpn. J. Appl. Phys.*, 35:2740, 1996.
43. Y. Furukawa, K. Kitamura, Y. Ji, G. Montemezzani, M. Zgonik, C. Medrano, and P. Günter. *Optics Letters*, 22:501, 1997.
44. M.H. Garrett, I. Mnushkina, Y. Furukawa, K. Kitamura, L.E. Halliburton, N.C. Giles, and C.D. Setzler. In *Proc. of PRM*, 1997, p. 295.
45. F. Jermann, M. Simon, and E. Krätzig. *J. Opt. Soc. Am. B*, 12:2066, 1995.
46. F. Jermann and E. Krätzig. *Appl. Phys. A*, 55:114, 1992.
47. T.R. Volk, S.B. Astaf'ev, and N.V. Razumovskiĭ. *Sov. Phys.: Solid State*, 37:583, 1995.
48. K. Polgár, Á. Péter, L. Kovács, G. Corradi, and Zs. Szaller. *J. Cryst. Growth*, 177:211, 1997.
49. K. Niwa, Y. Furukawa, S. Takekawa, and K. Kitamura. *J. Cryst. Growth*, 208:493, 2000.
50. M.H. Li, Y.H. Xu, R. Wang, X.H. Zhen, and C.Z. Zhao. *Cryst. Res. Technol.*, 36:191, 2001.
51. G. Malovichko, V. Grachev, and O. Schirmer. *Appl. Phys. B, Lasers & Optics*, 68:785, 1999.
52. S.C. Abrahams and P. Marsh. *Acta Cryst. B*, 42:61, 1986.
53. B.C. Grabmaier, W. Wersing, and W. Koestler. *J. Cryst. Growth*, 110:339, 1991.
54. N. Iyi, K. Kitamura, F. Izumi, J.K. Yamamoto, H. Asano T. Hayashi, and S. Kimura. *J. Solid State Chem.*, 101:340, 1992.
55. Y.S. Luh, M.M. Fejer, R.L. Byer, and R.S. Feigelson. *J. Cryst. Growth*, 85:264, 1987.
56. D.H. Jundt, M.M. Fejer, and R.L. Byer. *IEEE J. Quantum Electron.*, QE-26:135, 1990.
57. G.I. Malovichko, V.G. Grachev, V.T. Gabrielyan, and E.P. Kokanyan. *Sov. Phys. Solid State*, 28:1453, 1986.
58. G.I. Malovichko, V.G. Grachev, L.P. Yurchenko, Y. Ya. Proshko, and E.P. Kokanyan V.T. Gabrielyan. *Phys. Stat. Solidi (a)*, 133:K29, 1992.
59. K. Kitamura, Y.K. Yamamoto, N. Iyi, S. Kimura, and T. Hayashi. *J. Cryst. Growth*, 116:327, 1992.
60. A.P. Wilkinson, A.K. Cheetham, and R.H. Jarman. *J. Appl. Phys.*, 74:3080, 1993.
61. N. Zotov, H. Boysen, F. Frey, T. Metzger, and E. Born. *J. Phys. Chem. Solids*, 55:145, 1994.
62. J. Blümel, E. Born, and T. Metzger. *J. Phys. Chem. Solids*, 55:589, 1994.
63. G.G. DeLeo, J.L. Dobson, M.F. Masters, and L.H. Bonjack. *Phys. Rev. B*, 37:8394, 1988.
64. H. Donnerberg, S.M. Tomlinson, C.R.A. Catlow, and O.F. Schirmer. *Phys. Rev. B*, 40:11909, 1989.
65. G.E. Peterson and A. Carnevale. *J. Chem. Phys.*, 56:4848, 1972.
66. H. Donnerberg, S.M. Tomlinson, C.R.A. Catlow, and O.F. Schirmer. *Phys. Rev. B*, 44:4877, 1991.
67. P. Lerner, C. Legras, and J.P. Dumas. *J. Cryst. Growth*, 3-4:231, 1968.
68. D. Redfield and W.J. Burke. *J. Appl. Phys.*, 45:4566, 1974.
69. W. Rossner, B. Grabmaier, and W. Wersing. *Ferroelectrics*, 93:57, 1989.
70. N. Iyi, K. Kitamura, Y. Yajima, S. Kimura, Y. Furukawa, and M. Sato. *J. Solid State Chem.*, 118:148, 1995.

71. S. Shimamura, Y. Watanabe, T. Sota, K. Suzuki, N. Iyi, K. Kitamura, T. Yamazaki, A. Sugimoto, and K. Yamagishi. *J. Phys.: Condens. Matter*, 8:6825, 1996.

72. T. Volk, B. Maximov, T. Chernaya, N. Rubinina, M. Wöhlecke, and V. Simonov. *Appl. Phys. B*, 72:647, 2001.

73. T. Chernaya, B. Maximov, T. Volk, N. Rubinina, and V. Simonov. *JETP Letters*, 73:103, 2001.

74. S. Sulyanov, B. Maximov, T. Volk, H. Boysen, J. Schneider, N. Rubinina, and Th. Hansen. *Appl. Phys. A*, 74 [Suppl.]:S1031, 2002.

75. L.J. Hu, Y.H. Chang, F.S. Yen, S.P. Lin, I-Nan Lin, and W.Y. Lin. *J. Appl. Phys.*, 69:7635, 1992.

76. R. Mouras, M.D. Fontana, P. Bourson, and A.V. Postnikov. *J. Phys.: Condens. Matter*, 12:5053, 2000.

77. D.A. Dutt, F.J. Feigl, and G.G. Deleo. *J. Phys. Chem. Solids*, 51:407, 1990.

78. K. Sweeney and L. Halliburton. *Appl. Phys. Lett.*, 43:336, 1983.

79. J. Ketchum, K. Sweeney, L. Halliburton, and A. Armington. *Phys. Letts. A*, 94:450, 1983.

80. O.F. Schirmer, O. Thiemann, and M. Wöhlecke. *J. Phys. Chem. Solids*, 52:185, 1991.

81. A. García-Cabanes, J.A. Sanz-García, J.M. Cabrera, F. Agulló-López, C. Zaldo, R. Pareja, K. Polgár, K. Raksányi, and I. Földvári. *Phys. Rev. B*, 37:6085, 1988.

82. G.K. Kitaeva, K.A. Kuznetsov, A.N. Penin, and A.V. Shepelev. *Phys. Rev. B*, 65:054304, 2002.

83. G.I. Malovichko, V.G. Grachev, O.F. Schirmer, and B. Faust. *J. Phys.: Condens. Matter*, 5:3971, 1993.

84. B. Faust, H. Müller, and O. Schirmer. *Ferroelectrics*, 153:297, 1994.

85. X. Feng, J. Ying, J. Liu, and Z. Yin. *Science in China (series A)*, 33:113, 1990.

86. O.F. Schirmer, H.J. Reyher, and M. Wöhlecke. Characterization of point defects in photorefractive oxide crystals by paramagnetic resonance methods. In F. Agulló-López, editor, *Insulating Materials for Optoelectronics. New Developments*, page 93. World Scientific Publishing, 1995.

87. C. Prieto and C. Zaldo. *Sol. St. Commun.*, 83:819, 1992.

88. M.G. Clark, F.J. DiSalvo, A.M. Glass, and G.E. Peterson. *J. Chem. Phys.*, 59:6209, 1973.

89. G.I. Malovichko, V.G. Grachev, E. Kokanyan, and O.F. Schirmer. *Phys. Rev. B*, 59:9113, 1999.

90. V.G. Grachev and G.I. Malovichko. *Phys. Rev. B*, 62:7779, 2000.

91. G.M. Salley, S.A. Basun, A.A. Kaplyanskii, R.S. Meltzer, K. Polgár, and U. Happek. *J. Lumines.*, 87:1133, 2000.

92. L. Rebouta, P.J.M. Smulders, D.O. Boerma, F. Agulló-López, M.F. daSilva, and J.C. Soares. *Phys. Rev. B*, 48:1993, 3600.

93. A. Lorenzo, H. Jaffrezic, B. Roux, G. Boulon, and J. García-Solé. *Appl. Phys. Lett.*, 67:3735, 1995.

94. A. Böker, H. Donnerberg, O.F. Schirmer, and X.Q. Feng. *J. Phys: Condens. Matter*, 2:6865, 1990.

95. S.A. Basun, A.A. Kaplyanskii, A.B. Kutsenko, V. Dierolf, T. Troster, S.E. Kapphan, and K. Polgár. *Phys. Solid State*, 43:1043, 2001.

96. G.A. Torchia, J.O. Tocho, and F. Jaque. *J. Phys. Chem. Solids*, 63:555, 2002.

97. G. Corradi, H. Söthe, J.-M. Spaeth, and K. Polgár. *J. Phys.: Condens. Matter*, 3:1901, 1991.

98. I.W. Park, Y.N. Choi, S.H. Choh, and S.S. Kim. *J. Korean Phys. Soc.*, 32:S693, 1998.

99. J.O. Tocho, J.A. Sanz-García, F. Jaque, and J. García-Solé. *J. Appl. Phys.*, 70:5582, 1991.
100. G. Lifante, F. Cussó, F. Jaque, J.A. Sanz-García, A. Monteil, B. Varrel, G. Boulon, and J. García-Solé. *Chem. Phys. Letters*, 176:482, 1991.
101. J.O. Tocho, E. Camarillo, F. Cussó, F. Jaque, and J. García-Solé. *Sol. St. Commun.*, 80:575, 1991.
102. J. García-Solé, T. Petit, H. Jaffrezic, and G. Boulon. *Europhys. Letters*, 24:719, 1993.
103. E. Montoya, A. Lorenzo, and L.E. Bausá. *J. Phys.: Condens. Matter*, 11:311, 1999.
104. C. Bonardi, C.J. Magon, E.A. Vidoto, M.C. Terrile, L.E. Bausá, E. Montoya, D. Bravo, A. Martin, and F.J. Lopez. *J. Alloys Compounds*, 323:340, 2001.
105. D. Bravo, A. Martin, and F.J. Lopez. *Sol. St. Commun.*, 112:541, 1999.
106. E. Krätzig and O.F. Schirmer. Photorefractive centers in electro-optic crystals. In P. Günter and J.-P. Huignard, editors, *Photorefractive Materials and Their Applications I*, number 61 in Topics in Appl. Phys., page 131. Springer-Verlag, 1988.
107. K. Buse, F. Jermann, and E. Krätzig. *Appl. Phys. A*, 58:191, 1994.
108. N.V. Kukhtarev, V.B. Markov, S.G. Odulov, M.S. Soskin, and V.L. Vinetskii. *Ferroelectrics*, 22:949, 1979.
109. H. Kurz, E. Krätzig, W. Keune, H. Engelmann, U. Gonser, B. Dischler, and A. Räuber. *Appl. Phys.*, 12:355, 1977.
110. K. Peithmann, J. Hukriede, K. Buse, and E. Krätzig. *Phys. Rev. B*, 61:4615, 2000.
111. V.E. Wood, N.F. Hartman, and C.M. Verber. *Ferroelectrics*, 27:237, 1980.
112. O. Althoff, A. Erdmann, L. Wiskott, and P. Hertel. *Phys. Stat. Solidi (a)*, 128:K41, 1991.
113. I. Kanaev, V. Malinovskii, and A. Pugachev. *Soviet Physics Solid State*, 27:1772, 1985.
114. D. Mahgerefteh and J. Feinberg. *Optics Letters*, 13:1111, 1989.
115. A. Motes and J.J. Kim. *J. Opt. Soc. Am. B*, 4:1379, 1987.
116. G.G. Valley. *Appl. Opt.*, 22:3160, 1983.
117. L. Holtmann. *Phys. Stat. Solidi (a)*, 113:89, 1989.
118. F. Jermann and J. Otten. *J. Opt. Soc. Am. B*, 10:2085, 1993.
119. M. Simon, F. Jermann, and E. Krätzig. *Opt. Mater.*, 3:243, 1994.
120. R. Gerson, J.F. Kirchhoff, L.E. Halliburton, and D.A. Bryan. *J. Appl. Phys.*, 60:3553, 1986.
121. R. Orlowski and E. Krätzig. *Ferroelectrics*, 26:831, 1980.
122. H. Wang, J. Wen, J. Li, H. Wang, and J. Jing. *Appl. Phys. Lett.*, 57:344, 1990.
123. J.C. Deng, J.K. Wen, Z.K. Wu, and H.F. Wang. *Appl. Phys. Lett.*, 64:2622, 1994.
124. S.I. Bae, J. Ichikawa, K. Shimamura, H. Onodera, and T. Fukuda. *J. Cryst. Growth*, 180:94, 1997.
125. T.R. Volk and N.M. Rubinina. *Phys. Stat. Solidi (a)*, 108:437, 1988.
126. T.R. Volk, M.A. Ivanov, F. Ya. Shchapov, and N.M. Rubinina. *Ferroelectrics*, 126:185, 1992.
127. E. Krätzig. *Ferroelectrics*, 21:635, 1978.
128. M. Wöhlecke and L. Kovács. OH^- ions in oxide crystals. *Critical Reviews Solid State Materials*, 26:1–86, 2001.
129. J.R. Herrington, B. Dischler, A. Räuber, and J. Schneider. *Sol. St. Commun.*, 12:351, 1973.
130. A. Gröne and S. Kapphan. *J. Phys. Chem. Solids*, 56:687–701, 1995.
131. Y. Watanabe, T. Sota, K. Suzuki, N. Iyi, K. Kitamura, and S. Kimura. *J. Phys.: Condens. Matter*, 7:3627, 1995.

132. Ma. J. de Rosendo, L. Arizmendi, J.M. Cabrera, and F. Agulló-López. *Sol. St. Commun.*, 59:499, 1986.
133. L. Kovács, K. Polgár, and R. Capelletti. *Cryst. Latt. Def. and Amorph. Mat.*, 15:115, 1987.
134. L. Kovács, I. Földváry, and K. Polgár. *Acta Phys. Hung.*, 61:223, 1987.
135. Y. Zhang, Y.H. Xu, M.H. Li, and Y.Q. Zhao. *J. Cryst. Growth*, 233:537, 2001.
136. Y. Kong, J. Deng, W. Zhang, J. Wen, G. Zhang, and H. Wang. *Phys. Letters A*, 196:128, 1994.
137. J.K. Yamamoto, K. Kitamura, N. Iyi, and S. Kimura. *J. Cryst. Growth*, 128:920, 1993.
138. J.M. Cabrera, J. Olivares, M. Carracosca, J. Rams, R. Müller, and E. Diéguez. *Advances in Physics*, 45:349, 1996.
139. U. Schlarb and K. Betzler. *Phys. Rev. B*, 48:15613, 1993.
140. U. Schlarb and K. Betzler. *Ferroelectrics*, 156:99, 1994.
141. D.H. Jundt. *Optics Letters*, 22:1553, 1997.
142. U. Schlarb and K. Betzler. *Phys. Rev. B*, 50:751, 1994.
143. D.E. Zelmon, D.L. Small, and D. Jundt. *J. Opt. Soc. Am. B*, 14:3319, 1997.
144. G.K. Kitaeva, I.I. Naumova, A.A. Mikhailovsky, P.S. Losevsky, and A.N. Penin. *Appl. Phys. B*, 66:201, 1998.
145. M. Nakamura, S. Higuchi, S. Takekawa, K. Terabe, Y. Furukawa, and K. Kitamura. *Jpn. J. Appl. Phys.*, 41:L49, 2002.
146. K. Kasemir, K. Betzler, B. Matzas, B. Tiegel, M. Wöhlecke, N. Rubinina, and T. Volk. *Phys. Stat. Solidi (a)*, 166/1:R7, 1998.
147. U. Schlarb, B. Matzas, A. Reichert, K. Betzler, M. Wöhlecke, B. Gather, and T. Volk. *Ferroelectrics*, 185:269, 1996.
148. E. Wiesendanger and G. Güntherodt. *Sol. St. Commun.*, 14:303, 1974.
149. F. Abdi, M. Aillerie, M. Fontana, P. Bourson, T. Volk, B. Maximov, S. Sulyanov, N. Rubinina, and M. Wöhlecke. *Appl. Phys. B, Lasers Opt.*, 68:795, 1999.
150. J.C. Deng, Y.F. Kong, J. Li, J.K. Wen, and B. Li. *J. Appl. Phys.*, 79:9334, 1996.
151. J.Q. Yao, W.Q. Shi, J.E. Millerd, G.F. Xu, E. Garmire, and M. Birnbaum. *Optics Letters*, 15:1339, 1990.
152. J.Q. Yao and T.S. Fahlen. *J. Appl. Phys.*, 55:65, 1984.
153. C. Chen, Y. Wu, A. Jiang, B. Wu, G. You, R. Li, and S. Lin. *J. Opt. Soc. Am. B*, 6:616, 1989.
154. D.A. Bryan, R.R. Rice, R. Gerson, H.E. Tomaschke, K.L. Sweeney, and L.E. Halliburton. *Opt. Eng.*, 24:138, 1985.
155. T.R. Volk, N.M. Rubinina, and A.I. Kholodnykh. *Kvantovaya Elektronika (Quantum Elektronics)*, 15:1705, 1988.
156. T.R. Volk, V.V. Krasnikov, V.I. Pryalkin, and N.M. Rubinina. *Kvantovaya Elektronika (Quantum Elektronics)*, 17:262, 1990.
157. M.M. Fejer, M.J.F. Digonnet, and R.L. Byer. *Optics Letters*, 11:230, 1986.
158. W.J. Kozlovsky, C.D. Nabors, R.C. Eckhardt, and R.L. Byer. *Optics Letters*, 14:66, 1989.
159. M. Bode, P.K. Lam, I. Freitag, A. Tunnermann, H.A. Bachor, and H. Welling. *Optics Commun.*, 148:117, 1998.
160. S. Schiller and R.L. Byer. *J. Opt. Soc. Am. B*, 10:1696, 1993.
161. S.J. Lin and T. Suzuki. *Optics Letters*, 21:579, 1996.
162. T.Y. Fan, A. Cordova-Plaza, M.J.F. Digonnet, R.L. Byer, and H.J. Shaw. *J. Opt. Soc. Am. B*, 3:140, 1986.

163. A. Cordova-Plaza, M.J.F. Digonnet, and H.J. Shaw. *IEEE Journal of Quantum Electronics*, QE-23:262, 1987.

164. J.K. Yamamoto, A. Sugimoto, and K. Yamagishi. *Optics Letters*, 19:1311, 1994.

165. J. Capmany, D. Jaque, J.A. Sanz-Garcia, and J. Garcia-Sole. *Optics Communications*, 161:253, 1999.

166. L. Lallier, J.P. Pocholle, M. Papuchon, M.P. DeMicheli, M.J. Li, Q. He, D.B. Ostrovsky, C. Grezes-Besset, and E. Pelletier. *IEEE Journal of Quantum Electronics*, 27:618, 1991.

167. S.J. Field, D.C. Hanna, D.P. Sheperd, A.C. Tropper, P.J. Chandler, P.D. Townsend, and L. Zhang. *Optics Letters*, 16:481, 1991.

168. M. Houe and P.D. Townsend. *J. Phys. D: Appl. Phys.*, 28:1747, 1995.

169. R.L. Byer. Quasi-phasematched nonlinear interactions and devices. *J. Nonlinear Optical Physics & Materials*, 6:549–592, 1997.

170. I. Camlibel. *J. Appl. Phys.*, 40:1690, 1969.

171. L. Goldberg, W.K. Burns, and R.W. McElhanon. *Appl. Phys. Lett.*, 67:2910, 1995.

172. A. Kuroda, S. Kurimura, and Y. Uesu. *Appl. Phys. Lett.*, 69:1565, 1996.

173. A. Harada and Y. Nihei. *Appl. Phys.Lett.*, 69:2629, 1996.

174. I.I. Naumova, N.F. Evlanova, O.A. Gliko, and S.V. Lavrishchev. *J. Cryst. Growth*, 181:160, 1997.

175. A. Harada, Y. Nihei, Y. Okazaki, and H. Hyuga. *Optics Letters*, 22:805, 1997.

176. J. Capmany, E. Montoya, V. Bermudez, D. Callejo, E. Diguez, and L.E. Bausa. *Appl. Phys. Lett.*, 76:1374, 2000.

177. K. Mizuuchi, K. Yamamoto, and M. Kato. *Electronic Lett.*, 32:2091, 1996.

178. S. Sonoda, I. Tsuruma, and M. Hatori. *Appl. Phys. Lett.*, 70:3078, 1997.

179. T. Sugita, K. Mizuuchi, Y. Kitaoka, and K. Yamamato. *Optics Letters*, 24:1590, 1999.

180. W.M. Young, R.S. Feigelson, M.M. Fejer, M.J.F. Digonnet, and H.J. Shaw. *Optics Letters*, 16:995, 1991.

181. W.M. Young, M.M. Fejer, M.J.F. Digonnet, A.F. Marshall, and R.S. Feigelson. *J. Lightwave Technology*, 10:1238, 1992.

182. V.A. Fedorov, Yu. N. Korkishko, F. Vereda, G. Lifante, and F. Cusso. *J. Cryst. Growth*, 194:94, 1998.

183. T. Kawaguchi, K. Mizuuchi, T. Yoshino, M. Imaeda, K. Yamamoto, and T. Fukuda. *J. Cryst. Growth*, 203:173, 1999.

184. R. Nevado and G. Lifante. *Appl. Phys. A, Mater. Sci. Process.*, 72:725, 2001.

185. E. Cantelar, R.E. DiPaolo, F. Cusso, R. Nevado, G. Lifante, W. Sohler, and H. Suche. *J. Alloys and Compounds*, 323:348, 2001.

186. R.E. DiPaolo, E. Cantelar, R. Nevado, J.A. Sanz-Garcia, M. Domenech, P.L. Pernas, G. Lifante, and F. Cusso. *Ferroelectrics*, 273:2607, 2002.

187. J.J. Xu, G.Y. Zhang, F.F. Li, X.Z. Zhang, Q.A. Sun, S.M. Liu, F. Song, Y.F. Kong, X.J. Chen, H.J. Qiao, J.H. Yao, and L.J. Zhao. *Optics Letters*, 25:129, 2000.

7

Photorefractive Effects in $KNbO_3$

Marko Zgonik,[1] Michael Ewart,[2] Carolina Medrano, and Peter Günter

Nonlinear Optics Laboratory, Swiss Federal Institute of Technology, ETH—Hönggerberg, 8093—Zürich (Switzerland)
[1] present address: Faculty of Mathematics and Physics, University of Ljubljana and Jozef Stefan Institute, Ljubljana (Slovenia)
[2] present address: wzw-optic AG, Balgach (Switzerland)

7.1 Introduction

Oxygen octahedra ferroelectrics are important materials because of their electrooptic and nonlinear-optic properties [1]. In this group, the perovskites $BaTiO_3$ and $KNbO_3$ crystals are especially promising materials. These two crystals are interesting for nonlinear-optical applications because of their large value of the spontaneous polarization at room temperature and the high packing density of oxygen octahedra, which are highly polarizable units [2]. These properties give rise to large electrooptic coefficients [3–5] and high nonlinear optical susceptibilities [6, 7].

The possibility of growing large single-domain crystals of potassium niobate [8] is also an advantage in applications in nonlinear optics. To date, extensive research has been devoted to the photorefractive properties of undoped and doped $KNbO_3$. Many different kinds of impurities have been investigated with the aim of enhancing the photorefractive effect. At first, iron-doped crystals were studied [9, 10]. The photorefractive effect in nominally undoped crystals is caused by iron impurities that occur in the niobium pentoxide component of the growth material. Additional doping with iron was investigated in order to characterize the charge-transport parameters and to determine the most favorable dopant level for photorefractive applications with continuous-wave visible light [11].

More recently, other dopants, such as Cu, Co, Mn, Ni, and Rh, have been investigated [12–17]. This research aimed at increasing the photorefractive sensitivity at visible wavelengths and especially at near-infrared wavelengths. Reeves et al. claimed that Fe is the best dopant for visible and near-infrared photorefraction [13]. The results reported in [12] show that Ni, Mn, and Rh are also very promising dopants.

When the iron impurities Fe^{3+} are reduced to the Fe^{2+} valence state, thus making the electrons the dominant charge carriers [3, 4, 18], the photorefractive sensitivity of $KNbO_3$ is dramatically increased. Large photorefractive sensitivity of 2 cm^{2}J^{-1} has also been measured in nominally undoped KNbO$_3$ when one is writing holograms with UV light, therefore relying on band-to-band excitation of charge carriers [19].

Section 7.2. of this book chapter contains a description of the intrinsic properties of KNbO$_3$ relevant to photorefraction. Section 7.3 contains a discussion of various extrinsic properties that can be modified by doping the crystal during growth, by postgrowth treatment or by ion implantation. The valence states and nature of iron defects in KNbO$_3$ are discussed in more detail. A definition of the term "reduction state" and the different methods that are used to reduce KNbO$_3$ are presented. In Section 7.4 some examples of the photorefractive effects in various spectral ranges are described and are followed by conclusions.

7.2 Intrinsic Properties of KNbO$_3$

7.2.1 Material Constants Relevant to Photorefraction

Potassium niobate (KNbO$_3$) is a ferroelectric oxide [10, 20, 21] with perovskite structure. The point group of the crystal at room temperature is orthorhombic [22, 23] *mm2* with the pseudocubic unit cell shown in Figure 7.1. The crystals are usually grown by the Czochralski method (top-seeded solution growth) [24]. The boules are grown from a KNbO$_3$ melt with a 1.5% surplus of K$_2$O [10]. The crucible material is platinum. Between the growth temperature at 1050°C and room temperature two phase transitions take place from the high-temperature

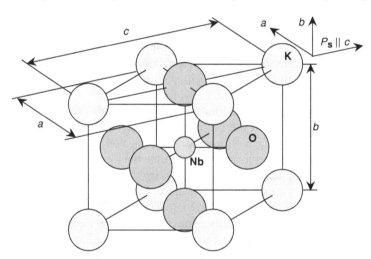

FIGURE 7.1. Pseudocubic unit cell of KNbO$_3$ at room temperature. The lattice constants are $a = 0.569$ nm, $b = 0.397$ nm, and $c = 0.573$ nm. The relative size of the spheres corresponds to the relative size of the ionic radii but the lattice constants' dimensions are drawn 2.5 times larger for clarity. The directions of the orthorhombic axes are shown together with the spontaneous polarization P_s.

FIGURE 7.2. Temperature dependencies of the principal refractive indices at a wavelength of 546 nm. From the high-temperature optically isotropic material the crystal becomes uniaxial in the tetragonal phase with the ordinary and extraordinary refractive indices n_o and n_e respectively. In the orthorhombic phase the crystal is biaxial with the refractive indices n_b, n_a, and n_c.

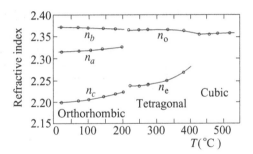

cubic phase. At 450°C the transition is from cubic to tetragonal, and at 203°C the tetragonal to orthorhombic transition occurs upon cooling. When heating KNbO₃, these transitions are at 218°C and 456°C respectively [22, 23].

Refractive indices of KNbO₃ were measured in the visible and near-IR spectral regions [20, 25]. The temperature dependencies of the principal refractive indices at a wavelength of 546 nm were determined by Wiesendanger [20] and are shown in Figure 7.2. A more complete determination of the dispersion (400 nm–3400 nm) and temperature dependence of the refractive indices in the orthorhombic phase can be found in [25]. Note that in the convention chosen the refractive indices obey $n_b > n_a > n_c$.

At room temperature the dielectric tensor of KNbO₃ has three independent elements [23]. The principal axis system is shown in Figure 7.1 and relates to the cubic system as follows: $a\|$ [−101], $b\|$ [010], and $c\|$ [101], where the positive c direction is parallel to the spontaneous polarization. Because KNbO₃ is an acentric crystal, second-order nonlinearities occur. Together with the anisotropic refractive indices, the optical nonlinearities make pure crystals good optical frequency converters. The second-order effects are responsible for electrooptic phase gratings as they occur in the photorefractive effect. The five independent elements of the electrooptic tensor are listed in Table 7.1, together with the other intrinsic material properties [26]. More details on the temperature dependence of these parameters can be found in [25, 27]. In addition, it is important to keep in mind that the refractive indices and the electrooptic tensor have all a normal dispersion, that is, they decrease in value from the visible to the infrared frequency spectrum [25, 27]. In Figure 7.3 the dispersion of the electro optic tensor elements at room temperature is shown.

For photorefractive applications one has to modify the crystals. The usual way is to dope them with various transition-metal ions or to control the stoichiometry of oxygen. The level of substitution is normally less than one impurity ion in every 1000 unit cells, and the oxygen vacancies are also present below this concentration range; therefore, we believe that the intrinsic properties are not changed appreciably, especially by judging them as arising from the basic properties of the oxygen-octahedra structure averaged over the whole sample.

TABLE 7.1. Basic set of the material parameters: dielectric (ε_{ij}^S), elastic stiffness (c_{ijkl}^E), piezoelectric stress (e_{ijk}), elastooptic (p_{ijkl}^E), and clamped electrooptic (r_{ijk}^S) tensor elements of KNbO$_3$. All materials constants are given at room temperature ($22°$C) at the wavelength $\lambda = 633$ nm, where the refractive indices are $n_a = 2.2801$, $n_b = 2.3296$, and $n_c = 2.1687$

Parameter	Value	Units
ε_{11}^S	34 ± 2	
ε_{22}^S	780 ± 22	
ε_{33}^S	24 ± 2	
c_{1111}^E	226 ± 3	10^9 N m^{-2}
c_{1122}^E	96 ± 5	10^9 N m^{-2}
c_{1133}^E	68 ± 7	10^9 N m^{-2}
c_{2222}^E	270 ± 3	10^9 N m^{-2}
c_{2233}^E	101 ± 3	10^9 N m^{-2}
c_{3333}^E	186 ± 5	10^9 N m^{-2}
c_{1313}^E	25 ± 0.3	10^9 N m^{-2}
c_{2323}^E	74 ± 0.8	10^9 N m^{-2}
c_{1212}^E	95.5 ± 1	10^9 N m^{-2}
e_{311}	2.46 ± 0.2	C m^{-2}
e_{322}	-1.1 ± 0.7	C m^{-2}
e_{333}	4.4 ± 0.3	C m^{-2}
e_{131}	5.16 ± 0.12	C m^{-2}
e_{232}	11.7 ± 0.5	C m^{-2}
p_{1111}^E	-0.20 ± 0.04	
p_{1122}^E	0.11 ± 0.03	
p_{1133}^E	0.55 ± 0.05	
p_{2211}^E	-0.13 ± 0.03	
p_{2222}^E	0.23 ± 0.03	
p_{2233}^E	0.16 ± 0.02	
p_{3311}^E	0.64 ± 0.11	
p_{3322}^E	0.15 ± 0.03	
p_{3333}^E	0.82 ± 0.04	
p_{1313}^E	0.45 ± 0.1	
p_{2323}^E	0.57 ± 0.15	
r_{113}^S	21 ± 2	10^{-12} m V^{-1}
r_{223}^S	7.1 ± 0.7	10^{-12} m V^{-1}
r_{333}^S	35 ± 2	10^{-12} m V^{-1}
r_{131}^S	27.8 ± 3	10^{-12} m V^{-1}
r_{232}^S	360 ± 30	10^{-12} m V^{-1}

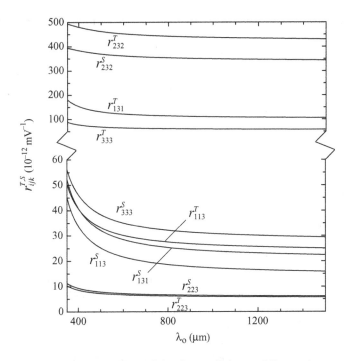

FIGURE 7.3. Wavelength dispersion of the free and clamped linear electrooptic coefficients of KNbO$_3$ crystal at room temperature.

The main effect of the doping is the shift of the ferroelectric phase transition temperatures. Close to the phase transition temperatures, the shift in T_C principally affects the dielectric permittivity (which diverges) and only partially the spontaneous polarization. Nevertheless, the dielectric permittivity and the spontaneous polarization as a function of $(T - T_C)$ do not change under the influence of the doping centers as long as the concentration remains small. In KNbO$_3$ the effect of doping on the phase transition temperatures is negligible. Consequently, no effect of doping on the linear electrooptic properties and other material constants from Table 7.1 can be expected around room temperature.

The band structure of KNbO$_3$ is characterized by a gap of 3.3 eV at room temperature [23] with the valence band formed by the O^{2p} orbitals and with the conduction band formed by Nb^{5d} orbitals [28–32]. The oxygen atoms form the octahedra that are typical of the perovskite structure. The oxygen atoms and especially their vacancies play a key role in the photorefractive properties of KNbO$_3$.

7.2.2 Geometry Dependence of the Electrooptic and Dielectric Constants

The photorefractive space-charge field E_{sc} and the response time depend on the dc dielectric constant ε^{eff}. The electrooptic index change Δn is determined by

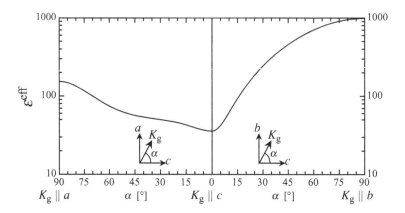

FIGURE 7.4. Effective dielectric constant $\varepsilon^{\mathrm{eff}}$ of KNbO$_3$ as a function of the angle between the grating and the crystal axes.

the scalar r^{eff}. Both $\varepsilon^{\mathrm{eff}}$ and r^{eff} depend on the grating direction, and the latter also depends on the beam polarizations and on the angle between the beams. In order to compare experimental observations in different geometries it is necessary to know the geometrical dependence of $\varepsilon^{\mathrm{eff}}$ and r^{eff}. In both cases the clamped coefficients must be known together with the piezoelectric and elastooptic tensors that determine the unclamped coefficients. Following the theory presented in the chapter "Space-Charge-Driven Holograms in Anisotropic Media" in Volume 1 of this series, the curves that are shown in Figures 7.4 and 7.5 were calculated with the basic tensor element values for KNbO$_3$ from Table 7.1. Figure 7.4 shows the effective low-frequency dielectric constant $\varepsilon^{\mathrm{eff}}$ for a sinusoidally modulated space-charge field with wave vector $\boldsymbol{K}_{\mathrm{g}}$.

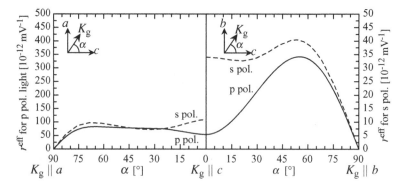

FIGURE 7.5. Effective scalar electrooptic coefficient r^{eff} of KNbO$_3$ as a function of the angle α between the grating wave vector $\boldsymbol{K}_{\mathrm{g}}$ and the crystal axes. The curves for s polarized light (dotted lines) are magnified by a factor of ten (right-hand scale) compared to r^{eff} for p polarized light (left-hand scale, solid line). The curve for p polarized beams is calculated for a crossing angle of $0°$.

The same theory also yields the geometrical dependence of r^{eff} with an example shown in Figure 7.5. For p polarized beams, r^{eff} is calculated for a crossing angle of zero degrees, which corresponds to the limit of almost-copropagating beams. More details on the calculation of r^{eff} can be found in the chapter "Space-Charge-Driven Holograms in Anisotropic Media" in Volume 1 of this series and in [26, 33, 34].

7.3 Doped KNbO$_3$

7.3.1 Doping of KNbO$_3$ with Fe

Impurities with ionization energies smaller than the gap energy provide the charge centers for photoexcitation. The main source of photosensitive centers are impurities that are introduced by intentional doping of the crystal melt, e.g., with Fe$_2$O$_3$. The polished Fe-doped crystals are yellowish in color. We will first present the known properties of Fe in KNbO$_3$, such as the dominant valence states and the segregation coefficient. The segregation coefficient describes the fraction of Fe in the growth melt that is incorporated into the crystal lattice. Both the valence and the nature of the impurity are important for understanding impurity states and their relation to the reduction state. By "impurity nature" we mean the type of defect that is formed by the impurity.

Iron is incorporated into the crystal lattice either by substituting for niobium (Nb^{5+}) or for potassium (K$^+$). Electron paramagnetic resonance experiments [35–37] show that the angular variation of the Fe^{3+} lines is well described by the Hamiltonian for an Fe^{3+} ion substituting a Nb^{5+}. In a perfectly ionic crystal (K$^+$Nb^{5+}O$_3^{2-}$) the question of charge compensation arises. One possibility is for an oxygen vacancy V$_O^{2+}$ to be formed near the Fe^{3+} ion such that the defect complex Fe^{3+}-V$_O^{2+}$ is formed [38]. The measurements reported by Michel-Calendini et al. [35] do not indicate such defects. However, those authors point out that their apparatus was unsuited to detecting the influence of the V$_O^{2+}$ on the Fe^{3+} resonance lines. Theoretical calculations on the geometrical microstructure of Fe^{3+}-V$_O^{2+}$ [39] show that this is an energetically favorable way in which charge neutrality is maintained. The Fe^{3+} ion moves away from the V$_O^{2+}$ ion by 0.03–0.04 nm relative to the center of the oxygen octahedra. As a result, the defect energy is decreased by 1.0–2.7 eV compared to when the Fe ion is centered in the oxygen octahedra. The range of displacements and energies is caused by the different assumptions as to which ions may move away from their positions in the ideal lattice.

That Fe^{x+}-V$_O^{y+}$ complexes, with $x = 2, 3, 4$ and $y = 1, 2$, have lower energies than isolated Fe^{x+} is also supported by theoretical calculations performed for cubic ABO$_3$ perovskites [40] in general and for the special case of SrTiO$_3$ [41]. Detailed molecular orbital calculations for SrTiO$_3$ [41] show that local charge compensation by a V$_O^{2+}$ center lifts the degeneracy of the Fe^{2+} and Fe^{3+} electronic states and lowers their energy.

TABLE 7.2. Ionic radii [45] of the
elements of K, Nb, and Fe. The values
were determined for crystals formed by
the elements

Element	Valence	Radius [nm]
K	1+	0.133
Nb	5+	0.069
	4+	0.074
Fe	3+	0.064
	2+	0.074

Iron substitution of K^+ in $KNbO_3$ is not supported by any experimental evidence. This is in contrast to the case of $KTaO_3$, for which EPR studies show
Fe substituting K [42, 43]. Theoretical calculations for $KNbO_3$ using muffin tin
potentials and a linear combination of atomic orbitals [44] indicate that Fe substitution of K is more favorable than substitution of Nb. However, the models used
in [44] are questionable. In fact, the theory predicts that the orthorhombic $KNbO_3$
lattice should be unstable. In order to compute the defect energies in $KNbO_3$ at
room temperature the lattice had to be stabilized artificially.

The ionic radii of the atoms of Fe-doped $KNbO_3$ indicate which are the most
likely substitutions. A comparison of the ionic radii of K, Nb, and Fe [45] is shown
in Table 7.2. Iron and Nb ions have similar radii, while the K ion is much larger.
Because of lattice stability and because the bonds should have a certain degree
of covalency in order to stabilize the ferroelectric phase [46], Nb substitution by
Fe is the most likely defect type in $KNbO_3$.

Iron impurities in ABO_3 crystals can occur as Fe^{2+}, Fe^{3+}, and Fe^{4+}. In $KNbO_3$,
Fe^{3+} has been observed directly with EPR [10, 47]. Optical studies of n-type
$KNbO_3$ that were correlated with EPR measurements infer the existence of Fe^{2+}
[11, 47]. The thermodynamic stability of Fe^{4+} in $KNbO_3$ has not been verified
experimentally. It is a stable species in $SrTiO_3$ [48, 49]. In $BaTiO_3$, Fe^{4+} is thought
to play a role in the photorefractive effect [50, 51].

The concentration of a dopant in a crystal is determined by the dopant level in
the growth melt and by the segregation coefficient. The segregation coefficient is
the ratio between the concentration of a dopant in the crystal and the concentration
of the dopant in the growth melt. The concentrations are generally given as atomic
ratios, i.e., parts per million (ppm). The segregation coefficient for Fe in $KNbO_3$
was determined when the first optical-quality crystals were grown in 1976 [10].
Plasma mass spectroscopy yielded a value of $5 \pm 1\%$. In the meantime, a number
of analyses have been performed on $KNbO_3$ crystals by a number of different
institutions but with inconsistent results. Newer, more careful, measurements
[52] show that the segregation coefficient is $1.7 \pm 0.5\%$ for Fe concentrations in
the growth melt ranging from 1000 at. ppm to 30,000 at. ppm. The experimental
method consisted in spark-source mass spectroscopy and secondary-ion-emission
mass spectroscopy. In this chapter we shall use this value when estimating the

absolute Fe concentration in KNbO$_3$ for a given doping level of the growth melt.

A segregation coefficient of 1.7% means that for a 1000 at. ppm Fe-doped melt the resulting crystals contain 17 at. ppm Fe. This translates into a total Fe concentration of $2.6 \cdot 10^{23}$ m^{-3}. The average separation of Fe impurities in such a crystal is 39 lattice constants, e.g., 22 nm, when viewed along the a direction.

7.3.2 Doping with Rh, Ni, Cu, Ce, Mn, and Co

Steady improvement in semiconductor lasers operating in the near-infrared spectral region has stimulated research for using photorefractive materials working at these wavelengths. Several attempts have been reported to extend the photorefractive response beyond the visible range in KNbO$_3$ crystals. Ma et al. [53] reported on doping experiments with different elements. They investigated the sensitivity of polycrystalline KNbO$_3$ powder in the near infrared and obtained particularly promising results when doping with nickel. In our laboratory we have followed the approach that all possible dopants should be tried to induce the appropriate electron energy levels in KNbO$_3$.

We grew single crystals doped with different concentrations of Rh, Ni, Cu, Ce, Mn, and Co ions [12, 14–17, 54]. Large good-quality strain-free crystals were successfully grown in most cases after several adjustments of the melt composition. In the case of Ce doping with concentrations higher than 1 ppm the growth failed. Despite the low dopant concentration in the melt, the crystals show strong photorefractive properties, which are comparable to 1000 atomic ppm Fe-doped crystals. We assign this observation to a higher segregation coefficient of Ce. Segregation coefficients for Mn and Rh doping were determined to be (5.2 $\pm$ 0.7)% and (3.1 $\pm$ 1)% [52]. From the oriented boules, rectangularly shaped crystals were cut and oriented, and the faces normal to the a- or b-axis were polished. The polished Rh-doped crystals are reddish-brown in color, the Ni-doped ones are dark green, the Cu-doped ones are light gray, and the Ce, Co, and Mn-doped crystals show light-orange coloration.

In order to characterize the samples, we performed optical, electrical, and photorefractive experiments. The optical absorption spectra were measured in the wavelength range between 300 and 2400 nm. In all the crystals there is an increase in the absorption coefficient α between 400 and 1000 nm. Values between 1 and 2 cm^{-1} at 488 and 515 nm for the Ce- and the Co-doped samples, respectively, are higher than typical values of $\alpha = 0.3$ cm^{-1} for moderately doped Fe:KNbO$_3$ samples. At 700 nm, absorptions of 0.15 and 0.2 cm^{-1} were observed for Ce- and the Co-doped samples, respectively. In the Ni-doped samples several absorption bands can be seen centered at 500, 630, and 900 nm. In Cu-doped crystals, the absorption bands are also present but less pronounced. The short-wavelength band edge moves toward longer wavelengths as the dopant changes from Mn, Rh, Fe, and Mn–Rh. Samples doped with Mn present two main absorption regions, one centered near 450 nm (2.75 eV, $\alpha = 1.2$ cm^{-1}), the other at 700 nm (1.77 eV, $\alpha = 0.25$ cm^{-1}). Figure 7.6 shows absorption spectra of a pure as grown KNbO$_3$

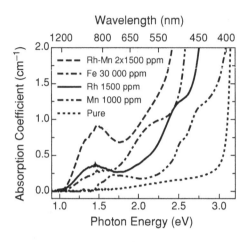

FIGURE 7.6. Absorption spectra of a pure as-grown $KNbO_3$ crystal and of crystals doped with Fe, Mn, Rh, and Mn-Rh. Spectra were measured with light polarized along the a-axis except for Rh-Mn, where the light was polarized parallel to the b-axis.

crystal and of several doped ones, which were found to be the most promising for photorefractive applications. As shown in the figure, the absorption band of Mn doped $KNbO_3$ is rather broad, well defined, and extends until 800 nm (1.55 eV), where the measured absorption coefficient is 0.15 cm^{-1}. Crystals doped with Rh present an enhanced absorption around 860 nm (1.44 eV). This absorption increases with the doping level, and for concentrations of 1500 ppm of Rh, an absorption coefficient of 0.4 cm^{-1} is measured with light polarized along the a axis. In crystals with a high concentration of Fe, values of $\alpha = 0.3$ cm^{-1} are measured at 860 nm. Mn–Rh-doped samples present a broad absorption band with a maximum absorption coefficient $\alpha = 1$ cm^{-1} near 830 nm (1.49 eV), as can also be seen in Figure 7.6.

7.3.3 Reduction State of Doped KNbO₃

The term "reduced" means that a large fraction of impurities and/or dopants in the crystals are in a valence state that is reduced by one unit charge relative to the untreated samples. For example, the dopant Fe in an untreated $KNbO_3$ crystal is mostly in the Fe^{3+} state, whereas in a reduced sample it is mainly present in the Fe^{2+} state. A change of impurity states leads to electrons as the main photoexcited charge carriers, while unreduced crystals show a hole-dominated charge transport. A detailed thermodynamic model of the reduction state of Fe-doped $KNbO_3$ can be found in [55].

Reduced $KNbO_3$ shows enhanced nonlinear properties, e.g., faster two-beam coupling and phase conjugation. In addition, such crystals show a pronounced nanosecond photorefractive response to short-pulse excitation [56] that could be applied to fast optical correlators. The speed of the photorefractive effects involved in applications such as optical computing (correlation and associative memory functions) is critical to the feasibility of such designs.

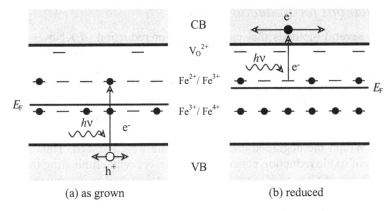

FIGURE 7.7. Ionization levels in an as-grown (a) and in a reduced (b) Fe-doped KNbO$_3$ crystal. The Fermi energy E_F is raised by introducing oxygen vacancies that release two electrons through thermal ionization. The doubly ionized vacancies are designated by V_O^{2+}. In reality, E_F is pinned to one of the impurity levels, but in the figure it is shifted slightly for clarity. Photoexcitation in the unreduced crystal generates holes in the valence band (VB), while electrons in the conduction band (CB) are generated in the reduced crystal.

Oxygen is present in KNbO$_3$ as O^{2-} with saturated bonds oriented along the octahedra edges. The double negative charge of the oxygen is present only in an ideally ionic crystal. Any covalent contribution to the lattice bonds will decrease the charge on the oxygen. When oxygen is removed, it forms O$_2$ molecules in the surrounding atmosphere and the crystal remains electrically neutral. The oxygen vacancies (V$_O$) act as shallow donors containing two electrons with typical thermal ionization energies on the order of $\approx$0.1 eV [50]. These localized states are formed by Nb^{5d}-V$_O$ complexes [40]. Figure 7.7 shows a schematic model of the ionization levels in an as-grown and in a reduced Fe-doped KNbO$_3$ crystal.

In Figure 7.7 only two possibilities for optical charge excitation are shown. This simplification is used to make the fundamental difference between reduced and unreduced Fe-doped KNbO$_3$ clear. If the photon energy is sufficient, multiple excitation channels may occur. For example, electrons may also be excited in unreduced or reduced crystals by ionizing a filled Fe^{3+}/Fe^{4+} level. Conversely, holes can also be generated in a reduced crystal. Other possible excitations include hole generation in unreduced Fe-doped KNbO$_3$ by exciting an electron into an empty Fe^{3+}/Fe^{4+} level.

The Fermi level is determined by the filling of electron states. The electrons are produced by thermal ionization of oxygen vacancies (V$_O$). The electrons are trapped by the lowest free orbital of Fe^{3+}, resulting in Fe^{2+}. At room temperature the oxygen vacancies V$_O$ are thermally ionized and form V$_O^{2+}$, so two electrons are available for reduction per vacancy. Thus, the removal of oxygen from Fe-doped KNbO$_3$ increases the concentration of Fe^{2+} and decreases that of Fe^{3+}.

7.3.4 Methods for Reducing KNbO₃

There are several methods that can be used for reduction of KNbO₃ crystals. These are electrochemical reduction [11, 18, 57], ion implantation [58–60], reduction by hydrogen loading [10], and high-temperature gas-induced reduction. A large body of knowledge on reducing BaTiO₃ has been gathered since the 1950s [61–66]. Those studies were aimed at optimizing the performance of BaTiO₃ capacitors and were mainly performed on ceramic or multidomain samples. At the time, capacitors degraded after a certain time due to the formation of conductive channels. Within the degraded areas the crystals were reduced. Thus the earliest studies of oxide reduction concerned themselves with eliminating this effect. From the point of view of photorefraction, crystal reduction is a positive effect. It is correlated with the transition from p-type to n-type conductivity and with an increase of the effective densities of the photoactive centers.

7.3.5 Electrochemical Reduction of KNbO₃

The knowledge that was acquired on BaTiO₃ capacitors became useful when KNbO₃ was developed in the 1970s as a nonlinear optical material. Electrochemical reduction relies on passing a current parallel to the polar axis through the crystal [10]. The crystal is heated close to the orthorhombic–tetragonal transition at 218°C and isolated from oxygen by placing it in a vacuum [11, 67] or in silicon oil [18]. Initially, for a given voltage the current through the crystal increases until, after about ten days, it stabilizes. As expected, the treated samples are n-type semiconductors with a corresponding increased photorefractive speed [57].

The physical processes involved in electrochemical reduction are not definitely known. One explanation is that the current passing through the crystal injects electrons that are trapped in deep traps [18]. A better explanation is that oxygen vacancies drift in the applied external field and accumulate near the cathode. Additional oxygen vacancies may be created when the sample is insulated from air at the elevated temperature. This model is compatible with the empirical fact that the crystals have to be insulated from oxygen during reduction. The high density of the accumulated oxygen vacancies would be destroyed by reoxidation from the surrounding air. An additional effect that occurs is that planar defects are induced by the electric current [68]. The planar defects are formed by two-dimensional crystal defects. They are semimetallic and give rise to channel formation [69, 70] and to large conductivities.

The big advantage of the electrochemical reduction treatment is that the reduction state of the optical quality samples can be monitored during treatment. As will be seen in the next chapter, the optimal degree of reduction for a given application is achieved by a tradeoff between a fast photorefractive response and a large two-beam coupling gain. The disadvantages arise from the unidirectional nature of the treatment and because of channel formation. The homogeneity of the optical samples is decreased after such treatment [71]. In addition, the reproducibility of the method was difficult to demonstrate because of the importance of the crystal domain structure. However, the very promising features of reduced KNbO$_3$ motivated further research into other methods of reduction.

7.3.6 Ion Implantation

Helium-ion implantation was first used to produce photorefractive waveguides in BaTiO$_3$ [72]. As part of the research into producing waveguides in KNbO$_3$, it was found that He$^+$ implantation reduced the crystals [73]. The He$^+$-implanted waveguides were n-type conductors, and the photorefractive effect was strong right up to near-infrared wavelengths. More recently, proton implantation has been developed as a reduction method that is ideally suited to waveguide applications at telecommunications wavelengths [58–60]. Strong photorefraction was observed at wavelengths as long as 1550 nm. The photorefractive sensitivity at 860 nm of proton-implanted waveguides approaches that of high-temperature gas-reduced rhodium-doped KNbO$_3$, as will be discussed later.

The current state of the art uses protons with energies of 3 MeV and doses of typically $3 \cdot 10^{20}$ m^2. The result is a layer with a thickness of typically 55 µm, which is terminated by a low-index barrier at the depth where the protons are stopped. The implanted crystals are reduced because the protons generate oxygen vacancies when they collide with the lattice oxygen atoms.

The advantage of reduction by proton implantation is that the reduction degree can be set by the implantation dose. Although research is continuing, there is no a priori reason why the waveguides should have inhomogeneous photorefractive properties. The disadvantages are twofold: implantation is currently unsuited to bulk crystal applications and the electrical properties of the waveguide are different from those of the rest of the crystal. As a result, the benefits of applying electric fields cannot be taken advantage of.

7.3.7 High-Temperature Reduction Methods

Two high-temperature reduction methods are available for KNbO$_3$: hydrogen loading and oxygen-vacancy creation in a reducing atmosphere. Hydrogen loading leads to OH$^-$ formation on the oxygen octahedra edges [74] with one electron available for impurity valence reduction per OH$^-$. Hydrogenated crystals have a blue color and show an increased dark conductivity. This reduction method was first applied to KNbO$_3$ by Flückiger [10], but no systematic data are available on the treated crystals. The hydrogen-loading approach to reduction has the advantage that many samples that were cut from a single boule can be treated in different ways. The main problem of treating single samples cut from boules are the phase transitions. The integrity of the samples after heating and cooling could be ensured by multiple heating coils to accurately control the thermal gradient during the phase transitions [75]. Because the treated crystals have to be poled in general, reduction in hydrogen is problematic because of the large dark conductivity. An advantageous tradeoff between the degree of reduction and the dark conductivity might be achieved. To this end a careful control of the hydrogen content would be necessary.

Reduction by oxygen-vacancy creation at high temperatures in a reducing atmosphere was first applied to photorefractive $BaTiO_3$ [76]. A theoretical point defect model that gave a good approximation of the observed photorefractive performance was developed by Wechsler and Klein [50]. However, the difficulties in reproducible doping of $BaTiO_3$ and the slow photorefractive response even of reduced crystals meant that interest in this line of research faded. Later, rhodium-doped $BaTiO_3$ was reduced and was found to have advantageous photorefractive properties at visible and near-infrared wavelengths [77–79].

The high-temperature gas-induced reduction that we use for $KNbO_3$ relies on producing oxygen vacancies that act as electron donors and reduce the valence of the impurities. The oxygen vacancies are produced at the surface of the crystal by exposing it at $900°C$ to an atmosphere with a low oxygen partial pressure. Diffusion causes the vacancies to spread across the whole volume. After the treatment the crystal is cooled down in the same atmosphere. The faster the cooling rate, the higher the effective treatment temperature. Because the reduction state decreases with decreasing treatment temperatures, the fastest possible cooling rates are desirable.

An advantage of the high-temperature gas treatment was found to be that it is ideally suited to the mass production of homogeneous bulk crystals. The photorefractive properties of reduced Fe-doped $KNbO_3$ are reproducible and homogeneous. An optimal effective trap density can be achieved. The photorefractive response time is shorter than in unreduced $KNbO_3$, although not as fast as that observed in electrochemically reduced crystals. The disadvantage is that the degree of reduction is limited by the requirement that the crystals be poled after treatment. This limits the maximal dark conductivity that can be allowed and also limits the photoconductivity and hence the speed of the photorefractive response.

From the point of view of crystal treatment with the aim of optimizing the photorefractive parameters, the theoretical model [55] predicts that the highest sensitivity is achieved for either extremely reduced or extremely oxidized crystals. Because the dark conductivity is large in such extreme cases, a suitable compromise must be found by choosing intermediate reduction states.

7.3.8 Postgrowth Reduction Treatment

This section describes the high-temperature gas-induced reduction treatment of $KNbO_3$. We chose to reduce the raw boules while they were still in the growth furnace because this eliminates the problem of carefully controlling the phase transitions. The treated boules are cooled through the phase transitions in a well-defined temperature gradient. The gradient is caused by the melt above which the boules are suspended [10]. The melt is hotter than the boule, so the phase transition starts at the top of the boule and moves down. After cooling to room temperature, optical samples can be prepared in the usual manner.

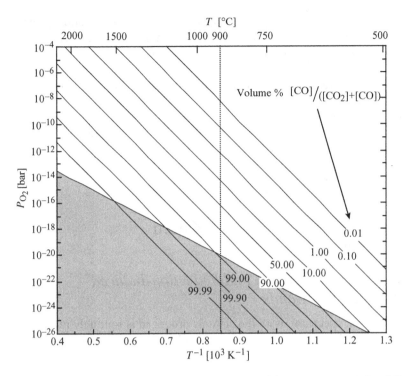

FIGURE 7.8. Oxygen partial pressure P_{O2} as a function of the temperature T for different CO:CO$_2$ ratios. The dashed line at 900°C indicates the treatment temperature that was used. In the shaded area carbon precipitation becomes significant.

The apparatus used for the reduction treatment consists of a growth oven and a gas control system. As well as the features necessary for growing KNbO$_3$ by the top-seeded-solution growth method, high-temperature gas-induced reduction needs controlled atmospheres and high postgrowth cooling rates. The oxygen partial pressure P_{O2} of the treatment gas is controlled by using a mixture of CO$_2$ and CO. This mixture has the advantage over pure Ar or H$_2$/Ar mixtures that its P_{O2} can be adjusted through the volume ratio of the two gases.

In the experimentally practicable temperature range of 750 to 950°C, P_{O2} can be varied from 10^{-6} to 10^{-20} bar. This allows a flexible control of P_{O2}. From an experimental point of view, it should be noted that the CO fraction in the mixture must be as small as 0.01% to achieve $P_{O2} \approx 10^{-6}$ bar at $T = 900°$C. A P_{O2} of 10^{-6} bar is the lower limit that is reached with purified Ar. Lower values of P_{O2} are reached with a higher relative concentration of CO. Figure 7.8 shows a plot of the oxygen partial pressure P_{O2} as a function of inverse temperature.

In the literature on gas-induced oxygen vacancies in BaTiO$_3$ [50, 76, 80], reduction starts for P_{O2} smaller than 10^{-2} bar, and completely reduced crystals are expected for $P_{O2} \approx 10^{-6}$ bar. Theoretical oxygen partial pressures P_{O2} as low as 10^{-18} bar can be reached at 900°C by using a mixture of CO/CO$_2$ that is 99% pure

CO. In the course of the experiments the $CO:CO_2$ ratio was gradually increased from 1:4000 up to 1:3. This corresponds to nominal values of $P_{O2} = 10^{-4}$ bar and $P_{O2} = 10^{-14}$ bar, respectively. The $KNbO_3$ boules were found to be reduced when a ratio of 4:30 ($P_{O2} = 10^{-13}$ bar) was used. In practice, the minimum oxygen partial pressure is limited by oxygen leaks from the air around the furnace.

Oxygen leaking into the oven and the effects of the cooling rate (as will be discussed later) mean that the effective oxygen partial pressure of the treatment is higher than that set by the $CO:CO_2$ ratio. The experimental results indicate that the effective oxygen partial pressure of a 4:30 mixture of CO/CO_2 in conjunction with a cooling rate of $355°C/hour$ is $\approx 10^{-6}$ bar.

7.4 Photorefractive Data on Reduced and Unreduced $KNbO_3$

7.4.1 Absorption Spectra and Reduction-Induced Absorption Bands

The absorption spectrum of all the crystals produced is routinely measured. The spectra of three reduced crystals with various Fe contents are shown in Figure 7.9.

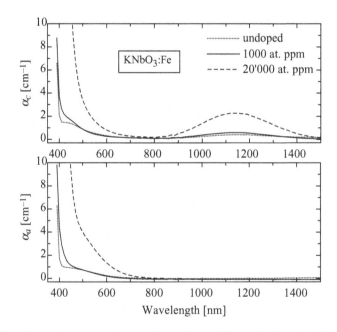

FIGURE 7.9. Absorption spectra for c and a polarized light of reduced $KNbO_3$ for three different Fe concentrations in the growth melt.

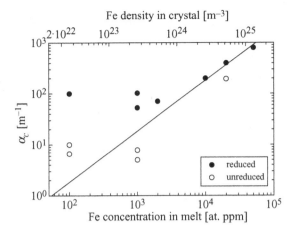

FIGURE 7.10. Absorption constant α_c at 488 nm for c polarized light in reduced and unreduced Fe-doped KNbO$_3$. The experimental errors are smaller than the points. The solid line indicates a linear dependence. Where more than one point is shown at a given dopant level it means that more than one boule was grown. This provides an estimate of the reproducibility of the current growth and reduction techniques.

Compared to unreduced KNbO$_3$, an additional absorption band is induced around 490 nm. The typical absorption band between 400 nm and 500 nm of reduced KNbO$_3$ [11, 9] is clearly visible. This is associated with an increase of the Fe^{2+} concentration at the expense of Fe^{3+}. In contrast to the earlier measurements, the crystals produced in these experiments also show reduction-induced absorption around 1200 nm. This interesting feature appears only for c polarized light and not for a polarized light (Figure 9). It does not lead to strong photorefraction at near-infrared wavelengths. Two-beam coupling experiments at 1064 nm imply that the photoconductivity at this wavelength is smaller than $10^{-11}\Omega^{-1}m^{-1}$. The absorption around 1200 nm must be due to an internal transition into an excited state of the Fe^{2+} impurities.

The absorption constant α_c at 488 nm is shown in Figure 7.10 as a function of Fe doping concentration [Fe^M]. The solid line describes the case in which α_c is strictly proportional to the intentional dopant level. Such a relationship does not hold in general. In reduced crystals when [Fe^M] $\geq$ 10,000 at. ppm a linear growth with the dopant level is approached. At lower dopant levels α_c is independent of [Fe^M] with $\alpha_c = 8 \pm 3$ m^{-1} for unreduced and $\alpha_c = 80 \pm 30$ m^{-1} for reduced crystals.

Figure 7.11 shows the effect of reduction on the absorption in Rh-doped KNbO$_3$. The inset provides a close-up view of the absorption at near-infrared wavelengths. The reduced Rh-doped crystal has less absorption at near-infrared wavelengths between 800 and 1100 nm, but its absorption in the visible part of the spectrum is increased by approximately a factor of ten.

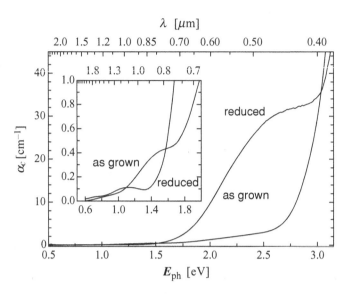

FIGURE 7.11. Absorption constants of Rh:KNbO$_3$ for c-polarized light, α_c, as a function of the photon energy E_{ph} (lower abscissa) and of the wavelength λ (upper abscissa). Note the smaller absorption of the reduced sample at near-infrared wavelengths.

7.4.2 Photorefractive Characterization Method

Characterization of the photorefractive properties is usually performed by observing two-beam intensity coupling in order to determine the effective trap densities N_{eff}, the charge diffusion lengths, and the dielectric response times. Photorefractive grating period and orientation are varied to get as much information as possible. The theory on photorefractive two-beam coupling is found elsewhere in this book series.

For beams entering the same crystal surface, a practical lower limit of the achievable grating period for a wavelength of 488 nm is $\Lambda_g = 0.28$ μm. Then the beams have an external full crossing angle of 132°, and if they are p polarized, they enter the crystal under the Brewster angle. The limit of $\Lambda_g = 0.28$ μm can be extended either by coupling the light into the crystal through prisms or by using index-matching fluid. In order to measure the effective trap density, the two-beam coupling gain must be limited by the trap-limited field E_q,

$$E_q = \frac{eN_{eff}}{2\pi\,\varepsilon_0\varepsilon^{eff}} \cdot \Lambda_g, \tag{7.1}$$

and not by the diffusion field E_D,

$$E_D = \frac{2\pi k_B T}{e} \frac{1}{\Lambda_g}. \tag{7.2}$$

FIGURE 7.12. Two-beam coupling geometry with antiparallel pump and probe beams used for measuring N^*_{eff}. The p-polarized beams enter the sample under the Brewster angle $\theta_B \approx 66°$ and so the grating wave vector is inclined by 66° relative to the c-axis, which is defined as parallel to the spontaneous polarization P_S. In this example the crystal is a b plate, which means that the polished faces are perpendicular to the b-axis.

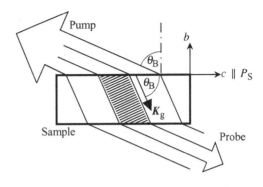

Here e denotes the elementary charge, ε_0 is the electric constant, ε^{eff} is the effective dielectric constant, k_B is the Boltzmann constant, and T is the absolute temperature. If $\Lambda_g = 0.28$ μm is the smallest measured grating period, the upper limit of N_{eff} (defined by $E_q = E_D$) is $N_{eff} = 2.5 \cdot 10^{22}$ m^{-3}. This limit is calculated under the assumption that the effective dielectric constant is $\varepsilon^{eff} = 35$.

Much shorter grating periods of $\Lambda_g = 0.11$ μm are reached for a wavelength of 488 nm with antiparallel beams [81]. Then the upper limit of measurable effective trap densities is $N_{eff} = 3.7 \cdot 10^{24}$ m^{-3}, when $\varepsilon^{eff} = 800$. In KNbO$_3$ crystals $\Lambda_g = 0.11$ μm is produced by p-polarized beams entering opposite crystal surfaces under the Brewster angle of 66°. A drawing of this geometry is found in Figure 7.12.

The signal beam is detuned by a few degrees from the pump path so that self-pumped phase conjugation of the pump does not distort the measurements. For crystals cut normal to the axes, the grating is tilted by 66° relative to the c-axis. In this geometry the effective trap density is calculated from the gain Γ_{exp} with

$$N^*_{eff} = \frac{\Gamma_{exp}\lambda_0\varepsilon_0\varepsilon^{eff}}{en^3 r^{eff}\Lambda_g}, \tag{7.3}$$

where λ_0 is the vacuum wavelength of the laser beams and n is the refractive index. The disadvantage of such "single-shot" measurements is that electron–hole competition is not detected. For this reason the N^*_{eff} determined from antiparallel two-beam coupling is understood as a lower limit unless bipolar charge transport is insignificant. When the single-charge-carrier, single-impurity-species model is accurate and when absorption gratings and losses are insignificant, then $N_{eff} = N^*_{eff}$.

The experimental values of N^*_{eff} as a function of [Fe^M] are shown in Figure 7.13. For [Fe^M] < 2000 at. ppm the effective trap density N^*_{eff} rises with [Fe^M]. At these low dopant levels the difference in N^*_{eff} between the unreduced and the reduced crystals is typically a factor of three to ten. When [Fe^M] > 2000 at. ppm the effective trap density increases only slowly with [Fe^M] in the reduced crystals. There N^*_{eff} tends to saturate at a value of $\approx 2 \cdot 10^{23}$ m^{-3}. This saturation

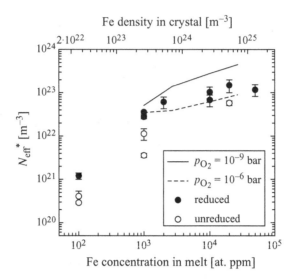

FIGURE 7.13. Effective trap density N_{eff}^*, which was measured with antiparallel beams in reduced and unreduced Fe-doped $KNbO_3$. The solid and dashed lines are the result of a numerical simulation of the thermodynamic point defect model from [54].

effect provides an estimate of the density of oxygen vacancies that are produced by the reduction treatment. For large Fe concentrations N_{eff}^* is limited by the density of charge donors. This in turn corresponds to twice the density of doubly ionized oxygen vacancies. Therefore the density of doubly ionised vacancies is $\approx 10^{23}$ m^{-3}.

The reduction-induced increase of N_{eff}^* is between a factor of two and three at $[Fe^M] = 20,000$ at. ppm. The lines shown in Figure 7.13 are the results of numerical solutions of the thermodynamic point defect model. The computed values of N_{eff} for the transition $Fe^{2+} \rightarrow Fe^{3+} + e^-$ are plotted at two treatment oxygen partial pressures P_{O2} and for a treatment temperature of 900°C. Two partial pressures are used so that an estimate of the effective oxygen partial pressure can be made. The ratio of 3:10 of the CO/CO_2 mixture corresponds to a nominal oxygen partial pressure of 10^{-14} bar. As was pointed out earlier in this chapter, oxygen leaks and the effect of the cooling process increase the actual oxygen partial pressure. The theory indicates that for a treatment temperature of 900°C the effective oxygen partial pressure is between 10^{-9} bar and 10^{-6} bar.

The photorefractive response times τ_{pr} at 1 W cm^{-2} are shown in Figure 7.14. In the antiparallel two-beam coupling geometry the photorefractive response time τ_{pr} is smaller than 1 s in general. The fastest time of 27 ms is observed in a highly reduced, nominally pure crystal. Figure 7.14 shows that the reduced crystals respond faster than the unreduced crystals if $[Fe^M] < 2000$ at. ppm. The larger the value of $[Fe^M]$, the slower the photorefractive response.

FIGURE 7.14. Photorefractive response time τ_{pr} measured at 1 W
cm^{-2} with antiparallel beams in
reduced and unreduced Fe-doped
KNbO₃. The solid line is a guide
to the eye and indicates a linear
dependence. The experimental errors
are within the point size.

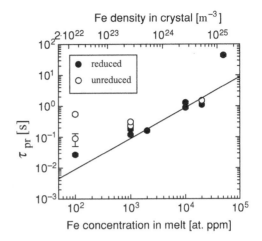

7.4.3 Charge Mobility Determination by Short-Pulse Experiments

The drift mobility of photoexcited charges in the two bands is considered to be an intrinsic material property. In addition to the band mobility, the trap limited mobility is also important and depends to some extent on crystal extrinsic properties. The difference between the two mobilities is caused by the presence of shallow levels into which the free charges may be trapped and thermally reexcited. This process may occur several times before the charges finally recombine into deep levels.

We have undertaken several experiments in order to determine the absolute values of the mobilities of electrons and holes in KNbO₃ [56, 82, 83]. Pulsed photorefractive experiments differ from the cw case in that charge transport is separated in time from the excitation process: charges diffuse in the dark. This fact simplifies the analysis of these measurements. Two crystal properties dominate the grating dynamics: the mobility and the lifetime of the photoexcited charges. Pulsed pump and probe techniques and continuous read-out of short-pulse-induced gratings were used to study the grating evolution over some microseconds, and this yielded information on shallow trap densities and, by varying the crystal temperature, on their thermal ionization energies.

An overview of the multiple grating dynamics that are observed in Fe-doped KNbO₃ is given in Figure 7.15. A sample that was reduced in CO/CO₂ is compared to an unreduced sample at two different temperatures. Both samples where grown from melts that contained 1000 at. ppm of Fe. At room temperature (upper graph in Figure 7.15) the diffraction efficiency in the reduced crystal shows two regimes: the initial nanosecond response and an additional buildup after 1 ms. The nanosecond response is caused by electron diffusion in the conduction band

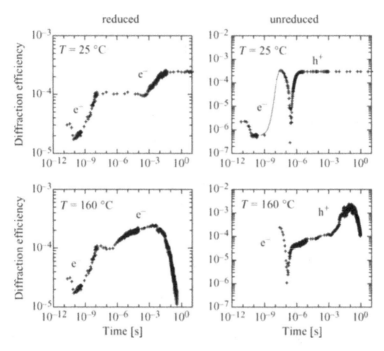

FIGURE 7.15. Grating dynamics after excitation by 75 ps pulses at 532 nm in 1000 at. ppm Fe-doped $KNbO_3$. The grating period was 2 μm and the writing energy density was 2.5 mJ cm^{-2}.

[82, 83]. This type of grating has previously been observed in electrochemically reduced $KNbO_3$. The electron lifetime of 4 ns means that at the grating period of 2 μm that was used for the measurements, diffusion in the conduction band is stopped before the electron distribution is averaged out. After 4 ns the electrons are trapped. Thermal reexcitation and further electron diffusion causes the millisecond buildup. When the temperature is increased (lower graph in Figure 7.15), the nanosecond response remains unchanged but the millisecond response becomes faster and the buildup starts after 1 μs at 160°C. The grating decays within one second.

   In the unreduced crystal the diffraction efficiency builds up in less than 50 ns, but the nanosecond response was not observed directly. After the initial buildup, the diffraction efficiency drops sharply to the detection limit after 0.3 μs. This is caused by electron–hole competition: both species are excited by the writing pulses. The initial grating builds up because of photoexcited electron diffusion, but the photoexcited holes generate a compensating space-charge field. The electrons have a higher mobility than the holes so they diffuse faster. The distribution of ionized electron donors causes the initial response. As the holes diffuse, the space-charge field generated by the ionized hole donors starts to build up. This field has the opposite sign to that of the ionized electron donors. Finally, the space-charge field caused by hole diffusion reaches an equal magnitude as that of the electron donors. At this point no net space-charge field is present and the diffraction efficiency vanishes. The holes continue to diffuse and a space-charge field with the opposite sign to that of the initial field builds up. This causes the diffraction efficiency to recover. The holes are trapped after ≈ 4 µs and do not diffuse any further at room temperature. The grating amplitude remains constant for more than 10 s. When the temperature is increased (lower graphs in Figure 15), the unreduced sample shows an additional grating buildup. At 160°C this starts after 10 ms. The grating finally decays after about 10 s.

Usually, electron–hole competition is inferred from cw two-beam coupling experiments. Pulsed experiments allow direct observation of simultaneous electron and hole excitation and transport, and the change of sign of the space-charge field amplitude is observed in real time.

We discuss here in more detail the measurements of the buildup-time constant of the hole, dominated grating over a range of grating periods. For charge carriers in a mobile state that has a limited lifetime, the time τ that they spend diffusing can be expressed as

$$\frac{1}{\tau} = \frac{1}{\tau_D} + \frac{1}{\tau_0}, \tag{7.4}$$

where τ_D is the diffusion time and where τ_0 is the lifetime of the photoexcited holes. The diffusion time is given by

$$\tau_D = \frac{e\Lambda_g^2}{4\pi^2 \mu_h k_B T}. \tag{7.5}$$

In order to measure the mobility μ_h it is necessary to determine the diffusion time τ_D.

The microsecond grating buildup that is observed was measured at grating periods ranging from 0.5 µm to 10 µm. The energy densities of the writing pulses were between 0.5 J m^{-2} and 1.0 J m^{-2}. At the energy density levels used for these experiments the grating buildup times are not affected by saturation effects, i.e., they do not depend on the writing fluence.

According to (7.4), the buildup time τ is independent of the grating period Λ_g when $\tau_D \gg \tau_0$. This corresponds to the case in which charge diffusion is stopped by the limited lifetime τ_0 before the free-charge distribution is homogeneous. At short grating periods, when $\tau_D \ll \tau_0$, the diffusion time determines the grating buildup time. In this case the free charges diffuse completely before their lifetime is over. In this limit the grating buildup time τ depends quadratically on the grating period Λ_g. The experimental values of τ are shown in Figure 7.16 together with a numerical fit to (7.4).

The buildup time of the space-charge electric field is 0.35 ± 0.05 µs at large grating periods. It decreases to 0.18 ± 0.03 µs at a grating period of 0.5 µm. The numerical fit with the hole lifetime τ_0 and the hole mobility μ_h as the free parameters results in $\mu_{h,\text{trap}} = (6 \pm 3) \cdot 10^{-7}$ m^{2}V^{-1}s^{-1} and lifetime $\tau_0 = (0.37 \pm 0.03)$ µs.

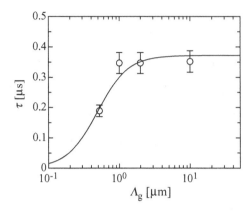

FIGURE 7.16. Buildup time of the space-charge electric field caused by hole diffusion. The solid line shows a numerical fit to equation (7.4).

7.4.4 Thermal Ionization Energies of Shallow and Deep Traps

By varying the temperature of the crystal between room temperature and $180°C$ the thermal activation energies for the different components of the space-charge field dynamics can be measured. In order to extract the time constants of the electric space-charge field, its dynamics were modeled as monoexponential functions of the form $(1 - \exp(-\frac{t}{\tau}))$ for the buildup and $\exp(-\frac{t}{\tau})$ for the grating decay.

In the unreduced $KNbO_3$ three different regimes occur, as explained and shown previously in Figure 7.15. These are the buildup on submicrosecond time scales and further buildup on millisecond time scales that is followed by the final decay after some milliseconds to seconds. We are interested here in the last two effects because they are caused by thermal reexcitation of trapped holes. The temperature dependence of the buildup and decay times at a grating period of 2 μm is shown in Figure 7.17. The time constants that are determined by diffusion of thermally excited charges can be described by an Arrhenius law. The straight lines in Figure 7.17 show the numerical fit of the thermal activation energies for the three time constants.

On the left side of Figure 7.17 an activation energy of 0.12 ± 0.02 eV is identified. This is the activation energy of the hole lifetime τ_0 in (7.4). The lifetime is determined by recombination into traps that have an activation energy of 0.12 eV. At a grating period of 0.52 μm the temperature dependence of the buildup time yielded the same activation energy. Because the diffusion time τ_{Di} contributes 50% of the total buildup time at this grating period, we conclude that the hole mobility has the same activation energy. Such a shallow trapping level for holes in $KNbO_3$ was also identified by temperature-dependent EPR measurements [84]. An activation energy on the order of 0.1 eV was reported and the level was identified as O^{2p} states.

The right side of Figure 7.17 shows the time constants of the final (ms/s) grating dynamics. The correspondence with an Arrhenius law is less pronounced than for the microsecond time constant. The buildup time seems to be dominated by an activation energy of 1.2 ± 0.2 eV at higher temperatures and by an energy

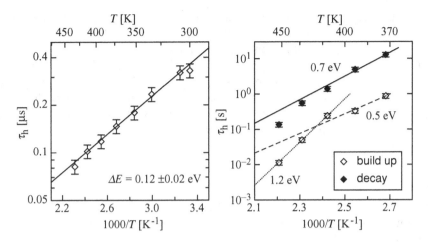

FIGURE 7.17. Buildup and decay times of the space-charge fields caused by hole diffusion in unreduced KNbO₃ at a grating period of 2 μm. The left graph shows the time constant of the space-charge field that compensates the initial electron diffusion grating. The right graph shows the slower buildup and final decay times of the grating.

of 0.5 ± 0.1 eV at lower temperatures. The measured decay times indicate an activation energy of 0.7 ± 0.1 eV. Because of the limited temperature range, a more definite resolution of the possible different activation energies was not possible. We therefore conclude that there are probably two levels acting as deep traps. The average activation energy of these levels is 0.8 ± 0.3 eV.

Temperature-dependent measurements were also carried out in reduced Fe-doped KNbO₃. The grating dynamics after the nanosecond buildup show an additional buildup that occurs on submillisecond time scales (see Figure 7.15). The final decay times are on the order of seconds at room temperature. Figure 7.18

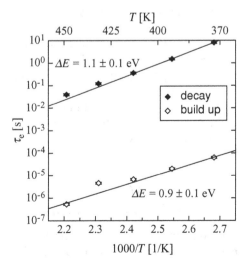

FIGURE 7.18. Buildup and decay times as a function of temperature in reduced Fe-doped KNbO₃. The solid lines show an Arrhenius law. The buildup corresponds to the second regime of the dynamics that are shown in Figure 7.15 and is caused by the diffusion of electrons that are thermally excited from deep traps.

shows the measured buildup and decay times as a function of the temperature. The measurements were carried out at a grating period of 2 μm and at writing energy densities where saturation effects do not occur.

The grating buildup and decay are caused by the diffusion of electrons that are thermally excited from their deep traps. The deep traps have an activation energy of 1.0 ± 0.1 eV. In highly reduced crystals that were prepared by electrochemical reduction a shallow trap level of 0.3 eV was identified [85]. This level leads to a submillisecond response. Its influence on the grating dynamics depends on the grating period and on the writing fluence.

7.4.5 Charge Mobility Anisotropy

In tetragonal barium titanate ($BaTiO_3$) the mobility of holes shows an anisotropy of about a factor of 20 between the two different crystallographic directions. Mahgerefteh [86] had earlier pointed out that this anisotropy has the same temperature dependence of the anisotropy of the dielectric permittivity, a fact that led us to consider a possible relation between the drift mobility and the dielectric permittivity. Several hypotheses have been examined to help us obtain an understanding of these phenomena, but none was able to present a complete and satisfactory explanation. Even the description of the charge transport by small-polaron hopping was not completely successful [86], although it was demonstrated that in $BaTiO_3$ this process is dominant [87].

We measured the anisotropy of the photoexcited charge-carrier mobilities in potassium niobate in order to understand better its photoconductive properties [88]. We determined the anisotropy of the mobility by a purely optical method. Details of this method are described in Chapter 2 of this book. In contrast to the direct photoconductivity measurements, our results are not influenced by the quality of contacts between crystal and electrodes or by surface conductivity, both of which are particularly critical in highly insulating materials like $KNbO_3$ and $BaTiO_3$.

The results for the mobility anisotropy for electrons are $\mu_a/\mu_c = 1.15 \pm 0.09$ and $\mu_b/\mu_c = 1.9 \pm 0.2$, while for holes they are $\mu_a/\mu_c = 1.05 \pm 0.06$ and $\mu_b/\mu_c = 2.9 \pm 0.3$ (please note that these mobility ratios are unfortunately exchanged in the abstract of [88] because of a typing error). These mobility ratios are valid for the trap-limited case because under the selected experimental conditions the typical relaxation times lie in the several-ms range. Measurements in several different samples had shown that the mobility ratios do not depend appreciably on the doping concentration. We explain this fact by noting that a concentration of the trapping centers on the order of 10^{-5} per unit cell influences the charge density $n(I)$ but only negligibly the charge-carrier mobility.

7.4.6 Charge-Mobility Determination by Interband Experiments

Additional information on the mechanism of the photoconductivity is obtained by measuring the dynamics of photorefractive gratings in the uv spectral region

TABLE 7.3. Comparison of the photorefractive parameters of Fe-doped KNbO$_3$ at 488 nm. The Debye screening lengths Λ_{Db} are computed from the effective trap densities N_{eff} and are given for a grating-wave vector parallel to the c-axis. L_D is the diffusion length, τ_{Di} is the dielectric relaxation time, σ_{phot} and σ_{dark} are the photoconductivity and the dark conductivity, respectively, and R is the ratio between the measured and theoretically predicted two-beam coupling parameter arising mainly from electron–hole competition. The largest observed sensitivities per incident and per absorbed unit energy density, S_{n1} and S_{n2}, respectively, are shown in the last two rows

[Fe] in melt state	none		1000 at. ppm		20,000 at. ppm	
	reduced	unreduced	reduced	unreduced	reduced	unreduced
$\alpha_c [m^{-1}]$	100 ± 10	6.6 ± 0.7	100 ± 10	7.8 ± 0.8	400 ± 50	200 ± 20
$N_{eff} [10^{21} m^{-3}]$	12 ± 2	0.29 ± 0.02	36 ± 2	3.6 ± 0.4	150 ± 50	58 ± 7
$\Lambda_{Db} [\mu m]$	1.3	2.6	0.23	0.73	0.11	0.18
$L_D [nm]$	120 ± 20	200 ± 20	62 ± 3	78 ± 6	76 ± 7	10 ± 5
$\tau_{Di} [s]$	0.01	0.3	0.08	0.8	0.1	2.7
$\sigma_{phot} [10^{-9} \Omega^{-1} m^{-1}]$	25 ± 2	0.40 ± 0.04	3.7 ± 0.1	0.37 ± 0.03	3.1 ± 0.3	0.11 ± 0.01
$\sigma_{dark} [10^{-9} \Omega^{-1} m^{-1}]$	<0.1	0.50 ± 0.08	<0.01	0.01 ± 0.005	0.3 ± 0.1	0.02 ± 0.01
R	0.4	0.8	0.85	0.6	0.85	0.8
$S_{n1} [10^{-9} m^2 J^{-1}]$	1600 ± 300	2.1 ± 0.2	44 ± 3	9 ± 1	23 ± 8	2.6 ± 0.4
$S_{n2} [10^{-11} m^3 J^{-1}]$	1600 ± 400	32 ± 5	44 ± 5	120 ± 30	6 ± 2	1.3 ± 0.3

[89]. This is done by changing the direction of the photorefractive grating and modifying other experimental conditions in a suitable way. The results allowed a direct comparison of trap-limited mobilities of electrons and holes in the b- and c-axis directions. They are $\mu_h^b / \mu_e^b = 2.0 \pm 0.4$ and $\mu_h^c / \mu_e^c = 1.25 \pm 0.15$, showing that trap-limited hole mobility is larger than trap-limited electron mobility.

7.4.7 Summary of Photorefractive Data

Table 7.3 summarizes the results of photorefractive characterization of Fe-doped KNbO$_3$. The high-temperature gas-reduction treatment increases the effective trap density N_{eff} three to ten times. The diffusion lengths L_D are not strongly influenced except for a dopant concentration of 20,000 at. ppm. The dielectric response times τ_{Di} decrease ten to thirty times as a result of the reduction treatment.

In the undoped, unreduced crystal σ_{dark} is comparable to σ_{phot} at 1 W cm^{-2}. For the most heavily doped crystals in Table 7.3 the experimental errors are large because the gain becomes unstable at low intensities. The high-temperature gas-induced reduction treatment increases the photoconductivity. The crystals are electron dominated, the photoconductivity is increased, and so is the effective trap density, but the dark conductivity is still low. A comparison of the photorefractive sensitivities in the last two rows of Table 7.3 summarizes the marked improvement of the photorefractive performance that is caused by the reduction treatment. The sensitivities per incident (S_{n1}) and per absorbed unit energy density (S_{n2}) were

TABLE 7.4. Photorefractive data for doped $KNbO_3$ at infrared wavelengths. The data for the visible light are added for comparison. Absorption constants α_c are determined for c-polarized light; N_{eff}^* are the effective trap densities. Because of the electron–hole competition, N_{eff}^* are lower limits. The dominant photoexcited charge-carrier types are given in parentheses; response times τ_{pr} are measured in the antiparallel beam geometry at an intensity of 0.75 W cm^{-2}. A photorefractive figure of merit S_{n2} is the photorefractive index change per absorbed unit energy

Dopant	λ [nm]	α_c [m^{-1}]	N_{eff}^* [10^{22} m^{-3}]	τ_{pr} [s]	S_{n2} [10^{-12} m^{3}J^{-1}]
Rh (reduced)	476	2880	12.3 (e$^-$)	0.8	2.8
3000 at. ppm	488	2790	13.7 (e$^-$)	1.0	2.7
	514	2340	9.6 (e$^-$)	0.3	7.9
	860	15	4.6 (h$^+$)[a]	4	67
	1064	11	2.6 (h$^+$)	400	0.6
Rh (as grown)	476	310	1.4 (e$^-$)	1.6	1.5
3000 at. ppm	488	280	2.4 (e$^-$)	2.0	2.4
	514	230	0.9 (h$^+$)	2.9	0.8
	860	38	1.6 (h$^+$)	1500[b]	0.006
	1064	15			
Rh (as grown)[c]	860	2	1 (h$^+$)	190	5.3
1500 at. ppm	1064	1	0.42 (h$^+$)	380	3
Fe[c]	514	110	2.4 (h$^+$)		
30,000 at. ppm	860	30	0.7 (h$^+$)	13	2.8
	1064	20	0.3 (h$^+$)	70	0.6
Mn[c]	860	16	0.08	3	1.9
1000 at. ppm	1064	9	0.08	25	0.5
Ni[c]	860	10	2.1	35	0.7
3000 at. ppm	1064	7	0.1	420	0.006
Mn-Rh[c]	860	30	2.9	184	1.1
1500	1064	30	2.3	250	0.4
Fe (H$^+$-impl.)[d]	860	70	0.33	0.4[e]	17[g]
1000 at. ppm	1064	90	0.38	0.6[f]	10[g]

[a] e$^-$ for $\Lambda_g > 0.47$ μm.
[b] Measured at 4 W cm^{-2}.
[c] Measured at 0.75 W cm^{-2}.
[d] Proton-implanted waveguide with a thickness of 55 μm.
[e] Measured at a grating period of 1.1 μm.
[f] Measured at a grating period of 1.3 μm.
[g] Estimated for the antiparallel beam geometry.

calculated for the largest observed two-beam coupling gains. An improvement of up to three orders of magnitude in S_{n1} is demonstrated. For S_{n2} the improvement is less dramatic because the absorption constants are increased by typically a factor of ten by the reduction treatment.

Photorefractive data on doped $KNbO_3$ with an emphasis on infrared sensitivity are given in Table 7.4. The Rh doping and reduction treatment is very useful, because it has a substantial effect on the photorefractive properties of KNbO$_3$. The photorefractive sensitivity at 860 nm (1.44 eV) is the highest in such crystals. It should be noted that the absorption at this wavelength is decreased by the reduction treatment but that the effective trap density increases. At 1.064 μm (1.16 eV) no significant gain could be measured in the as-grown crystal, whereas the reduced sample shows an appreciable response. Electron–hole competition is observed in the reduced Rh-doped crystal at a wavelength of 860 nm. For counterpropagating beams, holes are the dominant photoexcited species, as is the case for the as-grown sample. But for grating periods Λ_g larger than 0.47 μm, electrons are the dominant species in the reduced crystal. For $\Lambda_g \geq 1$ μm the space-charge-field amplitude calculated from the measured gain is 60 % of the diffusion field, as is expected in the presence of electron–hole competition. The data presented on 1500 at. ppm doped Rh:KNbO$_3$ have values somehow in between the values for as-grown and reduced material, as can be judged from the previous data in Table 7.4. The cause of such variations is unintentional electrochemical reduction during poling of samples, which has to be considered for precise determination of the reduction state.

Iron-doped KNbO$_3$ also shows a photorefractive response at near-infrared wavelengths. In an as-grown sample optimally doped with Fe the values at 860 nm are $N_{eff}^* = 0.7 \cdot 10^{22}$ m^{-3} and $\tau_{pr} = 13$ s, and at 1.064 μm They are $N_{eff}^* = 0.3 \cdot 10^{22}$ m^{-3} and $\tau_{pr} = 70$ s. For both wavelengths, holes are the dominant photoexcited species. The photorefractive sensitivity of the reduced Rh-doped crystal at 860 nm is 20 times larger than that of KNbO$_3$:Fe (3%), while at 1.064 μm the sensitivities are equal.

Apart from Rh and Fe, the most promising dopants are Mn, Ni, and Rh-Mn, with their data also listed in Table 7.4. The Mn-doped crystal has a relatively fast response time and small absorption constants but lacks sufficiently large effective trap densities N_{eff}^*. The Ni-doped crystal at 860 nm has a large N_{eff}^* and a small absorption constant but the response time is 100 times slower than that of reduced Rh-doped KNbO$_3$. Double doping with Mn and Rh produces large N_{eff}^*, but the response time is slow.

The proton-implanted waveguide has the fastest response times of less than a second [58–60]. The values compiled in Table 7.4 were calculated from the published values that were measured with an intensity of 200 W cm^{-2} at a grating period of 1.1 μm and 1.3 μm. It was assumed that the response time scales linearly with the intensity, and the effect of the grating period was ignored. The time constants were measured with the grating wave vector tilted by 45° relative to the c-axis in the bc-plane. The photorefractive sensitivity S_{n2} at 860 nm is of the same order of magnitude as that for the reduced Rh-doped crystal. The effect of the ten-times-faster response time is offset by the higher absorption constant and by the lower effective trap density. At 1064 nm, S_{n2} is larger in the proton-implanted crystal because the time constant is much faster than that of reduced Rh-doped KNbO$_3$. The advantage of the proton-implanted crystals is that the sensitivity extends up to wavelengths of at least 1.55 μm [58–60]. At 860 nm, however, Rh-doped KNbO$_3$ has a larger sensitivity because of its ten-times-larger effective trap density.

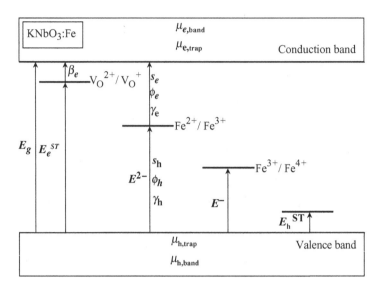

FIGURE 7.19. Schematic of the ionization levels and charge excitation and transport parameters for Fe-doped $KNbO_3$. The values of all the parameters are given in Table 7.5.

7.4.8 Material Parameters of Charge Transport in KNbO₃

A number of material parameters governing the charge transport in Fe-doped $KNbO_3$ at visible wavelengths have been determined and are summarized in this section. For a detailed description of various parameters one should consult other chapters of this book series. Figure 7.19 shows a schematic of the level structure of Fe-doped $KNbO_3$ and shows all the known physical parameters. The values of all the parameters are listed in Table 7.5.

The thermal ionization energies E^{2-}, E^-, and $E_{ST,h}$ are from temperature-dependent grating dynamics measurements [55]. The band-gap energy was measured by Wiesendanger [23], and the shallow electron level was identified by Biaggio et al. [82].

The trap-limited hole mobility $\mu_{h,trap}$ was measured by observing the grating-period dependence of the short-pulse-induced grating rise time. Similar measurements on a nanosecond time scale yielded the electron band mobility $\mu_{e,band}$ [56]. The ratio of the trap-limited mobilities of holes and electrons was measured in weakly Fe-doped $KNbO_3$ crystals [88] and yields the value of the trap-limited electron mobility $\mu_{e,trap}$. The upper limit of the hole band mobility $\mu_{h,band}$ is estimated from the nanosecond photorefractive response to pulsed uv illumination [83].

The excitation cross section for electrons s_e was determined by observing saturation effects in short-pulse photorefractive measurements in electrochemically reduced $KNbO_3$ [55]. From this, the excitation cross section for holes s_h is estimated by taking into account the relative absorption constants and quantum

TABLE 7.5. Material parameters of Fe-doped KNbO₃. The thermal ionization energies are given relative to the valence band edge

Parameter	Value	Remarks
E_G	3.3 eV	band-gap energy
$E_{ST,e}$	3.0 eV	V_O^{2+}/V_O^+ shallow trap level for electrons, observed in strongly el.-chem. red. crystals
E^{2-}	(2.3 ± 0.2) eV	Fe^{2+}/Fe^{3+} level
E^-	(0.8 ± 0.3) eV	Fe^{3+}/Fe^{4+} level
$E_{ST,h}$	(0.12 ± 0.02) eV	shallow trap for holes or small polaron level
$\mu_{e,band}$	$(5 \pm 1) \cdot 10^{-5}$ m²V⁻¹s⁻¹	electron band mobility
$\mu_{e,trap}$	$\approx 5 \cdot 10^{-7}$ m²V⁻¹s⁻¹	trap limited electron mobility, from $\mu_{h,trap}$ together with $\mu_{e,trap}/\mu_{h,trap}$ from [83]
$\mu_{h,band}$	$<5 \cdot 10^{-6}$ m²V⁻¹s⁻¹	hole band mobility
$\mu_{h,trap}$	$(6 \pm 2) \cdot 10^{-7}$ m²V⁻¹s⁻¹	trap-limited hole mobility
s_e	$(7 \pm 2) \cdot 10^{-3}$ m²J⁻¹	electron excitation probability rate for $Fe^{2+} + h\nu \rightarrow Fe^{+3} + e^-$
s_h	$\approx 2 \cdot 10^{-4}$ m²J⁻¹	hole excitation probability rate for $Fe^{3+} + h\nu \rightarrow Fe^{+2} + h^+$ estimated from absorption together with s_e
ϕ_e	0.065 (<0.14, >0.04)	quantum efficiency for $Fe^{2+} + h\nu \rightarrow Fe^{+3} + e^-$
ϕ_h	0.015 (<0.07, >0.008)	quantum efficiency for $Fe^{3+} + h\nu \rightarrow Fe^{+2} + h^+$
β_e	2000 s⁻¹	thermal excitation rate for electrons in the V_O^{2+}/V_O^+ shallow-trap level
γ_e	$\approx 1 \cdot 10^{-17}$ m⁻³s⁻¹	electron rec. cross section for $Fe^{+3} + e^- \rightarrow Fe^{2+}$, estimated from $\mu_e \tau_R$ together with $[Fe^{3+}]$ in 1000 at. ppm Fe-doped crystals
γ_h	$\approx 6 \cdot 10^{-16}$ m⁻³s⁻¹	hole recombination cross section for $Fe^{+2} + h^+ \rightarrow Fe^{2+}$, estimated from $\mu_h \tau_R$ together with $[Fe^{2+}]$ in 1000 at. ppm Fe-doped crystals

efficiencies. The quantum efficiencies for hole and electron excitation, ϕ_h and ϕ_e, were measured by comparing the diffusion lengths to the photoconductivities in a number of crystals [55]. The thermal excitation rate β_e of the shallow trap level V_O^{2+}/V_O^+ is determined in [82]. The recombination cross sections γ_e and γ_h are estimated from the product of the mobility and recombination time together with the acceptor densities that are calculated from the effective trap densities.

7.5 Conclusions

The list in Table 7.5 shows that a very comprehensive knowledge of the fundamental physical parameters that govern the photorefractive performance of Fe-doped KNbO₃ at visible wavelengths is available. It is now possible to give precise estimates of how a given crystal will perform and also to define the fundamental limits that can be achieved with KNbO₃. Doped KNbO₃ crystal can be grown in large size and good optical quality. The methods for controlling the valence state of dopants in a reproducible and homogeneous way are developed.

The new dopants and preparation of crystals have allowed the extension of the photorefractive applications of $KNbO_3$ into the near-infrared spectrum [12, 16, 90, 91]. Especially the high-temperature reduction treatment of Rh-doped $KNbO_3$ increased the photorefractive sensitivity at 860 nm by a factor of 30 in comparison with the as-grown material and therefore accelerating the photorefractive response time.

Ion implantation using MeV protons was successfully used for preparing $KNbO_3$ for special photorefractive applications where a relatively thin active photorefractive layer is sufficient, as in optical waveguides [59, 60]. Samples subjected to ion implantation show a consistent and reproducible reduction state which can be varied by adjusting the irradiation dose. Very high photorefractive gains up to 34 cm^{-1} at a wavelength of 514.5 nm were achieved in ion-implanted $KNbO_3$ and the fastest buildup time of 34 µs at an intensity of 200 W cm^{-2} (corresponding to a power of 2.5 mW in the interaction region).

UV light of a wavelength between 350 and 365 nm can also be used for efficiently writing photorefractive gratings with sub-mm thickness. Photon energy of this light lies slightly above the band gap of $KNbO_3$. The light is therefore absorbed and the effect is useful for applications involving visible or infrared light as the information carrier. Due to the faster dynamics with respect to conventional photorefraction, interband effects are very interesting for applications requiring a large parallel processing speed or a relatively quick switching time, as discussed in the chapter "Band-to-Band Photorefraction" of this book series. Screening of the externally applied electric field was also used to write reconfigurable waveguides for telecommunication wavelengths.

Photorefractive $KNbO_3$ was intensively used for short-pulse experiments, both in the visible and in the near-uv spectral band. It is interesting to note that the photorefractive grating can be written by a single pulse of sub-ns duration if sufficient charge-carrier density is produced in the active volume. This favors the uv excitation, where the photoexcitation process is much more efficient.

Photorefractive reflection gratings are important in phase conjugation devices, and $KNbO_3$ is the material of choice because of the large effective electrooptic coefficient for such photorefractive interactions. The advantages of reflection gratings are a faster response time—if the writing process is diffusion dominated—and a larger than diffusion-limited space-charge electric field in materials with sufficiently high trap densities as achieved in $KNbO_3$. Compact and stable reflection-grating-based phase conjugators were achieved in this way in Fe-doped crystals [92, 93, 94]. Strong two-beam coupling of counterpropagating beams has also attracted much attention, since it produces interesting spontaneous pattern formation. The first observation of hexagonal pattern by Honda [95] was followed by the work of several groups that was able to manipulate the dynamics and the spatial properties of the patterns [96–98].

Volume holograms can also be fixed in pure and Fe-doped $KNbO_3$ through the screening of the primary space-charge field. At elevated temperatures of around 100°C the compensating species was identified as the hydrogen ion [99]. Fixing at room temperature can be achieved by the use of ferroelectric domains [100]. A grating is recorded while the crystal is simultaneously depoled. By repoling the crystal, the presence of a latent grating is revealed.

The potential of KNbO₃ crystals for applications in all-optical image manipulation, data processing, beam reshaping and amplification, phase conjugation, etc. has therefore been shown. However, the commercialization of such devices is still a task for the near future.

References

1. M.E. Lines, A.M. Glass. *Principles and Applications of Ferroelectrics and Related Materials*, Clarendon Press, Oxford, (1977).
2. M. DiDomenico, Jr. and S.H. Wemple. *J. Appl. Phys.* **40**, 720–734 (1969).
3. P. Günter. *Phys. Rep.* **93**, 199–299 (1982).
4. P. Günter, J.P. Huignard. *Photorefractive Materials and Their Applications I + II*, Springer-Verlag, Berlin, (1989).
5. M. Zgonik, P. Bernasconi, M. Duelli, R. Schlesser, P. Günter, M.H. Garrett, D. Rytz, Y. Zhu, and X. Wu. *Phys. Rev. B* **50**, 5941–5949 (1994).
6. J.-C. Baumert, P. Günter, and H. Melchior. *Opt. Comm.* **48**, 215–220 (1983).
7. I. Biaggio, P. Kerkoc, L.-S. Wu, P. Günter, and B. Zysset. *J. Opt. Soc. Am. B* **9**, 507–517 (1991).
8. J. Hulliger, R. Gutmann, and H. Wuest. *J. Cryst. Growth* **128**, 897–902 (1993).
9. P. Günter. *Ferroelectrics* **22**, 671–674 (1978).
10. U. Flückiger, H. Arend. *J. Cryst. Growth* **43**, 406–416 (1978).
11. P. Amrhein. *Photorefractive Spatial Light Modulation in KNbO₃ Crystals*, Ph.D. thesis Nr. 9670, ETH Zürich (1992).
12. C. Medrano, M. Zgonik, I. Liakatas, P. Günter. *J. Opt. Soc. Am. B* **12**, 1152–1165 (1996).
13. R.J. Reeves, M.G. Jani, B. Jassemnejad, R.C. Powell, G.J. Mizell, W. Fay. *Phys. Rev. B* **43**, 71–82 (1991).
14. C. Medrano, M. Zgonik, N. Sonderer, C. Beyeler, S. Krucker, J. Seglins, H. Wuest, P. Günter. *J. Appl. Phys.* **76**, 5640–5645 (1994).
15. C. Medrano, M. Zgonik, I. Liakatas, M. Ewart, P. Günter. *Asian J. of Physics* **4**, 1–14 (1995).
16. M. Ewart, R. Ryf, C. Medrano, H. Wüest, M. Zgonik, P. Günter. *Opt. Lett.* **22**, 781–783 (1997).
17. M. Zgonik, C. Medrano, M. Ewart, H. Wüest, P. Günter. *Opt. Eng.* **37**, 1930–1935 (1995).
18. E. Voit, M.Z. Zha, P. Amrhein, and P. Günter. *Appl. Phys. Lett.* **51**, 2079–2081 (1987).
19. G. Montemezzani, P. Rogin, M. Zgonik, and P. Günter. *Phys. Rev. B* **49**, 2484–2502 (1994).
20. E. Wiesendanger. *Ferroelectrics* **1**, 141–148 (1970).
21. P. Günter. *J. of Appl. Phys.* **48**, 3475–3477 (1977).
22. A.W. Hewatt. *J. Phys. C* **6**, 2559–2572 (1973).
23. E. Wiesendanger. *Ferroelectrics* **6**, 263–281 (1974).
24. E. Voit, C. Zaldo, P. Günter. *Opt. Lett.* **11**, 309–311 (1986).
25. B. Zysset, I. Biaggio, P. Günter. *J. Opt. Soc. of Am. B* **9**, 380–386 (1992).

26. M. Zgonik, R. Schlesser, I. Biaggio, E. Voit, J. Tscherry, P. Günter. *J. Appl. Phys.* **74**, 1287–1297 (1993).

27. P. Bernasconi, M. Zgonik, P.Günter. *J. Appl. Phys.* **78**, (1995).

28. L. Douillard, F. Jollet, C. Bellin, M. Gautier, J.P. Duraud. *J. Phys. Cond. Mat.* **6**, 5039–5052 (1994).

29. F.M. Michel-Calendini, P. Pertosa, G. Metrat. *Ferroelectrics* **21**, 637–639 (1978).

30. F.M. Michel-Calendini, H. Chermette. *Ferroelectrics* **25**, 495–508 (1980).

31. F.M. Michel-Calendini, L. Castet. *Ferroelectrics* **13**, 367–430 (1976).

32. P. Pertosa, F.M. Michel-Calendini. *Phys. Rev. B* **17**, 2011–2020 (1978).

33. G. Montemezzani, M. Zgonik. *Phys. Rev. E* **35**, 1035 (1997).

34. G. Montemezzani. *Phys. Rev. A* **62**, 053803 (2000).

35. F.M. Michel-Calendini, M. Peltier, F. Micheron. *Solid State Commun.* **33**, 145–150 (1980).

36. E. Possenriede, O.F. Schirmer, H.J. Donnerberg, B. Hellermann. *J. Phys.: Cond. Mat.* **1**, 7267–7276 (1989).

37. E. Siegel, W. Urban, K.A. Müller, E. Wiesendanger. *Phys. Lett.* **53A**, 415–417 (1975).

38. E. Siegel, K.A. Müller. *Phys. Rev. B* **19**, 109–120 (1979).

39. H. Donnerberg. *Phys. Rev. B* **50**, 9053–9062 (1994).

40. F.M. Michel-Calendini. *Ferroelectrics* **37**, 499–502 (1981).

41. F.M. Michel-Calendini. *Solid State Commun.* **52**, 167–172 (1984).

42. D. Rytz, U.T. Höchli, K.A. Müller, W. Berlinger, L.A. Boatner. *J. Phys. C* **15**, 3371–3379 (1982).

43. R. Gonzalez, M.M. Abraham, L.A. Boatner, Y. Chen. *J. Chem. Phys.* **78**, 660–664 (1982).

44. M. Exner. *Computer-Simulation von extrinsischen Defekten in Kaliumtantalat- und Kaliumniobat-Kristallen*, Dissertation Universität Osanabrueck, Verlag Shaker, Aachen (1994).

45. R.C. Weast, ed. *Handbook of Chemistry and Physics*, CRC Press, Ohio (1972).

46. R.E. Cohen, H. Krakauer. *Phys. Rev. B* **42**, 6416–6423 (1990).

47. E. Voit. *Photorefractive Properties and Applications of KNbO₃*, Ph.D. thesis. Nr. 8555, ETH Zürich (1988).

48. K.W. Blazey, O.F. Shirmer, W. Berlinger, K.A. Müller. *Solid State Commun.* **16**, 589–592 (1975).

49. O.F. Schirmer, W. Berlinger, K.A. Müller. *Solid State Commun.* **16**, 1289–1292 (1975).

50. B.A. Wechsler, M.B. Klein. *J. Opt. Soc. Am. B* **5**, 1711–1723 (1988).

51. H. Kroese, R. Scharfschwerdt, O.F. Schirmer, H. Hesse. *Appl. Phys. B* **61**, 1–7 (1995).

52. K. Kitamura. NIMS, 1-1 Namiki, Tsukuba-shi, 305 Japan, *private communication*, (1996).

53. X. Ma, Z. Xing, D. Shen. Photorefractive Materials, Effects, and Devices Topical Meeting, Beverly MA, July 29–31, Technical digest, contribution PD1 (1991).

54. R. Ryf, G. Montemezzani, P. Günter. *J. Opt. A: Pure Appl. Opt.* **3**, 16–19 (2001).

55. M. Ewart. *Reduced KNbO₃ for Photorefractive Applications*, Ph.D. thesis Nr. 12484, ETH Zürich (1998).

56. M. Ewart, I. Biaggio, M. Zgonik, P. Günter. *Phys. Rev. B* **49**, 5263–5273 (1994).

57. A. Krumins, P. Günter. *Appl. Phys.* **19**, 153–163 (1979).

58. S. Brülisauer, D. Fluck, P. Günter, L. Beckers, C. Buchal. *J. Opt. Soc. Am. B* **13**, 2544–2548 (1996).

59. S. Brülisauer, D. Fluck, P. Günter, L. Beckers, C. Buchal. *Opt. Comm.* **153**, 375–386 (1998).
60. D. Fluck, S. Brülisauer, P. Günter, C. Buchal, L. Beckers. *Nuclear Inst. & Meth. in Phys. Res.*: Section B-Beam interactions with materials and atoms **148**, 678–682 (1999).
61. A. Onton, V. Marrello, AIP-Conference-Proceedings (USA), no. 31, pp. 320–325, (1976).
62. H.-J. Hagemann, D. Hennings. *J. Am. Cer. Soc.* **64**, 590–594 (1981).
63. N.-H. Chan, R.K. Sharma, D.M. Smyth. *J. Am. Cer. Soc.* **64**, 556–562 (1981).
64. N.-H. Chan, R.K. Sharma, D.M. Smyth. *J. Am. Cer. Soc.* **65**, 167–170 (1982).
65. H.M. Chan, M.P. Harmer, D.M. Smyth. *J. Am. Ceram. Soc.* **69**, 507–510 (1986).
66. R. Waser. *J. Am. Ceram. Soc.* **71**, 58–63 (1986).
67. C. Medrano, E. Voit, P. Amrhein, P. Günter. *J. Appl. Phys.* **64**, 4668–4673 (1988).
68. K. Szot, W. Speier, W. Eberhardt. *Appl. Phys. Lett.* **60**, 1190–1192 (1992).
69. K. Szot, F.U. Hillebrecht, D.D. Sarma, M. Campagna, H. Arend. *Appl. Phys. Lett.* **48**, 490–492 (1986).
70. K. Szot, J. Keppels, W. Speier, K. Besocke, M. Teske, W. Eberhardt. *Surface Science* **280**, 179–184 (1993).
71. P. Amrhein, P. Günter. *J. Appl. Phys.* **71**, 2006–2011 (1991).
72. K.E. Youden, S.W. James, R.W. Eason, P.J. Chandler, L. Zhang, P.D. Townsend. *Opt. Lett.* **17**, 1509–1511 (1992).
73. M. Zha, D. Fluck, P. Günter, M. Fleuster, C. Buchal. *Opt. Lett.* **18**, 577–579 (1993).
74. A. Förster, H. Hesse, S. Kapphan, M. Wöhlecke. *Solid State Comm.* **57**, 373–375 (1986).
75. G. Metrat, A. Deguin. *Ferroelectrics* **13**, 527–529 (1976).
76. S. Ducharme, J. Feinberg. *J. Opt. Soc. Am. B* **3**, 283–293 (1986).
77. B.A. Wechsler, M.B. Klein, C.C. Nelson, R.N. Schwartz. *Opt. Lett.* **19**, 536–538 (1994).
78. J.Y. Chang, S.H. Duan, C.Y. Huang, C.C. Sun. *Appl. Phys. B-Lasers and Opt.* **68**, 827–831 (1999).
79. J.Y. Chang, C.Y. Huang, R.R. Yueh, H.W. Pan, C.H. Lin, C.C. Sun. *Jpn. J. Appl. Phys. Part 1* **42**, 5140–5144 (2003).
80. A. Lahlafi, G. Godefroy, P. Jullien. *J. Opt. Soc. Am. B* **10**, 1276–1286 (1993).
81. R. Ryf, M. Zgonik, P. Günter. *Nonlinear Optics* **19**, 79–91 (1998).
82. I. Biaggio, M. Zgonik, P. Günter. *J. Opt. Soc. Am. B* **9**, 1480–1487 (1992).
83. M. Ewart, M. Zgonik, P. Günter. *Opt. Comm.* **141**, 99–106 (1997).
84. E. Possenriede, B. Hellermann, O.F. Schirmer. *Solid State Commun.* **65**, 31–33 (1988).
85. I. Biaggio, M. Zgonik, P. Günter. *Opt. Comm.* **77**, 312–317 (1990).
86. D. Mahgerefteh, D. Kirillov, R.S. Cudney, G.D. Bacher, R.M. Pierce, J. Feinberg. *Phys. Rev. B* **53**, 7094–7098 (1996).
87. H. Ihrig, D. Hennings. *Phys. Rev. B* **17**, 4593–4599 (1978).
88. P. Bernasconi, I. Biaggio, M. Zgonik, P. Günter. *Phys. Rev. Lett.* **78**, 106–109 (1997).
89. P. Bernasconi, G. Montemezzani, I. Biaggio, P. Günter. *Phys. Rev. B.* **56**, 12196–12200 (1997).
90. Y. Ding, H.J. Eichler, Z.G. Zhang, P.M. Fu, D.Z. Shen, X.Y. Ma, J.Y. Chen. *J. Opt. Soc. Am. B* **13**, 2652–2656 (1996).
91. Z.G. Zhang, Y. Ding, H.J. Eichler, P.M. Fu, G. Zhou, J.X. Tang, D.Z. Shen, X.Y. Ma, J.Y. Chen. *Chinese Phys. Lett.* **14**, 103–105 (1997).

92. Q. Byron He, S. Campbell, P. Yeh, X. Ma, D. Shen. *Appl. Opt.* **33**, 4320–4322 (1994).
93. T. Honda, H. Matsumoto. *Appl. Opt.* **33**, 4475–4479 (1994).
94. K. Nakagawa, M. Zgonik, P. Günter. *J. Opt. Soc. Am. B* **14**, 839–845 (1997).
95. T. Honda. *Opt. Lett.* **18**, 598–600 (1993).
96. J.O. Dimmock, P.P. Banerjee, F.L. Madarasz, N.V. Kukhtarev. *Opt. Comm.* **175**, 433–446 (200).
97. M. Schwab, C. Denz, A.V. Mamaev, M. Saffman. *J. Opt. B: Quantum Semiclass. Opt.* **3**, 318–327 (2001).
98. O. Sandfuchs, F. Kaiser, M.R. Belic. *J. Opt. Soc. Am. B* **18**, 505–513 (2001).
99. X.L. Tong, M. Zhang, A. Yariv, A. Agranat. *Appl. Phys. Lett.* **69**, 3966–3968 (1996).
100. R.S. Cudney, P. Bernasconi, M. Zgonik, J. Fousek, P. Günter. *Appl. Phys. Lett.* **70**, 1339–1341 (1997).

8

Photorefractive Properties of BaTiO₃

Marvin B. Klein

Hughes Research Laboratories, 30011 Malibu Canyon Road, Malibu, CA 90265, USA.
Present address: Lasson Technologies, Inc. 6059 Bristol Parkway, Culver City, CA 90230,
USA, mbklein@lasson.com

Barium titanate (BaTiO₃) was one of the first ferroelectric materials to be discovered, and also one of the first to be recognized as photorefractive [1]. The particular advantage of BaTiO₃ for photorefractive applications is the very large value of its electrooptic tensor component r_{42}, which in turn leads to large values of grating efficiency, beam-coupling gain, and four-wave mixing reflectivity. As an example, four-wave mixing reflectivities as large as 20 have been observed in BaTiO₃ [2], with no electric field applied to the crystal. Such large reflectivities are particularly desirable in phase conjugate resonator applicattions, where to date BaTiO₃ has been the material of choice.

The first observation of the photorefractive effect in BaTiO₃ was reported by Townsend and LaMacchia in 1970 [1]. This report came quite soon after the first reports of optical damage [3] and holographic storage [4] in LiNbO₃. In the work of Townsend and LaMacchia [1], simple sinusoidal holograms were written at 514.5 nm in melt-grown samples of BaTiO₃, and were read out at 632.8 nm. The diffraction efficiency and response time of the holograms were measured as a function of several experimental parameters, and a model was developed to explain some of the results. In 1972, Micheron and Bismuth [5] reported diffraction efficiency measurements on flux-grown samples doped with Fe and Ni. It was noted that Fe was more effictive than Ni, and an optimum Fe concentration was determined. Between 1972 and 1980, the photorefractive effect in BaTiO₃ received little or no attention, except for the studies of the bulk photovoltaic effect (using uniform illumination) by Koch et al. [6, 7]. The lack of progress on BaTiO₃ during this time interval was probably due to the sustained interest in LiNbO₃ for holographic memory applications and to the scarcity of BaTiO₃ samples with high optical quality.

In 1980, a renewal of interest in BaTiO₃ was generated by the work of Feinberg et al. [8] and Krätzig et al. [9], in which many of the favorable features of this material for real-time applications were demonstrated, and several important material

This chapter has been reproduced from: (P. Günter and J.P. Huignard, Eds.), Photorefractive Materials and Their Applications I, Chapter 7 (Topics in Applied Physics, Vol. 61) Springer Verlag, 1988

parameters were measured. Since 1980, a large number of papers on the fundamental properties and applications of BaTiO$_3$ have been published. For a description of device applications, see *Topics in Applied Physics*, Volume 62, Chapters 4
and 5.

At present, the advantages of BaTiO$_3$ for phase conjugate resonators and other
applications requiring large signal levels are widely recognized. At the same
time, many of the physical, optical, and dielectric properties of this material are
not widely understood. The purpose of this chapter is to review the important
properties of BaTiO$_3$, especially as they relate to the photorefractive effect. The
photorefractive effect can be a powerful technique in itself for making physical
measurements, and this feature will be emphasized. In Section 8.1, we discuss
the crystal growth, and lattice and domain structure of BaTiO$_3$, and describe the
dielectric and electrooptic properties of this material. A description of defect and
impurity centers in BaTiO$_3$ is given in Section 8.2, with special emphasis on
the photorefractive centers. Section 8.3 reviews the band-transport model of the
photorefractive effect, and its application to the particular case of BaTiO$_3$. In
Section 8.4 we review the photorefractive measurements of material properties in
BaTiO$_3$. Other characterization studies of photorefractive crystals are described in
Section 8.5. Techniques for optimizing the photorefractive properties of BaTiO$_3$
are described in Section 8.6, and our conclusions are given in Section 8.7.

8.1 Basic Properties and Technology

8.1.1 Crystal Growth

The early experiments on ferroelectricity in BaTiO$_3$ were performed on ceramic
samples. The first practical technique developed for the growth of single crystals
was the KF-flux or Remeika technique [10]. In this technique, BaTiO$_3$ powder
is dissolved in molten KF. The mixture is then cooled to $\approx$900 °C, the saturation
temperature of the mixture. Upon further cooling, spontaneous crystallization
occurs at the bottom of the crucible, producing single crystals in the form of
platelet pairs attached to each other along one edge at an angle of 39 °. The resulting crystals resemble butterflies, and the platelets are sometimes called "butterfly
wings." This unusual morphology is due to (111) twinning, and is characteristic of
KF-flux-grown crystals. While the Remeika crystals were relatively easy to grow
and their quality was satisfactory for the early measurements on BaTiO$_3$, the use
of these crystals for applications was limited by scattering due to flux inclusions
and absorption due to substitution of K and F for the Ba and O ions in the lattice.

A major advance in BaTiO$_3$ crystal growth technology was the development of
the top-seeded solution growth (TSSG) technique in the late 1960s [11]. In this
technique, a solution of BaO and TiO$_2$ was used, with TiO$_2$ added in excess of the
amount necessary for stoichiometry. The purpose of the excess TiO$_2$ was to lower
the melting point of the solution from a value of $\approx$1620 °C (in stoichiometric
BaTiO$_3$) to $\approx$1400 °C (with approximately 65 mol% TiO$_2$ in the melt). This eases
the requirements on furnace and crucible, but also avoids the high-temperature

hexagonal phase of stoichiometric BaTiO$_3$, which is deleterious to the production of high-quality crystals. The actual crystal growth process is performed in air, and involves the slow cooling of the solution by $\approx$40 °C, and the simultaneous pulling of the crystal from the melt. Single-crystal boules grown by the TSSG technique are typically 2–3 cm in diamter, and polished, poled samples from these boules (typically $0.5 \times 0.5 \times 0.5$ cm^3) are generally clear and free of visible defects. The only significant disadvantage of the TSSG technique is that it is incongruent, and is thus characterized by a slow growth rate.

8.1.2 Lattice Structure

At the growth temperatures used for the TSSG technique, the crystals that nucleate are cubic, with point group $m3m$ (symmetry O_h). Upon cooling, BaTiO$_3$ transforms to a tetragonal, ferroelectric phase (point group $4mm$, symmetry C_{4v} at $T_c \approx$ 130 °C, where T_c is the Curie temperature. The exact value of T_c in a given crystal depends on the concentration of impurities and other defects. Values of T_c a few degrees above 130 °C have been reported in pure, melt-grown crystals [12, 13], while values closer to 120 °C are measured in flux-grown crystals [14, 15].

The tetragonal phase of BaTiO$_3$ is the stable one at room temperature, and is of primary interest for applications. At $\approx$9 °C a transition to an orthorhombic phase (point group $mm2$) occurs. This phase transition is of some practical concern. Crystals cooled too rapidly through this phase transition are subject to cracking, and thus excessive cooling during shipment and handling must be avoided. At -90 °C a third and final transition to a rhombohedral phase (point group $3m$) occurs.

BaTiO$_3$ is a member of the perovskite family of ABO$_3$ compounds, which includes other well-known materials such as KNbO$_3$, KTaO$_3$, PbTiO$_3$, and SrTiO$_3$. The structure of BaTiO$_3$ (and other perovskites) in the cubic phase is a simple one (Figure 8.1), with Ba^{2+} ions at the cube corners, Ti^{4+} ions at the body centers, and O^{2-} ions at the face centers. The structure can be considered as a rigid grouping of oxygen octahedra linked at their corners by shared oxygen ions. The Ti^{4+} ions thus lie at the center of each octahedron, while the Ba^{2+} ions lie outside the octahedra. Many of the dielectric and optical properties of BaTiO$_3$ are determined by the characteristics of the basic TiO$_6^{8-}$ octahedron. For example, the energy band structure of BaTiO$_3$ near the band edge is determined by

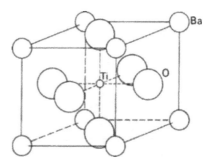

FIGURE 8.1. BaTiO$_3$ unit cell. The size of the ions is reduced for clarity.

titanium $3d$ orbitals (which are responsible for the low-lying levels of the conduction band) and oxygen $2p$ orbitals (which are responsible for the upper levels of the valence band). Furthermore, the dominant transverse optical (TO) mode in $BaTiO_3$ (which largely determines the dielectric properties in the cubic phase) consists of a vibration of the Ti^{4+} ion with respect to the TiO_6^{8-} octahedron.

The cubic–tetragonal phase transition near $130\,°C$ is a displacive one, the most important change being a shift of the Ti^{4+} ion from the center of the octahedron toward an oxygen ion at one of the face centers of the cubic unit cell. This shift produces a spontaneous polarization in the direction of motion. The polar nature of $BaTiO_3$ leads to a variety of well-known properties, including the piezoelectric effect, the pyroelectric effect, the electrooptic effect, and second harmonic generation. The reversibility of the polarization in $BaTiO_3$ leads to its well-known ferroelectric properties.

8.1.3 Domains and Poling

In the tetragonal phase of $BaTiO_3$ the spontaneous polarization may be oriented along any of the six pseudocubic $\langle 001 \rangle$ directions. Different regions of the crystal may polarize in each of these directions, each region of uniform polarization being referred to as a domain. In an as-grown crystal, the multidomain state that appears on cooling generally produces no net polarization. Therefore, such crystals show very small, if any, piezoelectric, pyroelectric, and electrooptic effects. These effects depend on the direction and sign of the polarization, and are thus averaged out to zero in a multidomain crystal. For this reason, it is important to find a procedure (called poling) that leads to a single domain state [16].

The domain structure of $BaTiO_3$ in the tetragonal phase is shown schematically in Figure 8.2. Two types of domains are found to exist: domains whose polarizations lie at $90°$ to each other, and domains with antiparallel polarizations. The

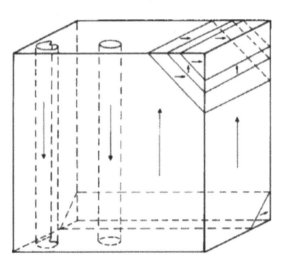

FIGURE 8.2. Domain structure of $BaTiO_3$ in the tetragonal phase. The arrows represent the direction of polarization.

former, called 90° domains, are readily seen when the crystal is placed between crossed polarizers, and may frequently be seen by the unaided eye. The 180° domains are more elusive, being observable only by etching or more elaborate techniques. Both types of domains must be eliminated in order to reduce scattering and produce useful electrooptic properties.

The poling process always involves the nucleation of new domains and the movement of domain walls. In this picture, the single domain state is achieved when nucleation is complete and when all walls are induced to migrate to the boundaries of the crystal, where they disappear. There are two physical effects that tend to impede the movement of domain walls, and thus make poling more difficult: (1) imperfections (e.g., inclusions, dislocations, strains) tend to pin walls, and (2) charges at the walls associated with local changes in polarization are transported very inefficiently in insulating crystals. In fact, it is known that conducting samples of BaTiO₃ (produced permanently by reduction [17] or temporarily by illumination [18]) are much easier to pole than insulating samples.

A number of related techniques are used at present for the poling of melt-grown samples of BaTiO₃. They all rely on a particular combination of electric field, heating, or an applied stress to produce a single-domain sample. In all cases, as-grown samples are first oriented, cut, and polished, with faces normal to the $\langle 100 \rangle$ family of axes in the cubic system. In the first technique, the 90° domains are removed at room temperature by applying uniaxial pressure alternately along two different directions. The resulting sample is left with only antiparallel 180° domains along the third direction. Following this mechanical poling process, an electric field is applied along the third direction at a temperature slightly below T_c. The sample is then cooled to room temperature, and the field is then removed. This step results in the removal of 180° domains, and produces a single-domain crystal. It should be noted that the prolonged application of an electric field at temperatures near T_c can produce electrochemical damage in the form of cracks at the positive face of the crystal. This damage may be related to the large mobility of oxygen vacancies, which are positively charged and move to the negative electrode [19, 20]. This results in the liberation of free oxygen at the positive electrode, with the cracking resulting from the loss of stoichiometry or the buildup of pressure in preexisting microcracks. The cracking could also be due to the lowering of the phase transition temperature (at the positive electrode) to a point below the temperature of the experiment. Regardless of the mechanism, the electrochemical damage can be controlled by reducing the electric field and cooling as fast as possible (except at the phase transition). However, these restrictions may not favorable for the elimination of all domains. Another mechanism for the electrochemical damage is the migration of metal ions into the crystal from the positive electrode (e.g., Ag^+ ions when silver paint is used for the contacts) into the bulk of the crystal.

A second poling technique involves cooling the sample through the cubic–tetragonal phase transition in the presence of an applied electric field [12]. This has the obvious advantage of eliminating the need for mechanical poling. However, there is a risk of cracking if the sample is cooled too rapidly through the phase transition. In addition, this technique may not eliminate all 180° domains [12, 13]. Recently Muser et al. [21] have improved upon this technique by first etching the sample to remove surface strains, which can pin domains. The sample is also held for an extended period of time at its highest temperature, in order that all charges associated with the original polydomain state can relax away.

A third poling technique involves the application of a large electric field to the sample at room temperature, with no previous mechanical poling. This technique is potentially the easisest one, and is commonly employed (with alternating fields) to study hysteresis loops in ferroelectrics [16]. However, it typically requires the application of large fields to overcome the sluggish motion of domain walls at room temperature.

After the completion of the poling process, it is important for many applications to determine the *degree of poling*. A useful figure of merit in this case is the *fractional poling factor F*, defined as the ratio of any second-order coefficient (e.g., electrooptic, nonlinear optic, piezoelectric, pyroelectric) in the sample in question to the same coefficient in a single-domain sample. It is difficult to determine F from a direct measurement of a particular second-order coefficient, first because many of the measurement techniques are inaccurate (especially when the use of electrical contacts is required), and second because disagreement frequently exists as to the value of the coefficient in a single-domain sample. One useful approach for determining (at least qualitatively) the degree of poling is direct observation of the domain structure. Lines and Glass [16] review a number of useful techniques for visualizing domains. One particularly promising technique for observation of 180° domains is etching of the c-faces of the sample [16]. Certain etchants (such as HF and HCl) act preferentially on the positive (or negative) c-faces, thus allowing direct visualization of the pattern of 180° domains on a c-face. This technique is more accurate for thin samples, in which 180° domains are more likely to extend completely to the opposite c-face.

8.1.4 Dielectric and Electrooptic Properties

The photorefractive performance of $BaTiO_3$ is significantly influenced by its dielectric and electrooptic properties. Specifically, the very large values of induced index change, beam-coupling gain coefficient, and degenerate four-wave mixing (DFWM) reflectivity are due to the large values of the electrooptic tensor components in $BaTiO_3$. Similarly, the relatively long time constants observed in $BaTiO_3$ are partially the result of the high values of dielectric constant in this material.

As indicated earlier, $BaTiO_3$ at room temperature is in a tetragonal phase, with uniaxial optical and dielectric behavior. The resulting values of dielectric constant are $\varepsilon_a = \varepsilon_b$ and ε_c. The magnitude and temperature dependence of ε_a and ε_c in the tetragonal phase are quite different [13], as shown in Figure 8.3. In order to understand this behavior, we must first consider the behavior of $BaTiO_3$ on cooling through the cubic–tetragonal phase transition. At temperatures above T_c, the Ti^{4+} ion in each unit cell is in a cubic environment, and is highly polarizable along all three cubic axes, resulting in a large value of dielectric constant. Upon cooling toward the Curie temperature, the Ti^{4+} ion experiences the formation of a double

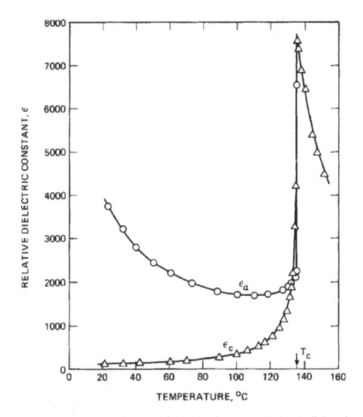

FIGURE 8.3. Temperature dependence of the low-frequency (1 kHz) dielectric constants in single domain BaTiO₃ (after [13]).

potential well along one particular cubic axis. At the phase transition, the depth of the double well has increased enough to "trap" the Ti^{4+} ion on one side or the other, leading to a macroscopic polarization, and a restricted polarizability for electric fields applied along the axis of polarization. However, for electric fields polarized along the a-axis (or b-axis), the Ti^{4+} ion remains free to vibrate and is highly polarizable. This leads to the large difference in the values of ε_a and ε_c as indicated in Figure 8.3. Note also in the figure that ε_c is resonant at the cubic–tetragonal transition at $T_c \approx 135\,°C$, consistent with the particular structural transformation at this temperature. On the other hand, the dielectric properties in the a-direction (or b-direction) are sensitive to the tetragonal–orthorhombic transition near $9\,°C$, where a second structural transformation occurs.

For a crystal of point group 4 mm, there are three independent components of the electrooptic tensor: $r_{113} = r_{223}, r_{333}$, and $r_{232} = r_{131}$ [22]. These coefficients relate the impermeability tensor $(1/n^2)_{ij}$ to the applied electric field E_k:

$$\Delta\left(\frac{1}{n^2}\right)_{ij} = \sum r_{ijk}E_k. \tag{8.1}$$

An alternative (and physically more fundamental) description uses the field-induced crystal polarization as the driving term. In this case the linear effect is described by

$$\Delta\left(\frac{1}{n^2}\right)_{ij} = \sum f_{ijk} P_k, \tag{8.2}$$

where the f_{ijk} are the linear polarization-optic coefficients, and the quadratic effect is described by

$$\Delta\left(\frac{1}{n^2}\right)_{ij} = \sum g_{ijkl} P_k P_l, \tag{8.3}$$

where the g_{ijkl} are the quadratic polarization-optic coefficients. In BaTiO$_3$, the coefficients defined in the above relations are related by

$$
\begin{aligned}
r_{13} &= f_{13}\varepsilon_0\varepsilon_c, \\
r_{33} &= f_{33}\varepsilon_0\varepsilon_c, \\
r_{42} &= f_{42}\varepsilon_0\varepsilon_a,
\end{aligned}
\tag{8.4}
$$

$$
\begin{aligned}
r_{13} &= 2g_{12}\varepsilon_0 P_s\varepsilon_c, \\
r_{33} &= 2g_{11}\varepsilon_0 P_s\varepsilon_c, \\
r_{42} &= g_{44}\varepsilon_0 P_s\varepsilon_a,
\end{aligned}
\tag{8.5}
$$

where P_s is the spontaneous polarization and ε_0 is the permittivity of free space [22]. In (8.4) and (8.5) we have used the contracted notation $r_{113} = r_{13}, r_{333} = r_{33}$, and $r_{231} = r_{42}$. The relation to the quadratic coefficients (8.5) is quite useful, for several reasons. First, the explicit dependence on P_s is consistent with the fact that $r_{ij} = 0$ in the cubic phase. Second, the g-coefficients are nearly invariant with temperature, frequency, and material for most oxide ferroelectrics [22]. Thus, the relations above are a form of Miller's rule [23], and show the explicit dependence of the electrooptic coefficients on the relative dielectric constant. The above relations clearly indicate why the coefficient r_{42} is so large in BaTiO$_3$: its magnitude is determined by the polarizability normal to the c-axis, and is thus proportional to ε_a. Thus, r_{42} is resonant at the tetragonal–orthorhombic transition [24] and r_{13} and r_{33} are resonant at the cubic–tetragonal phase transition [24].

The similarity in temperature dependence of r_{42} and ε_a in tetragonal BaTiO$_3$ has been noted by Johnson and Weingart [24], who plotted these two values together, as shown in Figure 8.4. The enhancement of r_{42} near the tetragonal–orthorhombic phase transition leads to an increase in the photorefractive beam-coupling gain Γ for the particular interaction geometry that exploits r_{42}. This enhancement in gain is sufficient to allow operation of a self-pumped phase-conjugate resonator at 840 nm [25]. It was shown in this work that the phase-conjugate resonator exceeded the threshold only when the sample temperature was less than 18 °C.

There is considerable variation among the values of dielectric constant reported in the literature for BaTiO$_3$ [26]. These variations are related to the experimental conditions (especially the quality of the contacts), and to composition variations

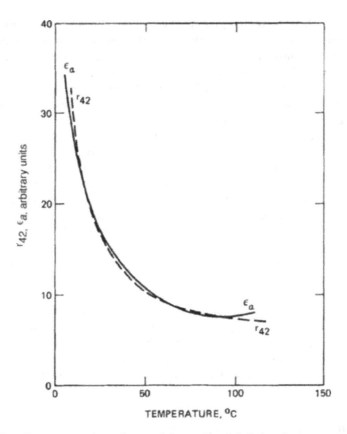

FIGURE 8.4. Temperature dependence of the unclamped dielectric constant ε_a and the electrooptic coefficient r_{42} in single-domain BaTiO₃ (after [24]).

from sample to sample (e.g., flux-grown vs. solution-grown samples). In the case of the electrooptic coefficients, the opposite problem exists: there is a scarcity of measured values, with few measurements reported for any component [27]. Furthermore, most measurements have been performed only on flux-grown crystals. In Table 8.1 we list selected values of the dielectric constant and electrooptic tensor

TABLE 8.1. Dielectric constants and electrooptic coefficients for single-crystal BaTiO₃ at 20 °C. The dielectric constant data are taken from measurements on solution-grown samples. All values of electrooptic coefficient (except the recent measurement [28]) were obtained using flux-grown samples. Reference numbers are given in square brackets after each value.

	ε_a	ε_c	r_{13} [pm/V]	r_{33} [pm/V]	r_{42} [pm/V]
Unclamped	3700 [12, 13]	135 [13]	24 [29]	80 [29]	1640 [24]
			19.5 [28]	97 [28]	
Clamped	2400 [12]	60 [13]	8 [30]	28 [30]	820 [31]

components in $BaTiO_3$. We have included values in Table 8.1 for both unclamped (constant stress) and clamped (constant strain) conditions. For an electric field applied uniformly throughout an unconstrained sample, the unclamped condition corresponds to applied frequencies below the bulk piezoelectric resonances of the sample, which can range between 1 MHz and 100 MHz. The clamped condition corresponds to applied frequencies above the piezoelectric resonances. At these high frequencies, all piezoelectric effects are frozen out, and thus do not contribute to the dielectric or electrooptic properties. In a mechanically constrained sample, the clamped coefficients are appropriate at all frequencies.

There has been some question over the years as to whether the dielectric constants and electrooptic coefficients used in models for the photorefractive effect should be the clamped or unclamped values. One might first expect that a sample uniformly illuminated throughout its volume by two interfering beams might be unclamped, while a sample illuminated in a portion of its volume might be considered as clamped (by the rigid mass of unilluminated crystal). However, in the illuminated portion, the strain or stress induced by the space charge electric field (via the piezoelectric effect) is spatially periodic along the direction of the space-charge field, and no average strain or stress is induced. Thus, the crystal can be considered as unclamped in the direction of periodicity. In recent beam-coupling experiments in $BaTiO_3$ [32, 33], the measured values of the electrooptic coefficient r_{13} in many samples were found to be larger than the literature value [30] for clamped conditions (Tables 8.1 and 8.3). This suggests that the appropriate coefficient for photorefractive measurements must be the unclamped value.

8.2 Band Structure and Defects

8.2.1 Intrinsic Band Structure

The electronic band structure of $BaTiO_3$ and other ABO_3 perovskite oxides has been extensively studied by Michel-Calendini et al. [34–40] and others [41]. Two basic calculational approaches have been used. The NLCAO (nonorthogonal linear combination of atomic orbitals) approach has allowed the calculation of the band structure of $BaTiO_3$ throughout the Brillouin zone, using $2p$ orbitals centered on the oxygen ions and the $3d$ orbitals centered on the titanium ions in the unit cell [34, 35, 41]. The self-consistent field–multiple scattering–$X\alpha$ (SCF–MS–$X\alpha$) technique is used primarily for electronic structure calculations at the zone center. Its specific advantage is that it allows the calculation of energy levels due to substitutional impurities, as well as the intrinsic electronic structure. Michel-Calendini and coworkers have used the SCF-MS-$X\alpha$ technique to calculate the electronic structure of pure $BaTiO_3$ [36, 37], as well as $BaTiO_3$:Fe [37–39] and $BaTiO_3$:Co [40]; their work will be followed in the rest of this discussion.

On the basis of the SCF–MS–$X\alpha$ calculations carried out on TiO_6^{8-} clusters, the following picture emerges of the intrinsic electronic states in cubic (O_h) and

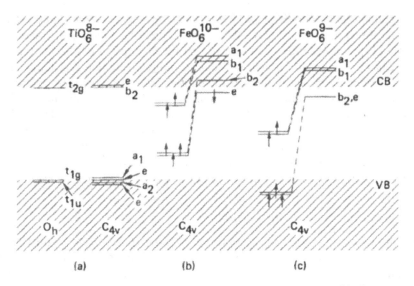

FIGURE 8.5. Schematic representation of the energy levels of $M^{n+}O_6^{(12-n)-}$ clusters in BaTiO₃: (a) TiO_6^{8-}, (b) FeO_6^{10-} (Fe^{2+}), and (c) FeO_6^{9-} (Fe^{3+}). In (b) and (c), the up-spin levels are located on the left side of each panel, and the down-spin levels are on the right. The levels in (a) and (c) have been calculated by the SCF–MS–Xα method [36, 39]. The levels in (b) are extrapolated from those of the $Fe^{2+}-V_0$ center [38]. The CB and VB correspond respectively to the conduction-band minimum (t_{2g} Ti 3d states) and the valence-band maximum (t_{1g} O 2p states of the TiO_6^{8-} cluster). Note the tendency of the higher valence states (e.g., Fe^{3+}) to lie lower in the band gap. Note also that Fe^{3+} can act either as a donor [occupied (a_1, b_1) states] or as an acceptor [empty (b_2, e) states].

tetragonal (C_{4v}) BaTiO₃. For both symmetries the top of the valence band is composed of oxygen 2p states, whereas the conduction bands are titanium 3d states. The states of the barium ion, on the other hand, lie well below the top of the valence band, and do not play an important role. In O_h symmetry the valence band edge can be decomposed into states with t_{1g} and t_{1u} symmetry, whereas the bottom of the conduction band has t_{2g} classification (Figure 8.5a). In this picture, the optical band gap results from t_{1u}–t_{2g} transitions, while the thermal gap is due to t_{1g}–t_{2g} transitions. As the symmetry is lowered from O_h to C_{4v}, the cubic levels split as well as shift. For example, in lower symmetry the levels (t_{1g}, t_{1u}) go over into (a_1, e; a_2, e), and (t_{2g}) into (b_2, e).

The optical gaps calculated by the Xα technique for undoped, tetragonal BaTiO₃ are 3.61 eV ($E\|\hat{c}$) and 3.56 eV ($E \perp \hat{c}$)[39]. The calculated values can be expected only approximately to predict the experimental values, for two reasons: (1) necessary approximations in the Xα approach limit its accuracy, and (2) all states in the conduction and valence bands are assumed to be discrete, so that transitions between them neglect broadening due to phonons, impurities, etc. The neglect of broadening should result in an overstimation of the band-gap energy.

The band-edge properties of $BaTiO_3$ can also be elucidated through the study of optical absorption. In a series of absorption studies of solution-grown samples, Wemple [42] and coworkers determined that the absorption edge of tetragonal $BaTiO_3$ has a significant Urbach tail, due to indirect (phonon-assisted) transitions. The optical gap was defined as the photon energy at which the absorption coefficient was 5×10^3 cm^{-1}. The resultant values at room temperature are 3.27 eV ($E \perp \hat{c}$) and 3.38 eV ($E \| \hat{c}$). Other optical measurements have yielded values of 3.1 eV [43] and 3.3 eV [44]. Subsequent absorption studies [45] have shown that doping with Fe^{3+} ions leads to a decrease in the band-gap energy, in proportion to the Fe concentration. This is likely due to the introduction of O $2p$–Fe $3d$ charge-transfer bands concentrated at energies just below the band-gap energy.

8.2.2 Vacancies and Impurities

There are two major types of defects in crystalline $BaTiO_3$ that could lead to deep levels in the band gap (and thus participate in the photorefractive effect): vacancies and impurities. It is always expected that impurities will be present in the 10–100 ppm range, unless extraordinary precautions are taken in purifying starting materials and controlling the growth environment. Vacancies are also commonly present, if only to compensate for the charge imbalance introduced by the impurities.

Of the three elements that make up the $BaTiO_3$ structure, the most volatile is oxygen. During crystal growth or processing at high temperatures, this species is the most likely to evaporate, thus creating oxygen vacancies, denoted by V_o. Since the oxygen ion in the $BaTiO_3$ lattice has a charge of $2-$, an oxygen vacancy has a charge of $2+$. Thus, an oxygen vacancy can trap one or two electrons, leading to donor levels in the $BaTiO_3$ band gap. The binding energy for the trapped electrons is thought to be quite small [44, 46–48], leading to shallow levels in the band gap. Berglund and Braun [46] estimate an energy depth of 0.2 eV for an O vacancy with one trapped electron and 0.025 eV for an O vacancy with two trapped electrons. At room temperature these levels are expected to be thermally ionized, and thus not significantly populated.

Vacancies in $BaTiO_3$ may also be produced from the incorporation of excess TiO_2 during solution growth [47, 49]. Both Ba and O vacancies are formed by the reaction

$$TiO_2 \rightarrow V_{Ba} + Ti_{Ti} + 2O_o + V_o, \tag{8.6}$$

where the subscripts on the right-hand side refer to the site in the lattice. According to the commonly used phase diagram for the BaO-TiO_2 system [50], the solubility of TiO_2 in $BaTiO_3$, at the typical growth temperature of 1400 °C is ≈ 1 mol%, leading to a very large concentration of Ba and O vacancies. More recent measurements [47, 49] have shown that the solubility of TiO_2 in $BaTiO_3$ is at most 100 ppm. Thus, the maximum concentration of O and Ba vacancies due to TiO_2 incorporation is 100 ppm. When acceptor impurities are also present (see below), the additional O vacancies required for charge compensation will cause a corresponding reduction in the number of Ba vacancies, through the equilibrium relation [51]

$$[V_{Ba}][V_o] = \text{const.}$$

Ba vacancies are deep acceptors [51], and could contribute to the photorefractive behavior of high-purity crystals.

We now wish to consider the possible impurities in BaTiO$_3$ and their relation to the photorefractive effect. Three major groups of impurities have been shown to exist in BaTiO$_3$ [52–54]: (1) calcium and strontium, (2) aluminum and silicon, and (3) transition metals. Calcium and strontium are in the same family of the periodic table as barium. They almost always occur with barium in nature, and are difficult to separate from it. Since calcium and strontium are isovalent with barium and substitute for barium in the BaTiO$_3$ lattice, they do not introduce any levels in the BaTiO$_3$ band gap, and thus cannot participate in the photorefractive effect.

The sources of aluminum and silicon impurities are the furnace walls and heating elements. However, aluminum and silicon possess only one stable valence state (Al^{3+} and Si^{4+}), and thus these elements cannot support the intervalence transfer, which is a key requirement of the photorefractive effect.

Transition metal impurities are ubiquitous in many oxide crystals, because of their abundance in nature, their chemical similarity to constituents of the compound in question (e.g., Ti in BaTiO$_3$), and possible contamination of starting powders from crucibles or metal utensils. The most likely transition metal impurities are Cr, Co, Ni, Fe, and Cu, and they are all expected to be present at levels as high as 50 ppm in BaTiO$_3$. Fe is the most abundant of the transition metals, and could be expected at still higher concentrations. It is believed that transition metal impurity ions (as well as Al^{3+} and Si^{4+}) substitute for Ti^{4+} in the BaTiO$_3$ lattice, due to the close match between their ionic radius and that of Ti^{4+}. In the case of Fe impurities in BaTiO$_3$, this has been proven in a number of separate studies [53, 55]. Due to the low binding energies of their 3d electrons, each of the transition metals can exist in several valence states, ranging typically between $+1$ and $+4$.

8.2.3 Charge Balance

When an impurity ion (with valence $+3$ or less) substitutes for Ti^{4+} in BaTiO$_3$, a charge imbalance is created. In the simplest case, the compensation for this imbalance is achieved through the creation of oxygen vacancies:

$$[M^{3+}] = 2[V_o^{2+}]_I, \tag{8.7}$$

where $[M^{3+}]$ is the concentration of a trivalent metal ion and $[V_o^{2+}]_I$ is the *impurity-related* concentration of oxygen vacancies. Note that the charge of the metal ion is expressed in relation to the isolated metal atom; the charge in the crystal is -1. In the simple model above, no free carriers are required for compensation, and the samples are expected to be highly insulating.

The above model does not account for the fact that the oxygen vacancy concentration is also influenced by the growth or processing environment of the crystal. At a given temperature there may be an excess or a deficiency of oxygen, compared with the amount required to produce $[V_o^{2+}]_I$. In either of these limits, free carriers are produced to maintain the charge balance:

$$[M^{3+}] + n = 2[V_o^{2+}] + p, \tag{8.8}$$

where n is the concentration of electrons, and p is the concentration of holes. The growth environment of commercial $BaTiO_3$ (grown by the top-seeded solution growth technique) [11] is oxidizing (i.e., there is excess oxygen), leading to a reduction in the number of vacancies below the value $[V_o^{2+}]_I$, and the creation of free holes. Thus, in as-grown crystals the transition metals are generally acceptor impurities, and the samples are p-type in-the dark [47, 56–58]. Conversely, samples processed at low-oxygen partial pressures requires the production of free electrons for charge balance, and are thus of n-type in the dark [47, 56–58].

In considering charge balance we must also account for the fact that transition metal dopants or impurities (say Fe) can change valence state, while other metal impurities (say Al) cannot. We may thus write

$$[Al^{3+}] + 2[Fe^{2+}] + [Fe^{3+}] + n = 2[V_o^{2+}] + p. \tag{8.9}$$

The presence of several valence states in the transition metals allows the charge balance to be maintained over a range of oxygen partial pressures without the creation of large numbers of free carriers. In this pressure range the major impact is the variation of the relative amounts of Fe^{2+} and Fe^{3+}, due to changes in $[V_o^{2+}]$. In p-type samples, the free-carrier concentration at room temperature is further reduced by trapping into acceptor levels, since these levels are thought to lie deep in the band gap [47, 56, 57, 59]. This leads to high resistivity values in samples grown or processed in air. By contrast, heavily reduced samples of $BaTiO_3$ are semiconducting (n-type) even at room temperature [47, 56, 60, 61], since the induced donor levels are shallow [47, 56] and are thus incapable of trapping electrons.

In addition to isolated metal impurities and fully ionized oxygen vacancies indicated in (8.9), we must also consider the possibility that association of ionic defects or trapping of free charges may lead to new species. Two specific cases must be considered. First, at high temperatures during crystal growth or processing, oxygen vacancies can associate with metal ions, due to Coulomb attraction. Assuming that this association occurs among nearest neighbors, the resulting structure in $BaTiO_3$ would be an $(M^{n+}O_5)(M^{n+}O_5)^{(10-n)-}$ cluster. This complex is called an $M^{n+}-V_o$ center. Such centers are known to exist in perovskites, but no consensus exists as to their importance. The concentration of these centers can be determined by electron paramagnetic resonance (EPR). According to recent EPR measurements of Fe^{3+} in melt-grown samples [52], the population of $Fe^{3+}-V_o$ centers is much less than that of isolated Fe^{3+}.

A second possible form of association is the trapping of one or two free electrons at oxygen vacancies. As indicated earlier, the resultant states are thought to be quite shallow [46–48], and are thus likely to be populated significantly only at low temperatures, or in heavily reduced samples.

8.2.4 Energy Levels of Fe^{2+} and Fe^{3+}

The electronic structure of a $3d$ ion embedded in BaTiO$_3$ depends on the oxidation state, the electronic ground term (spin configuration) of the impurity, and the symmetry of the local electrostatic crystalline field. Michel-Calendini and coworkers [36–39] have used the spin-unrestricted SCF–MS–Xα technique to calculate the molecular orbital (energy level) diagram of FeO$_6^{(12-n)-}$ (Fe^{n+}) clusters with O_h and C_{4v} symmetries. These calculations provide information on the location of the empty and occupied Fe impurity states in the TiO$_6^{8-}$ band gap. The pertinent energy level diagrams for the FeO$_6^{10-}$ (Fe^{2+}) and FeO$_6^{9-}$ (Fe^{3+}) clusters are shown in Figures 8.5b and c [36, 38, 39]. In order to gain a better understanding of the photorefractive mechanisms in BaTiO$_3$, we wish to relate the absolute levels given in Figure 8.5 with the relevant ionization energies. The connection is easily seen from the following equation for the Fe^{2+}/Fe^{3+} system:

$$\{Fe^{3+}, (VB)^m, L^n\} \rightarrow \{Fe^{2+}, (VB)^{m-1}, L^{n+1}\}.$$

This equation describes the change in formal charge state that occurs during the conversion of Fe^{3+} ($e_\uparrow^2 b_{2\uparrow} a_{1\uparrow} b_{1\uparrow}$) to Fe^{2+} ($e_\uparrow^2 b_{2\uparrow} a_{1\uparrow} b_{1\uparrow} e_\downarrow$) by transferring an electron from the valence band (the fully occupied valence band has m electrons) to an $e_\downarrow$ impurity state (hole emission). Here L refers to the one-electron states $b_{1\uparrow}, e_\downarrow$, etc.; $\downarrow$ and $\uparrow$ refer to up- and down-spin electron spin states; and VB denotes the valence band. From this it follows that the energy of the ionization level above the valence band is given by

$$E(Fe^{2+}/Fe^{3+}) = E_T\{Fe^{2+}, (VB)^{m-1}, L^{n+1}\} - E_T\{Fe^{3+}, (VB)^m, L^n\},$$

where $E_T\{\cdots\}$ represents the total energy of the lowest multiplet of the corresponding charge state.

Calculated values of the ionization energies of transition metal impurities in BaTiO$_3$ are not available. Instead, we can gain experimental information on the energy levels from the thermogravimetric data of Hagemann and coworkers [44, 54]. The resulting ionization energy level diagram for BaTiO$_3$: Fe is given in Figure 8.6. Also included are the levels due to oxygen vacancies with one or two trapped electrons [44, 46]. The notation in Figure 8.6 is now consistent with that used by the semiconductor community. In this picture, the ionization energy of Fe^{2+} (leading to Fe^{3+} and a free electron) is 3.1–2.3 = 0.8 eV. Similarly, the energy required to convert Fe^{4+} to Fe^{3+} through the generation of a free hole is 0.5 eV. In Figure 8.6 we have assumed that the Fermi level is just below the Fe^{2+}/Fe^{3+} level, leading to an admixture of Fe^{2+} and Fe^{3+} (with [Fe^{3+}] > [Fe^{2+}]), and a negligible population of Fe^{4+}. Note that the energy positions determined by this and other approaches based on thermal activation may not be equal to the energies for *optical* transitions. Note also that the Fe energy level structure in

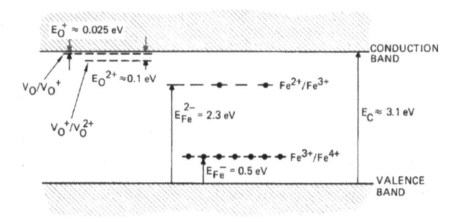

FIGURE 8.6. Energy levels of oxygen vacancies [44, 46] and Fe^{n+} ions [44, 54] in $BaTiO_3$. The band-gap energy of 3.1 eV is the value adopted Hagemann [44].

Figure 8.6 is clearly a simplification of the calculated structure given in Figure 8.5. Nevertheless, it is adequate to describe most thermal and gravimetric measurements on $BaTiO_3$. We shall see below that it is also adequate to describe most photorefractive measurements.

In the semiconductor framework of Figure 8.6, the populations of Fe^{2+}, Fe^{3+}, and Fe^{4+} can be related to the Fermi energy E_F by

$$\frac{[Fe^{4+}]n}{[Fe^{3+}]} = 2N_C \exp\left[-\frac{(E_C - E_{\overline{Fe}})}{k_B T}\right] \tag{8.10}$$

and

$$\frac{[Fe^{3+}]n}{[Fe^{2+}]} = \frac{1}{2}N_C \exp\left[-\frac{(E_C - E_{Fe}^{2-})}{k_B T}\right], \tag{8.11}$$

where N_C is the density of states in the conduction band and E_C is the energy of the conduction band edge. Equation (8.10) is simply the law of mass action for

$$Fe^{3+} \rightleftarrows Fe^{4+} + e^-, \tag{8.12}$$

while (8.11) is the corresponding relation for

$$Fe^{2+} \rightleftarrows Fe^{3+} + e^-. \tag{8.13}$$

8.3 Band-Transport Model

8.3.1 Energy-Level Model

In order to develop a model for grating formation in a photorefractive material, it is first necessary to postulate a system of centers in the band gap that can generate

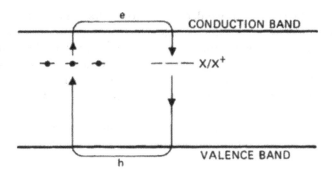

FIGURE 8.7. Energy-level model for the photorefractive effect. Electrons are photoionized from filled states X and recombine at empty states X^+; holes are photoionized from level X^+ and recombine at X.

carriers by photoionization, and trap carriers by recombination. Chapters 3 and 4 of this volume describe several energy-level models that have been postulated in the past. The most frequently used model assumes a single photorefractive species that can exist in two ionization states. This model was originally proposed by Chen [62] and Peterson et al. [63] for LiNbO$_3$ with Fe dopants or impurities. In the particular case of BaTiO$_3$ (with Fe impurities), this model is sufficient to explain steady-state beam-coupling measurements in as-grown crystals [32, 33], as well as certain features of their transient response [64]. A two-species model in which O Vacancies are also considered to participate in charge creation and trapping has also been proposed for BaTiO$_3$ [33]. This model differs from the single-species model with regard to the performance of reduced crystals (Section 8.6).

The energy-level model used for our ensuing discussion is shown in Figure 8.7. We assume that a single species X, which can exist in two valence states (X and X^+), is responsible for the photorefractive properties of BaTiO$_3$. In the case of iron-doped LiNbO$_3$, X corresponds to Fe^{2+} and X^+ corresponds to Fe^{3+}. In commercial samples of BaTiO$_3$, we also assume that the Fe^{2+} and Fe^{3+} are dominant [52], since Fe impurities (primarily in the form of Fe^{3+} ions) are consistently present at large concentrations. However, it should be noted that the presence of Fe^{2+} has not been proven directly, and in fact, the photorefractive species in as-grown samples could be Fe^{3+} (level X) and Fe^{4+} (level $X+$). The advantages of the Fe^{3+}/Fe^{4+} model are (1) the resulting location of the Fermi level below midgap is consistent with the p-type behavior of as-grown samples, and (2) the low hole ionization energy for Fe^{4+} is consistent with the observation of hole-dominated photoconductivity to 1.1 μm. In any case, the conclusions of the model described below are substantially unchanged for both Fe^{2+}/Fe^{3+} and Fe^{3+}/Fe^{4+}.

The representation of X/X^+ as a discrete, unique state is clearly a simplification of the Fe^{2+} and Fe^{3+} level structure given in Figures 8.5b and c. As indicated above, this simplification does not appear to affect the utility of the model, except perhaps in the interpretation of certain transient measurements [64, 65], to be discussed in Section 8.4.2.

In our energy-level model, we denote the concentration of X by N, and the concentration of X^+ by N^+. Other states, which are optically inactive, provide overall charge compensation within the crystal (Section 8.2.3). It is important to note that electrons or holes (or both) can contribute to the charge transport in BaTiO$_3$. For electron transport, state X is a donor, or "filled" state, and state X^+ is an ionized donor, or "empty" state. For hole transport, state X^+ is an acceptor or "filled" state, and state X is an ionized acceptor, or "empty" state. We further note that the sign of the space-charge, field is opposite for the two charge carriers. This changes the direction of beam coupling, and allows a measurement of the dominant photocarrier.

8.3.2 Transport and Rate Coefficients

In association with the energy-level model and notation given above, we must also define rate coefficients for the individual levels, and transport coefficients for electrons and holes. The important coefficients are (1) the photoionization cross section, (2) the recombination rate coefficient, (3) the dark generation rate, and (4) electron and hole mobility. In the single-trap-species model, the photoionization cross section for level X (thus creating an electron) is s_e, while the cross section for hole creation from level X^+ is s_h. The recombination rate coefficient for electrons at centers X^+ is γ_e, and the corresponding coefficient for holes at centers X is γ_h. The dark creation rate for electrons (from centers X) is β_e, and the corresponding coefficient for holes (from centers X^+) is β_h. Finally, the mobility for electrons (or holes) is μ_e (or μ_h). It must also be kept in mind that the photoionization cross sections and mobilities for BaTiO$_3$ at room temperature are anisotropic; the cross sections depend on the polarization orientation of the incident radiation with respect to the c-axis (Section 8.5.1), and the mobilities depend on the direction of transport with respect to the c-axis (see below).

The photoionization cross sections for states X and X^+ are related to the total absorption coefficient α by

$$\alpha = \alpha_P + \alpha_{NP}, \tag{8.14}$$

or

$$\alpha = N s_e + N^+ s_h + \alpha_{NP}, \tag{8.15}$$

where α_p is the photorefractive contribution and α_{NP} is the nonphotorefractive contribution. In as-grown samples of BaTiO$_3$, $N \ll N^+$ [52] and thus

$$\alpha_P \approx N^+ s_h. \tag{8.16}$$

If we define the quantum efficiency as $\Phi = \alpha_P/\alpha$, then

$$\alpha = N^+ s_h / \Phi. \tag{8.17}$$

The recombination-rate coefficients are related to the carrier recombination time for electrons (τ_{Re}) and holes (τ_{Rh}) by

$$\tau_{Re} = \frac{1}{\gamma_e N^+} \tag{8.18}$$

and

$$\tau_{Rh} = \frac{1}{\gamma_h N}. \qquad (8.19)$$

The mobility for holes and electrons is related to the conductivity σ by

$$\sigma = \sigma_e + \sigma_h = ne\mu_e + pe\mu_h, \qquad (8.20)$$

where σ_e, σ_h are the electron and hole conductivities. The accurate measurement of mobility in insulating oxides is particularly difficult. In semiconductors, the Hall effect and the four-point probe technique can be used to measure the carrier concentration and conductivity, respectively, yielding accurate values of Hall mobility, via (8.20). In as-grown BaTiO$_3$, with a typical resistivity value of 10^{10}–10^{12} Ωcm, Hall measurements are not possible, and conductivity measurements are more difficult. In order to obtain meaningful Hall data for BaTiO$_3$, workers have been forced to measure reduced samples with much lower resistivities. Another approach to the measurement of mobility is the drift technique, in which carriers are injected from one surface of a plane-parallel sample, and their transit time across the sample is measured. This approach is limited by the small signals involved and the reliance on an accurate model of charge trapping and transport in the sample.

In Table 8.2 we have collected literature values for mobility in BaTiO$_3$. In most cases, the mobilities were determined by Hall measurements on reduced samples, or drift time measurements on as-grown samples. In a few cases, the measured thermal emf was used to determine the free-carrier concentration. There is a large spread in the data, with measured values ranging from 3×10^{-4} to 1 cm^2/Vs. Note also that although the mobility in BaTiO$_3$ at room temperature is known

TABLE 8.2. Room-temperature mobility measurements in BaTiO$_3$.

Reference	Value [cm^2/Vs]	Identification	Type of experiment	Sample
Electron mobility				
[66]	2	μ_e	Hall	KF flux (reduced)
[61]	0.1	μ_e	Hall	KF flux (reduced)
[67]	0.5	μ_e	Hall, drift	Ceramic
[68][a]	0.07	μ_e	Thermal emf	Ceramic
[60]	0.13	$\mu_{e\parallel}$	Hall	TiO$_2$ solution (reduced)
	1.0	$\mu_{e\parallel}$		
[69]	5×10^{-4}	μ_e	Hall	KF flux (reduced)
[70]	0.2	μ_e	Thermal emf	KF flux (reduced)
[71]	0.5	μ_e	Hall	Ceramic
	0.6	μ_e	Thermal emf	
[72]	0.6	μ_e	Hall	Verneuil (polydomain, reduced)
[73]	$(3\pm) \times 10^{-3}$	μ_e	Drift time	TiO$_2$ solution (as-grown)
Hole mobility				
[74]	$(3 \pm 1) \times 10^{-4}$	μ_h	Drift time	KF flux (as-grown)
[75]	0.02–0.03	μ_h	Drift time	TiO$_2$ solution (as-grown)
[76]	>0.5	$\mu_{h\parallel}$	Photorefractive	TiO$_2$ solution (as-grown)

[a] Extrapolated from measurement at 1100 °C.

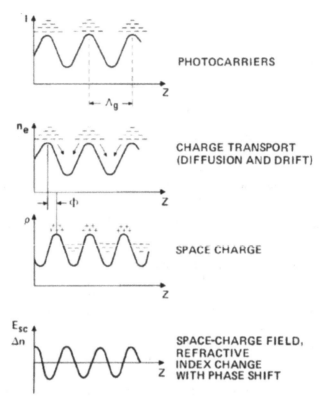

FIGURE 8.8. Gratings in a photorefractive material. The periodic irradiance pattern results from the interference of two waves in the material.

to be highly anisotropic [60], little attention has been paid to this property. One important exception is the photorefractive measurement by Tzou et al. [76], to be discussed later.

8.3.3 Grating Formation

The band-transport model [77–80] is commonly used to describe grating formation in a photorefractive material. Under the influence of periodic illumination (Figure 8.8), electrons (or holes) are optically excited from filled donor (or acceptor) sites to the conduction (or valence) band, where they migrate to dark regions in the crystal by drift or diffusion before recombining into an empty trap. The transported charges result in an ionic space-charge grating that is, in general, out of phase with the incident irradiance. The periodic space charge is balanced by a periodic space-charge electric field in accordance with Poisson's equation. This space-charge field modulates the refractive index through the electrooptic effect. If no electric field is applied to the crystal (as is generally the case in experiments with $BaTiO_3$), then diffusion alone leads to a phase shift of $\pi/2$ between the incident irradiance and the space-charge field. This shifted grating plays an important role in many device applications of BaTiO$_3$.

A mathematical description of the grating formation process for a single charge carrier was given in its most complete form by Vinetskii, Kukhtarev, and coworkers [77–79] (see also [80]). For the case in which both charge carriers play an important role, the early work of Orlowski and Krätzig [81] has been developed and expanded by Valley [82] and by Strohkendl et al. [83], leading to solutions in the steady-state and the transient regimes. See Chapter 3 for a review of the band-transport model. An important result of the steady-state two-carrier model is that the space-charge field (with no applied field and negligible photovoltaic effect) is purely imaginary, and can be written as

$$\delta E = im E_{sc} = im \frac{k_B T}{e} \frac{K}{1 + K^2/K_s^2} \bar{\sigma}(K), \qquad (8.21)$$

where m is the fractional modulation, K is the grating wave number, and K_s is the Debye screening wave vector, given by

$$K_s = \left(\frac{e^2 N_E}{\varepsilon \varepsilon_0 k_B T} \right)^{1/2}. \qquad (8.22)$$

In (8.22), ε is the relative dielectric constant and N_E is the effective number of empty traps, defined as

$$N_E = \frac{N N^+}{N + N^+}. \qquad (8.23)$$

The factor $\bar{\sigma}(K)$ in (8.21) accounts for the relative contribution of electrons and holes to the photorefractive performance of the sample. It is given by [83]

$$\bar{\sigma}(K) = \frac{C - 1}{C + 1}, \qquad (8.24)$$

in which

$$C = \frac{s_h N^+ (K^2 + K_e^2)}{s_e N (K^2 + K_h^2)}. \qquad (8.25)$$

The quantities K_e^{-1} and K_h^{-1} are the diffusion lengths (also known as the hopping range) for electrons and holes:

$$K_e^{-1} = \left(\frac{k_B T \mu_e}{e \gamma_e N^+} \right)^{1/2}, \quad K_h^{-1} = \left(\frac{k_B T \mu_h}{e \gamma_h N} \right)^{1/2}. \qquad (8.26)$$

For small values of $K (K \ll K_e, K_h)$, $\bar{\sigma}$ is independent of K, and can be written in terms of the photoconductivity of each carrier,[1]

$$\bar{\sigma} = \frac{\sigma_h - \sigma_e}{\sigma_h + \sigma_e} = \frac{\mu_h p - \mu_e n}{\mu_h p + \mu_e n}. \qquad (8.27)$$

---

[1] Note also that for large values of $K (K \gg k_e, K_h)$, $\bar{\sigma}$ is again independent of K and can be written in terms of the *absorption coefficient* for each carrier [82] : $\bar{\sigma} = (\alpha_h - \alpha_e)/(\alpha_h + \alpha_e)$.

When the photoconductivity for holes is dominant, $\bar{\sigma} = +1$, whereas $\bar{\sigma} = -1$ when electron photoconductivity dominates. In a beam-coupling experiment, this change in sign corresponds to a change in coupling direction. For the case in which the photoconductivity values for the two carriers are nearly equal (and in the absence of an external or photovoltaic field), the net space-charge field (and thus the beam-coupling gain) is small, being limited only by grating-dependent correction terms [82, 83]. In such crystals, the space-charge field (and thus the beam coupling gain) can actually change sign with grating period [32] (Section 8.4.3).

It is useful to determine the trap populations at the compensation point, using our two-level model. We start by writing the compensation condition

$$\sigma_h = pe\mu_h = ne\mu_e = \sigma_e, \qquad (8.28)$$

or

$$\frac{p}{n} = \frac{\mu_e}{\mu_h}. \qquad (8.29)$$

The free-carrier concentrations n and p can be related to the trap concentrations N and N^+ by the rate equations

$$s_e I N = \gamma_e n N^+ \qquad (8.30)$$

and

$$s_h I N^+ = \gamma_h p N, \qquad (8.31)$$

where I is the irradiance. These two equations may be combined to yield

$$\frac{N^+}{N} = \left(\frac{\gamma_h s_e p}{\gamma_e s_h n}\right)^{1/2}. \qquad (8.32)$$

Finally, by combining with (8.29), we find the condition at the compensation point as

$$\frac{N^+}{N} = \left(\frac{\gamma_h s_e \mu_e}{\gamma_e s_h \mu_h}\right)^{1/2}. \qquad (8.33)$$

Note that we are concerned here with the compensation of the *photoconductivity*; dark conductivity is neglected. In the dark, the rate equations (8.30) and (8.31) would be replaced by

$$\beta_e N = \gamma_e n N^+ \qquad (8.34)$$

and

$$\beta_h N^+ = \gamma_h p N, \qquad (8.35)$$

leading to a different condition for compensation. Note also that the values of free-carrier density and irradiance in the above discussion are spatially averaged values.

It may be seen from the above discussion that if one can adjust the ratio of trap concentrations to be equal to the value given in (8.33), the photorefractive gain (in the absence of an applied or photovoltaic field) can be reduced to a very small value. This could be important for applications in which the photorefractive effect is undesirable. Furthermore, the measurement of N^+/N at the compensation point gives a value of the ratio

$$R = \frac{\gamma_h S_e \mu_e}{\gamma_e S_h \mu_h}, \tag{8.36}$$

and thus provides information regarding transport coefficients. Techniques for varying $N^+/N = [Fe^{3+}]/[Fe^{2+}]$ are discussed in Section 8.6.

The transient response of a photorefractive grating when both electrons and holes contribute to the transport of charge has been analyzed by Valley [82] and Strohkendl et al. [83] (see Chapter 3). In the absence of an applied or photovoltaic field, the erasure rate for our single-species model is given by

$$\gamma = \gamma_{die} \frac{1 + (K/K_s)^2}{1 + (K/K_e)^2} + \gamma_{dih} \frac{1 + (K/K_s)^2}{1 + (K/K_h)^2}, \tag{8.37}$$

where γ_{die} and γ_{dih} are the dielectric relaxation rates for electrons and holes:

$$\gamma_{die} = \frac{\sigma_e}{\varepsilon \varepsilon_0} = \frac{n e \mu_e}{\varepsilon \varepsilon_0}, \quad \gamma_{dih} = \frac{\sigma_h}{\varepsilon \varepsilon_0} = \frac{p e \mu_h}{\varepsilon \varepsilon_0}. \tag{8.38}$$

The dielectric relaxation rates contain contributions from dark carriers and photocarriers. By substituting the rate equation solutions for the photocarrier concentrations n and p (8.30 and 8.31) into (8.38), we may write

$$\gamma_{die} = \frac{\sigma_{de}}{\varepsilon \varepsilon_0} + \left(\frac{e \alpha_{Pe} I}{\varepsilon \varepsilon_0} \right) \mu_e \tau_{Re} \tag{8.39}$$

and

$$\gamma_{dih} = \frac{\sigma_{dh}}{\varepsilon \varepsilon_0} + \left(\frac{e \alpha_{Ph} I}{\varepsilon \varepsilon_0} \right) \mu_h \tau_{Rh}, \tag{8.40}$$

where σ_{de}, σ_{dh} are the dark conductivities for electrons and holes, and α_{Pe}, α_{Ph} are the photorefractive absorption coefficients for electrons and holes. Note that the dielectric relaxation rates vary linearly with intensity in this model. It should also be noted that the dielectric relaxation rates and the diffusion rates are anisotropic, through their dependence on μ and ε. The application of (8.37) to the specific case of BaTiO$_3$ will be discussed in Section 8.4.2.

8.4 Physical Measurements Using the Photorefractive Effect

Aside from the many device applications of photorefractive materials, the photorefractive effect may also be used to great advantage for measuring fundamental material properties. The specific advantage of the photorefractive effect for such measurements is that it can be used to generate internal electric fields without the use of contacts. This allows the measurement of electrical and electrooptic properties without concern for the quality and characterization of the contacts.

In our discussion below, we will describe techniques for steady-state and transient photorefractive measurements in BaTiO$_3$. A more detailed description of techniques in the transient (and short-pulse) regime is provided in Chapter 6. We will then discuss the specific measurements that have been made in BaTiO$_3$.

8.4.1 Steady-State Measurements

One useful technique that provides information on the magnitude and sign of the steady-state space-charge field (and thus a number of important photorefractive parameters) is the measurement of the beam-coupling gain Γ, given by

$$\Gamma = 2\pi n^3 \frac{F r_{\mathrm{ang}} E_{\mathrm{sc}}}{\lambda}, \tag{8.41}$$

where n is the refractive index, F is the fractional poling factor, and r_{ang} is the appropriate combination of electrooptic tensor components and angular and polarization factors for a fully poled crystal. If we substitute for E_{sc} in (8.41), we obtain

$$\Gamma = \frac{2\pi n^3}{\lambda} \frac{k_{\mathrm{B}} T}{e} r_{\mathrm{eff}} \frac{K}{1 + K^2/K_{\mathrm{s}}^2}, \tag{8.42}$$

or

$$\Gamma = \frac{(2\pi)^2 n^3}{\lambda} \frac{k_{\mathrm{B}} T}{e} r_{\mathrm{eff}} \frac{\Lambda}{\Lambda^2 + \Lambda_0^2}, \tag{8.43}$$

where

$$r_{\mathrm{eff}} = F \bar{\sigma} r_{\mathrm{ang}}, \tag{8.44}$$

and

$$\Lambda_0 = l_{\mathrm{s}} = \frac{2\pi}{K_{\mathrm{s}}} = \left(4\pi^2 \varepsilon \varepsilon_0 \frac{k_{\mathrm{B}} T}{e^2 N_{\mathrm{E}}} \right)^{1/2} \tag{8.45}$$

is the Debye screening length. In including $\bar{\sigma}$ as part of the effective electrooptic coefficient, we follow the approach of the previous studies of steady-state beam coupling in BaTiO$_3$ [33, 52], which neglected the K-variation of $\bar{\sigma}$. This assumption is strictly true only for $K \ll K_{\mathrm{e}}$, K_{h}, or when one photocarrier is dominant ($\bar{\sigma} = \pm 1$).

In (8.42) [or (8.43)] there are two material parameters that can be measured directly in a beam-coupling experiment: r_{eff} and l_{s} (or K_{s}). We may then obtain $F \bar{\sigma}$ from r_{eff}, provided r_{ang} is known, and $N_{\mathrm{E}} = N N^+/(N + N^+)$ from l_{s} (or K_{s}), provided ε is known. The most direct means of determining these parameters is

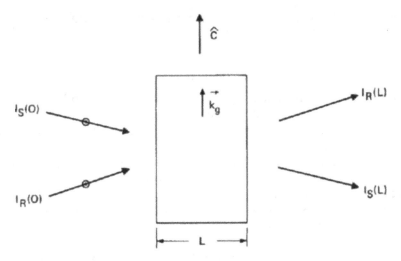

FIGURE 8.9. Crystal orientation and beam notation for beam-coupling measurements. Both beams are s-polarized.

to fit the function given by (8.42) or (8.43) to the data of Γ versus K (or Λ), with r_{eff} and K_s (or l_s) as variables. An alternative approach [32] is to recast (8.42) as

$$\frac{K}{\Gamma} = \frac{e\lambda}{2\pi n^3 k_B T r_{\text{eff}}} \left(1 + \frac{K^2}{K_s^2}\right). \tag{8.46}$$

Thus, if r_{eff} is independent of K, the experimental data plotted as K/Γ vs. K^2 should lie on a straight line, whose intercept yields r_{eff}, and whose value of slope/intercept yields K_s. In the more general case (accounting for the K-dependence of σ), the data may still retain an approximate straight-line behavior, but the interpretation is more complicated [83].

The photorefractive gain is commonly measured using the geometry shown in Figure 8.9. A first step in all measurements is to orient the c-axis of the crystal to provide gain for the signal beam I_S. The knowledge of this orientation, along with an independent electrical measurement of the polarity of the c-axis, yields the sign of the dominant charge carriers [8, 81]. For the beam notation shown in Figure 8.9 and the crystal orientation that gives gain for the signal beam, the transmission of the signal beam is given by

$$\frac{I_S(L)}{I_S(0)} = \frac{[I_S(0) + I_R(0)]\exp[(\Gamma - \alpha)L]}{I_R(0) + I_S(0)\exp(\Gamma L)}. \tag{8.47}$$

For negligible depletion of the pump wave $[I_S(0)\exp(\Gamma L) \ll I_R(0)]$, (8.47) reduces to

$$I_S(L) = I_S(0)\exp[(\Gamma - \alpha)L]. \tag{8.48}$$

The quantity typically measured in the laboratory is the effective gain [84] γ_0, defined by

$$\gamma_0 = \frac{I_S(L) \text{ with reference wave}}{I_S(L) \text{ without reference wave}}. \tag{8.49}$$

When depletion of the reference wave can be neglected, $\gamma_0 = \exp(\Gamma L)$. Thus, the measurement of γ_0 provides a direct means of determining Γ.

The measurement of beam-coupling gain as a function of grating period has been used by several authors [8, 32, 33] to measure the sign of the dominant charge carrier, the Debye length, and the effective electrooptic coefficient r_{eff} in as-grown samples of BaTiO$_3$. The effective number of empty traps is then obtained from the Debye length, using (8.45) and a given value of dielectric constant. The factor $F\bar{\sigma}$ is obtained from r_{eff}, using (8.44) and a given value of the electrooptic coefficient. Finally, the factor $\bar{\sigma}$ can be obtained from $F\bar{\sigma}$, using an estimated value of the fractional poling factor F.

As indicated earlier, all measurements to date [8, 32, 33] have ignored the K-dependence of $\bar{\sigma}$, an assumption that is accurate only in samples with small values of the diffusion length, for both carriers. Strohkendl et al. [83] have shown that this assumption can lead to values of N_E that can differ by a factor of 3 from those obtained from a fit to the data using the full K-dependence of $\bar{\sigma}$.

The steady-state beam-coupling data for BaTiO$_3$ are collected in Table 8.3. We have also included in Table 8.3 values of Debye length and effective trap density

TABLE 8.3. Measurements of Debye length and effective electrooptic coefficient in BaTiO$_3$.

Crystal	Sign of dominant photo-carrier	Debye length $\Lambda_0 = l_s$ [µm]	Effective no. of empty traps N_E [$\times 10^{16}$ cm^{-3}]	Effective EO coefficient r_{eff} [pm/V]	$F\bar{\sigma} = \dfrac{r_{eff}}{r_{ang}}$	$\bar{\sigma}$
G/L[a]	+	0.46	4.4[f]	+7.3	+0.30[i]	+0.33[k]
R1[a]	+	0.34	8.1[f]	+9.5	+0.40[i]	+0.44[k]
R2[a]	−	0.57	2.8[f]	−7.9	−0.33[i]	−0.37[k]
GB3[a]	+	0.39	6.1[f]	+9.7	+0.41[i]	+0.46[k]
K2[a]	+	0.32	8.7[f]	+12.0	+0.50[i]	+0.56[k]
SC[a]	+	0.47	4.2[f]	+8.3	+0.35[i]	+0.39[k]
GB4[a]	+	0.55	3.0[f]	+4.2	+0.18[i]	+0.20[k]
ROCKY[b]		1.17	0.6[g]	2.0	0.14[j]	0.14[l]
MMD[b]		1.28	0.5[g]	3.4	0.24[j]	0.24[l]
HOP[b,c]	+	0.66	1.9[g]	+5.0	+0.36[j]	+0.36[l]
CAT[b,d]	+	0.36	6.2[g,b] 6.0[g,d]	+9.8	+0.70[j]	+70[l]
SWISS[b]	+	0.44	4.2[g]	+12.5	+0.89[j]	0.89[l]
DOYLE[b]	+	0.45	4.1[g]	+13.4	+0.96[j]	+0.96[l]
--------[e]		2.8	2.9[h]			

[a] [32], $\lambda = 442$ nm. [d] [64], $\lambda = 515$ nm. [g] $\varepsilon = \varepsilon_{33} = 150$. [j] $r_{ang} = r_{13} = 14$ pm/V.
[b] [33], $\lambda = 515$ nm. [e] [86], $\lambda = 497$ nm. [h] $\varepsilon = \varepsilon_{11} = 4200$. [k] $F = 0.9$.
[c] [8], $\lambda = 515$ nm. [f] $\varepsilon = \varepsilon_{33} = 168$. [i] $r_{ang} = r_{13} = 24$ pm/V. [l] $F = 1.0$.

determined from measurements of response time as a function of grating period [64, 86].

The measured values of N_E given in Table 8.3 vary between 0.5×10^{16} cm^{-3} and 87×10^{16} cm^{-3}. We noted earlier that the measured value of N_E depends on the chosen value of dielectric constant. Uncertainties in this parameter lead to comparable uncertainties in the value of N_E. It should also be noted that in as-grown samples of BaTiO$_3$, $N \ll N^+$ [52], thus yielding [through (8.23)] $N_E \approx N = [\text{Fe}^{2+}]$.

The values of $\bar{\sigma}$ given in Table 8.3 are dependent on the assumed values for r_{ang} and F. Unfortunately, there is disagreement as to the values of electrooptic coefficients for a perfectly poled (single domain) crystal, and the fractional poling in a given sample. In most beam-coupling measurements, the grating normal is along $\hat{c}$ and the optical polarization is along $\hat{a}$ (or $\hat{b}$), thus coupling to the electrooptic tensor component r_{13}. This avoids the complicating effects of beam fanning [87]. Until very recently, the only reported measurement of the unclamped r_{13} in a single-domain sample was the value $r_{13} = 24$ pm/V (see Table 8.1). However, the accuracy of this value is open to question, due to interference from nearby piezoelectric resonances. Recently, Ducharme et al. [28] have reported a value $r_{13} = 19.5$ pm/V at 1 kHz. This smaller value leads to larger values of $F\bar{\sigma}$, although there is still a variation of a factor of ≈ 7 among the measured samples, indicating a significant variation in the admixture of electron and hole conductivity (and/or the fractional poling).

A number of approaches can be used to determine the fractional poling F in a given sample [16]. Ducharme and Feinberg [33] have measured r_{eff} after poling at two different voltages and obtained the same value, thus indicating that the sample was free of 180° domains ($F = 1$). On the basis of etching studies, Klein and Schwartz [52] concluded that $F \gtrsim 0.9$ in their samples.

It should be noted that there is at least one report of a BaTiO$_3$ sample with anomalous beam-coupling behavior [32]: the magnitude of the steady-state gain was very small, and the sign of the gain changed as a function of grating period (Figure 8.10). It was theorized in [32], "that the photoconductivity was nearly compensated in this sample, leading to the appearance of a weaker effect of unknown origin." This is consistent with the later calculations of Valley [82] and Strohkendl et al. [83], who have shown that the change in sign can be explained by the weak dependence of $\bar{\sigma}$ on Λ (or K), in the limit that the electron and hole diffusion lengths cannot be neglected in comparison with the grating period.

8.4.2 Transient Measurements

An expression for the erase rate when both electrons and holes contribute to the photoconductivity was given in Section 8.3.3 (8.37) When $K \ll K_e$, K_h, (8.37) reduces to

$$\gamma = \gamma_{\text{di}}\left[1 + \left(\frac{K}{K_s}\right)^2\right], \tag{8.50}$$

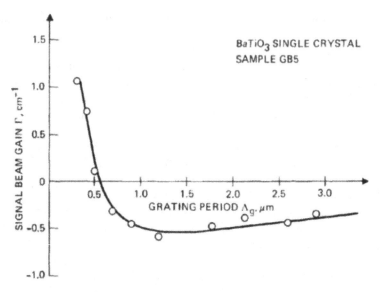

FIGURE 8.10. Beam-coupling gain as a function of grating period for the anomalous crystal GB5. The beams were p-polarized in this experiment, to exploit the larger electrooptic coefficient r_{33}.

where

$$\gamma_{di} = \gamma_{die} + \gamma_{dih}. \tag{8.51}$$

Thus, a measurement of γ vs. K at a fixed intensity yields values for K_s and γ_{di} [8, 64]. One may then obtain N_E from K_s, using (8.22). This measurement complements the separate determination of K_s from steady-state beam-coupling measurements.

By comparison with (8.39) and (8.40), the dielectric relaxation rate may be written as [64]

$$\gamma_{di} = \gamma_{dark} + \gamma_{light}, \tag{8.52}$$

where γ_{light} is a function of intensity that is zero when $I = 0$. Thus, a measurement of γ_{di} as a function of intensity yields γ_{dark} from the low-intensity limit, as well as the intensity dependence of γ_{light}. Assuming that the dark relaxation rate is impurity activated from a single center (with thermal activation energy E_0), then

$$\gamma_{dark} \approx \exp(-E_0/k_B T), \tag{8.53}$$

According to our two-level model, see (8.39) and (8.40), we expect that $\gamma_{light} \approx I$. This simple linear dependence is not generally observed in transient experiments (see below).

As discussed in Section 8.4.2, the measurement of decay rate vs. intensity (in the limit of small K) yields γ_{di} as a function of intensity. This in turn allows the measurement of the intensity dependence of γ_{light} in (8.52). In general, the

TABLE 8.4. Intensity dependence of decay rate, thermal activation energy, and dark resistivity in as-grown BaTiO$_3$ at room temperature.

Reference	Exponent x ($\gamma_{light} \sim I^x$)	Thermal activation energy E_0 [$\gamma_{dark} \sim \exp(-E_0/k_B T)$] [eV]	Dark resistivity [Ω cm]
[1]	0.5–1.0	0.7	
[8]	0.8		
[9]	0.6–0.9[a]		
[64, 88]	0.8[b]		
[65]	0.5		
[64]	0.62	0.98	10^{11}
	0.71 (40 °C)		
[33]	0.67 (oxidized)	0.97 (oxidized)	
	1.0 (reduced)		

[a] Exponent x determined from photoconductivity measurement.
[b] Exponent x determined from scaling of erase energy in short-pulse regime.

intensity variation of γ_{light} has been found to be of the form $\gamma_{light} \approx I^x$. In the limit of zero intensity, the dark activation energy E_0 and the dark resistivity can also be measured. In Table 8.4 we list measured values of the exponent x, thermal activation energy, and dark resistivity, determined from measurements of decay rate in BaTiO$_3$. We have also included data on the intensity variation of the photoconductivity, as well as the intensity variation of the write energy in a short-pulse experiment.

It is clear from Table 8.4 that as-grown BaTiO$_3$ is characterized by a sublinear variation of erase rate with intensity. This is important for short-pulse applications; with nanosecond pulses, the measured write time is an order of magnitude longer than that predicted from a linear intensity variation [88]. The sublinear response is in disagreement with the simple model leading to (8.39) and (8.40). This behavior suggests that other shallow levels may play an intermediate role in the trapping of electrons or holes [89]. Note also that the intensity variation becomes linear in reduced samples. This is consistent with the shallow position of the donor levels (V_o^+ or Fe^{2+}), which precludes the presence of more shallow intermediate levels.

In a recent transient and short-pulse study of as-grown BaTiO$_3$, Tzou et al. [76] have measured the ratio $\mu_{h\|}/\mu_{h\perp}$ and the recombination time τ_{Rh} for holes. The key to this experiment was the use of the grating obtained from counterpropagating beam directions to access larger experimental values of grating wave vector K. The importance of this limit is that K is now comparable to K_h in the parallel and perpendicular directions, and thus K_h can be obtained from a fit to the data. From the measured values of K_h, the values $\mu_{h\|}\tau_{Rh} = 5 \times 10^{-10}$ cm^2/V and $\mu_{h\perp}/\mu_{h\|} = 18 \pm 7$ were obtained, assuming $\varepsilon_\perp/\varepsilon_\| = \varepsilon_a/\varepsilon_c = 25$. In order to obtain individual values of $\mu_{h\|}$ and $\mu_{h\perp}$, the recombination time must be measured. This can be obtained from a time-resolved measurement of the grating formation in a short-pulse experiment [90]. Using 8-ns pulses, it was found that τ_{Rh} was less than 1 ns. This leads to the result $\mu_{h\|} > 0.5$ cm^2/Vs, which is consistent with estimates made earlier by Valley [90]. The measured value of hole mobility is larger than most reported values of μ_h in $BaTiO_3$ (see Table 8.2). It is possible that the sample measured contained a significant electron contribution to the photorefractive effect, so that the measured mobilities would also represent an admixture of the electron and hole mobility. Nonetheless, this is the first photorefractive measurement of the mobility in $BaTiO_3$; further measurements should help reduce the uncertainty in the correct value for photorefractive calculations.

8.5 Other Measurements in Photorefractive Crystals

While the physical and optical properties of $BaTiO_3$ have been studied for many years, only recently have attempts been made to study samples with known photorefractive properties, in order to understand the physical origins of the photorefractive effect. In this section we will describe such measurements on commercial as-grown samples, and the reasoning that leads to the conclusion that Fe is the dominant photorefractive species in $BaTiO_3$ [52].

8.5.1 Optical Absorption Coefficient Measurements

In Section 8.3.2 it was noted that the absorption coefficient for as-grown $BaTiO_3$ can be written as $\alpha = N^+ s_h / \Phi$. Thus, the measurement of α in a number of crystals yields a relative measurement of $N^+ \approx [Fe^{3+}]$, provided that the quantum efficiency is constant. In Figure 8.11 we show the measured absorption coefficients as a function of wavelength for four different $BaTiO_3$ crystals [52]. We see that the absorption coefficient is anisotropic, as expected from the uniaxial symmetry of $BaTiO_3$ at room temperature. Note also that three of the four crystals show a broad, featureless "tail" extending from the fundamental absorption edge near 3500 Å. One sample (GB3) does show a more prominent peak at 5000 Å, indicating that the impurity content is different in this crystal. Crystal R2 is also anomalous in the sense that the absorption anisotropy is reversed in sign from the other crystals. This effect may be related to the electron photoconductivity in this crystal. In Table 8.5 we present the value of α at 4416 Å for eight samples (with $E \perp \hat{c}$), in order to allow a relative determination of the density of filled traps.

8.5.2 Impurity Identity and Concentrations

Secondary-ion mass spectroscopy (SIMS) has been used to determine the impurities in two different crystalline $BaTiO_3$ samples [52]. These measurements, while only qualitative, give a sensitive indication of all impurities. The elements observed were H, Li, Na, K, C, Al, Si, Ca, Sr, Ni, Mn, Cr, and Fe. Of the elements listed, the first seven do not exist commonly in more than one valence state, and thus cannot support the intervalence transfer required for the photorefractive species. Ca and Sr substitute readily for Ba in the $BaTiO_3$ structure; however, they are isovalent with Ba, and thus produce no energy level in the $BaTiO_3$ band gap.

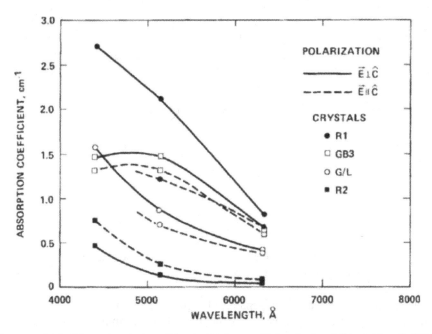

FIGURE 8.11. Spectral variation of the absorption coefficient in the visible for four BaTiO$_3$ samples.

The remaining species are transition metals and could contribute to the photore-fractive effect.

In later measurements [52], six different photorefractive crystals (R1, R2, G/L, GB3, 1334, and BW1) were analyzed for impurities by spark emission spec-troscopy. A semiquantitative determination of all metals was carried out, along with a quantitative measurement of transition metals. All samples contained Al and Si at concentration levels in the 50–500 ppm range. As indicated above, these

TABLE 8.5. Absorption coefficient at 442 nm (for $E \perp \hat{c}$), iron impurity concentration, and Fe^{3+} concentration (by EPR) in as-grown single domain BaTiO$_3$.

Crystal	Absorption coefficient [cm^{-1}]	Iron impurity concentration [ppm]	Fe^{3+} concentration by EPR [10^{18} cm^{-3}]
G/L	1.6	120	2.0
R1	2.7	147	
R2	0.5	72	1.9
GB3	1.5	55	
BW1	3.5	51	4.7
1334	2.4	138	5.6

For Fe, 100 ppm $\approx 5 \times 10^{18}$ cm^{-3}.

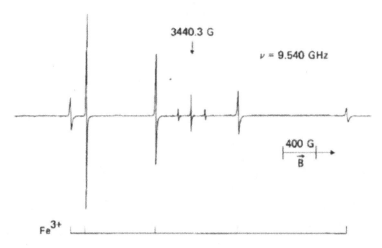

FIGURE 8.12. EPR spectrum observed with the steady magnetic field directed along the [010] axis [$B \perp \hat{c}$ and in the (100) plane].

species are not expected to participate in the photorefractive effect. Among the transition metals, iron was consistently the most abundant impurity, and the only element occurring consistently at concentrations greater than 50 ppm. The predominance of Fe impurities has previously been noted in $BaTiO_3$ [54]. The values of iron concentration for each crystal are included in Table 8.5.

8.5.3 EPR Measurements

EPR spectroscopy is a powerful technique for determining the presence and concentration of a variety of paramagnetic point defects and impurities. Furthermore, in the case of impurities, EPR can be used to identify the valence state. The disadvantage of EPR is that it is not equally sensitive to impurities in all possible valence states [52]. In the case of $BaTiO_3$, experiments are typically performed at room temperature, in order to avoid the tetragonal–orthorhombic phase transition near 9 °C. Thus, only those species with long spin-lattice relaxation times (and thus negligible line broadening) can be observed. For the important case of Fe impurities, we note that Fe^{3+} is observable at room temperature, but Fe^{2+} and Fe^{4+} are not.

In [52], the EPR spectra of four different crystals of $BaTiO_3$ were measured at room temperature. In Figures 8.12 and 8.13 we show representative spectra from crystal R2. The spectra were taken with the steady magnetic field directed along the [010] axis [$B \perp \hat{c}$ and in the (100) plane]. The spectrum shown in Figure 8.12 exhibits five intense resonances, which is characteristic of a $^6S_{5/2}$ magnetic ion (Fe^{3+}) in a noncubic electrostatic crystalline field [91]. One of the unique features of this spectrum as well as spectra at other angles is the sharpness of the lines. This reflects the high microscopic quality (i.e., reduced number of structural imperfections and reduced strain) of the. crystals studied. In Figure 8.13 we show

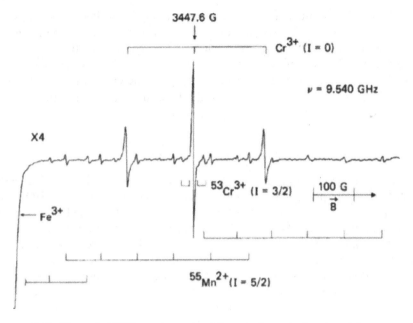

FIGURE 8.13. Expanded EPR spectrum about the $g \approx 2$ spectral region at the same magnetic field orientation as in Figure 8.12.

an expanded spectrum about the $g \approx 2$ spectral region at the same magnetic field orientation. We have indicated in this figure the resonance transitions that were assigned to the even isotopes ($I = 0$) 50, 52, and 54 and the odd isotope 53 ($I = 3/2$, 9.55% abundance) of Cr^{3+}, along with the odd isotope ($I = 5/2$) 55 of Mn^{2+}. Note that every spectral feature in Figure 8.13 has been identified.

The concentration of Fe^{3+} in several different $BaTiO_3$ crystals was determined by EPR [52] and the results are included in Table 8.5. By comparison, the concentrations of Cr^{3+} and Mn^{2+} were found to be at least two orders of magnitude lower than that of Fe^{3+}. No signals were observed in any of the samples from singly ionized oxygen vacancies. The detection limit for this measurement was $\approx 10^{16}$ cm^{-3}.

The angular dependence of the EPR spectra of crystal R2 was studied in detail. From the angular behavior in the plane perpendicular to the polar axis $\hat{c}$ of the five major transitions of the Fe^{3+} ion, it was established that the sites containing the paramagnetic species have a fourfold axis of symmetry (C_{4v}). It was further concluded that this symmetry resulted from a tetragonally distorted crystalline structure, rather than from the influence of a neighboring oxygen vacancy. The local site giving rise to the EPR spectrum of Fe^{3+} with C_{4v} symmetry in tetragonal $BaTiO_3$ is assumed to be associated with a tetragonally distorted FeO_6^{9-} cluster. This reasoning is based on the fact that if the local C_{4v} paramagnetic site were due to local compensation at a nearest-neighbor oxygen site, i.e., FeO_5^{7-} cluster ($Fe^{3+} - V_o$), then one would expect to observe a strong axial EPR spectrum, which is not consistent with the observations of [52]. Further, for the case of an $Fe^{3+} - V_o$ center, it is expected that at certain orientations of the magnetic field, three sets of axial resonances should be observed, each set having a unique tetragonal axis. This arises because there are three equivalent oxygen sites in the O_h FeO_6^{9-} unit from which an oxygen vacancy can be created to form the C_{4v} FeO_5^{7-} cluster. Again this was not observed, which supports the view that the iron resonances are associated with tetragonally distorted FeO_6^{9-} octahedra.

8.5.4 Correlation of Measured Parameters

The EPR and spectroscopic measurements described above indicate that Fe is the most abundant photoactive impurity in commercial samples of $BaTiO_3$. The importance of Fe for the photorefractive properties of $BaTiO_3$ can be determined [52] by seeking a correlation between the total iron concentration (from emission spectroscopy), the concentration of Fe^{3+} (by EPR), the relative concentration of filled traps (from the absorption coefficient), and the concentration of empty traps (from photorefractive beam coupling). In Figure 8.14 we plot the parameters given

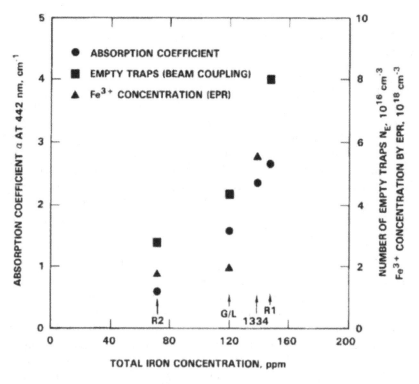

FIGURE 8.14. Absorption coefficient, effective concentration of empty traps, and Fe^{3+} concentration vs. total iron concentration for four samples of $BaTiO_3$. The samples are identified at the bottom of the plot.

above for each of four crystals. We see a clear correlation of the data, in the sense that all measured parameters increase monotonically with total Fe concentration. Ideally, a straight-line behavior should be observed for each of the three plotted parameters, but the accuracy here is limited by experimental uncertainties, as well as possible variations in the oxidation/reduction state or quantum efficiency among our crystals.

To summarize the measurements of [52], we have

$$N = [Fe^{2+}] \approx (2\text{--}9) \times 10^{16} \text{ cm}^{-3},$$
$$N^+ = [Fe^{3+}] \approx [Fe] \approx (2\text{--}8) \times 10^{18} \text{ cm}^{-3},$$

and

$$\frac{[Fe^{3+}]}{[Fe^{2+}]} \approx 40\text{--}120.$$

Note that the large ratio of $[Fe^{3+}]$ to $[Fe^{2+}]$ is consistent with the oxidizing growth environment of BaTiO$_3$ when the top-seeded solution growth technique [11] is used. It is also consistent with data on BaTiO$_3$: Fe ceramics [53, 54] processed in a similar atmosphere, for which Fe^{3+} is determined to be the dominant species.

A plot of electron and hole conductivity vs. the $[Fe^{2+}]/[Fe^{3+}]$ ratio provides a useful means for understanding the relative contribution of holes and electrons to the photorefractive effect in as-grown BaTiO$_3$ [52]. Following the approach of Orlowski and Krätzig (for LiNbO$_3$) [81], we define normalized electron and hole conductivities as

$$\bar{\sigma}_e = \frac{\sigma_e}{\sigma_e + \sigma_h} \tag{8.54}$$

and

$$\bar{\sigma}_h = \frac{\sigma_h}{\sigma_e + \sigma_h}. \tag{8.55}$$

In the limit of small diffusion lengths these parameters are related to the normalized differential conductivity [defined in (8.27)] by

$$\bar{\sigma} = \bar{\sigma}_h - \bar{\sigma}_e. \tag{8.56}$$

From the analysis of Section 8.3.3, we may write the normalized conductivities as

$$\bar{\sigma}_e = \frac{x^2 R}{1 + x^2 R} \tag{8.57}$$

and

$$\bar{\sigma}_h = \frac{1}{1 + x^2 R}, \tag{8.58}$$

where $x = N/N^+ = [Fe^{2+}]/[Fe^{3+}]$, and R is defined in (8.36). We can thus plot the normalized conductivity for BaTiO$_3$ if we know the value of the ratio R for the Fe^{2+} and Fe^{3+} states.

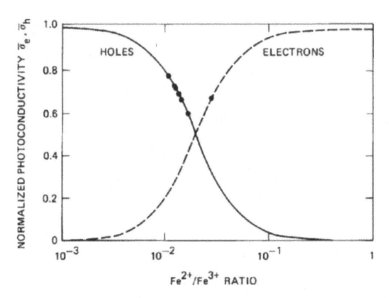

FIGURE 8.15. Relative contribution of electrons and holes to the photoconductivity in BaTiO$_3$, assuming $R = 2500$. The experimental points are taken from the beam-coupling data, using $\bar{\sigma} = \bar{\sigma}_h - \bar{\sigma}_e$.

For BaTiO$_3$, little information is available regarding the values of recombination rate coefficient and photoionization cross section for electrons and holes. Measured values of the electron and hole mobilities for BaTiO$_3$ (given in Table 8.2) differ widely. While the existing data on photorefractive crystals provide no direct information regarding these parameters, the condition $R \gg 1$ is most consistent with the data. This may be explained with reference to Figure 8.15, where we have plotted the normalized conductivity for BaTiO$_3$ [using (8.57) and (8.58)], with $R = 2500$. We have also included data points for each of the crystals studied in [52], using the values of $\bar{\sigma}$ determined from the beam-coupling measurements (Table 8.3). We see that all crystals lie in the transition region between hole and electron photoconductivity. With the assumed value of the ratio R, the compensation point corresponds to $x = R^{-1/2} = [Fe^{2+}]/[Fe^{3+}] = 0.02$. If we had assumed a much larger or smaller value for the ratio R, the resulting values of $x = [Fe^{2+}]/[Fe^{3+}]$ (as read from Figure 8.15) would be inconsistent with the values obtained using the individual measured values of $[Fe^{2+}]$ and $[Fe^{3+}]$. Finally, we note that crystal R2, with dominant electron photoconductivity, is still characterized by the condition $x \ll 1$, thus justifying the earlier approximation that $N_E \approx N = [Fe^{2+}]$ in this crystal.

8.6 Optimization of Photorefractive Properties

As indicated earlier, the major advantage of BaTiO$_3$ for device applications is the very large value of its electrooptic tensor component r_{42}. The major limitation of this material (in its as-grown form) is its relatively long response time. Typical values of the response time at an intensity of 1 W cm^{-2} are 0.1–1.0 s [1, 8, 64, 65], while for many applications response times less than 1 ms are desirable.

By analogy with earlier studies of LiNbO$_3$ [79, 85, 92], it has been suggested that chemical reduction should lead to an improvement in the photorefractive properties of as-grown BaTiO$_3$ [33, 52]. As discussed earlier, a reducing atmosphere is one with a low partial pressure of oxygen. When an as-grown sample is heated in a reducing atmosphere, the concentration of oxygen vacancies increases, leading to a reduction in the valence state of iron impurities, along with an increase in the population of free electrons (8.9). The major experimental question is whether a significant conversion to Fe^{2+} can be created without a significant increase in the free electron concentration (and thus the conductivity) in the sample.

8.6.1 Influence of Oxidation/Reduction

In order to understand the influence of reduction on the photorefractive properties of BaTiO$_3$, we have used the single-species, two-carrier model [82, 83] to calculate the absolute value of the beam-coupling gain Γ (for a 0.7 μm grating period) and the photorefractive response time τ (also for $\Lambda = 0.7\,\mu$m) as a function of the reduction ratio $N/N^+ = [\text{Fe}^{2+}]/[\text{Fe}^{3+}]$. The values of Γ and τ are calculated from (8.42) and (8.37), respectively, using $\tau = \gamma^{-1}$. In our calculation, we assumed $R = 2500$ (corresponding to a compensation point $N/N^+ = 2 \times 10^{-2}$), $\lambda = 500$ nm, a total dopant/impurity concentration $(N + N^+)$ of 5×10^{18} cm^{-3}, and $r_{\text{ang}} = r_{33} = 80$ pm/V. In order to plot the response time, the curve was scaled to a value of 1.1 s at $N/N^+ = 8 \times 10^{-3}$ and $I = 0.25$ W cm^{-2}, corresponding to the measured value in our experiment discussed below.

The results of our calculation are given in Figure 8.16. We see that the gain changes sign at the compensation point, and reaches a broad maximum at $N/N^+ = 1$, where the effective trap density is the largest. On the other hand, the response time is largest (least favorable) at the compensation point, and decreases rapidly for larger or smaller values of reduction ratio. This suggests that in a reduced crystal with $[\text{Fe}^{2+}]/[\text{Fe}^{3+}] = 1$–10, the gain can be optimized, *and* the response time can be reduced to a value on the order of 1–10 ms. Note that the inclusion of nonzero diffusion lengths in the calculation leads to a less rapid decrease in the response time at large values of reduction ratio. The influence on the gain is noticeable only at the compensation point.

The plots in Figure 8.16 suggest an explanation for the observation by Ducharme and Feinberg [33] that the gain changed sign and increased in magnitude in a reduced crystal, but the response time (at ≈ 1 W cm^{-2}) did not change appreciably. We assume first that their as-grown sample was nearly compensated, with a typical value of reduction ratio $x \approx 10^{-2}$. If the reduction treatment increased $[\text{Fe}^{2+}]/[\text{Fe}^{3+}]$ only by a factor of 4–6, then our single-species model indicates that the gain would increase, with only a slight reduction in response time.

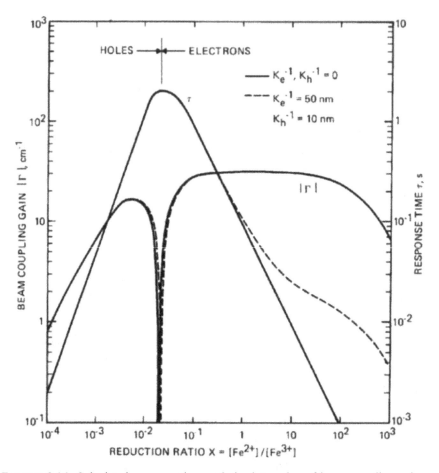

FIGURE 8.16. Calculated response time and absolute value of beam-coupling gain vs. reduction ratio for $\Lambda = 0.7$ μm in $BaTiO_3$.

The above calculations are based on the assumption that a single species (iron impurities or dopants) contributes to the photorefractive effect. This is entirely reasonable in as-grown crystals, in which we expect all oxygen vacancies to be fully ionized. However, in reduced crystals the increased concentration of free electrons will increase the rate of recombination with doubly ionized vacancies, leading to an increase in singly ionized vacancies, which are absorbing and could contribute to the photorefractive effect.

Two approaches have been used to calculate the photorefractive properties of a material with two separate optically active species [33, 82]. In both cases, each of the two species (labeled donor, with total concentration N_D, and acceptor, with total concentration N_A) can exist in two forms: filled and unfilled, or neutral and ionized. The model of Ducharme and Feinberg [33] assumes that the minority species is fully ionized and is present only for charge compensation. This model is equivalent to the single-species model of Kukhtarev [77–79] applied separately to the regions $N_D > N_A$ and $N_D < N_A$. The model predicts that the number of photorefractive charges (equal to the effective density of empty traps N_E in our notation) should go to zero at the compensation point $N_D = N_A$, whereas the single-species model predicts a maximum at this point. The model of Valley [82] assumes that the donor species contributes only to electron transport, while the acceptor species contributes only to hole transport. This model reduces to the basic model of Kukhtarev [77–79] when either of the two species is fully ionized. It also predicts that the density of empty traps should go to zero at the compensation point.

The best means of determining which of the above models applies to reduced BaTiO$_3$ is to measure the response time and Debye length over a wide range of oxidizing and reducing conditions. The measurements of Ducharme and Feinberg [33], while more limited in the. range of oxygen partial pressure, do indicate a clear minimum in the number of photorefractive charges at or near the compensation point, which supports the importance of a two-species model.

We have recently studied the photorefractive and electrical properties of a reduced crystal of BaTiO$_3$, in which the oxygen partial pressure was varied over 12 orders of magnitude [93, 94]. The crystal (labeled G/L) had been previously characterized in its as-grown state [52], yielding $[Fe^{2+}] \approx 4.4 \times 10^{16}$ cm^{-3} and $[Fe^{3+}] \approx (4 \pm 2) \times 10^{18}$ cm^{-3}. The beam-coupling response time (see below) at an intensity of 0.25 W cm^{-2} and a grating period of 0.7 μm was 1.1 s.

The crystal was subjected to three heat treatments. In each case the sample was heated to 1000 °C, held for ≈40 h, and then cooled to room temperature. A mixed atmosphere of CO and CO$_2$ was maintained throughout each experiment, with gas mixtures chosen to provide oxygen partial pressures in the range 10^{-4}–10^{-12} atm at 1000 °C. Following each heat treatment, photorefractive beam-coupling measurements were performed to determine the sign of the photo-carriers and the rise or decay time of the signal beam when the pump beam was chopped. Our results are given in Table 8.6. The response time measured via beam coupling is not generally equal to the fundamental material response time [95]. Nevertheless, our response time data should indicate the qualitative behavior of the material response time as a function of oxygen partial pressure.

The data presented in Table 8.6 are consistent with the model described earlier. As the oxygen partial pressure is decreased, the response time first increases

TABLE 8.6. Resistivity, sign of photocarriers, and beam-coupling response time of BaTiO$_3$ crystal (sample G/L) subjected to several reduction treatments.

Treatment CO/(CO + CO$_2$)	P_{O_2} at 1000 °C [atm]	Resistivity [Ω cm]	Sign of photocarriers	Response time [s]
as-grown	0.2	$\sim 10^{12}$	+	1.05
0.0	3×10^{-5}	$\sim 10^{12}$	+	1.75
0.001	8×10^{-9}	$\sim 10^{12}$	−	0.90
0.10	6×10^{-13}	10^6	−	0.1

and then decreases, as the compensation point is passed. At the same time, the dominant photocarrier changes from holes to electrons. Overall, a decrease in response time by a factor of ≈ 10 was achieved. However, under the most reducing conditions, the resistivity of the crystal decreased dramatically. Although the crystal was free of $90°$ domains, electrical poling was only partially successful, due to the limited poling field that could be applied. We note also that we did not observe any change in T_c as a result of the heat treatment. This invariance in T_c differs from the results of Ducharme and Feinberg [33], who observed a decrease in T_c of $6°C$ when an as-grown crystal was treated at $650°C$ in 10^{-6} atm of oxygen.

From our response time measurements (Table 8.6) and our theoretical plot of τ versus $[Fe^{2+}]/[Fe^{3+}]$ (Figure 8.16), we can estimate the ratio $[Fe^{2+}]/[Fe^{3+}]$ as a function of oxygen partial pressure. We find that the induced change in $[Fe^{2+}]/[Fe^{3+}]$ is relatively small, varying by less than a factor of 50 over a variation of nearly 12 orders of magnitude in oxygen partial pressure. The relative stability of Fe^{3+}, i.e., its resistance to conversion to Fe^{2+}, is a known property of $BaTiO_3$. In particular, in [53] negligible conversion of Fe^{3+} to Fe^{2+} was observed at oxygen partial pressures as low as 10^{-22} atm. Only in a pure H_2 environment (10^{-26} atm oxygen) was conversion of $\approx 50\%$ achieved. However, under these heavy reduction conditions, the crystals are highly conductive and opaque. We believe that the stability of Fe^{3+} in $BaTiO_3$ is a result of the shallow position of the Fe^{2+} levels (Figure 8.5b). Any attempted conversion to Fe^{2+} is inhibited by thermal ionization of this species, leading to conversion back to Fe^{3+}, along with the generation of free electrons:

$$Fe^{2+} \rightarrow Fe^{3+} + e^-. \tag{8.59}$$

The equilibrium densities of Fe^{2+}, Fe^{3+}, and n determined from this reaction are related in (8.13). We have also pointed out [52] that OH^- and H^+ ions tend to stabilize Fe^{3+}. This emphasizes the importance of avoiding the introduction of hydrogen during growth, and processing with other reducing agents, such as CO/CO_2.

There are a number of other approaches to inducing electron photoconductivity in $BaTiO_3$, while maintaining high resistivity. One approach would be to dope the crystals with transition metals other than iron. Here we must consider two cases: acceptor dopants (metals with valence states lower than $4+$) and donor dopants (metals with valence states greater than $4+$). For acceptor dopants, we require that the energy levels of the reduced species be substantially deeper than those of Fe^{2+}. Possible candidates are Cr and Mn, which are known to be more easily reduced than Fe in $BaTiO_3$ ceramics [54]. In fact, Mn is known to have deeper levels in the $BaTiO_3$ band gap than Fe [44]. However, even in this case, the ability to produce insulating, n-type behavior may be limited by the shallow nature of the compensating oxygen vacancies.

A second approach is donor doping, which offers the advantage of electron conductivity without the need for a reducing heat treatment. In donor-doped ceramics, charge compensation is by free electrons and metal ion vacancies, with the latter being favored at higher dopant concentrations [51, 96]. In single crystals, the required dopant concentration for metal ion compensation has not been determined. Metal ion-compensated samples are expected to be insulating, since the metal ion (acceptor) levels are deep in the band gap [51], and few oxygen vacancies are present.

8.7 Conclusions

Since the "rediscovery" of BaTiO$_3$ in 1980, this material has become the material of choice in applications requiring large gain and/or four-wave mixing reflectivity. At the same time, BaTiO$_3$ has developed a reputation for being slow, and thus unsuited for certain applications. However, all evidence suggests that the slow speed of BaTiO$_3$ is a specific property of this material in its as-grown state, and is not a fundamental property. The calculations from the previous section show that reduced samples of BaTiO$_3$ should have response times below 1 ms, with little or no decrease in beam-coupling gain. Of course the challenge is to obtain reduced samples without sacrificing the high resistivity obtained in as-grown samples.

In order to further characterize commercial samples of BaTiO$_3$, it would be useful to subject a given sample to heat treatments over a wide range of oxygen partial pressures (as done by Kurz et al. [85] for LiNbO$_3$), and correlate the changes in EPR signals and absorption coefficient with changes in the photorefractive behavior. Mössbauer or low-temperature EPR experiments should also be performed in order to measure the concentration of Fe^{2+} and Fe^{4+} ions. These measurements would serve to test the energy-level models discussed earlier, and confirm the identification of Fe as the photorefractive species.

Acknowledgments. I would like to acknowledge helpful discussions with R.N. Schwartz, G.C. Valley, B.A. Wechsler, D. Rytz, R. Pastor, R.A. Mullen, S. Ducharme, and J. Feinberg.

References

1. R.L.Townsend, J.T. LaMacchia: *J. Appl. Phys.* **41**, 5188 (1970).
2. J. Feinberg, R.W. Hellwarth: *Opt. Lett.* **5**, 519 (1980).
3. A. Ashkin, G.D. Boyd, J.M. Dziedzic, R.G. Smith, A.A. Ballman, J.L. Levinstein, K. Nassau: *Appl. Phys. Lett.* **9**, 72 (1966).
4. F.S. Chen, J.T. LaMacchia, D.B. Fraser: *Appl. Phys. Lett.* **13**, 223 (1968).
5. F. Micheron, G. Bismuth: *J. de Phys.* **33**: Suppl. 4, Colloq. 2, 149 (1972).
6. W.T.H. Koch, R. Munser, W. Ruppel, P. Wurfel: *Solid State Commun.* **17**, 847 (1975).
7. W.T.H. Koch, R. Munser, W. Ruppel, P. Wurfel: *Ferroelectrics* **13**, 305 (1976).
8. J. Feinberg, D. Heiman, A.R. Tanguay, Jr., R.W. Hellwarth: *J. Appl. Phys.* **51**, 1297 (1980).

9. E. Krätzig, F. Welz, R. Orlowski, V. Doormann, M. Rosenkranz: *Solid State Commun.* **34**, 817 (1980).

10. J.P. Remeika: *J. Am. Ceram. Sot.* **76**, 940 (1954).

11. V. Belrus, J. Kalinajs, A. Linz, R.C. Folweiler: *Mater. Res. Bull.* **6**, 899 (1971).

12. S.H. Wemple, M. DiDomenico, Jr., I. Camlibel: *J. Phys. Chem. Solids* **29**, 1797 (1968).

13. I. Camlibel, M. DiDomenico, S.H. Wemple: *J. Phys. Chem. Solids* **31**, 1417 (1970).

14. W.J. Merz: *Phys. Rev.* **91**, 513 (1953).

15. F. Jona, G. Shirane: *Ferroelectric Crystals* (Macmillan, New York 1962).

16. M.E.Lines, A.M. Glass: *Principles and Applications of Ferroelectrics and Related Materials* (Clarendon, Oxford 1977).

17. M. DiDomenico, Jr., S.H. Wemple: *Phys. Rev.* **155**, 539 (1967).

18. V.M. Fridkin, A.A. Grekov, N.A. Kosonogov, T.R. Volk: *Ferroelectrics* **4**, 169 (1972).

19. J. Rodel, G. Tomandl: *J. Mater. Sci.* **19**, 3515 (1984).

20. R.C. Pastor: unpublished.

21. H.E. Muser, W. Kuhn, J. Albers: *Phys. Status Solidi* A**49**, 51 (1978).

22. S.H. Wemple: "Electrooptical and Nonlinear Optical Properties of Crystals," in *Applied Solid State Science*, ed. by R. Wolfe, Vol. 3 (Academic, New York 1972), pp. 263–383.

23. R.C. Miller: *Appl. Phys. Lett.* **5**, 17 (1964).

24. A.R. Johnston, J.M. Weingart: *J. Opt. Soc. Am.* **55**, 828 (1965).

25. M. Cronin-Golomb: Paper ThT3, CLEO '85, Baltimore, Md.

26. W.R. Cook, H. Jaffe: "Piezoelectric, Electrostrictive and Dielectric Constants, and Electromechanical Coupling Factors of Piezoelectric Crystals," in *Landolt-Börnstein*, Group III, Vol. 11, ed. by K.-H. Hellwege, A.M. Hellwege, (Springer, Berlin, Heidelberg 1979) pp. 287–470.

27. W.R. Cook, R.F.S. Hearmon, H. Jaffe, D.F. Nelson: "Piezooptic and Electrooptic Constants," in *Landolt-Börnstein*, Group III, Vol. 11, ed. by K.H. Hellwege, A.M. Hellwege (Springer, Berlin, Heidelberg 1979) pp. 495–670.

28. S. Ducharme, J. Feinberg, R.R. Neurgaonkar: IEEE J. QE-**23**, 2116 (1987).

29. I.P. Kaminow: *Appl. Phys. Lett.* **7**, 123 (1965); Erratum, *Appl. Phys. Lett.* **8**, 54 (1966).

30. I.P. Kaminow: *Appl. Phys. Lett.* **8**, 305 (1966).

31. A.R. Johnston: *Appl. Phys. Lett.* **7**, 195 (1965).

32. M.B. Klein, G.C. Valley: *J. Appl. Phys.* **57**, 4901 (1985).

33. S. Ducharme, J. Feinberg: *J. Opt. Soc. Am.* B**3**, 283 (1986).

34. F.M. Michel-Calendini, G. Mesnard: *J. Phys.* C**6**, 1709 (1973).

35. P. Pertosa, F.M. Michel-Calendini: *Phys. Rev.* B**17**, 2011 (1978).

36. F.M. Michel-Calendini, H. Chermette, J. Weber: *J. Phys.* C**13**, 1427 (1980).

37. F.M. Michel-Calendini: *Ferroelectrics* **37**, 499 (1981).

38. P. Moretti, F.M. Michel-Calendini: *Ferroelectrics* **55**, 219 (1984).

39. F.M. Michel-Calendini, L. Hafid, G. Godefroy, H. Chermette: *Solid State Commun.* **54**, 951 (1985).

40. F.M. Michel-Calendini, P. Moretti: *Phys. Rev.* B**27**, 763 (1983).

41. L.F. Mattheiss: *Phys. Rev.* B**6**, 4718 (1972).

42. S.H. Wemple: *Phys. Rev.* B**2**, 2679 (1970).

43. C. Gahwiller: *Phys. Kondens. Mater.* **6**, 269 (1967).

44. H.-J. Hagemann:. Ph.D. thesis, Rheinisch-Westfälische Technische Hochschule Aachen (1980).

45. G. Godefroy, C. Dumas, P. Lompre, A. Perrot: *Ferroelectrics* **37**, 725 (1981).

46. C.N. Berglund, H.J. Braun: *Phys. Rev.* **164**, 790 (1967).

47. N.H. Chan, R.K. Sharma, D.M. Smyth: *J. Am. Ceram. Soc.* **64**, 556 (1981).
48. G.V. Lewis, C.R.A. Catlow: *J. Phys. Chem. Solids* **47**, 89 (1986).
49. R.K. Sharma, N.H. Chan, D.M. Smyth: *J. Am. Ceram. Soc.* **64** 448 (1981).
50. D.E. Rase, Rustum Roy: *J. Am. Ceram. Soc.* **38**, 102 (1955).
51. J. Daniels, K.H. Hardtl: *Philips Res. Rep.* **31**, 489 (1976).
52. M.B. Klein, R.N. Schwartz: *J. Opt. Soc. Am.* B**3**, 293 (1986).
53. H.J. Hagemann, A. Hero, U. Gonser: *Phys. Status Solidi* A**61**, 63 (1980).
54. H.J. Hagemann, D. Hennings: *J. Am. Ceram. Soc.* **64**, 590 (1981).
55. E. Siegel, K.A. Muller: *Phys. Rev.* B**20**, 3587 (1979).
56. S.H. Wemple: "Electrical Contact to *n*- and *p*-Type Oxides," in *Ohmic Contacts to Semiconductors*, ed. by B. Schwartz (Electrochemical Society, New York 1968).
57. N.H. Chan, R.K. Sharma, D.M. Smyth: *J. Am. Chem. Soc.* **65**, 167 (1982).
58. N.G. Eror, D.M. Smyth: *J. Solid State Chem.* **24**, 235 (1978).
59. J. Daniels: *Philips Res. Rep.* **31**, 505 (1976).
60. C.N. Berglund, W.S. Baer: *Phys. Rev.* **157**, 358 (1967).
61. S. Ikegami, I. Ueda: *J. Phys. Soc. Jpn.* **19**, 159 (1964).
62. F.S. Chen: *J. Appl. Phys.* **40**, 3389 (1969).
63. G.E. Peterson, A.M. Glass, T.J. Negran: *Appl. Phys. Lett.* **19**, 130 (1971).
64. S. Ducharme, J. Feinberg: *J. Appl. Phys.* **56**, 839 (1984).
65. D. Rak, I. Ledoux, J.P. Huignard: *Opt. Commun.* **49**, 302 (1984).
66. K. Kawabe, Y. Inuishi: *Jpn. J. Appl. Phys.* **2**, 590 (1963).
67. P. Gerthsen, R. Groth, K.H. Hardtl: *Phys. Status Solidi* **11**, 303 (1965).
68. H. Veith: *Z. Angew. Phys.* **20**, 16 (1965).
69. D.L. Ridpath, D.A. Wright: *J. Mater. Sci.* **5**, 487 (1970).
70. E.V. Bursian, Y.G. Girshberg, E.N. Starov: *Sov. Phys.-Solid State* **14**, 872 (1972).
71. P. Gerthsen, K.H. Hardtl, A. Csillag: *Phys. Status Solidi* A **13**, 127 (1972).
72. A.M.J.H. Seuter: *Philips Res. Rep., Suppl.* **3** (1974).
73. J.P. Boyeaux, F.M. Michel-Calendini: *J. Phys.* C **12**, 545 (1979).
74. G.A. Cox, R.H. Tredgold: *Phys. Lett.* **11**, 22 (1964).
75. L. Benguigui: *Solid State Commun.* **7**, 1245 (1969).
76. C.-P. Tzou, T.Y. Chang, R.W. Hellwarth: *Proc. SPIE* **613**, 58 (1986).
77. V.L. Vinetskii, N.V. Kukhtarev: *Sov. Phys.-Solid State* **16**, 2414 (1975).
78. N.V. Kukhtarev: *Sov. Tech. Phys. Lett.* **2**, 438 (1976).
79. N.V. Kukhtarev, V.B. Markov, S.G. Odulov, M.S. Soskin, V.L. Vinetskii: *Ferroelectrics* **22**, 949, 961 (1979).
80. G.C. Valley, M.B. Klein: *Opt. Eng.* **22**, 704 (1983).
81. R. Orlowski, E. Krätzig: *Solid State Commun.* **27**, 1351 (1978).
82. G.C.Valley: *J. Appl. Phys.* **59**, 3363 (1986).
83. F.P. Strohkendl, J.M.C. Jonathan, R.W. Hellwarth: *Opt. Lett.* **11**, 312 (1986).
84. J.P. Huignard, A. Marrakchi: *Opt. Commun.* **38**, 249 (1981).
85. H. Kurz, E. Krätzig, W. Keune, H. Engelmann, U. Gonser, B. Dischler, A. Rauber: *Appl. Phys.* **12**, 355 (1977).
86. N.V. Kukhtarev, E. Krätzig, H.C. Kulich, R.A. Rupp, J. Albers: *Appl. Phys.* B **35**, 17 (1984).
87. J. Feinberg: *J. Opt. Soc. Am.* **72**, 46 (1982).
88. L.K. Lam, T.Y. Chang, J. Feinberg, R.W. Hellwarth: *Opt. Lett.* **6**, 475 (1981).
89. A. Rose: *Concepts in Photoconductivity and Allied Problems* (Robert Krieger, New York 1978).
90. G.C. Valley: *IEEE J.* QE-**19**, 1637 (1983).

91. A.W. Hornig, R.C. Rempel, H.E. Weaver: *J. Phys. Chem. Solids* **10**, 1 (1959).
92. D.L. Staebler and W. Phillips: *Appl. Opt.* **13**, 788 (1974).
93. M.B. Klein: Paper FQ1, CLEO '86; San Francisco, Ca.
94. B.A. Wechsler, M.B. Klein, D. Rytz: *Proc. SPIE* **681**, 91 (1986).
95. J.M. Heaton, L. Solymar: *Opt. Acta* **32**, 397 (1985).
96. H.M. Chan, M.P. Harmer, D.M. Smyth: *J. Am. Ceram. Soc.* **69**, 507 (1986).

Additional References

Bacher, G.D., Chiao, M.P., Dunning, G.J., Klein, M.B., Nelson, C.C., and Wechsler, B.A.: "Ultralong Dark Decay Measurements in $BaTiO_3$," *Opt. Lett.* **21**, 18 (1996).

Cudney, R. and Kaczmarek, M.: "Light-Induced Removal of 180° Ferroelectric Domains in $Rh:BaTiO_3$," *Opt. Exp.* **7**, n9, 323–328 (2000).

Garrett, M.H., Chang, J.Y., Jenssen, H.P., and Warde, C.: "High Beam-Coupling Gain and Deep- and Shallow-Trap Effects in Cobalt-Doped Barium Titanate, $BaTiO_3$:Co," *J. Opt. Soc. Am. B* **9**, 1407 (1992).

Motes, A., Kim, J.J.: "Beam Coupling in Photorefractive $BaTiO_3$ Crystals," *Opt. Lett.* **12**, 199 (1987).

Smirl, A., Valley, G.C., Mullen, R.A., Bohnert, K., Mire, C.D., Boggess, T.F.: "Picosecond Photorefractive Effect in $BaTiO_3$," *Opt. Lett.* **12**, 501 (1987).

Rytz, D., Klein, M.B., Mullen, R.A., Schwartz, R.N., Valley, G.C., and Wechsler, B.A.: "High-Efficiency Fast Response in Photorefractive $BaTiO_3$ at 120 °C," *Appl. Phys. Lett.* **52**, n21, 1759–1761 (May 23, 1988).

Rytz, D., Stephens, R.R., Wechsler, B.A., Keirstead, M.S. and Baer, T.M.: "Efficient Self-Pumped Phase Conjugation at Near-Infrared Wavelength Using Cobalt-Doped $BaTiO_3$," *Opt. Lett.* **15**, 1279 (1990).

Rytz, D., Wechsler, B.A., Garrett, M.H., Nelson, C.C., and Schwartz, R.N.: "Photorefractive Properties of $BaTiO_3$:Co," *J. Opt. Soc. Am. B* **7**, 2245 (1990).

Wechsler, B.A., Klein, M.B.: "Thermodynamic Point Defect Model of $BaTiO_3$ and Application to the Photorefractive Effect," *J. Opt. Soc. Am.* B, **5**, 1711 (1988).

Wechlser, B.A., Klein, M.B., Nelson, C.C., and Schwartz, R.N.: "Spectroscopic and Photorefractive Properties of Infrared-Sensitive Rhodium-Doped Barium Titanate," *Opt. Lett.* **19**, 536 (1994).

9

Space-Charge Waves in Sillenites: Rectification and Second-Harmonic Generation

M.P. Petrov and V.V. Bryksin

Physico-Technical Institute, Russian Academy of Sciences, St. Petersburg 194021, Russia

This paper consists of two parts. The first contains a short review of the main characteristics and holographic phenomena typical of sillenites, and the second part is devoted to new phenomena associated with space-charge waves, namely to second-harmonic generation and rectification of these waves.

9.1 Major Characteristics of Sillenites as Holographic Materials

9.1.1 General Properties of Sillenites

Sillenites are reversible holographic materials, efficient photoconductors. They exhibit a strong electrooptic effect, excellent piezoelectric properties, and a relatively high Faraday effect. Owing to these features, sillenites find wide application in dynamic holography and interferometry, optical information processing, optical sensors of electric and magnetic fields, and piezoelectric devices. As reversible holographic materials, sillenites offer numerous advantages (a fast enough response time of $1–10^{-3}$ s, photosensitivity in red–blue light, and a high-enough diffraction efficiency, sensitivity, and resolution), which makes them attractive both for use in practice and experimental scientific investigations. The simple (cubic) structure of crystals facilitates experiments and theoretical analysis considerably. In addition, crystals of excellent optical quality and large size (Figure 9.1) are commercially available at a reasonable price.

9.1.1.1 Crystallographic Form

Crystals of the sillenite family were discovered by L.G. Sillen [1] in 1937. The common chemical formula is $Bi_{12}MO_{20}$, where M is the metal ion (typically with a real or effective valence equal to four). Many compounds of this family have been synthesized. However, only single crystals in which M is Si, Ge, or Ti have

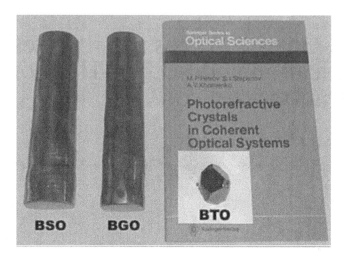

FIGURE 9.1. Single crystals of BSO, BGO, and BTO grown by the Czochralski technique. The weights of the boules: BSO, 2500 g; BGO, 2400 g; BTO, 120 g.

been grown of the size and quality acceptable for applications. (Below we use the abbreviations BSO, BGO, and BTO for $Bi_{12}SiO_{20}$, $Bi_{12}GeO_{20}$, and $Bi_{12}TiO_{20}$, respectively). The crystals of large size are grown mostly by the Czochralski technique. Sillenites belong to the space group I23 (point group 23). The body-centered cubic crystalline unit cell contains two chemical formulae. The overall structure of the crystal may be described in terms of the 7-oxygen coordinated Bi polyhedra, which share corners with other identical Bi polyhedra and with MO_4 tetrahedra. The five nearest oxygen atoms around Bi form an incomplete octahedral arrangement; the two remaining oxygen atoms form longer Bi-O bonds. (A detailed description of the sillenite structure and, in particular, the structure of BGO was given in [2]). In real crystals, the occupancy factor of M cations is less than unity [3].

9.1.1.2 Optical Properties and Photoactive Centers

Nominally pure crystals are transparent in the red and near infrared (0.4–7.5 μm) [4, 5]. The absorption coefficient α is on the order of 0.03–0.05 cm^{-1} for BSO at $\lambda = 632.8$ nm [6], but the crystals exhibit a rather strong light absorption in the yellow and green regions; the edge of BTO absorption is shifted toward the red region [7] ($\alpha = 1.5$ cm^{-1} at $\lambda = 0.63$ μm). Absorption spectra for BSO and BGO were studied in [8] for different melt compositions. For the stoichiometric composition, the absorption coefficient α in BSO changes from 10 cm^{-1} to 1.25 cm^{-1} at variation of λ from 450 to 500 nm, and in BGO this coefficient changes from 20 cm^{-1} to 2.6 cm^{-1} at the same variation of λ. Figure 9.2 shows the optical absorption spectrum of BTO. The band-gap edge of BGO and BSO is about (3.2 ± 0.5) eV [5, 8, 9]. There is a broad absorption shoulder [10] starting approximately from 2.6 eV. In the dark, sillenites are nearly dielectrics (the dark conductivity is

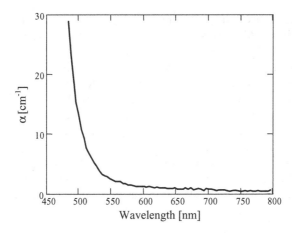

FIGURE 9.2. Absorption spectrum of BTO (courtesy of F. Rahe). The crystal was grown at the Ioffe Physical Technical Institute, St. Petersburg, Russia.

on the order of 10^{-12}–10^{-15} ohm^{-1}cm^{-1} [4, 8, 11, 12]. However, under illumination by light corresponding to the range of absorption, the photoconductivity can be several orders of magnitude higher than the dark conductivity. The crystals can easily be doped with Al, Ga, Mn, Cr ,Ca, Mg, Sr, Sn, Mo, Nd, and some other dopants, which can result in strong variations of conductivity and photoconductivity and can change the crystal color. A detailed review of doped crystals was presented in [13,14]. Absorption characteristics and photoconductivity, as well as photorefractive parameters, can differ considerably depending on the crystal's origin (see, for instance, comparison of various materials in [6]). It is commonly accepted that the photoconductivity has an electronic nature (BTO [15]), whereas the dark conductivity is associated with the hole conductivity. Doping with a sufficiently high amount of Al and Ga changes the photoconductivity from n type to p type [16, 17, 18]. A detailed description of the photoinduced phenomena can be found in [13]. There are numerous publications concerning the electron–hole competition in holographic recording (see, e.g., the review paper [19] and remarks in [7] concerning BTO). However, other models, which do not include bipolar conductivity, also exist. In [20], a model describing long-term storage of grating by formation of a "grating of photosensitivity" instead of a space-charge grating in a specially prepared (preliminarily illuminated during 30 minutes at $\lambda = 633$ nm) BSO crystal was suggested. In spite of the long history and numerous investigations, the nature of the light-induced charge transport (which is a vital component of the photorefractive process) is a vexed question. Comparison of nominally pure and doped crystals shows that the fundamental role is likely to be played by the intrinsic defects rather than impurity centers, although impurity centers can essentially modify the optical spectrum and carrier mobility of the crystals. In [18], a model (which is one of the most popular now) in which it was assumed that the centers responsible for the light absorption leading to the photorefractive effect are the intrinsic defects arising when atoms of bismuth replace atoms in the M sites was suggested. Such defects were called "antisite defects." This model was confirmed in many subsequent works (see, e.g., [8, 21, 21a]. The location of the Bi^{3+} ion on a Ge or Si site is associated with a hole localized at the oxygen neighbors. This covalent defect $Bi^{3+} + h$ is considered to be responsible for the absorption shoulder starting from 2.6 eV [22]. In [21], the assumption was made that $Bi^{3+} + h$ acts as a filled trap and $Bi^{4+} + h$ acts as an empty trap. These defects play the role of deep traps for the free-charge carriers and provide formation of the volume space-charge distribution in accordance with the spatial distribution of the incident light intensity. The efficient trap concentration N_A in sillenites is on the order of 10^{16} cm^{-3} (see Table 9.1). A relatively low trap density limits the amplitudes of space charge field gratings by the so-called saturation field

$$E_q = eN_A/\varepsilon\varepsilon_0 K. \tag{9.1}$$

Here K is the wave number of the grating, $K = 2\pi/\Lambda$, Λ is the grating spacing, and ε is the dielectric constant of the crystal. In addition to deep traps, sillenites have shallow traps, which can play a very important role in the photoconductivity of the materials (for instance, they provide the effect of photoconductivity quenching [23]). These traps form a complicated energy-level structure in the interval 0.18–1.2 eV (below the conductivity band). In accordance with [13], some of these traps can be associated with bismuth defects in the volume of the material (0.5–0.6 eV), and others can be due to some oxygen vacancies in the surface states (0.18–0.32 eV).

9.1.1.3 Electrooptical, Piezoelectric, and Photogalvanic Properties

Since sillenites belong to the cubic group (but without a center of inversion), there are only three nonzero electrooptic coefficients, and they are equal to one another ($r_{41} = r_{52} = r_{63}$). With no applied field, the optical indicatrix is a sphere. Refractive indices and electrooptic coefficients for different crystals are presented in Table 9.1. It is important to note that there is a systematic discrepancy between the electrooptic coefficients obtained by interferometric measurements and in holographic experiments. No commonly adopted explanation of this fact is known. Sometimes the possibility that the hole grating compensates the electron one is considered [7]; sometimes it is supposed that refractive index inhomogeneities reduce the modulation degree inside the sample [24]. If holographic experiments are performed in the presence of an applied electric field, one more factor can be very important. As is often discussed, the internal field in sillenites is lower (for instance, by a factor of 0.6 [25]) than the expected internal field corresponding to the applied voltage. This is due to various screening effects caused, for instance, by the influence of contacts (or contact areas) or inhomogeneous illumination of the sample [26]. Reduction in the internal field results in the renormalization of the calculated efficient electrooptical coefficients to smaller values.

In addition to the electrooptic effect, the 23-point group permits natural activity and piezoelectric effect in the sillenites. Typically, the optical activity plays a

TABLE 9.1. Major characeristics of the sillenites.

	BSO	BGO	BTO
Unit cell, Å	10.1043 [3]	10.455 [1]	10.177 [133]
	10.102 [4]		10.178 [136]
ε	56 [5]	40 [5]	47 [134]
			at $f = 1$ KHz
n_0	2.54 [5]	2.54 [5]	2.25 [134]
	at $\lambda = 633$ nm,	at $\lambda = 633$ nm,	2.68
	2.52–2.53 [25]	2.5476 [133]	at $\lambda = 488$ nm,
	at $\lambda = 633$ nm,	at $\lambda = 632.8$ nm	2.56 [133]
	2.87–2.51▲ [5]		
	at $\lambda = 400$–700 nm		at $\lambda = 630$ nm
Faraday effect	0.25 min Oe^{-1} cm^{-1} [4]	0.26 min Oe^{-1} cm^{-1} [4]	4.3 deg T^{-1}mm^{-1}
	4 deg T^{-1} mm^{-1} [137]	4.2 deg T^{-1}mm^{-1} [137]	[137]
	at $\lambda = 600$ nm	at $\lambda = 600$ nm	at $\lambda = 600$ nm
	7.1 deg T^{-1} mm^{-1} [137]	7.5 deg T^{-1} mm^{-1} [137]	6.9 deg T^{-1}mm^{-1}
	at $\lambda = 500$ nm	at $\lambda = 500$ nm	[137]
			at $\lambda = 500$ nm
ρ, deg/mm	60.2 [135]	60.9 [135]	6.3 [134]
	at $\lambda = 450$ nm,	at $\lambda = 450$ nm,	5.3 [7]
	20.5 [135]	19.5 [135]	at $\lambda = 630$ nm,
	at $\lambda = 650$ nm,	at $\lambda = 650$ nm,	2.3 [7]
	21.2 [25]	25 [4]	at $\lambda = 1060$ nm,
	at $\lambda = 630$ nm,	23 [137]	6.3 [136]
	25 [4]	at $\lambda = 600$ nm,	at $\lambda = 633$ nm,
	23 [137]	102 [137]	11.9 [136]
	at $\lambda = 600$ nm	at $\lambda = 500$ nm,	at $\lambda = 514.5$ nm,
			7 [137]
			at $\lambda = 600$ nm,
			42 [137]
			at $\lambda = 400$ nm,
			6.49 [26]
			at $\lambda = 630$ nm
Band gap, eV	3.25 [11,16]		3.1 [138]
Trap density, cm^{-3}	4×10^{15} [6]	6.4×10^{15} [6]	1.2×10^{15} [7]
	6.3×10^{16} [139]	1.1×10^{16} [143]	6.2×10^{16} [143]
	1.2×10^{16} [140]		
	3.4×10^{16} [141]		
	2×10^{15} [142]		
	3.4×10^{16} [143]		
$\mu\tau$, cm^2/V	1.4×10^{-7} [48]	1.2×10^{-7} [48]	3×10^{-7} [121]
	2×10^{-6} [121]		
	$(1.2$–$0.8) \times 10^{-6}$ [107]		
μ cm^2/Vs	$(1.6$–$2.5) \times 10^{-2}$ [121]	2×10^{-2} [105]	6×10^{-2} [121]
	$(13$–$80) \times 10^{-2}$ [107]	1.2×10^{-2} [146]	2×10^{-2} [147]
	550×10^{-2} [144]	3×10^{-2} [152]	
	340×10^{-2} [145]	14×10^{-2} [148]	
	24×10^{-2} [131]		
	2.9×10^{-2} [16]		

(continued)

TABLE 9.1. (*cont.*)

	BSO	BGO	BTO
Electrooptic coefficient r_{41}, pm/V	4.4 [25] stress free 3.7 [25] clamped 4.31 [149] 5 [5] 2.1–2.9* [8]	3.4 [150] 1.4–2.8* [8] 3.4 [5]	1.4* [7] at $\lambda = 1060$ nm 5.3 [136] at $\lambda = 633$ nm 5.17 [134] 4.75* [148] 5.75 [26]
Piezoelectric coefficients d_{14}, C/N d_{111}, C/N	40.5×10^{-12} [151] 23.4×10^{-12} [151]	33.9×10^{-12} [151] 19.6×10^{-12} [151]	48×10^{-12} [151] 27.8×10^{-12} [151]

▲ the data were not confirmed by other authors
*the data were obtained from holographic experiments

detrimental role since it reduces diffraction efficiency [27]. However, for a specific geometry (reflective grating with wave vector $\mathbf{K}$//[111] the existence of optical activity has a positive aspect since it allows the use of orthogonally polarized beams [28].

The piezoelectric effects in sillenites have attracted considerable attention (see, e.g., [29–32]). One way in which piezoelectric properties of the crystal manifest themselves is the change in the refractive index of the crystal through a chain of interactions: the space-charge field gives rise to strain via the piezoelectric effect, and the strain causes a variation in the refractive index via the photoelastic effect. In a standard situation, when an electric field is applied along a crystallographic axis, the piezoelectric effect results in small corrections in the electrooptic effect. However, in specific cases the piezoelectric effect can be more important. For instance, rather important changes in two-beam coupling can be caused by the piezoelectric effect for the grating orientations with intermediate angles between the [001] and [$\bar{1}$10] axes [31]. The other way is that the piezoelectric effect can result in a deformation and small changes in the orientation of the sample surface due to space charge fields arising during holographic recording. Light diffraction from photoinduced surface piezoelectric gratings was studied in the spatial light PRIZ modulator [33, 34] and in BTO crystals [35]. The light deflection caused by the piezoelectric reorientation of the crystal surface due to a spatially homogeneous electric field originating from spatial rectification of space-charge waves was revealed in [36, 37].

Sillenites exhibit the photogalvanic (or photovoltaic) effect [38, 39]. However, photovoltaic fields here are rather weak (on the order of 10^{-2} V/cm) because of a high photoconductivity.

9.1.1.4 Most-Popular Electrooptic Configurations

In this section, we ignore, for the sake of simplicity, natural optical activity of sillenites (a more thorough analysis can be found, e.g., in [26, 27, 46]).

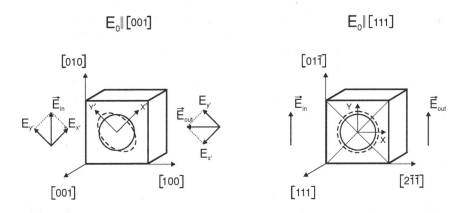

(a) Longitudinal geometry

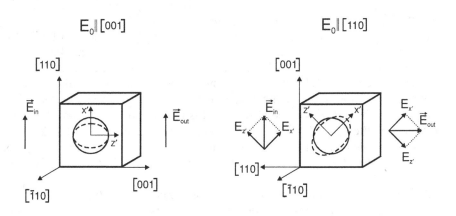

(b) Transverse geometry

FIGURE 9.3. Typical geometries of the electrooptic effect in sillenites. Here E_0 is the applied electric field. To find the refractive index variations in holographic gratings, the space-charge field E_{sc} is used instead of E_0.

a. Longitudinal effect (Figure 9.3a). In this case, the light propagates along the electric field (an applied field or a space-charge field). Most often, the electric field is oriented along one of the crystallographic [100], [010], or [001] axes. Let us assume that $E//[001]$. In this case the optical indicatrix shape is transformed from a sphere into an ellipse with the principal axes oriented along the [001] (z-axis), [$1\bar{1}0$] (x-axis) and [110] (y-axis). Correspondingly, the refractive indexes are $n_z = n_0$, $n_x = n_0 - r_{41}n_0^3 E_0/2$ and $n_y = n_0 + r_{41}n_0^3 E_0/2$. If the incident readout light has a vertical polarization, it can be represented in the crystal as a superposition of two eigenmodes, one of which is polarized along x and the other along y. They experience the difference in phase retardation $\Delta \Phi = 2\pi (n_y - n_x) T/\lambda = 2\pi r_{41} n_0^3 U$, where T is the sample thickness and U is the applied voltage. The half-voltage in this case has a minimal value $U_{\lambda/2} = \lambda/(2r_{41} n_0^3)$. At the output, a horizontally polarized component of the light wave emerges with the intensity $I_{out} = I_{inp} \sin^2 \Delta \Phi/2$, where I_{inp} is the input intensity (the optical activity of the crystal is neglected here). Since this component is orthogonally polarized with respect to the input light, it can easily be selected by an analyzer of polarization placed at the output (a more detailed description can be found, e.g., in [26]). The longitudinal configuration is efficiently used for studying electrooptic effects (measurements of electrooptic coefficients). It was used in the spatial light modulator PROM [34, 40–42] for image recording and for incoherent-to-coherent image transformation. Potentially, it can also be used for holographic experiments [43, 44]. Sometimes, one more longitudinal geometry is used when the electric field is oriented along the [111] axis [28]. In this case the optical indicatrix has the shape of an ellipsoid of revolution around the [111] axis, and the variation of refractive index is $\Delta n = r_{41} n_0^3 E/(2\sqrt{3})$ for the light propagating along [111] for any polarization orthogonal to [111]. Correspondingly, $n_z = n_0 - r_{41} n_0^3 E_0/\sqrt{3}$, $n_x = n_0 + r_{41} n_0^3 E_0/2\sqrt{3}$, and $n_y = n_0 + r_{41} n_0^3 E_0/2\sqrt{3}$. An analysis of reflective gratings for different crystallographic orientations in sillenites can be found in [45]. One of the drawbacks of the longitudinal configuration is the necessity to use transparent electrodes when recording is performed in an applied external field [41, 42]. If a refractive index grating is formed by recording with counter-propagating beams (the space-charge field is parallel to light propagation), the diffraction efficiency of the holograms can be rather poor, $\eta = 0.14\%$ [28]. Perhaps this is because of a low saturation field E_q at high spatial frequencies (on the order of hundreds of V/cm at spatial frequencies $\Lambda^{-1} = K/2\pi$ on the order of 10^4 mm^{-1}).

b. Transverse configuration. This is the most popular geometry for holographic investigations. In this configuration, light propagates orthogonally to the electric field (applied or space-charge field). The most widely used crystal cuts for this geometry are (110), $(\bar{1}10)$, and $(1\bar{1}0)$. They are absolutely equivalent from the point of view of electrooptic applications. Let us consider, for example, the $(\bar{1}10)$ cut. There are two most important specific cases. In the first one (Figure 9.3b), the grating wave vector **K** and electric field are oriented along [001], light propagates along [$\bar{1}10$], and light polarization is along [110]. Here only one eigenmode participates in diffraction, and the diffracted light polarization is the same as the incident light polarization. This is a very important (favorable) moment for two-beam interaction (energy exchange). The refractive index variation amplitude is $\Delta n = -r_{41} n_0^3 E_{sc}/2$, where E_{sc} is the space-charge field.

In the second case (Figure 9.3b), the wave vector **K** (and the space charge field) is oriented along [110] and light propagates along [$\bar{1}10$]. The input light polarization is along [001]. Here, there are two eigenmodes with the light polarization oriented at ± 45 degrees with respect to [001]. For one of these modes, the grating will have a refractive index variation amplitude $\Delta n = r_{41} n_0^3 E_{sc}/2$, and for the other mode the refractive index variation amplitude is $\Delta n = -r_{41} n_0^3 E_{sc}/2$. The incident light diffracts from these two gratings, and since the gratings are phase shifted by π, the diffracted light waves are in antiphase. At the output, their superposition is the wave with the polarization oriented perpendicular to the incident light polarization. So this geometry provides a rotation of the diffracted light polarization by 90 degrees and prevents two-wave interaction (energy exchange). In spite of the fact that the half-wave voltage differs by a factor of two for the isotropic and anisotropic geometries, the diffraction efficiency is the same for both configurations. Similar results can be obtained for the other equivalent cuts, i.e., (110) and ($1\bar{1}0$).

Note that the maximal amplitude of the refractive index grating ($\Delta n = r_{41} n_0^3 E_{sc} \sqrt{3}$) in cubic photorefractive crystals can be achieved with (for instance) the (110) cut at $\mathbf{K}//[1\bar{1}1]$ and light polarization parallel to $\mathbf{K}$ [34] (the same is valid for the other equivalent cuts).

The transverse configuration in which E_{sc} is orthogonal to [111] is analyzed, for instance, in [34]. Detailed analysis of the polarization properties of light diffraction in sillenites can be found in [46]. The transverse geometry is used in the majority of holographic experiments. It was also used (both the (110) and (111) cuts) in the spatial light modulator PRIZ [47, 34]. Note that in the PRIZ modulator the external electric field was applied along the light propagation direction, but this field was needed to provide the carrier drift rather than the electrooptic effect. The light-phase modulation in the PRIZ occurred due to the space-charge field gratings having wave vectors orthogonal to the light propagation. These gratings originated from inhomogeneous distribution of space charge in the crystal when the modulator was illuminated by spatially inhomogeneous recording light. Extremely important specific features of the modulator were the ability to suppress a homogeneous background of images (image contouring), since homogeneous illumination did not produce transverse fields and the ability to perform suppression of the static component (or selection of the time-varying components) of images. Now this function is called novelty filtering.

9.1.2 Holographic Experiments Using Sillenites

9.1.2.1 Space-Charge Field in the Steady-State Regime

The first holographic experiments with BSO and BGO were described in [48], and those with BTO were reported in [49]. For sillenites, both conventional mechanisms (diffusion and drift) of holographic recording are efficient. For the steady-state regime of recording and under the condition that the contrast ratio of the recording interference pattern satisfies $m \ll 1$, the space-charge field $E_{sc} = -m(iE_D + E_0)$, where $E_D = K k_B T^1/e$, E_D is the diffusion field, k_B is the Boltzmann constant, and T^1 is the temperature. For $\Lambda = 1$ μm and room temperature, $E_D \approx 1.6$ kV/cm. As was mentioned above, the absolute value of E_{sc} is limited by the so-called saturation field E_q [50]. For the realistic values $N_A = 1.2 10^{16}$ cm^{-3}, $\Lambda = 1$ μm, and $\varepsilon = 56$, the saturation field has magnitude $E_q = 7$ kV/cm [51]. If $E_D \gg E_q$ and $E_0 \gg E_q$, then $E_{sc} \approx -imE_q$, and in the arbitrary case the modulus of the space-charge field is $E_{sc} = m\{(E_0^2 + E_D^2)/[(1 + E_d/E_q)^2 + (E_0/E_q)^2]\}^{1/2}$. The formulae given above were obtained for $m \ll 1$, which means that a linear approximation is used, since any terms proportional to m^2 (and higher powers) were ignored. If m is comparable to unity, higher harmonics of the gratings with wave numbers $K_p = pK$ and amplitudes $E_p = (-\frac{m}{1+\sqrt{1-m^2}})^{|p|}(E_0 + i\frac{|p|}{p}E_D)$, where $p \neq 0$, emerge [52]. Here E_p means the complex amplitude of the corresponding Fourier components of the space-charge field rather than the amplitude of the cosine grating.

Let us comment on the physical factors determining the most important holographic characteristics of sillenites.

9.1.2.2 Diffraction Efficiency

Diffraction efficiency (η) is determined here as the ratio between the intensities of the diffracted (first diffraction order) and readout beams. Usually, relatively thick (with a thickness of 1–10 mm) samples are used in holographic studies of sillenites, while in spatial light modulators (SLM) the thickness is less. The requirement for thin gratings $T\lambda/n_0\Lambda^2 \ll 1$, where T is the crystal thickness in the light propagation direction, has to be met. So the SLM thickness is typically less than 1 mm. For a thick (volume, Bragg) transmission hologram, the maximum diffraction efficiency is [53]

$$\eta = \sin^2(\pi n_1 T/\lambda \cos\theta_0), \tag{9.2}$$

where θ_0 is half of the angle between the recording beams inside the crystal (here it is the Bragg angle), and n_1 is the amplitude of the refractive index grating. Let us recall that here the optical activity is ignored. A typical experimental magnitude of η is 0.1–10% (an exception is [6], where $\eta = 95\%$ was obtained in low-conductivity BGO at $m \approx 1$ and under the applied field $E_0 = 14$ kV/cm). One of the limiting factors is the optical activity [27], which does not allow the efficient use of the sample thickness in the transmission geometry. Another factor, which is important at high spatial frequencies, is the limited magnitude of the space charge field E_{sc} because of a relatively low trap concentration and hence low E_q. For BTO, where the optical activity is several times lower than those in BSO and BGO, the diffraction efficiency can reach 30% [54]. To minimize the role of the optical activity, the orientation of the incident light polarization is often selected to be slightly rotated to compensate the polarization plane rotation caused by optical activity and to provide a proper polarization orientation at the midpoint of the crystal thickness.

For a reflection thick hologram [53]

$$\eta = \coth^{-2}(\pi n_1 T/\lambda \cos\theta_0). \tag{9.3}$$

However, sillenites are rarely used for recording of such holograms mostly because of too-high spatial frequencies (on the order of 10^4 mm^{-1}), which leads to a low saturation field E_q.

For thin gratings (which are typical of SLM) [55],

$$\eta = J_1^2(2\pi n_1 T/\lambda) \approx (\pi n_1 T/\lambda)^2. \tag{9.4}$$

Here, J_1 is the Bessel function of the first order and first kind. For thin gratings the diffraction efficiency is typically less than 1%.

9.1.2.3 Sensitivity

Sensitivity of holographic materials can be determined in different ways [55–58, 34]. One has to be careful in the exact estimation of sensitivity from the published data because of different definitions of sensitivity used in the literature. Roughly speaking, the inverse sensitivity of sillenites (the amount of the recording optical energy needed to have a 1% diffraction efficiency) is in the interval 10^{-2}–10^{-4} J/cm^2 for holographic recording and 10^{-4}–10^{-6} J/cm^2 for spatial light modulators.

9.1.2.4 Resolution

The situation with definitions of resolution of sillenites is the same as for other holographic materials, since there is no universal criterion for this characteristic. Roughly, the resolution of sillenites is on the order of hundreds of mm^{-1} for complicated pictures (it is limited by the dynamic range of the material) and can reach 10^3 mm^{-1} in the case of recording of a single grating (the limitations come from E_q, since $E_q \approx \nu^{-1}$, where $\nu = 1/\Lambda$ is the spatial frequency). The resolution of spatial light modulators (PROM, PRIZ) is limited to about 10 mm^{-1} (the limitation is determined by purely electrostatic effects in the modulators, since the amplitude of the space-charge grating is inversely proportional to ν when $\nu > 1/T$).

9.1.2.5 Speed of Recording and Storage Time

If we use the simplest consideration and assume that the drift (and diffusion) length is much less than the grating spacing, the speed of recording τ_{sc}^{-1} is determined by the Maxwell relaxation time, so

$$\tau_{sc}^{-1} = \tau_M^{-1} = \sigma/\varepsilon\varepsilon_0 = \sigma_d/\varepsilon\varepsilon_0 + \sigma_I/\varepsilon\varepsilon_0, \tag{9.5}$$

where σ_d is the dark conductivity, $\sigma_I = e\mu n_{av}$ is the photoconductivity, μ is the mobility, and n_{av} is the average density of the photoexcited carriers. In a more general case [34],

$$\tau_{sc} = \tau_M[(1 + K^2 L_D^2)^2 + K^2 L_0^2]/(1 + K^2 L_D^2), \tag{9.6}$$

where L_D and L_0 are the diffusion and drift length, respectively. Under standard experimental conditions τ_M is on the order of 10^{-2}–10^{-4} s. The speed of grating erasure is determined by the same relaxation parameters and can be controlled by the intensity of illumination. The storage time is determined by dark conductivity.

The situation with sillenites can be more complicated if holes contribute to the recording and erasure processes. A more detailed discussion of the problem can be found in the review paper [59].

9.1.2.6 Wave Mixing (Wave Amplification) Effects

These effects result from the fact that in dynamic media, readout of the recorded grating occurs simultaneously with the recording process. If the diffracted beams have the same polarization as the recording beams, then an interference of recording and diffracted beams takes place inside the sample. The interference is constructive or destructive depending on the phase relationship between the diffracted beam and one of the recording beams that propagates in the same direction inside the sample. Because of diffraction from the phase grating, the diffracted beam has a $\pi/2$ shift with respect to the recording beam. If the grating has the same position as the interference pattern (a nonshifted grating), no additional phase shift between the diffracted and recording beams arises, and there is no change in the intensity of the resultant beam. However, if the grating is shifted in space by $\pi/2$ (or $-\pi/2$) compared with the interference pattern (which happens in the diffusion mechanism of recording, for example), the net shift of the diffracted beam will be π or zero, and the resultant intensity (superposition of the diffracted and recording beams) will be higher or lower than for the nonshifted grating. So enhancement (or suppression) of one of the beams (or energy exchange) and, as a consequence, variation in the contrast ratio m occurs. Then the conditions for grating recording along the light propagation direction change because of variations in m. The theory of these effects has been developed in detail (see, for instance, corresponding chapters in as well as in [31, 34] and corresponding references therein). In the simplest situation (no loss, no pump-beam depletion, a $\pi/2$ shift of the grating, the grating of the transmission type), amplification of one of the beams can be described by a simple exponential:

$$I(z) = I_0 \exp(\Gamma z). \tag{9.7}$$

Here, I_0 is the incident light intensity of the beam to be enhanced and Γ is the gain factor per unit length, $\Gamma = 2\pi n_0^3 r_{\mathrm{eff}} |E_{\mathrm{eff}}|/(\lambda \cos \theta_0)$, where r_{eff} is the effective electrooptic coefficient, which depends on the selected crystal orientation (for instance, for sillenites with the (110) cut and the transverse configuration with $\mathbf{K}//[001]$, $r_{\mathrm{eff}} = r_{41}$), $E_{\mathrm{eff}} = \mathrm{Im}(E_{\mathrm{sc}}/m)$ is the imaginary component of the acting field (for instance, $E_{\mathrm{eff}} = E_{\mathrm{D}}$ for the diffusion mechanism). The E_{sc}/m ratio is taken for the front side of the crystal. In sillenites, the conventional mechanisms of recording provide a rather low $\Gamma \approx 1$ cm^{-1}. Moreover, the crystal thickness cannot be used efficiently because of the optical activity, so enhancement is rather poor. However, the situation is different if the nonstationary recording technique is used. We shall return to this question below.

9.2 Space-Charge Waves

9.2.1 A Brief Historical Review

Space-charge waves (SCW) are eigenmodes of spatial and temporal oscillations of the charge density. In principle, they can exist in various physical systems, for instance, in clystrons, in high-temperature plasma, and in plasma in semiconductors. They can be waves of free electrons [60] or the recombination waves in semiconductors [61]. In this section, we discuss the SCW associated with the process of trap recharging in semi-insulating crystals (to which the sillenites belong) when the crystal is placed in an external electric field. In photorefractive crystals, the trap recharging is an intrinsic process playing a crucial role in the transient processes at holographic recording. So if the necessary conditions for the excitation of the trap-recharging waves are met, these waves can be detected and studied by holographic means. In most cases, the processes of hologram buildup and decay have a purely relaxational character. However, in [62] it was shown that when the carrier drift length is longer than the diffusion length and longer than the grating spacing, these processes can acquire an oscillatory form. For recording of the simplest gratings, the oscillations were shown to be in the form of waves, which afterwards were named SCW. Before [62] appeared, theoretical and experimental investigations of SCW had been performed in semiconductors with no relation to photorefractive phenomena. The notion of SCW under consideration was introduced in [63–65], where the theory of these waves and some mechanisms of SCW instabilities were proposed. In [63–65], the term "waves of spatial charge exchange" was used instead of the term "SCW."

For freely propagating SCW (along the x-axis), the electric field of SCW can be presented as

$$E_w(t, x) = E_w \exp(-t/\tau_r) \cos(k_w x - \Omega_w t), \tag{9.8}$$

where $\tau_r \approx \tau_M d^2$ is the relaxation time, k_w and Ω_w are the wave number and eigenfrequency of SCW, respectively, $d = k_w L_0$, and it is assumed that $d \gg 1$, $L_0 > L_D$, and $E_0 \ll E_q$. Under these assumptions the parameter d is the "quality factor" of SCW. Typical magnitudes of SCW parameters in sillenites are as follows: τ_M is on the order of 10^{-3} s (at light intensities on the order of $100 \, \text{mW/cm}^2$), $d = 1$–6, $\Omega_w/2\pi$ is on the order of 10–100 Hz, and $k_w = 2\pi/\Lambda_w$, where $\Lambda_w \approx 5$–50 μm.

SCW exhibit some interesting properties; for instance, their dispersion law has an unusual form (when the conditions $d \gg 1$, $L_0 > L_D$, and $E_0 \ll E_q$ are satisfied) [62–65]

$$\Omega_w = 1/(\tau_M k_w L_0) = 1/(\tau_M d). \tag{9.9}$$

As a result, the group and phase velocities are oppositely directed. The decay does not depend on frequency; it is determined mostly by the Maxwell (dielectric) relaxation time τ_M and the quality factor of SCW that depends on the carrier lifetime. SCW propagate in the direction (the direction of phase velocity) opposite to that of the applied field. The electric field of SCW is shifted by $\pi/2$ with respect to the interference pattern exciting the SCW if attenuation is neglected. Sometimes SCW are regarded as new quasiparticles in solids, and they were even named spastrons [66]. A comprehensive historical review of SCW investigations up to 1995 is presented in [31]. Below we briefly present only major points.

During a long period only a few papers [67, 68] concerning experimental investigations of SCW were published. This was mostly because of the absence of sufficiently efficient experimental techniques of SCW excitation and detection at that time. The situation changed radically when researchers working with photorefractive phenomena realized that SCW can be relatively easily excited and detected by the holographic technique and that the sillenites are extremely suitable crystals for such investigations. Experimentally, the first unambiguous detection of SCW in photorefractive crystals was performed in [69], where the technique of moving grating proposed in [70, 71] was used for SCW excitation. It was expected [69] that if the wave vector of the moving interference pattern and speed of propagation of the pattern coincided with k_w and with the speed of SCW propagation $v_w = \Omega_w/k_w$, the resonance SCW excitation would occur. Indeed, a strong resonance enhancement of the diffracted beam in BSO for the definite speed of the interference pattern motion was observed in [69]. To prove that the resonance enhancement of the diffracted light is associated with the SCW excitation rather than beam-coupling enhancement, a special experimental geometry was selected in which diffraction was accompanied by polarization rotation that prevented two-beam coupling. To be on the safe side, a rather thin sample (1 mm) was used to reduce two-beam coupling effects. So the detected resonance could be explained only by excitation of SCW. In this case the light diffraction occurs from the moving grating, whose amplitude at resonance is much higher than far from the resonance. (In [69] the term "running gratings" was used instead of SCW). Among other experiments proving the existence of SCW, the paper [72], where the direct detection of the SCW free motion and decay was demonstrated, should be mentioned. The obtained parameters were the velocity of SCW propagation $v_w = 200$ µm/s, $\tau_r = 100$ ms, and $\Omega_w/2\pi = 10$ Hz at the light intensity ($\lambda = 514$ nm) $I_0 = 10$ mW/cm^2, $\Lambda_w = 20$ µm, and $E_0 = 6$ kV/cm. From these data it follows that $\tau_M \approx 3$ ms and $d \approx 6$.

In the next years, several theoretical papers concerning SCW excitation in compensated semiconductors [73] and SCW instabilities under different conditions [74–77] were published. A short description of SCW (in sillentes) was presented in [34]. Starting from 1988, interest in nonlinear interactions of SCW grew sharply. In [78], spatial subharmonic instabilities were discovered in BSO during illumination of the crystal by a moving interference pattern. After that, a large number of experimental and theoretical papers were published on this and related topics (see, e.g., [79–96]). In 1997, self-generation and direct visualization of SCW in BSO was demonstrated in [97].

During the last several years other nonlinear effects have been studied in sillenites both theoretically and experimentally. They are overall and spatial rectification of SCW, doubling of wave vectors, and second-harmonic generation.

Below we describe these effects. Before presenting some theoretical results, we briefly describe the most popular technique of SCW excitation, since the theoretical analysis is usually closely connected with the selected method of SCW excitation and detection.

9.2.2 Basic Methods of SCW Excitation

9.2.2.1 Moving-Grating Technique

Initially, the idea to use a moving grating arose to provide a $\pi/2$ phase shift between the grating and interference pattern and to increase the two-wave mixing gain (Γ) even for the drift mechanism of recording [70] (see the detailed theoretical analysis [98, 99]). In this case the crystal is illuminated by two beams, the position of one of which is shifted in some way, for instance, by a rotating mirror. Then the interference pattern is described as

$$W(z, t) = W_0\{1 + m \cos[K(x - vt)]\} = W_0[1 + m \cos(Kx - \Omega t)]. \qquad (9.10)$$

Here, W_0 is the summary intensity of the beams, v is the speed of the interference pattern, K is the wave number of the interference pattern, and $\Omega = Kv$. The same result (a moving interference pattern) can be obtained if the frequency of one of the beams is detuned from the frequency of the other beam by the frequency Ω. This can be achieved by a moving mirror due to the Doppler effect (this technique is most popular (Figure 9.4)). The frequency Ω is called the detuning frequency (and this method of SCW excitation is referred to sometimes as the "detuning method"). When the interference pattern moves, the space-charge grating cannot follow the interference pattern without a delay, since formation of the space-charge grating needs some time. So the grating position is shifted with respect to the position of the interference pattern. Formally, this means that the space-charge field grating acquires an imaginary part even for the drift mechanism. If the external field is applied to the crystal, the selection of a proper velocity of the interference pattern motion can provide the imaginary part of the space-charge field comparable to the applied field, which can be much higher than the diffusion field. This is the main idea of how to enhance the two-wave interaction (to increase Γ). The diffracted beam enhancement at a definite speed of the interference pattern motion was demonstrated for the first time in BSO

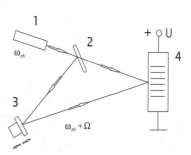

FIGURE 9.4. A simplified version of the setup for the SCW excitation by a moving interference pattern. 1, laser; 2, beam splitter; 3, moving mirror; 4, crystal.

in [70]. It is interesting that the conditions of the resonance enhancement of two-beam coupling coincide with the conditions of the resonance excitation of SCW: $K = k_w$ and $v = v_w$. Therefore, to provide the conditions under which only SCW excitation takes place, it is necessary to suppress two-beam coupling. This can be achieved by selecting the geometry where diffraction from the refractive index grating occurs with the polarization plane rotation and, of course, by using thinner samples to reduce absolute enhancement even for high Γ [69]. From the time the moving-grating technique was suggested, it has been used in many experiments, including parametric excitation of SCW subharmonics. To detect the SCW, diffraction of the recording or probe beam is used.

9.2.2.2 The Method of Alternating Applied Field

This method was initially developed to increase two-beam coupling. The first theoretical analysis of this mechanism and some experimental data were presented in [100, 101]. Since then the principles of recording in the oscillating applied field have been described in many publications (see, e.g., [102–104, 31]). The principle of recording in an alternating field is rather simple. Let us consider a square-wave alternating field. During one half-period of the field oscillations the electrons move in one direction as in the usual drift mechanism, but during the second half-period the electrons move in the opposite direction. If switching of the field from one direction to the other is fast enough (the grating is not erased during the switching time, this requirement is satisfied in the case of the square-wave applied field) and if the switching period is longer than the carrier lifetime (but shorter than τ_M), the space-charge grating grows and finally reaches some saturation level. This grating exhibits interesting features. From the symmetry point of view, it is equivalent to the diffusion grating, and the space-charge field grating is shifted by $\pi/2$ relative to the interference pattern, but the amplitude of this grating can be much higher than for the diffusion mechanism because of the drift mechanism that actually forms the grating. One more striking advantage of such a recording mechanism can be seen for the long-drift approximation when $d = KL_0 > 1$ (and $L_D < L_0$). For this situation and $m \ll 1$, the space-charge field $E_{sc} = -imE_0d$ if $E_0d \ll E_q$ and $E_{sc} = -imE_q$ if $E_0d \gg E_q$. So in this case the static space-charge field grating is additionally enhanced by a factor of d as compared with the grating field in the conventional drift mechanism, when a static electric field is applied, and a corresponding increase in the gain factor takes place because the grating is shifted. In the case of a sinusoidal alternating field, the effect is weaker.

Above, we mentioned only data concerning the formation of a static steady-state imaginary space-charge field. In a more detailed theory [104, 31], it was shown that an oscillating (both imaginary and real) space-charge field arises in addition to the static space-charge field. Moreover, in [105] it was experimentally shown using the alternating field technique that there exist high-frequency resonances of the gain factor (Γ) (on the scale of ten kilohertz) in BGO. They were attributed to the high-frequency branch of space-charge waves (this branch will be discussed at the end of the paper). In [106, 107], the AC field technique was also used, in combination with the oscillating interference pattern method, for detecting the high-frequency SCW branch. However, in [106, 107] the measured value was the amplitude of the alternating current flowing through the crystal rather than the diffracted beam intensity (or Γ). If we compare the advantages and drawbacks of the two methods (moving grating and AC field), one can say that the important advantage of the AC field technique is the reduction of various screening effects, which are very important in the case of application of the DC field. The AC field method also has a significant advantage for the two-beam coupling enhancement because of the nonresonance character of the formation of the static imaginary field. (A discussion of these two methods from the point of view of gain enhancement can be found in [108].) In [109], the method of SCW excitation including application of both DC and AC fields was proposed. In this case the crystal can be illuminated uniformly or by a static interference pattern. However, this technique has not yet found wide use. Though the AC applied field technique is well developed at present, a more comprehensive theoretical analysis is desirable.

9.2.2.3 Oscillating Grating Method

This technique is not a new one. It has been used for scientific investigations and practical applications for many years. It has proved to be very convenient from the experimental point of view for SCW investigations. In addition, it provides detection of new phenomena that do not exist when excitation is achieved by other methods, such as a moving-grating technique. This technique also facilitates greatly detection of second-harmonic generation.

In the oscillating grating method, the crystal is usually illuminated with two beams, one of which is phase-modulated. A typical setup for the optical excitation and detection of SCW is shown in Figure 9.5a,b. It includes a laser for the formation of the required interference pattern (often it is a Nd: YAG or an Ar-ion laser). After a beam splitter, an electrooptical phase modulator is installed. For some measurements, an additional (probe) beam (mostly from a He-Ne laser, but sometimes from a semiconductor laser) is used. To measure the current, a loading resistance is included in the electric circuit.

The recording laser beams form the interference pattern $W(x, t)$ oscillating near a midpoint

$$
\begin{aligned}
W(x, t) &= W_0[1 + m \cos(Kx + \Theta \cos(\Omega t))] \\
&\approx W_0 + W_0 m \cos(Kx) - \frac{1}{2} W_0 m \Theta \sin(Kx + \Omega t) \quad (9.11) \\
&\quad - \frac{1}{2} W_0 m \Theta \sin(Kx - \Omega t).
\end{aligned}
$$

Here W_0 is the average light intensity, Θ is the amplitude, and Ω is the frequency of phase modulation. The condition $\Theta \ll 1$ was used in the expansion of the initial expression into a series. The second term in (9.11) is the static interference

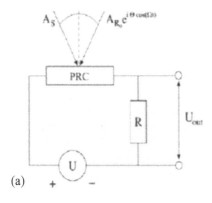

(a)

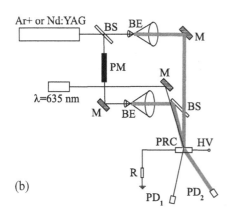

(b)

FIGURE 9.5. (a) A simplified version of the setup for SCW excitation by an oscillating interference pattern and for the detection of rectification effects. (b) Experimental setup (schematically) for SCW excitation by an oscillating interference pattern and for detection of the non-Bragg diffraction orders of the probe beam and one of the recording beams. PRC is the photorefractive crystal; R is the real loading resistor; HV is the high-voltage source; PM is the phase modulator; BE is a beam expander; BS is a beam splitter, M is a mirror, PD_1 and PD_2 are photodetectors for detecting the diffracted probe and recording beams, respectively. In this scheme, the recording laser is an argon–ion laser $\lambda = 514$ nm and the source of the probe beam is a semiconductor laser diode at $\lambda = 635$ nm.

pattern, whereas the last two terms describe two running interference patterns moving in opposite directions. In (9.11) the terms proportional to m^2 and Θ^2 are omitted. However, for some calculations these terms will be taken into account. In accordance with (9.11), three gratings of the space-charge field (and refractive index) will be recorded: one is a static grating and two gratings are moving ones.

Resonance SCW excitation occurs when K and Ω of one of the moving interference patterns coincide with k_w and the eigenfrequency Ω_w of the SCW [52]. Starting from this point, one can say that this technique is equivalent to the technique of detuning described above, with the only difference that the contrast ratio m is replaced by $m\Theta$. This is quite true. However, in the oscillating interference pattern method the static grating plays a very important role. Let us consider the diffraction process (for simplicity, we use a probe beam with the wavelength differing from the wavelength of recording beams). The probe beam will diffract (in the same direction) simultaneously from the static and two running gratings, and in the general case, three diffracted beams will interfere. Since the resonance grating has a higher amplitude, let us consider only diffraction from the static and the resonance moving grating. The diffracted light from the moving grating is frequency-shifted by Ω, and thus combination of diffraction from the static and dynamic grating results in the light intensity oscillating with a beating frequency Ω at the output. This is important from the experimental point of view since detection of the alternating output signal provides a high signal-to-noise ratio and greatly facilitates the process of measurements. The detected signals do not depend on the applied field direction since the alternating interference pattern contains two components propagating oppositely to each other. This is one of the specific features of the method of an oscillating interference pattern with respect to the method of detuning (moving grating). The next feature is more delicate. When we discuss nonlinear effects, we shall see that in the method of alternating interference pattern the moving space-charge grating can interact with the static space-charge grating and thereby result in an alternating current in the outside circuit and homogeneous oscillating electric field in the crystal. The generation of the oscillating current in the outside circuit is of interest not only from the scientific but also from the applied point of view, since it allows compact and high-sensitivity sensors of phase-modulated laser beams to be built. In the detuning method there is no static grating, and hence there is no alternating current.

9.2.3 SCW Rectification and Second-Harmonic Generation

9.2.3.1 The Linear Regime of Space-Charge Oscillations

When a crystal is illuminated by the oscillating interference pattern described by (9.11), electrons are photogenerated from traps into the conduction band with the photogeneration rate given by

$$g(x, t) = g_0[1 + m \cos(Kx + \Theta \cos \Omega t)], \tag{9.12}$$

where $g_0 = \alpha \beta W_0 / \hbar \omega_{\mathrm{ph}}$, α is the absorption coefficient, β is the quantum efficiency of electron excitation, and ω_{ph} is the optical frequency.

As a result, an inhomogeneous carrier density $n(x, t)$ and inhomogeneous electric field $E(x, t)$ arise in the crystal. These processes are described by a standard system of equations [50, 52, 110, 111]:

$$\frac{\partial n(x, t)}{\partial t} + \frac{n(x, t) - \tilde{n}_0}{\tau} = g_0[1 + h(x, t)] - \frac{\varepsilon}{4\pi e} \frac{\partial^2 E(x, t)}{\partial x \partial t}, \tag{9.13}$$

$$\frac{\varepsilon}{4\pi} \frac{\partial E(x, t)}{\partial t} + j(x, t) = I(t), \tag{9.14}$$

$$j(x, t) = en(x, t)v(E(x, t)) + eD(E(x, t))\frac{\partial n(x, t)}{\partial x}. \tag{9.15}$$

Here

$$h(x, t) = m \cos(Kx + \Theta \cos \Omega t), \tag{9.16}$$

$\tilde{n}_0$ is the steady-state density of free electrons, $j(x, t)$ is the current density in the crystal, $I(t)$ is the normalized current density in the outside circuit, and $v(E(x, t))$

and $D(E(x, t))$ are the drift velocity and diffusion coefficient, respectively. In the general case, $v(E(x, t))$ and $D(E(x, t))$ are nonlinear functions of the applied field. Equations (12–16) take into account the Poisson equation, and they are correct when there is a large enough number of free traps, and the system of traps is far from saturation.

Quite a general solution of these equations is given in [111]. In the consideration presented below, we make some simplifications for the sake of simplicity. We ignore the diffusion process ($D = 0$) since in the experiments the relationship $E_0 \gg E_D$ was satisfied; we take $\tilde{n}_0 = 0$ since dark conductivity is neglected and $\partial n/\partial t = 0$ since an adiabatic approach is used. In addition, we assume the drift velocity to be $v = \mu E(x, t)$, thus admitting that there is a linear relationship between the electric field and current, i.e., Ohm's law is satisfied (at the end of the paper we shall discuss the situation in which these simplifications are not used). Note that the approximation $\partial n/\partial t = 0$ means that we automatically omit from consideration the so-called high-frequency branch of SCW. We shall return to this question below. Equations (12–16) can be solved in both the linear and nonlinear approximations. In the linear case, it is postulated that $m \ll 1$ and all terms proportional to m^2 are ignored. Then the following expression for the space-charge field can be obtained for $\Theta < 1$:

$$E_{sc}(x, t) = E_{sc}(\text{static}) + E_{sc}(\text{running})$$
$$= \frac{m}{2} E_0 \left\{ -\exp(iKx) - \frac{i\Theta}{2} \frac{1}{1 - \omega d + i\omega} \exp[i(Kx - \Omega t)] + c.c. \right\}.$$

$$(9.17)$$

Here, $\omega = \Omega \tau_M$ and $d = L_0 K$ (in a more general case $d = L_0 K (1 - m^2)^{1/2}$ [52]). Equation (9.17) does not include the waves that propagate opposite to SCW, i.e., the nonresonant waves with phase factors $\exp[\pm i(Kx + \Omega t)]$, although these waves were taken into account in calculations. In the approximation used ($E_D/E_0 \ll 1$ and $E_0/E_q \ll 1$), d is the quality factor of SCW. If the diffusion and space-charge field saturation are taken into account [85], the expression for the quality factor has another form:

$$Q = (1/d + E_D/E_0 + E_0/E_q)^{-1}. \qquad (9.18)$$

If $Q > 1$, we deal with wave processes. When $Q < 1$, the processes are purely relaxational, and there are no eigenmodes of oscillations. As mentioned above, a typical quality factor in sillenites is not higher than 5–6. Detailed experimental investigation of the quality factor was performed in [113] for BSO. It was found that $Q \approx 3.1$ for $E_0 = 8.3$ kV/cm and grating spacing of 20 μm.

A very important characteristic of SCW is their dispersion law. In the approximation used, when diffusion and trap saturation are ignored and the quality of the SCW resonance is high enough ($d \gg 1$) and $m \ll 1$, the dispersion law has the form given by (9.9). Then the phase velocity is $v_{ph} = (\tau_M k_w^2 L_0)^{-1}$, and the group velocity is $v_{gr} = -(\tau_M k_w^2 L_0)^{-1}$.

For SCW excitation, condition (9.9) has to be satisfied, so the soundest proof that in the experiment the SCW are indeed excited is fulfillment of condition (9.9).

The oscillating field (9.17) exhibits the resonance dependence on the modulation frequency, and the resonance occurs when $\Omega = \Omega_R \approx 1/(\tau_M d)$, i.e., just when the conditions for SCW excitation are satisfied. (More precisely, $\Omega_R = 1/[\tau_M(d^2 + 1)^{1/2}]$.)

To analyze the diffraction effects, we use the approximation of a thin hologram where non-Bragg diffraction orders can arise (it is a simpler case from the experimental point of view). Then we have to decide what diffraction order is more suitable for SCW investigations. In [52, 112], experiments with two diffraction orders of the recording beams were analyzed. The first one occurs when the Bragg order is detected. In this case the resonance effects can be masked by two-wave interactions between the diffracted beam and one of the recording beams. The second case is that of non-Bragg diffraction. For the non-Bragg orders of diffraction, there are no recording beams propagating in the same direction as the diffracted beam, and so there is no usual two-beam coupling. This is a more favorable case for SCW study. However, there is an alternative. If a probe (nonrecording) beam is used, then Bragg diffraction is quite suitable for measurements. The oscillating intensity amplitude of the output beam including the first non-Bragg order in the case of detection of the recording beam diffraction is [52, 112]

$$W_{osc} = \frac{2W_{r,P}\eta\Theta\omega d}{\sqrt{1 + 2\omega^2(1 - d^2) + \omega^4(1 + d^2)^2}}. \tag{9.19}$$

The same expression is valid for the oscillating intensity amplitude of the output beam including the Bragg order when the probe beam is used. Here, $W_{r,P}$ is the intensity of the probe beam (or of one of the recording beams), and η is the diffraction efficiency for the static grating. In the case that the transverse configurations mentioned above ((110), ($\bar{1}10$), and ($1\bar{1}0$) crystal cuts) are used, $\eta \approx (\pi n_0^3 r_{41} m E_0 T/2\lambda)^2$. Here the optical activity is ignored because of a relatively small sample thickness T. It is important to underline that in such an experiment the output oscillating light intensity is proportional to the amplitude of the oscillating space-charge field, i.e., to the SCW amplitude. This is because the output intensity is determined by the superposition of the beam diffracted from the static grating and of the beam diffracted from the moving grating (the phase of this beam is oscillating at Ω).

Detailed experimental investigations of the resonance SCW excitation by an alternating interference pattern and verification of the dispersion law of SCW were performed in [114]. BTO and BGO crystals with the (110) cut were used. The electric field and grating wave vector were oriented along [$1\bar{1}0$]; the light polarization was along [001]. For recording, the second harmonic of a Nd:YAG laser was used; the light intensity was changed in the interval 25–400 mW/cm^2. Figure 9.6 shows examples of the frequency dependence of the non-Bragg diffraction order intensity for BTO. The fitting parameters $d = 1$ and $d = 3.3$ correspond to two values of the wave numbers of the gratings with $K = 113$ mm^{-1} and $K = 378$ mm^{-1},

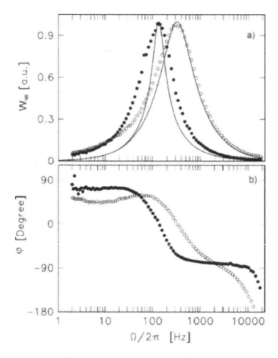

FIGURE 9.6. Frequency dependencies of amplitude W_{osc} (a) and phase φ (b) of the intensity of the beam diffracted from the BTO sample into the first non-Bragg order for two values of the wave number K (o: $K = 113$ mm^{-1}; •: $K = 378$ mm^{-1}). The total light intensity is $I_0 = 225$ mW/cm^2, the contrast ratio $m = 0.2$, the applied electric field $E_0 = 9.4$ kV/cm, and $\Theta = 0.5$ rad. The solid lines (a) correspond to calculations according to (9.19). Data are taken from [114].

and $\tau_M = 0.35$ ms. It can be seen from the figure that a good signal-to-noise ratio can be achieved, and this allows one to find the position of the resonance maximum with high accuracy. It is important that the resonance position not be sensitive to misalignments of the setup at its reconfiguration when the resonance position is measured as a function of the angle between recording beams (or as a function of K). This provides a very good opportunity to check whether the measured dispersion law corresponds to the SCW dispersion law. One of the main sources of possible experimental errors here is the inhomogeneous illumination of the sample, since the resonance depends on the light intensity propagating through the crystal, which, in turn, depends on the sample thickness. However, as shown by estimates, at the selected thickness of the samples (0.7 mm) [114] one can neglect the intensity inhomogeneity along the direction of light propagation. One more important source of errors can be inhomogeneity of the internal electric field, but this factor contributes mainly to inhomogeneous broadening of the resonance peak rather than its position.

Figure 9.7a shows the resonance frequency position as a function of wave number for BTO and BGO. One can see quite a good agreement between the theoretical curve ($\Omega_w \approx 1/k_w$) and experimental data, which proves that the dispersion law given by (9.9) is valid in our case. An additional proof of the validity of (9.9) is given in Figure 9.7b, where the position of the resonance minimum in the overall rectification effect as a function of Λ is shown. As will be shown later, the position of this minimum corresponds to the resonance excitation of SCW.

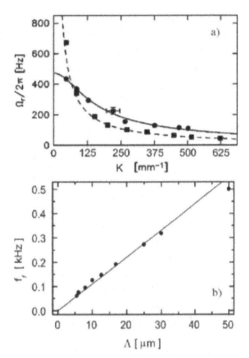

FIGURE 9.7. (a) Resonance frequency $\Omega_r/2\pi$ as a function of the wave number K for a BTO and a BGO sample [114]. The solid and dashed curves are fits of the theoretical expression for the resonance frequency to the experimental data. The light intensity is $W_0 = 225 \text{ mW/cm}^2$ and the applied field is $E_0 = 9.4 \text{ kV/cm}$ for BTO ($\bullet$) and $W_0 = 340 \text{ mW/cm}^2$ and $E_0 = 9.1 \text{ kV/cm}$ for BGO ($\blacksquare$). The contrast ratio is $m = 0.2$ for both crystals. (b) Frequency position of the dc current minimum (resonance frequency f_f) in BSO as a function of grating spacing Λ ($E_0 = 8 \text{ kV/cm}, m = 0.43$, $W_0 = 130 \text{ mW/cm}^2$, $\Theta = 1 \text{ rad}$). The solid line corresponds to a function $\propto \Lambda$ [116].

9.2.3.2 Nonlinear Regime of Space-Charge Oscillations

Nonlinear SCW interactions arise due to the nonlinearity of the holographic recording mechanism in the crystals under consideration. From the formal point of view the nonlinearity of the recording process is associated with the product $n(x, t)E_{sc}(x, t)$ in the expression describing the current density $J(x, t)$. For the approximation used, $J(x, t) = e\mu n(x, t)[E_0 + E_{sc}(x, t)]$. Since $E_{sc}(x, t)$ is the space-charge field arising in the crystal during the charge-grating formation process, one can say that the nonlinearity originates from the fact that the space-charge field affects the space-charge grating formation. The nonlinearity can be characterized by the ratio $E_{sc}(x, t)/E_0$, and it becomes very important when the space-charge field is comparable to the applied external field.

Let us consider in more detail the product $n(x, t)E_{sc}(x, t)$. For the method of SCW excitation by an alternating interference pattern and for $m \ll 1$, the charge density can be represented as a sum

$$n(x, t) = n_{av} + n_{st} \exp[iKx] + n_{run} \exp[i(Kx - \Omega t)] + \text{c.c.} \tag{9.20}$$

Here we again ignore the waves ($\exp[\pm i(Kx + \Omega t)]$), the indexes "st" and "run" mean static and moving components, and the quantities n_{st}, n_{run}, as well as $E_{sc}(x, t)$ (see (9.17)) are proportional to m. In fact, for the SCW excitation mechanism under consideration, the quantities n_{run} and E_{sc} (running) are also proportional to some other parameters, such as Θ (see (9.17)).

For very low m, the product $n(x, t)E_{sc}(x, t)$ can be ignored, but when m is on the order of unity, this product cannot be neglected, and corrections in the current magnitude proportional to m^2 have to be taken into account. These corrections result in a renormalization of n_{st}, n_{run}, E_{sc} (static), and E_{sc} (running) and lead to new terms in the expression for current density that describe interaction of different space-charge gratings.

1. The term proportional to $m^2 \exp[i2Kx]$ results in the appearance of the second (and even higher) spatial harmonics of the static grating. These harmonics make an additional contribution to the higher-order diffraction beams. The theory [52] predicts also the nonlinear excitation of the SCW of the type $\exp[i(pk_w x - \Omega_w t/p)]$, where p is an integer when $\Omega = \Omega_w/p$ [52]. The intensity of these waves reduces with increasing p.
2. The terms containing multipliers of the type $\exp[\pm i(Kx - \Omega t)] \exp[\pm i(Kx - \Omega t)]$ in the product $n(x, t)E_{sc}(x, t)$ describe the interaction of moving gratings resulting in doubling of K and Ω (second-harmonic generation) or cancellation of time and coordinate dependencies of the corresponding current components (SCW overall rectification), i.e., appearance of an additional DC current.
3. The terms containing multipliers of the type $\exp[\pm iKx] \exp[\pm i(Kx - \Omega t)]$ describe the interaction of moving and static gratings and result in doubling of K or elimination of the dependence on coordinate (spatially homogeneous term). In the case of doubling of K there is an opportunity for the temporal subharmonic of SCW to be excited (i.e., excitation of the waves with $\exp[i(2k_w x - \Omega_w t/2)]$ when $\Omega = \Omega_w/2$).

In the case that the interaction includes two initial gratings, the resulting grating is proportional to m^2. In this sense, we are dealing with the second-order nonlinearity of SCW interactions. The interaction of two moving gratings formally recalls the second-order nonlinear interaction of optical waves that results, for instance, in optical second-harmonic generation and optical rectification. At the same time, the effects of the nonlinear interaction of a moving grating with a static grating resulting in changes in the absolute value of the wave vector without variation of the eigenfrequency have no direct analogues in optics, since in optics they are prohibited by conservation of momentum. Sometimes we shall use the term "SCW scattering from a static grating" for this nonlinear interaction of a moving and static grating, but it should be underlined that it is rather a conditional terminology.

Below, we use the term "overall rectification" for the process associated with interaction of moving gratings and resulting in a constant contribution to the DC current, and the term "spatial rectification" for the process of scattering of a moving grating from a static one that results in a spatially homogeneous, but oscillating in time, contribution to the current density.

We do not discuss here effects of parametric excitation of spatial subharmonics of SCW, since these processes were described in detail in the publications mentioned above, and this question is out of the scope of the present paper.

9.2.3.3 Second-Harmonic Generation

The phase factor of a moving grating is described as $\exp[iKx - i\Omega t]$. So when two equivalent moving gratings interact, the resulting phase factor is equal to the product of the phase factors of the initial gratings, and in this particular case it is equal to $\exp[i2Kx - i2\Omega t]$. The appearance of such a phase factor is an indication of second-harmonic generation. It is important to note that second-harmonic generation in this case necessarily includes forced waves; since the dispersion law is violated in this process. It follows from the dispersion law (9.9) that if the initial space-charge wave has k_w and Ω_w, then the wave with the doubled wave vector $(2k_w)$ has to have the eigenfrequency $\Omega_w/2$. In the opposite case, if a new wave has a doubled frequency $(2\Omega_w)$, then it must have wave number equal to $k_w/2$. So simultaneous doubling of the frequency and wave vector of the SCW eigenmode will not result in the creation of a new eigenmode. Nevertheless, if the process of second-harmonic generation includes participation of forced waves (noneigenmodes), the limitations imposed by the dispersion law are removed. Two mechanisms of second-harmonic generation can be mentioned here. The first occurs when two forced charge waves (noneigenmodes of the SCW) interact and give rise to the second harmonic that is an eigenmode of SCW, and the second mechanism occurs when two eigenmodes interact and give rise to the second harmonic that is not an eigenmode. Both mechanisms were studied theoretically [110] and observed experimentally [115, 116]. To detect the second harmonic of SCW, one has to detect the second diffraction order and perform measurements at a doubled frequency (2Ω) compared with the modulation frequency Ω. Figure 9.8a,b shows the signal at 2Ω for the second non-Bragg diffraction order in BGO. Three maxima are observed in the general case. However, only two of them are due to second harmonic generation (the left and the right peaks), whereas the midpeak originates from frequency doubling of the signal oscillating at Ω by the photodetector. The low-frequency peak corresponds to the excitation of two forced waves and subsequent generation of the second harmonic that is an SCW eigenmode. The high-frequency peak is connected with the initial excitation of SCW eigenmodes and as a consequence, generation of the second harmonic that is a forced wave. The exact mathematical expression for the output signal is rather cumbersome [110, 116], and we do not present it here. However, it is possible to make some predictions qualitatively. For instance, the dependence of the output signal on m and θ and frequency position of the resonance maxima can be predicted. The output beam is the superposition of several diffracted beams propagating in the same direction. One of them is the beam diffracted from the second static harmonic. Its amplitude is proportional to m^2. The second one is the beam diffracted from the second harmonic of the moving grating. Its amplitude is proportional to $m^2\theta^2$, since the second harmonic amplitude is proportional to the product of the amplitudes of two initial moving gratings each of which is proportional to $m\theta$. As a result, the intensity of the output signal is proportional to $m^4\theta^2$. Concerning the positions of the resonances, it can easily be found that the first (low-frequency) maximum of the output signal is at the modulation

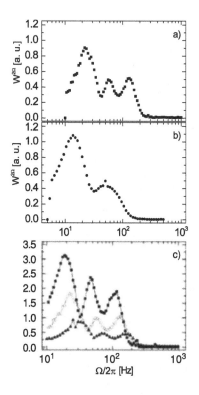

FIGURE 9.8. Output signal $W^{2\Omega}$ at 2Ω as a function of modulation frequency $\Omega/2\pi$ for two wave numbers K in the experiments for the detection of second harmonic generation. (a) BGO; $E_0 = 8$ kV/cm, $m = 0.43$, $W_0 = 130$ mW/cm^2, $\Theta = 1$ rad; $K = 483$ mm^{-1}; (b) $K = 897$ mm^{-1}. (c) Output signal $W^{2\Omega}$ at 2Ω as a function of modulation frequency $\Omega/2\pi$ in the experiments for the detection of second-harmonic generation in BGO for different applied fields E_0. $m = 0.43$, $W_0 = 130$ mW/cm^2, $\Theta = 1$ rad; $K = 483$ mm^{-1}; $\blacksquare$-$E_0 = 10$ kV/cm, $\bullet$-$E_0 = 8$ kV/cm, $\blacktriangle$-$E_0 = 6$ kV/cm. The lines are guides to the eye. Data are taken from [116].

frequency $\Omega = \Omega_w/4$ (the frequency of detection is $\Omega_w/2$) and the high-frequency maximum is at the modulation frequency $\Omega = \Omega_w$ (the frequency of detection is $\Omega = 2\Omega_w$). Experimental results qualitatively agree with this conclusion.

A simplified theoretical expression for the output signal (at 2Ω) when $d \gg 1$ and attenuation is ignored has the form

$$J_{2,2}(t) = \frac{1}{4}\eta I_R m^2 \Theta^2 \cos(2\Omega t)$$

$$\times \left[\frac{2}{(1 - \omega^2 d^2)(1 - 4\omega^2 d^2)} - \frac{1 + 8\omega^2 d^2}{(1 - 4\omega^2 d^2)(1 - 16\omega^2 d^2)}\right.$$

$$\left. - \frac{1 + 9\omega^2 d^2}{(1 - \omega^2 d^2)^2(1 - 16\omega^2 d^2)}\right]. \tag{9.21}$$

Here I_R is the intensity of one of the recording beams. Equation (9.21) does not include the corrections for the internal field and d caused by screening effects. The corresponding expression including such correction can be found in [110, 116]. The experiments with the second-harmonic generation demonstrate important advantages of the oscillating-gratings technique since the detected optical signal is proportional to the instant magnitude rather than the square of the modulus of the excited SCW field amplitude, and it can be detected at the frequency of the moving grating that provides light diffraction. This is because there is a background of the light diffracted from the static grating with wave number $2K$.

9.2.3.4 Overall Rectification

This effect arises when two phase-conjugated SCW with the phase factors $(\exp[iKx - i\Omega t]$ and $\exp[-iKx + i\Omega t])$ interact. As a result of this interaction, an additional current component arises that is proportional to the product of phase-conjugated multipliers, and therefore it is time- and coordinate-independent (which means that the total DC current flowing through the crystal changes). This effect is reminiscent of rectification of light waves in nonlinear optics due to the second-order nonlinearity when a static polarization arises under illumination of a proper material by sufficiently intense light [117]. In optics, this effect is typically rather weak; the nonlinearity here is characterized by the ratio E_{ph}/E_{at} (where E_{ph} and E_{at} are the field amplitude of the optical wave and the electric field in the electronic shell of the atom, respectively), and typically this ratio is much less than unity. For SCW, the characteristic parameter is the square of the contrast ratio m, which can be comparable to unity, and therefore, the effects can be very high. Above, we described the origin of the overall rectification rather formally, as resulting from multiplication of phase-conjugated waves. However, the fact that we discuss the product $n(x, t)E_{sc}(x, t)$ and analyze the new contributions to the homogeneous current density means that we consider the interactions between waves of the free-carrier charge density and waves of the space-charge field. Therefore, in more physical terms, the effect of rectification implies the onset of the DC current when the field grating (formed mostly by trapped charges) and the grating of free-charge carriers move in the same direction synchronously, and the phase shift between them is not equal to $\pi/2$.

The calculations performed in [110] have shown that the DC current flowing through the crystal depends on Ω due to the SCW overall rectification. The expressions for variations in the DC current density versus Ω have the following forms:

a. for arbitrary Θ,

$$I_0(\omega) = \frac{\sigma E_0}{1 + q}\left\{1 - \frac{m^2}{2(1 + q)}\sum_{l=-\infty}^{\infty}\frac{J_1^2(\Theta)}{(1 - \omega d_q l)^2 + \omega^2 l^2}\right\}$$

$$= \frac{\sigma E_0}{1 + q}\left\{1 - \frac{m^2}{2(1 + q)\omega}\mathrm{Re}\int_0^{\infty} d\phi \times \exp\left(-\frac{2\phi}{1 - i\omega d_q}\right)J_0(2\Theta\sin\phi)\right\}$$

(9.22)

b. for $\Theta < 1$,

$$I_0(\omega) = \frac{\sigma E_0}{1 + q}\left[1 - \frac{m^2}{2(1 + q)} + \frac{m^2\Theta^2}{8(1 + q)}\right.$$

$$\left. \times\left(2 - \frac{1}{(1 - \omega d_q)^2 + \omega^2} - \frac{1}{(1 + \omega d_q)^2 + \omega^2}\right)\right].$$

(9.23)

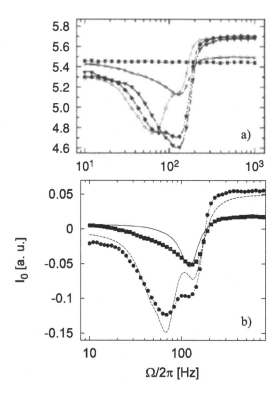

FIGURE 9.9. (a) Overall rectification in BGO. DC current as
a function of phase modulation
frequency for $E_0 = 8$ kV/cm,
$W_0 = 130$ mW/cm^2, $m = 0.43$,
and $K = 630$ mm^{-1} for different
Θ. ■- $\Theta = 0$, ●-$\Theta = 0.2\,\pi$,
$\leftrightarrow - \Theta = 0.6\pi$, ◆$- \Theta = 0.8\pi$,
○$- \Theta = 1.0\,\pi$. (b) Comparison
between theory (lines), (9.22), and
experiment (symbols) for overall
rectification in BGO. $\Theta = 0.2\,\pi$
(solid   line,   ●), $\Theta = 0.9\pi$
(dashed line, $\leftrightarrow$). The fitting
parameters are $d_q = 3.6 \pm 0.5$,
$\tau_M = 3.4 \cdot 10^{-4}$ s, and $q = 0.7$.
Data are taken from [116].

Here, the parameter q characterizes the difference between the real field inside
the crystal E_{int} and calculated applied field $E_0 = U/L$, so $E_{\text{int}} = E_0/(1+q)$,
where U is the applied voltage and L is the distance between the electrodes,
$d_q = d/(1+q) = K\mu\tau E_0/(1+q)$. The parameter q is introduced here for the
first time. In fact, so far we have not discussed experiments in which this factor
could play a crucial role and, in addition, could be directly measured. However,
the measurements of overall rectification provide an instrument of direct determination of the parameter q, and as will be seen later, the internal field reduction is
quite strong and typical of sillenites. Moreover, it plays a crucial role in the experimental detection of the spatial rectification effect by electrooptic or piezoelectric
measurements. Note that in the theoretical analysis of many other experiments
performed earlier, this parameter can lead only to the renormalization of the
internal field magnitude.

Figures 9.9a,b,c and 9.10 and 9.11 show the dependencies of the DC current on the modulation frequency for BGO, BSO [116], and BTO [118]. (The
BTO sample is different from that used for the measurements shown in Figure
9.6.) The theoretical data reasonably agree with the experimental results. As can
be seen from Figure 9.9, the measured variations in the DC current caused by
overall rectification reach 20%, so it is a giant effect. The fitting parameters are
$d_q = 3.6(\Lambda = 13.1$ μm$)$, $q = 0.7$, $\tau_M = 3.410^{-4}$ s, and $\mu\tau = 1.610^{-7}$ cm^2/V

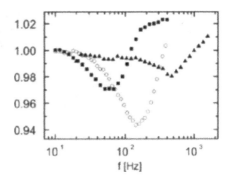

FIGURE 9.10. Overall rectification in BSO [116]. DC current I0 (in arbitrary units) as a function of phase modulation frequency f for different grating spacings ($\leftarrow$: $\lambda = 50$ μm, •: $\lambda = 12.6$ μm, ■: $\lambda = 5.5$ μm; $E_0 = 8$ kV/cm, $W_0 = 130$ mW/cm^2, $m = 0.43$, $\Theta = 1$ rad.

for BGO; $d_q = 1.9 (\Lambda = 50$ μm), $q = 0.8$, $\tau_M = 210^{-4}$ s, and $\mu\tau = 3.4 10^{-7}$ cm^2/V for BSO; and $d_q = 1.6$ ($\Lambda = 10$ μm), $q = 2.1$, $\tau_M = 5\ 10^{-4}$ s, and $\mu\tau = 0.9\ 10^{-7}$ cm^2/V for BTO.

The magnitudes of the product $\mu\tau$ are in reasonable agreement with the literature data (Table 9.1). It is important to note that the reduction in the internal electric field obtained from measurements of rectification ($E_{int} = E_0/(1 + q) = 0.59\ E_0$ for BGO and $E_{int} = 0.52\ E_0$ for BSO) is surprisingly close to the data reported in [25], where it was found that $E_{int} = 0.6\ E_0$ for BSO. This can mean that the factor $q \approx 0.7$ is quite a typical value for BSO and BGO, and many data obtained from measurements in a permanent external field must be corrected by a factor $1/(1 + q) \approx 0.6$. One of the interesting differences between the experimental data for BGO and BTO is a poorer signal-to-noise ratio and a much higher q for BTO. This means that the real electric field inside the BTO crystal is lower than that in BGO and much lower (approximately one-third) than the expected value of E_0. The main reason for this are screening effects caused by the nonohmic character of the electrodes in BTO. It is an important conclusion since it implies that the investigation of the overall rectification is of practical interest for analyzing the electrode quality and estimation of the internal field in photorefractive materials and even in semiconductors.

It is interesting to note that the effect of overall rectification can be detected also when SCW are excited by a moving interference pattern rather than an

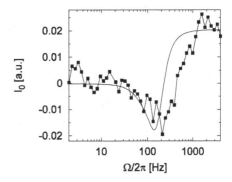

FIGURE 9.11. Overall rectification in BTO [118]. DC current as a function of phase modulation frequency for $E_0 = 9$ kV/cm, $W_0 = 344$ mW/cm, $m = 0.57$, $\Theta = 0.94$, and $K = 630$ mm^{-1}. The measured frequency dependence of the output signal I_0 measured at the loading resistor (■) is compared with the theory (9.22) (solid line). The fitting parameters are $d_q = 1.6$, $\tau_M = 5 \cdot 10^{-4}$ s, and $q = 2.1$.

oscillating one. The corresponding theory has been published elsewhere [119]. The expression for the DC current in this case has the form

$$
\begin{aligned}
\delta I_0 &= -\frac{\sigma E_0}{2(1+\rho\sigma)}\left(\frac{m}{\tau\tau_M}\right)^2 \frac{1+\Omega\tau_M E_D/E_0}{|\Omega-\Omega_1|^2|\Omega-\Omega_2|^2}, \\
\Omega_1 &\cong -K\mu E_0 - i\left(\tau^{-1}+\tau_M^{-1}+K^2 D\right), \\
\Omega_2 &\cong \frac{1}{\tau\tau_M}\frac{1}{K\mu E_0+i(\tau^{-1}+\tau_M^{-1}+K^2 D)}.
\end{aligned}
\tag{9.24}
$$

This expression is derived under rather general conditions when diffusion processes ($D \neq 0$) and fast oscillations of photoexcited carriers ($\partial n/\partial t \neq 0$) are taken into account. One can see from (9.24) that the DC current has two resonances. One of them is the so-called low-frequency resonance. It corresponds to the minimum of the current, which occurs at the detuning frequency equal to the frequency of SCW excitation Ω_w. The nature of this minimum is the same as that of the DC minimum described by (22, 23). The second resonance in (9.24) (the so-called high-frequency resonance) corresponds to the DC current maximum at a frequency that corresponds to the high-frequency branch of SCW (Ω_{wh}). The origin of this branch will be discussed later. The effect of overall rectification should not be confused with the effect of DC current generation by a moving interference pattern when no electric field is applied (see, for instance, [120]), since in the last case no space-charge waves are excited as eigenmodes of charge oscillations. However, from a more general point of view, the overall SCW rectification and DC current variations in zero external field have a common nature, since in both cases the DC current variation arises when the photoexcited carrier grating moves together with the space-charge field grating, and the phase shift between these two gratings is not equal to $\pi/2$.

9.2.3.5 Spatial Rectification

Spatial rectification results from the interaction of SCW (for instance, with the phase factor $\exp[iKx - i\Omega t]$) with a static charge grating (with the phase factor $\exp[-iKx]$). The interaction is described by the product of the SCW and the field of the static grating, and this product is proportional to the phase factor $\exp[-i\Omega t]$. As a result of the interaction, an electric current arises, homogeneous in space but oscillating in time. In other words, the origin of the homogeneous alternating current is as follows. If, for instance, there is a standing grating of electric field $\cos(Kx)$ and a moving grating of the electric charge $\cos(Kx - \Omega t)$, then at the moment when the phase shift between the gratings is zero, a homogeneous current appears, but at the moment when the phase shift is 90 degrees, the current disappears. When the moving grating propagates along the static one, periodical oscillations of the homogeneous current in the crystal and in the outside circuit arise. To describe the alternating current, we can use the expression [111]

$$I(t) = \sigma E_0 \mathrm{Re}\left\{ i \frac{m^2 \Theta}{2\tau\tau_M} \left[\frac{1 + i\Omega\tau_M + \Omega\tau_M E_D/E_0}{(\Omega + \Omega_1)(\Omega + \Omega_2)} - \frac{1 + i\Omega\tau_M - \Omega\tau_M E_D/E_0}{(\Omega - \Omega_1^*)(\Omega + \Omega_2^*)} \right] \exp(i\Omega t) \right\}, \quad (9.25)$$

which is very similar to the expression obtained in [121] if we make the assumption that $d \gg 1$.

It is necessary to note that the existence of an alternating homogeneous current does not mean automatically that there is a homogeneous oscillating electric field in the crystal. If the crystal is connected to an ideal source of voltage (with zero internal resistance), no variations in the voltage or electric field will result unless a loading resistor is installed in the circuit. However, it was found that experimentally we nearly always could detect oscillations of the internal electric field by a probe beam via the electrooptic effect [122, 123] or via the self-deflection effect [36, 37], even with no real loading resistor in the circuit. So the experiments unambiguously demonstrate the importance of effects that play the same role as an effective loading resistor. These effects are associated with screening phenomena due to the nonohmic character of the electrodes connected to the crystal and inhomogeneous illumination of the crystal (along the x-axis).

The next expression describes the amplitude of the homogeneous alternating internal field in the crystal [124] when the existence of screening effects, inhomogeneous illumination along the x-axis, internal resistance of the voltage source, and real loading resistance is taken into account by the parameter $q = R_L/R_c$, where R_c is the resistance of the crystal and R_L is the efficient loading resistor including all the factors mentioned:

$$E_{RSC} = E_{int} m^2 \Theta \omega d \left[\frac{\sqrt{\frac{\omega^2}{4} + 1}}{\sqrt{\omega^2 + (q^{-1} + 1)^2} \cdot \sqrt{1 + 2\omega^2(1 - d_q^2) + \omega^4(1 + d_q^2)^2}} \right].$$
$$(9.26)$$

Here, $E_{int} = E_0/(1 + q)$. In the case of the source of voltage, E_0 is a fixed value and $E_{RSC} \to 0$ if $q \to 0$. In the case of the source of current (I_s), $q \to \infty$, $E_{int} = I_s R_c/L$, so E_{int} is independent of q and E_{RSC} also does not depend on q since $(q^{-1} + 1)^2 \approx 1$.

Equation (9.26) predicts a resonance dependence very similar to the resonance dependences given by (9.17) and (9.19). This is quite obvious, since the process of spatial rectification is associated with the space-charge waves, the resonance excitation of which is described by (9.17) and (9.19).

In the experiments reported in [122, 123], the electrooptic effect in an oscillating homogeneous internal field was employed (there was no light diffraction). The crystal (BGO) was illuminated with a polarized probe beam. Figure 9.12 shows the output signal (the intensity of the probe beam after transmission through

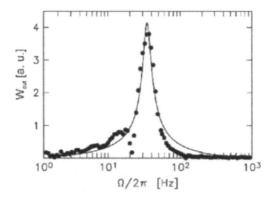

FIGURE 9.12. Spatial rectification in BSO [122]. Measurements were performed using the optical technique of detection of homogeneous oscillating space-charge field E_{RSC} ((9.25)) via the electrooptic effect. A probe beam was used, the polarization state of the probe beam was modulated by an oscillating homogeneous space charge field, and the output signal was the light intensity after the polarizer was placed in front of the photodetector. A detailed description of the setup for such measurements was given in [118]. The dependence of output signal W_{out} on phase modulation frequency $\Omega/2\pi$ for BGO. The total intensity of the recording beams is $W_0 = 60$ mW/cm^2, the contrast ratio is $m = 0.23$, the intensity of the probe beam (a HeNe laser, $\lambda = 633$ nm) is $W_P = 12$ mW/cm^2, the applied electric field is $E_0 = 7.5$ kV/cm, and the wave number is $K = 125$ mm^{-1}. The solid line shows the theoretical frequency dependence according to (9.26).

the crystal and the analyzer of polarization) as a function of phase-modulation frequency of the recording beam [122]. The polarization orientation was selected to provide a linear dependence of the output intensity on the polarization state variations caused (via the electrooptic effect) by variations in the internal field. The magnitude of the internal alternating field in this particular experiment was estimated as 50–60 V/cm. The theory of these effects is presented in [124]. It also gives a more general expression taking into account the effects of order m^4 that are associated with spatial rectification of the space-charge waves having a phase factor $\exp[i(2k_w x - \Omega_w t/2)]$. These waves are excited because of formation of the second spatial harmonic $(2k_w)$ due to the nonlinearity under consideration and oscillations of the interference pattern at $\Omega = \Omega_w/2$. This model explains the weak low-frequency peak in Figure 9.12.

The effect of spatial rectification can be detected directly, by measuring the alternating current flowing through the crystal. This technique is easier to perform, and sometimes it is more sensitive than measurements through the electrooptic effect. Figure 9.13 shows an example of current measurements in BTO. Note that the first measurements of this type were performed in [125]. The effect of spatial rectification is closely connected with the effect of "photo emf" or "nonsteady-state photoelectromotive force (photo emf)" discovered in [126, 127] and described in more detail in [128, 129]. The photo emf effect is also associated with the interaction of static gratings and moving gratings in the case that the

FIGURE 9.13. Spatial rectification in BTO [118]. Symbols • show the dependence of the output signal at the photodetector in measurements by the optical technique of detection of homogeneous oscillating space-charge field E_{RSC} via the electrooptic effect. Symbols ■ show the frequency dependencies of the output voltage measured at the loading resistor. $E_0 = 9$ kV/cm, $W_0 = 344$ mW/cm, $m = 0.57$, $\Theta = 0.94$, and $\Lambda = 10$ μm. The solid line is calculated through (9.26) using the fitting parameters $d_q = 1.6$, $\tau_M = 5 \cdot 10^{-4}$ s, and $q = 2.1$.

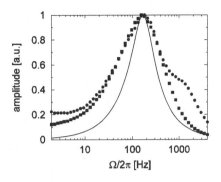

sample is illuminated by an oscillating interference pattern. The photo emf effect can exist even if there is no applied field. However, in zero external field and when the grating formation occurs through the diffusion mechanism, no SCW eigenmodes can be excited. Only the forced waves and the relaxational charge oscillations take place in this case. So there are no resonance frequencies and no dispersion law. When the electric filed is applied and the condition Q (or d) >1 is met, then eigenmodes of charge oscillations can be excited, and the effect of photo emf is overcome by the effect of SCW spatial rectification. Since spatial rectification is a resonance effect, it provides a much stronger alternating current (and higher sensitivity of the corresponding devices), but in a narrower frequency bandwidth.

9.2.3.6 Spatial Doubling

Spatial doubling arises when SCW (for instance, with the phase factor $\exp[iKx - i\Omega t]$) scatter from a static charge grating (with the phase factor $\exp[iKx]$). As a result of scattering, a space-charge wave with a doubled wave vector (or with the phase factor $\exp[i2Kx - i\Omega t]$) arises. This wave can be detected by measuring oscillations (at Ω) of the second-order diffraction beam intensity. The theoretical and experimental data proving the existence of this effect (as well as the first preliminary data on overall rectification and second-harmonic generation of SCW) were published in [52, 115]. Figure 9.14 shows the light intensity of the second non-Bragg diffraction beam that arises due to diffraction from the moving grating with wave number $2K$ oscillating at Ω. The maximum corresponds to excitation of the eigenmode with the wave number $2k_w$ and eigen frequency $\Omega_w/2$. The right shoulder is due to excitation of the forced wave with wave number $2k_w$ and frequency Ω_w. The curve shows the theoretical calculation [52, 115] in accordance with

$$
\begin{aligned}
J_{2,1}(t) = {} & 4\eta I_R \Theta d\omega [(\omega^2 + 4)(\omega^2 + 1)]^{1/2} \\
& \times [1 + 2\omega^2(1 - d^2) + \omega^4(1 + d^2)^2]^{-1/2} \\
& \times [1 + 2\omega^2(1 - 4d^2) + \omega^4(1 + 4d^2)]^{-1/2} \cos(\Omega t). \quad (9.27)
\end{aligned}
$$

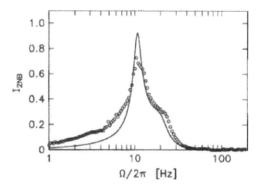

FIGURE 9.14. Intensity of the second-order non-Bragg diffraction beam detected at Ω as a function of Ω for BGO. A Nd:YAG laser at $\lambda = 532$ nm was used. Experimental parameters are $\Theta = 0.5$, $E_0 = 8$ kV/cm, $k/2\pi = 20$ mm^{-1}, light intensity $I \approx 40$ mW/cm^2. The solid curve is the amplitude of $J_{2,1}$ as a function of Ω calculated in accordance with (9.27).

9.2.4 High-Frequency Branch of SCW

So far, we have mostly discussed the SCW associated with trap recharging on the assumption of a quasistationary electron distribution in the conduction band $(\partial n/\partial t = 0)$. This means that we ignored all temporal processes with characteristic times shorter than τ_{M}. However, if we do not make the assumption $\partial n/\partial t = 0$, the solution of (9.12)–(9.16) leads to one more branch of the space-charge oscillations that is not directly associated with traps and that corresponds to wave oscillations of free carriers. The theory of these oscillations was developed many years ago (see, for instance, [59, 64, 65] and the review paper [130]). The modern history of this problem is described in [31]. The dispersion law for these waves has a more conventional form:

$$\Omega_{\text{wh}} = k_{\text{wh}} v = k_{\text{wh}} \, \mu E_0. \tag{9.28}$$

Here Ω_{wh} is the eigenfrequency of the high-frequency branch of waves and k_{wh} is their wave number. The relaxation rate $1/\tau_{\text{Rh}}$ for these waves in the case of sillenites is determined mostly by the recombination time. However, in the general case it depends also on diffusion and differential conductivity.

These high-frequency SCW can also be excited in the optical way, and the resonance excitation occurs when K and Ω of the interference pattern coincide with Ω_{wh} and k_{wh}. To detect the resonance, different techniques can be used. For instance, in [106] two-wave mixing gain measurements were used to detect the resonance in the 10 kHz range under an AC applied field in BGO. (The theory of high-frequency SCW in square wave and sine wave alternating fields is presented in [25].) In [131], the so-called holographic time-of-flight technique was used to determine the electron mobility in BSO. This technique allows one to measure the speed of propagation of the grating of the photoexcited free carriers and to prove the existence of high-frequency SCW. In [121], the resonance was detected by measuring the current in the outside circuit when the crystal (BSO) was illuminated by an alternating interference pattern and the external DC fields (4–10 kV/cm) were applied, and in [107, 108] a similar resonance was revealed when the crystal was illuminated by an alternating interference pattern and an AC field was applied. The measurements of the dependencies of the resonance frequency on external field and on wave number agree with the dispersion law described by (28). The detection of the high-frequency waves provides the method of direct measurements of the carrier mobility. The data are presented in Table 9.1. It is necessary to note that there is a large difference (approximately within two orders of magnitude) between the mobilities obtained using different techniques. The reasons for this are not quite clear yet. One possible explanation is that in some experiments shallow traps play an important role, and hence a reduced value of μ is detected, while in other experiments [145] the influence of shallow traps is eliminated. Another explanation based on more correct analysis of the resonance frequency measurements is presented in more recent publication [153].

9.2.5 Crystals with Negative Differential Conductivity

So far, we have discussed the most conventional situations in which the volt–ampere characteristics of a material agree with Ohm's law. However, in semiconductors one can meet quite often a volt–ampere characteristic with a falling section (the section with negative differential conductivity). Usually this happens at relatively high applied fields when the carrier mobility becomes a function of the applied field. In semiconductor quantum-well structures such a situation can arise even at relatively moderate (on the order of several kilovolts) fields due to the Stark effect [132]. The question arises, what happens with SCW in such a situation? In [111], this question was investigated theoretically for a quite general case when diffusion and mobility depend on applied field ($D(E_0)$, $\mu(E_0)$) but without taking into account the trap-saturation effect. The theory shows that many useful results can be obtained simply by introducing a negative Maxwell (dielectric) relaxation time τ_M. The negative τ_M means that charge carriers will tend to move in the area with a higher charge density, which results in the increase of SCW amplitude and SCW instabilities. This situation looks quite probable for semiconductors. However, it is doubtful that it can be realized in sillenites because too-high voltages are required.

Acknowledgments. The authors would like to thank Prof. Dr. E. Krätzig for his invaluable contribution to the investigations presented in this chapter. The authors also deeply appreciate the contributions of Dr. S. Wevering, H. Voght, Dr. A. Gerwens, F. Rahe, C. Ruter, Dr. A. Paugurt, Dr. V. Petrov, and Dr. P. Kleinert. The work was supported by the Russian Foundation for Basic Research (Grant 06-02-16170).

References

1. L.G. Sillen, Arkiv Kemi. *Mion. Geol.* 12A (18), 1 (1937); thesis, University of Stockholm (1940); B. Aurivillius and L.G. Sillen. Nature **155**, 305 (1945).
2. S.C. Abrahams, P.B. Jamieson, and J.L. Bernstein. *J. of Chem. Phys.* **47** (10) 4034 (1967).

3. S.C. Abrahams, J.L. Bernstein, and C. Svensson. *J. of Chem. Phys.* **71**, 788 (1979).

4. G.M. Safronov, V.N. Batog, Yu.I. Krassilov, et al. *Izvestia AN USSR, Neorg. Materiali* (in Russian) **6**, n2, 284 (1970).

5. R.E. Aldrich, S.L. Hou, and M.L. Harvill. *J. Appl. Phys.* **42**, 493–494 (1971).

6. J.P. Herriau, D. Roias, J.P. Huignard, J.M. Bassat, and J.C. Launay. *Ferroelectrics*, **75**, 271–279 (1987).

7. S.G. Odulov, K.V. Sherbinin, and A.N. Shumeljuk. *JOSA* **B11**, n9, 1780 (1994).

8. H. Vogt, K. Buse, H. Hesse, E. Kratzig, and R.R. Garcia. *J. Appl. Phys.* **90**, n7, 3167 (2001).

9. A.A. Reza, D.B. Senulene, V.A. Beliaev, and E.I. Leonov. *Letters J. Techn. Phys.* (in Russian), **5**, n8, 465–469 (1979).

10. H.J. Reyher, U. Hellwig, and O. Thieman. *Phys. Rev.* B47, 5638 (1993).

11. M. Peltier, F. Micheron. *J. Appl. Phys.* **48**, n9, 3683–3690 (1977).

12. SH.M. Efendiev, V.E. Bagaev, R.A. Aliev, E.R. Mustafaev. *Phys. Stat. Sol.(a)* **109**, 345 (1988).

13. V.K. Malinovskii, O.A. Gudaev, V.A. Gusev, and S.I. Demenko. "Photoinduced phenomena in sillenites" (1990) (in Russian). Press "Nauka," Siberian division.

14. L. Arismendi, J.M. Cabrera, and F. Agullo-Lopez. *Inter. J. Optoelectronics* **7**, 149 (1992).

15. J.A. Kilner, J. Drenan, P. Dennis, B.C.H. Steele. *Solid. Sta. Ionics* **5**, 527 (1981).

16. S.L. Hou, R.B. Lauer, and R.E. Aldrich. *J. Appl. Phys.* **44**, 2652–2658 (1973).

17. B.C. Grabmaier and R. Obershmid. *Phys. Status Solidi* A**96**, 199 (1986).

18. R. Obershmid. *Phys. Status Solidi* A**89**, 263 (1985).

19. K. Buse. *Appl. Phys.* B**64**, 391 (1997).

20. A.A. Kamshilin, O. Kobozev, A.I. Grachev, and P.M. Karavaev. *J. Opt. Soc. Am.* B**19**, n2, 202–207 (2002).

21. B. Briat, H.J. Reyher, A. Hamri, N.G. Romanov, J.C. Launay, and F. Ramaz. *J. Phys.: Condens.* Matter **7**, 6951 (1995).

21a. H. Vogt, E. Krätzig. Determination of the Diffusion Length of Charge Carriers in Nonstoichiometric Sillenite-Type Crystals by the Technique of Non-Steady Photocurrents. *J. Appl. Phys.* **94**, 2507 (2003).

22. H.J. Reyher, U. Hellwig, and O. Thieman. *Phys. Rev.* B47, 5638 (1993).

23. A.A. Kamshilin and M.P. Petrov. *Sov. Phys. Solid. State* **23**, 1811–1814 (1981).

24. U. van Stevendaal, K. Buse, H. Malz, H. Veenhuis, and E. Krätzig. *JOSA* B**15**, 2868 (1998).

25. A. Grunnet-Jepsen, I. Aubrecht, and L. Solymar. *J. Opt. Soc. Am.* B **12**, n5, 921 (1995).

26. J.P. Wilde, L. Hesselink, S.W. McCahon, M.B. Klein, D. Rytz, and B.A. Wechsler. *J. Appl. Phys.* **57**, n5, 2245 (1990).

27. B.I. Sturman, D.J. Webb, R. Kowarschik, E. Shamonina, and K.H. Ringhfer. *J. Opt. Soc. Am.* B**11**, n9, 1813–1819 (1994).

28. N.V. Kukhtarev, B.S. Chen, P. Venlateswartu, G. Saalamo, and M. Klein. *Opt. Commun.* **104**, 23 (1993).

29. V.V. Shepelevich, S.M. Shandarov, and E.A. Mandel. *Ferroelectrics* **110**, 235–249 (1990).

30. V.V. Shepelevich, S.F. Nechiporko, A.E. Zagorskiy, N.N. Egorov, Hu Yi, K.H. Ringhofer, E. Shamonina, and V.Ya. Gaivoronsky. *Ferroelectrics* **266**, 305–333 (2002).

31. L. Solymar, D.J. Webb, and A. Grunnet-Jepsen. *The Physics and Applications of Photorefractive Materials*. Clarendon Press, Oxford (1996).
32. G. Pauliat, P. Mathey, and G. Roosen, *J. Opt. Soc. Am.* **B8**, 1942–1946 (1991).
33. A.M. Bliznetsov, M.P. Petrov, and A.V. Khomenko. *Sov. Tech. Phys. Lett.* **10**, 463 (1984).
34. M.P. Petrov, S.I. Stepanov, and A.V. Khomenko. *Photorefractive Crystals in Coherent Optical Systems*. Springer-Verlag, Berlin, Heidelberg (1991).
35. S.I. Stepanov, N. Korneev, A. Gervens, and K. Buse, *Appl. Phys. Lett.* **72**, n8, 879–881 (1998).
36. M.P. Petrov, A.P. Paugurt, and V.V. Bryksin. *JETP Letters*. **70**, n4, 260–264 (1999).
37. M.P. Petrov, A.P. Paugurt, V.V. Bryksin, S. Wevering, B. Andreas, E. Kratzig. *Appl. Phys. B* **69**, 341–344 (1999).
38. M.P. Petrov, A.I. Grachev. *JETP Lett.* **30**, 18–21 (1979).
39. A.I. Grachev and M.P. Petrov. *Ferroelectrics* **43**, 181–184 (1982).
40. S.L. Hou, D.S. Oliver. *Appl. Phys. Lett.* **18**, 325 (1971).
41. J. Feinleb and D.S. Oliver. *Appl. Opt.* **11**, 2752 (1972).
42. D. Cassasent, F. Caimi, A. Khomenko. *Appl. Opt.* **20**, 4215–4220 (1981).
43. S. Mallick, M. Miteva, and L. Nikolova, *J. Opt. Soc. Am.* **B14**, n5, 1179 (1997).
44. E.Yu. Ageev, S.M. Shandarov, S.Yu. Veretennikov, A.G. Mart'anov, V.A. Kartashev, A.A. Kamshilin, V.V. Prokof'ev and V.V. Shepelevich. *Kvantovaia Elektronika* (in russian) **31**, n4, 343–345 (2001).
45. A.G. Mart'ianov, S.M. Shandarov, and R.V. Litvinov. *Fiz. Tverdogo Tela* (in Russian), **44**, n6, 1006–1010 (2002).
46. A. Marrakchi, R.V. Johnson, and A.R. Tanguay, Jr. *J. Opt. Soc. Am.* **B3**, 321–336 (1986).
47. M.P. Petrov, A.V. Khomenko, M.V. Krasin'kova, V.I. Marakhonov, and M.G. Shlyagin. *Sov. Phys.-Tech. Phys.* **26**, 816 (1981).
48. J.P. Huignard, F. Micheron. *Appl. Phys. Lett.* **29**, 591 (1976).
49. T.G. Pencheva and S.I. Stepanov. *Sov. Phys.-Solid State* **24**, 1214–1216 (1982).
50. N.V. Kukhtarev, V.B. Markov, S.G. Odulov, M.V. Soskin, and V.L. Vinetskii. *Ferroelectrics* **22**, 949–960 (1979).
51. J.P. Huignard and P. Günter. Optical processing using wave mixing in photorefractive crystal, 205–274, in *Photorefractive Materials and Their Applications II*, edited by P. Günter and J.P. Huignard, Springer-Verlag, Berlin–Heidelberg (1989).
52. V.V. Bryksin and M.P. Petrov. *Phys. Solid State*, **40**, n8, 1317–1325 (1998).
53. H. Kogelnik, *Bell Syst. Techn. J.* **48**, 2909 (1967).
54. A.A. Kamshilin, S.V. Miridonov, M.G. Miteva and E.V. Mokrushina. *J. Techn. Fys.* (in Russian) **59**, 113 (1989).
55. R.J. Colier, C.B. Burrckhardt, and L.H. Lin. *Optical Holography*. Academic Press, New York (1971).
56. P. Günter and H.J. Eichler. Introduction to photorefractive materials. Electrooptic and photorefractive materials, ed. by P. Günter, *Springer Proc. Phys.* **18**, 206–2028, Springer Verlag, Berlin, Heidelberg (1987).
57. G. Hamel de Montchenault, Bloiseaux, and J.P. Huignard. *Electr. Lett.* **22**, n19, 1030–1032 (1986).
58. J.P. Huignard and P. Günter, Optical processing using wave mixing in photorefractive crystal, 205–274, in *Photorefractive Materials and Their Applications II*, edited by P. Günter and J.P. Huignard. Springer-Verlag, Berlin–Heidelberg (1989).

59. G.C. Valley and J.F. Lam, Theory of photorefractive effects in electrooptic crtystals (75–98), in *Photorefractive Materials and Their Applications I*, edited by P. Günter and J.P. Huignard, Springer-Verlag, Berlin–Heidelberg (1988).

60. B.K. Ridley. *Proc. Phys. Soc.* **86**, 637–645 (1965).

61. O.V. Konstantinov and V.I. Perel'. *Soviet Phys. Solid State* **6**, n11, 2691–2696 (1965).

62. N.V. Kukhtarev. *Sov. Tech. Pys. Lett* **2**, 438–440 (1976).

63. R.F. Kazarinov, R.A. Suris, B.I. Fuks. *Sov. Phys. Semicon.* **6**, 500 (1972).

64. R.F. Kazarinov, R.A. Suris, B.I. Fuks. *Sov. Phys. Semicon.* **7**, 102 (1973).

65. R.F. Kazarinov, R.A. Suris, B.I. Fuks. *Sov. Phys. Semicon.* **7**, n4, 480–486 (1973).

66. L. Solymar, D.J. Webb, and A. Grunnet-Jepsen. *Progr. Quant. Electr.* **18**, 377–450 (1994).

67. B.I. Fuks , M.S. Kagan, R.A. Suris, N.G. Zhdanova. *Phys. Status Sol. (a)* **40**, k61 (1977).

68. N.G. Zhdanova, M.S. Kagan, R.A. Suris, B.I. Fuks. *Sov. Phys. JETP* **74**, 189 (1978).

69. S.I. Stepanov, V.V. Kulikov, and M.P. Petrov. *Sov. Tech. Phys. Lett.* **8**, 229–230 (1982), *Opt. Commun.* **44**, 19–23 (1982).

70. J.P. Huignard and A. Marrakchi. *Opt. Commun.* **38**, 249–254 (1981), *Opt. Lett.* **6**, 622 (1981).

71. P. Günter. **41**, 83–88 (1981).

72. G. Hamel de Montchenault, B. Loiseaux, and J.P. Huignard. *Electr. Lett.* **22**, n19, 1030–1032 (1986).

73. V.N. Alimpiev and I.R. Gural'nik. *Sov. Phys. Semicond.* **18**, 978–980 (1984).

74. V.N. Alimpiev and I.R. Gural'nik. *Sov. Phys. Semicond.* **20**, 512–514 (1986).

75. A.S. Furman. *Sov. Phys. Solid State* **29**, 617–622 (1987).

76. A.S. Furman. *Ferroelectrics* **83**, 41–53 (1988).

77. A.S. Furman. *Sov. Phys. JETP* **67**, 1034–1038 (1988).

78. S. Mallick, B. Imbert, H. Ducollet, H. Herriau, and J.P. Huignard. *J. Appl. Phys.* **63**, 5660–5663 (1988).

79. D.J. Webb and L. Solymar. *Opt. Commun.* **74**, 386 (1990).

80. L.B. Au, L. Solymar, and L.H. Ringhofer. *Subharmonics in BSO.* In Proceedings of the topical conference on phtotrefractive materials, effects and devices II, Aussois, France, 87–91 (1990).

81. J. Takacs, M. Schaub, and L. Solymar. *Opt. Commun.* **91**, 252 (1992).

82. J. Takacs and L. Solymar. *Opt. Lett.* **17**, 247 (1992).

83. B.I. Sturman, M. Mann, and K.H. Ringhofer. *Opt. Lett.* **17**, 1620 (1992).

84. B.I. Sturman, M. Mann, and K.H. Ringhofer. *Appl. Phys.* **A55**, 235 (1992).

85. B.I. Sturman, M. Mann, J. Otten, and K.H. Ringhofer. *J. Opt. Soc. Am.* **B10**, 1919 (1993).

86. C.H. Kwak, M. Shamonin, J. Takacs, and L. Solymar. *Appl. Phys. Lett.* **62**, 328 (1993).

87. J. Richter, A. Grunnet-Jepsen, J. Takacs, and L. Solymar. *IEEE J. Quantum Electron.* **30**, 1645 (1994).

88. H.C. Pedersen and P.M. Johanssen. *Opt. Lett.* **19**, 1418 (1994).

89. T.E. McClelland, D.J. Webb, B.I. Sturman, and K.H. Ringhofer. *Phys. Rev. Lett.* **73**, 3082 (1994).

90. B.I. Sturman, T.E. McCleland, D.J. Webb, E. Shamonina, and K.H. Ringhofer. *J. Opt. Soc. Am.* **B12**, n 9, 1621–1627 (1995).

91. P. Buchhave, S. Lyuksyutov, M. Vasnetsov, and C. Heyde. *J. Opt. Soc. Am.* B**13**, 2595–2601 (1996).
92. H.C. Pedersen and P.M. Johanssen. *Phys. Rev. Lett.* **77**, 3106 (1996).
93. B.I. Sturman, Maria Aguilar, F. Agullo-Lopez, and K.H. Ringhofer. *Phys. Rev. E* **55**, n5, 6072–6083 (1997).
94. H.C. Pedersen, P.M. Johanssen, and D.J. Webb. *J. Opt. Soc. Am.* B**15**, 1528–1532 (1998).
95. P.M. Johanssen, H. Pedersen, E.V. Podivilov, and B.I. Sturman. *J. Opt. Soc. Am.* B**16**, 103 (1999).
96. E. Nippolainen, A.A. Kamshilin, V.V. Prokofiev, and T. Jaaskelainen. *Appl. Phys. Lett.* **78**, n7, 859–861 (2001).
97. S.F. Lyuksyutov, P. Buchhave, and M.V. Vasnetsov. *Phys. Rev. Lett.* **79**, 67 (1997).
98. G.C. Valley. *J. Opt. Soc. Am.* B **1**, 868–873 (1984).
99. P. Refreigier, L. Solymar, H. Rajbenbach, and J. P. Huignard. *J. Appl. Phys.* **58**, 45–57 (1985).
100. S.I. Stepanov and M.P. Petrov. *Sov. Tech. Phys. Lett.* **10**, 572–573 (1984).
101. S.I. Stepanov and M.P. Petrov. *Opt. Comm.* **53**, 292–295 (1985).
102. S.I. Stepanov and M.P. Petrov. Nonstationary holographic recording for efficient amplification and phase conjugation, 262–289, in *Photorefractive Materials and Their Applications I*, edited by P. Günter and J.P. Huignard, Springer-Verlag, Berlin–Heidelberg (1988).
103. K. Walsh, A.K. Powell, C. Stace, and T.J. Hall. *J. Opt. Soc. Am.* B **7**, 288–303 (1990).
104. A. Grunnet-Jepsen, C.H. Kwak, I. Richter, and L. Solymar. *J. Opt. Soc. Am.* B**11**, n1, 124–131 (1994).
105. G. Pauliat, A. Villing, J.C. Launay, and G. Roosen. *J. Opt. Soc. Am.* B**7**, 1481–1486 (1990).
106. M. Brushinin, V. Kulikov, and I. Sokolov. *Phys. Rev.* B **67**, 075202 (2003).
107. M.A. Brushinin, V.V. Kulikov, and I.A. Sokolov. *OSA Trends in Optics and Photonics*, v. 87, *Photorefractive Effects, Materials, and Devices*, 150–158 (2003); *Phys. Rev.* B**65**, 245204 (2002).
108. S.L. Sochava, E.V. Mokrushina, V.V. Prokof'ev, and S.I. Stepanov. *JOSA* B **10**, 1600–1604 (1993).
109. V.A. Kalinin, L. Solymar. *Appl. Phys. Lett.* **68** (2), 167–169 (1996); *Appl. Phys. Lett.* **69** (27), 4265–4267 (1996).
110. V.V. Bryksin, M.P. Petrov. *Phys. Solid State* **44**, n10, 1869–1879 (2002).
111. V.V. Bryksin, P. Kleinert, and M.P. Petrov. *Physics of the Solid State* **45**, n11, 2044–2052 (2003).
112. M.P. Petrov, V.M. Petrov, V.V. Bryksin, A. Gervens, S. Wevering, and E. Krätzig, *JOSA* B**15**, n7, 1880–1888 (1988).
113. H.C. Pedersen, D.J. Webb, and P.M. Johansen, *JOSA* B**15**, n10, 2573–2580 (1998).
114. M.P. Petrov, V.V. Bryksin, V.M. Petrov, S. Wevering, and E. Krätzig. *Phys. Rev.* A**60**, 2413 (1999).
115. M.P. Petrov, V.V. Bryksin, S. Wevering, and E. Krätzig. *Appl. Phys.* B**73**, 669–703 (2001).
116. M.P. Petrov, V.V. Bryksin, H. Vogt, F. Rahe, E. Krätzig. *Phys. Rev.* B**66**, 085107 (2002).
117. M. Bass, P.A. Franken, J.F. Ward, G. Wenreich. *Phys. Rev. Lett.* **9**, 446 (1962).

118. M.P. Petrov, V.V. Bryksin, F. Rahe, C.E. Ruter, and E. Krätzig. *Opt. Commun.* **227**, 183 (2003).
119. V.V. Bryksin, P. Kleinert, and M.P. Petrov. *Physics of the Solid State*, to be published.
120. U. Haken, M. Hundhausen, and L. Ley. *Phys Rev.* B**51**, n16, 10579–10590 (1995).
121. I.A. Sokolov and S.I. Stepanov. *JOSA* B**10**, n8, 1483–1488 (1993).
122. M.P. Petrov, A.P. Paugurt, V.V. Bryksin, S. Wevering, and E. Krätzig. *Phys. Rev. Lett.* **84**, 5114–5117 (2000).
123. M.P. Petrov, A.P. Paugurt, V.V. Bryksin, S. Wevering, and E. Krätzig. *Optical Materials* **18**, 99–102 (2001).
124. V.V. Bryksin, M.P. Petrov. *Phys. Solid State* **42**, n10, 1854–1860 (2000).
125. S. Manssurova, S. Stepanov, N. Korneev, C. Dibon. *Optics Commun.* **152**, 207 (1998).
126. M.P. Petrov, S.I. Stepanov, and G.S. Trofimov. *Sov. Tech. Phys. Lett.* **12**, 379–381 (1986).
127. G.S. Trofimov and S.I. Stepanov. *Sov. Phys. Solid State* **28**, 1559–1562 (1986).
128. M.P. Petrov, I.A. Sokolov, S.I. Stepanov, and G.S. Trofimov. *J. Appl. Phys.* **68**, 2216 (1990).
129. S. Stepanov. *Handbook of Adv. Electronic and Photonic Materials and Devices*, ed. by H.S. Nalwa. Vol.2: *Semiconductor Devices*. p. 202 (2001).
130. A.F. Volkov and Sh.M. Kogan. *Sov. Phys.-Usp.* **11**, 881 (1969).
131. J.P. Partanen, J.M. Jonathan, and R.W. Hellwarth. *Appl. Phys. Lett.* **57**, 2404–2406 (1990).
132. L. Esaki and T. Tsu. *IBM J. Res. Dev.* **14**, 61 (1970); V.V. Bryksin, Yu.A. Firsov. *JETP* **61**, 2373 (1971) (in Russian).
133. A.A. Ballman, H. Brown, P.K. Tien, and R.J. Martin. *J. Cryst. Growth* **20**, 251 (1973).
134. A.J. Fox and T.M. Bruton. *Appl. Phys. Lett.* **27**, 360–363 (1975).
135. S.C. Abrahams, C. Svensson, and A.R. Tanguay, Jr. *Sol. St. Commun.* **30**, 293–295 (1979).
136. V.V. Prokofiev, J.F. Carvalho, J.P. Andreeta, N.J.H. Gallo, A. Hernandes, J. Frejlich, A.A. Freschi, P.M. Garcia, J. Maracabia, A.A. Kamshilin, and T. Jaaskelainen. *Cryst. Res. Technol.* **30**, n2, 171 (1995).
137. A. Feldman, W.S. Brower, and D. Horowitz. *Appl. Phys. Lett.* **16**, 201 (1970).
138. M.L. Barsukova, V.A. Kuznetsov, A.N. Lobachev, and Yu.V. Shaldin. *J. Cryst. Growth* **13/14**, 530 (1972).
139. P. Acioly, M. dos Santos, P.M. Garcia, and J. Frejlich. *J. Appl. Phys.* **86** (1), 247 (1989).
140. J.P. Huignard, J.P. Kerian, G. Rivet, and P. Gunter. *Opt. Lett.* **5**, 102 (1980).
141. P.M. Garcia, L. Cescato, and J. Frejlich, *J. Appl. Phys.* **66**, 47 (1989).
142. L. Boutsikaris, S. Malis, and N.A. Vainos. *JOSA* B**15**, 1042 (1998).
143. V.M. Petrov, S. Wevering, M.P. Petrov, and E. Krätzig. *Appl. Phys.* B**68**, 73–76 (1999).
144. S.L. Sochava, K. Buse, and E. Krätzig. *Phys. Rev.* B**51**, 4684 (1995).
145. I. Baggio, R.W. Hellwarth, and J.P. Partanen. *Phys. Rev. Lett.* **78**, 891 (1997).
146. I.S. Zakharov, P.P. Akinfiev, P.A. Petukhov, V.M. Skorikov. *Izvestia Vuzov, Physica* (in Russian), n3, 121 (1978).
147. A.M. Plesovskikh and S.M. Shandarov. *Fiz. Tverdogo Tela* (in Russian) **44**, n1, 57–61 (2002).

148. S. Stepanov, S.M. Shandarov, and N.D. Khat'kov. *Sov. Phys. Solid. State* **29**, 1754 (1987).
149. P. Bayvel, M. McCall, and R.V. Wright. *Opt. Lett.* **13**, 27 (1988).
150. P.V. Lenzo, E.G. Spencer, and A.A. Ballman. *Appl. Opt.* **5**, 1688 (1966).
151. V.I. Chmyrev, L.A. Skorikov, and M.I. Subbotin. *Inorg. Mater.* **19**, 242 (1983).
152. V.M. Fridkin, B.N. Popov, and K.A. Verkhovskaya. *Sov. Phys.-Solid State* **20**, 730 (1978).
153. M. Petrov, V. Bryksin, A. Emgrunt, M. Imlau, and E. Kratzig. *JOSA* **B22**, n7, 1529–1537 (2005).

10

Photorefractive Effects in $Sn_2P_2S_6$

Alexander A. Grabar,[1] Mojca Jazbinšek,[2] Alexander N. Shumelyuk,[3] Yulian M. Vysochanskii,[1] Germano Montemezzani,[2,4] and Peter Günter[2]

[1] Institute of Solid State Physics and Chemistry, Uzhgorod National University, 88000 Uzhgorod (Ukraine) **agrabar@univ.uzhgorod.ua**
[2] Nonlinear Optics Laboratory, Swiss Federal Institute of Technology, 8093 Zurich (Switzerland) **nlo@phys.ethz.ch**
[3] Institute of Physics, National Academy of Sciences, 03650 Kiev (Ukraine) **shumelyuk@iop.kiev.ua**
[4] Present Address: Laboratoire Matériaux Optiques, Photonique et Systèmes (LMOPS, CNRS UMR 7132), University of Metz and Supélec, 57070 Metz (France)

10.1 Introduction

Tin hypothiodiphosphate ($Sn_2P_2S_6$) is a very promising material for photorefractive applications in the red and near-infrared wavelength range. The main advantage of $Sn_2P_2S_6$ as an infrared-sensitive photorefractive material is its high beam-coupling gain combined with relatively fast response, about two orders of magnitude faster than in the conventionally used photorefractive material $BaTiO_3$ doped with rhodium. Compared to oxide ferroelectrics, which are wide-band-gap dielectrics, $Sn_2P_2S_6$ crystals are characterized by a narrower band gap (2.3 eV) and a higher photorefractive sensitivity, and are transparent in the range of 530–8000 nm.

The growth and the structure of $Sn_2P_2S_6$ were first reported by Carpentier and Nitsche [1, 2], and Dittmar and Schäfer [3] in 1974. The optical properties were first investigated by Gurzan et al. in 1977 [4] and the thermooptic properties by Grabar et al. in 1984 [5]. Until the first observation of photorefraction by Grabar et al. in 1991 [6], $Sn_2P_2S_6$ was mainly investigated owing to its interesting ferroelectric properties that are combined with semiconducting properties, and the fact that the phase transition changes with external fields and chemical composition. The interest in $Sn_2P_2S_6$ for photorefractive applications increased considerably after the work of Odoulov et al. in 1996 [7, 8], where photorefraction at 1.06 μm wavelength after preillumination of the crystal was demonstrated. Interband photorefraction in the blue-green spectral range was also observed [9]. In 2001, a new "brown" crystal modification was obtained [10] that yields relatively fast self-pumped phase-conjugation response in the near infrared [11]. More recently, doped crystals have been also reported to exhibit advantageous photorefractive properties [12].

Many of the optical, dielectric, and photorefractive properties of $Sn_2P_2S_6$ are not yet completely understood and the variation of material parameters reported is still very broad. The purpose of this chapter is to review the photorefractive and related properties of this crystal and a few applications demonstrated to date based on photorefractive $Sn_2P_2S_6$. Defect centers that are responsible for photorefraction in $Sn_2P_2S_6$ are not yet clearly identified. However, we consider various crystal types with respect to their characteristic properties. We distinguish "type I yellow" nominally pure crystals that exhibit a pronounced electron–hole competition and are sensitive to preillumination, "type II yellow" nominally pure crystals without considerable competition and preillumination effects, modified "brown" crystals with a variation of nonstoichiometric defects, and various types of intentionally doped crystals. In Section 10.2 we discuss the crystal growth, the lattice and domain structures, the electronic structure, light-induced effects, as well as the optical, dielectric, and electrooptic properties of $Sn_2P_2S_6$. In Section 10.3 we describe the basic photorefractive characteristics of the above-mentioned crystal types, the influence of an external electric field on photorefraction, beam-fanning effects, and main features of interband photorefraction. $Sn_2P_2S_6$ is particularly interesting for phase conjugation applications in the red and near infrared. In Section 10.4 we discuss the self-pumped phase conjugation characteristics of various crystal types. Finally, some conclusions are given in Section 10.5.

10.2 Physical Properties

$Sn_2P_2S_6$ belongs to a large family of $M_2P_2S(Se)_6$ crystals (M = Pb, Fe, Zn, Sn, Hg, Eu, Cd . . .) [13, 14]. It is a proper ferroelectric undergoing a second-order phase transition at $T_0 = 337\,K$ with a point symmetry change $2/m - m$ [1, 2, 15]. Basic material properties of $Sn_2P_2S_6$ are listed in Table 10.1.

10.2.1 Crystal Growth

The optical quality and the photorefractive properties of $Sn_2P_2S_6$ crystals critically depend on the crystal growth technique. First data on the growth, structure, and properties of $Sn_2P_2S_6$ and some structural analogue crystals were reported in [1, 3], where also the growth technique was described. Generally, single crystals can be grown by one of two alternative methods, i.e., vapor-transport or Bridgman. The crystals grown using the vapor-transport technique are characterized by a good optical quality, whereas the Bridgman's crystals are not transparent. They are, however, well adapted for piezoelectric [20, 26] or pyroelectric [26–28] applications. In addition, a "soft" method of $Sn_2P_2S_6$ synthesis at ambient temperature in water solution was described [29, 30], but in such a way that only crystalline powder could be obtained. The data presented in this chapter were obtained predominantly with crystals grown by the vapor-transport method. In

TABLE 10.1. Basic physical properties of $Sn_2P_2S_6$ at room temperature. For details and other tensor components please refer to a specific section or given references

		Reference
Spontaneous polarization P_s	15 $\mu C/cm^2$	[2, 16, 17]
Transparency range	530 nm–8000 nm	[4]
Main refractive indices[a]	$n_1 = 3.0256$	[18]
	$n_2 = 2.9309$	
	$n_3 = 3.0982$	
Dielectric constant ϵ_{11}	$230 - 300$	[12]
Electrooptic coefficient r_{111}^T [a]	174 pm/V	[19]
Piezoelectric coefficient d_{111}	244 pC/N	[20]
Density ρ	$3.54 \cdot 10^3$ kg/m^3	[3]
Elastic constant C_{1111}	$4.2 \cdot 10^{10}$ N/m^2	[21]
Pyroelectric coefficient p_1	$7 \cdot 10^{-4}$ C/(m^2 K)	[22]
Coercive field E_c	$7.5 \cdot 10^4$ V/m	[2]
Heat capacity C_p	240 J/(mol K)	[23]
Thermal conductivity λ_1	0.5 J/(s m K)	[24]
Nonlinear optical susceptibility[b]	$\chi_{111}^{(2)} = 24$ pm/V	[25]
	$\chi_{1111}^{(3)} = 17 \cdot 10^{-20}$ m^2/V^2	[25]

[a] at 632.8 nm
[b] at 1907 nm

this technique high purity stoichiometric amounts of the constituent elements are sealed into an evacuated quartz tube together with a transport agent (usually I_2). The tube is placed in a horizontal furnace with the temperature gradient between "hot" ($\approx 630°C$, evaporation) and "cold" ($\approx 600°C$, crystallization) zones. The average growth time is about 2–3 weeks, and an optically perfect single crystal boule with a volume of about 2 cm^3 can be obtained as a result. As alternative transporters, SnI_2 or SnI_4 compounds can also be used. A transporter gas replacement leads to variations in the growth speed and crystallization conditions due to the difference in the transport dynamics and the decomposition energy, and hence the crystal stoichiometry. This is usually manifested in the sample color varying from light-yellow to orange or even light-brown tone. A simplified scheme of the vapor-transport growth of $Sn_2P_2S_6$ is shown in Figure 10.1.

A substantial enhancement of the photorefractive properties was reported in [10] with modified "brown" $Sn_2P_2S_6$ crystals. These samples were obtained using SnI_2 transport gas along with some other technological variations in the growth regime: slightly increased temperature gradient and overall growth temperature decreased by about 80°C. Because no dopants were introduced in the initial compound, these samples most probably differ from the conventional "yellow" samples by the concentration of intrinsic defects. This is indirectly confirmed by the fact that $Sn_2P_2S_6$ Bridgman's crystals grown with the initial compound strongly deviating from stoichiometry (up to 10% deficit of phosphorus with respect to the stoichiometric composition) are similar (brown) in color. This variation can be caused by a higher decomposition energy of SnI_2 in comparison with SnI_4,

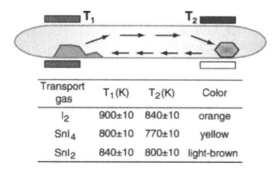

Transport gas	T_1(K)	T_2(K)	Color
I_2	900±10	840±10	orange
SnI_4	800±10	770±10	yellow
SnI_2	840±10	800±10	light-brown

FIGURE 10.1. Scheme of the vapor transport growth of $Sn_2P_2S_6$ [1]. High-purity stoichiometric amounts of the constituting elements are sealed into an evacuated quartz tube together with a transport gas. The tube is placed in a horizontal furnace with the temperature gradient between hot–evaporation (T_1) and cold–crystallization (T_2) zones. Several transport agents can be selected to grow $Sn_2P_2S_6$ by this technique: I_2, SnI_2, or SnI_4. The choice of the transport gas affects the growth speed and the stoichiometry level of the grown crystals.

leading to the creation of additional vacancies in the tin sublattice. However, the presence of other kinds of noncontrolled impurities cannot be excluded.

Among the methods commonly used for improving the photorefractive parameters, doping is usually the most productive and controllable. Previous investigations show that partial substitutions of Sn by metals (Pb, Fe, In, Cu, Cd, and others) do not change substantially the optical absorption and the color of $Sn_2P_2S_6$, indicating that such impurities do not create levels in the band gap. Replacements of S by Se give solid solutions $Sn_2P_2(Se_xS_{1-x})_6$ with the absorption edge shifted toward the red [1, 29], but this is accompanied by a significant decrease of the phase-transition temperature. As is shown later (Section 10.2.3), the electron states that determine the band-gap edges in $Sn_2P_2S_6$ are formed mainly by sp-hybridized orbitals of sulfur and phosphorus. The absorption edge is therefore expected to be most sensitive to the variations in the S sublattice. As a chalcogen element, Te is a natural candidate for doping $Sn_2P_2S_6$. Unlike many others, this element is also well suited for growth by the vapor-transport technique at the temperatures used for obtaining $Sn_2P_2S_6$ (near 630°C [1]). The samples of doped $Sn_2P_2S_6$ were grown from the initial compound of stoichiometric polycrystalline $Sn_2P_2S_6$ and tellurium in the regimes close to those used for obtaining nominally pure $Sn_2P_2S_6$ crystals. Red-colored single crystals of good optical quality with typical linear dimensions of about 5–7 mm were obtained in such a way [12].

10.2.2 Crystal Structure and Structural Phase Transitions

The crystal symmetry of $Sn_2P_2S_6$ at ambient temperature is monoclinic (point group m) [1–3]. The crystal structures of the ferroelectric and paraelectric phases were described in two crystalline coordinate systems and, consequently, in two

FIGURE 10.2. Structure of $Sn_2P_2S_6$. The symmetry plane is parallel to the plane of the figure. The unit cell, coordinate system, and the direction of the spontaneous polarization $\vec{P}_s$ are indicated.

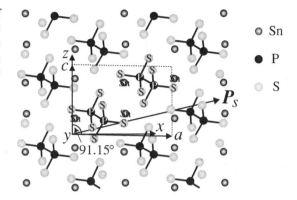

different space groups: Pc – $P2_1/c$ [1,31] and Pn – $P2_1/n$ [3], with two formula units per cell ($Z = 2$). The second one is used more often and is preferable due to the closeness to the orthorhombic structure: at room temperature the unit cell parameters chosen in the Pn space group are $a = 9.375$ Å, $b = 7.488$ Å, $c = 6.513$ Å, $\beta = 91.15°$ [3]. The spontaneous polarization P_s is about 15° off the crystallographic a-axis. The directions of the crystallographic axes and P_s are shown in Figure 10.2.

We use a Cartesian coordinate system with z-axis parallel to the crystallographic c-axis, y-axis normal to the mirror plane, and x-axis normal to y and z (see Figure 10.2). The positive direction of the x-axis is chosen according to the IEEE standard on piezoelectricity so that the piezoelectric constant d_{111} is positive, the $+z$-axis so that it coincides with the $+c$-axis, and $+y$ so that xyz is a right-handed system. Note that recent measurements [18] show that in this system the piezoelectric constant d_{333} is negative and is therefore not according to the piezoelectric standard, but we keep here the system used in most of the previous publications because it is closer to the crystallographic abc system.

$Sn_2P_2S_6$ crystals and their structural analogues include two main structural elements: metal (Sn) ions and (P_2S_6) groups (anions), representing two trigonal PS_3 pyramids connected with P-P bonds. According to x-ray data [2,31–33], at room temperature (ferroelectric phase) the mean positions of the Sn ions are shifted by about 0.3 Å along the polar direction from their centrosymmetric positions in the paraelectric phase. This feature, along with the softening of the optical mode observed in Raman [34–36] and neutron scattering [37] spectra, were the reasons for which this transition was considered of displacive type. This interpretation is supported also by the fact that the Curie constant is large, $(0.5–0.7) \cdot 10^5$ K [38], which is typical of the displacive phase transformations. There is, however, also evidence for an order–disorder type of phase transition. A pronounced central peak observed in Brillouin [39] and neutron [37] scattering spectra, as well as a high entropy of the phase transition (≈ 8.6 K^{-1}mol^{-1}), obtained in calorimetric measurements [40], suggests rather a transition of the order–disorder type. Mössbauer spectra studies [41] have shown that the shift of the tin ions is accompanied by their ordering in a local double-well potential. Similar results follow from the precise refinement of the x-ray scattering patterns carried out for $Sn_2P_2Se_6$ [42]. From the above-mentioned experimental data and the theoretical analysis [43] it follows that in $Sn_2P_2S_6$ the ferroelectric phase transitions should be treated as a crossover from the displacive type to the order–disorder one.

10.2.3 Electronic Structure and Light-Induced Effects

Photorefraction is based on defect energy levels of electrons inside the band gap of the material. Light of sufficient photon energy can excite electrons from these levels to the conduction band, or holes to the valence band, or cause direct band-to-band excitations of charge carriers, and thus change the material properties. The most direct consequence is a change in the material conductivity; i.e., the material is photoconductive. Charges can be also retrapped at trap sites of different energy than initially, and thus change the absorption and consequently photorefractive characteristics of the material.

The electronic band structure of $Sn_2P_2S_6$ was investigated experimentally by x-ray photoelectron spectroscopy (XPS) and soft x-ray fluorescence spectroscopy (SXF) [44]. By comparing the most intensive x-ray photoemission lines it was found that vapor-transport nominally pure $Sn_2P_2S_6$ samples exhibit a deviation from stoichiometry. Namely, the $Sn(3d_{5/2})$:$P(2p)$:$S(2p)$ ratio was found to be 19:21:60 instead of the stoichiometric 20:20:60, indicating the possibility of intrinsic defects as Sn- and S- vacancies. Studying both XPS and SXF spectra, it was shown that the valence band consists mainly of five resolvable bands between 3.3 eV and 14.5 eV. Band structure and density of states calculations were also performed by the Full-Potential Linearized Augmented Plane-Wave (FLAPW) method (see Figure 10.3). It was also confirmed that $Sn_2P_2S_6$ can be viewed as an ionic crystal, built of Sn^{2+} ions with high ionicity and $(P_2S_6)^{4-}$ fragments. Note that data indicating that the Sn ion is in the +2 state with very high ionicity (0.8) were also obtained independently by Mössbauer spectroscopy [41]. Within the (P_2S_6) group, P–P and P–S bonds are largely covalent and characterized by sp-hybridization. These states dominate in the center of the valence band, where some influence of Sn $5s$ states is also present. Using the band-structure calculations, the valence zone bands were assigned with the generative electronic states, and the electron density distribution was calculated, all with very good agreement between theoretical calculations and experimental data.

Light-induced charge transfer in nominally pure yellow $Sn_2P_2S_6$ was investigated by means of optical absorption and electron paramagnetic resonance (EPR) spectroscopy and their combination [45]. Light-induced metastability at room temperature is observed in light-induced optical absorption measurements. EPR data support the identification of holes as dominating charge carriers. For excitation with energies exceeding the band gap (2.5 eV) at low (10 K) and ambient (298 K) temperatures, EPR reveals that the following processes are likely to occur. A hole is captured at one of the two different Sn^{2+} sites, creating a $5s^1$ state, i.e., Sn^{3+}. Upon light excitation with the energy of 1.4 eV ($\lambda = 885$ nm) the hole is

FIGURE 10.3. The calculated partial density of states of Sn and P, split into *s* states and *p* states (from [44]).

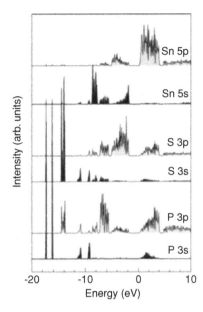

transferred to another Sn^{2+} center. The corresponding band in the light-induced absorption spectra remains for several days. Upon illumination with 2.0 eV photons (600 nm) the hole transfers to another center, which can be associated with S^-. In addition, the optical absorption spectra reveal the release of numerous carriers under interband excitation and the trapping at several different levels in the band gap. In samples previously irradiated by white light, the band-to-band excitation at the temperature 10 K produces a wide absorption band in the range of 1–2.5 eV, which disappears within a few minutes.

Light-induced effects were manifested also in some other properties. For example, a light-induced change of the dark electric conductivity was observed [46]. The DC electric conductivity measured along the nonpolar [010] direction was increased by two orders of magnitude at room temperature after sample preillumination with white light. The low-resistant state remains evident for several days and can be eliminated by sample annealing. These two states differ not only in the values of conductivity but also in the dependence of the DC conductivity on the temperature. In the high-resistant state one finds the usual exponential Arrhenius-like dependence, whereas in the low-resistant state the dark DC conductivity is found to be proportional to $T^{1/4}$ [46], which is inherent to a model of hopping conductivity with the variable hopping range.

Since the above-described defects are intrinsic and therefore not limited in quantity, light-induced sensitization is an effective way to improve the holographic performance. On the other hand, this leads to an instability of the photorefractive parameters. Note that a significant influence of a preillumination was observed in the photorefractive properties of nominally pure yellow $Sn_2P_2S_6$ crystals, as will be described in Section 10.3.1.

The comparative studies of the photovoltaic effect in both ceramic (sintered from the powder) and single crystals were performed in [47]. At white xenon lamp illumination of $2.1 \, W/cm^2$ the measured photovoltaic current was $25 \, nA/cm$. The voltage–current curve measurement under the light irradiation gives a photovoltaic field of $6 \, V/cm$, which is several orders of magnitude lower than that observed in perovskite-type crystals. This permits one to neglect the photovoltaic contribution to the charge transport processes in photorefractive experiments.

Experimental investigations of the non-steady-state photoelectromotive force effect (photo-EMF) in $Sn_2P_2S_6$ crystals were carried out in [48] for illumination at $\lambda = 633 \, nm$. The photoconductivity was found to be dominated by holes, with a diffusion length of photocarriers of $L_D \approx 1 \, \mu m$ and a characteristic photoconductivity-relaxation time $\tau_{ph} = 0.5$–$1.5 \, ms$ in the investigated light-intensity region (0.1–$0.01 \, mW/mm^2$). These experiments indicate that the diffusion length is shorter than the Debye screening length $l_s \approx 3 \, \mu m$ extracted from the experimental dependencies of the photo-EMF signal on the spatial frequency. This is in contradiction with the data obtained from photorefractive measurements [8]. The origin of this discrepancy is at the present point not completely clear. It might be supposed that the assumption of a pure monopolar photoconductivity taken in [48] is not exactly fulfilled for the crystals used.

10.2.4 Domain Structure and Poling

Usually as-grown $Sn_2P_2S_6$ samples are polydomain. They can, however, easily be poled by slow cooling from the deep paraelectric phase ($\approx 360 \, K$) down to room temperature under a DC electric field of about $1 \, kV/cm$ ($1 \, K/min$ and $0.5 \, K/min$ in the vicinity of the phase transition ≈ 335–$340 \, K$). As follows from symmetry considerations, domains appearing below the ferroelectric phase transition are connected by the symmetry elements, which are lost at the phase transformation. In the case of $Sn_2P_2S_6$ where the $2/m - m$ transition occurs, only two types of $180°$ domains with P_s vector lying in the (010)-plane should appear. Such domains are indistinguishable in the polarizing microscope, which complicates their direct observations. First studies of the domain pattern on the crystal surfaces were carried out using a liquid-crystal technique [49]. Another technique is based on the photorefractive properties and exploits the difference of the two-beam gain sign in opposite domains, which makes possible a 3D scanning of the polarization state over the sample using a narrow beam [50], or to get a 2D pattern of relatively large domains in a thin plate [51]. It is also possible to obtain an "optical" hysteresis loop dynamically as the field dependence of the beam coupling gain during sample repolarization with a low-frequency (50–150 Hz) electric field [6].

Observation of the optical hysteresis loop [2, 16, 17] was used for determining the spontaneous polarization P_s ($\approx 15 \, \mu C \, cm^{-2}$ at 293 K) and its variation with temperature (see Figure 10.5). A great influence of the prehistory (relaxation in the ferroelectric phase, surface state, preirradiation) on the switching processes was observed, which was partially manifested in domain fixing by the internal fields after sample relaxation [52,53]. It was found that a single-domain state in Sn$_2$P$_2$S$_6$ can be fixed by producing a photoelectret state. To reach it, the illuminated sample should be kept at about 400 K under a field of about 1600 V/cm for more than 1 hour. In this state the unipolar state is usually preserved even after a short-time heating above the phase transition due to the presence of a bias field [54].

A simple method permitting one to check the Sn$_2$P$_2$S$_6$ sample unipolarity was proposed in [55]. It was revealed that domain walls in these crystals are able to reflect light that is easily observed in the direction perpendicular to the incident beam. This effect, called directional light scattering, is not observed in the single-domain state and in the paraelectric phase and can be used as a tool for testing the unipolarity. The scattering pictures are usually not uniform: dark and bright parts correspond to single- and polydomain parts, separated by the planar defects (growth planes). This distribution correlates well with the domain patterns obtained by alternative techniques [51], proving its relation to the ferroelectric domains structure. The peculiarities of the scattering suggest that the light reflecting planes are charged domain walls, i.e., walls nonparallel to the polar vector. Studying this reflection, it was established that the P_s direction makes an angle of about 8–10° with these planes. Due to the semiconductor characteristics of Sn$_2$P$_2$S$_6$, a large part of the surface charge is screened by the charge carriers to create local fields near the wall with a thickness on the order of the Debye screening length, i.e., of order 1 μm, which is comparable with the light wavelength. The orientation of the reflecting planes is apparently determined by the minimum of the elastic contribution to the total energy of the domain walls [55]. Such a configuration of the domain structure in Sn$_2$P$_2$S$_6$ correlates well with the patterns obtained by surface chemical etching [56], where wedge-like pictures are observed on the xy surface, and rhombi and triangles on the polar yz-cut, as shown in Figure 10.4.

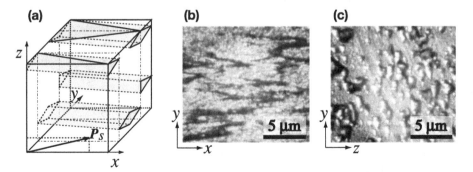

FIGURE 10.4. (**a**) Domain structure scheme of Sn$_2$P$_2$S$_6$ in the ferroelectric phase. Domain walls can be parallel to the symmetry plane (indicated by the dash-dotted lines) or inclined (shaded planes). Photograph of the etched surface perpendicular to the z-axis (**b**) and perpendicular to the P_s vector (**c**) of a polydomain Sn$_2$P$_2$S$_6$ single crystal.

10.2.5 Dielectric Properties

The complex dielectric permittivity of $Sn_2P_2S_6$ was studied in a wide frequency range using different techniques: impedance measurements at low frequencies [38], resonator (waveguide) technique at high frequencies [57], and the quasi-optic (complex transmission) in the submillimeter wave range [58]. The last two methods provided information about the lattice contribution and permitted us to measure the soft-mode parameters. A great influence of the measuring conditions was observed, such as material of the contacts, sample prehistory, and its domain state, influencing the correct determination of the dielectric constant ϵ. In [59] the thickness dependence of the dielectric permittivity was studied, and the formation of near-electrode layers with changed dielectric parameters was demonstrated.

As is well known, domain walls contribute significantly to the dielectric constant in the ferroelectric state of a polydomain sample. The increase of the total area of the walls leads to the rise of the domain contribution to ϵ as we approach the phase-transition temperature T_0. Since at sample cooling from the paraelectric state at $T < T_0$ the number of domain walls exceeds that at sample heating from the deep ferroelectric phase, ϵ is also higher at sample cooling. The temperature dependence of ϵ for $T > T_0$ obeys the Curie–Weiss law $\epsilon(T) = C/(T - T_0)$ with the Curie–Weiss constant C varying from $0.65 \cdot 10^5$ K to $0.70 \cdot 10^5$ K in nominally pure yellow samples. An example of the temperature dependence of the dielectric constant is presented in Figure 10.5.

The dielectric constant depends on the crystal type. At similar conditions (at room temperature and low frequency 1 kHz), the dielectric constant ϵ_{11} was about 230 in yellow $Sn_2P_2S_6$ samples, about 300 in brown samples, and about 280 in Te-doped samples of different concentration of Te in the initial compound.

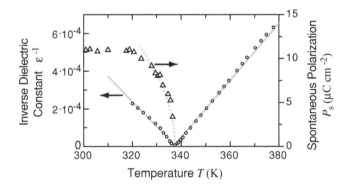

FIGURE 10.5. Inverse dielectric constant ϵ^{-1} (circles, left scale) along the x-axis as a function of the temperature T at low frequency (1 kHz). The solid lines are according to $\epsilon = C/|T - T_0|$, with $T_0 = 337$ K, $C = 0.66 \cdot 10^5$ K above T_0 and $C = 0.75 \cdot 10^5$ K below T_0 (replotted from [57]). Temperature dependence of the spontaneous polarization P_s (triangles, right scale) is also shown (replotted from [60]).

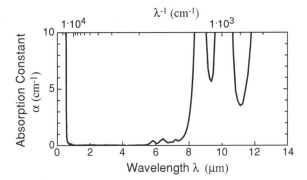

FIGURE 10.6. Absorption spectrum of Sn$_2$P$_2$S$_6$ at room temperature for nonpolarized light propagating along the z-axis.

Dielectric properties were studied also in photorefractive grating experiments [61]. The data on ε can be extracted from the decay of the space-charge grating (decay time is governed by dielectric relaxation, so that $\tau \propto \varepsilon \varepsilon_0 / \sigma$) or from the measurements of Debye screening length $\ell_s = \sqrt{\epsilon_{\text{eff}} \epsilon_0 k_B T / e^2 N_{\text{eff}}}$. Both these techniques were used to evaluate the anisotropy of the dielectric permittivity, $\varepsilon_{11}/\varepsilon_{33} \approx 3.7$–$3.9$ [61]. This value is in good agreement with the direct measurements reported in [62].

10.2.6 Linear and Nonlinear Optical Parameters

Sn$_2$P$_2$S$_6$ crystals are transparent between 530 and 8000 nm with a low absorption coefficient α in the transparency region (see Figure 10.6). The variations of the absorption edge with temperature and hydrostatic pressure were shown to follow the Urbach rule [63,64]. The spectral dependence of the absorption coefficients for yellow, brown, and Te-doped crystals measured on z-cut samples are presented in Figure 10.7. A wide additional absorption "shoulder" that arises in brown Sn$_2$P$_2$S$_6$ crystals leads to a significant rise of the absorption coefficient; for instance, at

FIGURE 10.7. Absorption spectra of nominally pure yellow, brown, and Te-doped Sn$_2$P$_2$S$_6$ crystals at room temperature for x-polarized light.

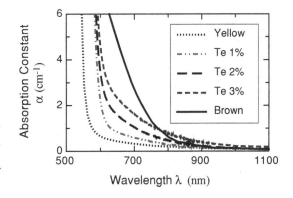

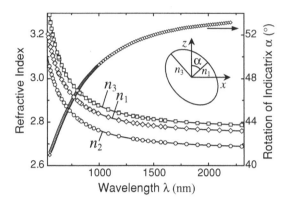

FIGURE 10.8. Dispersion of the refractive index main values (left scale) and the optical indicatrix azimuth (right scale) of $Sn_2P_2S_6$ crystals at $T = 295$ K [18].

He-Ne laser wavelength ($\lambda = 633$ nm) it reaches 5.7 cm^{-1} for the x-polarized, and 4.5 cm^{-1} for the y-polarized light. The absorption spectra of Te-doped crystals also clearly show an additional absorption band that is found to increase with the Te concentration. Note that the absorption constants in the near-infrared in doped crystals with lower Te concentration remain quite low and are of similar magnitude as those of nominally pure samples and lower than in brown crystals.

Due to the monoclinic symmetry, $Sn_2P_2S_6$ crystals are optically biaxial media and therefore characterized by the three main values of the refractive index: n_1, n_2, n_3. The refractive index dispersion was measured using the ordinary wedge method [15] and the minimum deviation method [18], and is presented in Figure 10.8. The $n(\lambda)$ curves are well approximated by a two-term Sellmeier oscillator model

$$n^2(\lambda) - 1 = \frac{S_1\lambda_1^2}{1 - (\lambda_1/\lambda)^2} + \frac{S_2\lambda_2^2}{1 - (\lambda_2/\lambda)^2} \qquad (10.1)$$

with the parameters listed in Table 10.2.

Temperature variations of the refractive indices were measured in [5] at 633 nm. Due to the monoclinic symmetry, the optical indicatrix rotates by varying the temperature or the light wavelength. Based on these data and using the temperature dependencies of the dielectric constant and of the spontaneous polarization, first evaluations of the electrooptic coefficients were done by considering the linear electrooptic effect as the quadratic Kerr effect being linearized by the polarization.

TABLE 10.2. Sellmeier coefficients for the dispersion of the refractive indices of $Sn_2P_2S_6$ at room temperature [18]

	S_1 (μm^{-2})	λ_1 (nm)	$E_1 = hc/\lambda_1$ (eV)	S_2 (μm^{-2})	λ_2 (nm)	$E_2 = hc/\lambda_2$ (eV)
n_1	93 ± 1	251 ± 4	4.93 ± 0.08	3.4 ± 1	428 ± 7	2.90 ± 0.05
n_2	91 ± 1	246 ± 6	5.04 ± 0.12	4.1 ± 1	407 ± 10	3.05 ± 0.07
n_3	92 ± 1	250 ± 4	4.95 ± 0.08	4.8 ± 1	440 ± 7	2.82 ± 0.05

TABLE 10.3. Electrooptic coefficients of Sn$_2$P$_2$S$_6$ measured at $\lambda = 633$ nm and room temperature, and estimated polarization-optic (PO) coefficients [19]

ijk	Unclamped EO coeff. r_{ijk}^T (pm/V)	Clamped EO coeff. r_{ijk}^S (pm/V)	Ratio r_{ijk}^S/r_{ijk}^T	PO coeff. f_{ijk} (m^2/C)
111	$+174 \pm 10$	$+50 \pm 5$	0.30 ± 0.02	0.089
221	$+92 \pm 8$	$+11 \pm 3$	0.12 ± 0.02	0.046
331	$+140 \pm 18$	$+42 \pm 10$	0.30 ± 0.09	0.072
131	$+25 \pm 15$	$+11 \pm 8$		0.012

The orientation of low-symmetry anisotropic crystals for optical measurements and applications is often quite a challenge. In Sn$_2$P$_2$S$_6$ only the y crystal axis (i.e., normal to the (010) symmetry plane) can be determined unambiguously on the basis of x-ray orientation. At room temperature the [100] direction that is close to the polar P_s vector is directed approximately at 45° to the main axes of the optical indicatrix at room temperature and $\lambda = 633$ nm. The variation of the indicatrix azimuth with the light wavelength was measured in [18] and is also depicted in Figure 10.8. In contrast to the definition of the coordinates in [18] in this review, the z-axis is reversed (see Section 10.2.2). The indicatrix rotation angle α, however, is still defined as the angle between the x-axis and the major dielectric axis as in [18].

The electrooptical (EO) coefficients of Sn$_2$P$_2$S$_6$ were measured by a direct interferometric technique for the most useful geometries, when the electric field is applied parallel to the crystallographic x-axis [19]. It was found that the room-temperature free EO coefficient r_{111}^T is the highest and reaches 174 pm/V at $\lambda = 633$ nm with a very weak dispersion in the wavelength range of 633–1300 nm. The ratio between free and clamped EO coefficients r_{ijk}^S/r_{ijk}^T was determined at $\lambda = 633$ nm by applying a fast-pulsed electric field instead of an AC field and measuring the temporal evolution of the electrically induced refractive index change. The measured EO coefficients and the ratios r_{ijk}^S/r_{ijk}^T are collected in Table 10.3. Note that the sign of r_{131} is reversed with respect to [19], where the $+z$-axis was defined such that the piezoelectric constant d_{333} was positive, and in our coordinate system it is negative, but $+z$ here coincides with $+c$ (see Figure 10.2). The temperature dependence of the EO coefficients near the structural phase transition is well described by the Curie–Weiss law with a peak value of $r_{111}^T = 4500$ pm/V (Figure 10.9).

10.3 Photorefractive Effects

In this section we review the basic photorefractive characteristics of different Sn$_2$P$_2$S$_6$ crystal types. Most of the experiments were performed with nominally pure "yellow" crystals. Yellow Sn$_2$P$_2$S$_6$ crystals are treated in Section 10.3.1, where the characteristic difference between Type I yellow and Type II yellow crystals is discussed, together with some interesting consequences of the out-of-phase

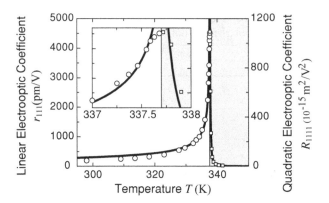

FIGURE 10.9. Temperature dependence of linear r_{111}^T ($T < T_0$) and quadratic R_{1111}^T ($T > T_0$, gray region) EO coefficient at $\lambda = 633$ nm. The theoretical lines are according to $r \approx C_1 (T_0 - T)^{-1/2}$ and $R \approx C_2 (T_0 - T)^{-2}$ with $C_1 = 1863 \sqrt{K}$pm/V, $C_2 = 57.2 \cdot 10^{-15}$ K²m²/V², and $T_0 = 64.7\,^\circ$C [19].

photorefractive grating formation in Type I yellow crystals. In Section 10.3.2 the basic photorefractive characteristics of modified "brown" crystals and doped crystals with some advantageous properties with respect to yellow $Sn_2P_2S_6$ crystals are discussed. The electric field influence on photorefractive grating formation and beam-fanning effects are treated in Sections 10.3.3 and 10.3.4 respectively. Illumination of $Sn_2P_2S_6$ in the blue-green spectral region causes band-to-band excitations. These are discussed in Section 10.3.5.

10.3.1 Photorefractive Effect in Nominally Pure Crystals

The nominally undoped ("yellow") as-grown crystals feature obviously nonlocal nonlinear response with a considerable coupling between the two recording beams. Rotation of the polar axis to 180° results in the inversion of the beam-coupling direction, with the gain factor changing its sign but keeping the same absolute value. This proves that the photorefractive grating is $\pi/2$-shifted with respect to the light fringes [7].

Further studies revealed two out-of-phase space-charge gratings in the steady state that develop with considerably different characteristic times. The formation of two gratings results in a pronounced transient peak in the dynamics of the beam coupling (Figure 10.10a) and in particular in the dynamics of the diffraction efficiency during the optical erasure of the grating (Figure 10.10b). When a virgin crystal starts to be exposed to two recording beams, the "fast" grating develops first and the intensity of the weak signal beam grows to its saturation level. Then the "slow" grating develops and the signal intensity gradually drops (Figure 10.10a). If the input signal beam is stopped when both gratings are saturated, the diffracted intensity decays quickly to zero value, reappears, reaches a maximum, and then

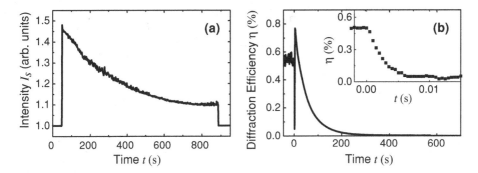

FIGURE 10.10. Manifestation of two out-of-phase space charge gratings in (**a**) Time evolution of the weak signal wave in a two-wave mixing experiment with total input light intensity $I = 30 \, W/cm^2$ and grating spacing $\Lambda = 2.3 \, \mu m$. The pump beam is turned on at $t = 48 \, s$ and off at $t = 887 \, s$. (**b**) Time evolution of the diffraction efficiency after turning off the signal beam at $t = 0 \, s$; grating spacing $\Lambda = 1.2 \, \mu m$. The inset shows the detailed behavior just after the signal wave is blocked. In these experiments light with $\lambda = 1.06 \, \mu m$ was used. The presented behavior is typical for Type I yellow $Sn_2P_2S_6$ crystals.

slowly disappears (Figure 10.10b) [7]. The zero efficiency within the decay curve indicates that the two gratings at a certain moment have the same amplitude but different signs (i.e., are π-shifted).

The buildup times of "fast" and "slow" gratings were extracted from the dynamics of the beam coupling. Their intensity dependencies appeared to be quite different (Figure 10.11). While the buildup rate for the fast grating scales linearly with the intensity, it is almost intensity independent for the slow grating within a wide intensity range. This suggested the formation of the "fast" grating via a photoinduced process and the "slow" grating mainly via a thermally induced process. From the direction of the beam coupling and from the measurements of the electromotive force (EMF) when the grating is being recorded by the vibrating fringe pattern [65, 66], it has been established that the "fast" grating is formed by positively charged carriers, while the "slow" grating is due to the motion of electrons. This is valid within the whole frequency range in which $Sn_2P_2S_6$ is sensitive, at least for nominally undoped crystals.

Since the out-of-phase grating makes the overall steady-state beam coupling smaller, several techniques have been proposed and implemented to avoid its formation or to diminish its amplitude. The first technique consists in cooling the crystal to slow down the thermally activated processes. Figure 10.12 shows the temporal development of the signal-beam intensity for several temperatures up to $-8°C$ [8, 67]. The smaller the temperature, the longer becomes the decay time of the "slow" grating and the smaller its amplitude. It is important that the amplitude of the "fast" grating remains unaffected within this temperature range. At $T = -20°C$ the amplitude of the "slow" grating becomes negligible compared to the amplitude of the "fast" grating.

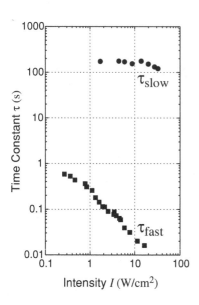

FIGURE 10.11. Buildup time of the fast and slow gratings versus the light intensity I at light wavelength $\lambda = 1.06\,\mu m$ for a Type I yellow $Sn_2P_2S_6$ crystal.

The other way to prevent the formation of the "slow" grating consists in the grating recording with moving fringes because of the frequency shift between the two recording waves. For a certain frequency detuning the "slow" grating is unable to follow a too-fast displacement of the fringes, and its amplitude is strongly reduced, while the "fast" grating is still unaffected by the fringe motion [67, 68]. The steady-state gain factor becomes practically equal to the transient gain for the frequency-degenerate case.

Another technique implements the active feedback during the grating recording [69, 70]. With the properly adjusted parameters of the electronic feedback loop the system finds itself the appropriate phase modulation of the signal wave, linear in time, that corresponds to the optimum frequency detuning.

As a consequence of the formation of two out-of-phase space-charge gratings with considerably different decay times, the gain spectrum (dependence of the gain factor on frequency detuning normalized to the decay rate of the fast grating)

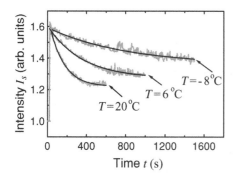

FIGURE 10.12. Dynamics of the signal beam intensity in a two-wave mixing experiment for different temperatures of the sample at light wavelength $\lambda = 1.06\,\mu m$, grating spacing $\Lambda = 6\,\mu m$, total intensity $I = 10\,W/cm^2$ for a Type I yellow $Sn_2P_2S_6$ crystal.

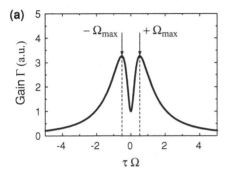
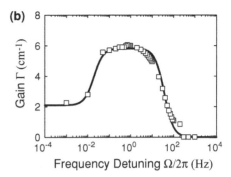

FIGURE 10.13. (**a**) Dependence of the steady-state gain factor Γ on frequency detuning in crystals with strong electron–hole competition (Type I yellow $Sn_2P_2S_6$). (**b**) The measured gain factor spectrum for nearly degenerate two-beam interaction in $Sn_2P_2S_6$ (open squares) at light wavelength $\lambda = 0.63\ \mu m$, total intensity $I = 3\ W/cm^2$, grating spacing $\Lambda = 1.8\ \mu m$. The full line is theoretical according to (10.2).

has two maxima and a deep detuning at zero (Figure 10.13a). The experimental results of the steady-state gain factor (Γ) versus frequency detuning are shown in Figure 10.13b. The equation describing the experimental data is given by [68]

$$\Gamma = \frac{\Gamma_h}{1 + (\tau_h \Omega)^2} - \frac{\Gamma_e}{1 + (\tau_e \Omega)^2}, \tag{10.2}$$

where Γ_h, Γ_e are the gain factors of the hole and electron gratings, and τ_h, τ_e are the time constants of the hole and electron gratings respectively. Such a gain spectrum corresponds to a nontrivial temporal dynamics of $Sn_2P_2S_6$-based coherent oscillators, which will be discussed in Section 10.4.2.

The amplitude and decay time of the "slow" grating depend also on the postgrowth crystal treatment. Being multidomain, the as-grown crystals possess nevertheless quite a high degree of poling, i.e., the total volume of domains with a certain orientation of the spontaneous polarization P_s differs significantly from the total volume of domains with the opposite orientation of P_s. Thus, even imperfectly single-domain, these crystals are suitable for grating recording and feature unidirectional beam coupling. As a rule, for such as-grown crystals the "slow" out-of-phase grating is well pronounced. The dynamics of the beam coupling and grating erasure shown in Figure 10.10 is typical just for these crystals, which we call Type I yellow $Sn_2P_2S_6$.

Depending on the aftergrowth poling procedure, the "slow" out-of-phase grating may either remain nearly the same as for as-grown samples or may become much slower and with much smaller amplitude. The poled crystals with nearly unobservable "slow" grating are Type II yellow $Sn_2P_2S_6$. Usually they also feature longer decay times of the "fast" grating and a smaller effective trap density, and are less sensitive to infrared recording [71].

The suppression of the "slow" grating improves the steady-state gain, increasing it to the peak value of the transient gain. To further increase the gain factor (i.e.,

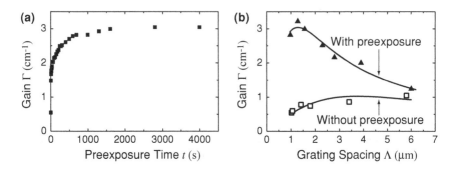

FIGURE 10.14. (a) Gain factor versus white light preexposure time at wavelength $\lambda = 1.06\,\mu m$ and grating spacing $\Lambda = 1.6\,\mu m$ for a Type I yellow crystal. (b) Grating spacing dependence of the gain factor Γ. Triangles and squares represent measurements with and without preexposure, respectively. The solid curves are theoretical according to (10.3).

space-charge amplitude) it is necessary either to apply an external electric field, which we discuss in Section 10.3.3, or to increase the effective trap density of the sample. The standard approach consists in deliberate doping of the sample (see Section 10.3.2), but to a certain extent this task can be fulfilled by illuminating the crystal with auxiliary light before or during recording.

An example of such "sensitization" of a $Sn_2P_2S_6$ sample of Type I to the recording at $\lambda = 1064\,nm$ is shown in Figure 10.14a, where the peak value of the gain factor is plotted as a function of the exposure time to incoherent white light from a halogen lamp. Before the measurement, the sample was kept in the dark for two months. Still in the dark it was mounted in the standard two-beam coupling scheme to measure the gain factor, and the first measurement at $t = 0$ was done. Then the sample was illuminated with white light for a certain time Δt. Illumination was stopped for a new gain measurement, switched on once more for Δt, stopped for measurements, and so on, repeatedly until saturation. Depending on the experimental conditions, the gain factor could be enhanced up to one order of magnitude [8].

The reason for this enhancement is in the light-induced modification of the effective trap density, as was proved by measurements of the Debye screening length ℓ_s (see Section 10.2.3). Figure 10.14b shows the grating spacing Λ dependencies of the gain factor for the virgin crystal (squares) and the crystal sensitized to saturation (triangles). Solid lines represent the corresponding theoretical curves according to the single-level single-carrier model [72, 73]

$$\Gamma = \frac{2\pi}{\lambda} \frac{\cos 2\theta_S}{\cos \theta_S} n^3 r^{\mathrm{eff}} E^{SC}, \qquad (10.3)$$

where we assume a symmetric incidence of the two beams polarized in the incidence plane, $2\theta_S$ is the angle between the beams inside the crystal and r^{eff} the effective electrooptic coefficient. The steady-state space-charge electric field in the case of diffusion-dominated charge transport is given by

$$E^{SC} = \frac{E_D E_q}{E_D + E_q},$$ (10.4)

where the diffusion electric field is $E_D = (k_B T/e)(2\pi/\Lambda)$ and the trap-density-limited field is

$$E_q = \frac{e}{\epsilon_{\text{eff}}\epsilon_0}\frac{\Lambda}{2\pi}N_{\text{eff}} = \frac{\Lambda}{2\pi\ell_s^2}\frac{k_B T}{e}.$$ (10.5)

Here k_B is the Boltzmann constant, T the absolute temperature, e the unit charge, ϵ_0 the electric constant, ϵ_{eff} the effective dielectric constant, N_{eff} the effective trap density, and $\Lambda = \lambda/(2n\sin\theta_S)$ the grating spacing.

It is clearly seen in Figure 10.14b that in the preilluminated crystal the gain factor is considerably larger and the largest gain is achieved at a smaller grating spacing than in the virgin crystal. Thus we can conclude that preillumination reduces the Debye screening length ℓ_s. We believe that this occurs because of the increase of the hole-donor density, which is caused by a light-assisted redistribution of the population among the energy levels of the intrinsic defects.

The sensitized state is not permanent but metastable: the decrease of the gain factor with time was detected when sensitized sample was kept in the dark [8]. By approximating the measured temporal dependence of the gain for the sample of Figure 10.14 by a single-exponential function we obtained ≈36 hours. This time can be considered as the characteristic time for the reestablishment of the equilibrium in the defect level population in Sn$_2$P$_2$S$_6$.

Apart from the sensitization with white light, a self-sensitization was observed that manifested itself in the intensity dependence of the gain factor that is a typical of photorefractive crystals. The measurements with He-Ne laser within the intensity range where photoconductivity largely dominates over the dark conductivity showed a gain, which was still not saturated. The grating spacing dependencies of the gain factor measured at different intensity levels revealed the change of the Debye screening length with the intensity. Figure 10.15 represents the intensity dependence of the effective trap density N_{eff} as obtained from the gain measurements and (10.3)–(10.5). One can see that there is still no saturation at

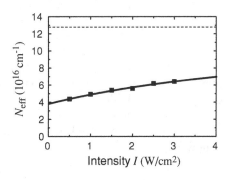

FIGURE 10.15. Intensity dependence of the effective trap density N_{eff} for a Type I yellow crystal at $\lambda = 0.63$ μm. Dots represent the experimental results, while the solid line is according to (10.6). The dashed line presents the saturated value of N_{eff} as expected from (10.6.)

$I = 3$ W/cm^2. These experimental data were analyzed phenomenologically by the dependence

$$N_{\text{eff}}(I) = N_{\text{eff}}(0) + \frac{\Delta N_{\text{eff}}}{1 + I^{\text{sat}}/I}. \tag{10.6}$$

With $\epsilon \approx 230$ (Section 10.2.5) we obtained $N_{\text{eff}}(0) \approx 3.8 \cdot 10^{16}$ cm^{-3}, $\Delta N_{\text{eff}} \approx 9 \cdot 10^{16}$ cm^{-3}, and $I^{\text{sat}} \approx 7$ W/cm^2. By extrapolating the dependence (10.6) to larger I we can expect further improvements of the gain for higher intensities [74].

It should be underlined that self-sensitization with red light differs qualitatively from sensitization with white light described above: red light does not produce a long-memory effect and $N_{\text{eff}}(I)$ returns to its equilibrium level $N_{\text{eff}}(0)$ within less than one minute. No hysteresis was observed when we measured the intensity dependence of the effective trap density with increasing and next with decreasing intensity. Similar behavior was reported for BaTiO$_3$ earlier [75] but with a much smaller ratio $\Delta N_{\text{eff}}/N_{\text{eff}}(0)$.

The study of the photorefractive response gives rich information about crystal properties. Apart from the effective trap density N_{eff}, such parameters as specific photoconductivity, diffusion length, and electrooptic constants have been estimated [76–78]. As was mentioned in Section 10.2.5, the anisotropy of dielectric properties have been evaluated in [61].

A gain factor of about 7 cm^{-1} was achieved in nominally undoped (Type I yellow) Sn$_2$P$_2$S$_6$ at $\lambda = 1064$ nm and preillumination, increasing to near 12 cm^{-1} at 633 nm. This is not much smaller than the gain in Co-doped and Rh-doped BaTiO$_3$ and quite sufficient for different wave-mixing experiments and for design of optical coherent oscillators (see Section 10.4). An obvious advantage of Sn$_2$P$_2$S$_6$ is its fast response with the decay time in the millisecond range. This is still slower than the microsecond-scale response of semiconductor photorefractives, but much faster than that of wide-band-gap ferroelectrics.

10.3.2 Photorefractive Properties in Modified and Doped Crystals

The photorefractive response is based on suitable donor and acceptor levels inside the material band gap, provided either by impurity levels or crystal nonstoichiometric defects. In the case of Sn$_2$P$_2$S$_6$ crystals, most photorefractive experiments were performed in nominally pure yellow samples. Doping with metals (Pb, In, Cu, Cd) was found not to change substantially the material parameters involved in photorefraction. As described in Section 10.2.1, modified brown crystals with presumably higher deviation from stoichiometry, and Te-doped crystals were recently obtained, with an increased absorption as shown in Figure 10.7. In this section we present the measured photorefractive parameters of these crystals that show significantly enhanced photorefractive response compared to nominally pure yellow Sn$_2$P$_2$S$_6$.

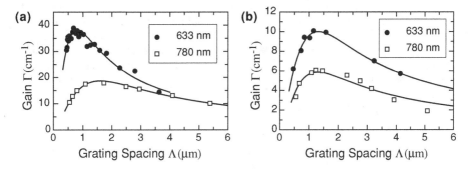

FIGURE 10.16. Grating spacing Λ dependences of the two-beam coupling gain coefficient Γ at wavelengths 633 nm and 780 nm for (**a**) brown $Sn_2P_2S_6$ and (**b**) Te doped $Sn_2P_2S_6$ containing 1% of Te in the initial compound. The solid curves are theoretical according to (10.3).

The main photorefractive parameters of modified brown and Te-doped crystals, the two-beam coupling coefficient Γ and the time constant τ, were measured for laser wavelengths of 633 nm and 780 nm as functions of the spacecharge grating spacing Λ and the light intensity [10–12]. In contrast to Type I yellow $Sn_2P_2S_6$, where the enhancement of the photorefractive sensitivity is substantially stimulated by white light and also by the He-Ne laser irradiation (Section 10.3.1), no influence of preillumination on the photorefractive sensitivity was observed in brown and Te-doped samples. This stability is a substantial advantage of these types of crystals.

Another advantage of brown and Te-doped crystals is the absence of any pronounced electron–hole compensation during the grating buildup, typical for Type I yellow $Sn_2P_2S_6$ (Figure 10.10). However, the time evolution of the amplified beams has a nonexponential behavior at relatively long exposure times. The magnitude of this slow contribution to photorefractive grating is rather small ($<10\%$) and changes sign with the grating spacing Λ.

Figure 10.16 shows the measured dependence of the photorefractive gain coefficient Γ on the grating spacing Λ. Modified brown crystals exhibit significantly increased gain, up to 38 cm^{-1} at 633 nm and 18 cm^{-1} at 780 nm. The time constants recalculated for the intensity of 1 W/cm^2 are about 4 ms at 633 nm, much faster than typically obtained in yellow samples (10–50 ms). Various Te-doped samples with the concentration of Te of 1%, 2%, and 3% in the initial compound, did not vary substantially in their photorefractive properties, although their absorption constants were significantly different (Figure 10.7). Maximal-gain values were increased from typical values in yellow samples (4–7 cm^{-1}) to about 10–12 cm^{-1}, still lower than in brown samples. The measured photorefractive response speed, however, was slightly faster in Te-doped samples than in brown samples.

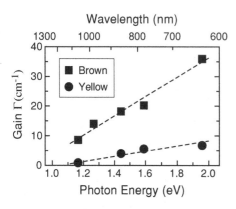

FIGURE 10.17. Photorefractive gain Γ at the optimal (Debye) grating spacing as a function of the photon energy or light wavelength λ in brown and, for comparison, in yellow (Type II) crystals without preillumination [80]. The dashed lines are guides for the eye.

The parameters characterizing photorefractive properties, the effective electrooptic coefficient r^{eff} and the effective trap concentration N_{eff}, were determined by approximating the $\Gamma(\Lambda)$ dependences (Figure 10.16) with the theoretical model (10.3). The effective number of traps is, as expected, increased with nonstoichiometry and doping. For example, at 780 nm the obtained N_{eff} increased from about $0.2 \cdot 10^{16}$ cm^{-3} in yellow $Sn_2P_2S_6$ to about $0.7 \cdot 10^{16}$ cm^{-3} in brown $Sn_2P_2S_6$ and $1.0 \cdot 10^{16}$ cm^{-3} in 1%Te:$Sn_2P_2S_6$ [12, 79]. Besides, r^{eff} is found to be considerably higher in brown samples, closer to the electrooptic coefficient r_{111}^T measured by the direct method (Table 10.3). The higher values of r^{eff} in brown samples are most likely due to the absence of electron–hole competition, resulting in high gain values. The photorefractive gain is still over 8 cm^{-1} in brown samples at the technologically important wavelength of 1.06 µm (see Figure 10.17).

The photorefractive response times τ were analyzed using the conventional relation $\tau^{-1}(\Lambda)$ in the single-level single-carrier model [72]. Both the diffusion length L_D and the Debye screening length ℓ_s parameters were extracted as the model parameters in the theoretical dependence $\tau^{-1}(\Lambda)$ [10]. The diffusion length L_D was shown to exceed the Debye screening length ℓ_s, which leads to the decreasing of the time constant with the space-charge period. This feature was also observed in yellow $Sn_2P_2S_6$ [8].

The above data show that the basic photorefractive properties of $Sn_2P_2S_6$ crystals can be effectively enhanced by modifications of the growth conditions as in the case of brown crystals, and the doping, i.e., with Te. Although the obtained gain values are not as high as in brown crystals, Te-doped crystals exhibit a faster response and lower absorption, which is also advantageous for applications. A summary of the typical photorefractive parameters of different crystal types is given in Table 10.4.

10.3.3 Drift-Driven Photorefraction

With an electric field applied to the sample during recording, the grating amplitude increases and the grating is shifted from/to the light fringes. The value of the shift depends on the field amplitude and on material parameters [72]. Therefore, the

TABLE 10.4. Typical photorefractive parameters of various Sn$_2$P$_2$S$_6$ crystals at two light wavelengths λ, without preillumination: α_x, absorption coefficient for x-polarized light; Γ_{max}, maximal two-wave mixing gain; τ, faster response time at a grating spacing of 1 μm and scaled to a light intensity of 1 W/cm^2; N_{eff}, effective trap density

Sn$_2$P$_2$S$_6$ sample	λ (nm)	α_x (cm^{-1})	Γ_{max} (cm^{-1})	τ (ms)	N_{eff} (10^{16} cm^{-1})
Yellow Type II	633	0.5	4–7	10–50	0.7
	780	0.2	2–5	100	0.2
Brown	633	5.7	38	4	2.5
	780	1.0	18	10	0.7
Te-doped (1%)	633	1.0	10	2.5	0.9
	780	0.4	6	7	1

application of an external field always results in the increase of the diffraction efficiency η and, to a certain extent, also of the two-beam coupling gain given by (10.3), where the space-charge field amplitude that is $\pi/2$ shifted to the light fringes changes to

$$E^{SC} = \frac{E_D(1 + E_D/E_q) + E_0^2/E_q^2}{(1 + E_D/E_q)^2 + E_0^2/E_q^2},$$

(10.7)

where E_0 is the externally applied field, and the diffusion electric field E_D and the trap-limiting field E_q are defined in Section 10.3.2. The limiting space charge field can be estimated using the relation (10.5) from the known data on the Debye screening length ℓ_s. With $\ell_s \approx 0.5$ μm [71] the limiting space charge field at ambient temperature is $E_q \approx 160\ \Lambda$ in V/cm if Λ is measured in microns. Thus for fringe spacings $\Lambda \geq 10$ μm one can expect a considerable enhancement of the gain factor when applying a moderate external field below the threshold of electric breakdown. These expectations were confirmed experimentally [81].

A standard two-beam coupling geometry was chosen with the light polarization in the plane of incidence and the grating vector $\boldsymbol{K}$ parallel to crystallographic direction x. The external field was applied in this direction too. A Ti-sapphire laser (TEM$_{00}$, $\lambda = 900$ nm) was used as a light source. The recording beam intensity was 1:600 (contrast $m = 0.08$) and total intensity in the sample was ≈ 20 W/cm^2.

A nominally undoped Sn$_2$P$_2$S$_6$ sample with dimensions $x \times y \times z = 3 \times 3 \times 4$ mm^3 that belongs to Type II Sn$_2$P$_2$S$_6$ crystals was used for this measurement [81, 82]. The secondary (out-of-phase) space charge grating is inhibited in these crystals, whereas their infrared sensitivity is smaller than in Type I crystals. The electric field enhancement of the gain factor was therefore of special importance for improving the performance of Type II crystals in the near infrared.

Spatial frequency dependencies of the gain factor are shown in Figure 10.18a for the applied field of 2 kV/cm (filled dots) and zero field (open squares). The dashed line is the fit to (10.3) and (10.7) with $E_0 = 0$ kV/cm. This fit gave the estimates for the parameters $\ell_s = 0.5$ μm and $r^{eff} = 130$ pm/V that were used for the calculation of spatial frequency dependencies of the gain factor with the

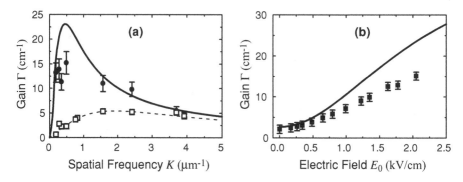

FIGURE 10.18. (**a**) Spatial frequency dependence of the gain factor Γ without (open squares) and with the applied field of 2 kV/cm (filled dots) at a wavelength $\lambda = 0.9 \, \mu m$ and total intensity $I = 20 \, \text{W/cm}^2$ for a Type II yellow crystal. Dashed line: the best fit of (10.3) and (10.7) to data with $E_0 = 0$. The solid line is calculated based on (10.3) and (10.7) with $E_0 = 2 \, \text{kV/cm}$ and the set of parameters extracted from the dependence with $E_0 = 0$. (**b**) Gain factor versus the applied field at the spatial frequency $K = 0.5 \, \mu m^{-1}$. Dots are the measured values, solid line is the expected dependence calculated for $K = 0.5 \, \mu m^{-1} (\Lambda \approx 12 \, \mu m)$.

applied field (solid line). Qualitatively, the calculated dependence is similar to the experimental one; it predicts, e.g., a shift of the maximum gain factor for relatively small spatial frequencies when the field is applied. At the same time the measured gain factor is smaller than calculated, especially at small spatial frequencies.

The electric field dependence of the gain factor is shown in Figure 10.18b for grating spatial frequency $K = 0.5 \, \mu m^{-1}$. The solid line represents the calculated dependence (10.3) plotted with the parameters $\ell_s = 0.5 \, \mu m$ and $r^{\text{eff}} = 130 \, \text{pm/V}$ mentioned above. Here too, a strong enhancement of the beam coupling is visible, but with a gain factor that is about 30% smaller than calculated. This discrepancy can be attributed to the inhomogeneity of the light intensity across the sample, which may lead to a reduction of the electric field in the illuminated region as compared to the applied field. The other reason for gain reduction may be related to the strong light-induced scattering. One can conclude from Figures 10.18a and 10.18b that an application of the external field strongly enhances the gain for large grating spacings. The gain factor with the external field is not only larger at the same spatial frequency (Figure 10.18a), but it becomes nearly three times larger than its optimum value at $E_0 = 0$. With the largest gain factor the signal beam is amplified 100 times in a 3-mm-thick sample.

An even higher gain factor might be expected for larger applied fields (the calculated dependencies for $K \leq 1 \, \mu m^{-1}$ are still not saturated for $E_0 = 2 \, \text{kV/cm}$). In practice, risk of electric breakdown does not allow one to apply more than 2 kV/cm to the available samples.

The data presented in Figures 10.18a and 10.18b were obtained when the external field was applied in the direction of spontaneous polarization. With the

FIGURE 10.19. Gain factor versus applied field during domain reversal at
a wavelength $\lambda = 0.9\,\mu m$, total intensity $I = 20\,W/cm^2$, and spatial frequency
$K = 0.5\,\mu m^{-1}$.

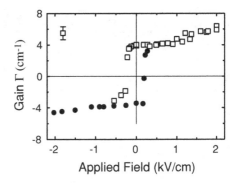

inverted polarity of the external field the repoling of the sample has been observed,
which resulted in the reverse direction of the beam coupling, i.e., the change of
the gain factor sign (Figure 10.19). A pronounced hysteresis has been observed
with the coercive field of about 200 V/cm. This proves that in Type II $Sn_2P_2S_6$
the coercive field can be smaller than the diffusion field, as distinct from SBN
crystals [83] and Type I $Sn_2P_2S_6$ crystals [6].

10.3.4 Photorefractive Beam Fanning and Light Scattering

Beam fanning, i.e., light scattering produced by the amplified noise photorefractive gratings, usually accompanies the laser beam propagation through relatively
thick samples of a photorefractive material with a large gain coefficient. This
effect is clearly observed in $Sn_2P_2S_6$, especially in brown and doped $Sn_2P_2S_6$.
The angular distribution of the scattering, observed on a screen placed behind
the sample, depends on the sample orientation and the light polarization relative
to the crystal axis. The scattered light appears after a certain time, in the case of
$Sn_2P_2S_6$ in about 10–100 msec [84]. When the light propagates along the z-axis,
the fanning looks like a diffuse light spot shifted from the initial laser beam toward the side corresponding to the positive gain. The greatest scattering occurs
when the light is polarized along the polar x-axis. At the intermediate light polarization (between x and y) the noise gratings, corresponding to the two different
polarization states, are mutually suppressed.

When the laser beam passes through a $Sn_2P_2S_6$ sample along the y-axis that
coincides with one of the main optical axes, the scattered light forms arc-like figures, similar to the conoscopic pictures observed in this direction (Figure 10.20).
The scattering intensity reaches a maximum when the incident light polarization makes an angle of about 45° with the plane of the optical main axes, i.e., for
x- or z-polarized light. In this case the mutually incoherent orthogonally polarized
waves form complementary noise gratings erasing each other, except in the directions where the Bragg condition is fulfilled for both wave polarizations. These
conditions of the synchronism between the wave vectors of the incident ($k^i_{1,2}$) and
scattered ($k^s_{1,2}$) light of both polarizations (1 and 2), and the space-charge grating

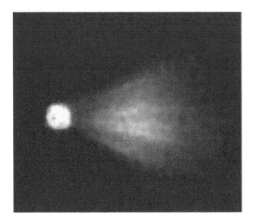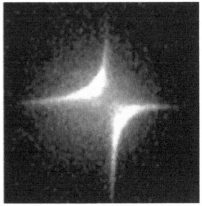

FIGURE 10.20. Photorefractive scattering (beam fanning) obtained with a He-Ne laser ($\lambda = 633$ nm) in a brown $Sn_2P_2S_6$ crystal for the polarization direction along the x-axis, and beams propagating along the z-axis (left) and the y-axis (right).

wave vector K, can be expressed as $k_1^i - k_1^s = k_2^i - k_2^s = K$. The calculated distributions of the scattering correlate well with the experimental ones [84].

10.3.5 Interband Photorefraction

Since $Sn_2P_2S_6$ has an absorption edge around 530 nm, blue-green light is already highly absorbed in the material. This opens the possibility to induce photorefractive gratings exhibiting faster response times by means of the interband photorefractive effect [85], where charge photoexcitation occurs via band-to-band transitions. Response times in the low-microseconds range can be observed in $Sn_2P_2S_6$ under cw blue illumination [86]. Under nanosecond pulsed excitation at the 532 nm wavelength the interband response is even slightly faster, which permits one to use this crystal for high-repetition-rate parallel processing tasks at this wavelength [9].

Interband photorefractive effects including examples involving $Sn_2P_2S_6$ are discussed in more detail in the chapter "Band-to-Band Photorefraction" in Volume I of this book series.

10.4 Optical Phase Conjugations and Self-Oscillations

Phase conjugation is an important process in several application schemes involving photorefractive materials, such as optical processing, laser beam manipulations, and image transforming. Self-pumped phase conjugation configurations that can be realized in high-gain photorefractive crystals have several advantages compared to usual four-wave mixing schemes due to the simplicity of optical alignment, the absence of certain coherence length limitations, and relatively low generation threshold.

Phase conjugation in the near infrared using $Sn_2P_2S_6$ crystals was demonstrated using a ring-cavity scheme [11, 79]. The phase-conjugate reflectivity, about 60% at 860 nm, was shown to be limited only by the transmission of the ring-cavity arrangement. Very fast phase-conjugate response using $Sn_2P_2S_6$ samples in the near infrared was demonstrated, on the order of 50 ms at 780 nm and 1 W/cm^2 [11], which is two orders of magnitude faster than in $Rh:BaTiO_3$. The characteristic properties of the self-pumped optical phase-conjugate mirrors based on $Sn_2P_2S_6$ are described in Section 10.4.1.

The formation of the "slow" out-of-phase photorefractive grating in Type I "yellow" $Sn_2P_2S_6$ results in a specific temporal dynamics of the self-pumped phase conjugator: the intensity of the conjugate wave features a high-contrast sinusoidal amplitude modulation, thus revealing the simultaneous excitation of two oscillation modes with different temporal frequencies. This phenomenon is described in Section 10.4.2.

10.4.1 Self-Pumped Optical Phase-Conjugate Mirrors

The experimental setup with an optimized configuration for the ring-cavity self-pumped phase conjugation in $Sn_2P_2S_6$ is depicted in Figure 10.21. Phase conjugation was demonstrated using Type II yellow and brown $Sn_2P_2S_6$ crystals. The crystals were coated with an around 100-nm-thick Al_2O_3 layer to additionally reduce the reflection losses. Figure 10.22 shows a typical experimental time evolution of the phase-conjugate reflectivity R, defined as the ratio between the measured intensity of the phase-conjugated wave 2 and the input wave 3. In the beginning the phase-conjugated signal emerges from the noise gratings of the scattered input beam; then it increases sharply and finally saturates at its maximum value. The reflectivity rise time is defined as the time in which the reflectivity rises from 10% to 90% of its saturation value. In the yellow sample the rise time is on the order of 5 s, and the maximum phase-conjugate reflectivity is 55%. In the brown sample the response is much faster, with a rise time of only about 20 ms,

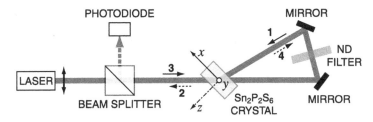

FIGURE 10.21. Experimental setup for ring-cavity self-pumped phase conjugation with $Sn_2P_2S_6$. All beams are polarized in the plane of the loop. The transmission grating is written by beam 3 with its self-diffracted beam 4 and by beams 1 and 2 counterpropagating in the loop.

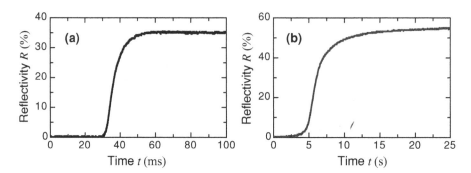

FIGURE 10.22. Time evolution of the phase-conjugate reflectivity R after turning on the input beam 3 using (**a**) brown $Sn_2P_2S_6$ of 2.4 mm thickness and (**b**) yellow $Sn_2P_2S_6$ of 9.7 mm thickness. The input beam is at 780 nm and has a power of 20 mW with a diameter of 0.8 mm (4 W/cm^2).

which is two orders of magnitude faster than in yellow $Sn_2P_2S_6$ and also than the typical response in Rh-doped $BaTiO_3$.

Besides the two-beam coupling mechanism and the initial spatial distribution of the incident light, the ring-cavity losses are of crucial importance for the self-pumped phase conjugation. We can define a loop transmission parameter T that accounts for the losses inside the cavity and the crystal face on the side of the cavity together with the absorption. To study the characteristics of the ring-cavity scheme, the cavity transmission was decreased additionally by adding neutral density (ND) filters into the ring as shown in Figure 10.21. The results of the dependence of the saturated reflectivity R on the loop transmission T are shown in Figure 10.23 for yellow and brown $Sn_2P_2S_6$ samples at 780 nm. By decreasing the transmission of the loop, the maximum reflectivity decreases until the threshold is reached. The measured results for the saturated reflectivity were

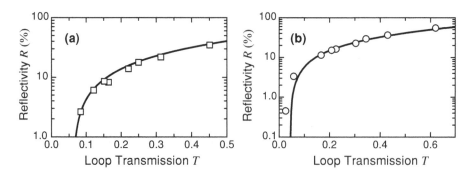

FIGURE 10.23. Measured saturated phase-conjugate reflectivity R as a function of the loop transmission T using (**a**) brown and (**b**) Type II yellow $Sn_2P_2S_6$. The input beam is at 780 nm and has a power of 20 mW with a diameter of 0.8 mm. Solid curves are theoretical with parameters (**a**) $\Gamma L = 4.9$ and $T_0 = 45\%$, and (**b**) $\Gamma L = 5.4$ and $T_0 = 62\%$.

FIGURE 10.24. Measured saturated phase-conjugate reflectivity R as a function of the incident intensity I_3 at 780 nm using brown Sn$_2$P$_2$S$_6$ (squares) and Type II yellow Sn$_2$P$_2$S$_6$ (circles). Solid curves are theoretical with parameters as in Figure 10.23 and considering additional uniform charge generation $I_\beta = 0.1$ W/cm^2 (brown Sn$_2$P$_2$S$_6$) and $I_\beta = 0.4$ W/cm^2 (yellow Sn$_2$P$_2$S$_6$).

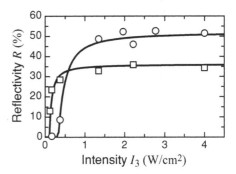

analyzed using a plane-wave solution of the coupled equations for the case of large coupling (depleted-pump) [87]. Comparing measured reflectivities to theoretical, two characteristic crystal parameters can be determined [11]: ΓL, the coupling strength, and T_0, the loop transmission without the ND filter. Theoretical curves in Figure 10.23 result in gain values of 20 ± 2 cm^{-1} for brown Sn$_2$P$_2$S$_6$ and 5.6 ± 0.5 cm^{-1} for yellow Sn$_2$P$_2$S$_6$ at 780 nm, which is in good agreement with two-wave mixing measurements results (Sections 10.3.1 and 10.3.2). A higher loop transmission T_0 (and consequently a higher phase-conjugate reflectivity) in the yellow sample is a result of lower absorption losses in this crystal. The rise time is not much affected by small additional losses, but increases considerably when the transmission threshold is approached [11].

Phase-conjugation performance is also affected by the input beam intensity. The response rate increased linearly with intensity in the measured intensity regime as expected for the two-beam coupling time constant [11]. The results of the saturated phase-conjugate reflectivity dependence on the input intensity are shown in Figure 10.24. The reflectivity increases for lower intensities above some threshold intensity and then it saturates at the maximum value limited by the ring-cavity characteristics. This kind of intensity thresholding was explained by the excitation of carriers due to the thermal effects and by background uniform illumination competing with the spontaneous buildup of the grating. The background uniform illumination I_β in the coupled-wave equations was taken into account to model intensity thresholding [11], presented with the solid line in Figure 10.24.

As already mentioned, the higher saturated reflectivity of yellow samples is attributed to a higher loop transmission. One can then expect even higher phase-conjugate reflectivities for longer light wavelengths λ, provided that the performance is not yet limited by the coupling strength, which decreases with λ. At 860 nm the measured saturated reflectivity in the brown sample was around 48% and in the yellow Sn$_2$P$_2$S$_6$ sample around 60%, considerably higher than at 780 nm (35% and 55% respectively). To check whether the reflectivities in the Sn$_2$P$_2$S$_6$ samples are also limited by the coupling strength, theoretical dependencies of the reflectivity as a function of the coupling strength were plotted in Figure 10.25 considering the obtained loop transmission parameters. Above the coupling-strength threshold the reflectivity increases with ΓL until it saturates at

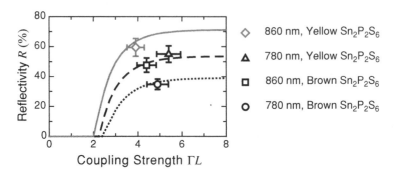

FIGURE 10.25. Theoretical dependencies of saturated reflectivity R on the coupling strength for loop transmissions $T = 0.45$ (dotted curve), $T = 0.62$ (dashed curve), and $T = 0.82$ (solid curve) and experimental points for brown $Sn_2P_2S_6$ sample of 2.4 mm thickness and Type II yellow $Sn_2P_2S_6$ sample of 9.7 mm thickness at 780 nm and 860 nm.

a value equal to the transmission of the loop reduced for the reflection losses at the input crystal surface. For the experimental points in Figure 10.25 the saturation of the reflectivity is already reached, and therefore the reflectivity is limited by the loop-cavity transmission and not by the coupling strength.

Self-pumped phase conjugation with even shorter response times than in brown $Sn_2P_2S_6$ and reflectivities limited by the transmission of the ring-cavity loop were observed also in Te-doped crystals [12]. Considering the above-presented thresholding behavior regarding the ring-cavity transmission, light intensity, and photorefractive coupling strength, one can optimize the phase-conjugate mirror parameters for each crystal type at a desired wavelength.

10.4.2 Phase Conjugation with Harmonic Intensity Modulation

The results on self-pumped conjugation described in the previous section are typical for $Sn_2P_2S_6$ crystals that possess only one dominant photorefractive grating, and the secondary, out-of-phase, grating is either strongly inhibited or does not exist at all (modified brown $Sn_2P_2S_6$, Type II yellow $Sn_2P_2S_6$). The gain spectrum, i.e., the dependence of the gain factor on frequency detuning, is Lorentzian for these crystals with the maximum gain at zero detuning $\Gamma^{max} = \Gamma(\Omega = 0)$ [68].

The formation of a pronounced "slow" out-of-phase grating, as described in Section 10.3.1, results in the splitting of the gain spectrum and the appearance of two maxima shifted symmetrically with respect to the pump frequency (see Figure 10.13a). In a self-pumped phase conjugator with open cavities, such as a ring-loop oscillator, semilinear oscillator, double-phase conjugate mirror, or Feinberg cat conjugator [88, 89], the frequency of the oscillation wave is imposed neither by boundary conditions, since one or two cavity mirrors are missing, nor by the frequency of the seed radiation that has a spectrum at least as wide as the reciprocal decay time of the photorefractive grating. It is also not likely that the cavity losses of the photorefractive oscillator change significantly within a few Hz

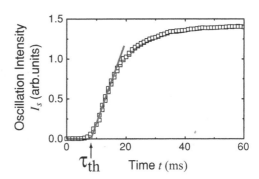

FIGURE 10.26. Dynamics of the intensity of the self-oscillation for a wavelength $\lambda = 633$ nm, pump intensity $I_p = 3$ W/cm^2, and grating spacing $\Lambda = 0.7\,\mu$m for Type I yellow $Sn_2P_2S_6$.

of the frequency range. This is why we believe that the main factor that imposes the oscillation frequency is the gain spectrum profile: the oscillation occurs at the frequency where the gain factor is the largest possible. With the two maxima in the gain spectrum of Figure 10.13a one can expect two-frequency coherent oscillation in Type I "yellow" $Sn_2P_2S_6$. This was confirmed experimentally with a He-Ne laser as a pump source [90].

The experimental setup was similar to that shown in Figure 10.21. The laser beam deflected down by the beam splitter was reflected back by an additional flat mirror in the direction of the photodiode. The distance between the additional mirror and the beam splitter was chosen in a way to ensure mutual coherence of the phase-conjugate wave and the reference wave coming to the detector. A shutter placed between the beam splitter and the additional mirror allowed us either to stop or to pass the reference wave to the detector.

A typical time evaluation of the conjugate wave intensity (reference wave is stopped) is shown in Figures 10.26 and 10.27. Similar to the dynamics shown in Figure 10.22 there is a certain delay time τ_{th} when the intensity of the oscillation wave is so small that it is comparable to the noise level. Finally, the phase-conjugate wave emerges from the noise and its intensity grows rapidly. As distinct

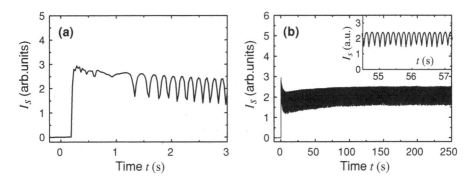

FIGURE 10.27. Temporal variations of the intensity of the self-oscillation for the incident pump intensity $I_p = 3$ W/cm^2 at a wavelength $\lambda = 0.63\,\mu$m and grating spacing $\Lambda = 0.9\,\mu$m for Type I yellow $Sn_2P_2S_6$.

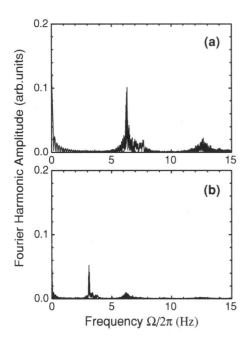

FIGURE 10.28. Fourier spectra of (a) temporal variation of the oscillation intensity and (b) beat frequency mark (oscillation wave plus a part of the pump wave).

from the data of Figure 10.22, however, the intensity of the phase conjugate wave in Figure 10.27 does not reach a steady-state value but oscillates permanently.

The Fourier transform of the intensity variation in saturation (i.e., for $t \gg t_{\mathrm{osc}}$) shows a well-defined peak (6.3 Hz for the experimental conditions of Figure 10.28a) that indicates the frequency separation between the two components of the oscillation spectrum and its second harmonic with 12.6 Hz.

When the pump beam is sent to the detector as a reference wave in addition to the oscillation wave, the beat frequency becomes two times smaller, about 3.1 Hz (Figure 10.28b). This fact proves that the two oscillation modes are shifted symmetrically with respect to the pump frequency, as one can expect from the shape of the gain spectrum (Figure 10.13). The frequency splitting depends on the pump intensity, which is quite natural for the crystals where the relaxation time depends on photoconductivity. At the same time, the frequency detuning grows with intensity not linearly but as a square root $\Omega \sim \sqrt{I}$. This fits well to the results of calculation of maxima separation in gain spectrum (see (10.2) of Section 10.3.1).

Taking into account that the amplitudes of the two out-of-phase gratings are roughly equal $\Gamma_s \approx \Gamma_f$ and the time constants of the slow grating are considerably smaller than those of the fast grating $\tau_s > \tau_f$, the equation for the frequency detuning reads

$$\Omega = \pm \sqrt{\frac{1}{\tau_f \tau_s}}. \tag{10.8}$$

With τ_s independent of intensity and $\tau_f \propto 1/I$ we get $\Omega \propto \sqrt{I}$.

It should be mentioned in addition that apart from the intensity, Ω depends also on the spatial frequency of the photorefractive grating [90]. It should also be mentioned that the presence of out-of-phase gratings in Type II "yellow" $Sn_2P_2S_6$ may in some cases result in a more complicated temporal dynamics of the oscillation wave. For a semilinear oscillator with two counterpropagating waves, for example, roughly sawtooth intensity modulations have been observed with regular shift to π of the phase of the oscillation wave in every new pulse [91].

10.5 Conclusion

In conclusion, $Sn_2P_2S_6$ crystals have been shown to be very promising photorefractive materials for the red and near-infrared spectral range. A big potential for further improvement of the relevant parameters by doping or deviation from stoichiometry still exists. The good photorefractive performance is due to a combination of large electrooptic coefficients, a sufficient amount of trapping sites, and a high photosensitivity. The last of these is related to the pronounced semiconductor properties of this crystal connected to the relatively narrow band gap (as compared to photorefractive oxide crystals). Particularly important is the significantly faster response speed of this material as compared to "classic" materials such as $BaTiO_3$, $KNbO_3$, and $Bi_{12}SiO_{20}$. These features permit one to use $Sn_2P_2S_6$ as an active medium for several dynamic holography applications, such as real-time interferometry and phase conjugation with short generation rise time and high efficiency. One may expect that additional progress in the performance of this crystal can be achieved by further improving the growth technology and by the directed search for new dopants.

Acknowledgments. The authors are very grateful to Prof. S. Odoulov for his contributions to the studies of the photorefractive effect in $Sn_2P_2S_6$ and the valuable help in preparing this chapter. We are also very indebted to Dr. I. Stoika and Dr. M. Gurzan for the development of the technology and the growth of excellent $Sn_2P_2S_6$ crystals. A.A. Grabar and Yu.M. Vysochanskii acknowledge partial support by a grant from the Swiss National Science Foundation.

References

1. C.D. Carpentier, R. Nitsche: *Mat. Res. Bull.* **9**, 401 (1974).
2. C.D. Carpentier, R. Nitsche: *Mat. Res. Bull.* **9**, 1097 (1974).
3. G. Dittmar, H. Schäfer: *Zs. Naturforsch.* **29B**, 312 (1974).
4. M.I. Gurzan, A.P. Buturlakin, V.S. Gerasimenko, N.F. Korda, V. Slivka: *Sov. Phys. Sol. State* **19**, 1794 (1977).
5. A.A. Grabar, Y.M. Vysochanskii, S.I. Perechinskii, L.A. Salo, M.I. Gurzan, V.Y. Slivka: *Sov. Phys. Sol. State* **26**, 2087 (1984).

6. A.A. Grabar, R.I. Muzhikash, A.D. Kostyuk, Yu. M. Vysochanskii: *Sov. Phys. Sol. State* **33**, 1314 (1991).

7. S.G. Odoulov, A.N. Shumelyuk, U. Hellwig, R.A. Rupp, A.A. Grabar: *Opt. Lett.* **21**, 752 (1996).

8. S.G. Odoulov, A.N. Shumelyuk, U. Hellwig, R.A. Rupp, A. Grabar: *J. Opt. Soc. Am. B* **13**, 2352 (1996).

9. R. Ryf, G. Montemezzani, P. Günter, A.A. Grabar, I.M. Stoika, Yu.M. Vysochanskii: *Opt. Lett.* **26**, 1666–1668 (2001).

10. A.A. Grabar, I.V. Kedyk, M.I. Gurzan, I.M. Stoika, A. Molnar, Y.M. Vysochanskii: *Opt. Commun.* **188**, 187 (2001).

11. M. Jazbinšek, G. Montemezzani, P. Günter, A.A. Grabar, I.M. Stoika, Yu.M. Vysochanskii: *J. Opt. Soc. Am. B* **20**, 1241 (2003).

12. A.A. Grabar, I.V. Kedyk, I.M. Stoika, Yu.M. Vysochanskii, M. Jazbinšek, G. Montemezzani, P. Günter: *OSA Trends in Optics and Photonics* **87**, 10 (2003).

13. W. Klingen, G. Eulenberger, H. Hahn: **55**, 229 (1968).

14. W. Brockner, R. Becker: *Zs. Naturforsch.* **42A**, 511 (1987).

15. Y.M. Vysochanskii, V.Y. Slivka: *Ferroelectrics of the Sn$_2$P$_2$S$_6$ family. Properties in vicinity of Lifshitz point* (Oriana-Nova, Lviv, 1994-in Russian).

16. M.M. Maior, V.P. Bovtun, M. Poplavko, Yu.B.M. Koperles, M.I. Gurzan: *Sov. Phys. Sol. State* **26**, 397 (1984).

17. V.N. Zhikharev, Yu.V. Popik, I.D. Seikovskii: *Cryst. Rep.* **41**, 855 (1996).

18. D. Haertle, A. Guarino, J. Hajfler, G. Montemezzani, P. Günter: *Opt. Express* **13**, 2047 (2005).

19. D. Haertle, G. Caimi, A. Haldi, G. Montemezzani, P. Günter, A.A. Grabar, I.M. Stoika, Y.M. Vysochanskii: *Opt. Commun.* **215**, 333 (2003).

20. Yu. M. Vysochanskii, M.I. Gurzan, M.M. Maior, E.D. Rogach, F.I. Savenko, V. Slivka: *Sov. Phys. Cryst.* **35**, 459, Translated from *Kristallografiya*. **35**, 784 (1990).

21. R.M. Yevych, S.I. Perechinskii, A.A. Grabar, Yu.M. Vysochanskii, V.Yu. Slivka: *Cond. Mat. Phys.* **6**, 315 (2003).

22. Yu. M. Vysochanskii, M.M. Maior, V.A. Medvedev, V.M. Rizak, V.Yu. Slivka, E.A. Sorkin: *Sov. Phys. Cryst.* **35**, 541, Translated from *Kristallografiya*. **35**, 918 (1990).

23. A.A. Vasilkevich, Yu.M. Vysochanskii, P.G. Ivanitskii, V.M. Rizak, I.M. Rizak, V.Yu. Slivka, V.I. Slisenko: *Sov. Phys. Solid State* **36**, 660, Translated from *Fiz. Tverd. Tela* **36**, 1205 (1994).

24. K. Al-Shufi, V.M. Rizak, I.M. Rizak, I.P. Prits, Yu.M. Vysochanskii: *Sov. Phys. Solid State* **35**, 1055, Translated from *Fiz. Tverd. Tela* **35**, 2122 (1993).

25. D. Haertle, M. Jazbinšek, G. Montemezzani, P. Günter: *Opt. Express* **13**, 3765 (2005).

26. M.M. Maior, M.I. Gurzan, Sh.B. Molnar, I.P. Prits, Yu.M. Vysochanskii: *IEEE Trans. Ultrason., Ferroelectr. and Freq. Contr.* **47**, 877 (2000).

27. M.M. Maior, Yu.M. Vysochanskii, I.P. Prits, Sh.B. Molnar, L.A. Seikovskaya, V. Slivka: *Sov. Phys. Cryst.* **35**, 767, Translated from *Kristallografiya*. **735**, 1300 (1990).

28. S.L. Bravina, N.V. Morozovsky: *Ferroelectrics* **118**, 217 (1991).

29. X. Bourdon, E. Prouzet, V.B. Cajipe: *J. Sol. State Chem.* **129**, 157 (1997).

30. X. Bourdon, V.B. Cajipe: *J. Sol. State Chem.* **141**, 290 (1998).

31. V. Voroshilov, Yu.M. Vysochanskii, A.A. Grabar, M.V. Potorij, I.P. Prits, V.M. Rizak, L.A. Sejkovskaja, V. Slivka, A.V. Vatsenko: *Ukr. J. Phys.* **35**, 71 (1990).

32. B. Scott, M. Pressprich, R.D. Willet, D.A. Cleary: *J. Sol. State Chem.* **96**, 294 (1992).

33. D.A. Cleary, R.D. Willett, F. Ghebremichael, M.G. Kuzyk: *Sol. State Commun.* **88**, 39 (1993).

34. Y.M. Vysochanskii, V. Slivka, A.P. Buturlakin, M.I. Gurzan, D.V. Chepur: *Sov. Phys. Sol. State* **20**, 49, Translated from *Fizika-Tverdogo-Tela*. **20**, 1990 (1978).
35. Yu.M. Vysochanskii, V.Yu. Slivka, B.M. Koperles, M.I. Gurzan, D.V. Chepur: *Sov. Phys. Sol. State* **21**, 864 (1979).
36. J. Hlinka, I. Gregora, V. Vorlicek: *Phys. Rev. B* **65**, 0643081 (2002).
37. S.W.H. Eijt, R. Currat, J.E. Lorenzo, P. Saint Gregoire, B. Hennion, Y.M. Vysochanskii: *Europ. Phys. J. B* **5**, 169 (1998).
38. M.M. Maior: *Phys. Sol. State* **41**, 1333 (1999).
39. A.I. Ritus, N.S. Roslik, Yu.M. Vysochanskii, A.A. Grabar, V. Slivka: *Sov. Phys. Sol. State* **27**, 1337 (1985).
40. K. Moriya, H. Kuniyoshi, K. Tashita, K. Ozaki, S. Yano, T. Matsuo: *J. Phys. Soc. Jpn.* **67**, 3505 (1998).
41. D. Baltrunas, A.A. Grabar, K. Mazeika, Yu. M. Vysochanskii: *J. Phys.: Cond. Matter* **11**, 2983 (1999).
42. R. Israel, R. de Gelder, J.M.M. Smits, P.T. Beurskens, S.W.H. Eijt, T. Rasing, H. van Kempen, M.M. Major, S.F. Motrija: *Zs. Krist.* **213**, 34 (1998).
43. J. Hlinka, T. Janssen, V. Dvorak: *J. Phys.: Cond. Matt.* **11**, 3209 (1999).
44. K. Kuepper, B. Schneider, V. Caciuc, M. Neumann, A.V. Postnikov, A. Ruediger, A.A. Grabar, Yu.M. Vysochanskii: *Phys. Rev. B* **67**, 115101 (2003).
45. A. Ruediger, O. Schirmer, S. Odoulov, A. Shumelyuk, A. Grabar: *Opt. Mater.* **18**, 123 (2001).
46. A.A. Grabar: *Ferroelectrics* **192**, 155 (1997).
47. Y.W. Cho, S.K. Choi, Y.M. Vysochanskii: *J. Mat. Res.* **16**, 3317 (2001).
48. I. Seres, S. Stepanov, S. Mansurova, A. Grabar: *J. Opt. Soc. Am. B* **17**, 1986 (2000).
49. Yu.M. Vysochanskii, M.M. Maior, S.I. Perechinskii, N.A. Tikhomirova: *Sov. Phys. Cryst.* **37**, 90 (1992).
50. A. Grabar, A. I. Bercha, V.Y. Simchera, I.M. Stoika: *Ferroelectrics* **202**, 211 (1997).
51. A. Grabar, I.V. Kedyk, I.M. Stoika, Yu.M. Vysochanskii: *Ferroelectrics* **254**, 285 (2001).
52. A.A. Bogomolov, V.V. Ivanov, E.L. Burdin, M.M. Mayor: *Ferroelectrics* **164**, 323 (1995).
53. A.A. Bogomolov, A.V. Solnyshkin, O.N. Sergeeva, S.V. Ershov, M.M. Major: *Ferroelectrics* **214**, 125 (1998).
54. V.M. Rizak, M.M. Major, Yu.M. Vysochansky, M.I. Gurzan, V. Slivka: *Ukr. J. Phys.* **33**, 1740 (1988).
55. A.A. Grabar: *Journal of Physics: Cond. Matt.* **10**, 2339 (1998).
56. D.I. Kaynts, A.A. Grabar, I.M. Gurzan, A.A. Horvat: *Ferroelectrics* **304**, 1017 (2004).
57. J. Grigas, V. Kalesinskas, S. Lapinskas, M.I. Gurzan: *Phase Transitions* **12**, 263 (1988).
58. A.A. Volkov, G.V. Kozlov, N.I. Afanas'eva, Yu.M. Vysochanskii, A.A. Grabar, V. Slivka: *Sov. Phys. Sol. State* **25**, 1482 (1983).
59. Yu.M. Vysochanskii, M.I. Gurzan, M.M. Maior, V.M. Rizak, V. Slivka: *Ukr. J. Phys.* **35**, 448 (1990).
60. A.P. Buturlakin, M.I. Gurzan, Yu.V. Slivka: *Sov. Phys. Sol. State* **19**, 1165 (1977).
61. A. Shumelyuk, D. Barilov, S. Odoulov, E. Keatzig: *Appl. Phys. B* **76**, 417 (2003).
62. V.M. Kedyulich, A.G. Slivka, E.I. Gerzanich, P.P. Guranich, V.S. Shusta, P.M. Lucach: *Uzhgorod University Bulletin, Phys. Ser.* **5**, 30 (1999).
63. I.P. Studenyak, V.V. Mitrovcij, G. Kovacs, O.A. Mykajlo, M.I. Gurzan, Yu. M. Vysochanskii: *Ferroelectrics* **254**, 295 (2001).
64. A.G. Slivka, E.I. Gerzanich, V.S. Shusta, P.P. Guranich: *Russ. Phys. J.* **42**, 796 (1999).

65. I. Seres, S. Stepanov, A. Shumelyuk, S. Odoulov, S. Mansurova: *Techn. Dig. Conf. Laser and Electro-Optics (CLEO)*, 104 (2000).
66. I. Seres, S. Stepanov, S. Mansurova, A. Grabar: *J. Opt. Soc. Am. B* **12**, 1986 (2000).
67. S. Odoulov, A. Shumelyuk, G. Brost, K.M. Magde: *Appl. Phys. Lett.* **68**, 3665 (1996).
68. A. Shumelyuk, S. Odoulov, G. Brost: *J. Opt. Soc. Am. B* **7**, 2125 (1998).
69. A. Shumelyuk, D. Barilov, S. Pavlyjuk, S. Odoulov: *Ukr. J. Phys.* **1**, 5 (2003).
70. A. Shumelyuk, S. Pavlyjuk, D. Barilov, S. Odoulov: *OSA Trends in Optics and Photonics* **62**, 601 (2001).
71. A. Shumelyuk, S. Odoulov, Yi Hu, E. Kraetzig, G. Brost: *Techn. Dig. Conf. Laser and Electro-Optics (CLEO)*, 171 (1998).
72. N.V. Kukhtarev, V.B. Markov, S.G. Odoulov, M.S. Soskin, V.L. Vinetskii: *Ferroelectrics* **22**, 949 (1979).
73. G. Montemezzani, M. Zgonik: *Phys. Rev. E* **35**, 1035 (1997).
74. A. Shumelyuk: *Techn. Dig. Conf. Laser and Electro-Optics (CLEO)*, 469 (2001).
75. G.A. Brost, R.A. Motes, J.R. Rotge: *J. Opt. Soc. Am. B* **9**, 1879 (1988).
76. S.G. Odoulov, A.N. Shumelyuk, A.A. Grabar: *Ukr. J. Phys.* **1-2**, 33 (1999).
77. M. Weber, G. von Bally, A. Shumelyuk, S. Odoulov: *Appl. Phys. B* **74**, 29 (2002).
78. M. Weber, F. Rickermann, G. von Bally, A. Shumelyuk, S. Odoulov: *Optik* **111**, 333 (2000).
79. M. Jazbinšek, G. Montemezzani, P. Günter, A.A. Grabar, I.M. Stoika, and Yu.M. Vysochanskii: *OSA Trends in Optics and Photonics* **87**, 190 (2003).
80. M. Jazbinšek, D. Haertle, G. Montemezzani, P. Günter, A.A. Grabar, I.M. Stoika, Yu.M. Vysochanskii: *J. Opt. Soc. Am. B* **22** (2005).
81. A. Shumelyuk, S. Odoulov, D. Kip, E. Kraetzig: *Appl. Phys. B* **72**, 707 (2001).
82. A. Shumelyuk, S. Odoulov, E. Kratzig, D. Kip: *Techn. Dig. Conf. Laser and Electro-Optics*, 69 (2000).
83. T. Woike, T. Wolk, U. Dorfler, R. Pankrath, L. Ivleva, M. Wöhlecke: *Ferroelectrics Lett.* **23**, 127 (1998).
84. A.A. Grabar, M.I. Gurzan, I.V. Kedyk, I.M. Stoika, Y.M. Vysochanskii: *Ferroelectrics* **257**, 245 (2001).
85. G. Montemezzani, P. Rogin, M. Zgonik, P. Günter: *Phys. Rev. B* **49**, 2484–2502 (1994).
86. G. Montemezzani, R. Ryf, D. Haertle, P. Günter, A.A. Grabar, I.M. Stoika, Yu.M. Vysochanskii: *Ukr. J. Phys.* **49**, 333 (2004).
87. M. Cronin-Golomb, B. Fischer, J.O. White, A. Yariv: *IEEE J. Quantum Electron.* **QE-20**, 12 (1984).
88. S. Odoulov, M. Soskin, A. Khizniak: *Optical Oscillators with Degenerate Four-Wave Mixing* (Harwood Academic Publishers, London 1989).
89. J. Feinberg, K. MacDonald: Phase-Conjugate Mirrors and Resonators with Photorefractive Materials. In: *Photorefractive Materials and Their Applications*, vol. 62, ed. by P. Günter, J.-P. Huignard (Springer, Berlin–Heidelberg–New York 1989) pp. 151–204.
90. A. Shumelyuk, S. Odoulov, G. Brost: *Appl. Phys. B* **68**, 959 (1999).
91. A. Shumelyuk, A. Hryhorashchuk, S. Odoulov: *OSA Trends in Optics and Photonics* **87**, 320 (2003).

11

Photorefractive Semiconductors and Quantum-Well Structures

D.D. Nolte, S. Iwamoto, and K. Kuroda

Photorefractive semiconductors have the advantages of high mobility, high-purity, single-crystal growth, controlled doping with defects, suitability to molecular beam epitaxy with quantum-confinement effects, and strong interband absorption. Bulk crystals have small linear electrooptic coefficients, but they have large quadratic coefficients when operating at wavelengths near the interband absorption. Many new aspects of photorefractive behavior have been discovered in semiconductors, including metastable gratings, magnetooptic effects, excitonic electrooptics, and negative differential resistance.

11.1 Bulk Photorefractive Semiconductors

11.1.1 Review of Early Literature

Research into photorefractive semiconductors was initiated separately but nearly simultaneously by Glass [1] and Klein [2] in 1984, followed by a determination of the sign of the dominant photocarrier by Glass in 1985 [3]. The first thorough and systematic study of a photorefractive semiconductor was on GaAs:Cr by Albanese and Steier [4, 5] in 1986 and a short paper on the intensity and temperature dependence of the photorefractive effect in GaAs [6]. Transient and ultrafast work was also begun in this year [7–9], in addition to the first work at 1.3 microns [10]. The first applications of photorefractive semiconductors appeared in 1987 with demonstrations of spatial light modulation [11], image transfer [12], and image processing [13]. Studies of enhancements and optimizations also were initiated in 1987 with studies of moving gratings and AC fields [14, 15], and polarization effects [16–18]. The first work on a photorefractive II-VI material also was published in 1987 [19]. In 1988, approximately 30 papers appeared in publication on various aspects of photorefractive semiconductors, including dynamics [20–28], electric fields [29, 30], absorption gratings [31, 32], cross-polarized coupling [33, 34], and four-wave mixing [25, 35, 36]. Since that time, photorefractive semiconductors have enjoyed steady interest in fundamental photorefractive phenomena as well as in applications.

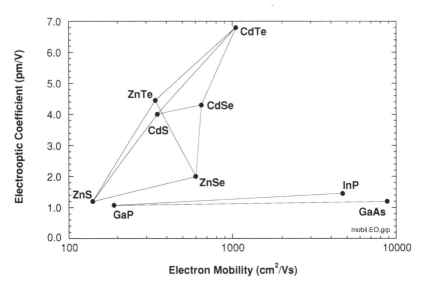

FIGURE 11.1. Electrooptic coefficient plotted against electron mobility for selected photorefractive semiconductors.

11.1.2 Semiconductor Properties

The chief interest in photorefractive semiconductors is the high carrier mobility that can be achieved even in semi-insulating materials. For instance, the electron mobility of GaAs is approximately 9000 cm^2/Vs and for InP is approximately 5000 cm^2/Vs. High mobilities translate to fast operation speeds. In GaAs, for instance, it is possible to attain grating response rates up to one MHz. However, the price paid for high mobility is a relatively small electrooptic coefficient. Both GaAs and InP, the highest-mobility semiconductors, have linear electrooptic coefficients of approximately 1 pm/V. Higher electrooptic coefficients are attainable in the II-VI semiconductors, but at the cost of mobility. Perhaps the best compromise between mobility and electrooptic coefficient is in CdTe. The electrooptic coefficient approaches 7 pm/V while retaining a relatively high electron mobility of 1000 cm^2/Vs. The electrooptic coefficients and mobilities of the most common group III-V and II-VI semiconductors are shown in Figure 11.1.

None of the common photorefractive semiconductors are semi-insulating as grown, except for liquid encapsulated Czochvalski (LEC) GaAs. Therefore, to render them semi-insulating, and hence able to support stored space-charge and space-charge electric fields, they must be doped with defects that produce deep energy levels within the band gap of the material. The most common dopants are the transition metals. The charge states of the transition metals within the III-V semiconductors InP, GaAs, and GaP are shown in Figure 11.2. Note that transition metal defects introduce donor levels lower in the band gap, but acceptor levels that are either resonant with the conduction band, or slightly below the band edge

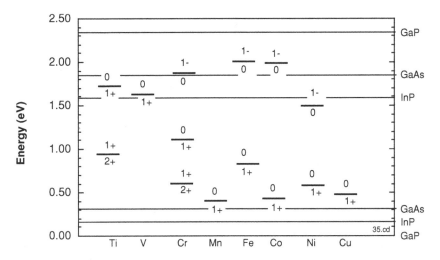

FIGURE 11.2. Transition metal energy levels in GaAs, InP, and GaP shown with the appropriate band-edge alignments.

for the heavier metals. The important metal Cr in GaAs:Cr has two donor levels among three charge states $(0/1 + /2+)$, which produces interesting photochromic effects [37, 38]. The Fe metal ion in InP:Fe has only a single deep donor level, but there is an internal excited state that can have an effect on the dynamic charge equilibrium under illumination [39, 40].

Of particular interest among photorefractive semiconductors is LEC GaAs, which is semi-insulating as grown through the role of the so-called EL2 defect [41–46]. This is likely an arsenic antisite defect (As_{Ga}), which produces two deep donor levels among the three charge states $(0/1+/2+)$. The donor states compensate as-grown shallow acceptors and pin the Fermi level midgap at $E_v + 0.7$ eV. The absence of highly diffusive transition metal dopants makes this material more stable to temperature, and can also lead to more homogeneous material properties. Therefore, this material has long been a mainstay of photorefractive semiconductor applications. Beyond its attractiveness for applications, it also exhibits interesting metastable behavior [47] that will be described in more detail in Section 11.1.4.

11.1.3 Optimization

Because of the small electrooptic coefficients of the semiconductors, considerable effort has been expended to produce the largest possible mixing in semiconductors. Electron–hole competition has long been understood as the principal effect that prevents any photorefractive material from achieving its theoretical maximum mixing efficiency [49, 50]. Complimentary gratings can be trapped on multiple defect levels that "screen" each other [48]. While selective growth and material processing can attempt to control the compensation levels, or remove parasitic

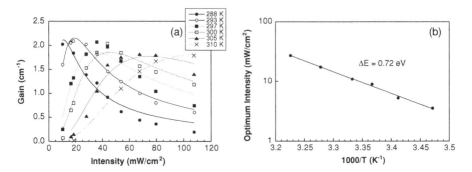

FIGURE 11.3. Demonstration of temperature-intensity resonance in InP. (a) Gain vs. intensity, and (b) optimum intensity vs. temperature [51, 52].

defects, this approach is difficult due to the complex interactions among defects and impurities in semiconductors during high-temperature growth and annealing. Therefore, it is more fruitful to try approaches that attempt to work with the given defect densities while trying to approach the theoretical maximum space-charge fields.

One such successful approach has been found for InP:Fe [51–53]. It was discovered that by balancing the optical emission of one charge type to one band with the thermal emission of the opposite charge type to the opposite band, a "resonance" condition could be set up that produced extremely large space-charge electric fields under DC applied fields. The temperature-intensity resonance in InP:Fe is shown in Figure 11.3. Figure 11.3a shows the photorefractive gain as a function of intensity for a selection of temperatures. Each curve shows a characteristic maximum when the resonance condition is achieved between photo- and thermal emission. The optimum intensity is plotted in Figure 11.3b as a function of $1/T$, showing the characteristic thermal fingerprint of the effect with a thermal activation energy of 0.72 eV, which is characteristic of the thermal emission of holes to the valence band. The resonance condition was extended to applications using moving gratings, AC fields, and band-edge effects [54–57]. An example of frequency-intensity resonance under AC fields in CdTe:V shows gains approaching 10 cm^{-1} achieved at moderate powers (5 mW/cm^2) and moderate fields (10 kV/cm).

11.1.4 Advanced Photorefractive Effects in Semiconductors

Advanced photorefractive effects in semiconductors refer to phenomena that go beyond the conventional properties of defects or refer to unusual conditions of fields and temperatures in photorefractive materials. These include (1) very low temperature (down to 4 K), (2) low-frequency Gunn-domain-like oscillations, (3) metastable defects and gratings, (4) magnetic fields, and (5) band-edge or near-resonance effects.

Deep-level photodiffractive spectroscopy (DLPS) [37, 38, 48] was developed as a defect spectroscopy related to deep-level transient spectroscopy (DLTS) [58, 59], but that concentrated on photorefractive and optical mixing properties of semiconductors. The spectroscopy is performed from liquid helium temperatures up to 400 K during which a nondegenerate four-wave mixing signal is monitored. The concentrations of nonequilibrium charge carriers on defect levels that are unoccupied in the dark, but populated under illumination, can be detected through the nonequilibrium screening of their complementary gratings and hence their effect on the four-wave mixing. In addition, by observing at which temperature the screening turns off, estimates of the level energies could be obtained. The experimental setup uses two laser wavelengths of 1.06 microns to write and 1.33 microns to read gratings in a crystal mounted in a liquid-helium cryostat. Using this system, an interesting low-temperature effect was observed in InP:Fe [60] that was similar to the so-called breathing mode oscillations that occur in semi-insulating GaAs under applied DC fields [61]. These effects are similar to Gunn oscillations but occur at much lower frequencies and only in semi-insulating materials. They appear as current oscillations in the external circuit in response to a DC applied field. Using DLPS, it was shown that the external current oscillations coincided with high-field domains that swept through the crystal. The effect was shown conclusively to be defect-mediated, and it ceased at temperatures above 80 K when the defect level in the lower half of the band gap could no longer support sufficient occupancy.

An important new phenomenon was discovered at low temperatures associated with metastable properties of intrinsic defects, specifically with the so-called EL2 defect in GaAs. It had been known that the EL2 defect could be photoquenched at low temperature in a manner that effectively removed the electrical activity of the defect [43, 62]. If the defect was the dominant compensating level, then photoquenching could render the material conducting by moving the Fermi level close to the valance band edge. If the photoquenching were performed under the spatially inhomogeneous illumination of photorefractive mixing, strong space-charge effects could be anticipated caused by Fermi-level modulation that would far exceed diffusion fields [47, 63, 64].

The nondegenerate four-wave mixing data from a GaAs:EL2 crystal are shown in Figure 11.4a. When the 1.06-micron write-beam illumination is turned on, a large transient diffraction is observed that tails off much more slowly. If the write beams are turned off at the height of the peak, the effect can last indefinitely (if the sample is kept cold), as shown in Figure 11.4b. When the 1.06 beam is turned on again at the end of the experiment, the diffraction efficiency is then quenched. The experimental data are understood through the view of the illumination as a form of modulation doping with coherent light [63, 64]. Fields exceed 2×10^4 V/cm, which is over an order of magnitude larger than possible with diffusion fields at that temperature and for that fringe spacing.

Another metastable defect that belongs to a broad class of defects across many materials in the so-called DX defect. The metastability in this defect is different than for EL2 in that a deep level is optically converted to a shallow level that then

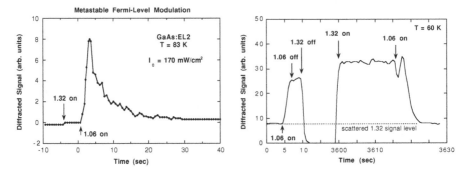

FIGURE 11.4. Fermi-level modulation in GaAs:EL2 showing the diffracted signal as a function of time for continuous illumination (left) and demonstrating optical memory (right) [47].

donates its electron to the conduction band. This metastability can be used during holography to generate plasma gratings [65–67].

Magnetic effects in photorefractive phenomena were investigated for the first time by Nolte et al. in two-wave mixing [68–70] and phase-conjugate mixing geometries [71, 72]. The diluted magnetic semiconductor CdMnTe has the large electrooptic coefficient of CdTe, but has strongly enhanced magnetooptic Faraday rotation for wavelengths near the band edge [73]. The Faraday rotation of the light polarization as it propagates through the cubic crystal produces interesting polarization and orientation effects that had not previously been considered. The two-wave mixing gain as a function of incident polarization is shown in Figure 11.5 for a <110> cut crystal with the grating vector along the <111> axis. The zero-field

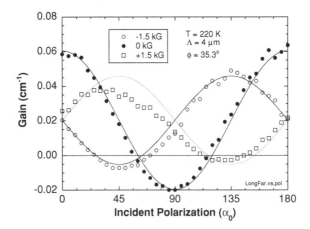

FIGURE 11.5. Gain in CdMnTe as a function of incident polarization for three magnetic field conditions [69].

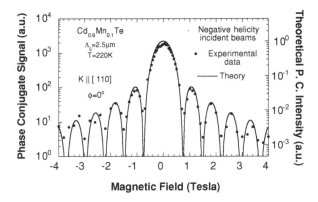

FIGURE 11.6. Phase-conjugate signal quenched with increasing magnetic field in CdM-nTe [74].

gain is shown together with the gains at ±1.5-kG magnetic fields, comparing experiment to theory [69].

An interesting question concerning magnetophotorefractive effects is the role that time-reversal symmetry plays in the generation of time-reversed light. It is well known that magnetic fields remove time-reversal symmetry of a semiconductor. What happens to phase-conjugate time-reversed light under the application of a magnetic field? Part of the answer is shown in Figure 11.6. The phase-conjugate signal for the grating vector aligned along the <110> axis is shown as a function of magnetic field [72]. In the four-wave mixing equations, there can be several contributions to the signal that have different polarizations. In the geometry used in Figure 11.6, one contribution is the time-reversed component, while the other contribution is not time reversed. The interesting point of this experiment is that the time-reversed signal is quenched as the magnetic field deviates from zero, as may have been anticipated intuitively. Conversely, the non-time-reversed signal is insensitive to the magnetic field. This may lead to the tentative conclusion that magnetic fields, as they break time-reversal symmetry, quench time-reversed light but not non-time-reversed light in four-wave mixing. However, this is not the complete story. Using a different polarization geometry, it was discovered that the magnetic field did just the opposite: it quenched the non-time-reversed light, while leaving the time-reversed light unaffected. The more general conclusion is therefore that magnetic fields distinguish between time-reversed and non-time-reversed light [72].

The final advanced photorefractive semiconductor property that is highlighted here is the resonant enhancement of the electrooptic effect through the quadratic effect associated with the direct band-to-band transition. The resonant refractive index change is virtually isotropic, and under sufficient electric field far exceeds the linear index change, especially very near the interband transition. A plot of the absorption change of GaAs in response to electric fields from 2 kV/cm up to 15 kV/cm is shown in Figure 11.7a. The associated refractive index change

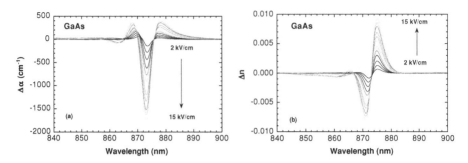

FIGURE 11.7. Quadratic electroabsorption (left) and electrorefraction (right) for a thin film of GaAs.

is shown in Figure 11.7b, obtained through the Kramers–Kronig transformation of the data in Figure 11.7a. Note the strong resonance structure at the bandedge associated with the exciton. In Figure 11.7a the absorption change approaches 2000 cm^{-1} under the highest field that a free surface can support before surface breakdown. Changes in the refractive index approach 0.3% of the total refractive index at the highest fields.

The enhanced quadratic electroabsorption effects decrease as the photon energy falls below the band edge. In this near-resonance regime, significant improvements in photorefractive gain are possible within crystals as thick as 1 mm. For the near-resonant enhancement in GaAs at the highest field of 14 kV/cm the gain exceeds the absorption, allowing net gain. Similar enhancements are seen in InP:Fe [55] using the temperature-intensity resonance as well to maximize the space-charge fields. Net gains approaching 20 cm^{-1} were possible in this crystal, which exceeds the gain in CdTe [57] that was based solely on the linear electrooptic coefficient.

11.2 Photorefractive Semiconductor Heterostructures

11.2.1 Early PRQW Literature Review

The first photorefractive quantum-well devices were fabricated from GaAs/AlGaAs quantum wells and demonstrated at Bell Labs in the summer of 1989 by Nolte, Glass, and Knox using four-wave mixing [75, 76]. These operated with transverse fields that produce the quantum-confined Franz–Keldysh effect (lifetime broadening) of quantum-confined excitons. Second-generation PRQW devices were fabricated at Purdue University in 1990, and were shown to respond to writing photon energies above the band gap [77] in four-wave mixing, and to have rich structure and behavior in two-wave mixing [78, 79]. The first longitudinal PRQW devices were based on CdZnTe/ZnTe multiple quantum wells sandwiched between nonsemiconductor dielectrics in a design similar to the PRIZ and PROM light valves [80–85], but with much thinner active layers of 1–2 microns as opposed to the thick millimeter layers of the light valves [86, 87].

The first detailed theoretical analyses of transport effects and spatial resolution in photorefractive quantum wells appeared in 1992 [88–90], as well as the first use of MBE-grown nonstoichiometric As-rich GaAs as a high-density optical storage medium [91]. The basic PRQW materials had been demonstrated by the end of 1993, including the longitudinal-field GaAs/AlGaAs devices [92, 93] and transverse-field InGaAs/GaAs devices [94]. In 1994, ten journal papers appeared on the topic of photorefractive quantum wells that included diffraction of ultrafast fsec optical pulses [95], optical microcavity effects [96, 97], broad-area longitudinal devices [98–100], and theoretical aspects of the devices [101–103].

11.2.2 Quantum-Well Electrooptic Properties

The strongest electrooptic response that can be achieved in a semiconductor arises from quantum confinement within semiconductor heterostructures. By confining electrons and holes within thin layers sandwiched between higher-band-gap materials, the Coulombic binding is enhanced, suppressing the strong thermal ionization that affects the unconfined excitons in GaAs. The classic structure for quantum confinement is a multiple-quantum-well structure with GaAs wells sandwiched between AlGaAs barriers. The GaAs well thickness is typically between 50 Å and 150 Å, which is comparable with the radius of the exciton. Well widths larger than this have weak quantum confinement, while widths smaller than this have increasing inhomogeneous broadening. Typical barrier concentrations tend to be around 30% Al, but significant quantum confinement can still be observed with Al concentration as low as 5%.

The absorbance of a thin film of GaAs is compared with the absorbances of several quantum-well structures with Al contents in the barriers of 5%, 10%, and 30% in Figure 11.8. The quantum confinement produces a shift of the band edge to higher photon energies for stronger confinement. In addition, stronger confinement produces recognizable excitonic resonant absorption peaks. There is almost no distinct excitonic absorption peak in the bulk film, but already at 5% Al content there is a noticeable resonance. By 10% Al content there is clear splitting of the heavy- and light-hole excitons, which becomes pronounced at 30% Al content.

The importance of sharply defined excitonic absorption peaks arises from the process of electroabsorption and the magnitudes of absorption change that can be achieved from the application of an electric field. The amplitude of an excitonic absorption peak is connected with the oscillator strength f of the transition and also with the line width. The peak absorption varies as

$$\alpha_p = f / \Gamma, \tag{11.1}$$

where Γ is the line width of the transition. When an electric field is applied to a quantum-confined exciton, the transition shifts and broadens. In the Franz–Keldysh (FK) effect, broadening is dominant, while in the quantum-confined Stark effect (QCSE) the energy shift is dominant. The maximum change in absorption

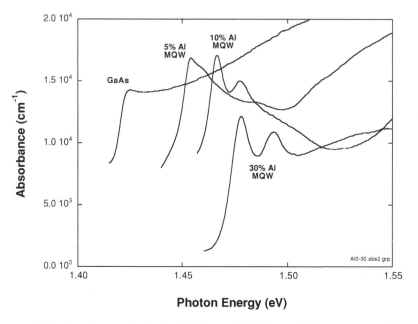

FIGURE 11.8. Absorbance of a GaAs thin film, and AlGaAs multiple-quantum well structures of increasing Al content with 100 wells each.

in response to the applied field in these two cases is given as

$$\Delta\alpha_p = \frac{f}{\Gamma^2}\Delta\Gamma, \quad \Delta\alpha_p = \frac{f}{\Gamma}\Delta E, \tag{11.2}$$

for the Franz–Keldysh and Stark effects respectively. Note that for the FK effect, the maximum electroabsorption depends on the inverse square of the transition line width. Therefore, even slight broadening of the transition (for example from well roughness in the quantum wells) can significantly lower the maximum electroabsorption. In (11.2) both ΔE and $\Delta\Gamma$ depend quadratically on the applied field.

The absorbance data for a quantum well under the FK effect and the QCSE are shown in Figure 11.9. In the FK effect, the electric field ionizes the excitons and broadens them primarily through lifetime broadening. In the QCSE the exciton shifts to lower energies in the band gap. The associated electroabsorption spectra are shown in Figure 11.10 for the FK effect up to 25 kV/cm and the QCSE up to 75 kV/cm. The FK fields are limited by surface breakdown of the samples, while the QCSE fields are limited by bulk breakdown. Fields as high as 150 kV/cm are possible in high-quality QCSE devices, while 25 kV/cm is close to the limit for the FK effect. Through the Kramers–Kronig relationship, any absorption change must have associated with it a refractive index change. The electrorefraction for the two electrooptic effects is shown on the right-hand axis in Figure 11.10. In both cases, the refractive index can be modulated by approximately 1%.

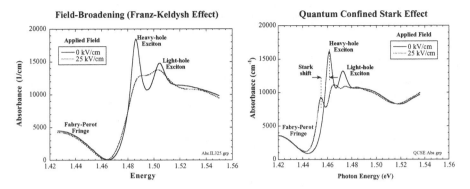

FIGURE 11.9. Franz–Keldysh (left) and quantum-confined Stark effects (right) in quantum-well structures.

The maximum electroabsorption for the FK and QCSE effects are compared in Figure 11.11 as functions of applied field. At low field, the Franz–Keldysh effect is an order of magnitude stronger than the quantum-confined Start effect. This is understood by considering the characteristic energy scales for each effect. The characteristic energy in the FK effect is the exciton binding energy of approximately 10 meV. The characteristic energy in the QCSE is the confinement energy in the quantum well, which can be in the range of 100 meV. It takes a higher field to affect the energies of the excitons when applying the field perpendicular to the well planes than when applying it parallel to the wells. Conversely, the FK effect saturates at lower fields and at lower values than the QCSE. The FK effect begins to saturate above 10–20 kV/cm, while the QCSE begins to saturate only above about 100 kV/cm. Absorption changes approaching 10,000 cm^{-1} are possible with the QCSE, while for the FK effect they usually can reach 6000 cm^{-1}. While the QCSE is ultimately the stronger electrooptic effect, in our experience we find that the fabrication and operation of FK-geometry devices is considerably easier and the devices are more robust. In addition, in practice the FK effect loses only a factor of 2 in peak performance relative to the QCSE. This tradeoff is acceptable

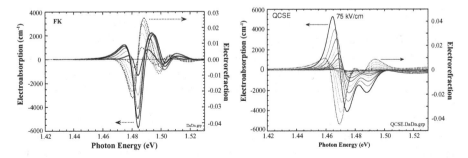

FIGURE 11.10. Electroabsorption and electrorefraction for the Franz–Keldysh effect (left) and the quantum-confined Stark effect (right).

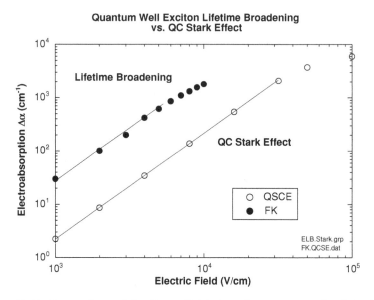

FIGURE 11.11. Comparison of the Franz–Keldysh and quantum–confined Stark electroabsorption as functions of applied electric field.

for most applications, and currently most PRQW devices and applications operate in the FK geometry.

11.2.3 Photorefractive Quantum-Well Classes

The thin-film nature of PRQW devices, combined with the layered semiconductor structures, defines three different PRQW geometries. Light can be incident either from the same side of the film (transmission geometries) or from opposite sides (reflection geometry). The electric field can be applied in the plane of the quantum well (transverse-field geometry) or perpendicular to the quantum-well planes (longitudinal geometries). These device classes are shown in Figure 11.12. There are two transmission geometries and one reflection geometry. There are two longitudinal-field geometries and one transverse-field geometry.

The characteristics of the photorefractive effect can be remarkably different in the different geometries and different from photorefractive effects in most bulk materials. The most striking difference from the well-known bulk photorefractive effects is the longitudinal transmission geometry in which the transport direction and the grating vector are at right angles. This makes this geometry primarily a transient geometry in which either the field or the light is pulsed. This is because the transporting charge spreads sideways and washes out the gratings in most steady-state situations [88]. The longitudinal reflection geometry is also a little unusual in that there may typically be only 10–20 fringes between the contacts, which may be too few to fully describe these gratings as volume

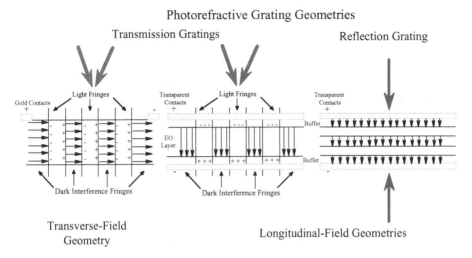

FIGURE 11.12. Photorefractive quantum-well (PRQW) geometries characterized by longitudinal or transverse field directions, and by transmission or reflection gratings.

gratings [104]. The transverse-transmission geometry behaves most like conventional transport photorefractive materials, but even in this geometry, surfaces [101, 105] and carrier nonlinearities [106, 107] have strong effects that produce unique behavior.

11.2.4 Two-Wave and Four-Wave Mixing in Franz–Keldysh PRQW

The main difference in mixing between thin films and bulk materials is the breakdown of Bragg selection in the Raman–Nath diffraction regime. The thin photorefractive gratings act as thin diffraction gratings that produce multiple diffraction orders. Hence, four-wave mixing is most usually performed in a forward-scattering geometry that is never Bragg-matched in the degenerate mixing case, and need not be Bragg-matched in the nondegenerate mixing case. Nevertheless, phase conjugation is still possible in such a forward-scattering geometry, and it has been used as the basis of all-order spectral dispersion compensation of fsec laser pulses [108]. Because of the diffraction fan-out, and because of the many mixtures that are possible (especially at high modulation depths [79]) among even two writing beams, the general situation is one of multiwave mixing. Only when specific effects are isolated do we speak of two-wave and four-wave mixing. Similarly, because electroabsorption and electrorefraction contribute nearly equally to mixing, it is not possible to talk about simple energy transfer between two beams, except for special wavelengths when the electroabsorption is zero. Hence, two-wave mixing is usually discussed in terms of cross-beam modulation.

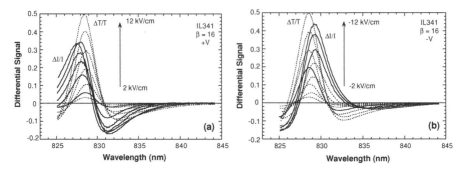

FIGURE 11.13. Cross-beam modulation and differential transmission for positive (a) and negative (b) fields on a transverse-field AlGaAs PRQW device [109].

Typical cross-beam modulation spectra, expressed as $\Delta I/I$ for the change in detected intensity, normalized by the high-field zero-mixing intensity, are shown in Figure 11.13 for both voltage polarities for $\beta = 16$. Also shown for comparison is the differential transmission $\Delta T/T$. The cross-beam modulation is not symmetric between $+V$ and $-V$, and does not track $\Delta T/T$ in any simple way. This is because of the photorefractive phase shift, which is neither 0 nor π, but is field dependent. Therefore at any given wavelength, or any given field, both the electroabsorption and electrorefraction contribute to the cross-beam modulation. From the two-wave mixing data, the phase shift and the internal grating efficiency ξ can be fit. These are shown in Figure 11.14. The internal efficiency is relatively low and field dependent. It is below 40% for all fields. The phase shift depends quadratically on field, increasing from 0 degrees (diffusion fields are negligible for all field values discussed here) and saturating around 70% of $\pi/2$ for fields higher than 6 kV/cm. The field value of 6 kV/cm is an important value in the physics of GaAs, corresponding to the intervalley transfer field associated with the Gunn effect in conducting samples. This is the field at which the electron transport in the

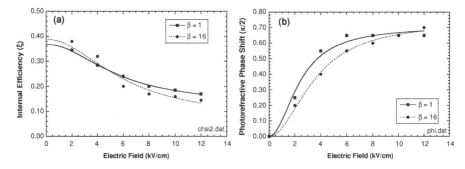

FIGURE 11.14. Internal efficiency (left) and photorefractive phase shift (right) as functions of applied electric field for transverse-transmission PRQW devices [109].

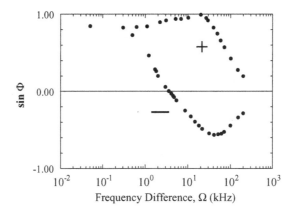

FIGURE 11.15. $\sin \phi$ as a function of the frequency offset between two write beams in a transverse-field device [112].

semiconductor becomes strongly nonlinear. The nonlinear transport is the origin of the strong field-dependent phase shift [106, 110, 111].

The photorefractive phase shift is a dynamic phase shift and is influenced by dynamic processes such as moving gratings. The sine of the phase shift is shown in Figure 11.15 as a function of frequency offset between the two writing beams for both DC field polarities. The characteristic response frequency for this device under these conditions is 20 kHz. When the photorefractive phase shift and the grating lag have the same sign, the phase shift approaches $\pi/2$. Conversely, when they have opposite sign, the phase shift can be driven to zero and even to change sign.

Another important phase associated with the quantum-confined excitons is a phase relating to the causal relationship between electroabsorption and electrorefraction. This so-called excitonic spectral phase is given by

$$\psi(\lambda) = \tan^{-1} \frac{\lambda}{4\pi} \frac{\Delta\alpha(\lambda)}{\Delta n(\lambda)} \qquad (11.3)$$

and is shown in Figure 11.16. The derivative of this phase with respect to angular frequency is approximately equal to the free-induction decay time of the excitons

$$\frac{\partial \psi}{\partial \omega} \approx T_2, \qquad (11.4)$$

which is 200 fsec for this multiple-quantum-well device. This phase is of more than only academic interest, because it contributes in a real manner to the phase of the diffracted field and hence contributes to the relative phase in two-wave mixing and homodyne detection [109].

One of the important features of photorefractive quantum wells is their ultralow-intensity nonlinear optical properties. Despite their fast grating response rates (greater than 1 kHz), they also can operate under extremely low light intensities

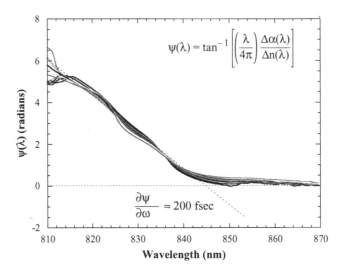

$$\psi(\lambda) = \tan^{-1}\left[\left(\frac{\lambda}{4\pi}\right)\frac{\Delta\alpha(\lambda)}{\Delta n(\lambda)}\right]$$

$$\frac{\partial\psi}{\partial\omega} \approx 200 \text{ fsec}$$

FIGURE 11.16. Excitonic spectral phase defined by the complex electroabsorption and electrorefraction of quantum-confined excitons [109].

because of their very large excitonic absorption. The saturation intensity is

$$I_{\text{sat}} = \frac{n_{\text{d}}}{\alpha}\frac{h\nu}{\tau},\tag{11.5}$$

where n_{d} is the dark carrier density, τ is the carrier lifetimes, α is the absorption, and $h\nu$ is the photon energy. Typical values for n_{d}, τ, and α are 1×10^8 cm^{-3}, 100 psec and 1×10^4 cm^{-1}, which give a saturation intensity in the range of 10 μW/cm^2. The cross-beam coupling measured at two wavelengths, one where only electroabsorption is acting and the other where only electrorefraction is acting, is shown in Figure 11.17 as a function of intensity. The gratings are fully develop for intensities above 10 μW/cm^2 as predicted. The effective n_2 coefficient for the PRQW devices is defined as

$$n_2 = \frac{\Delta n}{I_{\text{sat}}},\tag{11.6}$$

which gives an effective coefficient $n_2 = 3000$ cm^2/W. This is the largest effective n_2 coefficient of any active holographic material. The high sensitivity of these holographic films have made possible the first depth-resolved holography of living tissue [113].

The response time of the photorefractive gratings is another important figure of merit for photorefractive recording. The slowest response rate of a photorefractive quantum well is comparable with that of bulk semiconductors. It is approximately 1 kHz under the lowest intensity. However, the response rate of the quantum wells increases rapidly with increasing intensity, as shown in Figure 11.18a, depending linearly on intensity up to at least 5 mW/cm^2. At 5 mW/cm^2, the response rate is

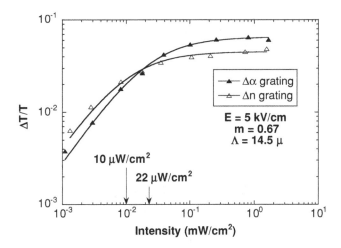

FIGURE 11.17. Intensity-dependent cross-beam modulation demonstrating the ultralow-intensity response of an AlGaAs PRQW device [114].

100 kHz, and the response time is approximately 2 microseconds at low fields. The increase in response time with increasing field is not as large as predicted from linear transport. The field dependence of the response time is shown in Figure 11.18b up to 12 kV/cm. The response time for linear transport is expected to rise quadratically with field. However, when nonlinear transport is considered, the response time remains small and not strongly field dependent. This is important for applications because the devices do not exhibit high-field slowing as is typical for many bulk photorefractive materials.

11.2.5 Microcavity Effects

The thin-film structure of the photorefractive quantum-well devices make them ideal candidates for microcavity enhancements [115, 116]. Semiconductor

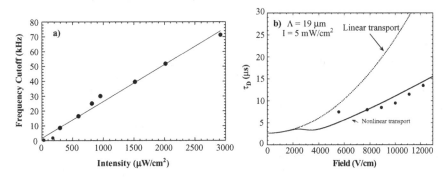

FIGURE 11.18. Response frequency and response time for transverse-field AlGaAs PRQW devices.

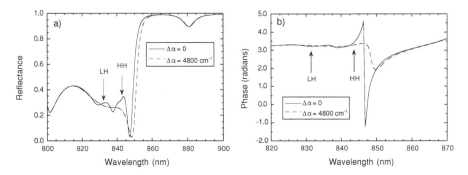

FIGURE 11.19. Calculated reflectance and phase for a photorefractive asymmetric Fabry–Perot structure under zero and maximum field [97].

microcavities are resonant Fabry–Perot structures that balance multiple passes with absorption per pass to increase the effective interaction length and hence the diffraction efficiency. Even simple PRQW devices attached to a glass sub-strate show mild Fabry–Perot enhancements when the semiconductor thickness is carefully controlled and the direction of the incident light is chosen appropriately [96]. Much stronger enhancements are possible when the PRQW is placed on a high-reflector substrate, placing the device into the balanced condition of an asymmetric Fabry–Perot structure [117–120].

The photorefractive properties of PRQW asymmetric Fabry–Perot structures (ASFP) were studied in detail as functions of device thickness and wave-length [97]. The calculated reflectance of a photorefractive ASFP is shown in Figure 11.19a as a function of wavelength at the optimal device thickness that bal-ances the amplitudes from the top and bottom surfaces of the semiconductor film. The reflection minimum occurs on the long-wavelength side of the heavy-hole exciton. The associated phase is shown in Figure 11.19b. For the Franz–Keldysh effect, a phase modulation of π is possible. The calculated reflected diffraction orders are shown in Figure 11.20. The solid curve is the high-field reflectance with no mixing, and the 0-order curve is the reflectance during mixing. The pres-ence of the π-phase grating quenches the reflectance. The first-order diffraction efficiency under these conditions is 1.5% into each of the first orders.

ASFP photorefractive devices were demonstrated experimentally by depositing a dielectric stack on a PRQW device to act as the high reflector [121]. The experimental results are shown in Figure 11.21. The output diffraction efficiency (defined as the ratio of diffracted to transmitted intensity) is plotted for three devices as functions of wavelength. The maximum output diffraction efficiency reached 200%. The input diffraction efficiencies (defined as the diffracted-to-incident intensities) reached 0.36% in one device operating relatively far below the heavy-hole exciton transition.

The concept of an asymmetric Fabry–Perot structure, with no dielectric layers on one side and a high reflector on the other, is easily extended to a resonant microcavity with reflectors on both sides. This increases the multiple passes and

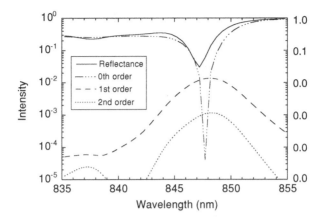

FIGURE 11.20. Calculated diffracted intensities in the diffraction orders corresponding to mixing under the conditions of Figure 11.19.

effective interaction length to provide maximum diffraction efficiency. Furthermore, by considering high-intensity pumping from the write beams, laser gain is achievable [122]. Calculations of the reflectance and diffraction efficiency of a semiconductor microcavity that is operating under amplified spontaneous emission is shown in Figure 11.22 as a function of wavelength for different cavity thicknesses that move the Fabry–Perot resonance through the band edge. Diffraction efficiencies should easily reach 10% under this condition, but the efficiency

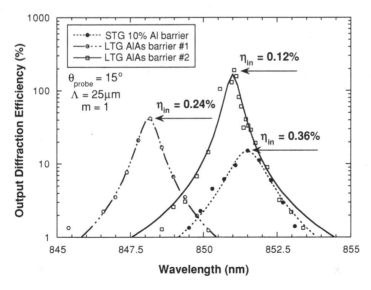

FIGURE 11.21. Experimental output diffraction efficiency for several photorefractive ASFP devices with their associated input diffraction efficiencies [121].

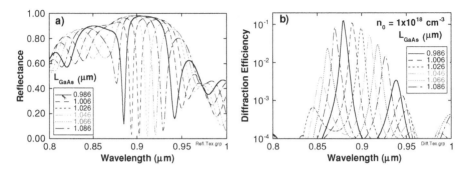

FIGURE 11.22. Calculated reflectance and diffraction efficiency for mixing in moderate-Q microcavities [122].

is highly sensitive to the thickness of the active layer. The operation of such a device shows interesting dynamics, including mode pulling as a function of carrier density, as shown in Figure 11.23a. The free-carrier condition that closely simulates a Franz–Keldysh effect in a PRQW is the density of 3×10^{17} cm^{-3} with a peak input diffraction efficiency of 2%. Experimental results from a holographic VCSEL structure that operates without net laser gain is shown in Figure 11.23b for comparison. The qualitative trends, including the mode pulling, are in good agreement.

It is important to consider why, in a resonant structure with multiple pass geometry, the diffraction efficiency is still only in the range of several percent and not much higher. This is due to the intrinsic absorptive nonlinearity on which the photorefractive quantum wells are based. In principle, no absorption-based holography can exceed approximately 4% diffraction efficiency. Even with multiple-pass microcavity geometries, this remains a general rule of thumb. The only means out of this limitation for semiconductor films is to operate under net laser gain (the simulations in Figures 11.22 and 11.23 have material gain, but not enough to

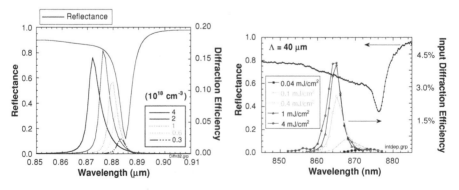

FIGURE 11.23. Diffraction efficiency as a function of excitation, showing mode pulling for a simulated structure (left) and experimentally (right) [122].

overcome cavity losses). To achieve net gain, cavities with much higher finesse are required, but these become much more sensitive to inhomogeneous mode pulling (under the inhomogeneous illumination of the write beams). Calculations show that a holographic VCSEL operating in the condition of net laser gain is limited to 30% diffraction efficiency. This may represent the fundamental limit of any exciton-based semiconductor dynamic holographic medium.

11.2.6 Devices at 1.06-Micron Wavelength

InGaAs/GaAs multiple quantum wells have an excitonic resonance at wavelengths longer than 0.9 μm [123, 124]. The radiation of the Nd:YAG laser belongs to this spectral region. Photorefractive MQW devices operating at 1064 nm would be quite useful for many applications [125]. For tailoring the excitonic transition to be resonant with a wavelength around 1064 nm, the fraction of indium in an InGaAs well should be greater than 0.2 [126], and was 0.25 in the experiment described below. With such a high indium fraction, however, a large lattice mismatch between InGaAs and GaAs causes a serious problem for crystal growth. To solve this problem, a strain-relief buffer layer was designed to have the weighted average lattice constant of the quantum-well and barrier layers so that the strain-related energy stored in the layers is minimized, and the strain is not accumulated in the thick MQW [127, 128].

Samples were grown by metal-organic vapor phase epitaxy on a Cr-doped semi-insulating GaAs substrate. The MQW consisted of 100 periods of 10-nm-thick InGaAs wells and 5-nm-thick GaAs barriers. Before the MQW was grown, a 200-nm-thick GaAs buffer, 100 periods of InGaAs(2nm)/GaAs(2nm) superlattice buffer, and a 600-nm-thick InGaAs strain-relief buffer were grown on the substrate. The fraction of indium in the strain-relief buffer was 0.164, which was adjusted to fit the lattice constant to the average lattice constant of the MQW. After the growth, the MQW was made semi-insulating by the irradiation of 3-MeV protons. The proton dose was 10^{15} cm^{-2}. Finally, Au-Ge alloy electrodes were deposited with a 1-mm spacing on the top of surface. The substrate was not removed because it is transparent to 1064-nm light.

The optical density and the differential transmission of a sample are shown in Figure 11.24. The excitonic resonance is clearly observed at around 1064 nm. Figure 11.25 shows the electroabsorption $\Delta\alpha$ under an electric field of 12 kV/cm. The electrorefraction Δn, derived from $\Delta\alpha$ using the Kramers–Kronig relation, is also plotted in Figure 11.25.

The photorefractive effects were studied by four-wave mixing in the Franz–Keldysh geometry using a Nd:YAG laser. In Figure 11.26, the diffraction efficiency η is plotted as a function of applied electric field under the conditions of a grating period $\Lambda = 8.2$ μm, and a total incident intensity $I = 6$ mW/cm^2. At an external electric field below 5 kV/cm, the diffraction efficiency is proportional to the fourth power of the applied field, as expected [114], and then it saturates at higher electric field. The differential transmission $\Delta T/T$ is also plotted in Figure 11.26. No saturation is observed in the differential transmission. The

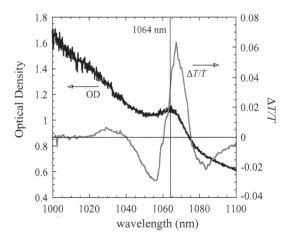

FIGURE 11.24. Absorption and differential transmission spectra of InGaAs/GaAs MQW designed for 1064 nm lasers.

saturation of the diffraction efficiency can be attributed to the nonlinear transport of electrons.

The photorefractive response time is plotted as a function of an applied electric field for incident intensities of 2.8, 21, 42, 88, and 375 mW/cm^2 in Figure 11.27. The response time decreases as the intensity increases, as expected. The field dependence of the response time is strongly affected by the transport mechanism of carriers. If the transport is linear in the electric field, the theory predicts a

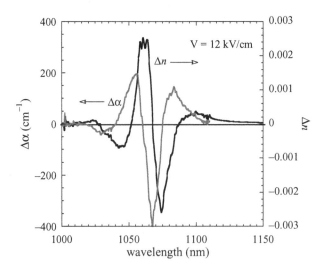

FIGURE 11.25. Electroabsorption $\Delta\alpha$ and electrorefraction Δn spectra of InGaAs/GaAs MQW designed for 1064-nm lasers under an electric field of 12 kV/cm.

FIGURE 11.26. Diffraction efficiency η and the differential transmission $\Delta T / T$ as a function of an applied field.

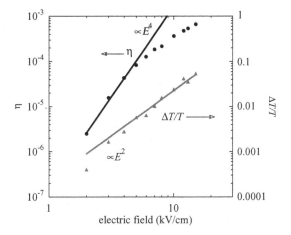

FIGURE 11.27. Photorefractive response time as a function of electric field for various incident intensity of light.

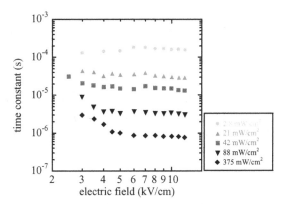

quadratic dependence (slowdown) of the response time on the external field. When the transport is nonlinear, however, the response time does not depend on the field strongly (see Figure 11.18). As shown in Figure 11.27, the response time is almost constant with field at low intensities, and it decreases with increasing field at higher intensities.

References

1. A.M. Glass, A.M. Johnson, D.H. Olson, W. Simpson, and A.A. Ballman. *Appl. Phys. Lett.* **44**, 948–950, 1984.
2. M.B. Klein. *Opt. Lett.* **9**, 350, 1984.
3. A.M. Glass, M.B. Klein, and G.C. Valley. *Electron. Lett.* **21**, 220, 1985.
4. G. Albanese, J. Kumar, and W.H. Steier. *Opt. Lett.* **11**, 650–652, 1986.

5. G. Albanese, J. Kumar, and W.H. Steier. *J. Opt. Soc. Am. a-Optics Image Science and Vision* **3**, P34–P34, 1986.
6. L.J. Cheng and A. Partovi. *Appl. Phys. Lett.* **49**, 1456, 1986.
7. J. Strait and A.M. Glass. *J. Opt. Soc. Am. B* **3**, 342–344, 1986.
8. J. Strait and A.M. Glass. *J. Opt. Soc. Am. B-Optical Physics* **3**, 342–344, 1986.
9. G.C. Valley, A.L. Smirl, M.B. Klein, K. Bohnert, and F. Boggess. *Opt. Lett.* **11**, 647–649, 1986.
10. J. Strait and A.M. Glass. *Appl. Opt.* **25**, 338–339, 1986.
11. L.J. Cheng, G. Gheen, T.H. Chao, H.K. Liu, A. Partovi, J. Katz, and E.M. Garmire. *Opt. Lett.* **12**, 705–707, 1987.
12. L.J. Cheng, G. Gheen, M.F. Rau, and F.C. Wang. *J. Appl. Phys.* **62**, 3991–3993, 1987.
13. G. Gheen and L.J. Cheng. *Appl. Phys. Lett.* **51**, 1481–1483, 1987.
14. J. Kumar, G. Albanese, and W.H. Steier. *Opt. Commun.* **63**, 191–193, 1987.
15. J. Kumar, G. Albanese, W.H. Steier, and M. Ziari. *Opt. Lett.* **12**, 120–122, 1987.
16. A. Partovi, E.M. Garmire, and L.J. Cheng. *Appl. Phys. Lett.* **51**, 299–301, 1987.
17. K. Walsh, T.J. Hall, and R.E. Burge. *Opt. Lett.* **12**, 1026–1028, 1987.
18. P. Yeh. *J. Opt. Soc. Am. B-Optical Physics* **4**, 1382–1386, 1987.
19. R.B. Bylsma, P.M. Bridengaugh, D.H. Olson, and A.M. Glass. *Appl. Phys. Lett.* **51**, 889–891, 1987.
20. V.N. Astratov, A.V. Ilinskii, and A.S. Furman. *Physica Status Solidi B* **150**, 611–615, 1988.
21. L.J. Cheng and A. Partovi. *Appl. Opt.* **27**, 1760–1763, 1988.
22. J. Dubard, A.L. Smirl, A.G. Cui, G.C. Valley, and T.F. Boggess. *Physica Status Solidi B-Basic Research* **150**, 913–919, 1988.
23. J.C. Fabre, J.M.C. Jonathan, and G. Roosen. *J. Opt. Soc. Am. B-Optical Physics* **5**, 1730–1736, 1988.
24. B. Imbert, H. Rajbenbach, S. Mallick, J.P. Herriau, and J.P. Huignard. *Opt. Lett.* **13**, 327–329, 1988.
25. I.C. Khoo and R. Normandin. *Appl. Phys. Lett.* **52**, 525–527, 1988.
26. A.L. Smirl, G.C. Valley, K.M. Bohnert, and T.F. Boggess. *IEEE J. Quant. Electron.* **24**, 289–303, 1988.
27. W.H. Steier, J. Kumar, and M. Ziari. *Appl. Phys. Lett.* **53**, 840, 1988.
28. G.C. Valley and A.L. Smirl. *IEEE J. Quant. Electron.* **24**, 304–310, 1988.
29. D.T.H. Liu, L.J. Cheng, M.F. Rau, and F.C. Wang. *Appl. Phys. Lett.* **53**, 1369–1371, 1988.
30. B. Mainguet. *Opt. Lett.* **13**, 657–659, 1988.
31. R.B. Bylsma, D.H. Olson, and A.M. Glass. *Opt. Lett.* **13**, 853, 1988.
32. L.M. Walpita. *J. Appl. Phys.* **63**, 5495–5499, 1988.
33. T.Y. Chang, A.E. Chiou, and P. Yeh. *J. Opt. Soc. Am. B-Optical Physics* **5**, 1724–1729, 1988.
34. L.J. Cheng and P. Yeh. *Opt. Lett.* **13**, 50–52, 1988.
35. B. Fischer and S. Weiss. *Appl. Phys. Lett.* **53**, 257–259, 1988.
36. M.B. Klein, S.W. McCahon, T.F. Boggess, and G.C. Valley. *J. Opt. Soc. Am. B-Optical Physics* **5**, 2467–2472, 1988.
37. D.D. Nolte, D.H. Olson, and A.M. Glass. *Appl. Phys. Lett.* **56**, 163–165, 1990.
38. D.D. Nolte and A.M. Glass. *Opt. Quant. Electron.* **22**, S47–S60, 1990.

39. P. Delaye, P.U. Halter, and G. Roosen. *Appl. Phys. Lett.* **57**, 360–362, 1990.
40. P. Delaye, P.U. Halter, and G. Roosen. *J. Opt. Soc. Am. B* **7**, 2268–2273, 1990.
41. A.D. Jonath, E. Voronkov, and R.H. Bube. *J. Appl. Phys.* **46**, 1754–1766, 1975.
42. G. Bastide, G. Sagnes, and C. Merlet. *Revue Phys. Appl.* **15**, 1517–1520, 1980.
43. G. Vincent, D. Bois, and A. Chantre. *J. Appl. Phys.* **53**, 3643–3649, 1982.
44. B.K. Meyer, J.-M. Spaeth, and M. Scheffler. *Phys. Rev. Lett.* **52**, 851–854, 1984.
45. M. Kaminska, M. Skowronski, and W. Kuszko. *Phys. Rev. Lett.* **55**, 2204–2207, 1985.
46. J. Lagowski, D.G. Lin, T.-P. Chen, M. Skowronski, and H.C. Gatos. *Appl. Phys. Lett.* **47**, 929–931, 1985.
47. D.D. Nolte, D.H. Olson, and A.M. Glass. *Phys. Rev.* **B40**, 10650–10652, 1989.
48. D.D. Nolte, D.H. Olson, and A.M. Glass. *Phys. Rev. Lett.* **63**, 891–894, 1989.
49. R. Orlowski and E. Krätzig. *Solid State Commun.* **27**, 1351, 1978.
50. F.P. Strohkendl, J.M.C. Jonathon, and R.W. Hellwarth. *Opt. Lett.* **11**, 312, 1986.
51. P. Gravey, G. Picoli, and J.Y. Labandibar. *Opt. Commun.* **70**, 190–194, 1989.
52. G. Picoli, P. Gravey, C. Ozkul, and V. Vieux. *J. Appl. Phys.* **66**, 3798, 1989.
53. G. Picoli, P. Gravey, and C. Ozkul. *Opt. Lett.* **14**, 1362, 1989.
54. B. Mainguet, F.L. Guiner, and G. Picoli. *Opt. Lett.* **15**, 938, 1990.
55. J.E. Millerd, S.D. Koehler, E.M. Garmire, A. Partovi, A.M. Glass, and M.B. Klein. *Appl. Phys. Lett.* **57**, 2776–2778, 1990.
56. J. Millerd, E. Garmire, and M.B. Klein. *Opt. Lett.* **17**, 100, 1992.
57. J.Y. Moisan, N. Wolffer, O. Moine, P. Gravey, G. Martel, A. Aoudia, E. Repka, Y. Marfaing, and R. Triboulet. *J. Opt. Soc. Am. B-Optical Physics* **11**, 1655–1667, 1994.
58. D.V. Lang. *J. Appl. Phys.* **45**, 3023–3032, 1974.
59. G.L. Miller, D.V. Lang, and L.C. Kimerling. "Deep level transient spectroscopy," in *Annual Reviews of Materials Science*. Palo Alto: Annual Reviews, 1977, 337–448.
60. D.D. Nolte, D.H. Olson, and A.M. Glass. *J. Appl. Phys.* **68**, 4111, 1990.
61. M. Kaminska, J.M. Parsey, J. Lagowski, and H.C. Gatos. *Appl. Phys. Lett.* **41**, 989, 1982.
62. M. Kaminska, M. Skowronski, J. lagowski, J.M. Parsey, and H.C. Gatos. *Appl. Phys. Lett.* **43**, 302–304, 1983.
63. D.D. Nolte and R.S. Rana. "Modulation Doping with Coherent Photons," presented at 20th Int. Conf. Phys. Semicond., Thessaloniki, Greece, 1990.
64. D.D. Nolte. *J. Appl. Phys.* **79**, 7514, 1996.
65. R.A. Linke, T. Thio, J. Chadi, and G.E. Devlin. *Appl. Phys. Lett.* **65**, 16, 1994.
66. R.L. MacDonald, R.A. Linke, J.D. Chadi, T. Thio, G.E. Devlin, and P. Becla. *Appl. Phys. Lett.* **19**, 2131, 1994.
67. R.L. MacDonald, R.A. Linke, G.E. Devlin, and M. Mizuta. 1995.
68. G.A. Sefler, E. Oh, R.S. Rana, I. Miotkowski, A.K. Ramdas, and D.D. Nolte. *Opt. Lett.* **17**, 1992.
69. R.S. Rana, E. Oh, K. Chua, A.K. Ramdas, and D.D. Nolte. *Phys. Rev. B* **49**, 7941, 1994.
70. R.S. Rana, E. Oh, K. Chua, A.K. Ramdas, and D.D. Nolte. *J. Lumin.* **60&61**, 56–59, 1994.
71. R.S. Rana, M. Dinu, I. Miotkowski, and D.D. Nolte. *Opt. Lett.* **20**, 1238, 1995.
72. M. Dinu, R.S. Rana, I. Miotkowski, D.D. Nolte, and S. Trivedi. "Magnetic Quenching of Time-Reversed Light in a Diluted Magnetic Semiconductor CdMnTe," presented at 23rd Int. Conf. of the Phys. of Semicond, Berlin, 1996.
73. E. Oh, A.K. Ramdas, and J.K. Furydna. *J. Lumin.* **52**, 183, 1992.
74. M. Dinu, I. Miotkowski, and D.D. Nolte. *Phys. Rev. B* **58**, 10435–10442, 1998.

75. A.M. Glass, D.D. Nolte, D.H. Olson, G.E. Doran, D.S. Chemla, and W.H. Knox. *Opt. Lett.* **15**, 264–266, 1990.
76. D.D. Nolte, D.H. Olson, G.E. Doran, W.H. Knox, and A.M. Glass. *J. Opt. Soc. Am.* **B7**, 2217, 1990.
77. D.D. Nolte, Q.N. Wang, and M.R. Melloch. *Appl. Phys. Lett.* **58**, 2067, 1991.
78. Q.N. Wang, D.D. Nolte, and M.R. Melloch. *Appl. Phys. Lett.* **59**, 256, 1991.
79. Q.N. Wang, D.D. Nolte, and M.R. Melloch. *Opt. Lett.* **16**, 1944, 1991.
80. S.G. Lipson and P. Nisenson. *Appl. Opt.* **13**, 2052, 1974.
81. R.A. Sprague. *J. Appl. Phys.* **46**, 1673, 1975.
82. B.A. Horwitz and F.J. Corbett. *Opt. Eng.* **17**, 353, 1978.
83. D. Casasent, F. Caimi, and A. Khomenko. *Appl. Opt.* **20**, 4215, 1981.
84. M.P. Petrov, A.V. Khomenko, M.V. Krasinkova, V.I. Marakhonov, and M.G. Shlyagin. *Sov. Phys. Tech.Phys.* **26**, 816, 1981.
85. D. Casasent, F. Caimi, M.P. Petrov, and A.V. Khomenko. *Appl. Opt.* **21**, 3846, 1982.
86. A. Partovi, A.M. Glass, D.H. Olson, G.J. Zydzik, K.T. Short, R.D. Feldman, and R.F. Austin. *Appl. Phys. Lett.* **59**, 1832–1834, 1991.
87. A. Partovi, A.M. Glass, D.H. Olson, G.J. Zydzik, K.T. Short, R.D. Feldman, and R.F. Austin. *Opt. Lett.* **17**, 655–657, 1992.
88. D.D. Nolte. *Opt. Commun.* **92**, 199, 1992.
89. M. Carrascosa, F. Agullorueda, and F. Agullolopez. *Applied Physics a-Materials Science & Processing* **55**, 25–29, 1992.
90. Q. Wang, R.M. Brubaker, D.D. Nolte, and M.R. Melloch. *J. Opt. Soc. Am. B-Optical Physics* **9**, 1626–1641, 1992.
91. D.D. Nolte, M.R. Melloch, J.M. Woodall, and S.E. Ralph. *Appl. Phys. Lett.* **61**, 3098, 1992.
92. A. Partovi, A.M. Glass, T.H. Chiu, and D.T.H. Liu. *Opt. Lett.* **18**, 906, 1993.
93. A. Partovi, A.M. Glass, D.H. Olson, G.J. Zydzik, H.M. Obryan, T.H. Chiu, and W.H. Knox. *Appl. Phys. Lett.* **62**, 464–466, 1993.
94. Q.N. Wang, D.D. Nolte, and M.R. Melloch. *J. Appl. Phys.* **74**, 4254–4256, 1993.
95. R.M. Brubaker, Q.N. Wang, D.D. Nolte, E.S. Harmon, and M.R. Melloch. *J. Opt. Soc. Am. B* **11**, 1038–1044, 1994.
96. K.M. Kwolek, M.R. Melloch, and D.D. Nolte. *Appl. Phys. Lett.* **65**, 385, 1994.
97. D.D. Nolte. *Opt. Lett.* **19**, 819–821, 1994.
98. W.S. Rabinovich, S.R. Bowman, R. Mahon, A. Walsh, G. Beadie, C.L. Adler, D.S. Katzer, and K. IkossiAnastasiou. Gray-scale response of multiple-quantum-well spatial light modulators, *J. Opt. Soc. Am.* **B-13**, 2235–2241, 1996.
99. C. De Matos, H. L'Haridon, J.C. Keromnes, G. Ropars, A. Le Corre, P. Gravey, and M. Pugnet, Multiple quantum well optically addressed spatial light modulators operating at 1.55 µm with high diffraction efficiency and high sensitivity, *J. Opt. A* **1**, 286–289, 1999.
100. C.S. Kyono, K. Ikossianastasiou, W.S. Rabinovich, S.R. Bowman, D.S. Katzer, and A.J. Tsao. *Appl. Phys. Lett.* **64**, 2244–2246, 1994.
101. L.F. Magana, F. Agullolopez, and M. Carrascosa. *Revista Mexicana De Fisica* **40**, 499–505, 1994.
102. L.F. Magaña, F. Agullo-Lopez, and M. Carrascosa. *J. Opt. Soc. Am. B* **11**, 1651, 1994.
103. S.L. Smith and L. Hesselink. *J. Opt. Soc. Am. B* **11**, 1878, 1994.
104. D.D. Nolte, I. Lahiri, and M.R. Melloch. *Opt. Lett.* **21**, 1888, 1996.

105. M. Aguilar, M. Carrascosa, F. Agullolopez, and L.F. Magana. *J. Appl. Phys.* **78**, 4840–4844, 1995.

106. R.M. Brubaker, Q.N. Wang, D.D. Nolte, and M.R. Melloch. *Phys. Rev. Lett.* **77**, 4249, 1996.

107. S. Balasubramanian, I. Lahiri, Y. Ding, M.R. Melloch, and D.D. Nolte. *Applied Physics B-Lasers and Optics* **68**, 863–869, 1999.

108. Y. Ding, A.M. Weiner, M.R. Melloch, and D.D. Nolte. *Appl. Phys. Lett.* **75**, 3255, 1999.

109. D.D. Nolte, T. Cubel, L.J. Pyrak-Nolte, and M.R. Melloch. *J. Opt. Soc. Am. B* **18**, 195–205, 2001.

110. Q.N. Wang, R.M. Brubaker, and D.D. Nolte. *Opt. Lett.* **19**, 822–824, 1994.

111. D.D. Nolte, S. Balasubramanian, and M.R. Melloch. *Opt. Mat.* **18**, 199–203, 2001.

112. S. Balasubramanian, I. Lahiri, Y. Ding, M.R. Melloch, and D.D. Nolte. *Appl. Phys. B* **68**, 863–869, 1999.

113. K. Jeong, L. Peng, J.J. Turek, M.R. Melloch, and D.D. Nolte. Fourier-domain Holographic Optical Coherence Imaging of Tumor Spheroids and Mouse Eye, *Appl. Opt.* **44**, 1798–1805, 2005.

114. Q.N. Wang, R.M. Brubaker, D.D. Nolte, and M.R. Melloch. *J. Opt. Soc. Am. B* **9**, 1626–1641, 1992.

115. C. Weisbuch, M. Nishioka, A. Ishikawa, and Y. Arakawa. *Phys. Rev. Lett.* **69**, 3314, 1992.

116. V. Savona, L.C. Andreani, P. Schwendimann, and A. Quattropani. *Sol. St. Commun.* **93**, 733, 1995.

117. M. Whitehead and G. Parry. *Electron. Lett.* **25**, 566–568, 1989.

118. R.H. Yan, R.J. Simes, and L.A. Coldren. *IEEE Phot. Tech. Lett.* **1**, 273, 1989.

119. J.F. Heffernan, M.H. Moloney, J. Hegarty, J.S. Roberts, and M. Whitehead. *Appl. Phys. Lett.* **58**, 2877, 1991.

120. D.S. Gerber, R. Droopad, and G.N. Maracas. *IEEE Phot. Tech. Lett.* **5**, 55, 1993.

121. K.M. Kwolek, M.R. Melloch, D.D. Nolte, and G.A. Brost. *Appl. Phys. Lett.* **67**, 736, 1995.

122. D.D. Nolte, K.M. Kwolek, C. Lenox, and B. Streetman. *J. Opt. Soc. Am. B* **18**, 257–263, 2001.

123. S. Iwamoto, H. Kageshima, T. Yuasa, M. Nishioka, T. Someya, Y. Arakawa, K. Fukutani, T. Shimura, and K. Kuroda. *Opt. Lett.* **24**, 321–323, 1999.

124. S. Iwamoto, H. Kageshima, T. Yuasa, M. Nishioka, T. Someya, Y. Arakawa, K. Kukutani, T. Shimura, and K. Kuroda. *J. Appl. Phys.* **89**, 5889, 2001.

125. S. Iwamoto, S. Taketomi, H. Kageshima, M. Nishioka, T. Someya, Y. Arakawa, K. Fukutani, T. Shimura, and K. Kuroda. *Opt. Lett.* **26**, 22–24, 1999.

126. S. Niki, C.L. Lin, W.S.C. Chang, and H.H. Wieder. *Appl. Phys. Lett.* **55**, 1339, 1989.

127. I.J. Fritz, D.R. Myers, G.A. Vawter, T.M. Brennan, and B.E. Hammons. *Appl. Phys. Lett.* **58**, 1608, 1991.

128. T.K. Woodward, T.S. II, D.L. Sivco, and A.Y. Cho. *Appl. Phys. Lett.* **57**, 548, 1990.

12

Recent Progress in Semiconductor Photorefractive Crystals

Konstantin Shcherbin

Institute of Physics, National Academy of Sciences, Prospect Nauki 46, 03650 Kiev (Ukraine) **kshcherb@iop.kiev.ua**

This chapter is meant to supplement previous extended reviews on photorefractive semiconductors. The emphasis is on enhancement of the photorefractive response with no external field. The advantages and a brief history of the study of photorefractive semiconductors are presented. Optimization of two-beam coupling and diffraction efficiency in cubic semiconductors by appropriate choice of the crystal orientation and polarization of the light beams is considered. Enhancement of the refractive index modulation by proper selection of the recording wavelength is demonstrated. It is shown that an understanding of the lightinduced charge transfer processes in semiconductors allows improvement of the photorefractive properties by auxiliary illumination at an appropriate wavelength. Selected applications of photorefractive semiconductors are demonstrated for phase conjugation, all-optical interconnect, and laser-based ultrasonics receiver. Generation of spatial subharmonics recently achieved in photorefractive semiconductors is also discussed.

12.1 Introduction

Photorefractive semiconductor crystals such as GaAs, CdTe, and InP are sensitive in the near infrared. Therefore these materials are well suited to the output radiation of many commercial solid-state and semiconductor lasers emitting in the near infrared and to the optical fibers possessing smallest dispersion and minimum losses just in this region of the spectrum. Another important advantage of semiconductors is the short response time as compared to the wide-band-gap photorefractive crystals. These features make bulk photorefractive semiconductors promising materials for fast holographic processing of optical data in the infrared region of the spectrum with telecommunications components available at present [1–3].

Smaller electrooptic constants of semiconductors leaking the symmetry center, as compared to those of ferroelectric crystals, result in smaller refractive index change and therefore in smaller gain and diffraction efficiency. To minimize these limitations the attention of researchers was focused initially on cadmium telluride [4] as a medium with the largest known electrooptic constant [2, 5, 6] among semiconductor crystals with lesser electrooptic effect. The experimental photorefractive recording in semiconductors was accomplished first, however, with GaAs and InP [7, 8] in 1984, while the first recording in CdTe was reported only in 1987 for vanadium-doped crystals [9]. Later, the photorefractive effect was observed in germanium-doped [10], vanadium-manganese-doped [11] and tin-doped [12] cadmium telluride. Germanium- and vanadium-doped crystals have demonstrated up to now the best photorefractive performance. That is why they are studied more carefully than other types.

Being the largest among all semiconductors, the electrooptic constant of cadmium telluride is still small to ensure the diffraction efficiency necessary for many practical applications. That is why different techniques are used to increase refractive index modulation. The methods well known for sillenites, such as moving grating recording in the dc-biased crystal [13, 14] or the recording of stationary grating in the presence of an ac field [14, 15] were also implemented with semiconductors [16–20]. Simultaneously, new techniques were developed that result from the distinctive features of semiconductors. They include temperature-intensity resonance of two-beam coupling in InP:Fe [21–23], near-band-edge enhancement of the photorefractive effect [24–26], and grating strengthening by means of auxiliary illumination [27, 28]. Using different enhancement techniques, the applications of semiconductors have been demonstrated for optical phase conjugation [29–31], design of optical interconnects [32, 33], laser-based ultrasonics detection with adaptive interferometers [34–39], etc. Studies of photorefractive properties have released in semiconductor crystals spectacular physical processes such as recording of metastable optical gratings [40–42], magneto-photorefractive phenomena in solid solutions of CdMnTe [43–45], photorefractive recording in quantum wells [46–50], generation of spatial subharmonics [51], all-optical beam steering, and optically controlled formation of solitons [52–54]. The shaping and stabilization of laser spectra was demonstrated with a photorefractive semiconductor used as an intracavity spectral filter [55–57].

Most of these techniques and related physical processes have been discussed in comprehensive books and review articles [1–3, 14, 58–60] including the chapter written by D.D. Nolte et al. for this volume [61]. In the present chapter, several ways of enhancement of the photorefractive response are described that are not related to the external field. This includes optimization of the recording geometry, selection of the wavelength, and optical sensitization of semiconductor for photorefractive recording. It is shown that an understanding of the photoinduced charge transfer processes in semiconductors allows for grating enhancement by auxiliary illumination at appropriate wavelengths. Selected applications of optimized semiconductors for phase conjugation, all-optical interconnect, and laser-based ultrasonic detection are demonstrated. The phenomena related to excitation of the space-charge waves, e.g., spatial subharmonics generation, recently achieved in photorefractive semiconductors are discussed.

12.2 Optimization of Semiconductors for Photorefraction

The refractive index grating in photorefractive crystals recorded via diffusion-driven transport (with no external field) is $\pi/2$-shifted with respect to the interference pattern formed by the recording beams [62]. The nonlinear response of this type leads to steady-state energy transfer from one beam to the other in the direction determined by crystal orientation and sign of the charge carriers. In undepleted pump-beam approximation the intensity of the transmitted signal beam is given by

$$I_S = I_{S0} exp\,(\Gamma\ell - \alpha\ell), \tag{12.1}$$

where I_{S0} is the input signal intensity, ℓ is the interaction length, Γ is the gain factor, and α is the absorption constant. The gain factor is directly proportional to the refractive index modulation Δn:

$$\Gamma = \frac{4\pi\,\Delta n}{\lambda\cos\theta'} = \frac{2\pi\,n_0^3 r_{eff} E_{SC}}{m\lambda\cos\theta'}, \tag{12.2}$$

where λ is the wavelength, θ' is the Bragg angle inside the crystal, $\Delta n = (1/2)n_0^3 r_{eff} E_{SC}/m$, n_0 is the refractive index, r_{eff} is the effective electrooptic constant that depends on the crystal orientation and light polarization, m is the fringe contrast and E_{SC} is the space-charge field resulting from the spatial charge redistribution. For diffusion process of recording, the space-charge field and the gain factor are defined as follows [62]:

$$E_{SC} = m\,\frac{k_B T}{e}\,\frac{\xi}{1 + l_s^2/\Lambda^2}, \tag{12.3}$$

$$\Gamma = \frac{2\pi\,n_0^3 r_{eff}}{\lambda\cos\theta'}\,\frac{k_B T}{e}\,\frac{\xi}{1 + l_s^2/\Lambda^2}, \tag{12.4}$$

where k_B is the Bolzman constant, T is the absolute temperature, e is the electron charge, Λ is the grating spacing, ξ is the electron–hole competition factor that accounts for compensation of the main grating by grating formed with carriers of different sign (this factor ranges from -1 to 1 and in general is spatial-frequency dependent [63, 64]), l_s is the Debye screening length,

$$l_s = \sqrt{\frac{4\pi^2\varepsilon\varepsilon_0 k_B T}{e^2 N_E}}, \tag{12.5}$$

ε and ε_0 are the dielectric constants of the material and vacuum, respectively, and N_E is the effective trap density. For one-center model $N_E = N_D^+(N_D - N_D^+)/N_D$, where N_D and N_D^+ are the full density of donors and the density of ionized donors (traps).

As one can see from equations (12.2) to (12.4), the nonlinear phase shift induced by photorefractive grating is inversely proportional to the wavelength. Therefore the gain factor and diffraction efficiency decrease with wavelength for fixed Δn.

Hence the photorefractive response of semiconductors in the infrared is reduced as compared to the materials with similar characteristics sensitive in the visible region of the spectrum. This unfavorable feature makes any possible improvement of the infrared sensitivity very important.

Equations (12.2)–(12.4) allow one to consider possible ways of photorefractive response enhancement. The refractive index and electrooptic constant are intrinsic characteristics of a crystal and therefore cannot be changed. The proper choice of interaction geometry only may be used to profit from the largest possible effective electrooptic constant. On the other hand, the Debye screening length and the electron–hole competition factor are the characteristics of a particular sample with given content and concentration of photorefractive centers. These properties therefore may be optimized. Various techniques may be used in addition to increase the space-charge field.

The refractive index modulation is proportional to the effective electrooptic constant r_{eff}, which is a component of the electrooptic tensor reduced to new Cartesian basis fitting to recording geometry [65]. The effective electrooptic constant, being a linear combination of the components of the principal electrooptic tensor, may appear to be larger than any component of the standard tensor $\hat{r}$. Therefore one way to optimize photorefractive response is to use recording geometry with the largest possible effective electrooptic constant. Some secondary effects like, e.g., the piezoelectric effect and contribution of absorption grating (see Section 12.2.1) may result in additional enhancement of the gain factor and diffraction efficiency in a specially selected geometry.

The second way to increase the refractive index change lies in enhancement of the space-charge field. The space-charge field depends on several factors. As follows from (12.3), the maximum of the space-charge field is reached at grating spacing $\Lambda = l_s$ for the diffusion process of recording. In the case of an external electric field the maximum is shifted from l_s to a larger grating spacing [13–15], but in both cases the smaller l_s is, the larger is the space-charge modulation that can be reached in the crystal.

The electron–hole competition reduces the space-charge field because the main grating is partially compensated by the grating formed by secondary charge carriers. Therefore, for effective photorefractive recording the crystal should possess a sufficiently large effective trap density, while the electron–hole competition should be small (the electron–hole competition factor modulus should be large). Both these requirements may be met by appropriate selection of the doping materials. The electron–hole competition in addition may be reduced by choosing an appropriate wavelength [66, 67] or by using special techniques of grating recording [21–23, 27, 28].

The external field also may be used to increase E_{SC}. It is not always easy, however, to apply a field of relatively high amplitude and frequency to fast semiconductors with relatively high conductivity of order of 10^{-7}– 10^{-9} $(\Omega$ cm$)^{-1}$. That is why any optimization of crystal properties and recording conditions is important to increase refractive index modulation in semiconductors.

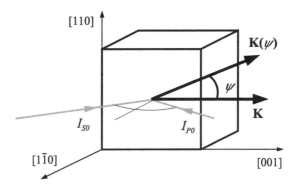

FIGURE 12.1. Two-beam coupling geometry; ψ is the angle between grating vector **K** and the [001]-axis.

12.2.1 Interaction Geometry

The effects of crystal orientation and beam polarization on photorefractive response are usually studied in the two-beam coupling configuration. A so-called "holographic" crystal cut [2, 3] is normally used for two-beam coupling experiments in cubic crystals with the input face parallel to the $(1\bar{1}0)$ crystallographic plane and the side faces corresponding to the (110) and (001) planes (see Figure 12.1). Initially the component r_{41} of the electrooptic tensor was exploited for grating readout in semiconductors [7–9]: Two beams I_{S0} and I_{P0} intersecting at an angle of 2θ in air entered the sample through $(1\bar{1}0)$ plane and recorded grating with grating vector **K** parallel to the [001]-axis (see Figure 12.1). The beams were s-polarized, i.e., along the [110]-direction. An early study of the effect of beam polarization on two-beam coupling constant in semiconductors [68] confirmed that this polarization is optimal for the grating vector aligned along the [001]-axis.

At the same time it was shown for sillenite crystals that neglecting optical activity, the maximum beam coupling is reached in cubic crystals for the grating orientation with **K** ∥ [111] and for light polarized along the same direction [14,69]. The first detailed study of the effect of grating orientation on two-beam coupling in semiconductors was reported for CdTe:Ge [10]. The gain factor was calculated and measured experimentally as a function of the angle ψ between the grating vector **K** and the [001]-direction (see Figure 12.1) for polarization vectors of the recording waves parallel and perpendicular to the plane of incidence. In the experiment, the sample mounted on a special holder was rotated around the normal to the input face and the gain factor was measured. Experimental dependences for s and p-polarized waves are shown by open circles in Figures 12.2a and 12.2b, respectively.

When the crystal is tilted to the angle ψ in the plane $(1\bar{1}0)$, the effective electrooptic constants for s- and p-polarized beams are as follows:

$$r_{eff}^{s} = -\cos\psi(\cos^{2}\psi - 2\sin^{2}\psi)\,r_{41}, \tag{12.6}$$

$$r_{eff}^{p} = -\frac{3}{2}\sin\psi\sin 2\psi\,r_{41}. \tag{12.7}$$

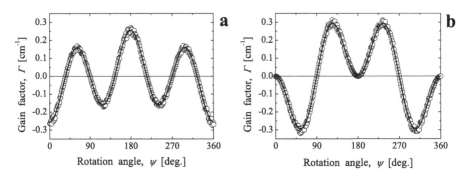

FIGURE 12.2. The gain factor versus angle ψ between the grating vector $\mathbf{K}$ and [001] direction measured in CdTe:Ge at $\lambda = 1.06$ μm. (**a**) recording light is s-polarized, i.e., perpendicularly to the plane of incidence; (**b**) p-polarized light (in the plane of incidence); experiment (circles) and the best fit (lines).

Equation (12.4) can be rewritten as $\Gamma = A(\Lambda) \times r_{eff}(\psi)$ with $A(\Lambda) = \text{const}$ for any particular grating spacing Λ. Solid lines in Figure 12.2 represent the best fit of the theoretical dependence of (12.4) to experimental data with $r_{eff}(\psi)$ given by (12.5), (12.6) and with $A(\Lambda) = 0.26$ as fitting parameter for both dependences. Excellent qualitative agreement between theoretical calculations and experimental data is evident.

For s-polarized waves the maximum of the effective electrooptic constant is equal to r_{41} and corresponds to the traditional recording geometry with $\mathbf{K} \parallel [001]$, while for p-polarization r_{eff}^p reaches $r_{eff}^p = \frac{2}{\sqrt{3}}r_{41}$ at $\mathbf{K} \parallel [111]$. Therefore the latter recording geometry ensures the largest refractive index modulation in cubic semiconductors. This conclusion was confirmed later on with InP:Fe [70].

The photorefractive crystals exhibit the piezoelectric effect as they leak the symmetry center. Therefore the space-charge field leads to elastic deformations, which in turn result in additional refractive index change [71,72]. It was shown that the photoelastic effect is important for ferroelectrics [71] and sillenites [72–75] as well as for semiconductors [76,77]. To consider the contribution of the piezoelectric effect to the refractive index modulation, an additional component should be added to the effective electrooptic constant [72]. The overall electrooptic constant should be rewritten as a sum $(r_{eff} + r_{eff}^{pz})$, where r_{eff}^{pz} is a particular effective electrooptic constant accounting for the piezoelectric contribution. This component also depends on the recording geometry.

One more phenomenon that modifies the beam coupling is a nonlinear absorption. The formation of the nonuniform space charge during grating recording is always accompanied by the appearance of the absorption grating [68, 78–80], since the ionized donor centers have an absorption constant different from that of the filled donor centers. For diffusion recording, the space-charge grating (and thus the absorption grating) is not shifted with respect to the interference fringes. The gain factor for weak unshifted absorption grating is $\Gamma_a = -\Delta\alpha$ [68]. For cubic photorefractive semiconductors with isotropic absorption the amplitude of the absorption grating does not depend on orientation of the grating vector. So, the gain factor from this contribution also does not depend on the grating vector orientation. In other words, the absorption grating does not change the shape of orientational dependences but introduces an offset opposite to the absorption modulation in maxima of interference fringes.

With the piezoelectric effect and nonlinear absorption taken into account, the expression for the gain factor is given by [77]

$$\Gamma = \frac{2\pi n_0^3 (r_{eff} + r_{eff}^{pz})}{\lambda \cos \theta'} \frac{k_B T}{e} \frac{\xi}{1 + l_s^2 / \Lambda^2} - \Delta\alpha. \qquad (12.8)$$

The influence of the piezoelectric effect and light-induced absorption on two-beam coupling in semiconductors was studied for gallium arsenide [76, 77] because this material is widely investigated and different physical characteristics including piezoelectric, elastic stiffness, and elastooptic constants are well known. Experimental dependences of the gain factor on grating vector orientation measured for CaAs:Cr [77] are shown in Figure 12.3 by open squares. The offset of the angular dependences, independent of the angle ψ, indicates the presence of the absorption grating and gives the amplitude of the absorption modulation $\Delta\alpha \approx -0.025$ cm^{-1} (for $\alpha \approx 1.1$ cm^{-1}). The peak-to-peak amplitudes of the dependences give the ultimate values of the refractive index modulation for s- and p-polarized light.

It should be noted that two largest extrema in Figure 12.3a ($\psi = 0°$, $\psi = 180°$) correspond to the traditional orientation with $\mathbf{K} \parallel [001]$ and s-polarization of the recording light. The effective electrooptic constant at this orientation is the handbook constant r_{41} for cubic crystals of the $\bar{4}3m$ point symmetry group. The tabulated value of the electrooptic constant already includes a photoelastic

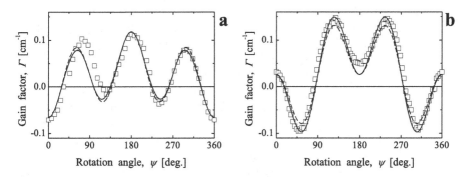

FIGURE 12.3. Gain factor versus the angle ψ between grating vector $\mathbf{K}$ and [001] direction measured in GaAs:Cr at $\lambda = 1.06$ μm. (a) s-polarized light; experiment (squares); best fit of (12.8) to experimental data with (solid line) and without (dashed line) photoelastic contribution. (b) p-polarized light; experiment (squares); calculation with the fitting parameters extracted from the data of Figure 12.3a with (solid line) and without (dashed line) photoelastic contribution.

contribution for this particular geometry and $r_{eff}^{pz}(\psi = 0°, 180°) = 0$. The effect of piezoelectricity is important for the other orientations of the grating vector. Therefore the factor $A(\Lambda)$ can be estimated from the gain factor measured at $\psi = 0°, 180°$ with previously estimated contribution of the absorption grating $\Delta\alpha \approx 0.025$ cm^{-1} taken into account.

The solid line in Figure 12.3a represents the best fit of the theoretical (12.8) to experimental data, taking into account the contribution of the amplitude grating as well as contributions of the photorefractive and photoelastic gratings. The standard values of the physical constants for GaAs collected in [77] are used. The solid line in Figure 12.3b is calculated with the values $\Delta\alpha$ and $A(\Lambda)$ estimated from the experimental data of Figure 12.3a. To show the relative contribution from different effects to the gain factor, the dashed lines in Figure 12.3 represent calculations with the photoelastic contribution omitted and with other fitting parameters the same.

As follows from (12.6) and (12.7), the gain factor should be $2/\sqrt{3} \approx 1.15$ times larger for grating vector and polarization of the light aligned along the [111]-direction (extrema in Figure 12.3b) as compared to the traditional beam-coupling orientation with $\mathbf{K} \parallel [001]$ and s-polarization of the light if the photoelastic contribution is omitted [10]. The improvement observed experimentally is, however, much larger, about 1.35 times. Considering the photoelastic effect perfectly explains this difference (compare dashed and solid curves in Figure 12.3b), thus indicating the importance of the photoelastic contribution for semiconducting GaAs.

As one can see, the absorption grating does not affect the shape of orientational dependences of the gain factor. It results only in the appearance of a permanent pedestal that is independent of the rotation angle. The situation is different for the diffraction efficiency. The dependence of the diffraction efficiency on rotation angle measured in GaAs:Cr for p-polarized light [77] is shown in Figure 12.4 by open squares. The dashed line corresponds to the intensity of the diffracted wave calculated with absorption grating not taken into account. This calculation predicts correctly the positions of the maxima but it cannot explain the obvious difference in their amplitudes. At the same time the calculation considering nonlinear absorption describes well the experimental data. The solid line in Figure 12.4 represents this calculation.

The strong influence of the absorption grating on orientational dependence of the diffraction efficiency is explained by the phase difference between different diffracted components. The diffracted wave may be decomposed into two components: the first component diffracted from the refractive index grating and second one diffracted from the absorption grating. The amplitude and the phase of the wave diffracted from the phase grating changes with rotation angle ψ, while the amplitude and the phase of the component diffracted from the amplitude grating are always the same, since the absorption of the crystal is isotropic. That is why two considered components are added either in phase or out of phase at different angles. The constructive or destructive interference of different contributions to the diffracted wave is the reason for the strong change in the shape of the orientational dependences of the diffraction efficiency [77]. It should be emphasized that the constructive interference may be effectively used

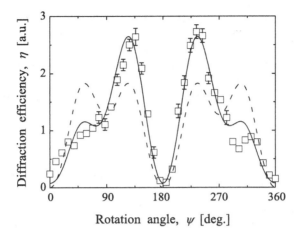

FIGURE 12.4. Diffraction efficiency versus angle ψ between grating vector **K** and [001] direction measured in GaAs:Cr at $\lambda = 1.06$ μm for p-polarized light; experiment (squares); calculation with the fitting parameters extracted from the data of Figure 12.3a with (solid line) and without (dashed line) contribution from absorption grating.

for optimization of the diffraction efficiency even in the case of small nonlinear absorption.

Thus the supplementary contributions from piezoelectric effect and nonlinear absorption may modify strongly photorefractive response in cubic semiconductors. Correct consideration of these contributions allows one to improve the performance of these materials noticeably. Such optimization may be important in the case of small diffraction efficiencies inherent to semiconductors with small electrooptic constants. At the same time, contributions from the secondary gratings are not so important for CdTe with much larger electrooptic constant. A later study shows the presence of light-induced absorption in CdTe (see Section 12.2.3), which is relatively small in many crystals in the vicinity of the operating wavelength $\lambda = 1.06$ μm and therefore does not affect the diffraction efficiency strongly.

Many physical characteristics of cubic semiconductors are isotropic, like the nonlinear absorption mentioned above.

The low-frequency dielectric permittivity is also isotropic; and so the Debye screening length does not depend on the recording geometry and it is constant for any particular sample. Therefore there is no need in optimization of the angle between recording beams at different recording geometry to maximize response by adjustment of grating spacing to the Debye screening length, as it is necessary to do for crystals with lower symmetry.

12.2.2 Optimization of Photorefractive Semiconductors by Appropriate Doping

It is required that photorefractive semiconductor exhibit relatively low dark conductivity to prevent a compensation of the light-induced grating by dark charge carriers. The crystal should also possess a certain amount of donor and trap centers necessary for spatial charge redistribution. Deep centers should lie in the band gap to ensure these properties pinning the Fermi level near the middle of the band gap and supporting photoinduced charge transport. Different intrinsic defects and impurities may serve as such photorefractive centers. Well-known examples of inherent defects that ensure photorefractive recording are the metastable EL2 centers in undoped GaAs crystals [1, 2]. Quite often, however, an appropriate dopant is added to a semiconductor during crystal growth or within a certain aftergrowth treatment. Correct selection of the dopant itself and doping concentration level ensures good photorefractive properties. Photorefractive recording has been reported for differently doped semiconductors such as GaAs:Cr [8, 16, 68], InP:Fe [8, 21], InP:Ti [81], CdTe:V [9], CdTe:Ge [10], CdTe:V,Mn [11], CdTe:Sn [12].

Many of these materials demonstrate high photorefractive performance at specific wavelengths. The high refractive index modulation achieved in the presence of an external electric field clears the way for practical applications [29–39, 58] including an industrial system for laser-based ultrasonics detection with InP:Fe [82]. It should be noted, however, that often it is difficult to apply high voltage to semiconductors because of their relatively high conductivity. The larger electrooptic constant of CdTe translates to a much-reduced requirement for the applied field. In addition, CdTe is the only photorefractive semiconductor known to be sensitive at 1.5 μm [83]. The latter wavelength is important because it corresponds to well-developed telecommunication components and also because 1.5 μm is in the eye-safe wavelength range. These two advantages make CdTe the most promising bulk photorefractive semiconductor for practical applications. Study of differenly doped CdTe crystals indicates that vanadium and germanium are suitable dopants that ensure high photorefractive performance of CdTe.

12.2.3 Electron–Hole Competition in Doped Semiconductors

The main factor limiting the refractive index change in both vanadium- and germanium-doped CdTe is electron–hole competition, which does not allow reaching the ultimate value of the diffraction efficiency and gain factor predicted by the theory. The electron–hole competition is still hardly controlled by doping optimization and therefore causes the reduction of the space-charge field in many crystals. When electron–hole competition is large, both types of carriers are effectively excited. The grating created by the main photocarriers is partially compensated by carriers of the opposite sign. In spite of such compensation one can succeed in finding such conditions of grating recording at which only one type of carrier dominates. In this case 90% of the theoretical limit of the gain factor may be reached, as has been demonstrated for CdTe:Ge [28].

The electron–hole competition manifests itself in different photorefractive properties. For example, the grating decay often is not exponential in crystals with bipolar conductivity. Moreover, for some samples a particular temporal evolution

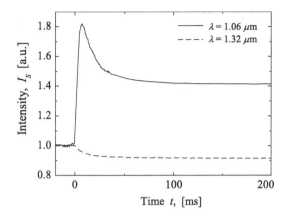

FIGURE 12.5. Temporal behavior of the signal beam intensity in two-beam coupling measured in CdTe:Ge sample at 1.06 μm (solid line) and at 1.32 μm (dashed line), the pump beam is switched on at $t = 0$.

of the signal beam is observed in two-beam coupling experiments [84, 85]. The signal-beam intensity measured in certain CdTe:Ge samples at 1.06 μm is shown by the solid line in Figure 12.5. In the beginning of recording, the intensity increases nearly 2 times in a few milliseconds at the total intensity level of order of 10 mW/cm^2. Then it goes down much more slowly and reaches the steady-state value. The experimental data prove the diffusion charge transport process for the whole period of grating recording. Thus the observed peak is related to the bipolar conductivity of the sample. First the grating is recorded by redistribution of carriers of one sign. Then the second grating is developed by carriers of opposite sign. The two gratings have different amplitudes and different decay times and partially compensate each other because they are out of phase.

The temporal behavior of the signal-beam intensity at 1.32 μm is monotonic (Figure 12.5, dashed line) in the same CdTe:Ge crystal. Moreover, the energy-coupling direction changes compared to that at 1.06 μm. The difference in the steady-state values and in the dynamics of the energy transfer at 1.06 μm and 1.32 μm suggests that the "fast" grating, which is well pronounced at 1.06 μm, disappears at 1.32 μm. The amplitude of the "slow" grating becomes smaller at 1.32 μm. This reduction is expected since the gain factor is inversely proportional to the wavelength and because of the reduction of the electrooptic constant and refractive index for larger wavelengths.

The described experimental observation indicates that two centers and two types of photoexcited carriers are responsible for the charge redistribution in CdTe:Ge at 1.06 μm. Study of photorefractive properties of CdTe:V gives the same conclusion [66]. The photoconductivity in CdTe changes with wavelength, which results in variation of photorefractive response. Even the type of photoconductivity may be different at different wavelengths. A change of main photogenerated carriers causes switching of the direction of energy transfer in two-beam coupling

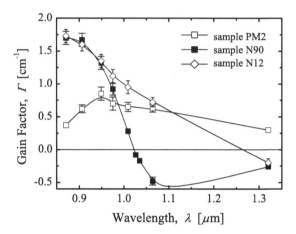

FIGURE 12.6. Two-beam coupling gain spectra measured in CdTe:Ge; experimental data for sample PM2 (open squares), sample N90 (black squares), and N12 (open diamonds); $2\theta = 67°$.

to the opposite one. Such behavior has been reported for both vanadium- [66] and germanium- [67, 85] doped crystals.

Wavelength Optimization

Generally, the refractive index change decreases for longer wavelength, since it is inversely proportional to the wavelength. In addition, the refractive index itself decreases with the wavelength increase. That is why it might be expected that the index change will decrease monotonically with the wavelength. It is not always the case, however, for photorefractive semiconductors because of the spectral dependence of the electron–hole competition.

The wavelength dependences of the two-beam coupling gain factor are shown in Figure 12.6 for three different CdTe:Ge samples by open squares (sample PM2), filled squares (sample N90), and open diamonds (sample N12). The gain factor varies smoothly with wavelength in the sample PM2, and its sign is always the same, while the gain factor changes sign with wavelength in the samples N12 and N90. Therefore the main photoexcited charge carriers that record the grating in the sample PM2 are always the same throughout the whole wavelength range studied, while they change with wavelength in two other samples.

An important experimental achievement is the high two-beam coupling gain factor $\Gamma = 1.7$ cm^{-1} reached in the samples N90 and N12 around 0.9 μm. This is the largest value ever reported for any photorefractive semiconductor without electric field. Such high photorefractive response can be profitable for applications with diode lasers operating at this wavelength range. At the same time, the electron–hole competition significantly reduces the space-charge field in these two crystals within the wavelength range studied, and the photorefractive response decreases to zero at the wavelength of electron–hole equilibrium. The zero-gain position varies from sample to sample. This variation is caused by the difference in relative concentrations of donor and trap centers formed in CdTe by germanium doping, which is still hardly controllable during the crystal growth. That is why a large scatter of photorefractive properties is observed at 1.06 μm with different CdTe crystals, and the photorefractive response of many samples is far from the theoretical limit.

At the same time, the observed variations of the photorefractive properties of CdTe are not as astonishing, since the electrical properties of the semiconductors depend strongly on the growth procedure and on impurity content. It is known, e.g., that for GaAs samples cut from different parts of the same ingot, not only does the beam coupling at 1.06 μm change quantitatively, but even the direction of energy transfer becomes opposite [86]. So even a semiconductor with well-developed growth technology like GaAs exhibits rather strong variation of the photorefractive properties because of differences in defect and impurity content. Therefore, in spite of the fact that CdTe:Ge and CdTe:V seem to be promising photorefractive materials, further optimization of crystal synthesis technique is necessary to improve the reproducibility of the crystals to the level necessary for industrial applications. One of the ways of such optimization is an appropriate codoping, which may introduce compensation levels for the secondary charge carriers and control the charge state of photorefractive centers.

Charge-Transfer Processes in CdTe at Different Wavelengths

Detailed spectroscopic study of cadmium telluride and identification of the impurity and defect centers that participate in space-charge formation are important both for better understanding of the background physical processes of photorefraction and for further improvement and optimization of the photorefractive recording. The study of photorefractive centers in CdTe:V [87,88] and CdTe:Ge [89,90] reveals that several impurity and defect centers are involved in the space-charge grating formation in both vanadium- and germanium-doped crystals. The complicated defect structure explains nontrivial photorefractive properties, which vary from sample to sample and differ with the wavelengths.

Pronounced light-induced absorption, i.e., the change of absorption under illumination, was detected in both materials [88–90]. The presence of light-induced absorption by itself indicates that at least two centers contribute to the charge redistribution. Indeed, if a one-center $Y^{0/+}$ participates in grating recording, the free electron excited from Y^0 may by trapped only by any Y^+. The center Y^0 from which the electron is excited becomes Y^+, while the center Y^+ that traps the electron becomes Y^0:

$$Y^0 + h\nu = Y^+ + e; \quad Y^+ + e = Y^0. \tag{12.9}$$

Thus the average relative concentration Y^0/Y^+ remains constant in this case and the average absorption does not change. Therefore light-induced absorption exists only if a part of the photoexcited carriers are trapped by a center another than $Y^{0/+}$.

The spectra of changes in absorption measured in CdTe:Ge at $T = 93$ K after consecutive illuminations by light with quantum energies 0.8 eV, 1.2 eV, 1.3 eV, and 1.5 eV [90] are shown in Figure 12.7. Altogether, four absorption bands

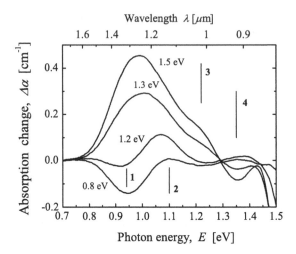

FIGURE 12.7. Spectra of the light-induced absorption measured in CdTe:Ge at $T = 93$ K after consecutive illuminations of the crystal by light with photon energies 0.8 eV, 1.2 eV, 1.3 eV, and 1.5 eV.

at energies of 0.94 eV (1.32 µm), 1.1 eV (1.13 µm), 1.22 eV (1.02 µm), and 1.35 eV (0.918 µm) can be distinguished from the data presented in this figure, while the most pronounced photoinduced absorption is clearly seen in the vicinity of 1 eV. The photoinduced absorption spectra differ from sample to sample, but the four characteristic bands mentioned above with different ratios of amplitudes are found in all tested CdTe:Ge samples. Rather strong photoinduced absorption up to 0.3 cm^{-1} has been detected also at ambient temperapture [89] at different wavelengths. Nonlinear absorption is, however, normally very small in CdTe:Ge at 1.06 µm.

The combination of the photorefractive, spectroscopic, magnetooptic, and electron paramagnetic resonance techniques [28, 89, 90] allows one to identify two centers each in two charge states that contribute to the charge redistribution in CdTe:Ge. These are neutral germanium Ge^0 and ionized germanium Ge$^+$ (Ge^{2+} and Ge^{3+} in ionic notation) with energies of optical excitations 1.35 eV and 0.94 ev, respectively, and a center of unknown origin X, which is present also in two charge states: X^0 (1.1 eV) and X$^-$ (1.22 eV).

A tentative model describing photoinduced charge-transfer processes in CdTe:Ge has been developed. The corresponding energy-level diagram is shown in Figure 12.8. Within this model the following photoexcitations take place under illumination:

$$Ge^0 + h\nu \rightarrow Ge^+ + e \quad \text{(band 4 } at \text{ 1.35 } eV \leftrightarrow 0.918 \text{ µm)};$$
$$Ge^+ + h\nu \rightarrow Ge^0 + h \quad \text{(band 1 } at \text{ 0.94 } eV \leftrightarrow 1.32 \text{ µm)};$$
$$X^0 + h\nu \rightarrow X^- + h \quad \text{(band 2 } at \text{ 1.1 } eV \leftrightarrow 1.13 \text{ µm)};$$
$$X^- + h\nu \rightarrow X^0 + e \quad \text{(band 3 } at \text{ 1.22 } eV \leftrightarrow 1.02 \text{ µm).} \quad (12.10)$$

FIGURE 12.8. Energy-level diagram describing photoexcitations in photorefractive CdTe:Ge; CB, conduction band, VB, valence band.

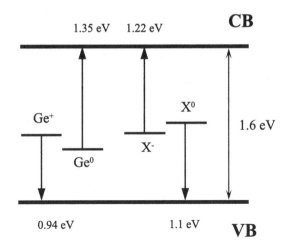

The relative concentrations of donor and trap centers formed by Ge-doping are different in every particular sample. The FWHM of any of the four revealed absorption bands (see Figure 12.7) is of the same order of magnitude as the frequency separation between them even at low temperature. Therefore one could expect that for any particular wavelength of the laser used for photorefractive recording, the absorbed photon may excite both a free electron to the conduction band and a free hole to the valence band from the $Ge^{+/0}$-center as well as from the center $X^{0/-}$. The relative concentrations of $Ge^{+/0}$ and $X^{0/-}$ centers in each charge state determine the type of photoconductivity and electron–hole competition at every wavelength. The most often used 1.06 μm radiation with quantum energy 1.17 eV falls just between all four bands (see Figure 12.7) and therefore all possible transitions with the excitation of electrons and holes occur. That is why a large spread of photorefractive properties is observed at this wavelength. The wavelength of electron–hole equilibrium also changes from sample to sample, since it depends strongly on the relative concentrations of both centers. Thus the variation of photorefractive properties of CdTe:Ge especially around 1 μm is explained by different contributions of the photorefractive centers formed by germanium doping to the charge redistribution.

Optical Sensitization of Photorefractive Semiconductors

The pronounced light-induced absorption detected in CdTe:Ge indicates that the two considered centers Ge and X interchange excitations when the sample is illuminated, i.e., the electron released from X^- can be trapped not only by X^0 but also by the Ge^+ center, etc. Such charge redistribution between different centers allows one to improve the photorefractive response and to increase the gain factor. Light with appropriate photon energy may redistribute charge carriers in the proper way to reduce the electron–hole competition at the recording wavelength.

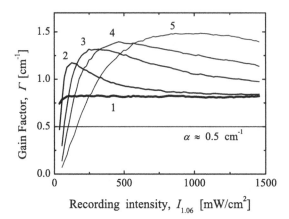

FIGURE 12.9. The intensity dependences of the gain factor measured in CdTe:Ge at 1.06 μm in the presence of auxiliary illumination at 1.32 μm. The intensities of 1.32-μm light are 0 mW/cm² (1); 80 mW/cm² (2); 370 mW/cm² (3); 640 mW/cm² (4); 1750 mW/cm² (5).

The data of Figure 12.9 illustrate the enhancement of photorefractive response achieved experimentally in the sample N12. The intensity dependences of the gain factor measured at 1.06 μm (1.17 eV) with additional 1.32 μm (0.94 eV) illumination of different intensity are shown in this figure. Similar dependences were obtained for 1.32 μm recording with additional illumination at 1.06 μm [85]. It follows from the data of Figure 12.9 that for any given intensity of the recording light a certain intensity of the sensitizing light can be found that maximizes the gain factor. Enhancement of the gain factor by a factor of 2 is achieved at both wavelengths with optimal auxiliary illumination. The largest gain factor $\Gamma \approx 1.5$ cm^{-1} measured at 1.06 μm with optimal 1.32 illumination reaches 90% of the theoretical limit calculated for the given grating spacing $\Lambda = 1$ μm with $r_{41} = 6.8$ pm/V [2] in the no-trap saturation limit.

The absorption constant of the studied crystal at 1.06 μm is $\alpha \approx 0.5$ cm. So the net gain of the signal wave (the gain factor larger than the absorption constant) is reached at $\lambda = 1.06$ μm with no auxiliary illumination. The sensitization technique allows for further improvement of this net gain: amplification of the signal beam overcoming both the absorption and the Fresnel losses is achieved. The temporal behavior of the signal intensity during grating recording is shown in Figure 12.10. The intensity measured behind the crystal is normalized to the input signal intensity. The data of Figure 12.10 demonstrate that net amplification of the input intensity is achieved in CdTe:Ge with no external field.

The enhancement of the diffraction efficiency by means of auxiliary illumination was first reported for gratings recorded in $Bi_{12}SiO_{20}$ at 0.488 μm with additional infrared incoherent illumination [91]. Later, the photorefractive response for pulsed infrared recording in $BaTiO_3$ was improved by pulse irradiation with green light [92] and for cw recording in $Bi_{12}TiO_{20}$ [93] and $Sn_2P_2S_6$ [94] with the simultaneous or preliminary incoherent visible illumination. In all these cases the improvement of photorefractive properties is based on appropriate influence on the relatively shallow traps available in the studied materials. The improvement in the photorefractive gain factor was reached also in InP:Fe at 1.06 μm with

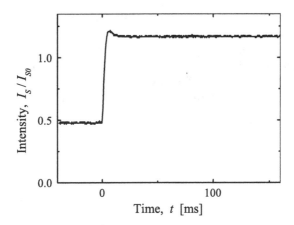

FIGURE 12.10. Temporal behavior of the signal-beam intensity I_S behind the CdTe:Ge sample normalized to the input intensity I_{S0} measured in two-beam coupling experiment at 1.06 μm with optimal 1.32-μm illumination the pump beam is switched on at $t = 0$.

incoherent illumination at the same wavelength [95] and in CdZnTe:V at 1.5 μm with additional illumination at 1.32 μm in the presence of a dc electric field [27].

For CdTe:Ge the effect of sensitization may be explained as follows. According to the tentative model describing charge transfer processes in CdTe:Ge (see Figure 12.8), both types of charge carriers are excited at each wavelength starting from the absorption edge to 1.5 μm. On the other hand, the main photoexcited charge carriers are electrons in the short-wavelength range, while generally the holes dominate at long wavelengths. At a certain wavelength a change in the type of photoconductivity may occur. In the studied sample this characteristic wavelength is between 1.06 μm and 1.32 μm, since the direction of energy transfer is different at these wavelengths. Therefore the photorefractive grating at 1.06 μm (1.17 eV) is recorded by electrons, while holes are the main carriers at 1.32 μm (0.94 eV).

Though the electrons are the main carriers at 1.06 μm, the holes are also excited and partially compensate the main grating. As one can see from the data of Figure 12.7, low-energy illumination decreases the absorption in the long-wavelength range at low temperature. Similar change exists at ambient temperature [89]. The decrease of absorption indicates by itself the reduction of photoconductivity. The excitation of holes corresponds to the low-energy absorption bands, whose amplitudes decrease under low-energy illumination. Therefore hole-conductivity decreases, which results in inhibition of the electron–hole competition for 1.06-μm recording, where electrons are the main photoexcited carriers.

The data of Figure 12.7 show that high-energy illumination results in increase of absorption in the long-wavelength range. This increase indicates that the hole conductivity becomes higher. Thus the electron–hole competition for 1.32-μm recording decreases under under 1.06-µm illumination, since the holes are the main carriers for 1.32-µm recording.

The technique of pseudo-3D presentation of light-induced absorption proposed and developed in [96] allows one to present in a clear way the main charge-transfer processes in CdTe:Ge [97] at different wavelengths. The photorefractive grating at 1.06 µm (1.17 eV) is recorded by photoelectrons excited mainly from Ge^0 centers. These electrons are trapped by Ge^+ and X^0 centers. At 1.32 µm the grating is formed by holes excited mainly from Ge^+ and trapped by Ge^0 and X^-. Thus the following charge redistribution exists:

$$Ge^0 + h\nu = Ge^+ + e, \quad X^0 + e = X^- \quad \text{at } 1.06 \text{ µm } (1.17 \, eV);$$
$$Ge^+ + h\nu = Ge^0 + h, \quad X^- + h = X^0 \quad at \ 1.32 \text{ µm } (0.94 \, eV). \quad (12.11)$$

Therefore the illumination at 1.06 µm improves recording at 1.32 µm, since it increases the concentration of Ge^+ ions, which are donors of the main photocarriers (holes). The positive effect of illumination at 1.32 µm is in a decrease of the Ge^+ concentration, which is the source of secondary carriers (holes) at 1.06-µm recording. Thus in both these cases the additional illumination inhibits the electron–hole competition at recording wavelength. The increase of the effective trap density was not detected experimentally but also may be important is some crystals.

A more general remark may be made. Let us assume that at wavelengths λ_e and λ_h photoexcitation of different charge carriers dominates and charge redistribution between different centers takes place. The illumination at λ_h will depopulate centers that are the donors of secondary carriers for grating recording at λ_e and/or increase the density of donors of the main carriers for recording at λ_e. A similar process with opposite charge-transfer direction will take place for grating recording at λ_h with supplementary illumination at λ_e. Consequently, the auxiliary illumination shifts the Fermi level in a proper direction, decreasing electron–hole competition at the recording wavelength. Therefore, the illumination at λ_h with a certain intensity should always increase the grating recorded at λ_e and vice versa. This prediction has an experimental proof. The positive effect of auxiliary illumination with appropriate wavelength was detected in all CdTe:Ge samples with different main charge carriers at different wavelengths. For example, grating enhancement was achieved in the sample N90 for grating recording at 0.91 µm with additional 1.06 µm illumination [67]. The change in the main carriers was reported also for CdTe:V [66]. Therefore the proper illumination of CdTe:V can reduce the electron–hole competition, which results in an increase in the photorefractive performance of the material.

To summarize, the optical sensitization technique reduces the electron–hole competition, which has not been done previously either by doping optimization or by special crystal growth procedures and aftergrowth treatment. The net gain of the signal wave is reached in CdTe:Ge crystal at $\lambda = 1.06$ µm, and the amplification of the input signal is achieved with no external field.

12.3 Applications of Optimized Photorefractive Semiconductors

12.3.1 Double-Phase-Conjugate Mirror in CdTe:Ge

The double-phase-conjugate mirror (DPCM) is a unique device that yields the conjugation of two mutually incoherent pump beams [3, 98, 99]. It is particularly interesting for bidirectional reconfigurable optical interconnects [100]. Photorefractive semiconductors have attracted much attention in this field because they are sensitive in the telecommunication wavelength range. DPCM has been built with all infrared-sensitive semiconductors, i.e., with InP:Fe [30, 32], GaAs:EL2 [101], CdTe:V [31], and CdTe:Ge [33]. The operation of DPCM requires high coupling strength $\Gamma l > 4$ [98]. That is why different techniques are used for grating enhancement.

In CdTe:Ge the oscillation in DPCM geometry was achieved at $\lambda = 1.06$ μm with relatively low external dc field at intensity of order a few milliwatts [33]. The use of a dc field is simpler in the case of semiconductors because ac fields with high frequency should be used to record a stationary grating in these fast materials. The grating recorded with a dc field is normally unshifted with respect to the interference fringes. The generation occurs at slightly shifted frequencies. This frequency detuning corresponds to the maximum gain of the moving grating recorded with the dc field [13, 14, 99].

In Figure 12.11 the phase-conjugate reflectivity R_{PC} is shown as a function of the applied field (a) and of the total pump intensity (b), measured with pump beams of equal intensities. R_{PC} is presented as the intensity ratio of the diffracted beam and contradirectional pump beam. When changing the amplitude of the external field or intensity, the coupling strength of the sample is changed. The threshold behavior of the oscillation is obvious with low threshold intensity about only 3 mW/cm² and low threshold field about 2 kV/cm.

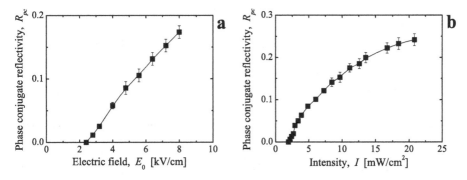

FIGURE 12.11. Phase-conjugate reflectivity of DPCM with CdTe:Ge at $\lambda = 1.06$ μm versus applied field (a) with total intensity $I = 11$ mW/cm² and versus total pump intensity (b) with field $E_0 = 8$ kV/cm; pump beam ratio $\beta = 1$.

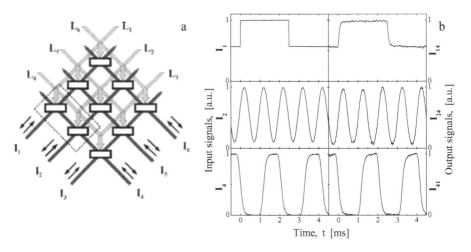

FIGURE 12.12. (**a**) Schematic representation of a 3×3 all-optical interconnect. (**b**) Temporal variation of the input and output signals for a 2×1 interconnect designed with two CdTe:Ge crystals at $\lambda = 1.06$ μm; I_1, I_2, and I_4 are the input signals; I_{41} I_{24}, I_{14} (I_{24}, e.g., stands for the signal emitted in 2 and measured at 4).

12.3.2 All-Optical Interconnect

Several architectures of an all-optical interconnect with DPCM as a basic element have been proposed, and their operations have been demonstrated [32, 33, 100]. The schematic representation of an $N \times M$-channel interconnect [33] with $N \times M$ optical switches controlled by $N + M$ auxiliary light beams is shown in Figure 12.12a for $N = M = 3$. The signals $I_1, \ldots, I_6$ are carried by the laser beams. The switches are the DPCMs in photorefractive crystals. When two laser beams approach the switch, a 3D phase grating develops inside the crystals and two relevant channels become automatically coupled. This coupling can be controlled with auxiliary light, which can adjust the crystal coupling strength $\Gamma \ell$ either below or above the threshold of oscillation. The auxiliary illumination increases the crystal conductivity and therefore reduces the space-charge field and $\Gamma \ell$. The reconfiguration of actual connections is controlled by switching on or off the erasing beams $L_1, \ldots, L_6$. If we intend to connect, e.g., channels I_1 and I_5, all erasing beams should be switched on except L_1 and L_5.

A one-line interconnect was built with $N = 1$ and $M = 2$ [33]. The two CdTe:Ge samples shown within the dashed-line window in Figure 12.12a connect the signals I_1, I_2, and I_4. The signal intensities can be modulated. DPCMs operate at 1.06 μm, while auxiliary radiation at 1.3 μm is used to increase strongly the photoconductivity of the samples and to erase partially a grating recorded by 1.06-μm radiation. With an external electric field of 8 kV/cm, about 3% of the input radiation is diffracted from the transmission grating in "open" position. Figure 12.12b presents the temporal variation of the input signals measured before the crystals and signals that come from the other channels. The one-to-one correspondence of the emitted and received signals is obvious. A switching time of less than 5 ms is achieved at intensities of the recording and erasing beams less than 100 mW/cm^2.

12.3.3 Laser-Based Ultrasonics Receiver with CdTe:Ge

Photorefractive materials are attractive for optical detection of ultrasonic surface displacements because of their adaptability to the speckle structure of the signal to be detected and to ambient environmental perturbations [102, 103]. In practice, only fast materials can ensure high performance of ultrasonic receivers because of the need to compensate for the full range of mechanical and optical disturbances that are encountered in the factory environment. Photorefractive semiconductors meet this requirement. Laser-based ultrasonic receivers have been demonstrated with GaAs:EL2 [37, 39], InP:Fe [34], CdTe:V [35], CdTe:Ge [38] as well as with quantum wells [104]. Three main adaptive interferometers used for ultrasonics detection may be distinguished. They are based on two-wave mixing [58, 102, 103], polarization self-modulation [105], and vectorial wave mixing [106].

The important characteristics of the adaptive interferometer are sensitivity and cutoff frequency, which determine minimal detectable displacement and range of adaptability to ambient environmental perturbations. The two-wave mixing adaptive interferometer with CdTe:Ge [38] operates at 1.06 μm as well as at 1.55 μm and demonstrates an attractive combination of these characteristics at low intensity of order 100 mW/cm^2. In Figure 12.13a the amplitude of intensity modulation ΔI_S normalized to the averaged transmitted signal intensity I_{SA} is plotted as a function of the modulation frequency measured at 1.06 μm. The cutoff frequency is about 1.3 kHz for a 6-kV/cm external field. At a lower field the cutoff frequency is slightly larger, while the signal amplitude is still acceptable for detection. Similar dependences measured at 1.55 μm are shown in Figure 12.13b. The cutoff frequency is lower because of the lower photoconductivity at the larger

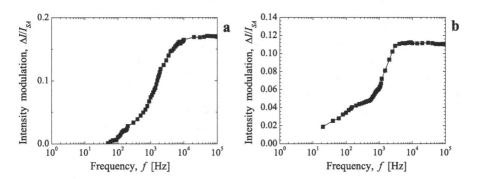

FIGURE 12.13. Frequency response of the two-wave mixing adaptive interferometer with CdTe:Ge; amplitude of output intensity modulation as a function of phase modulation frequency measured at $\lambda = 1.06$ μm (a) and at $\lambda = 1.55$ μm (b); $I = 140$ mW/cm^2, $E_0 = 6$ kV/cm, phase modulation amplitude $\Delta\varphi = 0.17$ rad.

wavelength, but it still nearly approaches 1 kHz. The shape of the tail at low frequency is related to complicated defect structure of CdTe:Ge.

The sensitivity of the adaptive interferometer is determined by the normalized detection limit δ_{lim}, which corresponds to the minimum detectable displacement with unity signal-to-noise ratio at 1 W light power with 1 Hz detection bandwidth [58, 103]. The adaptive interferometer with CdTe:Ge demonstrates high limiting sensitivity $\delta_{lim} \approx 1.2 \times 10^{-7} \text{nm} \sqrt{\text{W/Hz}}$ at both 1.06 μm and 1.55 μm for an applied field of 8 kV/cm. This sensitivity is only three times worse as compared to a classical interferometer operating with the same measuring system [38, 103]. The interferometer with CdTe:Ge is adaptive, however, and demonstrates excellent frequency response at low intensity.

12.4 Spatial Subharmonics in a Photorefractive Semiconductor

The generation of spatial subharmonics in photorefractive crystals is related to excitation of gratings with grating vectors $\mathbf{K}/2$, $\mathbf{K}/3$, and $\mathbf{K}/4$, etc., which are fractional to the grating vector $\mathbf{K}$ of the principal grating recorded by two incident light beams [107, 108]. The appearance of subharmonics is initiated by the space-charge waves (SCW), which may be effectively excited in a material with large mobility–lifetime product in the presence of an external electric field (see the chapter on space-charge waves written by B. Sturman [109]). From this point of view the semiconductor crystals may be regarded as well suited for the generation of subharmonics because they normally exhibit large values of the mobility–lifetime product. Moreover, the theory of SCW was developed first just for semiconductors [110]. Nevertheless, the spatial subharmonics were for a long time observed only in sillenites [107, 111–113].

To achieve generation of spatial subharmonics the semiconductor should meet the following requirements. The crystal should exhibit high resistivity to support high voltage, but at the same time it should possess large mobility–lifetime product $\mu\tau$ necessary for the excitation of SCW. These two requirements are contradictory, and therefore a compromise should always be found. In addition, the mobility–lifetime product may be overestimated for particular deeply compensated high-resistant photorefractive semiconductors. The semiconductor should also demonstrate good photorefractive properties simply for registration of the diffraction from the subharmonic grating. Selection of a crystal meeting all these requirements allowed us to achieve generation of spatial subharmonics in a photorefractive semiconductor, i.e., in CdTe:Ge [51, 114].

The spatial subharmonics are generated in CdTe in the presence of a nearly square-shaped ac field with frequency 700 Hz. Nearly far-field patterns taken from the screen placed at a distance of 1 m behind the sample are shown in Figure 12.14 for different amplitudes of the field. Subharmonic $\mathbf{K}/2$ appears for the field amplitude E_{AC} exceeding 2.2 kV/cm. The larger the field, the higher the order of the subharmonics generated.

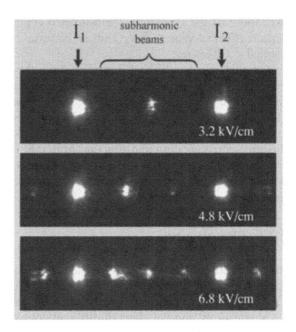

FIGURE 12.14. Spatial subharmonics in CdTe at $\lambda = 1.06\,\mu\text{m}$; the intensity distribution of the recording (I_1 and I_2) and subharmonic beams for different amplitudes of the ac field.

The excitation of subharmonics with different fractional grating vectors occurs at different threshold fields that depend on the spatial frequency of the principal grating. The experimental subharmonic "phase diagrams" that define the range of grating spacing and amplitude of the applied field where subharmonics can be excited are shown in Figure 12.15 by dots. Solid lines are calculated according to the SCW theory [108] for different subharmonics, all with the same set of crystal parameters [51]. From the best fit of the theory to experimental data the mobility–lifetime product of free carriers $\mu\tau = 10^{-10}\,\text{m}^2/\text{V}$ and the effective trap concentration $N_E = 4 \times 10^{20}\,\text{m}^{-3}$ are estimated. Apparently, the theoretical model describes qualitatively well the threshold behavior of subharmonic

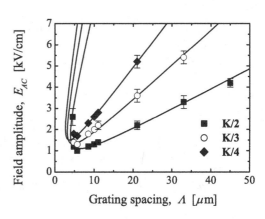

FIGURE 12.15. Threshold ac field amplitude E_{AC} for different subharmonic gratings as a function of the fringe spacing Λ of the principal grating measured experimentally (dots) and calculated according to the theory (lines) with the same set of crystal characteristics for different subharmonics.

excitation. At the same time, the estimated crystal characteristic correspond to certain effective values of N_E and $\mu\tau$ because the one-center model cannot be applied to CdTe:Ge.

Although spectacular and important from the fundamental physics point of view, the generation of spatial sabharmonics is generally an undesirable effect for practical applications because it results in energy loss of the pump beams [115, 116]. At the same time, spatial subharmonics may be used for estimation of crystal parameters [51] and for amplification of a seeding beam propagated in the direction corresponding to diffraction from the subharmonic grating [117, 118]. On the other hand, the SCW may be effectively used for optical-wave amplification when their spatial frequency is adjusted to the spatial frequency of the principal grating [14, 109, 116, 118, 119].

12.5 Conclusions

Photorefractive recording in infrared-sensitive semiconductors was reported first for GaAs and InP because these materials were available in semi-insulating form. Crystal-growth technology of GaAs and InP is well developed for microelectronics applications, and the effect of different dopants is well studied. The crystals with photorefractive response approaching the theoretical limit can be reproduced quite reliably now. Cadmium telluride has a great potential thanks to the larger electrooptic constant, as compared to GaAs and InP. Growth technology of the photorefractive CdTe is not so well developed, however. A spread of data is observed for different CdTe crystals, and the photorefractive response is reduced in many samples. That is why for industrial applications there is a need for development of growth technology of optimized photorefractive CdTe crystals with reproducible and spatially uniform properties. At the same time, a nearly theoretical limit of the gain factor and diffraction efficiency was reached in CdTe, and a net amplification of the signal wave in two-beam coupling was achieved using various optimization techniques. Different applications with selected germanium- or vanadium-doped crystals confirm the advanced prospects of this photorefractive semiconductor.

Acknowledgments. The author is grateful to S. Odoulov for helpful discussions. CdTe:Ge crystals were grown at Chernovtsy National University, Chernovtsy, Ukraine by Z.I. Zakharuk, I.M. Rarenko, P.M. Fochuk, and O.E. Panchuk.

References

1. A.M. Glass, J. Strait: The photorefractive effect in semiconductors. In: *Photorefractive Materials and Their Applications*, vol. 61, ed. by P. Günter, J.-P. Huignard (Springer, Berlin–Heidelberg–New York, 1989) 237–262.
2. D.D. Nolte: Photorefractive semiconductors. In: *Progress in Photorefractive Nonlinear Optics*, ed. by K. Kuroda (Taylor & Francis, London–New York, 2002), 9–45.

3. L. Solymar, D.J. Webb, A. Grunnet-Jepsen: *The Physics and Applications of Photorefractive Materials* (Clarendon Press, Oxford 1996).

4. V.L. Vinetsky, N.V. Kukhtarev, S.G. Odoulov, M.S. Soskin: USSR Invention Sertificat # 603276, Class G03H1/00, December 21, 1977, Published in Bull of Invention # 43, 250 (1978).

5. K. Tada, M. Aoki: *Jpn. J. Appl. Phys.* **10**, 998 (1971).

6. D.T.F. Marple: *J. Appl. Phys.* **35**, 539 (1964).

7. M.B. Klein: *Opt. Lett.* **9**, 350 (1984).

8. A.M. Glass, A.M. Jonson, D.H. Olson, W. Simpson, A.A. Ballman: *Appl. Phys. Lett.* **44**, 948 (1984).

9. R.B. Bylsma, P.M. Bridenbaugh, D.H. Olson, A.M. Glass: *Appl. Phys. Lett.* **51**, 889 (1987).

10. S.G. Odoulov, S.S. Slussarenko, K.V. Shcherbin: *Sov. Tech. Phys. Lett.* **15**, 417 (1989).

11. R.N. Schwartz, C.-C. Wang, S. Trivedi, G.V. Jagannathan, F.M. Davidson, Ph.R. Boyd, U. Lee: *Phys. Rev. B* **55**, 15378 (1997).

12. K. Shcherbin, V. Volkov, V. Rudenko, S. Odoulov, A. Borshch, Z. Zakharuk, I. Rarenko: *Phys. Stat. Sol. (a)* **183**, 337 (2001).

13. J.P. Huignard, A. Marrakchi: *Opt. Commun.* **38**, 249 (1981).

14. S.I. Stepanov, M.P. Petrov: Nonstationary holographic recording for efficient amplification. In: *Photorefractive Materials and Their Applications*, vol. 61, ed. by P. Günter, J.-P. Huignard (Springer, Berlin–Heidelberg–New York, 1989) 263–289.

15. S.I. Stepanov, M.P. Petrov: *Opt. Commun.* **53**, 292 (1985).

16. B. Imbert, H. Rajbenbach, S. Mallick, J.P. Herriau, J.P. Huignard: *Opt. Lett.* **13**, 327 (1988).

17. J. Kumar, G. Albanese, W.H. Steier, M. Ziari: *Opt. Lett.* **12**, 120 (1987).

18. M.B. Klein, S.W. McCahon, T.F. Boggess, G.C. Valley: *J. Opt. Soc. Am. B* **5**, 2467 (1988).

19. M. Ziari, W.H. Steier, P.M. Ranon, M.B. Klein, S. Trivedi: *J. Opt. Soc. Am. B* **9**, 1461 (1992).

20. J.-Y. Moisan, N. Wolffer, O. Moine, P. Gravey, G. Martel, A. Aoudia, E. Rzepka, Y. Marfaing, R. Triboulet: *J. Opt. Soc. Am. B* **11**, 1655 (1994).

21. P. Gravey, G. Picoli, J.Y. Labandibar: *Opt. Commun.* **70**, 190 (1989).

22. G. Picoli, P. Gravey, C. Ozkul, V. Vieux: *J. Appl. Phys.* **66**, 3798 (1989).

23. G. Picoli, P. Gravey, C. Ozkul: *Opt. Lett.* **14**, 1362 (1989).

24. A. Partovi, A. Kost, E.M. Garmire, G.C. Valley, M.B. Klein: *Appl. Phys. Lett.* **56**, 1089 (1990).

25. A. Partovi, E.M. Garmire: *J. Appl. Phys.* **69**, 6885 (1991).

26. J.E. Millerd, S.D. Koehler, E.M. Garmire, A. Partovi, A.M. Glass, M.B. Klein: *Appl. Phys. Lett.* **57**, 2776 (1990).

27. P. Pogany, H.J. Eichler, and M. Hage Ali: *J. Opt. Soc. Am. B* **15**, 2716 (1998).

28. K. Shcherbin, F. Ramaz, B. Farid, B. Briat, H.-J. von Bardeleben: *OSA Trends in Optics and Photonics* **27**, 54 (1999).

29. R.B. Bylsma, A.M. Glass, D.H. Olson, M. Cronin-Golomb: *Appl. Phys. Lett.* **54**, 1968 (1989).

30. V. Vieux, P. Gravey, N. Wolffer, G. Picoli: *Appl. Phys. Lett.* **58**, 2880 (1991).

31. G. Martel, N. Wolfer, J.Y. Moisan, P. Gravey: *Opt. Lett.* **20**, 937 (1995).

32. N. Wolffer, P. Gravey: *Appl. Phys. B* **68**, 947 (1999).

33. K. Shcherbin, S. Odoulov, Z. Zakharuk, I. Rarenko: *Opt. Mater.* **18**, 159 (2001).

34. P. Delaye, A. Blouin, D. Drolet, L.-A. de Montmorillon, G. Roosen, J.-P. Monchalin: *J. Opt. Soc. Am. B* **14**, 1723 (1997).

35. L.A. de Montmorillon, I. Biaggio, Ph. Delhaye, J.C. Launay, G. Roosen: *Opt. Commun.* **129**, 293 (1996).

36. E . Raita, A.A. Kamshilin, O. Kobozev, A. Shumelyuk: *OSA Trends in Optics and Photonics* **62**, 344 (2001).

37. A. Blouin, J.P. Monchalin: *Appl. Phys. Lett.* **65**, 932 (1994).

38. M.B. Klein, K. Shcherbin, V. Danylyuk: *OSA Trends in Optics and Photonics* **87**, 483 (2003).

39. B. Campagne, A. Blouin, L. Pujol, J.-P. Monchalin: *Rev. Scient. Instr.* **72**, 2478 (2001).

40. D.D. Nolte, D.H. Olson, A.M. Glass: *Phys Rev B* **40**, 10650 (1989).

41. R.A. Linke, T. Thio, J. Chadi, G.E. Davlin: *Appl. Phys. Lett.* **65**, 16 (1994).

42. D.D. Nolte: *J. Appl. Phys.* **79**, 7514 (1996).

43. G.A. Sefler, E. Oh, R.S. Rana, I. Miotkowski, A.K. Ramdas, D.D. Nolte: *Opt. Lett.* **17** (1992).

44. R.S. Rana, E. Oh, K. Chua, A.K. Ramdas, D.D. Nolte: *Phys. Rev. B* **49**, 7941 (1994).

45. R.S. Rana, M. Dinu, I. Miotkowski, D.D. Nolte: *Opt. Lett.* **20**, 1238 (1995).

46. D.D. Nolte, D.H. Olson, G.E. Doran, W.H. Knox, A.M. Glass: *J. Opt. Soc. Am. B* **7**, 2217 (1990).

47. Q. Wang, R.M. Brubaker, D.D. Nolte, M.R. Melloch: *J. Opt. Soc. Am. B* **9**, 1626 (1992).

48. D.D. Nolte, I. Lahiri, M.R. Melloch: *Opt. Lett.* **21**, 1888 (1996).

49. L. Peng, P. Yu, D.D. Nolte, M.R. Melloch: *Opt. Lett.* **28**, 396 (2003).

50. S. Iwamoto, S. Taketomi, H. Kageshima, M. Nishioka, T. Someya, Y. Arakawa, K. Fukutani, T. Shimura, K. Kuroda: *Opt. Lett.* **26**, 22 (2001).

51. K. Shcherbin: *Appl. Phys. B* **71**, 123 (2000).

52. M. Chauvet, S.A. Hawkins, G.J. Salamo, M. Segev, D.F. Bliss, G. Bryant: *Opt. Lett.* **21**, 1333 (1996).

53. T. Schwartz, Y. Ganor, T. Carmon, R. Uzdin, S. Shwartz, M. Segev, U. El-Hanany: *Opt. Lett.* **27**, 1229 (2002).

54. S. Shwartz, M. Segev, U. El-Hanany: *Opt. Lett.* **29**, 760 (2004).

55. N. Huot, J.M. Jonathan, G. Pauliat, P. Georges, A. Brun, G. Roosen: *Appl. Phys. B* **69**, 155 (1999).

56. A. Godard, G. Pauliat, G. Roosen, P. Graindorge, P. Martin: *Opt. Lett.* **26**, 1955 (2001).

57. A. Godard, G. Pauliat, G. Roosen, E. Ducloux, J.-L. Ayral: *OSA Trends in Optics and Photonics* **87**, 483 (2003).

58. L.A. de Montmorillon, Ph. Delhaye, G. Roosen: Photorefractive Interferometer for ultrasound detection. In: *Progress in Photorefractive Nonlinear Optics*, ed. by K. Kuroda (Taylor and Francis, London–New York 2002) 213–282.

59. D.D. Nolte, M.R. Melloch: Photorefractive quantum wells and thin films. In: *Photorefractive Effects and Materials*, ed. by D.D. Nolte (Kluwer Academic Publishers, Boston–Dordrecht–London 1995) 373–451.

60. J.E. Millerd, E. Garmire, A. Patrovi: Near-resonant photorefractive effects in bulk semiconductor. In: *Photorefractive Effects and Materials*, ed. by D.D. Nolte (Kluwer Academic Publishers, Boston–Dordrecht–London 1995) 311–372.

61. D. Nolte, S. Iwamoto, K. Kuroda: Photorefractive semiconductors and multiple quantum well structures. In: *Photorefractive Materials and Their Applications*, vol. 2, ed. by P. Günter, J.-P. Huignard (Springer, Science + Business Media, Inc., New York 2007) pp. xxx–xxx.

62. N.V. Kukhtarev, V.B. Markov, S.G. Odoulov, M.S. Soskin, V.L. Vinetskii: *Ferroelectrics*, **22**, 949 (1979).

63. M.B. Klein, G.C. Valley: *J. Appl. Phys.* **57**, 4901 (1985).

64. F.P. Strohkendl, J.M.C. Jonathan, R.W. Hellwarth: *Opt. Lett.* **11**, 312 (1986).

65. A. Yariv, P. Yeh: *Optical Waves in Crystals* (John Wiley & Sons, New York 1984).

66. Ph. Delhaye, L.A. de Montmorillon, I. Biaggio, J.C. Launay, G. Roosen: *Opt. Commun.* **134**, 580 (1997).

67. K. Shcherbin, O. Shumelyuk, S. Odoulov, E. Krätzig: *OSA Trends in Optics and Photonics* **73**, 202 (2002).

68. K. Walsh, T.J. Hall, and R.E. Burge: *Opt. Lett.* **12**, 1026 (1987).

69. V.V. Shepelevich, E.M. Khramovich: *Opt. Spectrosc.* **65**, 240 (1988).

70. J. Strait, J.D. Reed, N.V. Kukhtarev: *Opt. Lett.* **15**, 209 (1990).

71. A.A. Izvanov, A.E. Mandel, N.D. Khatkov, S.M. Shandarov: *Optoelectronics* **2**, 80 (1986).

72. S.I. Stepanov, S.M. Shandarov, N.D. Khat'kov: *Sov. Phys.- Solid State* **29**, 1754 (1987).

73. P. Günter, M. Zgonik: *Opt. Lett.* **16**, 1826 (1991).

74. G. Pauliat, P. Mathey, G. Roosen: *J. Opt. Soc. Am. B* **8**, 1942 (1991).

75. S. Shandarov: *Appl. Phys. A* **55**, 91 (1992).

76. R. Litvinov, S. Shandarov: *J. Opt. Soc. Am. B* **11**, 1204 (1994).

77. K. Shcherbin, S. Odoulov, R. Litvinov, E. Shandarov, S. Shandarov: *J. Opt. Soc. Am. B* **13**, 2268 (1996).

78. A.V. Alekseev-Popov, A.V. Knyaz'kov, A.S. Saikin: *Sov. Tech. Phys. Lett.* **9**, 475 (1983).

79. G.A. Brost, R.A. Motes, J.R. Rotge: *J. Opt. Soc. Am. B* **5**, 1879 (1988).

80. R.M. Pierce, R.S. Cudney, G.D. Bacher, J. Fainberg: *Opt. Lett.* **15**, 414 (1990).

81. D.D. Nolte, D.H. Olsen, E.M. Monberg, P.M. Bridenbaugh, A.M. Glass: *Opt. Lett.* **14**, 1278 (1989).

82. Two wave mixing non distructive evaluation probes, Tecnar Automation Ltée, 1321 Hocquart Street, St-Bruno, QC, Canada J3V 6B5, Canada.

83. A. Partovi, J. Millerd, E.M. Garmire, M. Ziari, W.H. Steier, S.B. Trivedi, M.B. Klein: *Appl. Phys. Lett.* **57**, 846 (1990).

84. K. Shcherbin: *Proc. SPIE* **3749**, 516 (1999).

85. K. Shcherbin, S. Odoulov, F. Ramaz, B. Farid, B. Briat, H.J. von Bardeleben, I. Rarenko, Z. Zakharuk, O. Panchuk, P. Fochuk: Characterisation of photorefractive CdTe:Ge. In: *Optics and Optoelectronics*, vol. 2, ed. by K. Singh (Narosa Dehra Dun 1998), 924–931.

86. M.B. Klein: *Techn. Dig. Conf. Photorefractive Effects, Materials and Devices* (PRM'89), 469 (1999).

87. J.C. Launay, V. Mazoyer, M. Tapiero, J.P. Zielinger, Z. Guellil, Ph. Delhaye, G. Roosen: *Appl. Phys. A* **55**, 33 (1992).

88. L.A. de Montmorillon, Ph. Delhaye, G. Roosen, H. Bou Rjeily, F. Ramaz, B. Briat, J.G. Gies, J.P. Zielinger, M. Tapiero, H.J. von Bardeleben, T. Arnoux, J.C. Launay: *J. Opt. Soc. Am. B* **13**, 2341 (1996).

89. B. Briat, K. Shcherbin, B. Farid, F. Ramaz: *Opt. Commun.* **156**, (1998).
90. B. Briat, F. Ramaz, B. Farid, K. Shcherbin, H.J. von Bardeleben: *Journ. Cryst. Growth* **197**, 724 (1999).
91. A. Kamshilin, M. Petrov: *Sov. Solid State Physics* **23**, 3110 (1981).
92. K. Buse, L. Holtman, E. Krätzig: *Opt. Commun.* **85**, 183 (1991),
93. S.G. Odoulov, K.V. Shcherbin: *Laser Physics* **3**, 1124 (1993).
94. S. Odoulov, A. Shumelyuk, U. Hellwig, R. Rupp, A. Grabar, I. Stoyka: *J. Opt. Soc. Am. B* **13**, 2352 (1996).
95. C. Ozkul, G. Picoli, P. Gravey, J. Le Rouzic: *Opt. Engineering* **30**, 397 (1991).
96. K. Krose, R. Schwarfschwerdt, A. Mazur, O. Schirmer: *Appl. Phys. B* **67**, 79 (1998).
97. K. Shcherbin, S. Odoulov, F. Ramaz, B. Farid, B. Briat, H.J. von Bardeleben, P. Delaye, G. Roosen: *Opt. Mater.* **18**, 151 (2001).
98. S. Weiss, S. Sternklar, B. Fischer: *Opt. Lett.* **114**, 12 (1987).
99. S. Odoulov, M. Soskin, A. Khyzhnyak: *Optical Coherent Oscillators with Degenerate Four-Wave Mixing* (Harwood Academic Publishers, London 1991).
100. S. Weiss, M. Segev, S. Sternklar, B. Fischer: *Appl. Opt.* **27**, 3422 (1988).
101. P.L. Chua, D.T.H. Liu, L.J. Cheng: *Appl. Phys. Lett.* **57**, 858 (1990).
102. R.K. Ing, J.P. Monchalin: *Appl. Phys. Lett.* **59**, 3233 (1991).
103. L.-A. de Montmorillon, P. Delaye, J.-C. Launey, G. Roosen: *J. Appl. Phys.* **82**, 5913 (1997).
104. D.D. Nolte, T. Cubel, L.J. Pyrak-Nolte, M.R. Melloch: *J. Opt. Soc. Am. B* **18**, 195 (2001).
105. A.A. Kamshilin, K. Paivasaari, M. Klein, B. Pouet: *Appl. Phys. Lett.* **77**, 4098 (2000).
106. A.A. Kamshilin, A.I. Grachev: *Appl. Phys. Lett.* **81**, (2002).
107. S. Mallick, B. Imbert, H. Ducollet, J.-P. Herriau, J.-P. Huignard: *J. Appl. Phys.* **63**, 5660 (1988).
108. B. Sturman, A. Bledowski, J. Otten, K. Ringhofer: *J. Opt. Soc. Am B* **10**, 1919 (1993).
109. B. Sturman: Space-Charge Wave Effects in Photorefractive Materials. In: *Photorefractive Materials and Their Applications*, vol. xx, ed. by P. Günter, J.-P. Huignard (Springer, Berlin–Heidelberg–New York 2004) pp. xxx.
110. R.F. Kazarinov, R.A. Suris, B.I. Fuks: *Sov. Phys. Semicond.* **7**, 480 (1973).
111. J. Takacs, M. Schaub, L. Solymar: *Opt. Commun.* **91**, 252 (1992).
112. J. Richter, A. Grunnet-Jepsen, J. Takacs, L. Solymar: *IEEE J. Quantum Electron.* **30**, 1645 (1994).
113. J. Takacs and L. Solymar: *Opt. Lett.* **17**, 247 (1992).
114. K. Shcherbin, S. Odoulov, V. Danilyuk: *OSA Trends in Optics and Photonics* **62**, 512 (2001).
115. A. Grunnet-Jepsen, L. Solymar, C.H. Kwak: *Opt. Lett.* **19**, 1299 (1994).
116. K. Shcherbin, V. Danylyuk, Z. Zakharuk, I. Rarenko, M.B. Klein: *Journ. Alloys Compounds.* **371**, 191 (2004).
117. D.R. Erbschloe, L. Solymar: *Appl. Phys. Lett.* **53**, 1135 (1988).
118. E.V. Podivilov, B.I. Sturman, H.C. Pedersen, P.M. Johansen: *Phys. Rev. Lett.* **85**, 1867 (2000).
119. M. Vasnetsov, P. Buchhave, S. Lyuksyutov: *Opt. Commun.* **137**, 181 (1997).

13

Amorphous Organic Photorefractive Materials

Reinhard Bittner and Klaus Meerholz*

* Institut für Physikalische Chemie, Universität zu Köln, Luxemburgerstr. 116, D-50939
Köln, Germany **klaus.meerholz@uni-koeln.de**

This chapter is structured in four main sections. First an introduction to the field of organic photorefractive (PR) materials will be given, which covers a brief historic overview as well as milestones achieved in the efforts made to improve the steady-state and dynamic performance of these promising new holographic materials.

Subsequently, in the theoretical section we will briefly summarize the basic framework of the well-established model of the PR effect in inorganic PR crystals, the Kukhtarev model [1]. The pecularities of the PR effect in polymers, such as charge transport and charge generation in organic materials, will be elaborated on in detail. This is followed by discussing Schildkraut's model [2, 3], which describes the photorefractive effect in amorphous organic PR materials, contrasted with Kukhtarev's model. The electrooptic properties of amorphous matrices doped with nonlinear optical chromophores and the mechanisms translating the PR space-charge field into a refractive-index modulation will be outlined.

In the third section, experimental techniques for investigating organic PR devices will be discussed. Representative examples for the general trends of the photorefractive key parameters, i.e., the PR refractive index modulation and the photorefractive gain as a function of various experimental parameters, will be presented. We will restrict ourselves to one material class derived from the first high-performance PR polymer.

Eventually a short survey of the most important amorphous organic PR materials known up to date will be given. Therefore material classes and the individual components of the materials will be systematically discussed rather than particular materials, the number of which nowadays has become too large to give a representative overview in the frame of this chapter.

We abstain from discussing potential applications of PR polymers, since many applications known for PR crystals could in principle apply to PR polymers as well, although not all conceivable applications are similarly suitable for the one or the other class of PR materials.

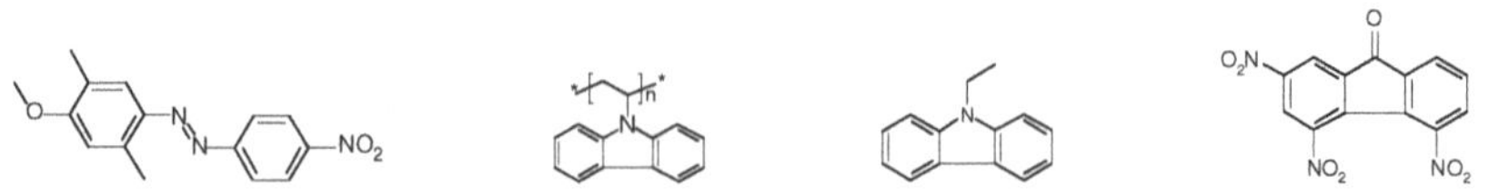

FIGURE 13.1. Components of the first high-performance PR composite [9].

13.1 Introduction

A Brief Historic Review

The photorefractive (PR) effect discovered in LiNbO$_3$ in 1966 [4] has been recognized as a reversible photoinduced refractive index change in materials combining photosensitivity, transport, and electrooptical effects. The PR effect has been intensively studied in inorganic crystals, the photorefractive behavior of which is nowadays fairly well understood. In contrast, photorefractive organic materials are a relatively new field, which still requires a lot of fundamental research in order to reach the level of understanding that allows effective tailoring and optimization of the materials.

The first PR polymer was reported in 1991 by Ducharme et al. [5]. It consisted of the electrooptic polymer bisphenol-A-diglycidylether (**bisA**) 4-nitro-1,2-phenylenediamine (**NPDA**) doped with diethylamino-benzaldehyde diphenyl-hydrazone (**DEH**) for providing photoconductivity. The nonlinear optical (NLO) chromophore **NPDA** acted also as sensitizer. In the following year, poly(N-vinylcarbazole) (**PVK**) was successfully introduced as the photoconductive host and used in combination with various chromophores [6–8].

The first high-performance PR polymer composite was presented in 1994 by Meerholz et al. [9]. The material consisted of **PVK** (hole-conducting polymer matrix), **ECZ** (plasticizer), **TNF** (sensitizer), and **2,5DMNPAA** (NLO chromophore). The chemical structures are shown in Figure 13.1. The most important feature of this composite was the low glass-transition temperature of $T_g \approx 5°C$, enabling for the first time full exploitation of the orientational enhancement mechanism [62]. The holographic characterization of this material yielded complete internal diffraction at an applied external field of 56 V/μm. Gain coefficients as large as $\Gamma_p = 220$ cm^{-1} were achieved at 90 V/μm applied field. However, the first high-performance PR polymer suffered from compositional instability. This problem could be improved significantly by using mixtures of NLO chromophores [10, 11] or by attaching larger brandied side groups than in the DMNPAA; see, e.g., [12–14].

Since these early days, the focus of research has been directed toward improving the index modulation amplitude of the recorded gratings and the recording speed. Due to the limited space in this book chapter, we can name only few examples and refer the reader to recent review articles [15–17] and the original literature for details.

Improvement of the Steady-State Performance

In an attempt to improve the grating strength, much work has been put into improving the chromophores [18]. Out of the long list of chromophores that have been developed, only a few are mentioned here, since they have gained some stardom: **DMNPAA** and **(F-)DEANST** [6] have already been mentioned above. They were typical electrooptical (EO) chromophores according to the initial belief that the Pockels effect would be the most relevant for translating the space-charge field into a refractive index modulation. After the orientational enhancement [62] effect was discovered, the optimization of the relation between first-order polarizability and ground-state dipole moment was focused on, leading at first to the colorless "chromophore" **2-BCNM** [19]. **ATOP** [20] and **DHADC-MPN** [21] are merocyanine dyes that exhibit large molecular figures of merit for chromophores to be utilized in PR polymers [18]. Lately, a clear trend toward using the di- and tricyanovinyle groups as acceptor groups is visible. In addition to maximizing the chromophore content, derivatives with glass-forming properties have been developed such that samples of pure chromophore (which then would also act as charge transporter) could be obtained. Details will be presented in Section 13.4.

The use of the novel chromophores often required use of longer wavelengths in the NIR. While 800 nm was easily achievable, it took some effort to extend the sensitivity beyond that. By using PbS nanoparticles, Prasad et al. were able to record holograms at 1300 nm [22]. Kippelen et al. reported on the use of two-photon sensitization [23]. This method has the inherent advantage that readout with a *cw* laser is nondestructive, independently of the intensity. However, this is overwhelmed by the necessity to use a femtosecond laser, which allows one to apply the very high light intensities required (on the order of GW/cm^2). The first results were obtained using 650-nm and 700-nm light; an extension to 1550-nm was reported recently [24].

Besides the chromophore and the chromophore density, a third parameter, which is of paramount importance for the index modulation amplitude, is the reduced temperature,

$$T_r = T - T_g. \tag{13.1}$$

This was investigated by varying the amount of plasticizer (which determines glass-transition temperature, T_g) or by changing the absolute temperature T (see below).

Improvement of the Recording Speed

The first report of a high-speed PR polymer utilized the buildup dynamics of the PR gain in order to discuss the PR response time [25]. The authors obtained a single response time of $\tau_g = 7.5$ ms for the buildup of the PR gain at 70 V/μm external field applied to a 100-μm-thick sample and 0.5 W/cm² total recording intensity.

The first report of a PR polymer composite exhibiting fast response in DFWM experiments has been provided by Herlocker et al. [26]. The authors reported response times of $\tau_1 = 4$ ms (weight factor 0.8) and $\tau_2 = 50$ ms (weight factor 0.2) for the buildup of the refractive index grating at a total light intensity of $I = 0.5$ W/cm^2 and $E = 95$ V/μm obtained from fitting the evolution of the refractive index modulation calculated from DFWM diffraction data to a biexponential associative growth function. Similarly, Wright et al. [27] reported a response time of 2 ms at $I = 1$ W/cm^2 and $E = 100$ V/μm by fitting their diffraction data using a stretched exponential function. Both materials were based on PVK.

A group of high-speed PR organic glasses was developed by Strohriegl et al. [28, 29, 30]. For **DR-DCTA** plasticized with **EHMPA** and sensitized with C_{60}, initial response times in DFWM experiments of down to $\tau = 0.5$ ms at 92 V/μm applied field, 10.8 W/cm^2 recording intensity at 645 nm operating wavelength were obtained.

Recently, Mecher et al. reported that preillumination can considerably increase the NIR recording speed [31]. The effect referred to as "gating" was found to result mainly from trap filling [32].

13.2 Physics/Theory

The PR effect occurs only if nonuniform illumination such as the interference pattern of two coherent laser beams is present, which distinguishes it from any other holographic recording mechanism. Upon illumination, mobile charges are generated, which migrate through drift or diffusion. Subsequently they are immobilized in some trapping sites, giving rise to the space-charge field E_{SC}. Due to the electrooptic effect, the space-charge field grating is translated into a refractive-index grating, the amplitude Δn of which, in absence of orientational enhancement (to be discussed later), is approximately proportional to the product of the effective electrooptic coefficient r_{eff} and space-charge-field amplitude $|E_{SC}|$. The index grating also possesses a phase shift φ_g relative to the incident interference pattern. The nonzero phase shift leads to coherent energy transfer between the writing beams due to self-diffraction. This phenomenon is often referred to as "two-beam coupling" (TBC) and is widely accepted as a fingerprint for photorefractivity. The recorded information can be erased by uniform illumination, opening the way to reversible storage of the optical interference fields, i.e., holograms.

The photorefractive effect in materials obeying band theory such as inorganic crystals (e.g., LiNbO$_3$) has been described by Kukhtarev et al. [1]. Even though, strictly speaking, this model does not apply to amorphous organic PR polymers, the basic concepts are nonetheless essential in understanding photorefractivity even in organic materials. Thus, the basics of this model will be described below, omitting features that do not apply to PR polymers.

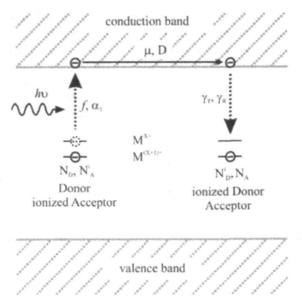

FIGURE 13.2. Energy-level scheme of Kukhtarev's model. The material exhibits a band structure characterized by delocalized energy levels. The majority carriers are assumed to be electrons, as is most common in inorganic PR materials. Each bar corresponds to an electronic level in an impurity center. If the upper level is occupied, it is a "donor" (or "generator"); otherwise, it is a "trap." Here D and μ are the carriers' diffusion coefficient and mobility. Within this model charge-carrier recombination (γ_R) and trapping (γ_T), and detrapping (α_T) and generation (f), respectively, are considered to be identical processes. Finally N_D and N_A are the number densities of the donor and acceptor sites of an impurity in the two redox states M^{x+} and $M^{(x+1)+}$. The material's energy structure is characterized by localized electronic levels, for which a Gaussian distribution is anticipated. The majority carriers are assumed to be holes, as is most common in organic PR materials. Each bar corresponds to an electronic level of a sensitizer or a trap. Here $f(E)$, $D(E)$, and $\mu(E)$ are the generation rate, the diffusion coefficient, and the mobility, respectively, each of which is field-dependent, γ_T and γ_R are the coefficients for trapping and (geminate) recombination respectively, and α_T is the detrapping rate. Finally, N_G, N_G^i, N_T, and N_T^i are the number densities of neutral and ionized sensitizers and traps, respectively.

13.2.1 The Kukhtarev Model for Inorganic Crystaline PR Materials

The so-called Kukhtarev model considers a crystalline semiconductor with a single impurity level in the band gap. Below we present the main features of the Kukhtarev model for an n-type photoconductor, which is the most common case. Figure 13.2 illustrates the model.

Donor impurities N_D may be photoionized (thermal charge generation shall be neglected) and thus donate an electron into the conduction band. Their ionized counterparts N_D^i are capable of capturing free electrons from the conduction band. Furthermore, there are acceptor impurities N_A, the concentration of which is typically much lower than that of the donors ($N_D \gg N_A$). These acceptor impurities are actually ionized donor impurities, which exist without illumination but are compensated by compensation charges in order to maintain charge neutrality. The compensation charges themselves do not contribute directly to the PR effect. Upon nonuniform illumination, charge carriers are excited from donor sites in the bright areas and diffuse or drift into the dark areas, where they get trapped by acceptor sites. This results in a spatially nonuniform space-charge distribution giving rise to the PR space-charge field, according to Poisson's equation (13.8).

Within the Kukhtarev model, charge generation is considered a single-step process, in which photon absorption transfers an electron into the conduction band with a constant photogeneration efficiency ϕ. In the conduction band the electron is delocalized, and its transport is characterized by a constant charge-carrier mobility μ and diffusion coefficient D. These parameters are assumed to be intrinsic, and they are related by the fundamental Einstein relation

$$D = \mu \frac{k_B T}{e_0} \tag{13.2}$$

where e_0 is the modulus of the elementary charge, and $k_B T$ is the thermal energy.

The model assumes a constant number of impurities and is characterized by the following set of fundamental equations:

$$\frac{\partial n_e}{\partial t} = \frac{\partial N_D^i}{\partial t} + \frac{1}{e_0} \frac{\partial J}{\partial x} \tag{13.3}$$

where n_e is the free electron density. The current density J is given by

$$J = e_0 n_e \mu E + e_0 D \frac{\partial n_e}{\partial t}, \tag{13.4}$$

where D and μ are related by (13.2). The rate equation for the ionized donors reads

$$\frac{\partial N_D^i}{\partial t} = f \left(N_D - N_D^i \right) - \gamma_R n_e N_D^i, \tag{13.5}$$

where γ_R is the recombination coefficient according to the Langevin theory [33], which equals the trapping coefficient γ_T:

$$\gamma_R = \gamma_T = \frac{\mu e_0}{\langle \varepsilon \rangle}, \tag{13.6}$$

with $\langle \varepsilon \rangle$ as the permittivity effective in the direction parallel to the grating vector $\vec{K}$.

Neglecting thermal charge generation, the photogeneration rate f of mobile charges may be formulated as

$$f = sI, \tag{13.7}$$

where I is the light intensity, and s is the cross section for photoexcitation of a charge carrier into the conduction band (i.e., includes the charge-generation efficiency). Here f is assumed to be identical to the detrapping coefficient α_T. Finally, the electrical field must satisfy Poisson's equation

$$\frac{\partial E}{\partial x} = -\frac{e_0}{\langle \varepsilon \rangle} \left(n_e - N_D^i + N_A \right). \tag{13.8}$$

As noted above, N_A is required to maintain charge neutrality, or in other words, if n_e is zero, the spatial averages of N_D^i and N_A are identical.

Steady-State Space-Charge Field in Inorganic PR Crystals

The set of nonlinear equations discussed in the preceding section can be solved analytically by means of linearization if a simple two-beam (sinusoidal) interference pattern with low contrast $m \ll 1$ of the light fringe pattern according to

$$m = \frac{2\sqrt{I_1 I_2}}{I_1 + I_2} \cos(2\theta_{int} \, p) \tag{13.9}$$

is assumed. Here, $I_{1,2}$ are the intensities of the recording beams, $p = 0$ ($p = 1$) for s- (p-) polarized writing beams, and $2\theta_{int}$ is the internal interbeam angle (see Figure 13.7). Thus, one obtains for the first-order Fourier component of the steady-state amplitude of the photorefractive space-charge field E_{SC}:

$$|E_{SC}| = m \sqrt{\frac{E_q^2 \left(E_0^2 + E_D^2 \right)}{(E_D + E_q)^2 + E_0^2}}, \tag{13.10}$$

where $E_0 = E_{ext} \cos \varphi$ (see Figure 13.7) is the projection of an externally applied dc field onto the grating wave vector $K = |\vec{K}|$. The field quantities E_D ("diffusion field," a field that can be achieved if diffusion is the only driving force leading to a nonuniform space-charge distribution) and E_q ("saturation field," a hypothetical field strength for the case that all traps were filled) are given by

$$E_D = \frac{K k_B T}{e_0} \tag{13.11}$$

and

$$E_q = \frac{e_0 N_A (N_D - N_A)}{K \langle \varepsilon \rangle N_D} \approx \frac{e_0 N_A}{K \langle \varepsilon \rangle}. \tag{13.12}$$

The phase shift φ_g between the interference pattern and $E_{SC}(x)$ is given by

$$\tan \varphi_g = \frac{E_D}{E_0} \left(1 + \frac{E_D}{E_q} + \frac{E_0^2}{E_D E_q} \right). \tag{13.13}$$

Dynamics of the Space-Charge Field in Inorganic PR Crystals

There are two main processes that determine the dynamics of space-charge-field **formation** in photorefractive crystals, namely the recombination of free-charge carriers (trapping) and the dielectric relaxation of the nonuniform charge distribution. Both processes take place with characteristic rates $1/\tau_r$ and $1/\tau_{di}$, respectively, which are

$$\frac{1}{\tau_r} = \gamma_R N_A \tag{13.14}$$

and

$$\frac{1}{\tau_{di}} = \frac{e_0 \mu n_{e,0}}{\langle \varepsilon \rangle}, \tag{13.15}$$

where $n_{e,0}$ is the zero-order density of mobile electrons, and τ_r corresponds to the lifetime of a free charge carrier. Under experimental conditions, which are typically observed in photorefractive crystals, the time scales for these two processes are significantly different, namely $1/\tau_{di} \ll 1/\tau_r$. In this case the dynamics of the PR space-charge field are described by a single exponential term:

$$E_{SC}(x,t) = |E_{SC}|[-\cos(Kx + \varphi_g) + \cos(Kx + \varphi_g + \omega_g t)\exp(-t/\tau_g)], \tag{13.16}$$

where E_{SC} and φ_g are given by (13.10) and (10.13), respectively, and τ_g and $1/\omega_g$ are the real and imaginary parts, respectively, of the complex time constant:

$$\tau_g = t_0 \frac{(E_\mu + E_D)^2 + E_0^2}{E_D(E_D + E_\mu) + E_q(E_D + E_\mu) + E_0^2}, \tag{13.17}$$

$$\omega_g = \frac{E_\mu E_0 + E_q E_0}{t_0\left((E_\mu + E_D)^2 + E_0^2\right)}, \tag{13.18}$$

where E_D and E_q are given by (13.11) and (13.12), respectively, and E_μ is commonly referred to as "drift field," given by

$$E_\mu = \frac{\gamma_R N_A}{K \langle \mu \rangle} = \frac{1}{K \langle \mu \rangle \tau_r}, \tag{13.19}$$

where $\langle \mu \rangle$ is the effective charge-carrier mobility in the direction parallel to the grating vector $\vec{K}$, and t_0 represents a characteristic relaxation time, including the dielectric relaxation rate

$$t_0 = \tau_{di} \frac{E_q}{E_\mu}. \tag{13.20}$$

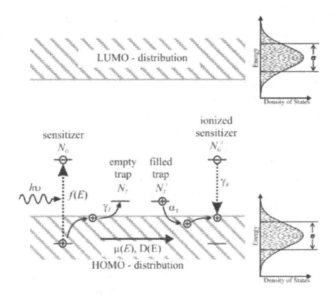

FIGURE 13.3. Energy-level scheme of Schildkraut's model.

For the time dependence of the space-charge-field **erasure** one obtains

$$E_{SC}(x, t) = -|E_{SC,i}| \cos(Kx + \varphi_g + \omega_g t) \exp(-t/\tau_g), \qquad (13.21)$$

with $E_{SC,i}$ as the initial space-charge-field amplitude prior to erasure.

Figures 13.2 and 13.3 illustrate the models developed by Kukhtarev for inorganic crystalline materials and by Schildkraut for organic amorphous materials which will be discussed in detail below. They constitute the simplest cases, i.e., further complications such as double doping in inorganic materials [34] or adding extrinsic traps to organics PR materials are not included.

13.2.2 Peculiarities of Amorphous Organic PR Materials

Many conclusions drawn from applying Kukhtarev's model to organic PR materials seem somehow reasonable. Therefore, this model is quite often used in the field of organic PR materials, in particular, for estimating the number density of PR traps (via (13.12) and (13.13)). However, amorphous organic materials differ substantially from inorganic crystals.

Charge-Carrier Generation

The direct photogeneration of free charge carriers by excitation of an electron from the *highest occupied molecular orbital* (HOMO) to the *lowest unoccupied molecular orbital* (LUMO) in a bulk organic semiconductor is mostly very inefficient, since recombination dominates strongly [35]. Thus, commonly a sensitizer (such as **TNF** shown in Figure 13.1) is added, which, in the case of a hole (electron) conducting polymer matrix, exhibits a significantly higher electron affinity (smaller ionization potential) as compared to the charge-transporting agent. In general, charge generation in organic solids first involves the creation of a bound electron–hole pair, which is separated by the thermalization length. In order to generate a free charge carrier, this pair must be split into individual charges. Due to the low dielectric constant in organic materials, the screening of the Coulomb interaction between electron and hole is small, and charge separation always competes with geminate recombination. By applying an electrical field, charge separation is enhanced, and hence the charge-generation process becomes field-dependent. The field dependence of the photogeneration efficiency $\phi(E)$ is generally described by the Onsager theory [36]. A good numerical approximation was given by Mozumder [37]:

$$\phi(h\nu, E, T) = \phi_0(h\nu)P(E, T). \tag{13.22}$$

Here, the primary quantum yield $\phi_0(h\nu)$, i.e., the fraction of absorbed photons resulting in bound thermalized electron–hole pairs, is considered to be independent of the applied field. It is multiplied by the probability $P(E, T)$ that recombination does not occur, where P is given by the following sum:

$$P = 1 - \frac{1}{2\xi} \sum_{n=0}^{\infty} A_n(\eta)A_n(\xi) \quad \text{with} \quad \xi = \frac{e_0 r_0 E}{2k_B T} \quad \text{and} \quad \eta = \frac{r_C}{r_0}, \tag{13.23}$$

with $A_n(x)$ given by the recursive formula

$$A_n(x) = A_{n-1}(x)\frac{x^n \exp(-x)}{n!} \quad \text{and} \quad A_0(x) = 1 - \exp(-x), \tag{13.24}$$

where r_0 is the thermalization length of the bound electron–hole pair and r_c is the Coulomb radius. For external fields in the range 10 V/μm–100 V/μm, which is typical for operating organic PR materials, $\phi(E)$ can be roughly approximated by a power law [3, 38] with typical values for the power-law exponent p of $p \approx 1.5$–3.5 [39], [51]:

$$\phi(E) = \phi_0 E^p. \tag{13.25}$$

Here, ϕ_0 is a constant to be determined experimentally.

Note that Onsager's model assumes a ballistic nature of the charge-generation process, i.e., the thermalization length and thus also the photogeneration efficiency strongly depend on the photon energy. This mechanism applies to a number of organic polymers (e.g., PPV [40] and polysilanes [41]). However, many organic materials do not show a wavelength dependence of the photogeneration efficiency [35], which contradicts the classical Onsager model. Therefore, based on Onsager's theory, models have been developed to account for this discrepancy, which basically differ in the details of the microscopic mechanism of the generation of the electron-hole pair [42, 43]. A model describing the charge photogeneration in PVK, for which all previous models have failed, is presented in [44].

Charge-Carrier Recombination

In inorganic PR crystals as well as in organic PR materials free charge carriers may recombine with charged generation centers (Figures 13.2 and 13.3) and the recombination coefficient follows the Langevin theory Equation (13.6).

However, in inorganic PR crystals and applying the low-light-intensity approximation (i.e., the number density of free charge carriers is negligibly small), the number density of recombination centers is constant and equals the density of compensating charges (i.e., N_A). In contrast, in organic PR polymers, the number density of recombination centers inherently depends on the electrical field and on the light exposure [45]. An experimental confirmation was given by Bäuml et al. [46].

Charge Transport

In amorphous organic materials without long-range order, the overlap of the molecular wave functions is poor, the formation of energy bands is inhibited, and charge carriers are located on **individual sites** (charge-transporting agents such as carbazole units shown in Figure 13.1) rather than dislocated in some band structure. Hence, charge transport requires the exchange of electrons between neighboring sites. The most commonly used model to describe charge transport in amorphous organic matrices is the hopping model developed by Bässler et al. [47]. It is assumed that carrier transport proceeds via hopping between spatially separated redox sites; i.e., the charge-transporting species are radical ions. The photoconductivity depends, among others, on the polarity of the matrix, on the average distance and mutual orientation between two hopping sites, and on the temperature. Transport in such amorphous organic solids is highly dispersive due to the wide distribution of the energy levels of the hopping sites and the distance and mutual orientations between two neighboring sites. Both aspects, energetic and spatial disorder, are modeled assuming Gaussian distributions for the density of states (DOS; Figure 13.3).

The field dependence of the hole mobility $\mu(E)$ is described by

$$\mu(\overline{\sigma}, \Sigma, E) = \mu_0 \exp\left[-\left(\frac{2\overline{\sigma}}{3}\right)^2\right] \begin{cases} \exp[C(\overline{\sigma}^2 - \Sigma^2)\sqrt{E}], & \Sigma \geq 1.5, \\ \exp[C(\overline{\sigma}^2 - 2.25)\sqrt{E}], & \Sigma \leq 1.5. \end{cases}$$
(13.26)

Here, μ_0 is the zero-field mobility, $\overline{\sigma} = \sigma + k_B T$ is a measure for the energetic disorder in the system, σ is the width of the energetic distribution functions, Σ is a parameter taking into account the spatial disorder, and C is an empirical parameter ($C = 2.9 \times 10^{-4}$ (cm/V)$^{1/2}$). For operating parameters typical of organic PR materials the field dependence of the charge-carrier mobility may be approximated by [48]

$$\mu(E) = \mu_i \exp[C'(\sqrt{E} - 1)] \qquad (13.27)$$

where μ_i is a constant and C' is an experimentally obtained characteristic parameter. Note that the Einstein relation is not valid for amorphous organic materials.

Charge-Carrier Traps

In inorganic PR systems there is a given and constant number of permanent traps; photogeneration and trapping are performed by the same chemical component in different redox states: "donors" M^{x+} and "acceptors" $M^{(x+1)+}$, respectively (see Figure 13.2). In amorphous organic systems the physical nature of the traps is not yet fully understood. Ionized sensitzers can be considered one kind of (charged) trap [45], the trapping behavior of which then follows the Langevin theory as for the case of charge-carrier recombination. On the other hand, there is very strong indication that in addition to the ionized donors there are also neutral trapping sites, for which the Langevin theory does not apply. Transport proceeds via hopping between the localized sites. Thus, each transport site may be considered as a potential intrinsic trap. In other words, traps are an inherent part of the transport manifold. Thus, one may speculate that these neutral traps are conformational traps close to the edge of the DOS (see Figure 13.3). Conformational traps are typically shallow traps, and charge carriers trapped therein may be detrapped thermally and then further participate in the overall process. Furthermore, the presence of some deep traps must be assumed as well. Charge carriers trapped therein cannot be detrapped thermally, but only by processes providing significantly higher energy than the thermal energy, e.g., by optical excitation.

Unlike in inorganic PR crystals, the number density of traps is inherently dependent on a manifold of physical conditions, e.g., electric field, light exposure [45], reduced temperature (13.1) and others. In particular, the trap density decreases as a function of the reduced temperature for $T_r < 0°C$, which has been attributed to slow collective motions of polymer-chain subunits inside the photoconducting polymer matrix below the glass transition point [49]. By that, the energy landscape of the conducting matrix is subject to a permanent flow, which on average reduces the lifetime of conformational traps and thus leads to a decrease in the effective trap density (see also Section 13.4.2).

Schildkraut et al. [2,3] were the first to include these aspects in a fairly general model to describe space-charge effects in disordered organic solids. A dynamic solution for the space-charge-field buildup in this model was given by Yuan et al. [50]. Schildkraut's model was advanced by Ostroverkhova et al. [51], also considering the buildup dynamics. Based on the original model, Cui et al. [52] considered the erasure dynamics of the space-charge field. In the following, Schildkraut's model will be presented including the advances and contributions enumerated above.

13.2.3 The Schildkraut Model for Amorphous Organic PR Materials

A schematic of Schildkraut's model [2,3] is depicted in Figure 13.3. In contrast to most inorganic systems, organic solids generally support hole transport. Mobile holes (number density n_h) are generated by photoexcitation of a so-called sensitizer (initial number density $N_{G,i}$), which must be an acceptor for the excess electron. Since the number density of sensitizer molecules is too small (typically 10^{19} cm^{-3}) to form a separate charge-transport manifold, electrons are assumed to be immobile, i.e., locally fixed on the sensitizer moieties. Accordingly, charged sensitizer moieties (number density $N_G^i(t)$) can act as recombination centers for mobile holes.

The most important deviation from Kukhtarev's model is that Schildkraut introduces an additional trap level (initial number density $N_{T,i}$, number density of filled traps $N_T^i(t)$), from which the carriers can be detrapped thermally (i.e., shallow traps). This leads to the following set of nonlinear equations in analogy to the equations (13.3) to (13.8):

$$\frac{\partial n_h}{\partial_t} = \frac{\partial N_G^i}{\partial t} - \frac{\partial N_T^i}{\partial t} - \frac{1}{e_0}\frac{\partial J}{\partial x}, \tag{13.28}$$

$$J = e_0 n_h \mu E - k_B T_\mu \frac{\partial n_h}{\partial t}, \tag{13.29}$$

$$\frac{\partial N_G^i}{\partial t} = f(N_{G,i} - N_G^i) - \gamma_R n_h N_G^i, \tag{13.30}$$

$$\frac{\partial N_T^i}{\partial t} = \gamma_T n_h (N_{T,i} - N_T^i) - \alpha_T N_T^i, \tag{13.31}$$

$$\frac{\partial E}{\partial x} = \frac{e_0}{\varepsilon}(n_h - N_G^i + N_T^i), \tag{13.32}$$

where $\gamma_T n_h$ is the trapping rate and α_T is the detrapping rate. The hole-generation rate is given by

$$f = \frac{\phi}{h\nu} s I, \tag{13.33}$$

where s is the cross section for light absorption, $h\nu$ is the photon energy, and ϕ is the generation efficiency for photoinduced charge carriers. Note that the second term of (13.29) assumes the validity of the Einstein relation (13.2), although it is known to be inapplicable in this case. However, it is the only known analytical relation between charge-carrier mobility and diffusion coefficient.

The above equations, which take into account shallow traps, do not differ substantially from those in crystalline photoconductors [53,54]. A fundamental difference appears only after the main photoconductivity parameters, namely $\phi(E)$ and $\mu(E)$, are introduced as field dependent in a semi-phenomenological way. In the following sections, solutions for the Schildkraut model will be discussed.

Steady-State Solution for the Space-Charge Field in Amorphous Organic Solids

By means of linearization, Schildkraut et al. derived analytical solutions for the zero- and first-order Fourier components of the free-charge-carrier density and the internal electric field from the set of nonlinear equations (13.28) to (13.32) together with (13.25) and (13.27). The number of variables in the model exceeds the number of independent equations, and thus additional assumptions are required in order to obtain an analytical solution for the PR space-charge field.

The system described by (13.28) to (13.32) shall be analyzed here by an approach different from those used in the current literature [55]. It was discussed before that Schildkraut's model is essentially based on the standard Kukhtarev model and introduces a trap level for holes as well as a field dependence for the photoconductivity parameters involved. For now, the field dependence of the photoconductivity parameters is set aside, and we shall discuss the impact of the introduction of the additional trap level T on the standard Kukhtarev model. Two limiting cases for the trap level are to be considered.

(A) **Shallow traps:** If thermal *detrapping* is high, the traps act as a reservoir for the mobile charge carriers. In this case the impact of the trap level on the photoconductivity can again be reduced to two limiting cases:

(A1) **Slight trap filling** ($n_h \ll N_T$): In this case (reservoir almost empty) the free carriers are easily trapped and detrapped, spending more time in the trap levels than contributing to the photoconductivity. This leads to an increase of the effective carrier lifetime and to a decrease (by the same factor) of the effective mobility as compared to their values in the case of no shallow traps. Since mobility and lifetime usually contribute as the product to the most important photoconductivity parameters (e.g., diffusion length, Debye screening length, average photoconductivity), the presence of shallow traps does not affect these parameters. The only manifestation of shallow traps will be a slower dynamics of the photoinduced charge carriers.

(A2) **Strong trap filling** ($n_h \gg N_T$): This limit (reservoir is full) is equivalent to the case of no shallow traps, and the effective lifetime and effective mobility tend to their real values. However, in the transient regime the photocurrent may be enhanced as compared to its steady-state value. Strong trap filling is hardly realized in organic amorphous photoconductors, where the number of shallow traps is typically very large compared to the number of intrinsic free carriers.

(B) **Deep traps:** The opposite limit is reached when *no detrapping* occurs. In this approximation the microscopic model of Schildkraut is identical to the standard Kukhtarev model, and the results for space-charge-field amplitude and phase are identical, only replacing donor and acceptor densities (N_D and N_A) by sensitizer and trap densities ($N_{G,i}$ and $N_{T,i}$), respectively, as was noticed by Schildkraut in his pioneering publication [2] (see also Figures 13.2 and 13.3). Note that the effective charge-carrier lifetime is reduced by deep traps.

Now the role of the field-dependent photoconductivity parameters can be analyzed by simply introducing $\phi(E)$ and $\mu(E)$. For now, the dependence on the external dc field will be arbitrary. Equations (13.28) to (13.32) are linearized in the usual way ($m \ll 1$), yielding for the first-order Fourier component of the space-charge field

$$|E_{SC}| = \frac{m E_q \sqrt{E_0^2 + E_D^2}}{\sqrt{\left(E_D + E_q(1 + F^{(1)}) + E_\mu \frac{\tau_r}{\tau_{di}}\right)^2 + \left(E_0 + \frac{E_D E_q}{E_0} F^{(2)}\right)^2}}, \tag{13.34}$$

where E_D, E_q, E_μ, τ_r, and τ_{di} are given by (13.11), (13.12), (13.19), (13.14), and (13.15), respectively, however, replacing N_A by $N_{T,i}$, N_D by $N_{G,i}$, and n_e by n_h. The dimension-less factors $F^{(1)}$ and $F^{(2)}$ are

$$F^{(1)} = \frac{\partial(\ln(\phi))}{\partial E} + \frac{E_\mu}{E_q} \frac{\tau}{\tau_{di}} \frac{\partial(\ln(\mu))}{\partial E}, \tag{13.35}$$

$$F^{(2)} = \frac{\partial(\ln(\mu))}{\partial E} - \frac{\partial(\ln(\phi))}{\partial E}, \tag{13.36}$$

reflecting the field dependence of the photoconductivity parameters.

The term $E_\mu \tau_r / \tau_{di}$ appears because a charge-carrier lifetime in the range of the dielectric relaxation time was assumed here. This term vanishes in the limit $\tau_r \ll \tau_{di}$, i.e., where the standard Kukhtarev equations were derived. If additionally the charge-carrier mobility and the photogeneration efficiency are constant, $F^{(1)}$ and $F^{(2)}$ vanish as well, and the amplitude of the space-charge field is described by the expression (13.10).

Provided that $\partial(\ln(\mu))/\partial E > 0$ and $\partial(\ln(\phi))/\partial E > 0$, which is generally fulfilled in amorphous organic photoconductors, the application of an external DC field will reduce the amplitude of the space-charge field as compared to a hypothetical material with identical energy levels, but with constant ϕ and μ.

The phase shift can also be expressed in terms of the field factors $F^{(n)}$:

$$\tan \varphi_g = \frac{E_0^3 + E_0 E_D \left(E_D + E_q + E_\mu \frac{\tau_r}{\tau_{di}}\right) + E_0 E_D F^{(3)} \left(E_\mu \frac{\tau_r}{\tau_{di}} + E_q\right)}{E_q(E_D^2 F^{(1)} - E_0^2(1 + F^{(2)}))}, \tag{13.37}$$

where

$$F^{(3)} = \frac{\partial(\ln(\mu))}{\partial E}. \tag{13.38}$$

Equations (13.34) and (13.37) can be compared with the results obtained from Schildkraut's model [2] with the specific dependence of $\phi(E)$ and $\mu(E)$ as discussed in the previous sections. Substituting (13.25) and (13.27) in (13.35) and (13.36), expressions are obtained that are very similar to those known from the

Schildkraut model reading (SI units):

$$|E_{SC}| = \frac{m\ell E_q \sqrt{E_D^2 + E_0^2}}{\sqrt{\left(E_D + E_q(1 + \ell p) + E_I\left(\frac{E_m}{E_0} + 1\right)\right)^2 + \left(E_0 + \frac{E_D E_q}{E_0}\left(\frac{E_m}{E_0} - \ell p\right)\right)}},$$
(13.39)

$$\tan \varphi_g = \frac{E_0^2 + E_D(E_D + E_q + E_I) + \frac{E_D E_m}{E_0}(E_I + E_q)}{\frac{E_D^2 E_q}{E_0}\left(\frac{E_m}{E_0} - \ell p\right) - E_0 E_q(1 + \ell p) - E_0 E_I\left(\frac{E_m}{E_0} + 1\right)}.$$
(13.40)

Here, p is the exponent of the power law (13.25). The only difference is the parameter ℓ, given by:

$$\ell = f_0 \frac{(N_{G,i} - N_{N,i} - n_0)}{\gamma_{R0} n_0 (n_0 + N_{T,i})},$$
(13.41)

which approaches unity for the limit of "no detrapping" (case B), i.e., where (13.39) was derived in [2]. The field quantities are defined as

$$E_D = \frac{k_B T}{e_0} K,$$
(13.42)

$$E_q = E_I \frac{\gamma_{R0}(n_0 + N_{T,i})}{f_0 + \gamma_{R0} n_0},$$
(13.43)

$$E_I = \frac{e_0 n_0}{\langle \varepsilon \rangle K} = E_\mu \frac{\tau_r}{\tau_{di}},$$
(13.44)

$$E_m = \frac{C'}{2} \sqrt{E_0 E_0}.$$
(13.45)

Here, γ_{R0}, f_0, and n_0 are the zero-order Fourier components of the recombination coefficient, the hole-generation rate, and the hole density, and $N_{G,i}$ and $N_{T,i}$ have been introduced with (13.30) and (13.31); C' has been introduced with (13.27), and all other parameters have been introduced with (13.10) to (13.13).

Dynamics of the Space-Charge Field in Amorphous Organic Solids

As discussed in the previous section, the introduction of shallow traps (case A) will increase the effective photocarrier lifetime. When the latter becomes comparable with the time scale of the experiment, in addition to the τ_g discussed in Section 13.2.1, which depended only on τ_{di}, there is another characteristic time proportional to the effective carrier's lifetime, which now influences the dynamics of the space-charge-grating **formation**. Thus the single exponential dynamics as derived for Kukhtarev's model ((13.16) and (13.21)) is expanded by a second exponential term. These simple qualitative considerations are supported by the analytical expression derived by Yuan et al. for the buildup dynamics of the PR space-charge field [56] and by the considerations of Cui et al. for the erasure of the space-charge field.

Based on Schildkraut's model, Yuan et al. derived an analytical expression for the buildup dynamics of the PR space-charge field [56]:

$$E_{SC}(x, t) = R_a \begin{Bmatrix} \cos(Kx + \varphi_a) - R_b \cos(Kx + \varphi_a + \varphi_b + \omega_1 t) \exp(-t/\tau_1) \\ -R_c \cos(Kx + \varphi_a + \varphi_c + \omega_2 t) \exp(-t/\tau_2) \end{Bmatrix}.$$

$$(13.46)$$

For the explicit form of the parameters $R_{a,b,c}$, $\varphi_{a,b,c}$, $\omega_{1,2}$, and $\tau_{1,2}$, reference is made to the original literature. Note that Yuan et al. derived two complex time constants of the system, where $1/\omega_{1,2}$ and $\tau_{1,2}$ are the imaginary and the real parts of these, respectively. According to (13.46), Schildkraut's model implies a very complicated behavior of the space-charge field as a function of time consisting of two exponentially decaying space-charge waves, which may additionally shift spatially with different speeds relative to a time-independent reference (first term in (13.46)). Thus, an oscillation of the space-charge field during buildup may be anticipated, which, however, has not been observed in polymers so far. Based on their framework, Yuan et al. carried out numerical simulations assuming various trapping conditions [50] and the absence of detrapping (case B; see above). Yet, the validity of the obtained numerical predictions requires experimental proof, which, in turn, would require reliable control over the trapping conditions in the experiment.

Ostroverkhova et al. [51] further developed Schildkraut's model by introducing two dedicated trap levels, both of which may be emptied thermally, i.e., assuming the limit of detrapping with slight trap filling (case A1; see above). The set of nonlinear equations (13.28) to (13.32) thus is expanded by a formal duplicate of (13.31), and (13.28) as well as (13.32) then contain two dedicated trap densities. Like Schildkraut et al., the authors applied the Fourier decomposition approach to find a solution for the expanded set of nonlinear equations. However, the time dependence of the zero-order Fourier parameters was also taken into account, which have been considered as changing instantaneously in all earlier investigations. The model thus modified can no longer be solved analytically, but must be solved numerically or semiempirically. The authors demonstrated a semiempirical solution for the grating buildup dynamics in a PR polymer determining several system parameters of the model directly by means of photophysical measurements (photocurrent transients) and comparison with Monte Carlo simulations of the zero-order Fourier components of the set of nonlinear equations. This way, detrapping and recombination rates and furthermore the products of trapping rates and trap densities were obtained. The photogeneration rate and the charge-carrier mobility were determined by xerographic discharge and time-of-flight experiments, respectively. The authors used the obtained results to numerically solve for the first-order Fourier component of the internal electrical field, i.e., the PR space-charge field, as a function of time. The results finally obtained showed good agreement with the experiment. For fitting their experimental data on the space-charge buildup dynamics the authors applied an associative biexponential fit function (13.95) [51].

Kulikovsky et al. expanded the above-described photophysical approach and thus managed to obtain also trapping rates and trap densities in a PR polymer composite separately [32].

Based on Schildkraut's model, Cui et al. considered the **erasure** of the PR space-charge field in polymers [57]. The authors assumed a distribution of the site energies for traps and charge-carrier (holes) generation sites as well as for the corresponding parameters involving the energetic depth of a site (detrapping rate, hole generation rate, etc.). By integration over all energy levels and, justified by physical arguments, applying the mean value theorem, Cui et al. obtained discrete values of the parameters subjected to an energy distribution, which are characteristic of the system under consideration. The zero-order Fourier components were assumed to change much faster than the first-order Fourier components and were thus considered constant in time. The authors finally derived an analytical expression for the decay of the PR space-charge field under uniform illumination, the one-dimensional complex form of which reads

$$
E_1(t) = -\frac{ie_0}{\langle \varepsilon \rangle K} \left\{ \begin{array}{l} R \exp(-t/\tau_1) + \dfrac{(DK^2 + iK\mu E_0)N_{T1,i}}{1/\tau_1 - 1/2} \exp(-1/\tau_2) \\[2ex] - \dfrac{(DK^2 + iK\mu E_0)N_{G1,i}}{1/\tau_1 - 1\tau_3} \exp(-1/\tau_3) \end{array} \right\},
$$

(13.47)

where E_0 is the projection of the applied electric field onto the grating wave vector K, D is the diffusion constant, e_0 is the elementary charge, $\langle \varepsilon \rangle$ is the effective permittivity of the medium, and R is a constant; $N_{T,i}$ and $N_{G,i}$ are the initial values of the first-order Fourier components of the trap density and the density of hole generators, respectively. The time constants $\tau_{1,2,3}$ are given by

$$
\frac{1}{\tau_1} = \frac{e_0\mu}{\langle \varepsilon \rangle}n_0 + DK^2 + iK\mu n_0,
$$

(13.48)

$$
\frac{1}{\tau_2} = f_0 + \gamma_R n_0,
$$

(13.49)

$$
\frac{1}{\tau_3} = \alpha_T + \gamma_T n_0,
$$

(13.50)

respectively. Here, n_0 is the zero-order hole density, f_0 and α_T are the hole photogeneration and the hole detrapping rate, and γ_R and γ_T are the geminate recombination and the hole-trapping coefficient, respectively. Note that the time constant τ_1 is complex and implies a space-charge-field component spatially shifting with respect to a time-independent reference (the second and third terms in (13.47)), i.e., an oscillation. This, however, has not been observed in PR polymers so far. Based on this empirical argument, Cui et al. proposed that the imaginary part in (13.48) may be omitted. Applying furthermore a low-light-intensity approximation (i.e., $f_0 \ll \gamma_R n_0$) and taking into account that the diffusion coefficient is typically small in polymers, Cui et al. proposed that the first two time constants should be very similar and experimentally barely distinguishable. Hence two decay-time constants should be anticipated for the PR space-charge field in polymers.

Another approach to the dynamics of the PR space-charge-field erasure might come from the analysis performed by Yuan et al. for the space-charge-field buildup [50]. Inverting the boundary conditions, which have been applied to derive (13.46), a similar expression is obtained, however, without the time-independent reference term (first term in (13.46)). This also yields two time constants and still implies oscillation effects, since the decay of the space-charge field is then described by two decaying space-charge waves shifting spatially with different velocities.

Note that this model may also be employed to describe the dark decay of the space-charge field. In this case, the detrapping rate α_T is correlated exclusively to thermal detrapping. Since f_0 and the zero-order hole density n_0 in the dark may be assumed to be negligibly small (provided there is no significant charge injection from the electrodes), the model suggests a single exponential dark decay behavior. However, this is contradicted by experimental results [58].

One should keep in mind that charge transport in amorphous systems is typically highly dispersive. Thus, it appears reasonable to assume that the time constants are not very well defined, but might be subject to some kind of distribution reflecting the energetic disorder [47]. The fact that the space-charge-field dynamics in the experiment usually cannot be fitted accurately (see Section 13.3.3), but only approximated by the number of exponential terms as discussed in this section, might support this assumption. However, it may also indicate that Schildkraut's model is still not sufficiently elaborated to reflect the actual situation of PR polymers.

Note finally that the actual dynamics of the PR index grating formation may additionally be determined by the poling dynamics of the EO chromophores of the material. This will be discussed in the following section.

13.2.4 Electrooptic Effects

In inorganic PR crystals the linear electrooptic (Pockels) effect translates the space-charge field into a refractive index modulation:

$$\Delta n = -\frac{n^3}{2} r_{\text{eff}} |E_{\text{SC}}|, \tag{13.51}$$

where r_{eff} is the effective electrooptic coefficient, a material-specific constant. Effects due to a Kerr mechanism do not play a role.

In contrast, in amorphous organic PR materials, both EO effects, Pockels and Kerr, contribute and depend on the externally applied field. In the following we will discuss the "oriented-gas model" as introduced by Kuzyk and Singer [59] and further developed by Wu [60] to describe the steady-state electrooptic properties

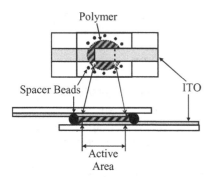

FIGURE 13.4. Typical device structure. Above: top view; below: side view.

of polymers. For the dynamics of chromophore orientation in an electric dc field, a more powerful model was developed by Binks et al. [61]. Finally, the "orientational enhancement mechanism" observable in low-glass-transition PR polymers and discovered by Moerner et al. [62] will be considered.

In order to introduce electrooptic (EO) effects into the material, noncentrosymmetric dipolar compounds (such as the **DMNPAA** shown in Figure 13.1), usually referred to as "chromophores" due to their colored appearance, are incorporated. However, for macroscopic second-order NLO effects the randomly distributed dipoles need to be oriented by poling in an external electrical field.

Therefore, typical photorefractive devices using organic amorphous materials as the active medium are sandwich structures (Figure 13.4) of the material between two glass slides coated with transparent electrodes (e.g., indium tin oxide, ITO) for the application of the electric field, the direction of which is then perpendicular to the device normal (z-direction).

The orientation of dipoles by an electric poling field E_p may be described by the so-called oriented-gas model introduced by Wu [60]. A Boltzmann distribution of the dipole angles in free space is assumed, i.e., any interaction with the matrix or other chromophores is neglected. It is important to note that the poling introduces birefringence into the material (see Figure 13.5, right). In order to take advantage of electrooptic effects, a tilted transmission geometry must be used, since for normal incidence the devices remain isotropic independently of the poling (see Figure 13.4, left). As a result, the sensed index modulation becomes polarization-dependent. In the following, we distinguish s-polarization (polarized "senkrecht" (German for perpendicular) to the plane of incidence) and p-polarization (polarized "parallel").

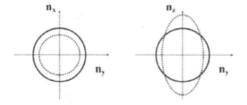

FIGURE 13.5. Refractive index ellipsoids before (solid line) and after poling (dashed line). Left side: top view; right side: side view.

In order to calculate the refractive index changes in a PR diffraction experiment (see Section 13.3.2, Figure 13.7), two limiting material classes are commonly distinguished (T_r according to (13.1)):

(M1) **"hard" materials** with "high" glass-transition temperature ($T_r > 60°C$) and

(M2) **"soft" materials** with "low" $T_g (T_r < 10°C)$ often introduced on purpose by adding a plasticizer (such as the **ECZ** shown in Figure 13.1).

However, there is no strictly defined borderline between these two limits. In both cases, poling must be performed close to T_g. Thus, for high-T_g polymers the temperature is elevated and the material is cooled after poling to room temperature (RT) with the external field applied, in order to preserve the orientational order of the chromophores. Unfortunately, over time, relaxation may occur. In contrast, low-T_g materials can be poled at room temperature (in situ), and the poling dynamics may significantly influence or even entirely determine the PR response dynamics of the material [63].

The Oriented Gas Model

The mean orientation of an assembly of molecular dipoles in an electric field can be described by a set of nth-order parameters A_n The nth-order parameter is defined as the Legendre polynomial of nth degree P_n averaged over the angular distribution of the dipoles with Θ as the angle enclosed by the direction of the applied field and the dipole axis [60]:

$$\langle P_n(\cos\theta)\rangle = A_n. \tag{13.52}$$

The second-order parameter A_2 directly correlates the first-order (i.e., the linear) polarizability of the molecular dipoles with the macroscopic birefringence of the dipole assembly, whereas the first- and third-order parameters A_1 and A_3 link the second-order (i.e., the first hyper-) polarizability to the macroscopic second-order (i.e., electrooptic) susceptibility. Thus, based on the microscopic molecular constants of the chromophores, the macroscopic changes of the linear and nonlinear refractive indices of polymers doped with NLO chromophores upon application of a poling field E_p (in the z-direction) can be derived.

The oriented-gas model yields for the **steady-state** induced birefringence (first-order index change $\Delta n^{(1)}$) and the second-order index change ($\Delta n^{(2)}$) sensed by light parallel or perpendicular, respectively, to the poling direction

$$\Delta n_Z^{(1)}(\omega) = \frac{2}{n}B, \tag{13.53}$$

$$\Delta n_{X,Y}(\omega) = -\frac{1}{2}\Delta n_Z^{(1)}(\omega), \tag{13.54}$$

$$\Delta n_Z^{(2)}(\omega) = \frac{3}{2n}C, \tag{13.55}$$

$$\Delta n_{X,Y}^{(2)}(\omega) = \frac{1}{3}\Delta n_z^{(2)}(\omega), \tag{13.56}$$

with

$$B = \frac{2NF_\infty}{45} \Delta\alpha u^2 \tag{13.57}$$

and

$$C = \frac{NF_\infty^2 F_0}{15} \beta u. \tag{13.58}$$

Here, u is the interaction energy of the molecular dipoles μ^* with the applied field E normalized by the thermal energy $k_B T$:

$$u = (\vec{\mu}^* \vec{E})/(k_B T) \tag{13.59}$$

Furthermore, ω is the circular frequency of the beam, and n is the bulk (average) refractive index; Z and X, Y denote the directions parallel and perpendicular to the poling direction, respectively, N is the number density of the chromophores, $\Delta\alpha = (\alpha_{||} - \alpha_\perp)$ is the anisotropy of the linear polarizability parallel and perpendicular to the chromophore molecular axis, respectively, and β is the first hyperpolarizability. F_0 and F_∞ are local-field correction factors taking into account the influence of the surrounding polymer matrix on the local electric field:

$$F_0 = (\varepsilon(\varepsilon_\infty + 2))/(2\varepsilon + \varepsilon_\infty), \tag{13.60}$$
$$F_\infty = (\varepsilon_\infty + 2)/3. \tag{13.61}$$

Here, ε is the static dielectric constant of the system, and ε_∞ is the dielectric constant at the frequency of the optical field involved.

In low-T_g PR materials (case M2; see above), the **orientational dynamics** of the chromophores in the polymer host matrix is of paramount importance for the overall dynamics of the PR response, since the possibility of in situ poling is responsible for the excellent performance of low-T_g PR polymers. The orientational dynamics of an assembly of molecular dipoles in a gas-like phase, i.e., which is not incorporated in a polymer host matrix, may also be described in terms of the oriented-gas model [60]. It has been assumed for a long time that this also applies in reasonable approximation for the case that the molecular dipoles are embedded in a polymer host matrix as far as T_g of the system is below or at least close to the room temperature (i.e., in low-T_g materials). However, Ribierre et al. have proven experimentally by comparing the electrooptic response behavior of a low-T_g PR polymer with the purely rheological response behavior of the system (dynamic shear compliance) that the poling dynamics of the molecular dipoles is dominantly ruled by the mechanical properties of the polymer matrix [64]. This renders the application of the oriented gas model to the poling dynamics of the chromophores in a PR polymer system purely phenomenological. Nevertheless, this approach has often been used for investigating the dynamic properties of low-T_g PR polymers [63, 65, 66]. Therefore, the latest developments in applying the oriented-gas model to the poling dynamics of chromophores embedded in an amorphous polymer matrix will be outlined in the following.

Binks et al. [61, 67, 68, 69] expanded the oriented-gas model to describe the poling and relaxation of molecular dipoles in media of arbitrary disorder by introducing a disorder parameter analogously to the Scher–Montroll formalism [70], a method to describe dispersive charge transport in conducting polymers. Hence, the authors described the time dependence of the rotational diffusion constant $D(t)$ by the following expression:

$$D(t) = D_0 t^{-1(1-\alpha)}, \tag{13.62}$$

where D_0 is the diffusion constant and $0 \leq \alpha \leq 1$ is a measure of the degree of disorder in the system.

By solving the rotational diffusion equation the authors derived an expression for the evolution of the order parameter A_2 as a function of time. A_2 is proportional to the birefringence of the system. For the case of turning on the poling field, it reads

$$A_2(t) = \frac{u^2}{u^2 + 15}\left[1 + \frac{\tau_1}{\tau_2 - \tau_1}\exp\left(-\frac{t^\alpha}{\alpha\tau_1}\right) - \frac{\tau_2}{\tau_2 - \tau_1}\exp\left(-\frac{t^\alpha}{\alpha\tau_2}\right)\right]. \tag{13.63}$$

Here, u is given by (13.59), and the time constants $\tau_{1,2}$ are given by

$$\tau_{1,2} = 1/[2D_0(2 \mp \sqrt{1 - u^2/5})]. \tag{13.64}$$

For complete order, i.e., $\alpha = 1$, (13.63) reduces to the biexponential form obtained by Wu [60] (neglecting terms correlated with orders of the modified Bessel function higher than 2, an approximation that has also been applied by Binks et al.):

$$A_2(t) = \frac{u^2}{u^2 + 15}\left[1 + \frac{\tau_1}{\tau_2 - \tau_1}\exp\left(-\frac{t}{\tau_1}\right) - \frac{\tau_2}{\tau_2 - \tau_1}\exp\left(-\frac{t}{\tau_2}\right)\right], \tag{13.65}$$

whereas for complete disorder, i.e., $\alpha = 0$, (13.63) becomes a power law:

$$A_2(t) = \frac{u^2}{u^2 + 15}\left[1 + \frac{\tau_1}{\tau_2 - \tau_1}t^{-\frac{1}{\tau_1}} - \frac{\tau_2}{\tau_2 - \tau_1}t^{-\frac{1}{\tau_2}}\right]. \tag{13.66}$$

For the case of turning off the poling field, Binks et al. obtained a power-law dependence, whereas Wu derived a monoexponential expression:

$$A_2(t) = A_{20}t^{-6D_0}, \tag{13.67}$$

$$A_2(t) = A_{20}\exp\left(-\frac{t}{6D_0}\right). \tag{13.68}$$

Note that the dynamics of the electrooptic contribution to the overall PR refractive index change would have to be described by the dynamic evolution of the first- and the third-order parameters $A_1(t)$ and $A_3(t)$, respectively. Thus the dynamics of the electrooptic contribution is more complicated, involving three time constants of significant amplitude.

Finally, in analogy to the remark regarding the formation dynamics of the space-charge field, one should keep in mind that a highly dispersive orientational dynamics in amorphous systems may result in a distribution for each time constant, which reflects small differences in the local environment of the individual chromophores. However, this applies only to the case that orientational dynamics plays a significant role for the photorefractive response.

Orientational Enhancement Mechanism

In photorefractive holographic experiments on high-T_g (case M1; see above) prepoled polymers, the refractive index changes originate solely from the EO effect due to the structural rigidity, which ideally does not allow for rotational diffusion of the chromophores. Similarly to inorganic PR crystals, the resulting index modulation is given by

$$\Delta n = -\frac{1}{2} n^3 r_{\text{eff}} E_{\text{SC}}, \tag{13.69}$$

$$r_{\text{eff}}^s = r_{13} \cos \varphi, \tag{13.70}$$

$$r_{\text{eff}}^p = r_{33} \cos \varphi \sin \alpha_1 \sin \alpha_2 + r_{13} \cos \varphi \cos \alpha_1 \cos \alpha_2 + r_{13} \sin \varphi \sin(\alpha_1 + \alpha_2). \tag{13.71}$$

Here, φ is the internal angle enclosed by the grating vector and the sample normal (z-axis) and $\alpha_{1,2}$ are the angles between each beam and the sample normal (Figure 13.7).

In contrast, polymers the T_g of which is only a bit larger or even lower than room temperature can be poled at room temperature in situ during holographic recording. The total local electric field E_T is the superposition of the uniform external field E_{ext} and the nonuniform internal space-charge field E_{SC} and thus, varies periodically in amplitude and direction (Figure 13.6, equation (13.72)). This leads to spatial variations of the direction of the local orientation and the degree of poling, and the refractive index change becomes quadratically proportional to the external electric field [62]. This square dependence has important consequences on the development of the PR grating in materials that are poled in situ, as

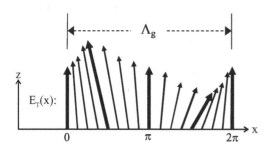

FIGURE 13.6. Illustration of the "orientational enhancement mechanism." See text for details.

illustrated by

$$
\begin{aligned}
E_T(x) &= [E_{ext} + E_{SC}\sin(Kx)]^2 \\
&= E_{ext}^2 + 2E_{ext}E_{SC}\sin(Kx) + [E_{SC}\sin(Kx)]^2 \\
&= E_{ext}^2 + 2E_{ext}E_{SC}\sin(Kx) + E_{SC}^2\left[\frac{1}{2} - \frac{1}{2}\cos(2Kx)\right] \\
&= \left[E_{ext}^2 + \frac{1}{2}E_{SC}^2\right] + 2E_{ext}E_{SC}\sin(kx) + \frac{1}{2}\cos(2Kx)E_{SC}^2.
\end{aligned}
\tag{13.72}
$$

Thus, besides the first-order "Bragg" grating (grating vector K, period Λ_g), which is proportional to the middle term in (13.72), an additional second-order grating (grating vector $2K$, period $\Lambda_g/2$) is formed in the material. This behavior is very different from the situation in PR crystals or permanently poled polymers, where gratings with higher spatial frequencies are solely Fourier components of a not ideally sinusoidal primary index grating. Shakos et al. [71] found that the $2K$ grating eats up about 13% of the total index modulation amplitude.

For the first-order refractive index changes one obtains for s- and p-polarization of the write beams, respectively,

$$
\Delta n_{K,s/p}^{(1)} = \frac{1}{n}\Delta n_Z^{(1)} E_{ext}E_{SC}G_{s/p}^{(1)},
\tag{13.73}
$$

$$
\Delta n_{K,s/p}^{(2)} = \frac{4}{3n}\Delta n_Z^{(2)} E_{ext}E_{SC}G_{s/p}^{(2)},
\tag{13.74}
$$

$$
\Delta n_{K,s/p} = \Delta_{K,s/p}^{(1)} + \Delta n_{K,s/p}^{(2)} = \left(\frac{B}{2n}G_{s/p}^{(1)} + \frac{2C}{n}G_{s/p}^{(2)}\right)E_{ext}E_{SC},
\tag{13.75}
$$

$$
G_s^{(1)} = -\cos\varphi,
\tag{13.76}
$$

$$
G_p^{(1)} = 2\cos\varphi\sin\alpha_1\sin\alpha_2 - \cos\varphi\cos\alpha_1\cos\alpha_2 + \frac{3}{2}\sin\varphi\sin(\alpha_1 + \alpha_2),
$$

$$
G_s^{(2)} = \cos\varphi,
$$

$$
G_p^{(2)} = \cos\varphi\cos\alpha_1\cos\alpha_2 + 3\cos\varphi\sin\alpha_1\sin\alpha_2 + \sin\varphi\sin(\alpha_1 + \alpha_2).
$$

$$
\tag{13.77}
$$

The described mechanism includes an enhancement of the "original" electrooptic contribution (such as in permanently poled materials) and the additional occurrence of a birefringence contribution due to a modulation of the linear polarization of the material. This phenomenon was first discussed by Moerner et al. as the so-called orientational enhancement mechanism [62].

The birefringence contribution dominates the Pockels contribution by about one order of magnitude [72]. The dominance of the birefringence contribution was experimentally proven with the first high-performance PR polymer [9], which exhibited gain coefficients with opposite sign for the two polarization, in accordance with (13.53), (13.54), and (13.75)–(13.77).

13.3 Experimental Techniques

In the following Section, first the procedure of material and device preparation will be described as well as the basic device characterization. This is followed by introducing the holographic techniques of degenerate four-wave-mixing (DFWM) and two-beam-coupling (TBC). Potential experimental problems such as the determination of the internal light conditions, the occurrence of multiple gratings, and beam fanning will be discussed. We will provide typical examples of steady-state and dynamic performance for materials related to the first high-performance PR polymer (Figure 13.1, [9]). Transmission ellipsometry (ELP) as an experimental technique for determining the efficiency and dynamics of the poling process will be discussed. Furthermore, by means of combination of transmission ellipsometry and DFWM, the PR space-charge field can be accessed directly.

13.3.1 Device Preparation and Simple Device Characterization

Typical organic PR devices consist of two glass slides coated with transparent electrodes (e.g., indium tin oxide, ITO) (Figure 13.5). Note that the thickness and the refractive index of the ITO affect the reflectivity of the various internal interfaces, which play an important role in the device performance (see below). In most cases mixtures of different components, so-called composites, are studied. To obtain a suitable mix, appropriate amounts of each component of a PR material are dissolved in a volatile solvent (e.g., methylene chloride, chloroform, or toluene). The solution is filtered and the solvent is then reevaporated. The material is homogenized mechanically by repeated melt-pressing. (Homogenization may be obsolete if the material is monolithic, i.e., consists of a single compound). The glass-transition temperature is usually determined by differential scanning calorimetry (DSC).

For device preparation some of the PR polymer is placed onto an electrode and heated until it softens. Spacers are placed around the material and the second electrode is pressed on top, yielding uniform films of a thickness given by the spacers. Typical thicknesses are 30–200 μm, which is a compromise between long interaction lengths and high Bragg selectivity on the one hand and viable externally applied absolute voltage for a desired electric field strength on the other. The active layer thickness d may be increased without sacrifice of electric field strength using stacks of several devices, so-called stratified volume devices [73, 74].

After the melt-press procedure the devices are cooled to room temperature. Note that the cooling procedure determines the amount of free volume enclosed in the system [75] and thus influences the orientational mobility of the chromophores. Fast cooling leads to a relatively large internal free volume facilitating fast orientation of the chromophores. However, the material then shows pronounced "physical aging," i.e., over time, it relaxes toward a thermodynamically more stable structure with less internal free volume, reducing the reproducibility of experimental results. If the majority of components are of low molecular weight, the resulting composites may be thermodynamically unstable (undercooled melts). Thus, too slow cooling may end up in phase separation or crystallization.

The absorption characteristics are measured in the solid film by UV/Vis/NIR spectroscopy. The shelf-life time of the devices based on PR polymer composites containing large amounts of low-molecular-weight compounds, which tend to crystallize, is tested by measuring the transmission of the devices (or the scattered light) as a function of time. In the regime $T_g > RT$, Arrhenius behavior is observed, which is ascribed to the thermally activated diffusion of the chromophore molecules [11]. The Arrhenius prefactor can be improved by using chromophore mixtures [10].

Note that in some cases, crystallization can be accelerated substantially in the presence of illumination and an applied electric field. This has been attributed to a reduced activation barrier for crystal formation when the chromophores are oriented.

13.3.2 Holographic Wave-Mixing Experiments: Setup and Data Acquisition

The holographic characterization of organic PR devices is typically done by means of four-wave-mixing (FWM) and two-beam-coupling (TBC) techniques. The experimental geometry is illustrated in Figure 13.7. Although these two types of experiments are in general discussed separately, commonly they are carried out simultaneously. Two equally polarized coherent laser beams ("writing beams" 1

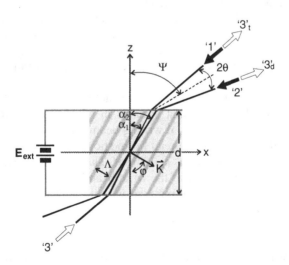

FIGURE 13.7. Illustration of the general setup configuration for holographic wave-mixing in PR polymer samples. d is the thickness of the polymer layer. The field polarity is $E > 0$.

and 2) of wavelength λ are overlapped in the sample with an angle 2θ to create a fringe pattern with period Λ. As already discussed above, experiments are performed in a tilted geometry (tilt angle ψ, defined outside the material). A typical geometry is $\alpha_1 = 50°$ and $\alpha_2 = 70°$ (external) [9], resulting in a tilt angle of $\psi = 60°$. For a bulk refractive index of $n = 1.70$ and a wavelength of 633 nm an interference grating fringe spacing of about $\Lambda = 3.1$ μm is obtained. The intensity of the probe beam 3 in DFWM experiments is typically several orders of magnitude weaker than the intensity of the writing beams. Often the probe beam is chosen with the polarization orthogonal to the writing beams. However, then the refractive index modulation sensed by FWM and TBC is different (see (13.75) through (13.77)).

As discussed above, an external voltage is applied to orient the chromophores. At the same time, charge-carrier generation and charge transport are improved (13.25)–(13.27). The direction of the electric field determines the sign of the PR phase shift φ_g (the other factors are the type of mobile carriers and the polarization of the writing beams; see below). Throughout this paper, we will use the convention $E_{ext} > 0$ if the anode faces the writing beams.

The tilted recording geometry (Figure 13.7) suffers from two kinds of problems:

(P1) Reflections of the two writing beams occur on the different optical interfaces of the multilayer heterostructure of the devices (air/glass1/ITO1/ organic/ITO2/glass2/air). In total, four interacting beams have to be considered [76]. The combination of each beam with all others eventually yields six competing gratings; two of these gratings are of transmission type, the remaining four of reflection type. The existence of the latter was shown by random-phase FWM [76], which shut off the two transmission holograms, and photo-EMF experiments [77]. Depending on the thickness and the refractive index of the glass substrates and the ITO contact layer, the reflectivity at the different interfaces and thus the relative strength of the gratings may vary strongly.

(P2) The cross section of the beams inside the active layer is elliptical, as illustrated in Figure 13.8. Note that the area in which both beams overlap is restricted to the cross section of writing beam 1 (assuming identical beam diameters). It is clear that the read beam should point to this overlap area in order to ensure ideal readout conditions. The illustration Figure 13.8 further shows that a major part of the write beam's 2 cross section inside the device simply illuminates the material without contributing to the optical interference field.

Both effects have a substantial impact on the grating contrast m, which determines directly the space-charge field and thus the index contrast Δn (see (13.10) together with (13.75)).

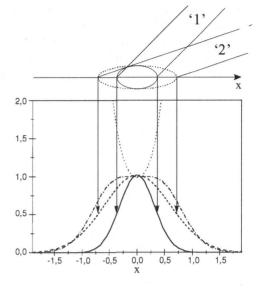

FIGURE 13.8. *Top:* Beam cross sections in the PR sample. *Bottom:* Intensity beam profiles in the propagation plane for beam 1 (solid line) and beam 2 (dashed line). Also shown are the beam intensity ratio b (dotted line) and the grating contrast m (dashed-dotted line).

Two-Beam Coupling (TBC)

The occurrence of beam coupling (coherent energy transfer between the writing beams) is considered proof of the PR nature of a holographic grating, since it can occur only if the grating is shifted relative to the incident interference pattern. In a TBC experiment the intensities of the two writing beams are monitored after they pass through the device. From these data the optical gain coefficient Γ can be calculated:

$$\Gamma(E) = \frac{1}{d}\left[\cos(\alpha_1) \ln \frac{I_1(E = 0)}{I_1(E)} - \cos(\alpha_2) \ln \frac{I_2(E = 0)}{I_2(E)} \right]. \tag{13.78}$$

The sign of Γ depends on the direction of the externally applied field and the polarization state of the writing beams.

The PR gain Γ is the real part of the complex coupling constant of the two-wave-mixing problem under consideration. Γ is related to the index modulation amplitude Δn and the phase shift φ_g:

$$\Gamma = \frac{2\pi \Delta n(\bar{e}_{p1}\bar{e}_{p2})}{\lambda_0 \cos \theta} \sin \varphi_g, \tag{13.79}$$

where 2θ is the internal intersection angle of the writing beams 1 and 2 (Figure 13.7) and $\bar{e}_{p1}\bar{e}_{p2}$ is the scalar product of their polarization unit vectors. The imaginary part of the coupling constant reads

$$\xi = \frac{\pi \Delta n(\bar{e}_{p1}\bar{e}_{p2})}{\lambda_0 \cos \theta} \cos \varphi_g \tag{13.80}$$

and introduces a phase delay between the writing beams, which, depending on the writing beam ratio, leads to bending or a tilt of the grating. The imaginary part ξ is barely considered in organic PR materials.

At this point, the tilted geometry introduces further systematic errors in obtaining the gain coefficient from the experimental data using (13.78):

(P3) Due to the inherent incomplete overlap of the writing beams (Figure 13.8) some portion X of one of the write beams can never be transferred to the other even for gigantic gain coefficients. This acts like an offset on the measured intensities:

$$\Gamma(E) = \frac{1}{d}\left[\cos(\alpha_1)\ln\frac{(1-X)I_1(E=0)}{I_1(E)-XI_1(E=0)} - \cos(\alpha_2)\ln\frac{I_2(E=0)}{I_2(E)}\right].$$
(13.81)

Thus, for $X > 0$, (13.78) yields too-small gain coefficients.

Other effects leading to erroneous gain coefficients are those affecting the particular beam intensities other than by coherent energy transfer:

(P4) Electric-field-dependent changes of the reflection and/or transmission coefficients (Fresnel equations) as a result of the field-induced index changes.
(P5) Electroabsorption phenomena [9]. Details will be discussed in Section 13.4.
(P6) Beam fanning [78, 79]. This does not occur in an ideal material without any scattering centers. In a real material it can be reduced to a minimum by selecting the polarity of the external electrical field such that the energy transfer occurs from beam 2 toward beam 1, i.e., $E > 0$ for s-polarization, $E < 0$ for p-polarization, respectively [78].

By reversing (13.79) the phase shift φ_g can be calculated. This requires knowledge of the refractive index modulation Δn for the same polarization and wavelength as the one applied in the TBC experiment (for the polarization anisotropy see (13.73) to (13.77)). An independent method to determine the phase shift is the moving-grating technique (MGT [80, 81]). However, it is valid only when the two beams impinge on the sample symmetrically with respect to the sample normal. In the tilted configuration, the analysis is more complicated, and additional approximations are required [82]. The phase shift measured by MGT in tilted configuration in general deviates quite substantially from the real value. A numerical solution was given by Grunnet-Jepsen et al. [83], and most recently Pedersen et al. presented an analytical model [84]. Reasonable results are obtained for a beam ratio of $b \approx 1$ and small gain coefficients ($\Gamma < 50$ cm^{-1}). Another independent method for measuring the PR phase shift is the AC phase-modulation technique [85, 86]. This technique is well established for PR crystals and has been adapted to the tilted geometry typically used for PR polymers by Liphardt et al. [87]. Reasonable results are obtained for $b > 100$ and $\Gamma < 50$ cm^{-1}.

Using (13.13) or (13.40) the saturation field and thus the number density of traps can be calculated from φ_g. Note that both (13.13) and (13.40) are derived from the underlying models assuming $m \ll 1$, a condition that is often not fulfilled.

Four-Wave Mixing (FWM)

In FWM experiments, index gratings are probed by a weak probe beam (3, $I_3 \ll I_{1,2}$) typically in a backward geometry (Figure 13.7). Degenerate ($\lambda_{1,2} = \lambda_3$) and nondegenerate ($\lambda_{1,2} \neq \lambda_3$) FWM must be distinguished. In both cases the probe beam must be appropriately adjusted to meet the Bragg condition. The intensities of the transmitted and the diffracted light are monitored to obtain the (external) diffraction efficiency, which is defined as

$$\eta = \frac{I_{3,\text{diffracted}}}{I_3}. \tag{13.82}$$

Kogelnik's coupled-wave model [88] can be used as a good approximation to calculate the total amplitude of the refractive index modulation Δn from the experimental FWM data. For slanted, lossy dielectric **transmission** gratings the model yields for the complex amplitude of the diffracted wave,

$$D(d) = -i \sqrt{\frac{\cos \alpha_1}{\cos \alpha_2}} \frac{\sin^2 \sqrt{v^2 - \zeta^2}}{1 - \zeta^2/v^2} \exp\left[-\frac{\alpha d}{\cos \alpha_1} + \zeta \right], \tag{13.83}$$

where v and ζ are

$$v = \frac{\pi \Delta n d}{\lambda_0 \sqrt{\cos \alpha_2 \cos \alpha_1}} \delta \tag{13.84}$$

and

$$\zeta = \frac{d}{4}\left(\frac{\alpha}{\cos \alpha_1} - \frac{\alpha + i\vartheta}{\cos \alpha_2} \right), \tag{13.85}$$

where $\delta = 1$ or $\delta = \cos(2\theta_{\text{int}})$ for s-polarized and p-polarized readout, respectively, where $2\theta_{\text{int}}$ is the internal angle between the transmitted beam and its diffracted portion. The angles α_1 and α_2 can be identified from Figure 13.7; α is the absorption coefficient, d is the layer thickness, Δn is the refractive index modulation, and λ_0 is the wavelength in free space. The parameter ϑ is a measure of the dephasing due to angular or wavelength Bragg mismatch and defined by

$$\vartheta = \Delta \theta_B K \sin(\psi - \theta_B) - \Delta \lambda \frac{K^2}{4\pi n}, \tag{13.86}$$

where θ_B is the Bragg angle and $\Delta \theta_B$ its absolute mismatch, $\Delta \lambda$ is the absolute wavelength mismatch, n is the bulk (i.e., the average) refractive index of the medium, and ψ can be identified from Figure 13.7 according to $\psi = \pi/2 - \varphi$. From (13.83), the (external) diffraction efficiency η is obtained according to

$$\eta = \frac{DD^*}{|\cos \alpha_1 / \cos \alpha_2|}. \tag{13.87}$$

For the geometry typically applied for FWM experiments on PR polymers, the slant of the grating can be neglected in reasonable approximation, which leads to a dramatic simplification of the analytical expression for the diffraction efficiency.

Note that (13.88) makes a compromise between complete neglect of the slant for the diffraction term and partial consideration for the absorption term:

$$\eta = \sin^2 \left(\frac{\pi \, \Delta n d}{\lambda \cos \psi} \right) \exp \left(- \frac{\alpha d}{\cos \psi} \right). \tag{13.88}$$

This model applies strictly only to linear gratings. Thus, s-polarized writing beams and positive external fields are typically used in FWM experiments for the evaluation of organic PR materials, because due to the polarization anisotropy, beam coupling is smaller than for p-polarized beams (see Section 13.2.4, (13.75)–(13.77)), and the nonuniformity of grating amplitude and phase throughout the device as well as beam-fanning effects are reduced to a minimum. Typically, readout is performed p-polarized, since the index modulation amplitude sensed by that polarization is larger ((13.75)).

In order to minimize the influence of the tilted geometry and its shortcomings concerning different beam path lengths and reflection at the various interfaces inside a PR polymer device, the internal diffraction efficiency is commonly considered. The slightly different absorption of the diffracted and the transmitted probe beams due to their slightly different optical path lengths are commonly neglected:

$$\eta_{\text{int}} = \sin^2 \left(\frac{\pi \, \Delta n d}{\lambda \cos \alpha_1} \right) \approx \frac{I_{3,\text{diffracted}}}{I_{3,\text{transmitted}} + I_{3,\text{diffracted}}}. \tag{13.89}$$

The maximum achieved internal diffraction efficiency very often significantly falls short of its theoretical value of unity. This has been commented on by many authors, and was attributed to various effects including restricted grating dimensions, absorption effects other than noted above, bending of the PR grating due to the imaginary part of the complex PR gain coefficient, and others. However, the actual reason for this effect is still not understood. At least hologram bending can be excluded from this list [89].

For reasons of comparison, it has become common practice to quote the field of maximum diffraction, $E_{\eta,\text{max}}$ (overmodulation field), as a measure of the performance of a material; the lower $E_{\eta,\text{max}}$, the higher the refractive index change for a given field strength and sample thickness. Note that only data that have been obtained applying identical setup geometries can be directly compared.

Kogelnik's coupled-wave model also provides a solution for the case of slanted lossy dielectric **reflection** gratings. In this case the complex amplitude of the diffracted wave is given by

$$D(d) = \sqrt{\frac{\cos \alpha_1}{\cos \alpha_2}} \, \frac{\sinh(i \nu \cosh \xi)}{\cosh(\xi + i \nu \cosh \xi)}, \tag{13.90}$$

where the parameter ξ is defined by the expression

$$\sinh \xi = \frac{\zeta}{i \nu}. \tag{13.91}$$

The parameters ν and ζ are defined by (13.84) and (13.85), respectively.

Equation (13.90) is dramatically simplified when the slant of the grating is neglected and negligibly small absorption is assumed, which finally yields for the diffraction efficiency

$$\eta = \tanh^2 \left(\frac{\pi \Delta nd}{\lambda_0 \cos \alpha_1} \right). \tag{13.92}$$

Note that there is no simple method to minimize the error made by neglecting absorption, as discussed before for transmission gratings.

Reflection gratings have barely been considered in the literature on organic PR materials up to now [90, 91].

13.3.3 Holographic Wave-Mixing Experiments: Results

Steady-State Performance

There are many degrees of freedom concerning material properties, experimental conditions, etc., that influence the steady-state performance of a PR polymer. This section gives an overview of the basic dependencies of the steady-state PR performance on the experimental conditions. Most importantly, the index-grating amplitude Δn and the gain coefficient Γ both strongly depend on the applied electric field E, the polarization of the beams, the grating contrast m (13.9), the total incident light intensity I, and the grating spacing Λ (the grating vector K). Furthermore, the internal tilt angle ψ plays a role.

We will give examples for typical low-T_g PR polymers derived from the first high-performance material (DMNPAA:PVK:ECZ:TNF; 50:36:13:1 %wt [9]) or corresponding composites using eutectic mixtures of DMNPAA and MNPAA [10]b]. The reduced temperature is $T_r \approx -15°C$ ($T_g \approx 5°C$). The setup has been described in the previous section. If not noted differently, the following conditions apply: external electric field $E = 90$ V/μm, s-polarized writing beams, power density $I \approx 1$ W/cm^2, contrast factor $m = 1$, p-polarized probe beam (power density ≈ 0.35 mW/cm^2).

Throughout the following section, the PR **refractive-index modulation** Δn was calculated from the experimental diffraction efficiency using (13.89). Δn depends approximately on the square of the external field (Figure 13.9) through (13.75), whereby it is anticipated that the PR space-charge field follows the externally applied field, which actually applies in reasonable approximation to typical low-T_g PR polymers. Hence, a simple power law can be assumed:

$$\Delta n_{s/p} \approx C_{s/p} E_{ext}^x, \tag{13.93}$$

where $C_{s/p}$ is a constant depending on the polarization of the light beams. Experimentallly, values for the exponent x between about 1.7 and about 2.2 have been found for PVK-based PR polymers [92]

Figure 13.10 shows a typical example of the dependence of the refractive index modulation as a function of the contrast factor m. According to (13.10) and

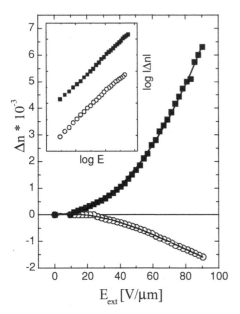

FIGURE 13.9. Field dependence of the PR index modulation for s- (open circles) and p-polarized (solid squares) probe beam. The sign of Δn was determined from TBC experiments. Inset: double logarithmic plot of the Δn moduli. The lines are a guide to the eye.

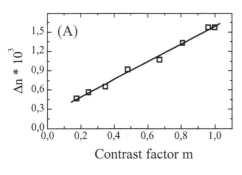

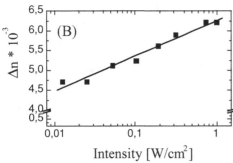

FIGURE 13.10. (A) Dependence of the PR index modulation on the contrast factor m. The total power density of the s-polarized writing beams was held constant and the probe beam was s-polarized. (B) Dependence of the refractive index modulation on the total incident intensity. The solid lines in both cases are linear fits.

FIGURE 13.11. Dependence of the refractive index modulation on the grating spacing and the grating vector (inset). The solid lines are linear fits.

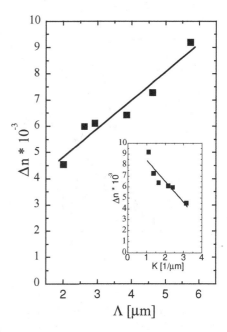

(13.39) together with (13.75), a linear dependency on the contrast factor m is expected and experimentally confirmed (Figure 13.10A). Note that the contrast factor changes as coherent energy exchange between the writing beams occurs due to TBC. This may result in a field asymmetry of the diffraction efficiency for opposite field directions if the zero field beam ratio is not unity. In particular for p-polarized writing beams, where the PR gain coefficient can become very large, significantly different histories of Δn as a function of the applied field may be observed for opposite field directions [76].

The total incident intensity and the grating vector (i.e., the grating spacing) enter into (13.75) via (13.39) to (13.44), or via (13.10) to (13.12) and (13.33). In any case, simple "first-sight predictions" as was possible for the dependence on E and m fail here. Figure 13.10(B) reveals an approximate logarithmic dependence of Δn on the incident intensity, whereas Δn scales approximately linearly with the grating spacing (Figure 13.11).

The refractive index modulation enters directly into the **diffraction efficiency** (e.g., (13.88)), a typical example of which is shown in Figure 13.12. The sine-square dependence leads to a periodic behavior of the diffraction efficiency as a function of the applied field. The externally applied field required to reach the first diffraction maximum ($E_{ext}(\eta_{max1})$) has become a commonly accepted measure for the steady-state performance of PR polymers [39,49]. The lower the value of $E_{ext}(\eta_{max1})$, the better the performance at a given field value. Naturally, a meaningful comparison of experimentally obtained $E_{ext}(\eta_{max1})$ values demands similar setup configurations.

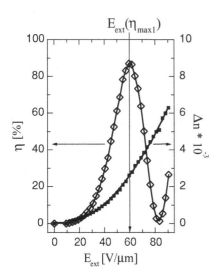

FIGURE 13.12. Typical field dependence of the FWM diffraction efficiency (Δn corresponds to Figure 13.9). The solid lines are a guide to the eye.

According to (13.79), the PR **gain coefficient** Γ is, in addition to geometrical factors, determined by the refractive-index modulation Δn as well as by the phase shift φ_g between the refractive index grating and the incident interference pattern. Therefore and in order to give an illustrative overview, mostly the general dependencies of the gain coefficient will be depicted below rather than the general dependencies of the PR phase shift. In all cases, the gain coefficient was calculated from experimental data by means of (13.78).

Figure 13.13 shows the field dependence of Γ as well as φ_g (inset). Furthermore, the general trend of Γ as a function of the grating contrast m is included. While the general features of Γ as a function of the external field apparently resemble the trend of Δn (Figure 13.9), from the inset it becomes clear that this is only approximately true for sufficiently high fields, where φ_g starts to level at roughly 20 degrees. This fact cannot be explained by means of the Kukhtarev model, which, for high fields, predicts according to (13.13)

$$\varphi_g \approx \arctan(E_0/E_q), \tag{13.94}$$

i.e., φ_g should approach $\pi/2$ when E_0 becomes large. The discrepancy is due to the field dependence of the limiting field E_q in organic materials, while it remains constant in inorganic PR crystals. Considering the dependence of Γ on the contrast factor m, it becomes obvious that φ_g significantly increases as m decreases, since Γ does not notably depend on m while Δn decreases (Figure 13.10A). This trend also holds for s-polarization.

Figure 13.14 shows the general trend of Γ as a function of the total incident intensity. Recalling Figure 13.10B reveals that the phase shift increases significantly as the total intensity decreases such that the general trend of Δn is reversed for Γ. Note that the logarithmic dependence on the intensity is nevertheless preserved.

FIGURE 13.13. Dependence of the
PR gain coefficients Γ_s (s-polarized
writing beams, open circles, $m =$
1.00) and Γ_p (p-polarized beams, solid
symbols) on the applied field. Γ_p is
furthermore shown for three different
contrast factors m (see text in plot). Inset: phase shift φ_g as calculated from
Γ_s using Δn as shown in Figure 13.9.
The solid lines are a guide to the eye.

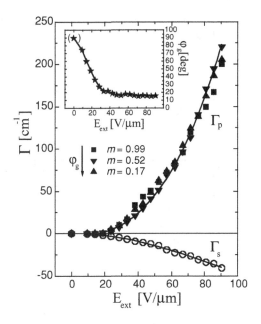

Figure 13.15 finally shows the dependence of Γ on the grating spacing and the
grating vector (inset). Comparison with Figure 13.11 reveals that φ_g increases as
the grating spacing decreases, or as the grating vector increases. Similarly, for the
intensity dependence the increase of φ_g inverts the trend for Γ as compared to Δn.

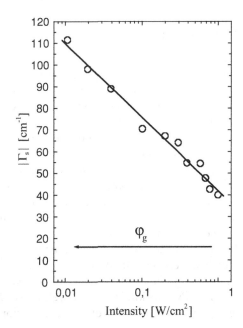

FIGURE 13.14. Dependence of the PR
gain coefficient Γ_s on the incident intensity. The arrow depicts the direction of
increasing φ_g. The solid line is linear fit.

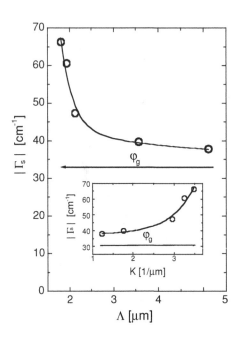

FIGURE 13.15. Dependence of the PR gain coefficients Γ_s on the grating spacing and the grating vector (inset). The arrow depicts the direction of increasing φ_g. The solid lines are a guide to the eye.

However, in this case the shape of the curve changes as well. Both findings can be explained by a reduction in the limiting field E_q ((13.12), or (13.43)). However, there is no simple mathematical function to describe the observed behavior, in contrast to most of the general trends described before.

Dynamic Performance

The theoretical models for the dynamic PR performance of polymers are rather complicated (see the theory section). As long as the dynamics of the material in question is clearly determined by the dynamic evolution of the PR space-charge field, Schildkraut's model and its derivatives may serve as a basis for interpreting experimental observations. However, as soon as the orientational dynamics takes over as the rate-limiting step, the rheological properties of the polymer matrix determine the dynamic performance [64]. In this regime, theoretical models of viscoelasticity may be employed to understand experimentally observed behavior. At the borderline between those regimes the dynamics of the PR effect resembles a mixture of processes of fundamentally different physical background, and a meaningful interpretation of experimental data becomes rather difficult.

The scope of this section is to give a general overview of the most obvious factors influencing the dynamic behavior of PR polymers. The discussion will mostly be restricted to low-T_g materials, where the buildup or the decay of the PR space-charge field can be expected to be the rate-limiting process.

Basically two kinds of photorefractive dynamics may be experimentally determined, which are the dynamic evolution of the FWM diffraction signal and the dynamic evolution of the TBC gain coefficient. The latter method is more problematic to experimentally investigate the dynamics of PR polymers, since the dynamic evolution of the gain coefficient includes the dynamics of both the refractive index grating and the PR phase shift. There is no theory covering the dynamic evolution of φ_g in any means, so that considering the dynamic evolution of the gain coefficient on its own remains purely phenomenological. In many cases, the TBC gain builds up faster than the FWM diffraction signal.

In most cases, the dynamics of the PR effect is tested by means of the FWM diffraction signal, which allows for the derivation of the evolution of the refractive index modulation as a function of time. According to (13.75), this is a direct measure of the dynamics of the PR space-charge field E_{SC}. It has become common practice to fit the obtained buildup curves for $\Delta n(t)$ to stretched exponential (13.95) or exponential associative (13.96) growth laws, respectively, whereby in the latter case mostly two exponentials have been used (i.e., $i = 2$ in (13.96)):

$$\Delta n(t) = A[1 - \exp(-t/\tau)]^{\beta}, \tag{13.95}$$

$$\Delta n(t) = \sum_i A_i[1 - \exp(-t/\tau_i)]. \tag{13.96}$$

Here, A and A_i are prefactors, τ and τ_i are the time constants, and β is the stretching parameter. In the case of decay curves for $\Delta n(t)$ most commonly multiexponential decay laws according to (13.97) have been applied:

$$\Delta n(t) = \sum_i A_i \exp(-t/\tau_i). \tag{13.97}$$

Although the analytical dynamic solutions of Schildkraut's model (see Section 13.2.3) actually suggest a more or less associative biexponential behavior, the fitting procedures should be regarded as phenomenological. There are still too many uncertainties as though one might try to assign the particular exponential terms to distinct physical processes. This applies the more if materials are examined that might be expected to involve orientational dynamics on similar time scales as found for the PR dynamics. In this case, the orientational dynamics must be tested independently, for example by ellipsometric techniques (see Section 13.3.4).

Another approach to the problem of understanding the dynamics of PR polymers has been proposed by Hofmann et al. [93]. They used a Laplace transformation formalism (Contin algorithm) to evaluate the dynamic behavior of a PR polymer, which yielded dedicated regimes of time constants subject to a dispersion. However, on the basis of the currently known theoretical models, this result cannot be interpreted conclusively.

First the dynamic performance will be discussed in **general terms**. The photorefractive effect is an integral effect, i.e., many charge carriers must be generated and redistributed in order to observe a photorefractive response. This statement of general validity leads to a set of **empirical rules** for the dynamic performance of organic PR materials, many of which also apply to PR crystals. It is clear that these are rules of thumb, which are strongly simplified and thus need not always be valid. In the following, the space-charge field buildup will be considered.

However, the space-charge field erasure follows similar trends whereby the arguments concerning trapping and detrapping must be exchanged.

The following issues may be anticipated to speed up the buildup of the PR space-charge field: there are experimental conditions like (1) a large internal tilt angle of the hologram to have maximum projection of the external field onto the grating wave vector, (2) small grating spacings for a short charge-carrier redistribution distance, and (3) a high recording intensity for carrier generation.

There are also material properties such as (4) high charge-carrier generation efficiency, (5) high mobility of the mobile charge carriers, and (6) efficient trapping that should be met. In organic PR materials, the points (4) and (5) imply a pronounced field dependence of the space-charge-field dynamics, since both the charge-carrier mobility and charge-carrier-generation efficiency are field-dependent, as discussed in the theory section. Finally, the charge carriers generated in the bright areas and subsequently redistributed must be efficiently immobilized when they arrive at the dark areas, which demands efficient trapping (point (6)).

Besides the simple counterparts of the above general statements on how to speed up the photorefractive response, the following material parameters may be anticipated to slow down the PR response: (7) easy detrapping, (8) large density of deep traps, and (9) strong absorption. Point (7) directly correlates to point (6) above, since easy detrapping phenomenologically may be considered a synonym for inefficient trapping. However, the physical background is different, since trapping efficiency is a question of the capture cross section for charge carriers and enters into the theoretical model as the trapping coefficient γ_T, whereas the ease of detrapping is a question of the trap depth and shows up in the theoretical models as detrapping rate α_T. A high density of deep traps (point (8)) reduces the charge-carrier mobility and thus effectively slows down the PR effect [94, 95]. (Please note that the dark decay is accelerated [58].) Finally, point (9) is somewhat ambiguous, since strong absorption may imply high charge-generation efficiency, which in fact accelerates the PR response. However, if absorption is too high, the incident light intensity drops too fast as a function of the penetration depth inside the material, which finally leads to a slowdown of the PR space-charge-field buildup.

In the following, **special issues concerning the dynamics of the PR effect** will be discussed that are not obviously covered by the general terms listed in the previous considerations. Figure 13.16 shows a set of dynamic FWM diffraction curves of a PVK-based low-T_g PR polymer obtained by applying three different **recording schemes** [96]:

(i)   The material is exposed to a nonuniform illumination pattern prior to switching on the external field that marks $t = 0$ for the evolution of the PR effect.

(ii) The material is held in the dark with the external field already applied long enough to reach quasi-steady-state parallel-plate-poling conditions. Then both writing beams are switched on simultaneously.

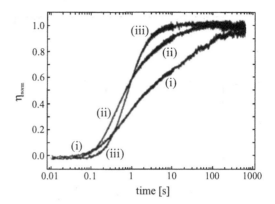

FIGURE 13.16. Dynamic buildup of the normalized FWM diffraction efficiency in a low-T_g ($T_r = -10°C$, (13.1)) PVK-based PR polymer for the recording schemes (i), (ii), and (iii) as described in the text. For more details reference is made to the original literature [96].

(iii) The material is parallel-plate poled as in (ii). However, it is additionally illuminated with one of the writing beams. Then $t = 0$ is defined as the time at which the second writing beam is switched on.

The depicted trends are not necessarily representative for any kind of low-T_g amorphous organic PR material. However, Figure 13.16 demonstrates the significance of the applied recording scheme. The essential differences in the listed recording schemes are the initial poling conditions and the initial density of free charge carriers.

Preorientation effects determine the orientational dynamics of the chromophores. In the above case (i), the material is initially statistically isotropic, while in the cases (ii) and (iii), the material is already parallel-plate poled, i.e., solely the reorientation according to the orientational enhancement mechanism must take place when the PR space-charge field builds up.

Preillumination effects play an important role in the recording scheme discussed above. However, their importance deserves a separate and more detailed consideration. Illuminating a PR polymer uniformly with the external field applied prior to writing the photorefractive grating may yield fairly different results depending on the employed photoconducting matrix.

In most photoconducting polymers the trap landscape changes due to preillumination. On the one hand, traps are formed from sensitizer moieties, which have provided a mobile charge carrier on illumination and are now capable of trapping another one. On the other hand, there is indication that deep traps in the system are activated that may not be correlated with sensitizer moieties. Both mechanisms may result in a retardation of the photorefractive response. This has been observed in a PMMA:DTNBI-based material [95] and in PVK-based PR polymers [97, 98] (Figure 13.17). Ostroverkhova et al. [51] analyzed this effect in terms of their modified Schildkraut model (see the theory section) and attributed the retardation of the PR response upon preillumination to an increase of the densities of ionized charge generators and filled traps, which, in accordance with Schildkraut's model, would lead to a deterioration of the PR dynamic performance. In general, in PR

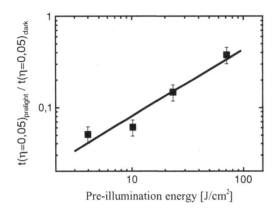

FIGURE 13.17. Relative retardation of the PR grating buildup due to preillumination in a low-T_g PVK-based PR polymer ($T_r = -8.5°C$, (13.1)) composed of 50%wt of a eutectic mixture of DMNPAA and MNPAA, 37%wt PVK, 12%wt ECZ, and 1%wt TNF [98].

polymers, the response dynamics of which is limited by charge-carrier mobility, preillumination may be anticipated to reduce the dynamic performance.

In contrast, in materials with response speed limited by the charge generation process, preillumination may significantly accelerate the buildup of the PR space-charge field. Mecher et al. preilluminated by a wavelength, where the utilized sensitizer showed a significantly better charge-generation efficiency than at the subsequently applied recording wavelength [31]. The authors referred to the above recording scheme as "gating," since the recording process is "gated" by the preillumination. Thus, by means of the preillumination a high density of uniformly distributed free charge carriers is provided inside the material prior to the PR grating recording process, which then are redistributed according to the nonuniform illumination pattern applied subsequently. The PR space-charge-field buildup could thereby be accelerated by up to a factor of 40 as compared to the buildup speed without preillumination. Kulikovsky et al. analyzed the phenomenon of "gating" the PR recording process in terms of photocurrent dynamics on the basis of the model proposed by Ostroverkhova et al. (see the theory section). The authors found that the gating effect can be quantitatively explained by deep trap filling [32].

Note finally that the existence of preillumination effects involving the optical activation of trapping sites [95] raises a problem for the evaluation of dynamic recording curves. If the trap landscape changes significantly as a result of illumination (and this is *not* restricted to preillumination), the time constant(s) for the grating build up, and possibly more factors depending on the particularly chosen fitting function (e.g., in the exponential prefactors multiexponential fitting) could depend on time.

The dynamics of **grating erasure** in PR polymers has been paid much less attention than the recording dynamics. However, it bears some surprises. Systematic investigations have been carried out only for PVK-based materials [52,98]. The erasure dynamics depends on the time the grating was recorded (Figure 13.18). As discussed before for the preillumination effects when recording a hologram,

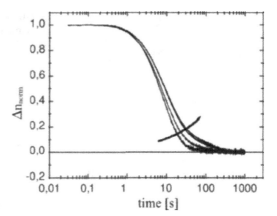

FIGURE 13.18. Time history for the erasure of the PR refractive index modulation in a PVK-based PR polymer as described in the caption of Figure 13.17 normalized to its value achieved at the end of recording. The arrow points in the direction of increasing recording time with $t_{rec} = 20$ s, 100 s, 500 s, and 3000 s. For $t_{rec} = 3000$ s, the grating is erased only slightly more slowly as compared to $t_{rec} = 500$ s.

this might be attributed to a change of the trap landscape during the recording process. In particular, there is strong indication that deep trapping sites are activated upon illumination during the recording process that are not charged sensitizers [99]. Only for recording times longer than the time typically regarded as the time required to reach steady state for the particularly investigated material does the erasure behavior stabilize (Figure 13.18). Even if the material is additionally doped with small amounts of extrinsic deep traps, the dependence of the grating erasure speed on the recording time of the hologram persists, and moreover, a local maximum of the grating strength during erasure is observed [98, 100].

The **dark decay** of holograms recorded in organic materials has also not been investigated much. The first systematic study performed by Bittner et al. [58] yielded a rather surprising result: after adding extrinsic traps to a PVK-based material that reduces the recording speed compared to the nondoped reference material due to trapping [94], the dark decay was accelerated. It was demonstrated that the dark decay was determined by the phase shift φ_g between the light-intensity grating and the recorded index grating when the recording was stopped (Figure 13.19). Here φ_g determines the gradual overlap of the distributions of electrons (localized at recombination centers) and holes (trapped somewhere in

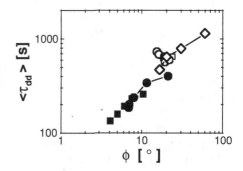

FIGURE 13.19. Averaged dark decay time constants for different recording conditions as a function of the corresponding PR phase shifts ϕ at the end of recording the holograms. Open symbols: no extrinsic traps; solid symbols: trap-doped material. For more details see [58].

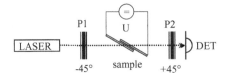

FIGURE 13.20. Experimental setup for transmission ellipsometry. P1, P2: polarizers; DET: detector; U: power supply.

the manifold of charge-transporting sites) if these are considered independently. Hence, the probability for a detrapped charge carrier to recombine should be larger if the phase shift is small.

Ostroverkhova et al. investigated the influence of the temperature on the dark decay time constant τ_{dd} [101]. They found that τ_{dd} was directly related to the thermal detrapping rate as implied by the theory (see the theory section).

Reflection Gratings

Recently, reflection gratings were observed in these materials [102]. At $E_{ext} = 50$ V/μm a diffraction efficiency of $\eta = 0.5$ and a gain of $\Gamma = 160$ cm^{-1} were obtained. The recording speed in the reflection geometry was about a factor 4–5 faster than in the typically utilized transmission geometry and otherwise identical conditions. This is attributed to the reduced grating spacing.

13.3.4 Determination of Electrooptic Activity

The EO properties of an organic PR material are usually investigated by transmission ellipsometry in the identical devices used for holographic recording [103, 104]. A sketch of the setup geometry is given in Figure 13.20.

The transmission of a setup as depicted in Figure 13.20 is given by:

$$T = \sin^2\left[\frac{2\pi d}{\lambda_0} (n_e \cos\alpha_{to} - n_o \cos\alpha_{te})\right], \tag{13.98}$$

where d is the sample thickness, λ_0 is the vacuum wave length, and n_o and n_e are the refractive indices sensed by the ordinary and the extraordinary portions of the linearly polarized light beam in the given geometry, respectively. Note, that n_e depends on the tilt angle, whereas n_o does not. The angles α_{to} and α_{te} are the internal angles enclosed by the ordinary and the extraordinary beam portions, respectively, and the sample normal.

The setup as depicted in Figure 13.20 has been widely used for measuring the parallel-plate-poling properties (i.e., poling dynamics and poling performance) of organic PR materials. The transmitted light intensity is a direct measure of the electric-field-induced birefringence of the material, and the apparent similarity between (13.98) and (13.89) suggests a direct experimental comparison of trends of η and T as a function of various parameters. For determining the parallel-plate-poling properties, the angles α_{to} and α_{te} are both often approximated by the (internal) tilt angle of the sample.

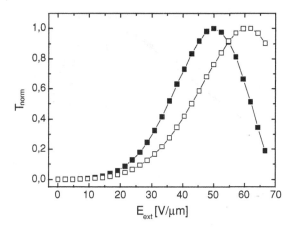

FIGURE 13.21. Normalized field-dependent ellipsometric transmission T_{norm} in the absence (open squares) or in the presence of a PR grating (closed squares). Material: 50%wt of a eutectic mixture of DMNPAA and MNPAA, 43%wt PVK, 6%wt ECZ, and 1%wt TNF.

The study of the poling behavior of the materials can be done independently or even during the holographic experiment. In the latter case, it is very advisable to use a wavelength where the PR material is essentially photoinsensitive to avoid effects on the holographic recording.

According to the orientational enhancement mechanism, the bulk-averaged birefringence of a parallel-plate-poled PR polymer enabling insitu poling changes if a holographic grating is written. Thus, from the ellipsometric transmission according to (13.98) with and without a holographic grating, the PR space-charge field can be determined directly and independently from holographic diffraction experiments [105, 106]. Figure 13.21 shows typical examples of ellipsometric transmission curves with and without a PR grating. The transmission maximum in the first case occurs at lower externally applied fields as compared to the latter case, since the PR space-charge field contributes to the overall birefringence. Figure 13.22 shows the PR space-charge field as a function of the applied field for a typical PR polymer composite determined according to the method described above.

13.4 Amorphous Photorefractive Materials

A variety of different approaches for developing amorphous organic photorefractive systems have been presented. The design concepts for the molecular moieties, providing the functions necessary for photorefractivity, i.e., photosensitive, photoconductive, and electrooptic properties, are universal to all material classes. Therefore, in the following we will first present the requirements of the individual components/functionalities. This is followed by a discussion of general material

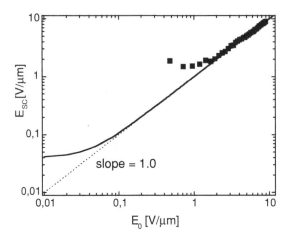

FIGURE 13.22. E_{SC} (solid squares) as determined for the material denoted in the caption of Figure 13.21 from the ellipsometric transmission in absence and in presence of a PR grating. Solid line: E_{SC} as calculated from Kukhtarev's model for $E_D = 0.04$ V/μm and $E_q = 100$ V/μm. Dashed line: projection of E_{ext} onto the grating wave vector K (i.e., E_0, saturation case for E_{SC}).

parameters, in particular the glass-transition temperature and absorption, as well as the potential cross influence between the components in the material (e.g., their relative electronic levels). Finally, the different material concepts will be briefly reviewed.

13.4.1 Requirements for the Individual Components

The following design rules apply to individual components. However, mutual cross influence cannot be excluded.

Semiconductor

There are no particular design rules known to date for organic charge-transport agents (CTA) to be used in PR materials. Unfortunately, empirical rules obtained from xerographic materials cannot be transferred directly, since trapping is undesirable there, but mandatory for photorefractivity. One generally distinguishes materials of two categories:

- **Materials with isolated charge-transport sites.** By far, most high-performance organic PR materials reported up to now utilize carbazole as the CTA, and among these, the PVK-based systems are by far the largest group. A comprehensive overview was given by Zhang et al. [107]. Other hole-conducting matrices that have been successfully used in the field of organic PR materials include triarylamine- (e.g., TPD-) containing side-chain polymers [25].
- **Conjugated materials** such as derivatives of the poly(phenylene-vinylene) (PPV) [108, 109].

Recently, copolymers of triarylamines and PPV have been successfully used in PR composites [31, 93]. Figure 13.23 shows some representative examples of

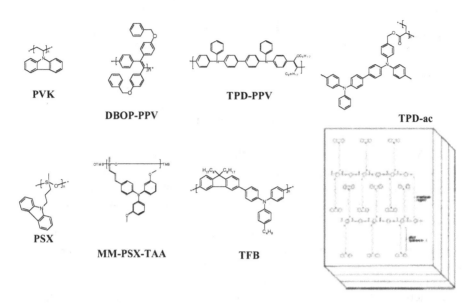

FIGURE 13.23. Chemical structures of semiconductive polymers used in PR-PC: **PVK, PSX** [111], **DBOP-PPV** [109], **MM-PSX-TAA** [112], **TPD-PPV** [31, 113], **TFB** [93], **TPD-ac** [25], and layered **PPT-Cz** (adapted from [114]).

matrix polymers. Please note that all compounds are hole conductors. They show maximum optical absorption at wavelengths shorter than 400 nm and thus are operable in the entire visible and NIR parts of the spectrum.

When it comes to selecting a suitable CTA for photorefractivity, first the mobility of pristine CTA is a good guess. Second, it has been established that a highly dispersive charge-carrier mobility (large width of the DOS, i.e., high σ; see Section 13.2.2) is beneficial for a fast and efficient space-charge-field formation [29, 30, 93, 110]. Please note in this context that the chromophore, the sensitizer, and both their contents will have a strong impact on the carrier mobility and on σ.

Furthermore, trapping is equally important for the formation of a space-charge field. Since conformational traps lie at the lower edge of the DOS, their effective concentration as well as their average depth scales with σ. This might explain why CTAs yielding a large width of the DOS tend to yield good performance.

Semiconductor Concentration

If a CTA is doped into an inert matrix, below a certain threshold there is no, or at best very limited, electrical conduction, since the CTAs are spatially separated too far for charge-carrier hopping. This threshold condition is not well defined, but corresponds to roughly 5–10%wt of a common CTA-like carbazole, DEH, or TPA derivatives. Above this threshold, charge transport becomes possible; the charge-carrier mobility increases with further increasing CTA concentration and finally levels usually significantly below 100%wt. In its basic framework,

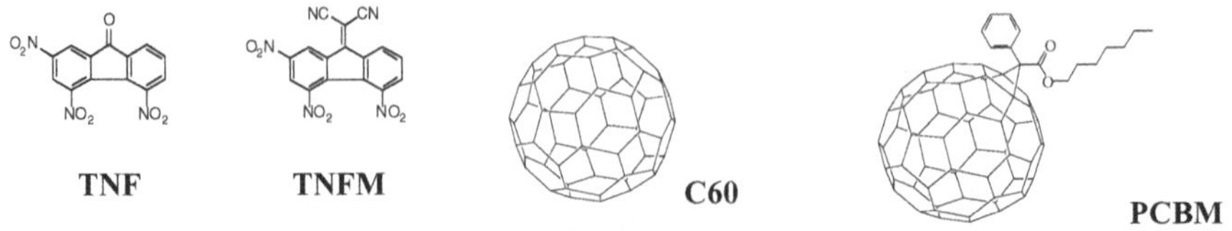

FIGURE 13.24. Chemical structures of various sensitizers used in organic PR materials.

the CTA concentration dependence of the charge-carrier mobility resembles a problem that might be approached applying percolation theory, as proposed by Silver et al. [115]. Another approach, the homogeneous lattice gas concept, was proposed by Gill [116]. However, both fall short of experimental reality, so that this issue remains a matter of empiricism.

It is important to point out that CTAs doped in low concentrations into an already electrically conducting matrix may act as extrinsic traps, the depth of which depend on the electronic levels of the doped CTA relative to the matrix CTAs [117] (see below). If the dopand acts as a trap, the charge-carrier mobility as a function of the dopand's concentration first reduces dramatically, reaches a local minimum, and then increases again when the dopand forms a second charge-transporting manifold. This behavior reflects correspondingly in the PR response speed [94].

Sensitizer

The task of the sensitizer is to provide the minimum degree of absorption necessary to generate charge carriers for the PR grating buildup. Since most of the CTAs used so far are hole conductors, the sensitizer must be an acceptor. Up to now, no systematic studies comparing different sensitizers have been performed.

The following types of sensitizers have been used in organic PR materials:

- **Fluorene derivatives** such as **TNF**, **TNFM** (both Figure 13.24; **TNFM** is also sometimes referred to as **TNFDM**). They are strong acceptors as reflected by the relatively low reduction potential (Table 13.2) and form charge-transfer (CT) complexes with many hole-conducting CTAs, in particular carbazole units, and even some chromophores [18, 118]. In these CT-complexes, partial electron transfer from the CTA (donor) to the sensitizer (acceptor) occurs already in the ground state.
- **Fullerenes** such as C_{60}, or its highly soluble derivative [6,6]-phenyl-C61-butyricacid-methylester (**PCBM** [119]; both Figure 13.24). Both are weaker acceptors than TNF and TNFM (Table 13.2).
- **Semiconductor nanoparticles** such as CdS [120], CdSe [121, 122] PbS, and HgS [22]. There seems to be a strong influence of the coating used to make the polar particles compatible with nonpolar polymer matrices.
- **Chromophores**: All typical EO chromophores contain donor and acceptor moieties. Thus, they can in principle act as sensitizer, an aspect that has been

accounted for by referring to "trifunctional" or fully functional chromophores such as **DHADC-MPN** [21] or **ATOP** [123] (see below). However, due to the necessity to include large amounts of chromophore to obtain a good PR material, in most cases this leads to strong absorption losses.

Few reports are known where a small amount of a chromophore was added as a sensitizer and was not intended to add to the overall EO response. One important example was given by Kippelen et al. [23]. They reported on the use of two-photon sensitization using chromophores.

The aforementioned sensitizers differ in their spectral sensitivity. Also, not every sensitizer is efficient with every CTA. However, for sensitizers in general there is a clear trend that the efficiency drops towards longer wavelength. The longest operating wavelengths reported are 1300 nm [22] and 1550 nm [24].

Sensitizer Concentration

There have been few reports on the influence of the sensitizer content [124, 125], both on PVK/TNFM. With increasing TNFM content the absorption of light by the sensitizer increases, and as a result, the number of photogenerated charge carriers increases, speeding up the recording process. In the steady state, the effective trap density is increased with increasing sensitizer content, leading to a stronger saturation field [98]. Thus, the space-charge field can reach the limiting value, i.e., the projection of the external field E_0, for sufficiently large sensitizer contents only. Thus, while the index modulation amplitude saturates for a given external field, the gain coefficient drops due to a reduced phase shift as a result of increased E_q, as becomes clear from (13.13). For the largest contents (3%wt) recording slows down [125].

Chromophore

Typical NLO dyes are D/A-type molecules with a permanent dipole moment μ_g. In a crude approximation their electronic ground and excited states, respectively, can be viewed as linear combinations of neutral and zwitterionic resonance structures (Figure 13.25, for reasons of simplicity, an arbitrary electronic π-system is replaced by a single double bond):

The linear combinations may be expressed in terms of a resonance parameter c allows one to classify the dyes from polyenes ($c^2 \approx 0$) over neutrocyanines ("cyanine limit,' $c^2 \approx 0.5$) to betaine-type structures ($c^2 \approx 1.0$) [72, 126]. This parameter is closely related to the parameter MIX introduced by Blanchard-Desce and Barzoukas [127] and the BLA parameter introduced by Marder et al. [128].

FIGURE 13.25. Limiting mesomeric structures of a typical push-pull substituted π-system.

For a two-center charge-transfer (CT) system the following expressions can be rigorously derived:

$$\mu_{ag} = c(1 - c^2)^{1/2}\Delta, \tag{13.99}$$

$$\Delta\mu = (1 - 2c^2)\Delta, \tag{13.100}$$

$$\mu_g = c^2\Delta, \tag{13.101}$$

where Δ is the difference between the dipole moments of the neutral and the zwitterionic states [72].

π-conjugated systems with a single donor-acceptor (DA) substitution usually exhibit one intense low-lying CT band. The static (extrapolated to infinite wavelength) second-order polarizability $\beta_0 = \beta(0; 0, 0)$ and the anisotropy of the first-order polarizability $\Delta\alpha_0 = \Delta\alpha(0; 0)$ of such systems can be approximated by the following two-level expressions [72, 129],

$$\Delta\alpha_0 = 2\mu_{ag}^2\lambda_{ag}/(hc_0), \tag{13.102}$$

$$\beta_0 = 6\mu_{ag}^2\Delta\mu\lambda_{ag}^2/(hc_0)^2, \tag{13.103}$$

where h is Planck's constant, c_0 the speed of light in vacuum, λ_{ag} the wavelength of the maximum of the molar decadic absorption coefficient ε, μ_{ag} the transition dipole moment, and $\Delta\mu$ the change of the dipole moment upon optical excitation within the CT band. The dependencies of $\Delta\alpha$, β, and μ on the mixing parameter are shown in Figure 13.26. Using the two-level model [129], the values for the operation wavelength $\beta = \beta(-\omega; \omega, 0)$ and $\Delta\alpha = \Delta\alpha(-\omega; \omega)$ can be calculated according to

$$\Delta\alpha(-\omega; \omega) = \Delta\alpha_0 \frac{\omega_{ag}^2}{\omega_{ag}^2 - \omega^2}, \tag{13.104}$$

$$\beta(-\omega; \omega, 0) = \beta_0 \frac{\omega_{ag}^2 (3\omega_{ag}^2 - \omega^2)}{3(\omega_{ag}^2 - \omega^2)^2}. \tag{13.105}$$

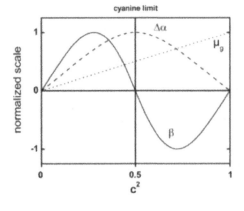

FIGURE 13.26. Normalized variations of the ground-state dipole moment μ_g (dotted line), the polarization $\Delta\alpha$ (dashed line), and the hyperpolarizability β (solid line) along the long molecular axis as a function of the two-center CT model mixing parameter c^2.

FIGURE 13.27. Normalized PR figures of merit as a function of the two-center CT model mixing parameter c^2. Solid line: FOM in high-T_g materials, which equals the electrooptic contribution. Dashed line: birefringence contribution; this is very similar to the FOM in low-T_g materials.

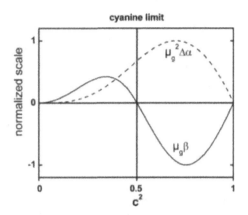

Close to the resonance (at ω_0 or λ_{ag}), a substantial enhancement can be achieved. However, in order to take advantage of the resonance enhancement, the absorption should still be small. Thus, a compromise known as the "nonlinearity/transparency tradeoff" has to be made. Chromophores exhibiting a rather sharp absorption edge such as cyanine and merocyanine dyes (e.g., ATOP, Figure 13.28 k) are clearly advantageous.

As already discussed in Section 2.4, two limiting material classes are commonly distinguished, with different molecular figures of merit (FOM) for the chromophores:

- In **"hard" materials** with "high" glass-transition temperature (valid for about $T_r > 60$ K) an appropriate FOM was defined as [72, 130]

$$F^{\text{Pockels}} = \frac{1}{M}[\mu_g \beta]. \tag{13.106}$$

- In **"soft" materials** with "low" T_g (valid for about $T_r < 10$ K) an appropriate FOM was defined as

$$F^{\text{Kerr}} = \frac{1}{M}\left[\mu_g \beta + \frac{2}{9}\frac{\mu_g^2 \Delta\alpha}{kT}\right] \approx \frac{2}{9M\,kT}[\mu_g^2\,\Delta\alpha]. \tag{13.107}$$

In most chromophores, the birefringence contribution dominates by about one order of magnitude over the electrooptic contribution [72].

These FOMs, displayed in Figure 13.27 can serve as design tools for chromophores. In both cases, the molar mass M was introduced as a measure of the molar volume. Note that there is no strictly defined borderline between these two cases.

The molecular parameters in (13.99)–(13.101) and the trends predicted by the above model calculations have been experimentally verified by optical (UV/vis) and electrooptic absorption measurements (EOAM) by Wortmann [72, 126, 130].

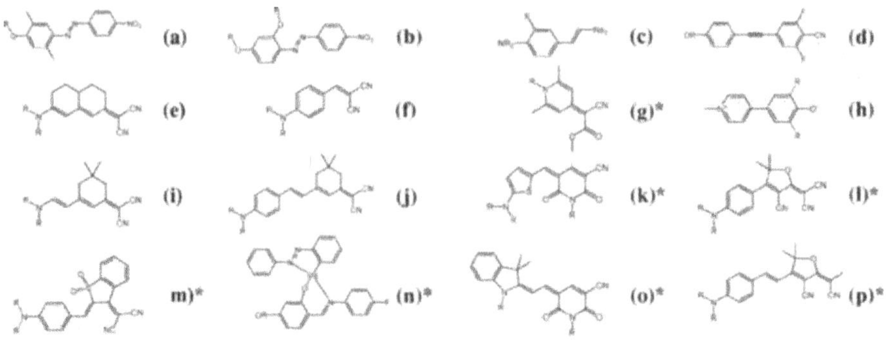

FIGURE 13.28. Chemical structures of chromophores used in PR composites; here R stands for an alkyl group; however, not all R are necessarily identical. All chromophores marked with * have been utilized as fully functional PR-LMWG's (see below). (a) **2,5 DMNPAA** (R = methyl) [8, 9], **BDMNPAB** (R = n-Bu [131]), **EHDNPB** (R = EtHEx [132, 133]), **HDMNPAB** (R = n-Hex [131, 133]), **ODMNPAB** (R = n-Oct [133]); (b) **1,3MNPAA** (R = methyl [10b]), **NPADVBB** (R = vinyl-tolyl [11]); (c) **F-DEANST** (R = Et [6]); (d) **FTCN** (R = $(CH_2)_2O(CH_2)_4$, [26]); (e) **DHADC-MPN** (R = n-Hex [21]); (f) **PDCST** (R = c-Hex), **7-DCST** (R = c-Hept), **AODCST** (R = $(CH_2)_2OCH_3$) [27, 134]); (g) **2-BCNE** [19]; (h) zwitterion (R = t-Bu [135]); (i) **DB-IP-DC** (R = n-Bu) [136]; (j) **Lemke-E** (R = Et [137]); (k) **ATOP-1** (R = n-Bu [20]); (l) **DCDHF-n** [138]; (m) methine dye (R = EtHex [139]); (n) **AZPON** [140]; (o) **IDOP** [18]; (p) **DCDHF-nV** [138d], (R = EtHex) [141].

Figure 13.24 shows a representation of the mostly used chromophores. Table 13.1 presents the respective FOMs. Starting from the early chromophores such as the relatively nonpolar **DMNPAA** and **(F-)DEANST**, there is a clear trend toward introducing stronger acceptor groups such as the dicyano-vinyl or even tricyano groups. For the conjugated system, simple alternating double bonds stabilized by nonconjugated bridging units proved most successful. Finally, the donor are alkoxy or dialkyamino groups in most cases. Even a formal zwitterion has been reported, however with limited performance [144]. Recently, there is a clear trend to vary the solubilizing alkyl groups such that the materials become glass-forming. These materials are marked by * in Figure 13.28. Following an empirical rule, long and branched groups are preferable, but there is no guarantee.

Chromophore Concentration

To optimize the EO effect, besides using a chromophore with large FOM, the chromophore content should be as high as possible. It has been found empirically

TABLE 13.1. Kerr figure of merit (FOM_{Kerr}) for some selected chromophores [18, 142, 143]

Chrom.	a	h	f	g	e	o	k
trivial name	DMNPAA	F-DEANST	PDCST	2-BCNE	DHADC-MPN	IDOP	ATOP
FOM	0.2	0.2	0.6	0.3	0.6	1.6	2.0

FIGURE 13.29. Illustration of the dimer-
ization of chromophores and the influence
of an applied electrical field.

that the maximum content of a chromophore that can be incorporated into a material depends on the alkyl content. For low chromophore content the PR index
modulation amplitude increases linearly [145]. However, due to the dipolar character of the chromophores, there is an equilibrium between "free" chromophores
exhibiting an EO response and dimers or "aggregates" [146], which due to their
centro-symmetry do not exhibit an EO response (Figure 13.29). Naturally, at low
content the chromophores are predominantly free, while at high content the number of aggregates increases. In some cases, this can lead to complete cancellation
of the EO effect. Due to the simultaneous increase of the dielectric constant the
equilibrium constant K is not really a constant, but decreases with increasing
chromophore content. Wortmann et al. found some indication that the application
of an electric field supports the dissociation of the dimers [147] (13.29).

We point out that the chromophore and its content have a substantial impact on
the photoconductivity parameters, reducing the carrier mobility μ by increasing
the width σ of the DOS distribution function (see Section 2.2, (13.26); [148]).
Furthermore, chromophore aggregates may constitute traps even though the single
chromophores may not.

The following scheme (Figure 13.30) summarizes the requirements that optimized multifunctional chromophores have to fulfill for photorefractive applications. While the molecular FOM F_{Kerr}^0 and the problem of dipolar aggregation
are currently well understood and quantitatively predictable as outlined in the
preceding chapter, other properties related to the influence of the dye on charge
transport and trapping properties remain less understood. This sets limitations on

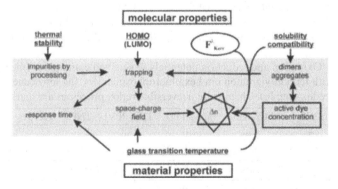

FIGURE 13.30. Interdependence between PR key processes under participation of multifunctional chromophores and their influence on refractvie index modulation Δn and the
response time (adapted from [18]).

an independent optimization of the index modulation amplitude by chromophore design and the speed of the material (response time) by the photoconductor.

Plasticizer

The glass-transition temperature T_g of any PR material as the material parameter determining the reduced temperature T_r ((13.1)) is essential for the PR performance, since the orientational enhancement effect can occur only in soft media. In order to reduce T_g and to let the orientational enhancement effect take place efficiently, very often plasticizers are added to PR polymer composites. The choice of a suitable plasticizer is first determined by the question of its compatibility with the polymer matrix. Second, it would be favorable if the plasticizer contributes constructively to the functionalities required for PR response, since otherwise it will only reduce the intrinsic performance parameters.

For carbazole-containing materials, N-ethylcarbazole (ECZ) has been widely used [8, 9]. In many other studies various phthalates (dioctylphthalate, DOP; diphenylphthalate, DPP; butyl-benzylphthalate, BBP) have been utilized. Please note that the chromophore will also act as a plasticizer. Spacer or solubilizing groups in the material may act as internal plasticizers.

The larger the content of the plasticizing compound, the lower the resulting T_g is.

13.4.2 General Material Parameters
Electronic Levels

The relative electronic levels of all the functional moieties involved is of paramount importance for the performance of the overall PR material.

The most important level is the energy of the charge-transport manifold. Since most CTA are hole conductors, it is the HOMO of the CTA. Ideally, all other components exhibit HOMO energies below the one of the CTA such that these compounds do not constitute traps and transport can take place as freely as possible. This condition is generally fulfilled for all sensitizers (Figure 13.24) and all potential plasticizers. However, often the HOMO of the chromophore is higher

TABLE 13.2. Oxidation and reduction potentials of selected compounds (measured in acetonitrile with tetrabutylammonium hexafluorophosphate vs. the ferrocene/ ferrocenium redox couple). Chemically irreversible redox processes are marked *

compound	ECZ	ECZ dimer	alkyl-TPD	TPD-PPV	TNF	TNFM	(C_{60}) PCBM
E_{ox}	0.78*	0.45	0.33	0.29	n.a.	n.a.	—
E_{red}	n.a.	n.a.	n.a.	n.a.	−0.85	−0.40	−1.00

compound	DMNPAA	PDCST	IDAP	2-BCNE	ATOP		
E_{ox}	1.21*	0.70	0.67	0.64*	0.50		
E_{red}	−1.23	—	—	—	—		

FIGURE 13.31. Diagram of energy levels for organic PR materials, where the respective functions
are provided by individual components. CTA =
charge transporting agent.

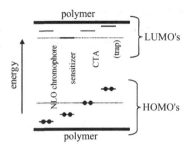

than that of the CTA (see Table 13.2) [45, 149]. Thus, the chromophore not only
constitutes a hole trap, but due to the high content may form an additional transport
manifold, which depending on the mobility can take over charge transport.

The LUMO energy of the sensitizer should be the lowest, to make sure that
an electron once generated remains localized (due to the low sensitizer content,
<1%wt, there is usually no electron transport). This condition is fulfilled by
all usual (hole-conducting) CTAs. There may be a conflict with chromophores
containing very strong acceptors whose LUMO is close to or even below the
LUMO of the sensitizer. In this case, electron transport may become the dominant
transport mechanism, even though a typical hole-CTA is present.

The situation is summarized in Figure 13.31. For an electron-conducting system, analogous relations will apply. However, HOMOs and LUMOs must be
interchanged in the above considerations.

The HOMO and LUMO positions are generally determined by cyclic voltam-
metry [150] assuming an offset between the electrochemical scale and the vacuum
scale of 4.8 eV [151]. Some values are given in Table 13.2. Note that redox potentials obtained in solution may differ from those in the solid state, since the
different physical conditions are afflicted with different molecular interactions
for the compound considered. However, the trends as determined in solution are
usually preserved in the solid state.

Glass-Transition Temperature

The reduced temperature T_r of the system (13.1) and thus the glass-transition
temperature of a material determines the steady-state value as well as the dynamics of the space-charge-field formation and, in low-T_g materials, the degree of
orientation and the orientational mobility of the EO chromophores.

The studies that have been devoted to this effect can be divided into two categories: first, a series of composites with varying amounts of plasticizer (ECZ
in PVK-based systems [39, 49, 63, 90]; carbazole-based system [152]) or chromophores (e.g., [153]) and otherwise identical composition was investigated.
Second, the absolute temperature T was varied [154, 155, 101]. It was found that
the reduced temperature (13.1) can serve as a universal point of reference to explain the performance changes in organic PR materials with a broad range of T_g's.
Overall, one can summarize the effects as follows:

- *Space-charge field E_{SC}* : As already discussed in the theory section, with increasing T_r, the detrapping of charges becomes more probable due to molecular motion of the trapping sites changing the trap depth. In turn, the number density of traps is reduced as confirmed by concomitant measurements of the PR gain coefficient [49], the dark conductivity increases [39, 49, 101, 155], and the power-law coefficient p, (13.25), decreases [39]. As a result, E_{SC} collapses for $T_r < 0$, whereby the exact critical reduced temperature $T_{r,c}$ becomes smaller when the chromophore content is reduced: it is $T_{r,c} \approx -5$ K for 50% of **2,5 DMPAA** [39, 49], while it is $T_{r,c} < -30$ K for 5–6%wt of **DEANST** [155]. The dark decay rate $1/\tau_{dd}$ also depends on T_r [58, 98]. Recently, it was demonstrated for DCDHF glasses (see Figure 13.28), that within a temperature window of only 10 K above the glass transition, the dark decay slows down by about two orders of magnitude [101], which may open a way for efficient thermal fixing of the space-charge field by decreasing the temperature.
- *EO coefficient r_{eff}* :With increasing T_r, the material's viscosity becomes reduced. In turn, the orientation of the dipolar chromophores becomes faster and simultaneously more efficient. As a result, both the effective EO coefficient r_{eff} and the orientation speed become more favorable. Both finally level for high T_r (low T_g).

Overall, since both, E_{SC} and r_{eff}, enter into the refractive index modulation Δn ((13.69)), there is an optimum steady-state performance for $T_g \approx RT$ [49].

Absorption

The absorption spectra of the components with respect to the desired operating wavelength are of significant importance. On the one hand, a minimum absorption must be provided in order to generate free charge carriers, which then give rise to the PR space-charge field. This minimum absorption is typically introduced in a PR polymer composite by means of a sensitizer added in small amounts to the system.

Please note that a PR polymer composite may be made faster in PR response by adding a strongly absorbing and/or a relatively large amount of sensitizer. Thus, a compromise must be found in this case as well.

Ideally, the other components of the system should not show notable absorption at the operating wavelength. Although this is a trivial demand, it often requires a tradeoff between the nonlinear properties of the NLO chromophore and its absorption coefficient at the operating wavelength of the system. Unfortunately, the chromophores showing high molecular optical nonlinearity usually also absorb strongly in the visible or even in the near infrared, which often prevents their application in PR polymer composites.

13.4.3 Material Concepts

One way to categorize the materials is to distinguish between multicomponent systems (composites) and monolithic systems. Commonly, a PR material is referred to as a composite if more than about 1–2%wt of the system is different from the main component. Otherwise, the system is regarded as monolithic.

Alternatively, according to the means to combine the functions necessary to observe the PR effect, four fundamentally different material concepts must be distinguished:

A. PR amorphous glasses (PR-AG), comprising
   1. sol/gel materials (PR-SG),
   2. fully functionalized polymers (PR-FFP),
   3. polymer composites containing LMW compounds (PR-PC),
   4. low-molecular-weight glasses (PR-LMWG).
B. Polymer-dispersed liquid crystals (PR-PDLC). They are a hybrid of the first and the fourth category.
C. Polymer-dissolved liquid crystals (PR-PDSLCs).
D. PR liquid crystals (PR-LC). They can be distinguished according to the liquid crystalline phase, i.e., nematic or ferroelectric.

We have listed the materials following roughly their mechanical properties, which in most cases is equivalent to decreasing the glass-transition temperature: while the PR-SG are "hard" materials, which do not exhibit a glass transition at all in many cases, the PR-LC are liquids (potential glass transition very well below room temperature). The transitions are smooth, and can be adjusted by adding a plasticizer.

Among these material classes, the first category, in particular PR-PC and PR-LMWG, are by far the most intensively studied. Similarly, there has been much work done on PR-LC.

(A1) Sol/Gel PR Materials (PR-SG)

Photorefractive sol/gel materials have been proposed by Chaput, Darracq, and Burzynski [156, 157, 158]. The materials consisted of a silica oxide backbone with carbazole moieties and an NLO chromophore covalently attached via alkyl spacer groups in order to introduce charge-transporting and nonlinear optical properties. For sensitization small amounts of TNF were added. The authors pointed out the excellent optical quality of this material class. More recently, Del Monte et al. reported on sol/gel materials that exhibited large gain even without applied electric field (Figure 13.32) [168]. PR-SG can be regarded as fully functionalized polymers with an inorganic backbone.

(A2) Fully Functionalized Polymers (PR-FFP)

Numerous monolithic PR polymers have been developed. Figure 13.32 shows some representative materials. A comprehensive overview about PR-FFP was given by Yu et al. [159]. The functional units are in principle identical to those on PR composites. The polymer backbone may be inert (**P1**, **P2**, **P7**, and **P8**) or functional (**P3–P6**, **P9**). The PR performance of PR-FFP is determined by the question of the extent to which the orientational enhancement effect can contribute to the overall PR effect. In general, up to now, the steady-state as well

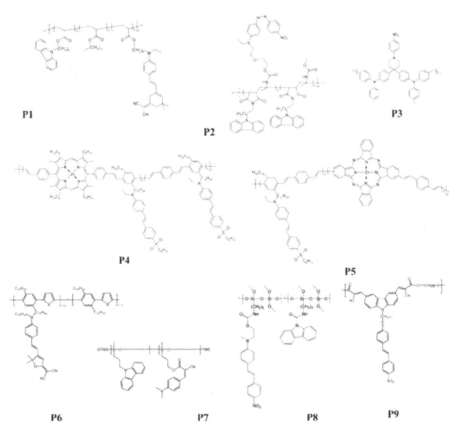

FIGURE 13.32. Chemical structures of PR-FFP and PR-SG: **P1** [162], **P2** [163], **P3** [164], **P4** and **P5** [165], **P6** [166], **P7** [167], **P8** [168], **P9** [169].

as the dynamic PR performance of all the known monolithic PR polymers is inferior to today's high-performance PR-PC, PR-LMWGs, and PR-PDSLCs.

One interesting limiting case is a high-T_g material with purely linear electrooptic response [109, 160]. Since the EO effect takes place instantaneously, such systems have been expected to be faster than systems showing orientational enhancement. Unfortunately, due to the relatively slow formation of the space-charge field, the high-T_g PR materials did not meet expectations.

(A3) Polymer Composites (PR-PC)

Most organic PR materials developed to date are polymer composites. PR polymer composites basically consist of a polymer host into which low-molecular-weight components are embedded. The polymer backbone may be inert or functional. In many cases, e.g., the case of PVK, functional moieties required for photorefractivity are covalently attached.

It is clear that this "mixing" technique has the inherent advantage that the systems can be tuned very easily by simply altering the composition or using different components. However, PR polymer composites are thermodynamically metastable systems, inherently suffering from possible phase separation and crystallization of the contained low-molecular-mass components. Accordingly, the thermodynamic stability of PR polymer composites will strongly depend on the miscibility of low-molecular-mass components with the polymer matrix. Thus, the compatibility of the polymer matrix and the low-molar-weight (LMW) compound (typically the NLO chromophore) is highly important. In most cases (preferably branched), aliphatic side groups attached are attached to the LMW compound.

It is clear that components added in only small amounts (e.g., the sensitizer) must at least be soluble in the system in the amount desired.

(A4) Low-Molecular-Mass Glasses (PR-LMWG)

"Low-molecular-mass glasses" (LMMG) consist of one or more low-molecular-mass components forming an amorphous glassy solid. Sometimes up to about 10%wt high-molecular-mass (polymeric) compounds are added to improve the dielectric properties. The majority component in PR-LMWGs are bifunctional molecules exhibiting electrooptic as well as charge-transport properties. This may be achieved in two ways:

1. Using a bifunctional moiety. This approach usually utilizes typical chromophores with long and branched alkyl side groups, which are supposed to perform as charge transporters as well. Examples are **2-BCNE**, **ATOP**, **DCDHF**, and **IDOP**. Please note that the NLO chromophore may change its optical properties when the molecule acts as a CTA.
2. By covalently binding transporting moieties with NLO chromophores so that these two properties are combined in one molecule but are decoupled by a suitable spacer group. This approach has the apparent advantage that both properties can be optimized independently of each other. Figure 13.33 shows

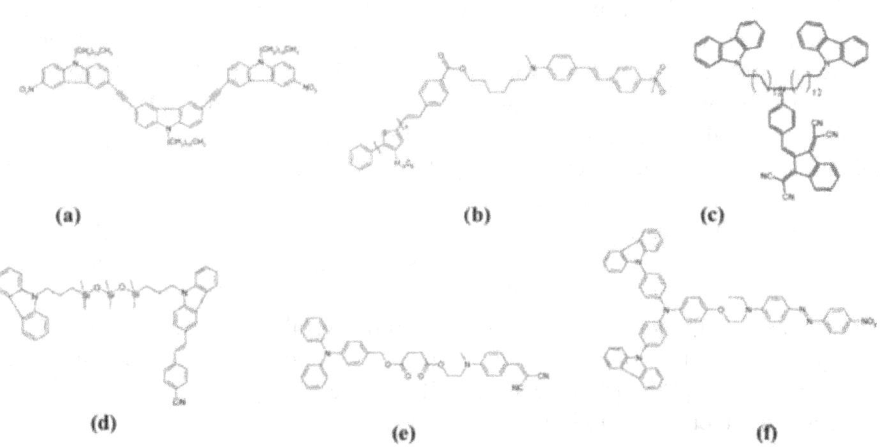

(a)   (b)   (c)

(d)   (e)   (f)

FIGURE 13.33. Chemical structures of PR-LMWG: (a) **TRC-1** [170] (b) **OT-1** [171], (c) [172], (d) **Si3(Cz-Cz/stilbene)** [173], (e) **TPA-DCVA** [174], (f) **DR-DCTA** [28].

five typical examples of this material class. The most extensively investigated class of PR-LMWGs are the DCTAs developed by Strohriegel et al. [28]. In particular, the compounds using disperse red-1 as the chromophore (Figure 13.28, Figure 13.32 f) shows excellent (i.e., fast) PR performance when the plasticizer is optimized [29, 161].

In all cases, a small amount of sensitizer was added. Depending on T_g, an additional plasticizer was also incorporated.

(B) Polymer-Dispersed Liquid Crystals (PR-PDLC)

In PR-PDLCs [175, 176], the liquid crystal is dispersed in a photoconducting polymer matrix (e.g., PVK) forming small droplets, which are in the mesophase. The polymer matrix decouples the droplets, which breaks the long-range orientational order and thus the droplets exhibit random director of their individual mesophase ("super-chromophore"). Ono et al. investigated the influence of the droplet size [177]. Smaller droplets allow for faster response times and larger index modulation.

This material class may be understood as an intermediate stage on the way from pure liquid crystals (PR-LC) to PR-PDSLCs; a direct comparison was reported in [178]. The basic operating conditions of PR-PDLCs are similar to those for PR-PCs and PR-LMWGs, i.e., a relatively strong electric field is necessary for operation. However, the inherent inhomogeneous character of a dispersion results in relatively poor optical quality (strong scattering), which makes these systems unfavorable in general. Except in [179] this approach has no longer been pursued after the initial work by Golemme and Ono.

(C) Polymer-Dissolved Liquid Crystals (PR-PDSLC)

A very promising approach to realizing photorefractive liquid-crystalline systems has been followed by Ono et al. They showed that mixing high- and low-molecular-weight liquid crystals leads to nematic phase systems operable at the low external electric dc fields characteristic for liquid-crystalline systems, but in the volume grating regime with grating spacings on the order of few microns [180]. The first system reported was composed of a typical commercially available liquid crystal and a side-chain liquid-crystalline polymer (SLCP) carrying a liquid crystal covalently attached to an acrylic polymer backbone by an alkyl spacer. The authors refer to their material class as polymer-dissolved liquid crystals (PDSLCs). The systems in general are doped with a small amount of common sensitizers (e.g., C_{60}). Optimization of these PDSLCs by varying the molar mass of the high-molar-mass component and using better liquid crystals (commercial name: E7) finally yielded composites showing fast response times while maintaining large gain coefficients [181, 182]. Please recall that the PR-PDSLCs reviewed above are photoconductive due to ionic diffusion and drift.

Recently, the concept has been extended by introducing "real" photoconduction to this material class. For this purpose a 1:1 copolymer carrying the same mesogenic group as SLCP as well as carbazole covalently attached to the acrylic backbone and TNF as sensitizer has been used at otherwise identical composition as described above [183, 184].

(D) Liquid-Crystalline PR Systems (PR-LC)

Photorefractive liquid-crystalline systems have been focused on as it has become clear that the PR effect in organic amorphous systems is significantly enhanced by orientational effects (see Section 13.2.4). At first sight this seems obvious, because liquid crystals are well known to exhibit large Kerr effects. Due to their collective orientational properties, this material class exhibits PR response at very low externally applied field, since the electrical poling in order to give rise to second-order nonlinearity is very easy. As in the case of PR-PDSLCs reviewed above, the photoconductive properties in PR-LC are due to ionic diffusion and drift, limiting the strength of the applied field. However, small fields are not favorable for charge transport and generation, both of which are essential for the PR effect. Using liquid-crystal cells faces moreover the problem that the inherent long-range orientational interaction requires very large grating spacing of several tens up to hundreds of microns to take place to a sufficient degree. For smaller grating spacings, the elastic energy counteracting a periodic modulation of the birefringence increases dramatically, eventually almost suppressing the PR response completely. The large grating spacings result in slow response times on the one hand and operation in the Raman–Nath regime (thin holographic grating) on the other. Furthermore, the spatial resolution of a hologram written at large grating spacings is unsatisfactory. The charge transport in these systems is due to diffusive motion of ions causing hydrodynamic turbulences if the applied fields are too high.

13.5 Conclusion

During the past decade amorphous organic photorefractive materials have matured to an established class of reversible holographic storage media offering a manifold of desirable properties. They feature good optical quality, high structural flexibility, good reproducibility, easy processing, and low cost. Synthetic efforts have led to a large contingent of materials offering high refractive index modulations, large gain coefficients, and high recording speed allowing the operation beyond video rate. The sensitivity range of PR polymers spans the visible far into the infrared (1500 nm). The initially critical shelf lifetimes of PR polymer devices could be considerably enhanced, and promising approaches to create thermodynamically stable materials have been proposed.

However, there are still some major drawbacks. The application of high electric fields is still indispensable, which is undesirable for many technical applications. The photorefractive properties of polymers exhibit a pronounced sensitivity against the illumination history, and the optical properties in general tend to degrade with time. Furthermore, the dark lifetime of holograms is still not satisfying, and up to now no convincing way has been found to replace the "quasi"-nondestructive readout of holograms by actual nondestructive readout. Finally, up to now no way has been found to optimize the steady-state and the dynamic performance simultaneously.

In order to meet these challenges, a detailed understanding of the PR process in polymers is required, the key parameters of which are the electronic properties of the photoconducting polymer matrix and the NLO response of the material. While the latter is currently fairly well understood, the theoretical explanation for the experimentally observed space-charge-field effects is still relatively diffuse, not allowing the reliable or even numerical prediction of the potential of some material on the basis of the structural properties. In particular, detailed insight into charge-carrier dynamics and trapping in organic PR materials is still missing.

Acknowledgments. We acknowledge contributions by Dr. S. Mansurova and F. Gallego-Gomez. We would also like to express our gratitude to all our collaborators. Finally, we would like to thank the Volkswagen (VW) foundation, the European Space Agency (ESA), and the Deutsche Forschungsgemeinschaft (DFG) for providing financial support for this research.

References

1. Kukhtarev N.V., Markov V.B., Odulov S.G., Soskin M.S., Vinetskii V.L. *Ferroelectrics* **22**, 949 (1979).
2. Schildkraut J.S., Cui Y. *J. Appl. Phys.* **72**, 5055 (1992).
3. Schildkraut J.S., Buettner A.V. *J. Appl. Phys.* **72**, 1888 (1992).
4. Ashkin A., Boyd G.D., Dziedzic J.M., Smith R.G., Ballmann A.A., Nassau K. *Appl. Phys. Lett.* **9**, 72 (1966).
5. Ducharme S., Scott J.C., Twieg R.J., Moerner W.E. *Phys. Rev. Lett.* **66**, 1846 (1991).
6. Donckers M.C.J.M., Silence S.M., Walsh C.A., Hache F., Burland D.M., Moerner W.E., Twieg R.J. *Opt. Lett.* **18**, 1044 (1993).
7. Orczyk M. E., Swedek B., Zieba J., Prasad P.N. *J. Appl. Phys.* **76**, 4995 (1994).
8. Kippelen, B., Sandalphon, Peyghambarian, N., Lyon, S.R., Padias, A.B., Hall Jr., H.K. *Electr. Lett.* **29**, 1873 (1993).
9. Meerholz K., Volodin B., Sandalphon, Kippelen B., Peyghambarian N. *Nature* **371**, 497 (1994).
10. (a) Meerholz K., Bittner R., De Nardin Y., Bräuchle C., Hendrickx E., Volodin B.L., Kippelen B., Peyghambarian N. *Adv. Mat.* **9**, 1043 (1997); (b) Meerholz K., De Nardin Y., Bittner R. *Mol. Cryst. Liq. Cryst.* **315**, 99 (1998).
11. (a) Hendrickx, E., Volodin, B.L., Steele, D.D., Maldonado, J.L., Wang, J.F., Kippelen, B., Peyghambarian, N. *Appl. Phys. Lett.* **71**, 1159 (1997); (b) Hendrickx, E., Wang, J.F., Maldonado, J.L., Volodin, B.L., Sandalphon, Mash, E.A., Persoons, A., Kippelen, B., Peyghambarian, N. *Macromol.* **31**, 734 (1998).
12. Cox A.M., Blackburn R.D., West D.P., King T.A., Wade F.A., Leigh D.A. *Appl. Phys. Lett.* **68**, 2801 (1996).
13. Smiley E.J., McGee D.J., Salter C., Carlen C.R. *J. Appl. Phys.* **88**, 4910 (2000).
14. Chun H., Moon I.K., Shin D.-H., Kim N. *Mol. Cryst. Liq. Cryst.* **370**, 107 (2001).

15. Meerholz, K., Kippelen, B., Peyghambarian, N. "Non-Crystalline Organic Photore-fractive Materials: Chemistry, Physics, and Applications," 571–632 in *Electrical and Optical Polymer Systems*, by D.L. Wise, G.E. Wnek, D.J. Trantolo, J.D. Gresser, and T.M. Cooper (eds.), World Scientific, 1998.
16. Zilker S.J. *ChemPhysChem* **1**, 72, (2000).
17. Ostroverkhova O., Moerner W.E. *Chem. Rev.* **104**, 3367 (2004).
18. Würthner F.,Wortmann R., Meerholz K. *ChemPhysChem* **1**, 17 (2002).
19. Lundquist, P.M., Wortmann, R., Geletneky, C., Twieg, R.J., Jurich, M., Lee, V.Y., Moylan, C.R., Burland, D.M. *Science* **274**, 1182 (1996).
20. Würthner, F., Wortmann, R., Matschiner, R., Lukaszuk, K., Meerholz, K., Bittner, R., De Nardin, Y., Bräuchle, C., Sens, R. *Angew. Chem. Int. Ed. Engl.* **36**, 2765 (1997).
21. Kippelen B., Marder S., Hendrickx E., Maldonado J.L., Guillemet G., Volodin B.L., Steele D.D., Enami Y., Sandalphon, Yao Y.J., Wang J.F., Röckel H., Erskine L., Peyghambarian N. *Science* **279**, 54 (1998).
22. Winiarz J.G., Zhang L., Park J., Prasad P.N. *J. Phys. Chem. B* **106**, 967 (2002).
23. (a) Blanche, P.A., Kippelen, B., Schülzgen, A., Fuentes-Hernandez, C., Ramos-Ortiz, G., Wang, J.F., Hendrickx, E., Peyghambarian, N. *Opt. Lett.* **27**, 19 (2002); (b) Kippelen, B., Blanche, P.A., Schülzgen, A., Fuentes-Hernandez, C., Ramos-Ortiz, G., Wang, J.F., Peyghambarian, N., Marder, S.R., Leclercq, A., Beljonne, D., Bredas, J.L. *Adv. Funct. Mat.* **12**, 615 (2002).
24. (a) Tay, S., Thomas, J., Eralp, M., Li, G., Kippelen, B., Marder, S.R., Meredith, G., Schülzgen, A., Peyghambarian, N. *Appl. Phys. Lett.* **85**, 4561 (2004). (b) Tay, S., Thomas, J., Eralp, M., Li, G., Norwood, R.A., Schülzgen, A., Yamamoto, M., Barlow, S., Walker, G.A., Marder, S.R., Peyghambarian, N. *Appl. Phys. Lett.* **87**, 171105 (2005).
25. Ogino K., Nomura T., Shichi T., Park S.-H., Sato H., Aoyama T., Wada T. *Chem. Mater.* **9**, 2768 (1997).
26. Herlocker J.A., Ferrio K.B., Hendrickx E., Guenther B.D., Mery S., Kippelen B., Peyghambarian, N. *Appl. Phys. Lett.* **74**, 2253 (1999).
27. Wright, D., Diaz-Garcia, M.A., Casperson, J.D., DeClue, M., Moerner, W.E., Twieg, R.J. *Appl. Phys. Lett.* **73**, 1490 (1998).
28. Hohle C., Hofmann U., Schloter S., Thelakkat M., Strohriegel P., Haarer D., Zilker S.J. *J. Chem Mater.* **9**, 2205 (1999).
29. Hofmann U., Grasruck M., Leopold A., Schreiber A., Schloter S., Hohle C., Strohriegl P., Haarer D., Zilker S.J. *J. Phys. Chem. B* **104**, 3887 (2000).
30. Grasruck M., Schreiber A., Hofmann U., Zilker S.J., Leopold A., Schloter S., Hohle C., Strohriegl P., Haarer D. *Phys. Rev. B* **60**, 16543 (1999).
31. (a) Mecher E., Gallego-Gómez F., Tillmann H., Hörhold H.-H., Hummelen J.C., Meerholz K. *Nature* **418**, 959 (2002); (b) Mecher E., Gallego-Gomez F., Meerholz K., Tillmann H., Hörhold H.H., Hummelen J.C. *Chem. Phys. Chem.* **5**, 277 (2004).
32. Kulikovsky L., Neher D., Mecher E., Meerholz K., Hörhold H.-H., Ostroverkhova O. *Phys. Rev. B* **69**, 125216 (2004).
33. Langevin P. *Ann. Chim. Phys.* VII, **28**, 433 (1906).
34. Buse, K., Adibi, A., Psaltis *Nature* **393**, 665–668 (1998).
35. Martin Pope, Charles E. Swenberg, "Electronic Processes in Organic Crystals and Polymers" 2nd edition, Oxford University Press **1999**.
36. Onsager, L. *Phys. Rev.*, **54**, 554, 1938.
37. Mozumder, A. *The J. of Chem. Phys.* **60**, 4300, 1974.
38. Khand K., Binks D.J., West D.P. *J. Appl. Phys.* **89**, 2516 (2001).

39. Däubler T.K., Bittner R., Meerholz K., Neher D. *Phys. Rev. B* **61**, 13515 (2000).
40. Cimrová V., Kmínek, I., Nešpurek S., Schnabel W. *Synth. Met.*, **64**, 271 (1994).
41. Barth S., Bässler H. *Phys. Rev. Lett.*, **79**(22), 4445 (1997).
42. Noolandi J., Hong K.M. *J.Chem. Phys.*, **70**(1), 3230 (1979).
43. Braun L.B. *J. Chem. Phys.*, **80**(9), 4157 (1984).
44. Cimrová V., Nešpurek S. *Chem. Phys.*, **184**, 283 (1994).
45. Grunnet-Jepsen A., Wright D., Smith B., Bratcher M.S., DeClue M.S., Siegel J.S., Moerner W.E. *Chem. Phys. Lett.* **291**, 553 (1998).
46. Bäuml, G., Schloter, S., Hofmann, U., Haarer, D. *Synth. Met.* **97**, 165 (1998).
47. Bässler H. *Phys. Stat Sol. B* **175**, 15 (1993).
48. Pautmeier L., Richert R., Bässler H. *Synth. Met.* **37**, 271 (1990).
49. Bittner R., Däubler T.K., Neher D., Meerholz K. *Adv. Mat.* **11**, 123 (1999).
50. Yuan B., Sun X., Hou C., Zhou Z., Jiang Y., Li C. *J. Appl. Phys.* **88**, 5562 (2000).
51. Ostroverkhova O., Singer K.D. *J. Appl. Phys*, **92**, 1727 (2002).
52. Cui Y., Swedek B., Cheng N., Zieba R., Prasad P.N. *J. Appl. Phys.* **85**, 38 (1999).
53. Tayebati P., Mahgerefteh D. *JOSA B*, **8**, 1053 (1991).
54. Bryushinin M.A., Sokolov I.A. *Phys. Rev B* **62**, 7186 (2000).
55. Mansurova S. personal communication.
56. Yuan B.H., Sun X.D., Hou C.F., Zhou Z., Jiang Y.Y., Li C. *J. Appl. Phys.* **88**, 5562 (2000).
57. Cui Y., Swedek B., Cheng N., Zieba R., Prasad P.N. *J. Appl. Phys.* **85**, 38 (1999).
58. Bittner R., Meerholz K., Steckman G., Psaltis D. *Appl. Phys. Lett.* **81**, 1 (2002).
59. Kuzuyk, M.G., Singer K.D., Zahn H.E., King L.A. *J. Opt. Soc. Am B* **6**, 742 (1989).
60. Wu J.W. *J. Opt. Soc. Am. B* **8**, 142 (1991).
61. Binks D.J., West D.P. *J. Chem. Phys.* **115**, 1060 (2001).
62. Moerner W.E., Silence S.M., Hache F., Bjorklund G.C. *J. Opt. Soc. Am. B*, **11**, 320 (1994).
63. Bittner R., Bräuchle C., Meerholz K. *Appl. Opt.* **37**, 2843 (1998).
64. (a) Ribierre J.-C., Cheval G., Huber F., Mager L., Fort A., Muller R., Méry S., Nicoud J.F. *J. Appl. Phys.* **91**, 1710 (2002); (b) Ribierre J.-C., Mager L., Fort A., Méry S. *Macromolecules* **36**, 2516 (2003).
65. Swedek B., Cheng N., Cui Y., Zieba J., Winiarz J., Prasad P.N. *J. Appl. Phys.* **82**, 5923 (1997).
66. Binks D.J., West D.P. *Appl. Phys. B* **74**, 279 (2002).
67. Binks D.J., Khand K., West D.P. *J. Appl. Phys.* **89**, 231 (2001).
68. Binks D.J., Khand K., West D.P. *J. Opt. Soc. Am. B* **18**, 308 (2001).
69. Binks D.J., West D.P. *Appl. Phys. Lett.* **77**, 11108 (2001).
70. Scher H., Montroll E.W. *Phys. Rev. B* **12**, 2455 (1975).
71. Shakos, J.D., West, D.P., Rahn, M.D., Khand, K. *J. Opt. Soc. Am. B* **17**, 373 (2000).
72. Wortmann, R., Poga, C., Twieg, R.J., Geletneky, C., Moylan, C.R., Lundquist, P.M., DeVoe, R.G., Cotts, P.M., Horn, H., Rice, J.E., Burland, D.M. *J. Chem. Phys.* **105**, 10637 (1996).
73. Johnson R.V., Tanguay A.R. *Opt. Lett.* **13**, 189 (1988).
74. Stankus, J.J., Silence, S.M., Moerner, W.E., Bjorklund, G.C. *Opt. Lett.* **19**, 1480 (1994).
75. Marcus R.A. *Ann. Rev. Phys. Chem.* **15**, 155 (1964).
76. Meerholz K., Mecher E., Bittner R., De Nardin Y. *J. Opt. Soc. Am. B*, **15**, 2114, 1998.
77. Bittner R., Meerholz K., Stepanov S. *Appl. Phys. Lett.* **74**, 3723 (1999).
78. Meerholz K., Bittner R., De Nardin Y. *Opt. Commun.* **150**, 205 (1998).

79. (a) Grunnet-Jepsen, A., Thompson, C.L., Moerner, W.E. *Science* **277**, 549 (1997); (b) Grunnet-Jepsen, A., Thompson, C.L., Moerner, W.E. *J. Opt. Soc. Am. B* **15**, 905 (1998).
80. Heerden P.J. *Appl. Opt.* **2**, 393 (1963).
81. Sutter, K. and Günter, P. *J. Opt. Soc. Am. B* **7**, 2274 (1990).
82. Walsh, C.A., Moerner, W.E. *J. Opt. Soc. Am.*, **9**, 1642 (1992).
83. Grunnet-Jepsen A., Thompson C.L., Moerner W.E. *Opt. Lett.* **22**, 874 (1997).
84. Pedersen H.C., Johansen P.M., Pedersen T.G. *Opt. Comm.* **192**, 377 (2001).
85. (a) Santos P.A., Cescato L., Frejlich J. *Opt. Lett.* **13**, 1014 (1988); (b) Garcia P.M., Cescato L., Frejlich J. *J. Appl. Phys.* **66**, 47 (1989).
86. Ilynikh P.N., Nestiorkin O.P., Zel'dovich B.Y. *Opt. Commun.* **80**, 249 (1991).
87. Liphardt M., Ducharme S. *J. Opt. Soc. Am. B* **15**, 2154 (1998).
88. Kogelnik H. *Bell Syst. Tech. J.*, **48**, 2909 (1969).
89. Bittner R., Bräuchle C., Meerholz K. *J. Inf. Rec.* **24**, 469 (1998).
90. (a) Bolink, H.J., Malliaras, G.G., Krasnikov, V.V., Hadziioannou, G. *J. Phys. Chem.* **100**, 16356 (1996); (b) Malliaras, G.G., Krasnikov, V.V., Bolink, H.J., Hadziioannou, G. *Pure Appl. Opt.* **5**, 631 (1996).
91. Kwon, O.P., Montemezzani, G., Gunter, P., Lee, S.H. *Appl. Phys. Lett.* **84**, 43 (2004).
92. Kippelen B., Meerholz K., Peyghambarian N., in "Nonlinear Optics of Organic Molecules and Polymers" (eds. Nalwa H.S., Miyata S.), CRC Press, **1997**.
93. Hofmann, U., Schreiber, A., Haarer, D., Zilker, S.J., Bacher, A., Bradley, D.D.C., Redecker, M., Inbasekaran, M., Wu, W.W., Woo, E.P. *Chem. Phys. Lett.* **311**, 41 (1999).
94. Malliaras G.G., Krasnikov V.V., Bolink H.J., Hadziioannou G. *Appl. Phys. Lett.* **66**, 1038 (1995).
95. Silence S.M., Bjorklund G.C., Moerner W.E. *Opt. Lett.* **19**, 1822 (1994).
96. Mecher E., Bittner R., Bräuchle C., Meerholz K. *Synth. Metals* **102**, 993 (1999).
97. Herlocker J.A., Fuentes-Hernandez C., Ferrio K.B., Hendrickx E., Blanche P.A., Peyghambarian N., Kippelen B., Zhang Y., Wang J.F., Marder S.R. *Appl. Phys. Lett.* **77**, 2292 (2000).
98. Bittner R., Ph.D. thesis, University of Cologne (2003).
99. Bittner R., in preparation.
100. Steckman G., Bittner R., Meerholz K., Psaltis D. *Opt. Comm.* **185**, 13 (2000).
101. Ostroverkhova O., Moerner W.E., He M., Twieg R.J. *Chem. Phys. Chem.* **4**, 732 (2003).
102. Gallego, F., Köber, S., Salvador, M., Meerholz, K. in preparation.
103. Sandalphon, Kippelen B., Meerholz K., Peyghambarian N. *Appl. Opt.* **35**, 2346 (1996).
104. Hendrickx E., Guenther B.D., Zhang Y., Wang J.F., Staub K., Zhang Q., Marder S.R., Kippelen B., Peyghambarian N. *Chem. Phys.* **245**, 407 (1999).
105. Gallego-Gómez F., Mecher E., Meerholz K. *Proc. SPIE* **5216**, 1 (2002).
106. Joo W.-J., Kim N.-J., Chun H., Moon I.K., Kim N., Oh C.-H. *J. Appl. Phys.* **91**, 6471 (2002).
107. Zhang, Y., Wada, T., Sasabe, H. *J. Mat. Chem.* **8**, 809 (1998).
108. Suh, D.J., Park, O.O., Ahn, T., Shim, H.K. *Opt. Mat.* **21**, 365 (2003).
109. Mecher E., Bräuchle C., Hörhold H.H., Hummelen J.C., Meerholz K. *Phys. Chem. Chem. Phys.* **1**, 1749 (1999).
110. Jakob, T., Schloter, S., Hofmann, U., Grasruck, M., Schreiber, A., Haarer, D. *J. Chem. Phys.* **111**, 10633 (1999).

111. (a) Zobel O., Eckl M., Strohriegl P., Haarer D. *Adv. Mat.* **7**, 911 (1995); (b) Schloter, S., Hofmann, U., Hoechstetter, K., Bauml, G., Haarer, D., Ewert, K., Eisenbach, C.D. *J. Opt. Soc. Am. B* **15**, 2560 (1998).

112. Wright, D., Gubler, U., Moerner, W.E., DeClue, M., Siegel, J.S. *J. Phys. Chem. B* **107**, 4732 (2003).

113. Hörhold, H.H., Tillmann, H., Raabe, D., Helbig, M., Elflein, W., Bräuer, A., Holzer, W., Penzkofer, A. *Proc. SPIE* **4105**, 431 (2001).

114. Kwon, O.P., Lee, S.H., Montemezzani, G., Günter, P. *J. Opt. Soc. Am. B* **20**, 2307 (2003).

115. Silver M., Risko K., Bässler H. *Philos. Mag. B*, **40**, 247 (1979).

116. Gill W.D. *J. Appl. Phys.* **43**, 5033 (1972).

117. Pai D.M., Janus J.F., Stolka M. *J. Phys. Chem.*, **88**, 4714 (1984).

118. Van Steenwinckel, D., Hendrickx, E., Persoons, A., Samyn C. *Chem. Mat.* **13**, 1230 (2001).

119. Hummelen J.C., Knight B.W., LePeq F., Wudl F. *J. Org. Chem.* **60**, 532 (1995).

120. (a) Winiarz, J.G., Zhang, L.M., Lal, M., Friend, C.S., Prasad, P.N. *J. Am. Chem. Soc.* **121**, 5287 (1999); (b) Winiarz, J.G., Zhang, L.M., Lal, M., Friend, C.S., Prasad, P.N. *Chem. Phys.* **245**, 417 (1999).

121. (a) Binks, D.J., West, D.P., Norager, S., O'Brien, P. *J. Chem. Phys.* **117**, 7335 (2002); (b) Binks, D.J., Bant S.P., West, D.P., O'Brien P., Malik M.A. *J. Mod. Opt.* **50**, 299 (2003).

122. Fuentes-Hernandez C., Suh D.J., Kippelen B., Marder S.R. *Appl. Phys. Lett.* **85**, 534 (2004).

123. Würthner F., Yao S., Schilling J., Wortmann R., Redi-Abshiro M., Mecher E., Gallego-Gomez F., Meerholz K. *J. Am. Chem. Soc.*, **123**(12), 2810 (2001).

124. Khand K., Binks D.J., West D.P., Rahn M.S. *J. Mod. Opt.* **48**(1), 93 (2001).

125. Van Steenwinckel, D., Hendrickx, E., Persoons, A. *J. Chem. Phys.* **114**, 9557 (2001).

126. Maslak P., Chopra A., Moylan C.R., Wortmann R., Lebus S., Rheingold A.L., Yap G.P.A. *J. Am. Chem. Soc.* **118**, 1471 (1996).

127. Blanchard-Desce M., Barzoukas M. *J. Opt. Soc. Am. B*, **15**, 302 (1998).

128. Marder S.R., Beratan D.N., Cheng L.T. *Science* **252**, 103 (1991).

129. Oudar J.L., Chemla D.S. *J. Chem. Phys.* **66**, 2664 (1977).

130. Moylan C.R., Wortmann R., Twieg R.J., McComb I.-H. *J. Opt. Soc. Am. B* **15**, 929 (1998).

131. Chen, Z.J., Wang, F., Huang, Z.W., Gong, Q.H., Chen, Y.W., Zhang, Z.J., Chen, H.Y. *J. Phys. D* **31**, 2245 (1998).

132. Cox, A.M., Blackburn, R.D., West, D.P., King, T.A., Wade, F.A., Leigh, D.A. *Appl. Phys. Lett.* **68**, 2801 (1996).

133. Rahn, M.D., West, D.P., Shakos, J.D. *J. Appl. Phys.* **87**, 627 (2000).

134. Diaz-Garcia, M.A., Wright, D., Casperson, J.D., Smith, B., Glazer, E., Moerner, W.E., Sukhomlinova, L.I., Twieg, R.J. *Chem. Mat.* **11**, 1784 (1999).

135. Fort, A., Müller, J., Cregut, O., Vola, J.P., Barzoukas, M. *Opt. Mat.* **9**, 271 (1998).

136. Chun, H., Moon, I.K., Shin, D.H., Kim, N. *Chem. Mat.* **13**, 2813 (2001).

137. (a) Hendrickx, E., Van Steenwinckel, D., Persoons, A., Samyn C., Beljonne D., Bredas J.L. *J. Chem. Phys.* **113**, 5439 (2000); (b) Hendrickx, E., Van Steenwinckel, D., Engels, C., Gubbelmans, E., van den Broeck, K., Beljonne, D., Bredas, J.L., Samyn, C., Persoons, A. *Materials, Science & Engineering C* **18**, 25 (2001); (c) Hendrickx E., Engels C., Schaerlaekens M., Van Steenwinkel D., Samyn C., Persoons A. *J. Phys. Chem. B* **106**, 4588 (2002).

138. (a) Wright, D., Gubler, U., Roh, Y., Moerner, W.E., He, M., Twieg, R.J. *Appl. Phys. Lett.* **79**, 4274 (2001); (b) Gubler, U., He, M., Wright, D., Roh, Y., Twieg, R., Moerner, W.E. *Adv. Mat.* **14**, 313 (2002); (c) Ostroverkhova, O., Gubler, U., Wright, D., Moerner, W.E., He, M., Twieg, R. *Adv. Funct. Mat.* **12**, 621 (2002); (d) Ostroverkhova, O., Moerner, W.E., He, M., Twieg, R.J. *Appl. Phys. Lett.* **82**, 3602 (2003).
139. Wang, L.M., Ng, M.K., Yu, L.P. *Appl. Phys. Lett.* **78**, 700 (2001).
140. Aiello, I., Dattilo, D., Ghedini, M., Golemme, A. *J. Am. Chem. Soc.* **123**, 5598 (2001); (b) Aiello, I., Dattilo, D., Ghedini, M., Bruno, A., Termine, R., Golemme, A. *Adv. Mat.* **14**, 1233 (2002).
141. (a) Hou, Z., You, W., Yu, L.P. *Appl. Phys. Lett.* **82**, 3385 (2003); (b) You, W., Hou, Z., Yu, L.P., *Adv. Mat.* **16**, 356 (2004).
142. Beckmann, S., Etzbach, K.-H., Krämer, P., Lukaszuk, K., Matschiner, R., Schmidt, A.J., Schuhmacher, P., Sens, R., Würthner, F. *Adv. Mat.* **11**, 536 (1999).
143. Wortmann, R., Glania, C., Kramer, P., Lukaszuk, K., Matschiner, R., Twieg, R.J., You, F. *Chem. Phys.* **245**, 107 (1999).
144. Fort, A., Mager, L., Müller, J., Combellas, C., Mathey, G., Thiébault, A. *Opt. Mat.* **12**, 339 (1999).
145. Meerholz, K., De Nardin, Y., Bittner, R., Wortmann, R., Würthner, F., Yao, S., Schilling, J., Wortmann, R., Redi-Abshiro, M., Mecher, E. *Appl. Phys. Lett.* **73**, 4 (1998).
146. Würthner, F., Yao, C., Debaerdemaeker, T., Wortmann, R. *J. Am. Chem. Soc.* **124**, 9431 (2002).
147. Wortmann R., Rösch U., Rdi-Abshiro M., Würthner F. *Angew. Chem. Int. Ed.* **42**, 2080 (2003).
148. Goonesekera, A., Ducharme, S. *J. Appl. Phys.* **85**, 6506 (1999).
149. (a) Hendrickx, E., Zhang, Y.D., Ferrio, K.B., Herlocker, J.A., Anderson, J., Armstrong, N.R., Mash, E.A., Persoons, A.P., Peyghambarian, N., Kippelen, B. *J. Mat. Chem.* **9**, 2251 (1999); (b) Herlocker, J.A., Fuentes-Hernandez, C., Ferrio, K.B., Hendrickx, E., Blanche, P.A., Peyghambarian, N., Kippelen, B., Zhang, Y., Wang, J.F., Marder, S.R. *Appl. Phys. Lett.* **77**, 2292 (2000).
150. Bard, A.J. and Faulkner, L.A. *Electrochemical Methods: Fundamentals and Applications* (Wiley, New York, 1984).
151. Gross, M., Müller, C.D., Bräuchle, C., Nothofer, H.G., Scherf, U., Neher, D., Meerholz, K. *Nature* **405**, 661 (2000).
152. Hwang, J., Seo, J., Sohn, J., Park, S.Y., *Opt. Mat.* **21**, 359 (2002).
153. Wu S.Z., Zeng F., Li F.X., Zhu Y.L., Zhao H.P. *J. Polymer Sci. B* **37**, 3302 (1999).
154. Van Steenwinckel, D., Hendrickx, E., Samyn, C., Engels, C., Persoons, A. *J. Mat. Chem.* **10**, 2692 (2000).
155. Swedek B., Cheng N., Cui Y., Zieba J., Winiarz J., Prasad P.N. *J. Appl. Phys.* **82**(12) 5923 (1997).
156. Chaput F., Riehl D., Boilot J.P., Cargnelli K., Canva M., Lévy Y., Brun A. *Chem. Mater.* **8**, 312 (1995).
157. Darracq B., Canva M., Chaput F., Boilot J.P., Riehl D., Lévy Y., Brun A. *Appl. Phys. Lett.* **70**, 292 (1997).
158. Burzynski R., Casstevens M.K., Zhang Y., Ghosal S. *Opt. Eng.* **35**, 443 (1996).
159. Yu L. *J. Polym. Sci. A* **39**, 2557 (2001).
160. Schloter S., Hofmann U., Höchstetter K., Bauml G., Haarer D., Ewert K., Eisenbach C.D. *J. Opt. Soc. Am. B* **15**, 2560 (1998).

161. Zilker, S.J., Hofmann, U., *Appl. Opt.* **39**, 2287 (2000).
162. Van Steenwinckel D., Engels C., Gubbelmanns E., Hendrickx E., Samyn C., Persoons A. *Macromolecules* **33**, 4074 (2000).
163. Hattemer E., Zentel R., Mecher E., Meerholz K. *Macromolecules* **33**, 1972 (2000).
164. Park S.-H., Ogino K., Sato H. *Synth. Met.* **113**, 135 (2000).
165. Wang Q., Wang L., Yu J., Yu L. *Adv. Mat.* **12**(13), 974 (2000).
166. You, W., Cao, S., Hou, Z., Yu, L. P. *Macromol.* **36**, 7014 (2003).
167. Bratcher, M.S., DeClue, M.S., Grunnet-Jepsen, A., Wright, D., Smith, B.R., Moerner, W.E., Siegel, J.S. *J. Am. Chem. Soc.* **120**, 9680 (1998).
168. Cheben, P., del Monte, F., Worsfold, D.J., Carlsson, D.J., Grover, C.P., Mackenzie, J.D. *Nature* **408**, 64 (2000).
169. Aoyama, T., Wada, T., Zhang, Y.D., Sasabe, H., Moritsuki, Y., Yonechi, Y., Koike, Y. *Polymer Journal* **33**, 718 (2001).
170. (a) Wang L., Zhang Y., Wada T., Sasabe H. *Appl. Phys. Lett.* **69**(6), 728 (1996); (b) Zhang, Y., Wang, L., Wada, T., Sasabea, H. *Appl. Phys. Lett.* **70**, 2949 (1997).
171. (a) Li W., Gharavi A., Wang Q., Yu L. *Adv. Mat.* **10**(12), 927 (1998); (b) Wang L., Ng M.-K., Yu L. *Phys. Rev. B* **62**(8), 4973 (2000).
172. Wang, Q., Wang, L., Saadeh, H., Yu, L. *Chem. Commun.*, 1689 (1999).
173. Mager L., Méry S. *MCLC* **322**, 21 (1998).
174. Ogino K., Park S.-H., Sato H. *Appl. Phys. Lett.* **74**(26), 3936 (1999).
175. Ono H., Kawatsuki N. *Opt. Lett.* **22**, 1144 (1997).
176. Golemme A., Kippelen B., Peyghambarian N. *Appl. Phys. Lett.* **73**, 2408 (1998).
177. Ono, H., Shimokawa, H., Emoto, A., Kawatsuki, N. *Polymer* **44**, 7971 (2003).
178. Ono, H., Shimokawa, H., Emoto, A., Kawatsuki, N. *J. Appl. Phys.* **94**, 23 (2003).
179. Van Steenwinckel D., Hendrickx E., Persoons A. *Chem. Mater.* **13**, 1230 (2001).
180. Ono H., Saito i., Kawatsuki N. *Appl. Phys. Lett.* **72**, 1942 (1998).
181. Ono H., Hanazawa A., Kawamura T., Norisada H., Kawatsuki N. *J. Appl. Phys.* **86**, 1785 (1999).
182. Ono H., Kawatsuki N. *J. Appl. Phys.* **85**, 2482 (1999).
183. Ono H., Kawamura T., Frias N.M., Kitamura K., Kawatsuki N., Norisada H., Yamamoto T. *J. Appl. Phys.* **88**, 3853 (2000).
184. Ono H., Kawamura T., Frias N.M., Kitamura K., Kawatsuki N., Norisada H. *Adv. Mat.* **12**, 143 (2000).

14

Organic Photorefractive Materials and Their Applications

Bernard Kippelen

School of Electrical and Computer Engineering, Georgia Institute of Technology, Atlanta, GA 30332 404 385-5163. `kippelen@ece.gatech.edu`

14.1 Introduction

During the past decade, polymers for photonic devices have matured considerably and numerous functionalities such as electrical, electrooptical, and light-emitting properties have been incorporated successfully into polymers through the intimate collaboration between scientists from optical sciences, physics, chemistry, and materials science. Photorefractive polymers belong to these new classes of materials with several functionalities that have emerged during the nineties. Building on the previous development of electrooptic and photoconducting polymers for optical switches and xerography, respectively, photorefractive polymers experienced a rapid development. Within ten years, they reached a performance level that makes them alternatives to the well-known inorganic photorefractive materials. Active multidisciplinary research efforts in this field are constantly providing new materials with improved performance and simultaneously a better understanding of the basic principles of photorefractivity in these complex materials. For instance, nanotechnologies are enabling hybrid approaches in which nanoscale inorganic materials can be mixed in an organic matrix; or self-assembly provides new means to control the morphology of new materials with improved properties.

In this chapter, we review the basic optical properties required to understand photorefractivity in organic materials, with an emphasis on the differences between inorganic and organic materials. In section two, we provide a review of the basic concepts of nonlinear optical and electrooptic properties of polymers. We discuss the molecular nonlinear optical properties at the microscopic level and provide models to relate them to macroscopic properties. We assume that the reader is familiar with bulk nonlinear and electrooptic properties [1–5]. Section 14.3 gives a description of the photoconducting properties of organic materials. One of the fundamental differences between organic and inorganic materials is the strong field dependence of the photogeneration efficiency of carriers and the field dependence of the charge mobility. Transport in amorphous organic semiconductors cannot be described by band-type models but can be reasonably well described by the disorder formalism, which gives an empirical expression for the field and temperature dependence of the mobility. Section 14.4 provides a brief review of the *Kukhtarev model* developed initially to describe photorefractivity in crystals and presents some of the more advanced models that incorporate the effects of the field dependence of the photogeneration efficiency and the mobility, as well as the effects of shallow and deep traps. The latter section describes also orientational photorefractivity, a process in which the refractive index of the material is modulated through the reorientation of the molecules rather than through the traditional Pockels effect. This form of orientational photorefractivity provides a means to achieve large refractive index modulations and consequently large diffraction efficiency, but has a dynamics that is limited by the orientational diffusion of the molecules. In Section 14.5, we describe the design of photorefractive polymers and illustrate some of their properties through selected examples of materials that have been developed in recent years. We do not give an exhaustive survey of the different materials reported in the literature, since such surveys can be found in several review articles [6–9]. The last section will describe some examples of applications that have been demonstrated so far with photorefractive polymers.

14.2 Molecular and Bulk Nonlinear Optics

14.2.1 Microscopic Theory: First and Second Hyperpolarizability

Like macroscopic systems, molecules exhibit second-order nonlinearities only if the centrosymmetry is broken. This can be achieved in conjugated molecules by connecting each end of the conjugated path with groups that have a different electronic affinity. In other words, the symmetry can be broken by deforming the π electron distribution by attaching a donor-like group at one end and an acceptor-like group at the other end as shown schematically in Figure 14.1. This class of molecules is referred to as *push-pull* molecules. Due to an excess of charge at the acceptor side, the molecule has a dipole moment in its ground state. The response of the molecule to an electric field, or its polarizability, depends strongly on the direction of the applied field with respect to the molecule: charge flow is favored toward the acceptor, while hindered toward the donor. This asymmetric polarization provides strong second-order nonlinear optical properties. On a molecular level, in the dipole approximation, the microscopic polarization components of a molecule under nonresonant excitation conditions can be written as

$$p_i = \mu_i + \sum_j \alpha_{ij} E_j + R^{(2)} \sum_{jk} \beta_{ijk} E_j E_k + R^{(3)} \sum_{jkl} \gamma_{ijkl} E_j E_k E_l + \cdots ,$$

$$(14.1)$$

A —— [π – conjugated bridge] —— D

FIGURE 14.1. Schematics of push-pull molecules for electrooptic and photorefractive applications.

where α_{ij}, β_{ijk}, γ_{ijkl} are tensor elements and E_j, E_k, E_l are components of the electric field. The subscripts i, j, k, l refer to Cartesian coordinates expressed in the frame of the molecule; μ_i is the ground-state dipole moment of the molecule. The linear polarizability is given by the tensor $\tilde{\alpha}$. The microscopic equivalent of the macroscopic susceptibility $\tilde{\chi}^{(2)}$ is described by $\tilde{\beta}$ and is called the *first hyper-polarizability*. Similarly, $\tilde{\gamma}$ is the second hyperpolarizability and is the molecular equivalent of the macroscopic susceptibility $\tilde{\chi}^{(3)}$. Note that $\tilde{\beta}$ is a *third-rank* tensor that is at the origin of *second-order* nonlinear effects, and that is called *first* hyper-polarizability. In (14.1), $R^{(n)}$ are degeneracy factors that depend on the nonlinear process and the frequencies of the electric fields involved in the interaction.

14.2.2 Two-Level Model

With the structural flexibility of organic compounds, molecules with increasing nonlinearity could be synthesized and incorporated into polymer matrices. Today, numerous electrooptic polymers exhibit electrooptic coefficients that are higher than that of the inorganic crystal of lithium niobate ($LiNbO_3$), which is a standard in the electrooptics industry with an electrooptic coefficient of 30 pm/V. These advances were possible through the simultaneous improvement of the magnitude of the nonlinearity at a molecular level and through a better understanding of intermolecular interactions between molecules in the polymer matrix. To pre-dict the nonlinear optical properties of push-pull molecules, Oudar and Chemla [10–13] proposed a simple two-level model in which they considered the in-tramolecular charge-transfer interaction between acceptor and donor, and the fact that β was governed not only by the ground-state electronic distribution but also by excited states. The hyperpolarizability was then taken to be the sum of two contributions:

$$\beta = \beta_{add} + \beta_{CT}, \qquad (14.2)$$

where β_{add} is the additive part of the substituents and their individual interaction with the conjugated π-electron system, and β_{CT} is the contribution from the intramolecular charge-transfer interaction between acceptor and donor through the conjugated connecting bridge. The charge-transfer part is approximated to a two-level model with a ground state (g) and an excited state (e) as illustrated in Figure 14.2. Within this two-level model the charge-transfer contribution to the

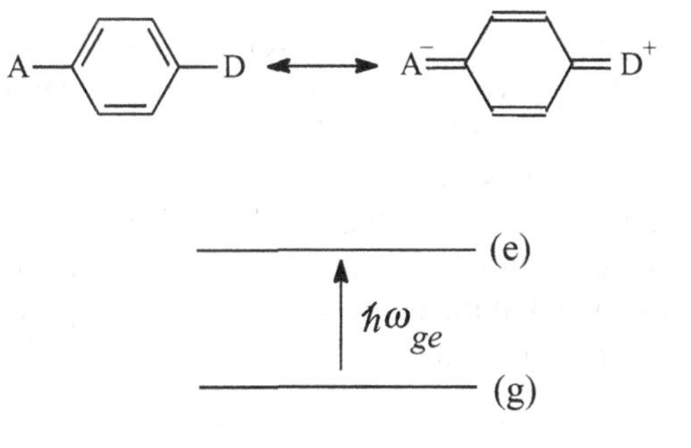

FIGURE 14.2. Ground state and first excited state of a push-pull molecule.

hyperpolarizability is given by

$$\beta_{CT}(\omega_3, \omega_2, \omega_1) = R\frac{e^2 f \Delta\mu_{ge}}{2m\hbar\omega_{ge}} \frac{\omega_{ge}^2(3\omega_{ge}^2 + \omega_1\omega_2 - \omega_3^2)}{(\omega_{ge}^2 - \omega_1^2)(\omega_{ge}^2 - \omega_2^2)(\omega_{ge}^2 - \omega_3^2)}, \quad (14.3)$$

where R is a degeneracy factor, e is the elementary charge, m the electron mass, ω_i are the frequencies of the optical waves involved in the second-order nonlinear process, $\hbar\omega_{ge}$ is the energy difference between the ground state (g) and the excited state (e), $\Delta\mu_{ge} = \mu_e - \mu_g$ is the difference in dipole moments between the excited state and the ground state, and f is the oscillator strength of the transition that is related to the transition dipole moment μ_{ge} through

$$f = \frac{2m}{\hbar^2}\hbar\omega_{ge}\mu_{ge}^2. \quad (14.4)$$

Substituting for $\omega_3 = 2\omega$, $\omega_1 = \omega_2 = \omega$, in (14.3) leads to the following expression for the charge-transfer contribution to the hyperpolarizability for a second-harmonic generation process:

$$\beta_{CT}^{SHG}(2\omega, \omega, \omega) = \beta^{SHG}(0)\frac{\omega_{ge}^4}{(\omega_{ge}^2 - \omega^2)(\omega_{ge}^2 - (2\omega)^2)} \quad (14.5)$$

with

$$\beta^{SHG}(0) = \frac{3e^2}{\hbar^2}\frac{\Delta\mu_{ge}\mu_{ge}^2}{\omega_{ge}^2} \propto \frac{(\mu_e - \mu_g)\mu_{ge}^2}{E_{ge}^2}, \quad (14.6)$$

where the superscript *SHG* refers to second-harmonic generation; $\beta^{SHG}(0)$ is called *the dispersion-free hyperpolarizability* since it does not contain any dependence on the frequency of the optical fields involved in the nonlinear interaction.

Similarly, starting from (14.3), the charge-transfer contribution to the hyperpolarizability of an electrooptic effect is obtained:

$$\beta_{CT}^{EO}(\omega, 0, \omega) = \beta^{EO}(0)\frac{\omega_{ge}^2(3\omega_{ge}^2 - \omega^2)}{3(\omega_{ge}^2 - \omega^2)^2}, \quad (14.7)$$

where the superscript *EO* refers to the electrooptic effect. When comparing values of the hyperpolarizability of different chromophores, the type of interaction (SHG or EO) should always be considered, as well as the wavelength at which it was measured. Note that due to the dispersion factors appearing in (14.7) and (14.5), the nonlinearity is enhanced when the wavelength of the experiment is close to the wavelength of the charge-transfer absorption band. The charge-transfer absorption band of the chromophore with energy $E_{ge} = \hbar\omega_{ge}$ is usually characterized by its corresponding wavelength $\lambda_{max} = hc/E_{ge}$.

FIGURE 14.3. Neutral, cyanine, and charge-separated form of a polyene molecule and corresponding bond-length alternation (BLA) value.

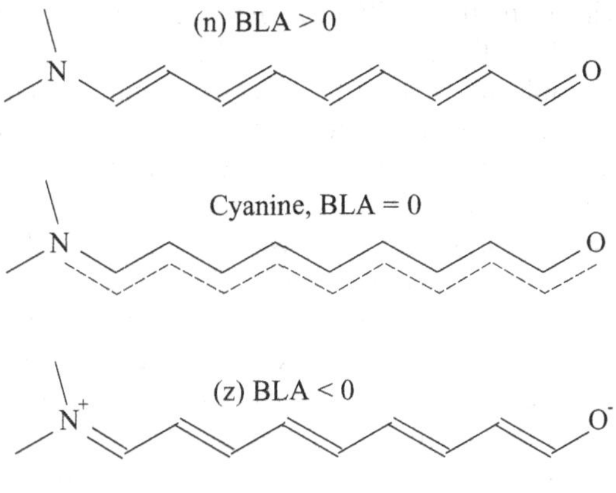

14.2.3 Bond-Length Alternation Model

While the two-level model guided the design of second-order nonlinear optical molecules for almost two decades, a model in which β is correlated to the degree of ground-state polarization was developed in recent years and has provided a way to synthesize molecules with larger nonlinearities [14]. The degree of ground-state polarization, or in other words the degree of charge separation in the ground state, depends primarily on the chemical structure (as for example, the structure of the π-conjugated system, or the strength of the donor and acceptor substituents), but also on its surroundings (as for example, the polarity of the medium). In donor–acceptor polyenes, this variable is related to a geometrical parameter, the bond-length alternation (BLA), which is defined as the average of the difference in the length between adjacent carbon–carbon bonds. To better understand the correlation between β and the degree of ground-state polarization of the molecule, it is illustrative to discuss the wave function of the ground state $|\psi\rangle$ in terms of a linear combination of the two limiting resonance structures:

$$|\psi\rangle = a|n\rangle + b|z >, \tag{14.8}$$

where $|n\rangle$ denotes the neutral form characterized by a positive BLA, and $|z\rangle$ the charge-separated form characterized by a negative BLA (since the double and single bond pattern is now reversed relative to the neutral form) (see Figure 14.3). The charge-separated form is also called *zwitterionic* or *quinoidal.* For substituted polyenes with weak donors and acceptors, the neutral resonance form dominates the ground-state wave function, and the molecule has a high degree of bond-length alternation. In other words, the coefficient a in (14.8) is much larger than b. With stronger donors and acceptors, the contribution of the charge-separated resonance form to the ground state increases, and simultaneously, BLA decreases. When the two resonance forms contribute equally to the ground-state structure, the molecule exhibits essentially no BLA, corresponding to a situation in which $a = b$ in (14.8). This zero BLA situation is called the *cyanine* limit, referring to the common structure of a cyanine molecule. Such cyanine molecules are known to be represented by two degenerate resonance forms, resulting in structures with virtually no BLA, as represented by the medium structure in Figure 14.3. Finally, if the charge-separated form dominates the ground-state wave function, the molecule acquires a reversed bond-alternation pattern ($a \ll b$). BLA is thus a measurable parameter that is related to the *mixing* of the two resonance forms in the actual ground-state structure of the molecule.

Correlation between the molecular β with the ground-state polarization and consequently with BLA was demonstrated through quantum-chemical calculations in polyene molecules [15][16]. In these studies the ground-state polarization was tuned by applying an external static electric field of varying strength and β was correlated with the average of the difference in the π-bond order between adjacent carbon–carbon bonds (more simply denoted the Bond-Order Alternation, BOA) that is qualitatively related, as BLA, to the *mixing* of the two resonance forms in the actual ground-state structure of the molecule. The π-bond order is a measure of the double-bond character of a given carbon–carbon bond (a BOA of about -0.55 is calculated for a donor–acceptor polyene with alternating double and single bonds). On going from the neutral polyene limit to the cyanine limit, β first increases, peaks in a positive sense for an intermediate structure, decreases, and passes through zero at the cyanine limit as shown in Figure 14.4. From that limit and going to the charge-separated resonance structure, β continues to decrease and thus becomes negative, peaks in a negative sense, and then decreases again (in absolute value) to become smaller in the charge-separated structure. Evidence for the proposed relationships was derived from a study of a series of six molecules that spanned a wide range of ground-state polarization and therefore BLA, that were examined in solvents of varying polarity by electric-field-induced second-harmonic generation (EFISH) experiments.

By applying this approach new molecules for electrooptic and photorefractive applications could be developed. For a review of the design of nonlinear molecules using the BLA model see for instance [17].

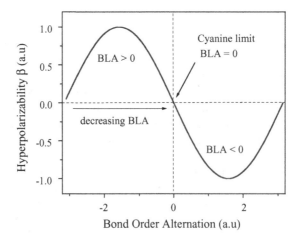

FIGURE 14.4. Magnitude of β as a function of bond-length alternation (BLA) and bond-order alternation (BOA).

14.2.4 Oriented Gas Model

14.2.4.1 Electrooptic Properties

So far, we have described the nonlinear optical properties of molecules on a microscopic level. Macroscopic nonlinearities in polymer films are obtained by doping or functionalizing the polymer with these molecules or molecular groups. To be electrooptic, a bulk material composed of nonlinear molecules must also lack an inversion center if the molecular β is to lead to a macroscopic nonlinearity. The method most widely used to impart noncentrosymmetry in noncrystalline systems is the *poled-polymer* approach. If dipolar nonlinear species are dissolved in a polymer and subjected to a large electric field, at or above the glass-transition temperature T_g of the polymer, the interaction of the dipole with the field causes the dipolar species to orient, to a certain extent, in the direction of the applied field. If the polymer is cooled back to the glassy state, with the field applied, then the field-induced noncentrosymmetric orientation can be frozen in place, yielding a material with a second-order optical nonlinearity. This process is referred to as *poling*.

To understand and optimize the electrooptic properties of polymers by the use of molecular engineering, it is of primary importance to be able to relate their macroscopic properties to the individual molecular properties. Such a task is the subject of ongoing research. However, simple descriptions based on the oriented gas model exist [18,19] and have proven to be in many cases a good approximation for the description of poled electrooptic polymers [20]. The oriented gas model provides a simple way to relate the macroscopic nonlinear optical properties such as the second-order susceptibility tensor elements expressed in the orthogonal laboratory frame {X,Y,Z}, and the microscopic hyperpolarizability tensor elements that are given in the orthogonal molecular frame {x,y,z} (see Figure 14.5).

Because of its simplicity, the oriented gas model relies on a large number of simplifications and approximations: (i) at the poling temperature the chromophores are assumed free to rotate under the influence of the applied field, and any coupling or reaction from the surrounding matrix is ignored; (ii) the chromophores have cylindrical symmetry and the only nonvanishing hyperpolarizability tensor element is β_{zzz}, where z is the symmetry axis of the chromophore; (iii) the permanent dipole moment μ of the chromophore is oriented along the z-axis of the molecule; (iv) the chromophores are assumed independent and noninteracting. Under this last approximation, the bulk response of a material is given by the sum of the responses from its molecular components. The relationship between the macroscopic polarization components in the laboratory frame {X,Y,Z} and the molecular components can be written as

$$P_I(t) = \frac{1}{V}\Big(\sum p_i(t)\Big)_I, \tag{14.9}$$

where V is the volume, the index I refers to the laboratory frame, and the index i to the frame attached to the molecule. Due to the large number of molecules, the summation in (14.9) is replaced by a thermodynamic average. The second-order

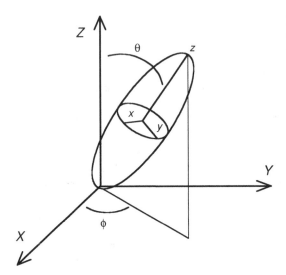

FIGURE 14.5. Laboratory frame $\{X, Y, Z\}$ and molecular frame $\{x, y, z\}$.

nonlinear susceptibility tensor elements can be written as [21]

$$\chi^{(2)}_{IJK} = N \langle a_{Ii} a_{Jj} a_{Kk} \, \beta^{*}_{ijk} \rangle_{IJK}, \qquad (14.10)$$

where N is the density of chromophores, and the a_{Ij} are the director cosines or the projection of the new axis I (in the laboratory frame) on the former axis j (in the molecular frame). The brackets $\langle . \rangle$ in (14.10) denote the orientational ensemble average of the hyperpolarizability tensor elements. The superscript * denotes that the local field correction factors have been included in β. By definition of a thermodynamic average, (14.10) can be rewritten as

$$\langle a_{Ii} a_{Jj} a_{Kk} \, \beta^{*}_{ijk} \rangle_{IJK} = \int a_{Ii} a_{Jj} a_{Kk} \, \beta^{*}_{ijk} \, G(\varphi, \theta, \psi) \, d\Omega, \qquad (14.11)$$

where $G(\varphi, \theta, \psi)$ is the normalized orientational distribution function, φ, θ, ψ are the Euler angles, and $d\Omega = d\psi \sin \theta d\theta d\varphi$ is the differential solid angle. In the following, an additional simplification is made: the distribution function $G(\varphi, \theta, \psi)$ is assumed independent of φ and ψ. This assumption is related to the approximation that the chromophores have cylindrical symmetry and that the second-order nonlinear optical properties depend mainly on the average polar angle θ of the molecules with the direction of the poling field (see Figure 14.5).

The normalized orientational distribution function $G(\varphi, \theta, \psi)$ is approximated to a Maxwell–Boltzmann distribution function, which can be written as

$$G(\Omega) \, d\Omega = \frac{\exp\left(-\frac{U(\theta)}{k_B T}\right) \sin \theta d\theta}{\int_0^{\pi} \exp\left(-\frac{U(\theta)}{k_B T}\right) \sin \theta d\theta}, \qquad (14.12)$$

where $U(\theta)$ is the interaction energy between the poling field and the dipole moment. This interaction energy is given by

$$U(\theta) = -\mu^* \cdot E_p - \frac{1}{2} p \cdot E_p \approx -x \, k_B T \cos\theta \quad \text{with } x = \frac{\mu^* E_p}{k_B T}, \qquad (14.13)$$

where μ^* is the modulus of the permanent dipole moment of the chromophore and is corrected for local field effects by the surrounding matrix, and E_p is the modulus of the poling field that is applied along the Z-axis of the laboratory frame. The second term in the right-hand side of (14.13) describes the reorientation of the molecule due to the induced dipole moment p. Since push-pull molecules have generally a high permanent dipole moment, it is a good approximation to neglect that second contribution. The force associated with the potential $U(\theta)$ is the torque exerted by the electric field on the molecule that tends to orient the molecule in the direction of the poling field. This torque tends to minimize the interaction energy, that is, to reduce the value of the angle θ. With the notation introduced in (14.13) the orientational average of any function $g(\theta)$ that depends solely on the polar angle θ through the interaction potential $U(\theta)$ can be written as

$$\langle g(\theta)\rangle = \frac{\int_0^\pi g(\theta)\exp(x\cos\theta)\sin\theta d\theta}{\int_0^\pi \exp(x\cos\theta)\sin\theta d\theta}. \qquad (14.14)$$

According to (14.10), the two independent tensor elements for poled polymers when Kleinman symmetry is valid are given by

$$\chi_{ZZZ}^{(2)} = N\langle(\hat{K}\cdot\hat{k})(\hat{K}\cdot\hat{k})(\hat{K}\cdot\hat{k})\rangle\beta_{zzz}^* = N\langle\cos^3\theta\rangle\beta_{zzz}^*, \qquad (14.15)$$

$$\chi_{ZXX}^{(2)} = N\langle(\hat{K}\cdot\hat{k})(\hat{I}\cdot\hat{k})(\hat{I}\cdot\hat{k})\rangle\beta_{zzz}^* = N\langle\cos\theta\sin^2\theta\cos^2\varphi\rangle\beta_{zzz}^*. \qquad (14.16)$$

Replacing the ensemble average by (14.14), (14.15), and (14.16) can be rewritten as

$$\chi_{ZZZ}^{(2)} = NL_3(x)\,\beta_{zzz}^*, \qquad (14.17)$$

$$\chi_{ZXX}^{(2)} = \frac{N}{2}\,(L_1(x) - L_3(x))\beta_{zzz}^*, \qquad (14.18)$$

where we have introduced the Langevin functions $L_n(x)$ of order n and with an argument x defined as

$$L_n(x) = \frac{\int_0^\pi \cos^n x \exp(x\cos\theta)\sin\theta d\theta}{\int_0^\pi \exp(x\cos\theta)\sin\theta d\theta}. \qquad (14.19)$$

The Langevin functions for $n = 1$ to 3 are given by

$$L_1(x) = \coth x - \frac{1}{x},$$

$$L_2(x) = 1 + \frac{2}{x^2} - \frac{2}{x}\coth x, \qquad (14.20)$$

$$L_3(x) = \left(1 + \frac{6}{x^2}\right)\coth x - \frac{3}{x}\left(1 + \frac{2}{x^2}\right).$$

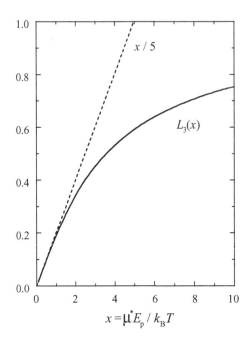

FIGURE 14.6. Langevin function $L_3(x)$ and its linear approximation.

When $x = \mu^* E_p / kT$ is smaller than unity, these functions can be approximated to

$$L_1(x) \approx \frac{x}{3}, \quad L_2(x) \approx \frac{1}{3} + \frac{2x^2}{45}, \quad L_3(x) \approx \frac{x}{5}. \tag{14.21}$$

The validity of this approximation for $L_3(x)$ for values of the argument $x < 1$ is illustrated in Figure 14.6. The condition $x < 1$ is satisfied in most of the experimental conditions. For instance, for a molecule with a dipole moment of $\mu^* = 15$ D (or 15×10^{-18} esu) and a poling field of $E_p = 100$ V/μm (corresponding to 3333.33 statvolt/cm) at a temperature of 450 K, the value for the argument is $x = 0.8$.

Thus, for small poling fields, when the linear approximation for the Langevin functions holds, the two independent susceptibility tensor elements given by (14.17) and (14.18) can be written as

$$\chi_{ZZZ}^{(2)} = N \frac{\mu^* E_p}{5 k_B T} \beta_{zzz}^*, \tag{14.22}$$

$$\chi_{ZXX}^{(2)} = N \frac{\mu^* E_p}{15 k_B T} \beta_{zzz}^*. \tag{14.23}$$

Note that within all the approximations made in this section we obtain the result $\chi_{ZXX}^{(2)} = \chi_{ZZZ}^{(2)}/3$.

14.2.4.2 Orientational Birefringence

Due to their rodlike shape, the chromophores possess a linear polarizability that is very different for directions of the optical field polarized parallel or perpendicular to the molecular axis. The poling process described in the previous section whereby the orientation of the molecules is changed by an external field leads, therefore, to birefringence in addition to the second-order nonlinear optical properties. In purely electrooptic polymers, where poling is initially achieved by a spatially uniform dc field, the poling induces a permanent birefringence that is spatially uniform and plays generally only a minor role during electrooptic modulation. In contrast, in photorefractive polymers with a glass-transition temperature close to room temperature, spatially modulated birefringence can be induced by the modulated internal space-charge field and can be of paramount importance, since it can lead to strong refractive index gratings.

The polarizability tensor of the molecule can be written in the molecular principal axes $\{x, y, z\}$ as

$$\tilde{\alpha} = \begin{pmatrix} \alpha_{xx} & 0 & 0 \\ 0 & \alpha_{yy} & 0 \\ 0 & 0 & \alpha_{zz} \end{pmatrix} = \begin{pmatrix} \alpha_{\perp} & 0 & 0 \\ 0 & \alpha_{\perp} & 0 \\ 0 & 0 & \alpha_{//} \end{pmatrix}, \tag{14.24}$$

where the subscripts $\perp$ and $//$ refer to the directions perpendicular and parallel to the main z-axis of the molecule, respectively. In the linear regime and on a microscopic level, an optical field at frequency ω interacting with the polymer film will induce the molecular polarization

$$p_i = \alpha_{ii} \, E_i(\omega), \tag{14.25}$$

where the index i refers to the different components in the frame of the molecule. The macroscopic polarization is given by the sum of the molecular polarizations. Since the molecules exhibit a strong anisotropic linear polarizability, the macroscopic polarization will depend on their average orientation. Since the average orientation can be changed by applying a poling field, the macroscopic polarizability can be continuously tuned by changing the value of the applied field. For a poling field applied along the Z-axis, the change in linear susceptibility along the poling axis between an unpoled and poled film is given by [22]

$$\Delta\chi_{ZZ}^{(1)} = N(\alpha_{//} - \alpha_{\perp})^* \left[\langle \cos^2 \theta \rangle - \frac{1}{3} \right], \tag{14.26}$$

where we have introduced the superscript * to include field-correction factors in the polarizability anisotropy. Recalling the relationship between the refractive index and the susceptibility $n^2 = 1 + 4\pi \chi$, the index change associated with a change in linear susceptibility is given by

$$\Delta n_{ZZ} = \frac{2\pi}{n} \Delta\chi_{ZZ}^{(1)}, \tag{14.27}$$

and therefore for an optical field polarized along the poling axis Z, the index change induced by the poling of the chromophores is given by

$$\Delta n_Z = \frac{2\pi}{n} N(\alpha_{//} - \alpha_\perp)^* \frac{2}{45} \left(\frac{\mu^* E_p}{k_B T}\right)^2,$$ (14.28)

where we have replaced the second-order Langevin function by its linear approximation (see (14.21)).

The same derivation can be made for an optical field polarized along a direction perpendicular to the poling axis and the refractive index change is given by

$$\Delta n_X^{(1)}(\omega) = \Delta n_Y^{(1)}(\omega) = -\frac{1}{2}\Delta n_Z^{(1)}(\omega).$$ (14.29)

Note that according to (14.28) and (14.29) the index change induced by the reorientation of anisotropic chromophores is quadratic in the poling field E_p and is therefore an orientational Kerr effect. The index change has also a quadratic dependence on the dipole moment. Equation (14.29) shows that the refractive index increases for an optical field polarized along the direction of the poling field and decreases for a polarization in a direction perpendicular to it.

14.3 Photoconducting Properties of Organic Materials

In this section we briefly summarize the physical processes of photogeneration of carriers and charge transport in organic amorphous semiconductors. Organic photorefractive materials differ from their inorganic counterparts in several ways: (i) the morphology of organic materials is nearly amorphous, while that of most inorganic crystals is highly ordered and periodic; (ii) the basic properties such as photogeneration and charge transport are strongly field-dependent; and (iii) transport in amorphous organic materials is described by a hopping mechanism through a manifold of localized states with energetical and positional disorder rather than by band-type models. In the simplest model, each molecule or hopping site can be approximated by a two-level model in which the lowest energy level is the highest occupied molecular orbital (HOMO) and the highest energy level is the lowest unoccupied molecular orbital (LUMO). When a molecule is in its ground state and neutral, the HOMO level is occupied by two electrons with opposite spin and the LUMO level is empty. When the molecule is optically excited, one electron is promoted from the HOMO level to the LUMO level, leaving a hole in the HOMO level (process 1 in Figure 14.7). Then the following processes can occur: (i) the electron recombines with the hole within the molecule (this process can be radiative, in which case the molecule is luminescent, or nonradiative); or (ii) an electron transfer reaction can take place with a neighboring molecule. In the latter case, that electron transfer reaction can take place with a molecule of the same type or with a molecule with a different electron affinity or ionization potential (sensitizer) (process 2 in Figure 14.7). The rate of this electron transfer reaction can be described by Marcus theory [23] and depends in part on the difference

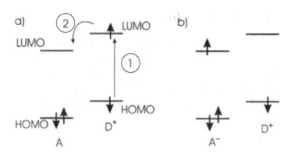

FIGURE 14.7. Schematic representation of the photogeneration of charges between donor- (D) and acceptor-like (A) molecules when the donor-like molecule is initially optically excited.

in energy between the initial and final states involved in the electron transfer reaction and the distance that separates them. In the case of Figure 14.7, this energy difference would be the energies of the LUMO levels (or electron affinities) between the D and A molecules. If the electron gets transferred efficiently, the products of such a reaction are two radical ions that are oppositely charged, namely the donor radical cation D^+ and the sensitizer radical anion A^-. The hole created on the radical cation D^+ can then propagate by electron transfer reactions of electrons that are in the HOMO levels of neighboring neutral donor-like molecules. In this case, the organic material is like a p-type semiconductor or has hole-transport properties. Likewise, if the concentration of sensitizer molecules is large enough, the electron that is in the LUMO level of the radical anion of the sensitizer radical anion A^- can be transferred to the LUMO level of neighboring neutral sensitizer molecules. This process gives n-type semiconducting or electron-transport properties to the material. As shown in Figure 14.8, the same photogeneration process can take place by initial optical excitation of the sensitizer molecule.

Photogeneration and transport in organic materials are subjects of intense research across several fields and application areas. These complex phenomena are not well understood yet in monolithic systems, and their understanding in multifunctional materials such as photorefractive polymers is an even bigger challenge because the different functional groups interact with each other. Such challenges are the subject of current and future research in photorefractive polymers. For completeness, we provide below a brief conceptual overview, and for more information we refer the reader to recent reviews in the field [24, 25].

FIGURE 14.8. Schematic representation of the photogeneration of charges between donor- (D) and acceptor-like (A) molecules when the acceptor-like molecule is initially optically excited.

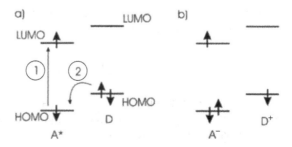

14.3.1 Photogeneration: The Onsager Model

The photorefractive space-charge buildup process can be divided into two steps: the electron–hole generation process followed by the transport of one carrier species. The creation of a correlated electron–hole pair (or exciton) after absorption of a photon can be followed by recombination. This process limits the formation of free carriers that can participate in the transport process and is therefore a loss for the formation of the space charge. These properties result in low photogeneration efficiency unless an electric field is applied to counteract recombination. The quantum efficiency for carrier generation is therefore strongly field-dependent and increases with the applied field. A theory developed by Onsager [26, 27] for the dissociation of ion pairs in weak electrolytes under an applied field has been found to describe reasonably well the temperature and field dependence of the photogeneration efficiency in some of the organic photoconductors [25].

Onsager theory is based on the assumption that the formation of an uncorrelated electron–hole pair involves the photogeneration of a correlated electron–hole pair, or exciton formation, followed by the formation of an intermediate charge-transfer state in which the electron and the hole are thermalized and separated by an average distance r_0. Then, one of the carriers can escape from its twin with a given probability if that distance is comparable to or higher than the so-called Coulomb radius. The Coulomb radius r_c is defined as the distance at which the Coulomb energy between a hole and an electron equals the thermal energy $k_B T$. It is given by

$$r_c = \frac{e^2}{4\pi \varepsilon_0 \varepsilon_r k_B T},$$
(14.30)

where e is the elementary charge and ε_r the relative dielectric constant. The dissociation probability is enhanced if an external field is applied to counteract geminate recombination. Since Onsager theories describe the probability of escape of an electron from its parent cation in the presence of an external field, these theories provide a good framework to describe the field-dependence of photogeneration in organic solids.

As illustrated in Figure 14.9, photogeneration in organic solids can be divided into three steps: (1) absorption of a photon and formation of an excited state; (2) thermalization and creation of a charge-transfer state; (3) dissociation. The overall photogeneration efficiency is therefore the product of two terms:

$$\phi(E) = \eta_0 \times \Omega,$$
(14.31)

where η_0 is the probability for the excited state to form a thermalized pair. It is often referred to as the primary quantum yield, Ω is the dissociation probability of a thermalized charge-transfer state.

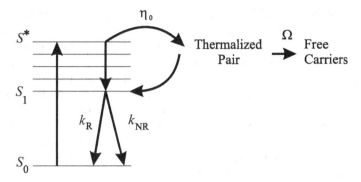

FIGURE 14.9. Schematic representation of photogeneration in organic solids as described by Onsager theory. Adapted from Noolandi and Hong (1979) [28].

An analytic expression for the photogeneration efficiency according to the Onsager model as derived by Mozumder [27] is given by

$$\phi(r_0, E) = \eta_0 \left(1 - \zeta^{-1} \sum_{n=0}^{\infty} A_n(\kappa) A_n(\zeta) \right), \qquad (14.32)$$

where $A_n(x)$ is a recursive formula given by

$$A_n(x) = A_{n-1}(x) - \frac{x^n \exp(-x)}{n!} \qquad (14.33)$$

and

$$A_0(x) = 1 - \exp(-x). \qquad (14.34)$$

In (14.32), $\kappa = r_c/r_0$ and $\zeta = e r_0 E / k_B T$.

While the Onsager model provides a good description of the field dependence of photogeneration in numerous organic systems, the values of the thermalization radius that are used in order to get the best fit to the experimental data are often too high to have a real physical meaning. In the absence of other models, Onsager's theory remains the best framework to describe photogeneration in organic sensitizers.

14.3.2 Charge Transport in Organics: The Disorder Formalism

After photogeneration of free carriers, the next step in the buildup of a photorefractive space charge is charge transport from brighter regions of the interference pattern, where charges are generated, to the darker regions, where they get trapped. In contrast to inorganic photorefractive crystals with a periodic structure, photorefractive polymers have a nearly amorphous structure. The local energy level of each molecule/moiety is affected by its nonuniform environment. Transport can no longer be described by band models but is attributed to intermolecular hopping of carriers between neighboring molecules or moieties [29–31].

In recent years, a model based on disorder due to Bässler and coworkers [32] describes the transport phenomena of a wide range of different materials and emerged as a solid formalism to describe the transport in amorphous organic materials [33–40]. This so-called disorder formalism is based on the assumption that charge transport occurs by hopping through a distribution of localized states with energetical and positional disorder. So far, most of the predictions of this theory agree reasonably well with the experiments performed in a wide range of doped polymers, main-chain and side-chain polymers, and in molecular glasses. However, when these photoconductors are doped with highly polar dyes such as photorefractive chromophores, significant differences have been observed that could not be rationalized within existing theories. More research is required to understand transport in photorefractive polymers.

In the Bässler formalism, disorder is separated into diagonal and off-diagonal components. Diagonal disorder is characterized by the standard deviation σ of the Gaussian energy distribution of the hopping site manifold (energetical disorder), and the off-diagonal component is described by the parameter Σ, which describes the amount of positional disorder. Results of Monte Carlo simulations led to the following universal law for the mobility when $\Sigma \geq 1.5$:

$$\mu(E, T) = \mu_0 \exp\left[-\left(\frac{2}{3}\frac{\sigma}{k_B T}\right)^2\right] \exp\left\{C\left[\left(\frac{\sigma}{k_B T}\right)^2 - \Sigma^2\right]E^{1/2}\right\}, \quad (14.35)$$

where μ_0 is a mobility prefactor and C an empirical constant with a value of 2.9×10^{-4} $(cm/V)^{1/2}$. Equation (14.35) is valid for high electric fields (a few tens of V/μm) and for temperatures $T_g > T > T_c$, where T_g is the glass-transition temperature and T_c the dispersive-to-nondispersive transition temperature. For a more detailed description the reader is referred to the original work by Bässler et al. [32] and the recent review by P. Borsenberger [25].

Numerous simulations and experimental work have shown that the width of the density of states σ in (14.35) is strongly dependent on the polarity of the polymer host and the dipole moment of the dopant molecule. It was found that the total width σ comprises a dipolar component σ_D and an independent Van der Waals component σ_{VdW} [40]:

$$\sigma^2 = \sigma_{VdW}^2 + \sigma_D^2. \quad (14.36)$$

Charge mobility in organic semiconductors is a property that is measured by a range of techniques that operate over different length scales and that require different sample geometries. The most common techniques are (i) the time-of-flight technique (TOF); (ii) the space-charge limited-current technique (SCLC) [41, 42]; (iii) the field-effect transistor technique [43]; and (iv) the pulse radiolysis time-resolved microwave conductivity technique (PR-TRMC) [44]. For thick photoconducting samples such as photorefractive polymers, the TOF technique is the most common. In the TOF geometry, the material is sandwiched between a transparent electrode (typically indium tin oxide (ITO)) and a metal electrode.

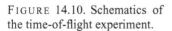

FIGURE 14.10. Schematics of the time-of-flight experiment.

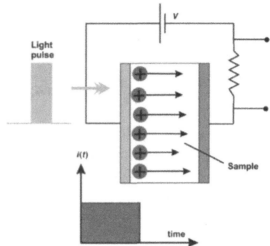

The sample is biased, with an appropriate field direction for transporting the carriers of interest, and mounted inside a temperature-controlled unit. As shown in Figure 14.10, a short laser pulse is transmitted through the ITO electrode and is strongly absorbed by the material. The excited states that are formed dissociate into electron/hole pairs. In hole (electron) transport materials, the injected holes (electrons) move toward the metal electrode under the influence of the applied electric field by drift. During their transit, the work produced by the displacement of the charges under the action of the electrical force in the sample leads to a transient electrical power in the circuit that manifests itself as a transient current. When the injected charges are collected, the transient current vanishes. By measuring the duration of the photocurrent, $\tau = L/v$, in a sample of thickness L, the drift velocity $v = \mu E$ can be calculated, from which mobility is derived. Because of dispersion in the arrival times of the carriers, the transient current is not a step function, as shown in the idealized Figure 14.10, but is typically followed by a tail. Typical transient currents measured in 20-μm-thick samples of 4,4′-bis(phenyl-m-tolylamino)biphenyl (TPD) doped into polystyrene (1:1 wt.%) are shown in Figure 14.11.

Typical values for charge mobility in amorphous photoconductors are in the range of 10^{-3} cm^2/Vs for holes and 10^{-4} cm^2/Vs for electrons. The smaller values observed for electron mobility are often correlated with the trapping effects induced by molecular oxygen in these materials. Note, however, that charge mobility in photorefractive polymers is expected to be significantly lower due to the high dipole moment of the dopant molecules. This high dipole moment according to (14.36) increases the dipolar contribution to the energetical disorder and consequently the overall width of the energetical distribution of the hopping site manifold, which leads to lower mobility.

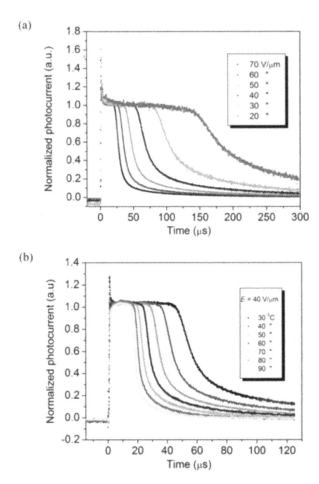

FIGURE 14.11. (a) Field dependence of the transient photocurrents measured in a guest/host sample of 4,4′-bis(phenyl-m-tolylamino)biphenyl (TPD) doped into polystyrene (1:1 wt. %). (b) Temperature dependence of the transient photocurrents measured in the same sample.

14.4 The Photorefractive Effect in Organic Materials

The first organic photorefractive materials were developed in the nineties [45, 46]. Since the first demonstration of photorefractivity in a polymer [47], numerous polymer composites have been developed, and the performance level of this new class of materials could rapidly be improved by several orders of magnitude, to a level where they compete with numerous inorganic crystals but are produced using completely different manufacturing techniques. Current materials can exhibit ms response time, can have refractive index modulation amplitude close to a percent, or two-beam coupling gain coefficients of 400 cm^{-1} as reviewed in [6].

Unfortunately, highest speed and dynamic range have not been combined yet in one composite.

The photorefractive effect as discovered and studied initially in inorganic materials [48–60] is described in great detail in other chapters of this book and has been the subject of several decades of productive research in the field. Therefore, in this section we will provide only the major results for the space-charge field of the traditional Kukhtarev model [59] and emphasize some of the refinements that have been made in more recent models to describe photorefractivity in organic materials [61–68]. Development of these new models was driven by the need to take into account significant differences in the physical properties of organic materials compared to their inorganic counterparts for which the Kukhtarev model was initially developed. Examples of such differences have been discussed in the previous section and include the field dependence of the photogeneration of carriers and that of the charge mobility. Other modifications include various aspects related to trapping mechanisms. In this section, we will also provide a description of orientational photorefractivity, which is responsible for large-refractive-index modulation amplitudes in the most efficient organic photorefractive polymers to date.

As in inorganic photorefractive materials, the different steps leading to the formation of a photorefractive grating in an organic material are summarized in Figure 14.12. They comprise (i) the absorption of light and the generation of charge carriers; (ii) the transport through diffusion and drift of electrons and holes over distances that are a fraction of the grating spacing, leading to separation of holes and electrons; (iii) the trapping of these carriers and the buildup of a space-charge field; (iv) and the modulation of the refractive index by the periodic space-charge field.

In most of the photorefractive crystals, this index modulation is produced via the electrooptic effect. We will see in this section that in most organic photorefractive materials studied to date it is produced by orientational Kerr effects or so-called orientational birefringence. Let us also recall that the nonlocal response of photorefractive materials, namely the dephasing between the refractive index modulation and the initial light distribution, is a signature of the photorefractive effect and is responsible for two-beam coupling in steady-state conditions in thick samples. Experimental techniques used to study the performance of organic photorefractive materials include laser-induced grating experiments such as degenerate four-wave mixing and two-beam coupling experiments [69].

14.4.1 The Kukhtarev Model

An expression for the internal space-charge field can be obtained through the Kukhtarev model [59], which was developed to describe photorefractivity in most inorganic materials. In this model, the photorefractive material is described by a band model. As for a traditional semiconductor, the material consists of conduction and valence bands separated by a band gap. In its simplest form, the model describes the transport of single-carrier species, and the band gap of the material

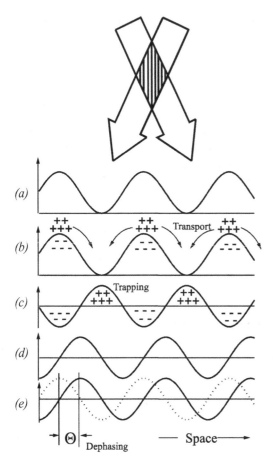

FIGURE 14.12. Illustration of the photorefractive effect. The overlap of two coherent laser beams creates an optical interference pattern (a). In the high-intensity regions, charge carriers are generated (b). One type of carrier is transported and trapped (c), creating an alternating space-charge field (d). The space-charge field induces a change in refractive index (e). The resulting index grating is phase shifted with respect to the initial light distribution.

contains localized energy levels that can be excited optically, promoting either holes in the valence band (VB) or electrons in the conduction band (CB). If one assumes that electrons are the only carriers that can be transported, under weak illumination conditions and negligible dark conductivity, the Fourier coefficient of the first-order space-charge field at steady state can be approximated to

$$E_1 = E_{sc} = -m \frac{E_q(i E_D + E_0)}{(E_q + E_D) - i E_0},$$

(14.37)

where m is the fringe visibility, E_0 is the applied field, E_D is the diffusion field, and E_q is the limiting space-charge field. The diffusion field is given by

$$E_D = \frac{2\pi k_B T}{\Lambda e},$$

(14.38)

where Λ is the grating spacing of the optical interference pattern. The limiting space-charge field is given by

$$E_q = \frac{2\Lambda e N_{eff}}{\varepsilon_r}, \tag{14.39}$$

where N_{eff} is the effective photorefractive trap density.

The space-charge field given by (14.37) is a complex number that is characterized by an amplitude and a phase given by

$$|E_{sc}| = m\left(\frac{(E_0^2 + E_D^2)}{(1 + E_D/E_q)^2 + (E_0/E_q)^2}\right)^{1/2} = m|E_m|, \tag{14.40}$$

$$\Theta = \begin{cases} \mathrm{arctg}\left[\frac{E_D}{E_0}\left(1 + \frac{E_D}{E_q} + \frac{E_0^2}{E_D E_q}\right)\right] & \text{for } E_0 \neq 0, \\ \frac{\pi}{2} & \text{for } E_0 = 0. \end{cases} \tag{14.41}$$

Since the space-charge field produces the refractive index modulation, the phase Θ of the complex field represents the phase shift between the refractive index modulation and the initial periodic light distribution that generated the space-charge field.

The internal periodic electric field described above will modulate the refractive index of an electrooptic material, and the resulting index modulation can be written as

$$\Delta n = \frac{1}{2}\left(\Delta \tilde{n} e^{i2\pi x/\Lambda} + c.c.\right), \tag{14.42}$$

where $\Delta \tilde{n}$ is a complex amplitude that contains the phase information of the space-charge field given by:

$$\Delta \tilde{n} = -\frac{1}{2}n^3 r_{eff}|E_{sc}|e^{i\Theta} \tag{14.43}$$

where r_{eff} is an effective electrooptic coefficient that depends on the orientation of the material and the polarization of the optical beams.

14.4.2 The Two-Trapping-Site Model

In recent years, the Kukhtarev model has been refined to provide a better description of photorefractivity in organic materials [61–68]. The most complete model to date was developed by Singer and Ostroverkhova and takes into account shallow and deep traps [68]. This model has provided a good description of the effects arising from the preillumination of some photorefractive polymers. Such effects include the deterioration of the response time of photorefractive polymers based on the matrix poly(vinylcarbazole) doped with low-ionization chromophores [68, 70] due to the accumulation of large densities of radical anions of the sensitizer, as well as optical gating effects reported in some composites that were conducive to improved response times [71, 72]. This model provides good insight into the

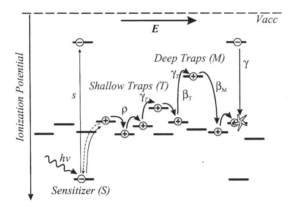

FIGURE 14.13. Schematic representation of the two-trapping-site model for the photore-fractive effect (redrawn from [68]).

influence of trapping mechanisms on the dynamics of the formation of the space-charge field. The main processes of the model are described in Figure 14.13. After a photon is absorbed by a neutral sensitizer molecule, an electron is transferred from the transport manifold to the sensitizer, and in turn a hole is injected into the transport manifold. Thermal ionization of the sensitizers and any photovoltaic effect are neglected. Holes hopping between sites in the transport manifold can either recombine, at a rate γ, with sensitizer anions to produce neutral sensitizer molecules or become trapped at rates γ_T and γ_M. The trapping sites are assumed to be of two types: shallow and deep. They each have a single-well defined ioniza-tion potential, smaller than the transport manifold. Thermally activated detrapping from shallow and deep traps is considered to occur at rates β_T and β_M, espectively. The set of nonlinear equations considered by the model are given by

$$\frac{\partial \rho}{\partial t} = \frac{\partial S^-}{\partial t} - \frac{\partial T^+}{\partial t} - \frac{\partial M^+}{\partial t} - \frac{1}{e}\frac{\partial J}{\partial x},$$

$$\frac{\partial E}{\partial x} = \frac{e}{\varepsilon_0 \varepsilon_r}(\rho + T^+ + M^+ - S^-),$$

$$\frac{\partial T^+}{\partial t} = \gamma_T(T - T^+)\rho - \beta_T T^+,$$

$$\frac{\partial M^+}{\partial t} = \gamma_T(M - M^+)\rho - \beta_M M^+,$$

$$\frac{\partial S^-}{\partial t} = sI(S - S^-) - \gamma S^- \rho,$$

$$J = e\mu\rho E - eD\frac{\partial \rho}{\partial x}, \tag{14.44}$$

where E is the electric field, ρ the free-charge (holes) density, J the current density; S, T, and M are the total density of sensitizers, shallow traps, and deep traps respectively; S^-, T^+, and M^+ the density of sensitizer anions and filled shallow and deep traps, respectively; s the photogeneration cross section; γ_T the shallow trapping rate; γ_M the deep trapping rate; γ the recombination rate; β_T and β_M are the thermal detrapping rates of the shallow and deep traps, respectively; I is the space-dependent irradiance; μ the charge mobility; D the diffusion coefficient; and e, ε_r, ε_o are the elementary charge, the dielectric constant, and the permittivity of free space, respectively.

14.4.3 Orientational Photorefractivity

Early photorefractive polymers [46, 47, 73–75] were designed to mimic inorganic crystals and semiconductors in which the refractive index modulation by the photorefractive space-charge field was due to the Pockels effect. In these materials, a spatially uniform electrooptic activity was obtained by a poling process prior to the formation of photorefractive gratings. However, as photorefractive polymers were developed through the guest host approach, the glass-transition temperature T_g of the resulting composite was lowered: the electroactive chromophore was acting as a plasticizer. The addition of plasticizers reduced T_g even further. In such plasticized materials, significant improvements in dynamic range could be observed [76–78]. If the operating temperature is close enough to T_g (around 10 degrees below), the dipolar chromophores are free to reorient in the total electric field, which is the superposition of the applied field and the spatially modulated space-charge field. For each coordinate along the grating (X direction in Figure 14.14a), the total field is changing both its amplitude and direction because the amplitude of the space-charge field is oscillating between its extrema $+E_{sc}$ and $-E_{sc}$. This results in a periodic orientation of the dipolar chromophores that orient in the total electric field as illustrated in Figure 14.14b. The periodic orientation of

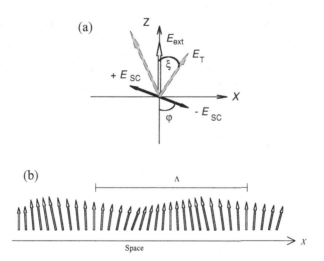

FIGURE 14.14. Schematic of the total poling field (a) and the spatially modulated orientation of the dipolar molecules in the sample (b).

anisotropic chromophores leads to a periodic refractive index modulation through orientational birefringence [79]. In this case, the refractive index modulation has an electrooptic and a birefringent contribution. The former is directly related to the second-order nonlinear optical properties of the chromophore, while the latter is due to the anisotropy of the linear polarizability. To maximize the electrooptic effect, a chromophore should have a high hyperpolarizability β and simultaneously a high dipole moment μ. A high dipole moment μ is required to obtain a good orientation of the chromophores during the poling process through the torque exerted by the field on the chromophore (see (14.39), (14.48)–(14.49)). To maximize the orientational birefringence, a high polarizability anisotropy $\Delta\alpha = (\alpha_{//} - \alpha_\perp)$ is desired. A high dipole moment μ is even more important in this case because the index change is quadratic in the dipole moment (see (14.54)). Based on the oriented gas model, one can define the following figure of merit for chromophores for orientational photorefractivity [80, 81]:

$$Q_{OP} = N_{max}(A(T)\Delta\alpha\mu^2 + \beta\mu),$$ (14.45)

where N_{max} is the maximum density of chromophores that can be doped or functionalized in the low-glass-transition-polymer composite before chromophore aggregation or interaction occurs, and $A(T)$ is a scaling factor. The subscript OP in Q_{OP} stands for *orientational photorefractivity*.

The derivation of expressions for the electrooptic and birefringent contributions to the refractive index modulation in the case of orientational photorefractivity is more complex than for traditional photorefractivity. The tensors for the linear polarizability and the first hyperpolarizability are diagonal in the frame attached to the chromophores. In the case of a spatially uniform poling with the poling field applied along the laboratory Z-axis, the molecular axis z usually coincides with the laboratory Z-axis after poling. Here, the poling direction is changing for each coordinate along the X direction (i.e., along the grating) (see Figure 14.14) and makes an angle $\xi(x)$ with the laboratory Z axis [79, 82]. By applying matrix transformations and by retaining only terms that have the spatial frequency K, the following expressions are derived for the different contributions to the refractive index modulation for each polarization of the reading beam. For the orientational birefringence contribution one gets

$$\Delta n^{(1)}_{K,s} = -\frac{2\pi}{n} B E_{ext} |E_{sc}| \cos\varphi,$$ (14.46)

$$\Delta n^{(1)}_{K,p} = \frac{2\pi}{n} B E_{ext} |E_{sc}| \left[2\cos\varphi \sin\alpha_1 \sin\alpha_2 - \cos\varphi \cos\alpha_1 \cos\alpha_2 \right.$$
$$\left. + \frac{3}{2} \sin\varphi \sin(\alpha_1 + \alpha_2) \right],$$ (14.47)

with

$$B = \frac{2}{45} N(\alpha_{//} - \alpha_\perp)^* \left(\frac{\mu^*}{k_B T} \right)^2,$$ (14.48)

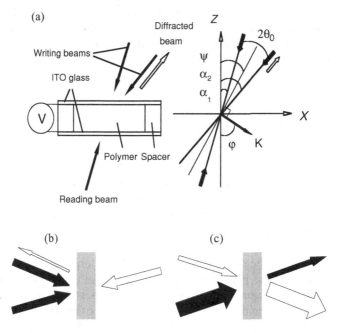

FIGURE 14.15. Schematic representation of the tilted configuration used for wave-mixing experiments in photorefractive polymers; (b) standard geometry for four-wave mixing experiments; (c) standard geometry for two-beam coupling experiments.

where the angles φ and α_i are defined in Figure 14.15. For the electrooptic contribution the expressions are

$$\Delta n_{K,s}^{(2)} = \frac{8\pi}{n} C E_{ext} |E_{sc}| \cos\varphi, \tag{14.49}$$

$$\Delta n_{K,p}^{(2)} = \frac{8\pi}{n} C E_{ext} |E_{sc}| [\cos\varphi \cos\alpha_1 \cos\alpha_2 + 3\cos\varphi \sin\alpha_1 \sin\alpha_2$$
$$+ \sin\varphi \sin(\alpha_1 + \alpha_2)], \tag{14.50}$$

with

$$C = \frac{N\beta^* \mu^*}{15 k_B T}. \tag{14.51}$$

Note that the field E_{ext} in (14.46), (14.47), (14.49), and (14.50) represents the total field applied between the electrodes. Its component along the grating vector (which is represented by E_0 in (14.37), (14.40), and (14.41)) is given by $E_0 = E_{ext} \cos\varphi$.

All the orientational photorefractive properties discussed in this section are contingent on the ability of the chromophores to orient at the operation temperature, that is, when the temperature is close to or above the glass-transition temperature T_g. Most efficient photorefractive polymers to date are based on such orientational photorefractivity. However, polymers with high T_g can be designed in which poling (orientation of the chromophores) is achieved at high temperature and frozen in the material, leading to a spatially uniform electrooptic coefficient. In this case, photorefractivity is described by the Pockels effect as in inorganic photorefractive crystals.

14.5 Organic Photorefractive Materials

14.5.1 Design of Photorefractive Polymers

As for inorganic materials, polymers must combine several key functionalities to exhibit photorefractive properties: weak absorption, transport of either positive or negative carrier species, trapping, and a field-induced refractive index modulation mechanism. Owing to the rich structural flexibility of organic molecules and polymers, these functionalities can be incorporated into a given material in many different ways. The organic materials can be amorphous like polymers, or exhibit liquid crystalline phases, or can consist of nanocomposite phases such as polymer-dispersed liquid crystals, or hybrid materials such as sol-gels. Photorefractivity has been evidenced in numerous classes of organic materials. With polymers, two main design approaches have been followed: (i) In the guest–host approach several materials with different functionalities are mixed into a polymer composite (see Figure 14.16a). A polymer is sometimes used to provide the mechanical properties to the film. But in some cases, molecular materials form amorphous glasses by themselves and there is no need for a polymer compound. In these molecular glasses, the molecules can exhibit several functionalities simultaneously. (ii) In the fully functionalized approach, different functional groups are attached as side groups to a polymer chain. The chain itself can in some cases incorporate one functionality (see Figure 14.16b). This second approach requires a much larger synthetic effort and has not been used as much as the guest–host approach. Both approaches have their strengths and weaknesses. A major limitation of the guest–host approach is the compatibility of the different compounds that form the photorefractive composite. This compatibility limits the number of combinations of sensitizers, and photoconductive and electroactive materials,

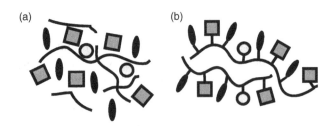

FIGURE 14.16. Schematics of different chemical designs for photorefractive polymers: (a) guest/host, (b) fully functionalized approach.

that can be tested. A significant problem in this approach is the phase separation between the different guest compounds that can occur over time. To be practical, a photorefractive material must combine good electrical and optical properties, but at the same time must exhibit high photostability under illumination, long shelf lifetime with high optical clarity, and good dielectric properties under extended bias of several kilovolts. Early guest–host photorefractive polymers had limited shelf lifetimes, but in recent years materials with shelf lifetimes of several years could be demonstrated. Nevertheless, shelf lifetime should always remain a major concern when new materials are designed and characterized.

As discussed earlier, an applied electric field must be applied to photorefractive polymers. This is achieved by fabricating a thick (≈ 100 μm) film of the polymer between two transparent conducting glass slides such as indium tin oxide (ITO) coated glass slides as shown in Figure 14.15. The sample must have high optical quality and the film must exhibit high dielectric strength such that fields of several tens of V/μm can be applied to the sample without causing any dielectric breakdown. Therefore, starting materials with high purity must be used when preparing photorefractive polymers. The purity of many commercial products is not good enough to prepare good samples, and additional purification steps should be carried out.

To write holograms efficiently in such samples, the geometry is such that the bisector of the two writing beams is tilted with respect to the normal of the sample. This way, the field applied to the two ITO-coated glass slides has a nonvanishing component along the grating vector K of the interference pattern. In this tilted configuration, the field applied to the sample acts as a drift source for the carriers (see Figure 14.15). The key basic parameters used to characterize the performance of photorefractive polymers are buildup time of the space-charge field, dynamic range (or refractive index modulation amplitude), and two-beam coupling coefficients. The last two parameters are measured by four-wave mixing and two-beam couplings experiments, respectively, and are schematically represented in Figures 14.15b,c.

14.5.2 Classes of Organic Photorefractive Materials

In this section we will describe a few examples of photorefractive polymer composites that illustrate the flexibility in the design of such materials. Organic photorefractive materials can be divided into several classes: (i) crystals, (ii) guest/host systems, (iii) fully functionalized polymers, (iv) glasses, (iv) sol-gels and organic/inorganic hybrid materials, (v) liquid crystals, (vi) polymer-stabilized and polymer-dispersed liquid crystals.

Figure 14.17 shows examples of chromophores used in photorefractive polymers. DMNPAA and DEANST are widely used in photorefractive polymers and have been initially selected for their electrooptic properties. DMNPAA exhibits also high polarizability anisotropy and was used in the composites in which overmodulation of the diffraction efficiency could be observed in 105-μm-thick samples at applied fields of 60 V/μm, with net gain coefficients of 200 cm^{-1} [78].

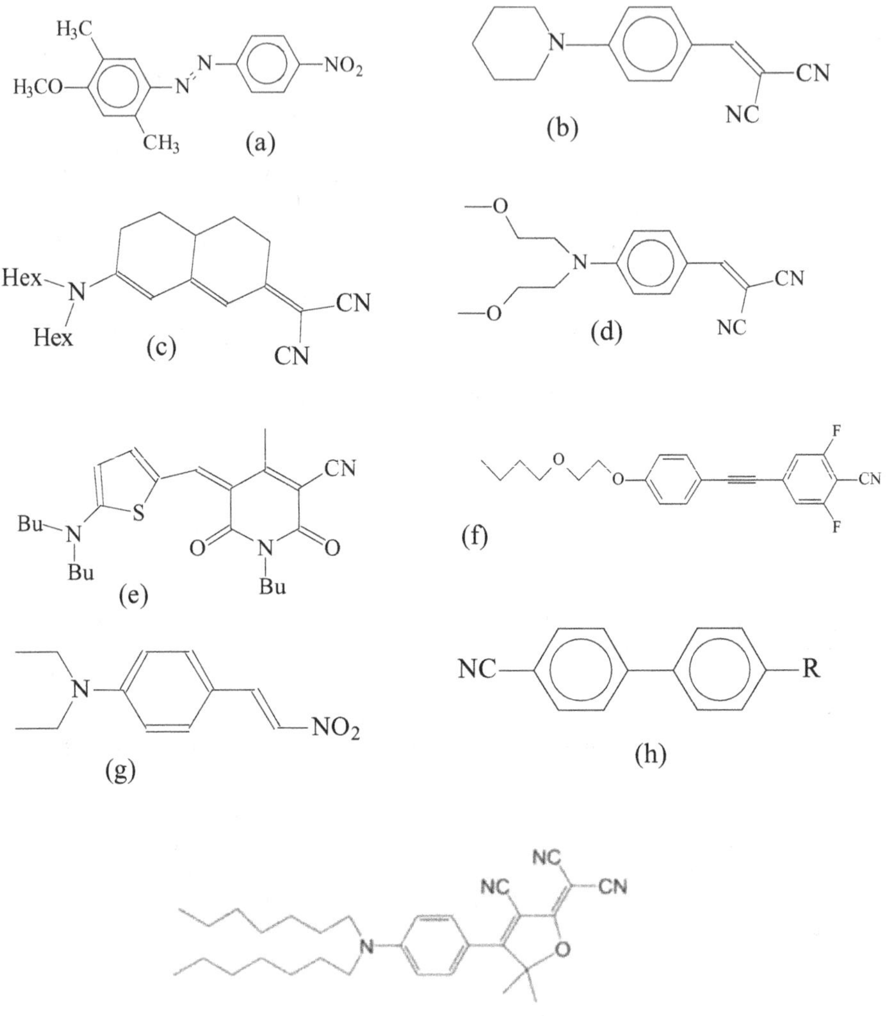

FIGURE 14.17. Examples of chromophores used in photorefractive polymers: (a) 2,5-dimethyl-4-(p-nitrophenylazo)anisole (DMNPAA); (b) 4-piperidinobenzylidene malononitrile (PDCST); (c) 2-N, N-dihexyl-amino-7-dicyanomethylidenyl-3,4,5,6,10-pentahydronaphthalene (DHADC-MPN); (d) 4-di(2-methoxyethyl)aminobenzylidene malononitrile (AODCST); (e) amino-thienyl-dioxocyano-pyridine (ATOP) derivative; (f) fluorinated cyano-tolane chromophore (FTCN); (g) Diethylamino-nitrostyrene (DEANST); (h) Cyanobiphenyl derivative R = C_5H_{11} (5CB), R = OC_8H_{17} (8OCB); (i) 2-dicyanomethylen-3-cyano-5,5-dimethyl-4-(4′-dihexylaminophenyl)-2,5-dihydrofuran (DCDHF-6).

FIGURE 14.18. Field dependence of the diffraction efficiency and gain coefficient measured at 633 nm in a 105-μm-thick sample of DHADC-MPN:PTCB:DIP:TNFDM (37.6:49.7:12.5:0.18 wt.%).

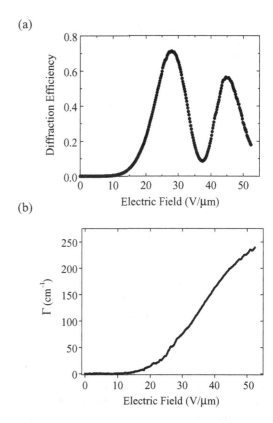

(a)

(b)

After the discovery of orientational photorefractivity, the design of chromophores was no longer driven by the electrooptic effect but rather by the orientational birefringence. DHADC-MPN and ATOP (chromophores (c) and (e) in Figure 14.17) are examples of molecules that are close to the cyanine limit and have a high figure of merit (see (14.45)) for orientational photorefractivity. Photorefractive polymers based on these materials exhibit high dynamic range at moderate applied field [83–85].

Figure 14.18 shows the diffraction efficiency and gain coefficient in a sample doped with DHADC-MPN that has a shelf lifetime of several years. Such samples stored for 6 days at 70°C did not show any signs of phase separation, illustrating that guest/host photorefractive polymers with high stability can be fabricated. However, despite their high dynamic range, polymers based on these chromophores have a rather slow response time (seconds). Photorefractive polymers with ms response times have been designed by using PDCST and AODCST (chromophores (b) and (d) in Figure 14.17) [86]. Such molecules have a fast orientational dynamics in a plasticized PVK matrix. The photorefractive speed in such samples is limited by the photoconducting properties of the composite.

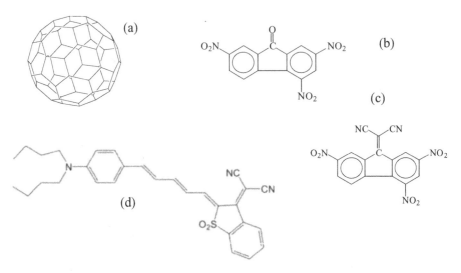

FIGURE 14.19. Examples of sensitizers: (a) fullerene C_{60}; (b) 2,4,7-trinitrofluorenone (TNF); (c) 2,4,7-trinitro-9-fluorenylidene-malononitrile (TNFDM); (d) 2-{5-[4-(Di-*n*-butylamino)phenyl]-penta-2,4-dienylidene}-3-(dicyanomethylidene)-2,3-dihydrobenzo-thiophene-1,1-dioxide (DBM).

These chromophores are colored and cannot be used across the full visible spectral region. Another chromophore that leads to ms response times when doped in a plasticized PVK matrix is FTCN (molecule (f) in Figure 14.18) [87]. This molecule is transparent in most of the visible spectrum but has lower efficiency compared with PDCST and AODCST. To fabricate photorefractive polymers containing transparent electroactive molecules, 5CB (molecule (h) in Figure 14.17), a molecule that is known for its liquid crystalline phase, has been used as a dopant into PVK [88]. High two-beam coupling coefficients (400 cm^{-1}) could be observed in PVK composites doped with the chromophore DCDHF-6 (molecule (i) in Figure 14.17) [89].

To tune the sensitivity of a photorefractive polymer to a particular spectral region, the composite is doped with a sensitizer. Examples of widely used sensitizers are shown in Figure 14.19. TNF was used in numerous polymers based on the photoconducting polymer PVK because mixtures of PVK/TNF were used in the first commercial photocopiers and were extensively studied in the literature [25]. In most cases, the sensitizer itself does not absorb the light but rather the charge-transfer complex that is formed between the sensitizer and the molecule that enables charge transport. This charge-transfer complex exhibits a new absorption band that is different from any of the absorption bands of the two compounds. When the polymer composite is p-type, that is, when the transport molecules have a donor-like character, the sensitizer is often a molecule with a strong acceptor character. In the case of an electron-transport matrix, the sensitizer can be a

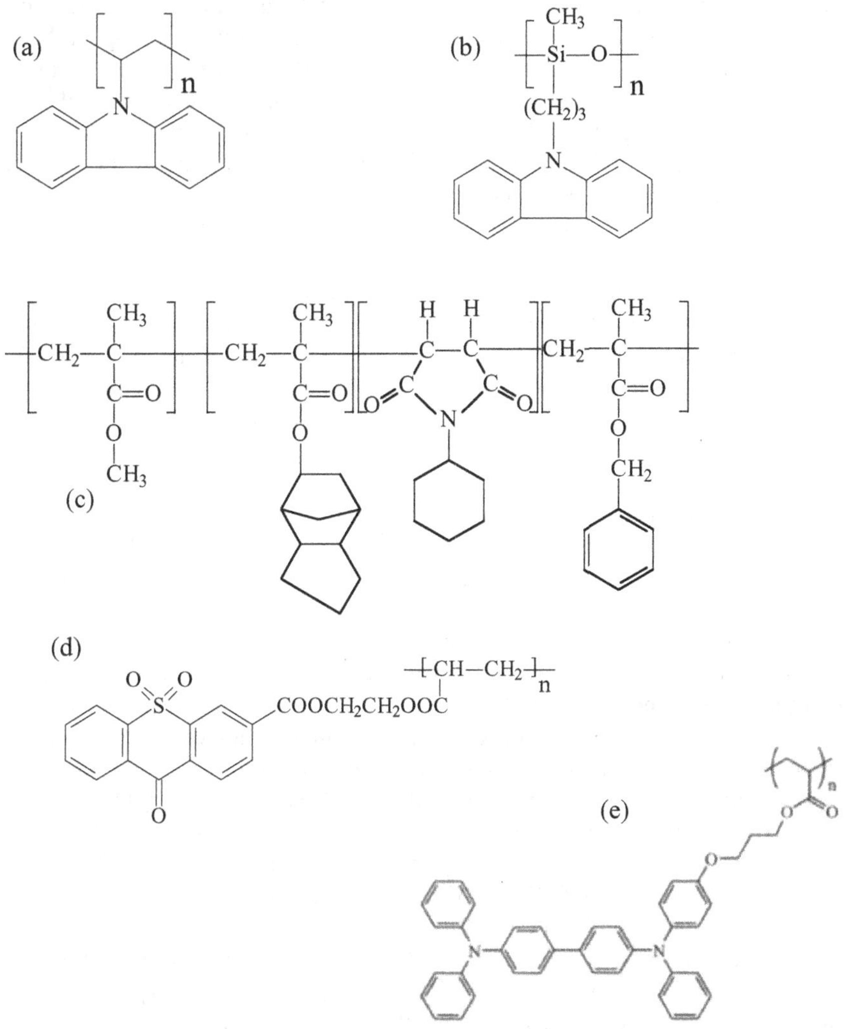

FIGURE 14.20. Examples of polymers used in photorefractive composites: (a) poly(vinylcarbazole) (PVK); (b) carbazole substituted polysiloxane (PSX); (c) poly(methylmethacrylate-co-tricyclodecylmethacrylate-co-N-cyclohexylmaleimide-co-benzyl methacrylate) (PTCB); (d) thioxanthene containing polymer (P-THEA); (e) polyacrylate polymer with side-chains of tetra-diphenyldiamine (TPD) groups (PATPD).

donor-like molecule. Charge-transfer complexes are good sensitizers for photorefractive applications because they have a high photogeneration efficiency. Efficiencies of 100% have been reported recently in charge-transfer composites formed between triphenyldiamine derivatives and C_{60} (molecule (a) in Figure 14.20) [90]. The central wavelength of the charge-transfer band can be tuned by either changing the electron affinity of the acceptor-like molecule or the ionization potential of the donor-like molecule that form the complex. While TNF and C_{60} provide sensitivity in the red part of the spectrum, in particular at 633 nm, which is widely used through the use of He:Ne lasers and semiconductor laser diodes, TNFDM enables near-infrared applications [84, 85, 91]. Since the solubility of C_{60} is limited, a more soluble derivative, PCBM, has also been utilized [92, 72]. A photorefractive polymer that uses the sensitizer DBM (molecule (d) in Figure 14.19) was reported recently to operate at 975 nm [93].

Examples of polymers used in photorefractive applications are shown in Figure 14.20. PVK and PSX [94–97] both contain the carbazole moiety, which has a donor-like character and provides for hole transport. Carbazole and some other hole-transporting moieties have been widely used in the field of photorefractive polymers. The polarity of carbazole makes PVK a good matrix for many of the guest electroactive molecules. The choice of a polymer matrix is often dictated by its compatibility with some of the other functional molecules that form the photorefractive composite. Hence, inert polymers have been used in some instances. PTCB (polymer (c) in Figure 14.20) has an excellent compatibility with DHADC-MPN (see Figure 14.17c) and good optical quality. The lack of transport properties, however, is responsible for the slow response time of such samples. PTCB is a copolymer that was deisigned to have low birefringence [98] and that is commercially available from Hitachi Chemical Co. Ltd. under the brand name OZ-1330. Bulk samples of photorefractive polymers based on DHADC-MPN and PTCB have been fabricated by injection molding [99]. Samples with good optical quality and high diffraction efficiency could be obtained, demonstrating the manufacturability of these materials in large volume at low cost using standard techniques such as injection molding. Other inert binders that have been used include poly(methylmethacrylate) (PMMA) [100, 101], poly(vinylbutyral) (PVB), polystyrene (PS), the polyimide Ultem [101], bisphenyl-A-polycarbonate (PC) [102], and poly(n-butyl methacrylate) (PBMA) [103]. Charge transport can also be provided by conjugated main-chain polymers such as poly(phenylene-vinylene) (PPV) derivatives [72, 92, 104]. Photorefractivity was recently reported in a polymer (P-THEA) with electron-transport properties [105]. The electron-transporting nature of the composite was confirmed by a reverse two-beam coupling compared with PVK-based materials. Electron transport was achieved through the thioxanthene unit (see polymer (d) in Figure 14.20).

In the photorefractive polymer composites described above, the value of the glass-transition temperature T_g is of great importance because it needs to be in a range where the electroactive chromophores can be reoriented by the photorefractive space-charge field. Adjustment of T_g is done though the addition of a plasticizer. Examples of plasticizers are shown in Figure 14.21. As for the polymer matrixes, plasticizers can be photoconducting or inert. ECZ (molecule (a) in Figure 14.21) is widely used with PVK. Inert plasticizers such as BBP (molecule (c) in Figure 14.21) have also been combined with PVK [106].

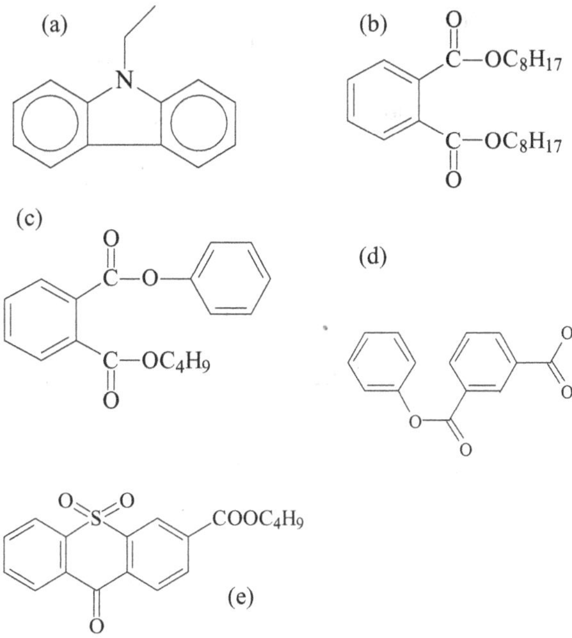

FIGURE 14.21. Examples of plasticizers: (a) N-ethylcarbazole (ECZ); (b) diisooctyl phthalate (DOP); (c) benzyl butyl phthalate (BBP); (d) diphenyliso phthalate (DIP); (e) 9-oxo-9H-thioxanthene-3-butyloxycarbonyl-10,10-dioxide (TH-nBu).

In an effort to optimize simultaneously the dynamic range and the response time in a given material, an attractive design concept is to have an amorphous glass of a multifunctional molecule that combines photosensitivity, transport, trapping, and electroactivity. This way the transport properties do not get diluted when a large amount of electroactive dopant is used, and vice versa. In addition, the material should be processable into thick films that have a good shelf lifetime. Photorefractive glasses were first developed with the bifunctional chromophore 2BNCM (molecule (a) in Figure 14.22) [107]. These chromophores combine high birefringence with transport properties, but unlike previous chromophores they do not crystallize. The glasses that are formed have high optical quality and long stability. Photorefractivity was observed in these glasses and the refractive index modulation amplitude was 1.5 times higher than that observed in DMNPAA-based photorefractive polymers. However, the speed of these materials was quite slow (on the order of a minute). Another monolithic approach based on multifunctional carbazole trimers (molecule (f) in Figure 14.22) has been reported [108]. The trimers have a low glass-transition temperature and form films at room temperature with good optical quality. Mixtures of a bifunctional carbazole trimer and TNF as a sensitizer showed a net gain of 76 cm^{-1} at an applied field of 30 V/μm. Infrared sensitivity at 780 nm was demonstrated in photorefractive glasses of molecules containing a tricyano-substituted furan group as

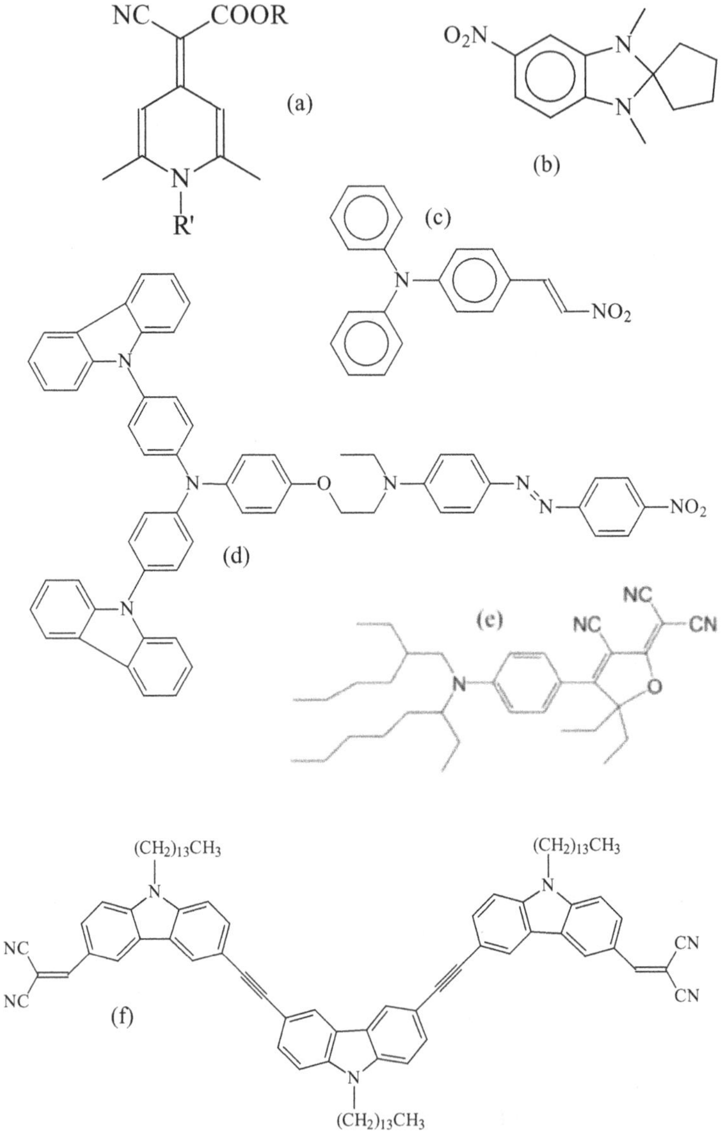

FIGURE 14.22. Examples of bifunctional chromophores for photorefractive applications: (a) N-2-butyl-2,6-dimethyl-4H-pyridone-4-ylidenecyanomethylacetate (2BNCM); (b) 1,3-dimethyl1-2,2-tetramethylene-5-nitrobenzimidazoline (DTNBI); (c) 4-(N,N'-diphenylamino)-(β)-nitrostyrene (DPANST) (d) 4,4'-di(N-carbazolyl)-4''-(2-N-ethyl-4-[2-(4-nitrophenyl)-1-azo]anilinoethoxy)-triphenylamine (DRDCTA); (e) monolithic material m06 with tricynaosubstituted furan group as acceptor; and (f) carbazole trimer.

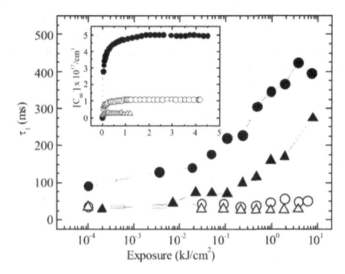

FIGURE 14.23. Exposure dependence of the dominant time constant (τ_1) at 48 V/μm in samples PATPD/ECZ/7DCST/C$_{60}$ (49.5/15/35/0.5 wt. %) (open triangles) and PATPD/ECZ/DBDC/C$_{60}$ (49.5/20/30/0.5 wt. %) (open circles), PVK/ECZ/7DCST/C$_{60}$ (49.5/15/35/0.5 wt. %)(filled triangles) and PVK/ECZ/DBDC/C$_{60}$ (49.5/20/30/0.5 wt. %) (filled circles). Total writing beam intensity 1.1 W/cm^2. Inset shows in situ exposure dependence of the C$_{60}$ radical anion concentration determined spectroscopically in the corresponding samples. DBDC stands for 3-(N,N-di-n-butylaniline-4-yl)-1-dicyanomethylidene-2 cyclohexene; 7DCST stands for 4-homopiperidino-benzylidenemalononitrile.

acceptor (see for instance molecule (e) in Figure 14.22 [109]). Other such derivatives exhibit high performance in the visible range [89]. Numerous reports [91, 110–113] and studies [114, 115] on photorefractive glasses can be found in the literature.

If bifunctional molecules do not form glasses, they can be incorporated into polymer binders. Examples of bifunctional molecules are shown in Figure 14.22. DTNBI (molecule (b) in Figure 14.22) was doped in PMMA and C$_{60}$ was used as a sensitizer. Diffraction efficiencies of 7%, subsecond grating growth times, and net two-beam coupling gain coefficients of 34 cm^{-1} were observed in such samples [100]. DPANST (molecule (c) in Figure 14.23) was doped into PBMA [103]. Recently, ms response times were reported in photorefractive glasses based on the bifunctional chromophore DRDCTA (molecule (d) in Figure 14.22) doped with the plasticizer DOP (molecule (b) in Figure 14.22) and C$_{60}$ as a sensitizer [110, 116].

Another approach to providing materials with long-term stability is to synthesize fully functionalized polymers. This route is more challenging than any of the guest/host approaches described so far and requires a major effort in chemical synthesis. It offers less flexibility than the guest–host approach. Several materials including conjugated polymers have been developed [117–122]. Other approaches are based on the combination of organic and inorganic materials such as in sol-gel approaches [123–125] and the use of inorganic semiconductor quantum dots as sensitizers [126–131]. Recently, hybrid photorefractive polymers were reported with grating buildup times of less than 100 ms and overmodulation of the diffraction efficiency at an applied field of 60 V/μm. These materials were sensitized with cadmium selenide (CdSe) quantum dots treated with the surfactant 4-methylbenzenethiol [132].

In the amorphous organic photorefractive materials discussed above, the electroactive molecules are assumed to be noninteracting. Their orientation in an electric field consequently requires high field values (typically a few tens of V/μm). In contrast, liquid crystals exhibit electroactivity at much lower fields. Photorefractivity in liquid crystals was first demonstrated by Rudenko and Sukhov in the nematic liquid crystal 5CB (see molecule (h) in Figure 14.17) doped with the dye rhodamine 6G [133]. Under excitation of an argon-ion laser, the dye dissolved in the liquid crystal undergoes reversible heterolytic dissociation and leads to the formation of ions. The drift and diffusion of these ions in the mixture leads to a space-charge field that reorients the axis (or director) of the liquid crystal. Since this orientational effect is a Kerr effect and is quadratic in the electric field, a dc field is applied to the sample. In contrast to photorefractive polymers, for which the applied voltage is several kilovolts, here the applied dc voltage required for efficient wave-mixing effects is approximately 1 V for a 100-μm-thick film [134]. High photorefractive gain was observed in nematic liquid crystal mixtures of 5CB and 8OCB doped with electron donor and acceptor molecules [135]. In this case, mobile ions were produced through photoinduced electron transfer reactions between the acceptor N-N′-di(n-octyl)-1,4,5,8-naphthalenediimide and the donor perylene. Due to the large coherent length of a liquid-crystalline phase, the resolution of these materials is limited. Therefore, most of the wave-mixing experiments were performed with writing beams intersecting in the sample with a small angle, leading to large values of the grating spacing. Consequently, most two-beam coupling experiments were performed in the thin-grating limit and the data were analyzed with coupled-wave theory for thick gratings, leading to large values of gain coefficients. These values cannot be compared with those measured in polymers in the thick-grating limit. To improve the resolution of photorefractive liquid crystals, polymer-stabilized nematic liquid crystals have been designed [136]. In this case, the addition of low concentrations of a polymeric electron acceptor creates an anisotropic gel-like medium in which higher resolution is achieved. Another approach that combines high speed, high resolution, and low field reorientation of a liquid crystalline phase is to use polymer-dispersed liquid crystals. In such materials the space-charge field is written in a photoconducting polymer matrix as in a traditional photorefractive polymer. This space-charge field reorients the director of the liquid crystal droplets dispersed in the polymer matrix and leads to an effective refractive index modulation [137, 138]. In 105-μm-thick films of PMMA doped with ECZ as a transport material, TNFDM as a sensitizer, and the liquid crystal mixture TL202 purchased from Merck, overmodulation of the diffraction efficiency at a grating spacing of 3 μm was observed at an applied field of 8 V/μm [139].

Recently, photorefractive polymers were successfully sensitized using nonlinear absorption. This method first proposed by von der Linde [140] with inorganic crystals involves the use of high-intensity writing beams for recording of gratings and low-light intensity for readout. This scheme allows for nondestructive readout, as was demonstrated recently in photorefractive polymers [141, 142]. Two-photon absorption in the sensitizer also allows for operation in the infrared. Using two-photon absorption in the sensitizer molecule DBM (molecule (d) in Figure 14.19), photorefractivity could be demonstrated at 1550 nm [143].

In many materials studied to date, little attention has been given to the stability of a given photorefractive polymer under continuous operation. As discussed in Section 4.2, the relative energies of the frontier orbitals of the transport molecule and the electrooptic chromophore play an important role in the buildup of radical ions of the sensitizer. As was shown in PVK-based materials sensitized with C_{60} and doped with low-ionization-potential chromophores, continuous exposure will lead to an accumulation of C_{60} radical anions [144], which gradually decreases the photoconductivity and slows down the dynamics of the buildup of the space-charge field. This deterioration of the response time can be minimized by using chromophores with higher ionization potential, as shown in [145]. However, such chromophores often lead to lower dynamic range. Another approach is to use a photoconducting matrix that has a lower ionization potential such as the polyacrylate polymer with side chains of tetra-diphenyldiamine (TPD) groups (PATPD) shown in Figure 14.20 (e). TPD groups have an ionization potential of 5.3 eV compared with 5.9 eV for carbazole. When doped with the chromophores 3-(N,N-di-n-butylaniline-4-yl)-1-dicyanomethylidene-2 cyclohexene (DBDC) and 4-homopiperidinobenzylidenemalononitrile (7-DCST) these materials exhibit operational stability for continuous exposures up to 4 kJ/cm^2 as shown in Figure 14.23 [146]. In contrast, when PVK is used as a matrix, the response time degrades rapidly. Figure 14.23 also shows that operational stability correlates with the density of C_{60} radical anions.

14.6 Applications

Photorefractive materials enable a wide range of optical applications, including reconfigurable interconnects, dynamic holographic storage, optical correlation, image recognition, image processing, and phase conjugation. Other chapters in this book provide numerous examples. In this section, we will briefly review examples of applications that have been implemented with photorefractive polymers.

Holographic storage was a strong driver during the early development of photorefractive polymers. Early demonstration of hologram recording and retrieval was performed in DMNPAA:PVK:ECZ:TNF samples and was made possible by the high diffraction efficiency measured in 105-μm-thick samples [147]. More detailed assessments of their potential for holographic storage were conducted with the holographic storage test-bed developed at IBM. In samples of the composite 2BNCM:PMMA:TNF, single digital data pages could be stored and retrieved using only 1 part in 10^6 of the total dynamic range [107]. High-contrast 64-kbit digital data pages could be stored in 130-μm-thick DTNBI:PMMA:C_{60} samples placed near the Fourier plane in a 4-f recording geometry. Due to the limited thickness of the sample, only single data pages were stored. A global absolute threshold resulted in a readout BER (Bit Error Rate) of 1.5×10^{-4} [148]. Using polysiloxane polymers substituted with carbazole side groups and doped with FDEANST and TNF, the IBM group demonstrated digital data recording at a density of 0.52 Mbit/cm^2 [94].

As a reconfigurable recording medium, photorefractive polymers have also been used for dynamic time-average interferometry [147], wave-front phase conjugation and phase doubling [149], and incoherent-to-coherent image conversion [149]. To demonstrate the technological potential of high-efficiency photorefractive polymers in device geometry, an all-optical, all-polymeric pattern-recognition system for security verification has been demonstrated [150]. In this system, the photorefractive polymer was used as the real-time optical recording and processing medium. In the proposed security verification system, documents were optically encoded with pseudorandomly generated phase masks and they were inspected by performing all-optical spatial correlation of two phase-encoded images in a photorefractive polymer. To authenticate the document it was compared with a master, which was an exact copy of the mask. The hologram written by the interference of a reference beam and a laser beam going through the test mask formed a holographic filter for the master mask. If the two phase patterns matched, light was strongly diffracted by the photorefractive polymer. The detection of that light provided a way to check the authenticity of the document that was tested. This system has many advantages: the use of a highly efficient photorefractive polymer as active material and its compatibility with semiconductor laser diodes keep the overall manufacturing cost to levels that are significantly lower than that of any previous proposed optical correlator. The system is fast because the processing is implemented optically in parallel. Furthermore, the high resolution of the photorefractive polymers allows the use of shorter focal length lenses in the 4-f correlator, thus making its design more compact compared with one using liquid crystal light valves. Figure 14.24 shows a photograph of a compact version of the system for security verification that was built.

While the previous application was based on real-time four-wave mixing effects, other applications rely on the strong coherent two-beam coupling effects observed in photorefractive polymers. Such applications include self-pumped phase-conjugation mirrors [151]. In this case, high optical gain is required and was obtained by stacking several photorefractive polymer samples of PDCST:PVK:BBP:C_{60}. A single-pass optical gain of 5 could be demonstrated. Beam-coupling in photorefractive polymers can also be used for optical limiting [152]. In this case, the application relies on the existence of amplified scattering, an effect referred to as beam fanning [153, 154].

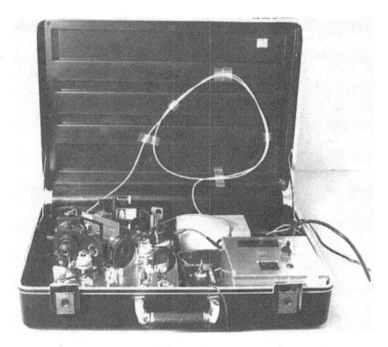

FIGURE 14.24. Photograph of a compact version of the optical correlator for security verification that uses photorefractive polymers.

The real-time holographic recording properties of photorefractive polymers have also been exploited in medical imaging applications by performing time-gated holography [85, 155]. In this method a hologram is formed by the temporal overlap in the photorefractive polymer of the reference pulse and the first-arriving (least-scattered) light from the stretched image-bearing pulse. The filtering of the useful photons from the scattered photons is thus achieved in real time without any need for digital processing. The image carried by the ballistic light can then be reconstructed in real time via diffraction of a probe beam (pulsed or continuous-wave). An example of a reconstructed image using this technique is shown in Figure 14.25. The imaging was performed in a degenerate four-wave mixing geometry using a femtosecond Ti:Sapp laser at 800 nm in the near infrared.

Other applications that have been demonstrated with photorefractive polymers include homodyne detection of ultrasonic surface displacements using two-wave mixing [156, 157], novelty filtering [158, 159], and edge enhancement [160, 161]. With the development of new photorefractive polymers with response times in the ms range, numerous optical processing techniques can be performed at video rates. However, all-optical processing techniques compete with computational methods. Therefore, it is important that photorefractive polymers exhibit faster response times in the future. Applications involving the formation of optical spatial solitons in organic photorefractive materials can be anticipated after the recent demonstration of such effects in photorefractive glasses [162].

FIGURE 14.25. Portion of an Air Force bar-target imaged through nine scattering mean free paths using low-coherence holographic time-gating in a photorefractive polymer in the near infrared (800 nm).

14.7 Conclusion and Outlook

Despite the high efficiency of existing materials and their response times that enable applications at video rates, much effort is still needed in the field before this class of optical materials reaches its full potential. Many of the early advances were made through a combination of different materials and established that photorefractive polymers were suitable for various real-time holographic applications that use low-cost, low-power laser diodes. However, a rational design that leads to controlled properties is required to push these materials to their ultimate performance level. Owing to the progress that has been made in the field of organic semiconductors, the future of photorefractive polymers looks bright. The ability to control the relative energy levels of the frontier orbitals of the different molecules doped into these polymer composites will play an important role in making materials that are stable under continuous operation. Current materials are amorphous or consist of liquid-crystalline phases. The development of new materials including supramolecular structures through nanofabrication technologies or hybrid materials that combine organic or inorganic nanoparticles will bring photorefractive polymers to new performance levels. Examples of materials that exhibit a nanoscale morphology are photorefractive composites that form layered photoconductive polymers [163–165]. Numerous material classes remain unexplored. Materials including transition metal complexes have been explored [166, 167], but many other material classes can potentially be investigated. It is likely that materials with microsecond response times will emerge with sensitivities extended to the telecommunication wavelengths. New applications for these materials will be generated as their performance, modeling, and manufacturing further advance.

Acknowledgments. This work was partially supported by NSF through an ECS research grant, a CAREER grant, and the STC program under agreement number DMR-0120967, and by the Defense Advanced Research Program Agency (DARPA). The author would like to acknowledge Profs. S.R. Marder and N. Peyghambarian, as well as the numerous students and scientists who collaborated with him over the years.

List of Symbols and Abbreviations

E	electric field amplitude
ω, ω_i	optical frequency
$\chi^{(2)}$	second-order nonlinear susceptibility
$R^{(n)}$	degeneracy factors
n	refractive index
r, r_{ij}	linear electrooptic coefficient, electrooptic tensor elements
μ, μ_i	dipole moment, dipole moment component
$\tilde{\alpha}, \alpha_{ij}$	linear molecular polarizability tensor, tensor components
$\tilde{\beta}, \beta_{ijk}, \beta_{zzz}, \beta$	first hyperpolarizability tensor, tensor components
$\tilde{\gamma}$	second hyperpolarizability tensor
β_{add}, β_{CT}	additive and charge-transfer contributions to the first hyperpolarizability
f	optical transition oscillator strength and lens focal length
$\Delta\mu_{ge}$	difference in dipole moment between excited state and ground state
$\hbar\omega_{ge}$	photon energy
μ_{ge}	transition dipole moment
$\beta(0)$	dispersion-free value of the first hyperpolarizability
$\hbar = h/2\pi$	Planck constant
E_{ge}	band gap energy
$|\psi\rangle, |n\rangle, |z\rangle$	wave functions
p_i	molecular polarizability component

$\{X,Y,Z\}$	laboratory Cartesian system of coordinates
$\{x,y,z\}$	molecular Cartesian system of coordinates
V	volume
a_{Ii}	director cosines
φ, θ, ψ	Euler angles
$d\Omega$	differential solid angle
$U(\theta)$	interaction energy
$G(\varphi, \theta, \psi), G(\Omega)$	Maxwell–Boltzmann distribution function
μ^*	corrected value of dipole moment
E_p	poling field
k_B	Boltzmann constant
T	temperature
$L_n(x)$	Langevin functions
r_c	Coulomb radius
η_0	probability to form a thermalized pair
Ω	dissociation probability
ϕ	photogeneration efficiency
I	optical intensity
Λ	grating spacing
ε_r	relative dielectric constant
$E_1, E_{sc}, E_0, E_m, E_{ext}$	dc or low-frequency electric field
E_D	diffusion field
E_q	limiting space-charge field
N_{eff}	effective photorefractive trap density
Θ	shift of the photorefractive space-charge field
$\Delta n, \Delta \tilde{n}$	refractive index modulation amplitude, complex refractive index modulation amplitude
r_{eff}	effective electrooptic coefficient
J	current density
S, T, M	total density of sensitizers, shallow traps, and deep traps, respectively
S^-, T^+, M^+	density of sensitizer radical anions, density of filled shallow and deep traps
γ_T, γ_M	shallow and deep trap trapping rates
β_T, β_M	shallow and deep traps detrapping rates
D	diffusion coefficient
$\mu(E, T)$	carrier mobility
μ_0	mobility prefactor
K	grating vector
$\sigma, \sigma_{VdW}, \sigma_D$	energetical disorder
Σ	positional disorder
T_g	glass-transition temperature
T_c	dispersive-to-nondispersive transition temperature
$\Delta\alpha$	polarizability anisotropy

Q_{OP} figure of merit for orientational photorefractivity
N_{max} maximum noninteracting chromophore density
ξ angle between the total field and the Z-axis

References

1. Franken PA, Hill AE, Peters CW, Weinreich G (1961). *Phys. Rev. Lett.* 7:118.
2. Bloembergen N (1964). *Nonlinear Optics*, Benjamin, New York.
3. Shen YR (1984). *The Principles of Nonlinear Optics*, Wiley, New York.
4. Boyd RW (1992). *Nonlinear Optics*, Academic Press, Boston.
5. Yariv A (1975). *Quantum Electronics*, Wiley, New York.
6. Ostroverkhova O, Moerner WE (2004). *Chem. Rev.* 104:3267.
7. Moerner WE, Grunnet-Jepsen A, Thompson CL (1997). *Ann. Rev. Mater. Sci.* 27:585.
8. Meerholz K, Kippelen B, Peyghambarian N (1998). In *Photonic Polymer Systems*, Marcel Dekker, New York.
9. Kippelen B, Peyghambarian N (1997). In *Sol-Gel and Polymer Photonic Devices, Critical Review of Optical Science and Technology*, CR 68, SPIE Optical Engineering Press, Bellingham.
10. Oudar JL, Chemla DS (1975). *Opt. Commun.* 13:164.
11. Chemla DS, Zyss J (1987). *Nonlinear Optical Properties of Organic Molecules and Crystals*, Academic Press, Orlando.
12. Zyss J (1994). *Molecular Nonlinear Optics: Materials, Physics and Devices*, Academic Press, Boston.
13. Kanis DR, Ratner MA (1994). *Chem. Rev* 94:195.
14. Marder SR, Beratan, DN, Cheng LT (1991). *Science* 252:103.
15. Gorman CB, Marder SR (1993). *Proc. Nat. Acad. of Sciences of U.S.A.* 90:11297.
16. Meyers F, Marder SR, Pierce BM, Brédas JL (1994). *J. Am. Chem. Soc.* 116:10703.
17. Marder SR, Kippelen B, Jen AKJ, Peyghambarian N (1997). *Nature* 388:845.
18. Williams DJ (1987). In *Nonlinear Optical Properties of Organic Molecules and Crystals*, Chemla DS, Zyss J, eds., Academic Press, New York.
19. Singer KD, Kuzyk MG, Sohn JE (1987). *J. Opt. Soc. Am. B* 4:968.
20. Burland DM, Miller RD, Walsh CA (1994). *Chem. Rev.* 94:31.
21. Yu YZ, Wong KY, Garito AF (1997). In *Nonlinear Optics of Organic Molecules and Polymers*, Nalwa HS, Miyata S, eds., CRC Press, Boca Raton.
22. Wu JW (1991). *J. Opt. Soc. Am B* 8:142.
23. Marcus RA (1993). *Rev. Mod. Phys.* 65:599.
24. Law KY (1993). *Chem. Rev.* 93:449.
25. Borsenberger PM, Weiss DS (1993). *Organic Photoreceptors for Imaging Systems*, Marcel Dekker, New York.
26. Onsager L (1938). *Phys. Rev.* 54:554.
27. Mozumder A (1974). *J. Chem. Phys.* 60:4300.
28. Noolandi J, Hong, KM (1979). *J. Chem. Phys.* 70:3230.
29. Mort J (1980). *Adv. in Physics* 29:367.
30. Scher H, Montroll EW (1975). *Phys. Rev. B* 12:2455.
31. Schmidlin SW (1977). *Phys. Rev. B* 16:2362.

32. Bässler H (1993). *Adv. Mat.* 5:662.
33. Borsenberger PM, Magin EH, Van der Auweraer M, De Schryver FC (1993). *Phys. Stat. Sol. (a)* 140:9.
34. Borsenberger PM, Bässler H (1994). *J. Appl. Phys.* 75:967.
35. Borsenberger PM, Detty MR, Magin EH (1994). *Phys. Stat. Sol. (b)* 185:465.
36. Borsenberger PM, Magin EH, Van der Auweraer M, De Schryver FC (1994). *Phys. Stat. Sol. (b)* 186:217.
37. Borsenberger PM, Gruenbaum WT, Magin EH (1995). *Phys. Stat. Sol. (b)* 190:555.
38. Borsenberger PM, Shi J (1995). *Phys. Stat. Sol. (b)* 191:461.
39. Borsenberger PM, Gruenbaum WT, Sorriero LJ, Zumbulyadis N (1995). *Jpn. J. Appl. Phys.* 34:L1597.
40. Borsenberger PM, Magin EH, O-Regan MB, Sinicropi JA (1996). *J. Polymer Sci. B: Polymer Phys.* 34:317.
41. Lampert MA, Mark P (1970). *Current Injection in Solids*, Academic, New York.
42. Murgatroyd PN (1970). *J. Phys. D: Appl. Phys.* 3:151.
43. Dimitrakopoulos CD, Malenfant PRL (2002). *Adv. Mater.* 14:99.
44. Hoofman RJO, de Haas MP, Siebbeles LDA, Warman JW (1998). *Nature* 392:54.
45. Sutter K, Günter P (1990). *J. Opt. Soc. Am. B* 7:2274.
46. Schildkraut JS (1990). *Appl. Phys. Lett.* 58:340.
47. Ducharme S, Scott JC, Twieg RJ, Moerner WE (1991). *Phys. Rev. Lett* 66:1846.
48. Ashkin A, Boyd GD, Dziedzic JM, Smith RG, Ballmann AA, Nassau K (1966). *Appl. Phys. Lett.* 9:72.
49. Chen FS (1967). *J. Appl. Phys.* 38:3418.
50. Chen FS (1969). *J. Appl. Phys.* 40:3389.
51. Amodei JJ (1971). *Appl. Phys. Lett.* 18:22.
52. Günter P (1982). *Phys. Rep.* 93:199.
53. Feinberg J (1983). In *Optical Phase Conjugation*, Fisher RA, ed., Academic Press, New York.
54. Günter P, Huignard JP (1988) (1989). *Photorefractive Materials and Their Applications*, Vols. I and II, Springer-Verlag, Berlin.
55. Nolte DD (1995). *Photorefractive Effects and Materials*, Kluwer, Boston.
56. Roosen G (1989). *Int. J. Optoelectron.* 4:459.
57. Nolte DD, Olson DH, Doran GE, Knox WH, Glass AM (1990). *J. Opt. Soc. Am. B* 7:2217.
58. Wang Q, Nolte DD, Melloch MR (1991). *Appl. Phys. Lett.* 59:256.
59. Kukhtarev NV, Markov VB, Soskin M, Vinetskii VL (1979). *Ferroelectrics* 22:949.
60. Yeh P (1993). *Introduction to Photorefractive Nonlinear Optics*, Wiley, New York.
61. Twarowski AJ (1989). *J. Appl. Phys.* 65:2833.
62. Schildkraut JS, Buettner AV (1992). *J. Appl. Phys.* 72:1888.
63. Schildkraut JS, Cui YJ (1992). *J. Appl. Phys.* 72:5055.
64. Cui YP, Swedek B, Cheng N, Zieba J, Prasad PN (1999). *J. Appl. Phys.* 85:38.
65. Yuan BH, Sun XD, Hou CF, Li Y, Zhou ZX, Jiang YY, Li CF (2000). *J. Appl. Phys.* 88:5562.
66. Yuan BH, Sun XD, Jiang YY, Hou CF, Zhou ZX (2001). *J. Mod. Opt.* 48:1161.
67. Yuan BH, Sun XD, Zhou ZX, Li Y, Jiang YY, Hou CF (2001). *J. Appl. Phys.* 89:5881.
68. Ostroverkhova O, Singer KD (2002). *J. Appl. Phys.* 92:1727.
69. Eichler HJ, Günter P, Pohl DW (1986). *Laser-Induced Dynamic Gratings*, Springer Verlag, Berlin.

70. Herlocker JA, Fuentes-Hernandez C, Ferrio KB, Hendrickx E, Zhang Y, Wang JF, Marder SR, Blanche PA, Peyghambarian N, Kippelen B (2000). *Appl. Phys. Lett.* 77:2292.
71. Kulikovsky L, Neher D, Meerholz K, Hörnhold HH, Ostroverkhova O (2004). *Phys. Rev. B* 69:125216.
72. Mecher E, Gallego-Gomez F, Tillmann H, Hörnhold HH, Hummelen JC, Meerholz K (2002). *Nature* 418:959.
73. Tamura K, Padias AB, Hall HK Jr., Peyghambarian N (1992). *Appl. Phys. Lett.* 60:1803.
74. Yu L, Chan W, Bao Z, Cao SXF (1993). *Macromolecules* 26:2216.
75. Kippelen B, Tamura K, Peyghambarian N, Padias AB, Hall HK Jr. (1993). *Phys. Rev. B* 48:10710.
76. Donckers MCJM, Silence SM, Walsh CA, Scott JC, Matray TJ, Twieg RJ, Hache F, Bjorklund GC, Moerner WE (1993). *Opt. Lett.* 18:1044.
77. Kippelen B, Sandalphon, Peyghambarian N, Lyon SR, Padias AB, Hall HK Jr. (1993). *Electron. Lett.* 29:1873.
78. Meerholz K, Volodin B, Sandalphon, Kippelen B, Peyghambarian N (1994). *Nature* 371:497.
79. Moerner WE, Silence SM, Hache F, Bjorklund GC (1994). *J. Opt. Soc. Am. B* 11:320.
80. Wortmann R, Poga C, Twieg RJ, Geletneky C, Moylan CR, Lundquist PM, DeVoe RG, Cotts PM, Horn H, Rice JE, Burland DM (1996). *J. Chem. Phys.* 105:10637.
81. Kippelen B, Meyers F, Peyghambarian N, Marder SR (1997). *J. Am. Chem. Soc.* 119:4559.
82. Kippelen B, Meerholz K, Peyghambarian N (1997). In *Nonlinear Optics of Organic Molecules and Polymers*, Nalwa HS, Miyata, eds., CRC Press, Boca Raton.
83. Hendrickx E, Volodin BL, Steele DD, Maldonado JL, Wang JF, Kippelen B, Peyghambarian N (1997). *Appl. Phys. Lett.* 71:1159.
84. Meerholz K, De Nardin Y, Bittner R, Wortmann R, Würthner F (1998). *Appl. Phys. Lett.* 73:4.
85. Kippelen B, Marder SR, Hendrickx E, Maldonado JL, Guillemet G, Volodin B, Steele DD, Enami Y, Sandalphon, Yao YJ, Wang JF, Röckel H, Erskine L, Peyghambarian N (1998). *Science* 279:54.
86. Wright D, Díaz-García MA, Casperson JD, DeClue M, Moerner WE, Twieg RJ (1998). *Appl. Phys. Lett.* 73:1490.
87. Herlocker JA, Ferrio KB, Guenther BD, Mery S, Kippelen B, Peyghambarian N (1999). *Appl. Phys. Lett.* 74:2253.
88. Zhang J, Singer KD (1998). *Appl. Phys. Lett.* 72:2948.
89. Wright D, Gubler U, Roh Y, Moerner WE (2001). *Appl. Phys. Lett.* 79:4274.
90. Hendrickx E, Kippelen B, Thayumanavan S, Marder SR, Persoons A, Peyghambarian N (2000). *J. Chem. Phys.* 112:9557.
91. Wurthner F, Yao S, Schilling J, Wortmann R, Redi-Abshiro M, Mecher E, Gallego-Gomez F, Meerholz K (2001). *J. Am. Chem. Soc.* 123:2810.
92. Mecher E, Brauchle C, Hornold HH, HUmmelen JC, Meerholz K (1999). *Phys. Chem. Chem. Phys.* 1:1749.
93. Eralp M, Thomas J, Tay S, Li G, Meredith G, Schülzgen A, Peyghambarian N, Walker GA, Barlow S, Marder SR (2004). *Appl. Phys. Lett.* 85:1095.
94. Lundquist PM , Poga C, DeVoe RG, Jia Y, Moerner WE, Bernal MP, Coufal H, Grygier RK, Hoffnagle JA, Jefferson CM, Macfarlane RM, Shelby RM, Sincerbox GT (1996). *Opt. Lett.* 21:890.

95. Schloter S, Hofmann U, Strohriegl P, Schmidt HW, Haarer D (1998). *J. Opt. Soc. Am. B* 15:2473.

96. Chun H, Moon IK, Shin DH, Song S, Kim N (2002). *J. Mater. Chem.* 12:858.

97. Wright D, Gubler U, Moerner WE, DeClue M, Siegel JS (2003). *J. Phys. Chem. B* 107:4732.

98. Iwata S, Tsukahara E, Nihei E, Koike Y (1997). *Appl. Opt.* 36:4549.

99. Herlocker JA, Fuentes-Hernandez C, Wang JF, Peyghambarian N, Kippelen B, Zhang Q, Marder SR (2002). *Appl. Phys. Lett.* 80:1156.

100. Silence SM, Scott JC, Stankus JJ, Moerner WE, Moylan CR, Bjorklund GC, Twieg RJ (1995). *J. Phys. Chem.* 99:4096.

101. Wortmann R, Poga C, Twieg RJ, Geletneky C, Moylan CR et al. (1996). *J. Chem. Phys.* 105:10637.

102. Burzynski R, Zhang Y, Ghosal S, Casstevens MK (1995). *J. Appl. Phys.* 78:6903.

103. Zhang Y, Ghosal S, Casstevens MK, Burzynski R (1995). *Appl. Phys. Lett.* 66:256.

104. Suh D, Park OO, Ahn T, Shim HK (2002). *Jpn. J. Appl. Phys.* Part 2 41:L428.

105. Okamoto K, Nomura T, Park SH, Ogino K, Sato H (1999). *Chem. Mater.* 11:3279.

106. Grunnet-Jepsen A, Thompson CL, Twieg RJ, Moerner WE (1997). *Appl. Phys. Lett.* 70:1515.

107. Lundquist PM, Wortmann R, Geletneky C, Twieg RJ, Jurich M et al. (1996). *Science* 274:1182.

108. Wang L, Zhang Y, Wada T, Sasabe H (1996). *Appl. Phys. Lett.* 69:728.

109. Hou Z, You W, Yu L (2003). *Appl. Phys. Lett.* 82:3385.

110. Schloter S, Schreiber A, Grasruck M, Leopold A, Kol'chenko M, Pan J, Hohle C, Strohriegl P, Zilker SJ, Haarer D (1999). *Appl. Phys. B* 68:899.

111. Wang LM, Ng MK, Yu L (2001). *Appl. Phys. Lett.* 78:700.

112. He M, Twieg RJ, Gubler U, Wright D, Moerner WE (2003). *Chem. Mater.* 15:1156.

113. Sohn J, Hwang J, Park SY, Lee GJ (2001). *Jpn. J. Appl. Phys.* Part 1 5A:3301.

114. Ostroverkhova O, Moerner WE, Twieg RJ (2003). *Chem. Phys. Chem.* 4:732.

115. Hwang J, Seo J, Sohn J, Park SY (2003). *Opt. Mater.* 21:359.

116. Hohle C, Hofmann U, Schloter S, Thelakkat M, Strohriegl P, Haarer D, Zilker SJ (1999). *J. Mater. Chem.* 9:2205.

117. Kippelen B, Tamura K, Peyghambarian N, Padias AB, Hall HK Jr. (1993). *Phys. Rev. B* 48:10710.

118. Zhao C, Park CK, Prasad PN, Zhang Y, Ghosal S, Burzynski R (1995). *Chem. Mater.* 7:1237.

119. Yu L, Chen YM, Chan WK (1995). *J. Phys. Chem.* 99:2797.

120. Li L, Chittibabu KG, Chen Z, Chen JI, Marturunkakul S, Kumar J, Tripathy SK (1996). *Opt. Commun.* 125:257.

121. Hendrickx E, Engels C, Schaerlaekens M, Van Steenwinckel D, Samyn C, Persoons A (2002). *J. Phys. Chem. B* 106:4588.

122. You W, Cao S, Hou Z, Yu L (2003). *Macromolecules* 36:7014.

123. Burzynski R, Casstevens MK, Zhang Y, Ghosal S (1996). *Opt. Eng.* 35:443.

124. Chaput F, Riehl D, Boilot JP, Cargnelli K, Canva M, Levy Y, Brun A (1996). *Chem. Mater.* 8:312.

125. Cheben P, del Monte F, Worsfold DJ, Carlsson DJ, Grover CP, Mackenzie JD (2000). *Nature* 408:64.

126. Winiarz JG, Zhang L, Lal M, Friend CS, Prasad PN (1999). *J. Am. Chem. Soc.* 121:5287.

127. Binks DJ, Bant SP, West DP, O'Brien P, Malik MA (2003). *J. Mod. Opt.* 50:299.

128. Binks, DJ, West DP, Norager S, O'Brien P (2002). *J. Chem. Phys.* 117:7335.
129. Binks DJ, Khand K, West DP (2001). *J. Opt. Soc. Am. B* 18:308.
130. Winiarz JG, Zhang L, Park J, Prasad PN (2002). *J. Phys. Chem.* 106:967.
131. Suh DJ, Park O, Jung H-T, Kwon MH, (2002). *Kor. J. Chem. Eng.* 19:529.
132. Fuentes-Hernandez C, Suh DJ, Kippelen B, Marder SR (2004). *Appl. Phys. Lett.* 85:534.
133. Rudenko EV, Sukhov AV (1994). *JETP Lett.* 59:142.
134. Khoo IC, Li H, Liang Y (1994). *Opt. Lett.* 19:1723.
135. Wiederrecht GP, Yoon BA, Wasielewski MR (1995). *Science* 270:1794.
136. Wiederrecht GP, Wasielewski MR (1998). *J. Am. Chem. Soc.* 120:3231.
137. Golemme A, Volodin BL, Kippelen B, Peyghambarian N (1997). *Opt. Lett.* 22:1226.
138. Ono H, Kawatsuki N (1997). *Opt. Lett.* 22:1144.
139. Golemme A, Kippelen B, Peyghambarian N (1998). *Appl. Phys. Lett.* 73:2408.
140. von der Linde D, Glass AM, Rodgers KF (1974). *Appl. Phys. Lett.* 25:155.
141. Blanche PA, Kippelen B, Schulzgen A, Fuentes-Hernandez C, Ramos-Ortiz G, Wang JF, Hendrickx E, Marder SR, Peyghambarian N (2002). *Appl. Phys. Lett.* 27:19.
142. Kippelen B, Blanche PA, Schulzgen A, Fuentes-Hernandez C, Ramos-Ortiz G, Wang JF, Peyghambarian N, Marder SR, Leclercq A, Beljonne D, Bredas JL (2002). *Adv. Funct. Mater.* 12:615.
143. Tay S, Thomas J, Eralp M, Li G, Kippelen B, Marder SR, Meredith G, Schulzgen A, Peyghambarian N (2004). *Appl. Phys. Lett.* 85:4561.
144. Grunnet-Jepsen A, Wright D, Smith B, Bratcher MS, DeClue MS, Siegel JS, Moerner WE (1998). *Chem. Phys. Lett.* 291:553.
145. Herlocker JA, Fuentes-Hernandez C, Ferrio KB, Hendrickx E, Zhang Y, Wang JF, Marder SR, Blanche PA, Peyghambarian N, Kippelen B (2000). *Appl. Phys. Lett.* 77:2292.
146. Fuentes-Hernandez C, Thomas J, Termine R, Meredith G, Peyghambarian N, Kippelen B, Barlow S, Walker GA, Marder SR, Yamamoto M, Cammack K, Matsumoto K (2004). *Appl. Phys. Lett.* In press.
147. Volodin BL, Sandalphon, Kippelen B, Kukhtarev NV, Peyghambarian N (1995). *Opt. Eng.* 34:2213.
148. Poga C, Lundquist PM, Lee V, Shelby RM, Twieg RJ, Burland DM (1996). *Appl. Phys. Lett.* 69:1047.
149. Volodin BL, Kippelen B, Meerholz K, Peyghambarian N, Kukhtarev NV, Caulfield HJ (1996). *J. Opt. Soc. Am. B* 13:2261.
150. Volodin BL, Kippelen B, Meerholz K, Javidi B, Peyghambarian N (1996). *Nature* 383:58.
151. Grunnet-Jepsen A, Thompson CL, Moerner WE (1997). *Science* 277:549.
152. Grunnet-Jepsen A, Thompson CL, Moerner WE (1997). *Mat. Res. Soc. Symp. Proc.* 479:199.
153. Grunnet-Jepsen A, Thompson CL, Twieg RJ, Moerner WE (1998). *J. Opt. Soc. Am. B* 15:901.
154. Meerholz K, Bittner R, De Nardin Y (1998). *Opt. Commun.* 150:205.
155. Steele DD, Volodin BL, Savina O, Kippelen B, Peyghambarian N (1998). *Opt. Lett.* 23:153.
156. Klein MB, Bacher GD, Grunnet-Jepsen A, Wright D, Moerner WE (1999). *Opt. Commun.* 162:79.
157. Klein MB, Bacher GD, Grunnet-Jepsen A, Wright D, Moerner WE (1999). *SPIE* 3589:22.

158. Goonesekera A, Wright D, Moerner WE (2000). *Appl. Phys. Lett.* 76:3358.
159. Hendrickx E, Van Steenwinckel D, Persoons A (2001). *Appl. Opt.* 40:1412.
160. Banerjee PP, Gad E, Hudson T, McMillen D, Abdeldayem H, Frazier D, Matsushita K (2000). *Appl. Opt.* 39:5337.
161. Ono H, Kawamura T, Kawatsuki N, Norisada H (2001). *Appl. Phys. Lett.* 79:895.
162. Cheng ZG, Azaro M, Ostroverkhova O, Moerner WE, He M, Twieg RJ (2003). *Opt. Lett.* 28:2509.
163. Kwon OP, Lee SH, Montemezzani G, Günter P (2003). *Adv. Funct. Mater.* 13:434.
164. Kwon OP, Lee SH, Montemezzani G, Günter P (2003). *J. Opt. Soc. Am. B* 20:2307.
165. Kwon OP, Montemezzani G, Günter P, Lee SH (2004). *Appl. Phys. Lett.* 84:43.
166. Peng Z, Gharavi AR, Yu L (1997). *J. Am. Chem. Soc.* 119:4622.
167. Aiello I, Dattilo D, Ghedini M, Bruno A, Termine R, Golemme A (2002). *Adv. Mater.* 14:1233.

15

Photosensitivity and Treatments for Enhancing the Photosensitivity of Silica-Based Glasses and Fibers

P. Niay,[1] B. Poumellec,[2] M. Lancry,[1] and M. Douay[1]

[1] Laboratoire de Physique, Atomes et Molécules (PHLAM), Unité Mixte de Recherche du Centre National de la Recherche Scientifique (CNRS, UMR 8523), Centre d'Etudes et de Recherches Lasers et Applications, Université des Sciences et Technologies de Lille, Bâtiment P5, 59655 Villeneuve d'Ascq, France
pierre.niay@univ-lille1.fr
[2] Laboratoire de Physico-Chimie de l'Etat Solide, Unité Mixte de Recherche du Centre National de la Recherche Scientifique (CNRS, UMR 8182), ICMMO (Institut de Chimie Moléculaire et des Matériaux d'Orsay), Université de Paris Sud-Orsay, Bâtiment 414, 91405 Orsay Cedex, France **Bertrand.Poumellec@lpces.u-psud.fr**

15.1 Introduction

Due to excellent physical and chemical properties, vitreous silica-based (v-SiO_2-based) glasses prove to be key materials in optical waveguides, metal-oxide semiconductors, and other optical elements. Thus, numerous elaboration methods are used to produce v-silica-based devices. For example, the techniques of vapor-phase deposition (e.g., flame hydrolysis deposition (FHD), modified chemical vapor deposition (MCVD), plasma-enhanced chemical vapor deposition (PECVD)) are common in the fabrication of most standard telecommunication silica fibers or planar waveguides [1–4]. Moreover, planar waveguides or thin films can be produced from silica-containing solid source by physical vapor-deposition techniques such as rf reactive sputtering deposition [5–6] and helicon-activated reactive evaporation (HARE) [7]. Waveguides can also be fabricated by means of electron [8] or ion (Si, Ge, Se, O, B, P, or H) implantation in pure silica [9–13]. Furthermore, other processes including sol-gel synthesis [14–15] and conventional melting of raw materials are routinely used to manufacture sol-gel fiber [15], glassy thin films, or multicomponent glass in which waveguides are realized by means of techniques such as ion exchange [16–17]. As a result of the large variety of elaboration processes, all the v-SiO_2-based glasses are different in terms of chemical composition and structural disorder [18]. Nonetheless, they share two properties, leading to important practical consequences. On the one hand, v-SiO_2-based glass exhibits *macroscopic inversion symmetry* and should not exhibit any even-order optical nonlinear effect ($\chi^{(2n)}$) such as the Pockels ($n = 1$) effect unless it was prepared by a poling technique that breaks the inversion symmetry [19–23]. On the other hand, all these glasses (and others such as chalcogenide and fluoride glasses) prove to be *photosensitive*, depending on both the exposure conditions of the glass to laser light and on their composition.

Photosensitivity is a new non elastic light–matter interaction that was discovered by Hill et al. in 1978 [24–25]. These authors launched a 50-mW TEM_{00} beam from an Ar^+ single-mode laser at 514 nm into a 1-m-long piece of germanosilicate fiber (*internal writing technique*). They observed that the optical power reflected from the input end of the fiber increased with exposure time until almost all the light was reflected. The effect was analyzed by assuming the photoinduced formation of a periodic refractive index change at the fiber core, i.e., a Bragg grating (BG). The important features of the phenomenon lie in the facts that the narrow-band filters are persistent [26] and a two-photon absorption process mediates their formation [27]. However, the details of the mechanism behind the formation of the BG were a matter for some speculation. Indeed, two types of model were proposed to account for the refractive index changes: A *local* one, in which the modification of the index is a function only of the local intensity of light, assumes the photoinduced formation or bleaching of color centers [28–29]. In contrast, the second one was a *photorefractive-like model* that involved carrier transport over macroscopic distances, leading to the photoinduced formation of longitudinal and transverse components of a space-charge field [30]. The double role of the field components was *to break the symmetry of the glass* and to induce a periodic change in index via the Pockels effect [30]. After some early objection to the relevance of a local model [31], various theoretical and experimental works have shown that it can accurately describe the early stages of BG growth [32–37], whereas convincing arguments were given to disregard the photorefractive-like model. Nearly at this time, R.G. Powels et al. reported a pioneering work in which four-wave mixing techniques were used to establish and probe refractive-index thick holograms in Eu^{3+}-doped silicate and phosphate glass plates [38–39]. The properties of the permanent gratings proved to be consistent with a local structural modification of the glass mediated by hot phonon localized around the Eu^{3+} ions [38].

Thus, one has to distinguish between *photorefractivity* and *photosensitivity*. The former word involves photoinduced conduction of carriers that leads to the formation of a space-charge field and thus to index changes via the Pockels effect in a noncentrosymmetric material. The latter is, in this paper, devoted to various photoinduced permanent changes in refractive index of centrosymmetric glass. From a practical point of view, the most important difference between photosensitivity and photorefractivity comes from the facts that any *uniform* (fringeless) exposure of a glass to convenient laser light leads to a significant index change and this change is permanent. Furthermore, exposing a fiber BG to uniform irradiation at room temperature does not necessarily erase the grating, as is observed for photorefractive grating, but on the contrary, may lead to an increase in the grating reflectivity [40–43]. On the other hand, it is worth noticing that exposing a glass to laser light may also induce other photoinduced phenomena that come together with the permanent change in refractive index and most often are closely related to the photosensitivity itself.

It was soon recognized that the *internal writing technique* suffered from two drawbacks for widespread applications of BG. First, such a grating can be used only close to the writing wavelength. Secondly, the photoinduced change in index proves to be too low ($<5\ 10^{-5}$) to write short ($\approx$ mm) highly reflective gratings. The *transverse holographic technique* proposed by G. Meltz and W.W. Morey helps in solving these issues [44]. Briefly, the method consists in exposing a single-mode germano-silicate fiber through the side of the cladding to a coherent UV-two-beam interference pattern with a wavelength λ_p selected to lie in the germanium oxygen-vacancy defect center (GODC) absorption band near 244 nm (see Figure 15.1). The UV exposure leads to the growth of a phase grating reflecting light around a Bragg wavelength λ_B given by

$$\lambda_B = 2n_{eff}\Lambda; \qquad \Lambda = \frac{\lambda_p}{2\sin(\alpha/2)}. \qquad (15.1)$$

In (15.1), Λ is the fringe spacing ($\Lambda \approx 0.2$ μm in the experiment by G. Meltz et al.), α is the crossing angle between the interfering beams, and n_{eff} is the effective index of the fundamental LP_{01} mode. Hence, by means of this technique and a proper choice for α and λ_p, one can now select the Bragg wavelength ($\lambda_B > n_{eff}\lambda_p$) almost anywhere in the transmission range of the fiber. Furthermore, the holographic technique proved to enhance the efficiency of the grating inscription [44]. K.O. Hill et al. then rapidly benefited from the tensorial nature of the change in refractive index (UV-induced birefringence) to write a polarization rocking filter within birefringent fiber using *the point-by-point side-writing technique* [45]. The method consists in periodically side-exposing a length W of the fiber core through an amplitude mask ($W = \Lambda/2$) in order to change the fiber birefringence with a grating period made equal to the fiber beat length ($\Lambda \approx 1$ cm). *The point-by-point side-writing technique* was later implemented to write long period gratings (LPG) that couple the fundamental guided mode to copropagating cladding modes [46].

However, the changes in refractive index obtained through Meltz and Morey's method saturated below a few 10^{-4} in most germano-silicate fibers, a figure even too low for most applications. Therefore, research efforts on photosensitivity were then intended (1) to enhance the photosensitivity of fiber and more generally that of v-SiO_2-based glasses, (2) to get a better understanding of the mechanisms of photosensitivity, and (3) to be in a position to assess and enhance the long-term stability of the photoinduced refractive index change. As a result of the progress in these fields, stable changes in refractive index in excess of 10^{-3} can routinely be obtained in most silica-based fibers, planar waveguides, and bulk samples. Such index changes allow for the fabrication of an outstanding number of passive optical devices including in-waveguide grating-based optical components and UV-written planar waveguides [47]. Photosensitivity also finds applications in trimming optical path difference in waveguide interferometers [48]. Advanced methods for the fabrication of these devices and specific applications of Bragg

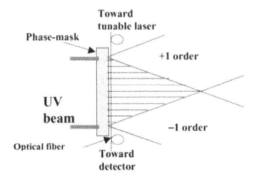

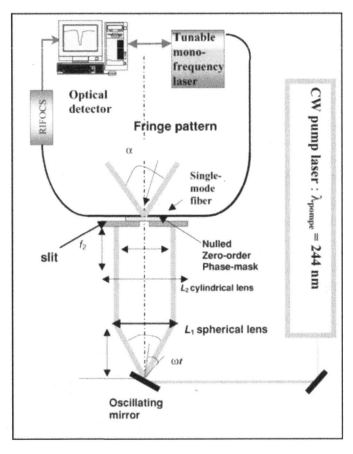

FIGURE 15.1. Typical experimental setup used for writing Bragg gratings: the beam from a CW UV laser impinges on a phase-mask, producing a fringe pattern. The fiber transmission spectrum can be monitored during or after the BG inscription by means of a frequency-tunable laser and a power-meter.

gratings are reviewed in recent books and in a special issue of the *Journal of Lightwave Technology* [49–52]. It is worth noting that photosensitivity can also be regarded as an undesirable phenomenon, as in the degradation of photolithography optics.

Our first purpose in this paper is to recall fundamental properties of photosensitivity in silica-based glasses with emphasis on the methods for measuring and enhancing this photosensitivity and the factors on which it depends. Secondly, after reviewing the photoinduced phenomena that come together with the permanent change in refractive index, we focus our interest on the main mechanisms of photosensitivity in germano-silicategermano-silicate glasses, keeping in mind that these mechanisms may also be relevant to other silica glasses. On the other hand, the refractive index change, like many other properties, exhibits a partial stability only. Indeed, exposure of a grating to increasing temperatures usually results in a decay of the refractive index modulation. However, an opposite behavior can sometimes be observed, illustrating the complexity of the photosensitivity mechanisms [40]. So our third purpose is to discuss the precautions that one has to take to predict the thermal stability of a BG.

15.2 Examples of Methods Used for Measuring Photoinduced Refractive Index Change

Recording the kinetics of refractive index change in the course of either writing or annealing experiments has a scientific as well as practical interest. Indeed, on the one hand, connecting changes in refractive index with those of other relevant quantities (for example, concentration of photoinduced species) is at the heart of an understanding of the photosensitivity mechanisms. Furthermore, measuring the initial slope of the plots that represent the growths of the index change versus exposure time for various optical power densities allows the number of steps involved in the process to be identified [53]. On the other hand, appropriate control of the spectral response of BG-based devices needs the index change to be tailored at every point along the grating. To meet this requirement, in turn, requires knowledge of the relation between this change and the parameters of the exposure such as optical power density and exposure time.

Numerous methods for measuring the photoinduced index change have been reported in the literature. These methods include the Z-scan technique [54], the measurement of the change in the deviation experienced by a probe beam impinging onto a UV-exposed preform-made prism [55], and the measurement of the numerical aperture of a directly written planar waveguide [47, 56]. Other classical methods such as ellipsometry [57], prism-coupling technique [58], refracted near-field method [59–60], and focusing [61–62] or coherent filtering techniques [62–63], have also been implemented for measuring photoinduced changes in index in exposed thin films, bulk plates, or waveguides. More accurate and sensitive methods use different configurations of bulk interferometers [64–65] or in-waveguide interferometers that include in-fiber [66–67] or in-planar waveguide [68] Mach–Zehnder interferometers, in-fiber Michelson interferometer [28], modal interferometer [69], and in-fiber [70–72] or in-waveguide Fabry–Perot interferometer [73]. On the other hand, specific methods for measuring the exposure-induced birefringence either in bulk, in fiber, or waveguide are reported in [74–76].

In the following, we discuss only the most common method so far used for measuring index changes in fibers: photoinscription and spectral characterization of in-fiber BG. It is worth noting that the inscription of BG has not only been extensively used for measuring refractive index changes in optical fibers but also in planar waveguides [77–79], in preform plates [80], and in bulk [17].

The method consists in measuring the spectral characteristics of a *uniform* BG during (or after) its inscription in the single-mode fiber under study. To this end, the fiber is transversely exposed to a uniform two-beam interference fringe pattern, as reported in [44]. The processes that have been used to produce the fringe patterns are reviewed in [49]. A typical experimental setup involving a nulled zero-order phase-mask to make the fringe pattern is displayed in Figure 15.1. By and large, the BG pitch Λ is usually selected so that λ_B is located in a spectral range where the exposure-induced loss can be neglected (no absorption grating). As shown in this figure, the transmission T of the fiber near λ_B is measured by means of a tunable single-frequency laser and a power-meter. Thus, the changes in both the BG first-order reflection coefficient ($R \approx 1 - T$), the grating width $\delta\lambda$, and the shift $\Delta\lambda$ in λ_B can be measured as a function of exposure time for various optical power densities at the fiber core.

To deduce the changes in refractive index from these data, one usually starts from several assumptions: a step-index fiber; periodicity of the exposure-induced change in refractive index along the fiber axis ($\overrightarrow{oz}$); uniformity of this change across the core. This last hypothesis has been discussed in [81]. It is worth noticing that nonuniformity of the index change within the core of a waveguide can induce effects such as form birefringence [82] and coupling to cladding modes. Thus, assuming further that the change in index is null in the fiber cladding, one can write the evolution of the refractive index $n(z)$ along z as a Fourier series expansion:

$$n(z, F, r) = (n_1 + \Delta n_{\text{mean}}(F)) + \Delta n_{\text{mod}}(F).\cos\left(\frac{2\pi z}{\Lambda}\right)$$

$$+\Delta n_{\text{mod2}}(F).\cos\left(\frac{4\pi z}{\Lambda}\right) + +; \quad 0 < z < L; \quad 0 < r < a,$$

$$n(z, r) = n_1; \quad z > L; \quad 0 < r < a; \quad n(z, r) = n_2; \quad r \geq a; \quad \forall z. \quad (15.2)$$

In (15.2), n_1 and n_2 are the refractive indexes respectively at the core and cladding before exposure, a is the core radius, L the grating length, $\Delta n_{\text{mean}}(F)$ the averaged change in index over the BG length, $\Delta n_{\text{mod}}(F)$ the amplitude of modulation responsible for the grating reflectivity to first order ($\lambda \approx \lambda_B$). As long as the measurement is restricted to the first-order spectrum, the series expansion is limited to the first two terms. In (15.2), (F) is for the exposure conditions such

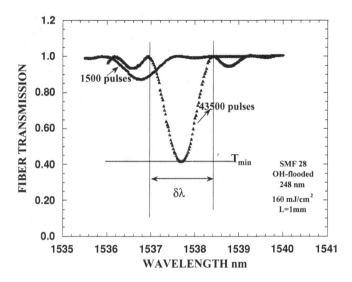

FIGURE 15.2. Samples of BG spectra during the inscription of 1-mm-long BG in an OH-flooded standard telecommunication fiber (SMF 28 Corning) by means of a KrF laser beam.

as λ_p or the cumulated fluence F_t. Standard formulas (15.3) can then be used to calculate Δn_{mod} and Δn_{mean} from the grating reflectivity R_{max} at $\lambda = \lambda_B$ (or its corresponding spectral width $\delta\lambda$) and the exposure-induced shift $\Delta\lambda$ in λ_B [49]. Figure 15.2 shows samples of BG transmission spectra and makes the above-mentioned quantities concrete.

$$R_{\text{max}} = \tanh^2\left(\frac{\pi\eta\Delta n_{\text{mod}} L}{\lambda_B}\right) = 1 - T_{\text{min}};$$

$$\delta\lambda = \frac{\lambda^2}{\pi n_{\text{eff}} L}\sqrt{\pi^2 + \left(\frac{\pi\eta\Delta n_{\text{mod}} L}{\lambda}\right)^2}, \tag{15.3}$$

$$\Delta\lambda = \frac{\lambda_B \Delta n_{\text{eff}}}{n_{\text{eff}}}; \quad \Delta n_{\text{eff}} \approx \eta\Delta n_{\text{mean}}.$$

In (15.3), n_{eff} is the effective index experienced by the fundamental mode. Assuming further that the parameters of the fiber do not change with exposure (or annealing) time, one can use (15.4) to calculate η, the fraction of the total optical power propagating along the core [83]:

$$\eta(V) = 1 - \frac{u^2}{V^2}\left[1 - \frac{K_0^2(v)}{K_1^2(v)}\right]; \quad v \approx 1.1428V - 0.996; \quad V = \frac{2\pi a\sqrt{n_1^2 - n_2^2}}{\lambda} \tag{15.4}$$

In (15.4), K_0, K_1, and V are modified Bessel functions of orders 0, 1 and the normalized frequency of the fiber, respectively [83].

In fact, accurately determining Δn_{mod} and Δn_{mean} from (15.3) requires some care in the design of the experiment and the analysis of the data. Firstly, one has to choose the BG length so as to avoid any saturation of the spectrum during the inscription and thereby to be in position to measure changes in R and λ_B to within a typical accuracy of 0.01 and 20 pm respectively. Secondly, as can be deduced from (15.3), one has to measure the actual length of the modulation along the fiber. An approach to solving this issue consists in using the side-diffraction technique in an improved version that allows the grating uniformity to be checked with a sensitivity of $\approx 10^{-6}$ in terms of modulation at a spatial resolution of ≈ 10 μm [84–87]. Thirdly, the hypothesis according to which η remains constant during exposure or annealing time is questionable each time Δn_{mean} becomes significant in front of $(n_1\text{-}n_2)$. In that case, as shown in [42], an accurate determination of both Δn_{mod} and Δn_{mean} requires the use of an iterative calculation. Fourthly, writing uniform BG with a view to establishing a relation between $n(z)$ and the local optical power density raises a more serious difficulty. Indeed, since Λ is usually lower than the spatial resolution of the optical detector, one can measure an averaged optical power only along the grating length. The details of the distribution of optical power along the fiber axis can hardly be deduced from this measurement, since the visibility of the fringe pattern remains unknown. This is especially true for the case that the fringe pattern is produced by means of a phase-mask, since the self-imaging Talbot effect may lead to a complicated nonsinusoidal distribution of power behind the mask [88–91].

In conclusion, the BG inscription-based method for measuring index change gives only a way to connect **effective values of Δn_{mod} and Δn_{mean} to mean optical power density ones**. An obvious procedure for overcoming this difficulty consists in using any one among the above-mentioned interferometer-based measurement methods in which an optical path is **uniformly** exposed to laser light. Yet the question of how to connect the data from the interferometer-based measurements to those deduced from inscription of BG remains a rather tricky issue [60, 92].

15.3 Examples of Methods Used for Increasing the Photosensitivity of V-Silica-Based Glass or Fiber to UV Laser Illumination

Pure silica glasses exhibit a poor photosensitivity to UV-laser light (for example, $\Delta n \approx 3\ 10^{-4}$ at accumulated fluence $F_t \approx 30\text{–}140$ kJ/cm^2; $\lambda_p = 193$ nm or 157 nm [93, 98]). Similar poor photosensitivity ($\Delta n_{mod} \approx 2 \times 10^{-5}$, $F_t \approx 100$ kJ/cm^2) is observed when a standard fiber is exposed to pump light at $\lambda_p > 193$ nm. Nevertheless, using uncommon exposure conditions such as a high fluence per pulse (650 mJ/cm^2) at a wavelength of 193 nm [100] or light at 157 nm yields an index change Δn of $\approx 1\text{–}5\ 10^{-3}$ ($F_t \approx 30$ kJ/cm^2) in this type of fiber (see Section 15.4) [103–104]. In contrast to what is observed with most UV lasers, exposing a pure silica sample to a femtosecond ($\approx$100–200 fs) laser beam at $\approx$800 nm yields strong permanent changes in index ($\approx$2–6 10^{-3}) [56, 102, 105]. Yet, up to now, the use of this type of laser for writing devices is only at an early stage of development. Therefore, for most applications, using UV lasers with a view to inducing index change of order a few 10^{-3} after a reasonable exposure time ($\Delta t \approx 15$ min) remains a key for fabricating most devices in silica-based samples. This, in turn, requires the glass to be photosensitized.

There is no unified theory of how to increase the photosensitivity of glass, because it depends on the chemical composition of the glass. However, for this purpose, researchers have followed two main strategies that are not exclusive. The first one consists in using *hydrogen treatment*, the second one in directly *manufacturing glass with intrinsic high photosensitivity*. Since the topics of photosensitivity enhancement have been recently reviewed in [107], we restrict the content of this section to a comparison between the efficiencies of some hydrogen treatments and to the presentation of some generalities about the manufacturing of highly photosensitive glass [108].

Two definitions can be used equally when referring to the photosensitivity of a glass: either *the change in index induced after exposure of the glass to a reasonable given accumulated fluence* or *the ratio of the index change to the cumulated fluence (i.e., the slope at the origin)*. The first definition understands that the fluence is chosen so that it corresponds to typical conditions of BGs inscription, whereas in the second definition, the value of the index change usually remains low ($\approx 10^{-4}$) to avoid any saturation phenomena. The methods described in the following usually enhance both the initial kinetics of index change and the change for a long exposure time. Accordingly, we do not distinguish further between the two definitions.

Hydrogen Treatments

The *high-pressure/low-temperature H$_2$loading* method consists in placing the sample in a high-pressure hydrogen (or deuterium) atmosphere (40–700 bars) at $\approx$ room temperature for a time long enough to allow for a nearly complete diffusion of the gas into the glass [109]. Up to now, this treatment seems to be the most efficient method for increasing the photosensitivity of a germano-silicate glasses. Examples of glass for which H_2 loading leads to a large photosensitivity and the corresponding values of the index changes can be found in [52, 107–108]. The highest changes ($\Delta n \approx 10^{-2}$) have been obtained with the Ge or N-doped silica glasses. On the other hand, the H_2 loading technique is also known to add some issues to the BG-based device technology. The first issue comes from unavoidable hydrogen out-diffusion during the writing. This phenomenon leads to drawbacks such as shift in the Bragg wavelength of LPG [110] or change in the photosensitivity of a planar waveguide with the writing time [111–112] (because unlike fibers, planar waveguides may have thin overcladding that favors rapid H_2 out-diffusion). Other common issues are a low thermal stability of the index change at moderate temperature ($\leq$200°C) [113] and UV-induced hydroxyl species formation [109]. The formation of OH species leads to a quenching of the luminescence of RE doping ions inserted in a laser host [113] and to a significant extra contribution to the propagation loss in the third optical window of optical communication ($\lambda \approx 1.53$ μm). An expensive solution for solving this last issue consists in using deuterium in place of hydrogen with a view to moving the OH species-related absorption peak from ≈ 1.39 μm to ≈ 1.9 μm [115].

Photolytic hypersensitization of H_2-loaded silica glasses and *low-temperature hypersensitization* of phospho-silicate glass are efficient means of enhancing the photosensitivity without incurring some disadvantages associated with the H_2 loading technique [111, 116–119]. *Photolytic hypersensitization* consists in using a uniform preexposure of a H_2-loaded sample for obtaining a permanent and controllable increase in the photosensitivity even after hydrogen out-diffusion from the sample [111, 116]. Indeed, since the gain in photosensitivity depends on the sensitization-cumulated fluence, hypersensitization locking offers a powerful tool for tailoring the photosensitivity profile along the fiber axis. *Low-temperature hypersensitization of H_2-loaded phospho-silicate fibers* does not require any post-exposure step: after H_2 out-diffusion, the fiber is permanently sensitized at 193 nm [119]. Following these treatments, stable strong gratings can be written in the glasses with reduced hydrogen-related species formation [112, 118–119].

OH flooding is a post-thermal sensitization of hydrogenated silica fibers using a high-temperature ($\approx 1000°$C) processing either by means of an oven or a CO_2 laser [120–121]. Sensitization remains effective after complete hydrogen out-diffusion [122]. Moreover, the technique has proved its efficiency to enhance the photosensitivity of pure silica fiber [123] and planar germano-silicate waveguides [124].

Figure 15.3 shows a comparison between the efficiencies of the three methods for enhancing the photosensitivity of a standard fiber to light at 193 nm. The H_2 *loading technique* leads to a gain in the photosensitization by a factor of ≈ 3, compared to other methods. Note that in this figure, the grating growth data have been corrected to account for the exposure- and sensitization-induced changes in η (see Section 15.2). On the other hand, it is worth noticing that the OH *flooding* technique may lead to a high increase in the core mean index ($\Delta n_{mean} \approx 4\ 10^{-3}$) and thus to a change of the single-mode fiber into a two-mode fiber.

Direct Fabrication of Glass with Intrinsic High Photosensitivity

For the purpose of manufacturing glass with intrinsic high photosensitivity, researchers have used different methods such as changing the concentration and the nature of the dopants in the silica glass and the technology involved in glass fabrication. To date, GeO_2 remains the most prominent dopant for highly photosensitive silica glass. One mechanism of photosensitivity of GeO_2-SiO_2 glass at $\lambda_p \approx 240$ nm is associated with the GODC-related absorption band centered at ≈ 242 nm (see Section 15.5). The intensity of this band is known to increase linearly with the concentration of GeO_2 under similar manufacturing parameters,

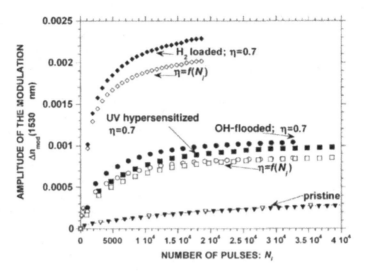

FIGURE 15.3. Growth of the refractive index modulation during the inscription of BG within sensitized and not-sensitized standard telecommunication fibers (SMF 28 Corning). The pump laser is an ArF laser (193 nm, 160 mJ/cm^2). The diamonds are for an H$_2$-loaded fiber, the squares for a UV hypersensitized fiber (2000 UV pulses and out-gassing), the circles for an OH-flooded fiber (1 s at 950°C and out-gassing). The triangles represent the growth in a pristine fiber. The full symbols correspond to data calculated at a fixed value of η, whereas the data in empty symbols were obtained through an iterative method to allow for the variation of η with treatment and exposure time.

since it is sensitive to oxydo-reducing conditions [125]. Accordingly, a first solution for making glass with intrinsic high photosensitivity could consist in doping the glass at a high concentration of germanium (≥ 10 mole %). Among the techniques used to deposit glass layers, PECVD allows accurate control of the glass composition. When implemented with hydride precursors such as silane (SiH$_4$) or tetra-ethoxy-silane (TEOS, Si(OEt)$_4$), the PECVD method can give rise to highly doped photosensitive glass layers ($\Delta n \approx 5\ 10^{-3}$, $F_t = 7$ kJ/cm^2, $\lambda_p = 193$ nm) [126–129]. Yet increasing the concentration of germanium in the core of a single-mode waveguide results in a reduction of the mode size and additional loss, for example when the device is spliced to a standard telecommunication fiber. An approach for solving this issue might consist in codoping the GeO$_2$-doped glass. Dopant such as boron leads to a decrease in the core index and thus allows the manufacturer to incorporate more GeO$_2$ while keeping constant the core-cladding index difference. As further advantages, the addition of boron to the core fiber enhances the UV-induced index change [130]. As a result, saturated modulation as large as $\approx 3.7\ 10^{-3}$ ($F_t = 15$ kJ/cm^2, $\lambda_p = 248$ nm) has been reported for gratings written in B-codoped germano-silicate thin films fabricated through the TEOS-PECVD technique [129]. Yet gratings in boron-codoped waveguides have much poorer high-temperature stability, compared to BG in not-codoped devices. Furthermore, B codoping and (or) use of conventional PECVD-forming

N-H species (N_2O is the oxidizing agent) can also give rise to excess loss at a wavelength near 1.5 μm. Accordingly, for the purpose of overcoming these difficulties, other techniques have been used, consisting either in doping silicate or in codoping germano-silicate (or phospho-silicate) glass with other dopants. For example, SnO_2, incorporated in silica glass [131], in phospho-silicate [132] or germano-silicate [133] fibers as a dopant through the MCVD process leads to photoinduced index changes ranging from $3 \ 10^{-4}$ to 1.4×10^{-3} after exposure to ≈ 0.6–20 kJ/cm^2 at 248-nm light. Similarly, BGs have been written in optical fibers with SiO_2 : SnO_2 : Na_2O cores and SiO_2 cladding that were produced by the rod-in-tube technique ($\Delta n \approx 6.2 \ 10^{-4}$, $F_t \approx 30$ kJ/cm^2, $\lambda_p = 248$ nm) [134]. Hydrogen-free silicon oxynitride technology has been implemented through the SPCVD (surface plasma chemical vapor deposition) process to incorporate nitrogen into silica or germano-silicate glass, leading to a new photosensitive material for fiber and waveguide optics. The 193-nm light-induced change in index reached $\approx 10^{-3}$ [135–136]. It is worth noticing that averaged negative index changes of up to -2×10^{-3} have been induced by writing a BG in planar germanium-free $N:SiO_2$ waveguides using light at 248 nm [137]. Similarly, following uniform exposure with light at 193 nm, large negative index changes have been measured in hydrogen-free GeO_2-SiO_2 thin glass film prepared either by the HARE process ($\Delta n = -0.006$, [138]) or by the conventional rf-sputtering method ($\Delta n = -0.14$, [139]).

In conclusion, UV-induced changes in index larger than 10^{-3} can be induced in most silica glass provided that one uses an appropriate photosensitization method or a glass with intrinsic photosensitivity. Each method must be investigated to account for the specifications of the photo-written device.

15.4 Factors on Which Depends the Photosensitive Response of Silica-Based Glasses

Most data displayed in the literature correspond to photoinduced changes in refractive index at a wavelength λ_{use} of ≈ 1.5 μm. In fact, the index changes depend on other numerous variables *that may be coupled*, adding to the complexity of studies dealing with the photosensitivity of glass. Schematically, the law of dependence can be expressed through the symbolic form

$$\overline{\overline{\Delta n}} = f(material, \ conditions \ of \ exposure, \ photosensitization, \ condition \ of \ use).$$

$$(15.5)$$

In (15.5), $\overline{\overline{\Delta n}}$ is the photoinduced change in the refractive index tensor. The word *material* includes factors such as the chemical composition of the glass, the elaboration process (governing the initial concentration of defects and the absorption coefficient at the pump wavelength), the physical form of the glass (fiber, preform plate, waveguide, thin film or bulk), and its thermal and mechanical history. The expression "*conditions of exposure*" represents factors such as the characteristics of the pump laser beam at the exposed sample, exposure time, elastic stress, and the temperature $T_{writing}$ of the sample during the exposure. The expression "*conditions of use*" represents parameters including the wavelength λ_{use} at which one takes advantage of the change in index, the temperature T_{use}, and time t_{use} at which the index change-based device is used and other conditions such as the electromagnetic or radioactive environment. T_{use} and t_{use} govern the spontaneous decay of the index change or its decay in the course of an accelerated aging experiment. For the purpose of illustrating the complexity of this multiple-variable dependence, in the following, we present some examples of how factors such as exposure time, λ_{use}, and the type of laser can affect the photoinduced index changes in *germano-silicate glasses or fibers*.

Types of UV-Photosensitivity According to Kinetics of Index Changes

Let us consider UV-induced inscriptions of BG within different optical germano-silicate fibers. A number of BG types have been distinguished, characterized by markedly different spectral and thermal behaviors. During a typical inscription, both the strength of the grating and the Bragg wavelength (BW) increase monotonically with exposure time. The last behavior is indicative of the induction of a positive change in the effective index of the fiber fundamental mode. This type of grating, referred to as a *type I BG*, is the most commonly fabricated BG and proves to be the least stable at high temperature of the type reported to date. With further exposure time, the type I grating reflectivity saturates, then decreases almost to zero before increasing again and eventually saturating [140]. At the time of the second spectrum formation, the BW slightly decreases, indicating a negative effective index change [140–143]. This behavior is shown in Figure 15.4, where both the reflectivity and the BW of BGs are plotted versus exposure time in the course of their inscriptions within high-NA germano-silicate pristine or annealed fibers. The grating formed during the second growth is much more stable than the type I grating [40]. These features have prompted us to call these gratings *type II A* gratings or negative gratings [140–141] to distinguish them from damage *type II* gratings written by one pulse at a high fluence (≈ 1 J/cm^2 [144]). The difference between the rates of formation of type II A and type I gratings depends on numerous parameters. For example, the influence of the parameter "glass thermal history" is illustrated in Figure 15.4 where the growth of the type II A grating in the annealed fiber looks significantly faster than that in the pristine counterpart. Other examples are displayed in [40, 145–147]. Although the formation of a type II A grating is likely correlated to that of a nonuniform index change across a plane perpendicular to the fiber axis [148], the underlying mechanism remains still not fully clarified. Up to now, the formation of type IIA gratings has been reported only in nonhydrogenated germano-silicate or oxynitride fibers or planar waveguides [99, 135, 149, 143].

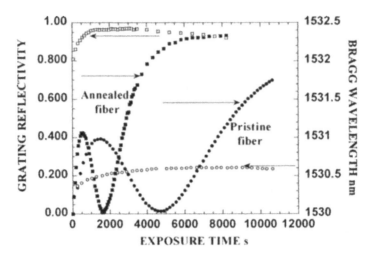

FIGURE 15.4. Growth of a type II A grating in high NA fibers (Ge; 28 mole %). Circles are for the inscription in a pristine fiber, squares represent the writing in an annealed fiber (3 hours at 1000 °C). Full and empty symbols represent the evolutions of the modulation and the Bragg wavelength respectively.

In B-Ge codoped (and in standard) hydrogenated fibers, significant behavior has been reported, with abnormal spectral evolution of regenerated BG following the erasure of the type I grating [150]. It contrasts with that of a type II A BG by exhibiting a large increase in the BW with increasing exposure [150]. Since the temperature sensitivity of this type of BG after annealing was lower than that of a type I grating, this new type of grating was designated *type I A*.

Spectral Dependence of the UV-Induced Index Changes: $\Delta n_{mod}(\lambda_{use})$

In the first report on the spectral dependence of the UV-induced index change, the change was detected by uniformly exposing one arm of an unbalanced Mach–Zehnder (MZ) fiber interferometer with 248-nm UV light [66]. From the measurement of the change in the spectral response of the MZ interferometer, the increase in the nonhydrogenated fiber core index was determined over the range 700 nm–1400 nm. It was found that the increase has an almost constant value over the analyzed spectral range [66]. This work was recently extended over the range 300 nm–2600 nm through the measurement of the refractive index spectral dependence of bulk N- or P- or Ge-doped silica prisms before and after exposure with either 193-nm or 248-nm radiation [55]. Figure 15.5 shows the spectral dependence of the UV exposure-induced refractive index changes in different samples (the P- or Ge-doped samples had been hydrogenated (140 atm) for different times, whereas the N-doped silica prism was nonhydrogenated). The interpretation of the plots is made complicated by the nonuniform absorption of the UV light

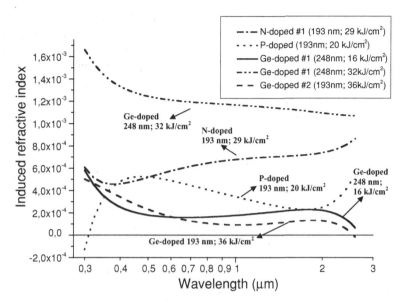

FIGURE 15.5. Spectral dependence of UV-induced changes of refractive index in N-doped, P-doped, or Ge-doped fibers. The P-doped and Ge-doped fibers were H_2 loaded at pressures that differ from sample to sample. The inscriptions have been carried out by means of either a KrF or an ArF laser (Courtesy of Dr. O. Butov and Prof. K.M. Golant).

across the prisms, the H_2 out-diffusion during exposure of P-doped or Ge-doped samples, and the dependence of the change on the conditions of process [55]. Yet it appears that the UV-induced spectra in the range 300 nm–2600 nm are not monotonic with wavelength in specific cases. On the other hand, the spectral response of the UV-induced change in index of the germano-silicate glasses looks flat over a large spectral range including the 1500-nm telecommunication window. The spectra look qualitatively different for different glass compositions and even for different dopant concentrations. From these features, it can be concluded that different electron and vibrational components of polarizability contribute to the overall index change, depending on the glass composition [55].

Dependence of the Index Changes in Germano-silicate Glasses on the Type of Pump Laser: $\Delta n_{mod}(laser)$

For a long time, photosensitivity of *nonhydrogenated germano-silicate* glasses has been associated with the UV-induced bleaching of the GODC band absorbing near 244 nm (see Section 15.5). The existence of this one-photon pathway is supported by the observation that the spectrum of the Ge-doped fiber type I photosensitive response mimics the excitation spectrum corresponding to the triplet state of the GODC defects [151]. Yet shorter-wavelength laser sources offer access to new absorption channels and thus alternative routes to writing

BG [152]. Indeed, provided one uses high-intensity 193-nm light from an ArF laser, strong gratings can be written not only in standard fibers [100] but also in GODC-free germanium-doped optical fiber [153]. Although residual one-photon absorption plays a role in explaining the details of the index change kinetics in the standard fiber, two-photon absorption across the glass band gap was the inferred prime cause of as large as $\approx 10^{-3}$ index change [100]. On the other hand, the F_2 excimer laser produces 7.9 eV photons that directly bridge the 7.6-eV absorption of low-GeO_2-doped silica glasses (5 mol%) and the tail of SiOH absorption [101, 156]. This allows access to strong single-photon-related photosensitivity in germano-silicate waveguides or in standard fibers, without the need for traditional enhancement techniques. Thus, exposing low-Ge-doped planar waveguides (3mol %) to light at 157 nm can induce change of index of $\approx 4 \times 10^{-3}$ [104]. Nevertheless, due to the high linear absorption coefficient of the GeO_2-SiO_2 glass at 157 nm, the shallow depth of penetration of the UV light (≈ 4 μm) into the doped glass leads to nonuniformity of the index change and thus to strong UV-induced form birefringence to the extent that the core of a waveguide is exposed to the radiation at 157 nm [104]. Furthermore, the lack of available phase-mask operating at 157 nm hampers the use of this radiation for writing optical fiber BG.

Fabrication techniques involving UV-photosensitivity are, for the most part, limited to the production of one- or two-dimensional structures [56]. The pulse duration of a femtosecond laser is on the order of ≈ 100 femtoseconds, making it easy to produce high peak-power densities such as ≈ 100 TW/cm^2 after focusing at the laser spot. As a result, the development of this type of near-infrared ($\lambda_p \approx 820$ nm) laser has prompted the investigation of multiphoton reaction-induced photosensitivity of glass that does not linearly absorb efficiently at the laser wavelength. Accordingly, it results in highly localized laser-induced effects including ablation, photo-structural changes, and consequently changes in index and birefringence. Since the process exhibits a highly nonlinear dependence on the laser power density, it allows for the fabrication of three-dimensional photonic devices. As a result, the focusing conditions have to be carefully adjusted so as to avoid damages within the glass and to keep the exposure-induced propagation loss at a low level [96, 104]. By so doing, good-quality uniform devices can be written in various glasses including silica and Ge-doped fibers [94] without the need for any photosensitization process.

The dependence of the index change on the pump wavelength for exposed H_2-*loaded germano-silicate* glass looks different from that for the nonhydrogenated counterpart. Indeed, large absorption changes in Ge-doped glass that has no initial absorption GODC band peaking at ≈ 242 nm can be produced by exposing the H_2-loaded sample to 248-nm as well as 193-nm light [152]. The kinetics of index change is merely faster at the beginning of the exposure and exhibits faster saturation for longer exposure time when the exposure is performed at 193 nm. Changes in index at saturation reach very similar values [99]. The photosensitivity (photosensitivity defined as the reciprocal of the accumulated fluence required to reach an index change of 3×10^{-4} in a 10 mol% Ge-doped fiber) increases almost exponentially with photon energy in the range 3.5 eV–5.4eV [151]. Similarly, a rapid LPG formation in a H_2-loaded standard fiber can be produced with 157-nm laser radiation [154]. The fluence (5 J/cm^2) required to write a 20-dB attenuation peak near 1590 nm with the 157-nm laser proves to be 250 times smaller in comparison with that with the 248-nm laser exposure [154]. The use of 157-nm radiation also offers further possibilities. For example, uniform preexposure of H_2-loaded Ge-doped glass with light at 157 nm at an accumulated fluence around 200 J/cm^2 leads to a permanent locking of the photosensitivity at 248 nm [155]. Thus strong and stable BG ($\Delta n_{mod} = 6 \ 10^{-4}$) can be written by exposing a 157-nm hypersensitized out-gazed standard fiber with a fringe pattern at 248 nm [157]. Only a slightly better modulation of 8×10^{-4} (yet at the expense of a lower stability) could be generated in the H_2-loaded fiber directly exposed to a fringe pattern at 248 nm from the same setup [155].

In conclusion, the fact that the photosensitivity of glasses depends on numerous correlated variables makes the statement of general rules for inscription difficult. This arises from quite different mechanisms leading to refractive index change. Obviously, one has to establish laws for peculiar experimental conditions. Whatever they are, it appears that the use of a femtosecond laser will provide a powerful tool in the future to write uniform devices including 3D structures without the need for any photosensitization process. The use of UV lasers in association with sensitization or hypersensitization processes looks more flexible for writing complex devices that need tailored index changes. Moreover, the shorter the UV pump wavelength, the faster the kinetics of index change.

15.5 Mechanisms of Photosensitivity in Germano-Silicate Glasses

The word "photosensitivity" by itself implies that the first step, at least, of the mechanisms involves the absorption of photons through either a one- or a two-photon process, depending on several parameters including UV-light power density and the glass attenuation at the pump wavelength. The absorption is followed by a noticeable modification of the absorption spectrum, i.e., growth of absorption bands and bleaching of others. This is indicative of migration and retrapping of electrons extracted from ionized atoms and/or ions. Light can also induce transport of atoms or ions leading to change in the glass structure and stress field. These processes are more or less efficient according to the chemical composition of the material, the treatment for increasing the photosensitivity, and the laser wavelength.

Several models have been proposed for explaining the glass photosensitivity in optical fibers and planar waveguides. In fact, they are made up of elementary mechanisms that never work alone but rather mix together, each explaining a part of the photoinduced index change only. Here, we do not have the space to discuss in detail all the actual processes according to the factors influencing the photosensitivity (see Section 15.4). Thus, after the presentation of the main elementary processes involved in the photosensitivity of germano-silicate glasses, we detail the mechanisms put forward for explaining the 5-eV UV-light-induced refractive index change in H_2-loaded Ge-doped SiO_2: the most important one from a practical point of view.

The color center model: Usually, UV light induces the bleaching of bands attributed to different GODCs (called GODC(II, A) in equation (15.6); "A" means that the defect contributes to the absorption spectrum) that absorb at 240–260 nm ($\approx$5 eV). After photon absorption, the excited defects change into defects (color centers called GODC(III, A) in equation (15.6)) that absorb at each side of the 5 eV absorption band (the structure of these defects is still an open question; see discussion in [158]). Color centers can also be produced through an intrinsic non-defect channel by exposing the nonhydrogenated glass to dense UV photon flux. The resulting two-photon-mediated band-to-band excitation generates pairs of electrons and holes that migrate to trapping centers forming defects such as GEC (germanium electron center, also called Ge(1)) and Ge(2) (structure not generally admitted) [159–160]. The defects absorb in the UV range. These changes in the absorption spectrum imply a change in the refractive index on account of the principle that both represent the response of the material to the electromagnetic wave. Thus, the knowledge of one allows the other to be computed using the Kramers–Kronig transformation [16]. First proposed as being the mechanism explaining the photoinduced refractive index change [161], the color center model appeared to give a weak contribution to the index modulation when silica glasses are not treated with hydrogen [162–164].

The permanent electric dipole model: This model is also based on the photoionization of bonds. The breaking of bonds by photon absorption leads to the formation of excited charges and ions. After the migration of the excited charge over a few interatomic distances, its retrapping can result in an electric dipole [29, 165]. The electrostatic field created therefore modifies the refractive index by the Kerr effect. Although the model predicts weak effects, it has been used for explaining some photoinduced birefringence [166].

The organized electron migration model: At the beginning of research on photosensitivity, the classical model of photorefractivity working in crystals like BSO and BGO was also proposed but disregarded rapidly (see Section 15.1). Indeed, in contrast to experimental observations, the model predicts that the grating period is two times smaller than that of the light fringe pattern [165].

The chemical species (ions or atoms) migration model: Laser irradiation can trigger the direct migration of chemical species [167–168], such as germanium [169]. This leads to the notion of chemical BG. Migration of hydrogen [170] or fluorine [171] has also been indirectly revealed through high-temperature annealing of BG written in H_2-loaded fibers. The ionic migration leads to a change in the distribution of dopant concentration in the fiber core and thus contributes to a refractive index change.

The stress field relaxation model: Bond-breaking is a way of excitation relaxation. This process can induce a stress relaxation [172]. To the contrary, after this process, a bond reconstruction can induce deformation of the glass structure and lead to a new stress field. These processes contribute to refractive index change through the photoelastic effect. It has been shown that UV-induced relaxation of a high-tensile stress in the core of Sn-Ge codoped optical fibers can lead to a change in the mean refractive index as large as 10^{-3} [173]. In contrast, in the case of nonhydrogenated noncodoped germano-silicate fibers, an increase in the axial stress has been observed [174–176].

The densification model: UV-induced densification is a mechanism that accounts for the axial stress increase by assuming the production of a local dense region after photon absorption. Although one can find some specific counterexamples, it is a common feature of silica glasses to densify when they are irradiated with various sources of electrons, ions, or photons including gamma and X-ray. This corresponds to changes in the glass structure [177–178] such as tetrahedron rotation, ordering, and coordinence change. We have analyzed and quantified this mechanism with regard to the elastic stress field associated with a heterogeneous densification [176]. Comparison of the results from the model with those deduced from the UV-exposure of nonhydrogenated Ge-doped glasses allows us to think that densification is in this case the dominant process [176, 179]. Finally, let us mention that this mechanism and thus kinetics are stress-dependent and self-limited [41, 164].

An example of a actual mechanism: *244 nm photoinduced refractive index change in* H_2*-loaded* Ge-*doped* SiO_2

As previously mentioned, the actual process is in fact a composition of several reactions. For the purpose of illustrating this point, we describe the state of our knowledge concerning the process now used worldwide for elaborating the Bragg gratings, i.e, the refractive index change when Ge-doped silica glasses hydrogen loaded are irradiated around 5 eV with a laser. Although other possibilities can be hypothesized (for example, see [180]), in [181] starting from a comprehensive survey of the literature, we have suggested a mechanism that is summarized by the following set of equations:

$$
ODC(II, A) \xrightarrow{1h\nu} ODC(II, A)^* \xrightarrow{1h\nu}
\begin{cases}
\xrightarrow{} ODC(III, A) + e' \\
Ge + e' \underset{1h\nu}{\overset{heat}{\rightleftharpoons}} Ge(1) \\
precursor + e' \underset{1h\nu}{\overset{heat}{\rightleftharpoons}} Ge(2) \xrightarrow{H_2, h\nu, fast} \\
\xrightarrow{slow} GeE' + e'
\end{cases}
\xrightarrow{1h\nu} ODC(II, A)
$$

$$Ge(2) + H_2 \xrightarrow{h\nu} GeH \, or \, GeH_2 \, or \, H(II) \xrightarrow{2h\nu} GeE' + H_2$$

$$TOR + 1/2 H_2 \xrightarrow{1h\nu} TOH, H_2O, RH \tag{15.6}$$

Equation (15.6) can be divided into three parts. The first one (at the top of the group) consists of a group of triggering equations that show that the absorption of photons constitutes the trigger of the mechanism. The second part displays serial reactions representing the propagation of the reaction catalyzed by hydrogen. The third part represents a nondefect channel one-photon reaction, working as an ending reaction partially.

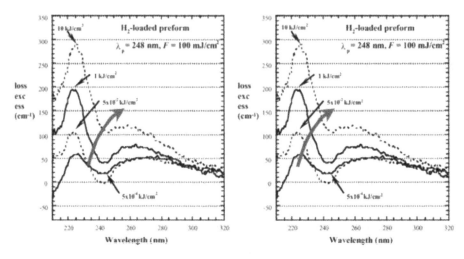

FIGURE 15.6. UV-photochromism of a H_2-loaded Ge-doped fiber preform core when exposed to a 248-nm pulsed laser (on the left) or a 244-nm CW laser (on the right). After Leconte [99].

Now let us detail each reaction more deeply to deduce how it affects the total change in refractive index.

The first group starts with a one-photon absorption by the GODC(II, A) defects. After absorbing another photon, the excited GODC(II, A) defect can turn either into a GeE' center or into a positively charged GODC(III, A) defect by releasing an electron. Then, electrons produced by the photoionization of GODCs can be trapped by Ge-atom-containing centers to form GEC paramagnetic defects such as Ge(1) and Ge(2). When the pump source is a laser, this last pathway proves to be more efficient than the first one and is reversible, giving back GODC(II, A) from GODC(III, A). Since GODC(II, A) absorbs around 5eV, the corresponding band is progressively bleached, whereas excess loss at 4.5 eV and 5.8 eV is created due to the increase in the concentration of GODC(III, A) defects (Figure 15.6). The ratio of the concentration GODC(III, A) over GODC(II, A) is thus linearly dependent on the power density, as observed in [182]. On the other hand, a small amount of GeE' is created, as can be observed by the growth of paramagnetic signal as a result of the irradiation [159]. These changes in the absorption spectrum account for a change in index, as we describe in the above-mentioned color center model. However, the equilibrium in the population of defects is rapidly established compared to the other part of the mechanism. Accordingly, once the glass is exposed to a fringe pattern, the photoinduced formation of these defects gives rise to a strong mean index but weak index modulation. We can also note that one of the photoinduced paramagnetic defects seems to play a specific role: Ge(2). It is observed to disappear when the glass is hydrogenated [183–184]. We suggest that Ge(2) changes into a hydrogen-containing defect under interaction with the hydrogen dissolved in the glass (Figure 15.7).

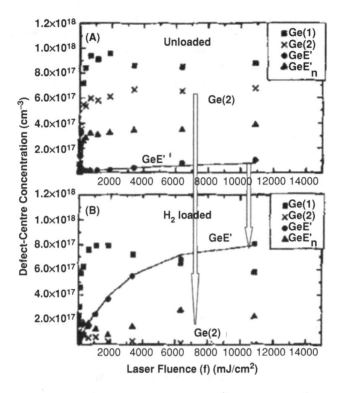

FIGURE 15.7. Paramagnetic defect-center concentration versus UV photon accumulated fluence (0–11 J/cm^2) in (A) unloaded and (B) H$_2$-loaded Ge-SiO$_2$ fibers [183] (courtesy of Dr. J. Friebele). From this experiment we suggest that Ge(2) produced without hydrogen is transformed further through reaction with hydrogen [183] as indicated by the arrows.

The second part of the reaction is composed of two successive reactions. The first is the Ge(2) reaction, with hydrogen under UV absorption leading to a hydrogen-containing species [184–185]. Then, under two-photon absorption, these species change into GeE' centers (defects that absorb near 6.3 eV [160]), releasing hydrogen. This part of the mechanism leads to several features: a dramatic increase in the GeE' concentration and thus a dramatic increase in the refractive index at the beginning of the exposure, as long as GODC(II, A) is not exhausted. Yet for longer exposure time when the absorption by GODC is completely bleached, both the 6.3-eV band (GeE') and the IR band at 2184 cm^{-1} (ascribed to GeH) keep on growing, indicating a second source of hydride species. It is worth noticing that the transformation of GeH into Ge(E') does not require free hydrogen to be released. This implies that it is possible to interrupt the reaction before the end, when the GeH concentration is at a peak. Thus, after H$_2$ out-gassing, it is possible to write BG without hydrogen. This is at the root of the glass hypersensitization process (see Section 15.3) [111].

Finally, the third part of the mechanism is a reaction corresponding to trapping of hydrogen into the glass host either into hydroxyl or hydride species. The first trap corresponds to a termination reaction, the second one to an additional source of GeH. Whereas the index growth supported by GeE' production seems to be limited by the germanium content only, the OH content obviously corresponds to the hydrogen content present at the moment of the reaction. The glass structure transformation due to achievement of new linkages and increase of the edge absorption is here the support of the refractive index change.

In conclusion, the example given above shows that the mechanism implies several elementary processes such as color centers transformation, electron migration, and stress relaxation. The mechanism can be hypothesized only from a large data set of experimental results coming from various investigation methods. At the same time, it presents interesting features leading to a hypersensitization process, water content mastering, and laser power density optimization. It is worth noticing that most H-bearing species are less stable than the GeE' one but able to be more rapidly produced. The optimization depends therefore on the type of targeted application.

15.6 Stability of Photoinduced Changes in Refractive Index

A full knowledge of the stability of photoinduced change in refractive index has both scientific and practical interest. For example, it can help in identifying the species responsible for the index change by looking for any correlation between temperature-induced decays of these species' concentration and that of the index change. On the other hand, most applications require long device lifetime and at least the possibility of forecasting possible spontaneous degradation of the photoinduced index change.

Since the topic of this paragraph has been reviewed in recent papers [13–111, 186–190], in the following we present only the principles of the main methods used for predicting the lifetime of a BG or for enhancing the thermal stability of a BG written in a standard telecommunication fiber. We also discuss some issues linked to the practical implementation of these methods. For a complete description of the techniques used to obtain gratings with intrinsic high stability, we refer the reader to references [40, 105, 107, 134, 144, 191–194].

Principle of the Method Used for Predicting the Lifetime of Photoinduced Change in Refractive Index and Problems Linked to This Prediction

In an attempt to predict the stability of the change in refractive index, one usually begins by performing accelerated aging experiments. There exist three methods for carrying out an accelerated test. The first approach consists in holding the

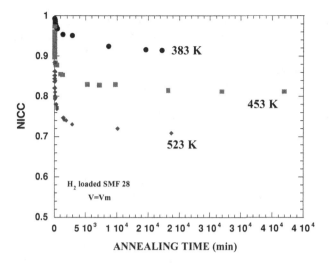

FIGURE 15.8. Isothermal decays of three identical uniform BG written in a H_2-loaded standard telecommunication fiber (SMF28, Corning).

temperatures of gratings at fixed values above room temperature and periodically recording the grating transmission spectrum (*isothermal annealing* experiments) [186]. For sake of illustration, Figure 15.8 displays the isothermal decay of the integrated coupling coefficients $NICC(t, T)$ corresponding to the decay of three gratings at three temperatures (383 K, 453 K, 523 K) [186]:

$$NICC(t, T) = \frac{\tanh^{-1} \sqrt{R(t, T)}}{\tanh^{-1} \sqrt{R(0.296K)}} \qquad (15.7)$$

In (15.7), R is the BG reflectivity, T is the fixed temperature at which the BG has been held for an annealing time t. Provided the BG is uniform and η does not depend on T, $NICC(t, T) \equiv \frac{\Delta n_{mod}(t,T)}{\Delta n_{mod}(0.296K)}$, where $\Delta n_{mod}(0.296$ K) and $\Delta n_{mod}(t, T)$ are the modulation at the beginning of the annealing and after annealing at T for t respectively. Similarly, one can define NBW for the normalized annealing-induced evolution of the photoinduced change in the mean refractive index:

$$NBW(t, T) = \frac{\Delta n_{mean}(t, T)}{\Delta n_{mean}(0.296K)}. \qquad (15.8)$$

NBW can easily be determined from the shifts experienced by the BW in the course of the writing and annealing experiments. In most experiment, both R and λ_B are measured at the temperature T of the isothermal annealing. In the following, we will explain why this practice can sometimes be a source of error.

The *true isochronal annealing* experiment consists in heating gratings at increasing temperature for a fixed time (a new grating is used for each annealing temperature). The *step isochronal annealing* is similar to the *true isochronal annealing* except that the same grating is used for each annealing step. On the condition that specific conditions are fulfilled [195], the step isochronal annealing provides a good approximation for the true isochronal annealing method [195] and is often used rather than this latter method, since it is less time-consuming.

The prediction of the grating lifetime from the raw data of the annealing experiments requires a proper choice of a physically stated model that would lead to a coherent extrapolation of the data beyond the time during which these data have been gathered. Moreover, given a model, one has to be in a position to check some basic statements of the model to get confidence in the predictions. For analyzing and predicting the decay of the BG reflectivity, it is usually assumed that the grating inscription populates a broad distribution of unstable species, those with low activation energy decaying faster than those with a high one [186]. Provided that the decay originates from only one first-order thermally activated reaction (B $\rightarrow$ A), it can be shown that there exists a boundary for the activation energy (the demarcation energy E_d) above which the population remains untouched and below which the species B are changed into A [186, 188]. Assuming further that the distribution is temperature-independent and that it exhibits slopes weak enough compared to $1/k_B T$ [188], one can show that the change in the population of species, after heating the grating at T for a time t, depends only on $E_d = k_B T \ln((k_0 t))$ (k_B is the Boltzmann constant and k_0 is a parameter) [186] [188]. On the other hand, the model assumes that the *refractive index modulation* is proportional to the B concentration [186]. Optimizing the collapse of all the degradation data into a single curve allows k_0 in the demarcation energy expression to be determined and leads to the notion of "master curve" (MC): $NICC(t, T) = f(E_d)$ or $NBW(t, T) = f(E_d)$ [186–187]. Figure 15.9 displays the MC corresponding to the NICC decays shown in Figure 15.8. The MC plot allows the user to predict

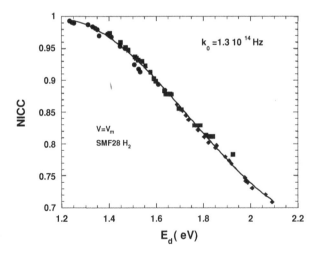

FIGURE 15.9. Normalized integrated coupling constant (NICC) as a function of the demarcation energy $E_d = k_B T \ln(k_0 t)$: master curve in (eV). The solid curve is a fit of the data to a third-order polynomial equation.

the grating lifetime, provided the anticipated conditions (t_{use}, T_{use}) of BG use correspond to a point on the MC that has actually been sampled during the annealing experiment. Furthermore, the shape of the distribution of activation energies can be extracted from the slope of the MC [186].

A key assumption in using these MCs is that the decay in the change of refractive index comes from only one reaction (one value for k_0 in the E_d expression above) [188]. The question of how to check this hypothesis has been recently discussed by one of the authors. To this end, he used an extension of the analysis made by T. Erdogan et al. [186] to a more complex situation [188]. For example, the cases of higher-order reaction, and parallel or serial reactions, have been investigated. Moreover, a link between the writing conditions and the grating stability has been established in the case that the kinetics of writing and bleaching can be described by means of a reversible reaction ($A \underset{Bleaching}{\overset{Writing}{\rightleftarrows}} B$)[195].

The main result of these studies, for our purposes, is the following analysis procedure: to check the hypothesis according to which the bleaching comes from one reaction, one first has to carry out isochronal and isothermal measurements for one quantity, NICC or NBW, and second to plot data versus $k_B T \ln(\Delta t_{isoc})$ for the first set and $k_B T_{isot} \ln(t)$ for the second set. If the shapes of the curves are the same (parallel curves), their collapse into an MC is possible by fitting k_0. The hypothesis of only one reaction is then fulfilled [195]. For the opposite situation, there are several parallel limiting reactions with different k_0 and different distribution. The analysis is thus more complex for achieving lifetime prediction. Comparing the MC deduced from NICC to that from NBW can then allow further conclusions to be deduced. The MCs can be different because the supports (species B) of the modulation and the mean index can eventually be different. For example, photoinduced diffusion of chemical species may give rise to change in the mean index but not to index modulation. In that case, we have to take care about estimation of the lifetime of the grating because the stability of the modulation and the mean index are not necessarily the same, and since the contrast may change on aging, it can hide the process visible on the NICC curve. Another tricky point is that the stability of the local refractive index change can depend on the strength of writing: the stronger the writing, the larger the stability [196]. Therefore, the stability of the UV-induced change in refractive index at the bright UV fringes can be higher than that at the dark fringes. This means that the stability is not uniform along the grating length (oz). The stability will be better at the bright fringe position than at the dark fringe one. The contrast of the grating ($C = \frac{\Delta n_{mod}}{\Delta n_{mean}}$) now changes in the course of the annealing experiment.

A further problem that would occur in analyzing the results of isothermal annealing experiments carried out on the reflectivity of BG written either in a Ge- or N-doped fiber comes from the fact that temperature-induced *reversible* changes in reflectivity are known to be significant in these fibers [197]. Since these reversible changes (or temperature-induced reversible change in λ_B) can spoil the analysis of the isothermal annealing experiments, it proves necessary to correct the raw data to account for these changes by means of relations similar to those established in [197].

Methods Used for Enhancing the Thermal Stability of BG Written in a Standard Telecommunication Fiber

A widespread method used for increasing the stability of a BG after its inscription consists in postannealing the BG for a time t_{anneal} at a temperature T_{anneal} above T_{use}. A coherent rationale of this method is formulated from the above-mentioned model [186]: the treatment wipes out the portion of the index change that would normally decay over the lifetime of the device and keeps only the stable portion of the index change [186]. This method proves to be effective for enhancing the stability of index changes in nonhydrogenated germano-silicate fibers. An alternative method for enhancing the stability of BG reflectivity consists in pre- or post-exposing the BG to uniform UV light [188–189]. The assumption according to which the stability depends on the writing conditions can be put forward to explain this observation as a blanket irradiation that increases the proportion of large-barrier energy sites at the grating place [189]. Yet, due to a large DC level of the mean index induced by the postexposure, the stabilization method increases the thermal stability of the grating reflectivity at the expense of a larger Bragg wavelength shift [190]. The hydrogen treatments (OH flooding, UV hypersensitization, low-temperature hypersensitization) also increase the stability of NICC and NBW compared to those in the pristine counterpart [198]. Yet since the above-mentioned treatments induce a large DC level, the absolute shifts in the BW are also larger in the treated fibers.

In conclusion, the stability of BG can be complex because mainly two quantities, NICC and NBW, define it. NICC and NBW can exhibit different stability if they are based on different decay reactions or if the decay reaction (even if unique) is related to the writing one. Thus, the stability is defined strongly by parameters such as the material composition, the optical power density, the characteristics of the laser used to illuminate the glass, the grating strength, and the contrast of the fringe pattern used for writing a grating. An obvious conclusion of this section is that some care must be taken in analyzing the stability of the UV-induced index change through the decay of a BG. More specifically, it proves necessary to have a good knowledge of the BG initial contrast and to perform accelerated testing on both the Bragg wavelength and the reflectivity. Unfortunately, a review of the scientific literature shows that more often than not, experiments have involved only the study of the reflectivity, making it difficult to draw a definitive conclusion about the validity of the lifetime prediction of the device, since both reflectivity and λ_B lifetime have to be certified.

15.7 Conclusion

The Bragg grating technology in silica waveguides and in related devices has revolutionized the fields of optical communications, fiber lasers, and optical fiber sensors [199]. It also has triggered numerous fundamental studies on the mechanisms of silica-based glass photosensitivity, which in turn have resulted in a better understanding of the phenomenon, leading to the conclusion that the mechanisms of photosensitivity in glass deeply differ from those responsible for photorefractivity in noncentrosymmetric crystals. This chapter has reviewed some pertinent aspects and trends of these 25 years of research on the fundamentals of photosensitivity that are likely to be of interest to the broader community. Thus, some advice given in the chapter (e.g., care should be taken when using a Bragg grating for measuring a refractive index change or when analyzing the stability of these index changes through accelerated aging experiments) are obviously not specific to silica-based glasses and are also available for gratings written in other glasses such as chalcogenide, fluoride, and phosphate glasses. Other aspects of photosensitivity, e.g., sensitivity of gratings to environmental effects such as pressure, elongation, exposure to ionizing radiation, use of gratings for various applications, are well documented in various textbooks and review papers [49, 200, 201] in which the reader can find copious useful information.

The present chapter highlights the role of hydrogen treatments for enhancing the photosensitivity of most silica glasses (except for oxynitride glasses), leading to index changes in excess of 10^{-3}. Although, one has to consider possible deleterious consequences of OH formation that result from the exposure of any H_2-loaded silica glass to UV light, this figure allows grating-based components to be easily fabricated. These sensitization techniques can also be efficiently used for the emerging class of optical silica fibers (e.g., air-silica microstructured fibers) on condition that both ends of the "holey fiber" be hermetically spliced to a standard single-mode fiber to minimize the effect of a rapid hydrogen out-diffusion from the solid core of the fiber to the air holes [202]. It also paves the way to the efficient fabrication of 3D optical band gaps through the photo inscription of Bragg gratings within all-solid photonic band-gap fibers fabricated by means of photosensitive materials.

After gathering useful information about the numerous correlated factors on which depends the photosensitive response of silica-based glasses, the chapter reviews the main basic mechanisms of photosensitivity in germanosilicate glasses. Due to the importance of the H_2-loading techniques for applications, a large part of this review is devoted to studying the specific case of the photosensitivity of H_2-loaded germanosilicate glasses under exposure to light at or around 244 nm since a widely used laser emits at this wavelength. The main conclusion is here on the role of the hydride or hydroxyl species for explaining the enhancement in the sensitized glass photosensitivity. Indeed, as the results of the transformation of the GeH into GeE' defects and the formation of hydroxyl species, the UV exposure leads to a large increase in the absorption in the UV range and thus to a large change in the refractive index as deduced from the Kramers–Kronig relationship. Currently, a great deal of research remains with a view to clarifying the mechanisms of photosensitivity to enlarge the range of applications still further. For example, one can take advantage of a mechanism involving an initial step of multiphoton absorption to tailor specific applications. Indeed, for some glasses, such nonlinearity can allow the UV-induced index modulation in a grating to be larger than the UV intensity modulation at the fringe pattern. This property can be used to enhance the efficiency of the grating inscription through lateral exposure of the sample each time the contrast of the fringe pattern is low, as is the case for microstructured fibers due to scattering by the microstructuration [203].

The remaining part of the chapter is devoted to a review of the methods used for predicting the long-term stability of Bragg gratings. The way to realize this prediction (more specifically, how the measurements should be achieved to be used afterwards for prediction) is clarified here. Nevertheless, the emergence of high-power fiber laser technology addresses the questions of enhancing the stability of standard gratings exposed to higher temperatures than those used in communication systems or to a high optical power density. Recent developments leading to high-temperature performance of Bragg gratings [107, 120, 204, and 205], in combination with advanced packaging technology, suggest that this should not be a problem. For example, the recent demonstration of chemical composition grating (periodic modulation of the chemical composition along the fiber axis) opens a useful path to high-temperature applications of fiber gratings.

List of Abbreviations

UV	Ultra-Violet
FHD	Flame Hydrolysis Deposition
MCVD	Modified Chemical Vapor Deposition
PECVD	Plasma Enhanced Chemical Vapor Deposition
HARE	Helicon Activated Reactive Evaporation
v-silica	vitreous-silica
TEM	Transverse Electric Magnetic
BG	Bragg Grating
GODC	Germanium Oxygen-vacancy Defect Center
LPG	Long Period Grating
F	Fluence
R	Reflectivity
TEOS	Tetra-Ethyl-Ortho-Silane
SPCVD	Surface Plasma Chemical Vapor Deposition
BW	Bragg Wavelength
NA	Numerical Aperture
MZ	Mach–Zehnder
GEC	Germanium Electron Center
NICC	Normalized Integrated Coupling Constant
NI_{mod}	Normalized refractive index modulation
NBW	Normalized mean refractive index

References

1. W.G. French, R.E. Jaeger, J.B. Mac Chesney, S.R. Nagel, K Nassau, A.D. Pearson. In *Optical Fiber Telecommunications*, S.E. Miller, A.G. Chynoweth eds. Academic Press, 233 (1979).

2. C.V. Poulsen, T. Storgaard-Larsen, J. Hübner, O Leistiko. *Proc. SPIE* vol. **2998**, 132 (1997).
3. E.M. Dianov, K.M. Golant, R.R. Khrapko, A.S. Kurkov, A.L Tomashuk, J. Light. *Technolo*. **13**, 1471 (1995).
4. M.V. Bazylenko, M. Gross, A. Simonian, P.L. Chu. *J. Vac. Sci. Technol.* **A14**, 336 (1996).
5. Z.Y. Yin, B.K. Garside. *Appl. Opt.* **21**, 4324 (1982).
6. J. Nishii, H. Yamanaka, H. Hosono, H. Kawazoe. *Appl. Phys. Lett.* **64**, 282 (1994).
7. A. Durandet, R. Boswell, D. Mc Kensie. *Rec. Sci. Instrum.* **66**, 2908 (1995).
8. S. Madden, M. Green, D. Barbier. *Appl. Phys. Lett.* **57**, 2902 (1990).
9. J. Albert, B. Malo, K.O. Hill, D.C. Johnson, J.L. Brebner, R. Leonelli. *Opt. Lett* **17**, 1652 (1992).
10. H. Hosono, J. Nishii, H. Kawazoe. *Technical Digest of the Topical Meeting on Photosensitivity and Non Linearity in Glass Waveguides.* 1995 O.S.A Technical Digest series **22**, 61 (1995).
11. P.J. Hughes, A.P. Knights, B.L. Weiss, S. Ojha. *Elec. Lett.* **36**, 427 (2000).
12. R.H. Magruder III, R.A. Weeks, R.A. Weller, R.A. Zuhr, D.K. Hensley. *J. Non-Cryst. Sol.* **239**, 78 (1998) and references therein.
13. J.L. Brebner, L.B. Allard, M. Verhaegen, M. Essid, J. Albert, P. Simpson, A. Knights. *Proc. SPIE* vol. **2998**, 122 (1997).
14. K.D. Simmons, G.I. Stegeman; B.G. Potter, J.H. Simmons. *Opt. Lett.* **18**, 25 (1995).
15. E.A. Mendoza, D.J. Fenell, R.A. Lieberman. *Proc. SPIE* **2288**, 62 (1994).
16. J.E. Roman, K.A. Winick. *Opt. Lett.* **18**, 808 (1993).
17. M. Ferraris, G. Motta, F. Ortenzio, G. Perrone, D. Pîrcàlàboiu, I. Montrosset, L. Cognolato. *Proceedings of the European Conference on Integrated Optics*, ECIO, Stockholm, 334/ Eth 10-1 (1997).
18. H. Hosono, Y. Ikuta, T. Kinoshita, K. Kajihara, M. Hirano. *Phys. Rev. Lett.* **87**, 175 501 − 1 − (2000).
19. Y. Sasaki, Y. Ohmori, *Appl. Phys.Lett.* **39**, 466 (1981).
20. U. Österberg, W. Margulis, *Opt. Lett.* **11**, 516 (1986).
21. R.A. Myers, N. Mukherjee, S.R.J. Brueck. *Opt. Lett.* **16**, 1732 (1982).
22. A.J. Iskushima, T. Fujiwara, K. Saito. *J. Appl. Phys.* **88**: Applied Physics Reviews, 1201 (2000).
23. G.P. Agrawal. *Non Linear Fiber Optics*, Optics and Photonics series, P.L. Liao, P.L. Kelley, I. Kaminow, eds; Academic Press, 435 (1995).
24. K.O. Hill, Y. Fujii, D.C. Johnson, B.S. Kawasaki. *Appl. Phys. Lett.* **32** 647 (1978).
25. B.S. Kawasaki, K.O. Hill, D.C. Johnson, Y Fujii. *Opt. Lett.* **3**, 66 (1978).
26. J. Stone. *J. Appl. Phys.* **62**, 4371 (1987).
27. D.K.W. Lam, B.K. Garside. *Appl. Opt.* **20**, 440 (1981).
28. F.P. Payne. *Elec. Lett.* **25**, 498 (1989).
29. J.P. Bernardin, N.M. Lawandy. *Opt. Com.* **79**, 194 (1990).
30. D.P. Hand, P. St. J. Russell. *Opt. Lett.* **15**, 102 (1990).
31. J. Bures, J. Lapierre, D. Pascale. *Appl. Phys. Lett.* **37**, 860 (1980).
32. S. LaRochelle, V. Mizrahi, G.I. Stegeman, J.E. Sipe. *Appl. Phys. Lett.* **57**, 747 (1990).
33. V. Mizrahi, S. LaRochelle, G.I. Stegeman. *Phys. Rev. A* **43**, 433 (1991).
34. C. Martijn de Sterke, Sunghyuck An, J.E. Sipe. *Opt. Com.* **83**, 315 (1991).
35. Sunghyuck An, J.E. Sipe. *Opt. Lett.* **16**, 1478 (1991).
36. B. Guo, D.Z. Anderson. *Appl. Phys. Lett.* **60**, 671 (1192).
37. J.P.R. Lacey, F.P. Payne, X.H. Zheng, L.M. Zhang. *Elec. Lett.* **28**, 1690 (1992).
38. E.G. Behrens, R.G. Powell. *Phys. Rev.* B **34**, 4213 (1986).

39. E.G. Behrens, F.M. Durville, R.G. Powell. *Phys. Rev.* B **39**, 6076 (1989).
40. P. Niay, P. Bernage, S. Legoubin, M. Douay, W.X. Xie, J.F. Bayon, T. Georges, M. Monerie, B. Poumellec. *Opt. Com.* **113**, 176 (1994).
41. P. Niay, P. Bernage, T. Taunay, M. Douay, E. Delevaque, S. Boj, B. Poumellec. *IEEE Phot. Technol. Lett.* **7**, 391 (1995).
42. D. Ramecourt, P. Niay, P. Bernage, I. Riant, M. Douay. *Elec. Lett.* **35**, 329 (1999).
43. S.H. Moffat, D. Grobnic, S.J. Mihailov, J. Albert, F. Bilodeau, K.O. Hill, D.C. Johnson. 1999 OSA TOPS **33**, *Bragg Gratings, Photosensitivity and Poling in Glass Waveguides*, 338 (2000).
44. G. Meltz, W.W. Morey, W.H. Glenn. *Opt. Lett.* **14**, 823 (1989).
45. K.O. Hill, F. Bilodeau, B. Malo, D.C. Johnson. *Elec. Lett.* **27**, 1548 (1991).
46. A.M. Vengsarkar, P.J. Lemaire, J.B. Judkins, V. Bhatia, T. Erdogan, J.E. Sipe. *J. Light. Technol.* **14**, 58 (1996).
47. V. Mizrahi, P.J. Lemaire, T. Erdogan, W.A. Reed, D.J. DiGiovanni, R.M. Atkins. *Appl. Phys. Lett.* **63**, 1727 (1993); M. Svalgaard, C.V. Poulsen, A. Bjarklev, O. Poulsen. *Elec. Lett.* **30**, 1401 (1994).
    G.D. Maxwell, B.J. Ainslie. *Elec. Lett.* **31**, 95 (1995).
48. R. Kashyap, G.D. Maxwell, B.J. Ainslie. *IEEE Photon. Tech. Lett.* **5**, 191 (1993).
49. R.Kashyap. *Fiber Bragg Gratings* Academic Press (1999).
50. A. Othonos, L. Kalli. *Fiber Bragg gratings: Fundamental and Applications in Telecommunications and Sensing* Artech House (1999).
51. I. Riant, P. Niay, B. Poumellec, M. Douay. Réseaux de Bragg photo-inscrits dans des fibres de télécommunication, chapitre 2 du traité Hermès edition Lavoisier, pp. 47–88 (2003) (in French).
52. Special issue of the *Journal of Lightwave Technology* on Fiber Gratings, Photosensitivity and Poling, *J. Light Technol.* **15** (1997).
53. G.C. Bjorklund, D.M. Burland, D.C. Alvarez. *J. Chem. Phys.* **73**, 4321 (1980).
54. Y. Watanabe, J. Nishii, H. Moriwaki, G. Furuhashi, H. Hosono, H. Kawazoe. *J. Non-Cryst. Sol.* **239**, 104 (1998).
55. O.V. Butov, K.M. Golant, A.L. Tomashuk. *Proc. XIX Int. Congr. Glass, Edinburgh, Phys. Chem. Glasses*, **43C**, X (2002).
56. D. Homoelle, S. Wielandy, A.L. Gaeta, N.F. Borelli, C. Smith. *Opt. Lett.* **24**, 1311 (1999).
57. S. Inoue, A. Nukui, K. Yamamoto, T. Yano, S. Shibata, M. Yamane, T. Maeseto. *Appl. Opt.* **37**, 48 (1998).
58. M.V. Bazylenko, D. Moss, J. Canning. *Opt. Lett.* **23**, 697 (1998).
59. K.I. White. *Opt. and Quant. Electron.* **11**, 185 (1979).
60. E. Brinkmeyer, D. Johlen, F. Knappe, H. Renner. *OSA TOPS 33, Bragg Gratings, Photosensitivity, Poling in Glass Waveguides*, 155 (2000).
61. D. Marcuse. *Appl. Opt.* **18**, 9 (1979).
62. K. Yamada, W. Watanabe, T. Toma, K. Itoh, J. Nishii. *Opt. Lett.* **26**, 19 (2001).
63. M. Born, E. Wolf. *Principles of Optics*, Pergamon Press (1959).
64. N.F. Borelli, C. Smith, D.C. Allan, T.P. Seward III. *J. Opt. Soc. Am.* B**14**, 1606 (1997).
65. T. Dennis, E.M. Gill, S.L. Gilbert. *Appl. Opt.* **40**, 1663 (2001).
66. B. Malo, A. Vineberg, F. Bilodeau, J. Albert, D.C. Johnson, K.O. Hill. *Opt. Lett.* **15**, 953 (1990).
67. E.M. Dianov, S.A. Vasiliev, A.A. Frolov, O.I. Medvedkov. *OSA Technical Digest, Bragg Gratings, Photosensitivity, Poling in Glass and Waveguides*, **17**, 175 (1997).

68. Y. Hibino, M. Abe, T. Kominoto, Y. Ohmori. *Elec. Lett.* **27**, 2294 (1991).
69. J. Canning, A.L.G. Carter. *Opt. Lett.* **22**, 561 (1997).
70. S. Legoubin, M. Douay, P. Bernage, P. Niay. *J. Opt. Soc. Am.* A**12**, 1687 (1995).
71. W.W. Morey, T.J. Bailey, W.H. Glenn; G. Meiltz. *Proc. Optical Fiber Conf. 1992*, 96 (1992).
72. P.L. Swart, M.G. Shlyagin, A.A. Chtcherbakov, V.V. Spirin. *Elec. Lett.* **38**, 1508 (2002).
73. T. Kitagawa, F. Bilodeau, B. Malo, S. Thériault, J. Albert, D.C. Johnson, K.O. Hill, K. Hattori, Y. Hibino. *Elec. Lett.* **30**, 1311 (1994).
74. T. Erdogan, V. Mizrahi. *J. Opt. Soc. Am.* B **11**, 2100 (1994).
75. D. Johlen, H. Renner, E. Brinkmeyer. *OSA TOPS* **33**, *Bragg Gratings, Photosensitivity, Poling in Glass Waveguides*, 182 (2000).
76. D. Moss, M. Ibsen, F. Ouellette, P. Leech, M. Faith, P. Kemeny, O. Leistiko, C.V. Poulsen. *Proc. Australian Conf. Opt. Fibre Tech.*, 333 (1994).
77. R. Kashyap, G.D. Maxwell, B.J. Ainslie. *IEEE Photonics Technol. Lett.* **5**, 191 (1993).
78. F. Bilodeau, B. Malo, J. Albert, D.C. Johnson, K.O. Hill, Y. Hibino, K. Hattori. *Opt. Lett.* **19**, 953 (1993).
79. J. Hübner, C.V. Poulsen, T. Rasmussen, L.U.A. Andersen, M. Kristensen. *OSA Technical Digest, Photosensitivity and Quadratic Nonlinearity in Glass Waveguides* **22**, 96 (1995).
80. W.X. Xie, P. Niay, P. Bernage, M. Douay, T. Taunay, J.F. Bayon, E. Delevaque, M. Monerie. *Opt. Com.* **124**, 295 (1996).
81. H. Renner. *OSA Tech. Digest, Bragg Gratings, Photosensitivity, Poling in Glass Waveguides 2001*, Stresa, paper BthC22 (2001).
82. A.M. Vengsarkar, Q. Zhong, D. Inniss, W.A. Reed, P.J. Lemaire, S.G. Kosinski. *Opt. Lett.* **19**, 1260 (1194).
83. L.B. Jeunhomme. *Single-mode Fiber Optics, Principles and Applications*, Optical Engineering series, M. Dekker, Inc. New York (1990).
84. P.A. Krug, R. Stolte, R. Ulrich. *Opt. Lett.* **20**, 1767 (1995).
85. D. Ramecourt, P. Bernage, P. Niay, M. Douay, I. Riant. *Appl. Opt.* **40**, 6166 (2001).
86. L.M. Baskin, M. Sumetsky, P.S. Westbrook, P.I. Reyes, B.J. Aggleton. *IEEE Phot. Tech. Lett.* **15**, 449 (2003).
87. F. Busque, X. Daxhelet, S. Lacroix. *Optical Fiber Conference Digest* (OFC 2003), **1**, 381 (2003).
88. K.O. Hill, B. Malo, F. Bilodeau, D.C. Johnson, J. Albert. *Appl. Phys. Lett.* **62**, 1035 (1993).
89. J.D. Prohaska, E. Snitzer, J. Winthrop. *Appl. Opt.* **31**, 3896 (1994).
90. P.E. Dyer, R.J. Farley, R. Giedl. *Opt. Com.* **115**, 327 (1994).
91. Z. Xiong, G.D. Peng, B. Wu, P.L. Chu. *J. Light. Technol.* **17**, 2361 (1999).
92. D. Ramecourt, Ph.D. thesis University of Lille, n° 2970 (2001).
93. N. Groothof, J. Canning, E. Buckley, K. Lyttikainen, J. Zagari. *Opt. Lett.* **28**, 233 (2003).
94. K.M. Davis, K. Miura, N. Sugimoto, K. Hirao. *Opt. Lett.* **21**, 1729 (1996).
95. Yamada, W. Watanabe, T. Toma, K. Itoh, J. Nishii. *Opt. Lett.* **26**, 19 (2001).
96. K. Sakuma, S. Ishikawa, T. Shikata, T. Fukuda, H. Hosoya. *In Optical Fiber Communication Conference* (OFC 2003), vol **2**, 445 (2003).
97. L. Sudrie, M. Franco, B. Prade, A. Mysyrowicz. *Opt. Com.* **171**, 279 (1999); *Opt. Com.* **191**, 333 (2001).

98. J. Zhang, P.R. Herman, C. Lauer, K.P. Chen, M. Wie. *Proceedings of SPIE* **4274**, 125 (2001).
99. B. Leconte. PhD thesis N° 2379 University of Lille (1998), in French.
100. J. Albert, B. Malo, K.O. Hill, F. Bilodeau, D.C. Johnson. *Appl. Phys. Lett.* **67**, 3529 (1995).
101. K.P. Chen, P.R. Herman, J. Zhang, R. Tam. *Opt. Lett.* **26**, 771 (2001).
102. S.J. Mihailov, C.W. Smelser, P. Lu, R.B. Walker, D. Grobnic, H. Ding, G. Henderson, J. Unruh. *Opt. Lett.* **28**, 995 (2003).
103. K.P. Chen, P.R. Herman, R. Taylor. *J. Light. Technol.* **21**, 140 (2003).
104. K.P. Chen, P.R. Herman, R. Tam. In *Proceedings on Bragg Grating, Photosensitivity and Poling in Glass Waveguides*. Stresa, paper BthA5 (2001).
105. E. Fertein, C. Przygodzki, H. Delbarre, A. Hidayay, M. Douay, P. Niay. *Appl. Opt.* **40**, 3506 (2001).
106. B. Poumellec, L. Sudrie, M. Franco, B. Prade, A. Mysyrowicz. *Opt. Express* **11**, 1070 (2003).
107. M. Fokine, PhD thesis, Department of Microelectronics and Information Technology, Stockholm, (2002).
108. A table that summarizes the main UV-induced index changes in silica fibers are available on request at marc.douay@univ-lille1.fr
109. P.J. Lemaire, R.M. Atkins, V. Mizzahi, W.A. Reed. *Elec. Lett.* **29**, 1191 (1993).
110. H. Kawano, H. Muentz, Y. Sato, J. Nishimae, A. Sugitatsu. *J. Light. Technol.* **19**, 1221 (2001).
111. M. Äslund, J. Canning, Y. Yoffe. *Opt. Lett.* **24**, 1826 (1999).
112. J. Canning, *Opt. Fiber Technol.* **6**, 275 (2000).
113. H. Patrick, S.L. Gilbert, A. Lidgard, M.D. Gallagher. *J. Appl. Phys.* **78**, 2940 (1995).
114. Y. Yan, A.J. Faber, H. de Wall. *J. Non-Cryst. Sol.* **191**, 283 (1995) and [103].
115. J. Stone. *J. Light. Technol.* **5**, 712 (1987).
116. G.E. Kohnke, D.W. Nightingale, P.G. Wigley, C.R. Pollock. In *Optical Fiber Communication Conference* (OFC99), paper PD-20 (1999).
117. J. Canning, M. Äslund, P-F. Hu. *Opt. Lett.* **25**, 1621 (2000).
118. M. Äslund, J. Canning. *Opt. Lett.* **25**, 692 (2000).
119. J. Canning, P-F. Hu. *Opt. Lett.* **26**, 1230 (2001).
120. M. Fokine, W. Margulis. *Opt. Lett.* **25**, 302 (2000).
121. R.M. Atkins, R.P. Espindola. US patent 5930420 (1999).
122. M. Lancry, PhD thesis, University of Lille, (2004).
123. J. Albert, M. Fokine, W. Margulis. *Opt. Lett.* **27**, 809 (2002).
124. C. Riziotis, A. Fu, S. Watts, R. Williams, P.G.R. Smith. In *Proceedings on Bragg Grating, Photosensitivity and Poling in Glass Waveguides*, Stresa, paper BThC31 (2001).
125. D.L. Williams, S.T. Davey, R. Kashyap, J.R. Armitage, B.J. Ainslie. *Proc. SPIE* **1513**, *Glasses for Optoelectronics* II, ECO4, the Hague, Netherlands, 158 (1991).
126. D. Moss, F. Ouellette, M. Faith, P. Leech, P. Kemeny, M. Ibsen, O. Leistiko, C.V Poulsen, J.D. Love, F.J. Ladouceur. *OSA Technical Digest on Photosensitivity and Quadratic Nonlinearity in Glass Waveguide* **22**, 120 (1995).
127. J. Canning, D.J. Moss, M. faith, P. Leech, P. Kemeny, C.V. Poulsen, O. Leistiko. *Electron. Lett.* **32**, 1479 (1996).
128. M. Svalgaard, C.V. Poulsen, A. Bjarklev, O. Poulsen. *Elec. Lett.* **30**, 1401 (1994).
129. H. Takeya, S. Matsumoto, K. YoshiaRa, M. Takabayashi, S. Miyashita, F. Uchiwara. *OSA TOPS* **33**, *Bragg Gratings, Photosensitivity, Poling in Glass Waveguides*, 272 (2000).

130. D.L. Williams, B.J. Ainslie, J.R. Armitage, R. Kashyap, R. Campbell. *Elec. Lett.* **29**, 45 (1993).
131. G. Brambilla, V. Pruneri, L. Reekie. *Appl. Phys. Lett.* **76**, 807 (2000).
132. L. Dong, J.L. Cruz, J.A. Tucknott, L. Reekie, D.N. Payne. *Opt. Lett.* **20**, 1982 (1995).
133. L. Dong, J.L. Cruz, L. Reekie, M.G. Xu, D.N. Payne. *IEEE Photon. Technol. Lett.* **7**, 1048 (1995).
134. G. Brambilla, V. Pruneri, L. Reekie, C. Contardi, D. Milanese, M. Ferraris. *Opt. Lett.* **25**, 1153 (2000).
135. E.M. Dianov, K.M. Golant, R.R. Khrapko, A.S. Kurkov, B. Leconte, M. Douay, P. Bernage, P. Niay. *Electron. Lett.* **33**, 236 (1997).
136. E.M. Dianov, K.M. Golant, V.M. Mashinsky, O.I. Medvedkov, I.V. Nikolin, O.D. Sazhin, S.A. Vasiliev. *Electron. Lett.* **33**, 1334 (1997).
137. D. Wiesman, J. Hübner, R. German, B.J. Offrein, G-L Bona, M. Kristensen, H. Jäckel. *OSA TOPS* **33**, *Bragg Gratings, Photosensitivity, Poling in Glass Waveguides*, 280 (2000).
138. R.A. Jarvis, J.D. Love, A. Durandet, G.D. Conway, R.W. Boswell. *Elec. Lett.* **32**, 550 (1996).
139. J. Nishii, H. Hosono. *OSA TOPS* **33**, *Bragg Gratings, Photosensitivity, Poling in Glass Waveguides*, 277 (2000).
140. W.X. Xie, P. Niay, P. Bernage, M. Douay, J.F. Bayon, T. Georges, M. Monerie, B. Poumellec. *Opt. Com.* **104**, 185 (1993).
141. L. Dong, W.F. Liu, L. Reekie. *Opt. Lett.* **21**, 2032 (1996).
142. M. Douay, W.X. Xie, P. Bernage, D. Pureur, P. Niay, B. Poumellec, L. Dong. *Proc. SPIE* 1997, vol. **2998**, 82 (1997).
143. J. Canning, D. Moss, M. Äslund, M. Bazylenko. *Elec. Lett.* **34**, 366 (1998).
144. J.L. Archambault, L. Reekie, P. St. J. Russel. *Electron. Lett.* **29**, 28 (1993).
145. T. Taunay, P. Niay, P. Bernage, M. Douay, W.X. Xie, D. Pureur, P. Cordier, J.F. Bayon, H. Poignant, E. Delevaque, B. Poumellec. *J. Phys. D Appl. Phys.* **30**, 40 (1997).
146. I. Riant, F. Haller. *J. Light. Technol.* **15**, 1464 (1997).
147. B. Poumellec, White Nights' Summer School on Photosensitivity in Optical Waveguides and Glasses (POWAG' 2002) Saint-Petersburg, Lecture 5 "Glass Photosensitivity," 2002.
148. D. Pureur, G. Martinelli, P. Bernage, P. Niay, M. Douay, M. Monerie. *J. Opt. Soc. Am.* A **14**, 417 (1997).
149. E.M. Dianov, K.M. Golant, V.M. Mashinsky, O.I. Medvedkov, I.V. Nikolin, O.D. Sazhin, S.A. Vasiliev. *Electron. Lett.* **33**, 1334 (1997).
150. Y. Liu, A.R. Williams, L. Zhang, I. Bennion. *Opt. Lett.* **27**, 586 (2002).
151. V. Grubsky, D.S. Starodubov, J. Feinberg. *Opt. Lett.* **24**, 729 (1999).
152. J. Albert, B. Malo, F. Bilodeau, D.C. Johnson, K.O. Hill, Y. Hibino, M. Kawachi. *Opt. Lett.* **19**, 387 (1994).
153. J. Albert, K.O. Hill, D.C. Johnson, F. Bilodeau, S.J. Mihailov, N.F. Borelli, J. Amin. *Opt. Lett.* **24**, 1266 (1999).
154. K.P. Chen, P.R. Herman, R. Tam, J. Zhang. *Electron. Lett.* **36**, 2000 (2000).
155. K.P. Chen, P.R. Herman, R. Tam. *IEEE Photon. Technol. Lett.* **14**, 170 (2002).
156. H. Hosono, Y. Ikuta. *Nucl. Instrum. and Methods in Phys. Res.* B 166-167-691 (2000).
157. K.P. Chen, X. Wei, P.R. Herman. *Electron. Lett.* **38**, 17 (2002).
158. B. Poumellec, M. Douay, J.C. Krupa, J. Garapon, P. Niay. *J. Non-Crystalline Sol.* **317**, 319 (2003).

159. J. Nishii, N. Kitamura, H. Yamanaka, H. Hosono, H. Kawazoe. *Opt. Lett.* **20**, 1184 (1995).
160. H. Hosono, M. Mizuguchi, H. Kawazoe, J. Nishii. *Japan. J. of Appl. Phys. Part 2—Letters* **35**, L234-L236 (1996).
161. P.S.J. Russell, D.P. Hand. *Electron. Lett.* **26**, 1846 (1990).
162. M. Douay, W.X. Xie, E. Fertein, P. Bernage, P. Niay, J.F. Bayon, and T. Georges., *Proceedings of the SPIE Meeting on Photosensitivity and Self-Organization in Optical Fibers and Waveguides* **2044**, Quebec, Canada, 88 (1993).
163. E. Fertein, PhD thesis No 1389, University of Lille (1995) in French.
164. T. Taunay, PhD thesis No 1935, University of Lille (1997) in French.
165. P.S.J. Russell, L.J. Poyntz-Wright, and D.P. Hand. *Proceedings of the SPIE meeting* **1373** *Fiber Laser Source and Amplifier* II, 126 (1990).
166. T. Meyer, P.A. Nicati, P.A. Robert, D. Varelas, H.G. Limberger, R.P. Salathé. *Opt. Lett.* **21**, 1661 (1996).
167. N.M. Lawandy. *Opt. Com.* **74**, 180, (1989).
168. A. Miotello, R. Kelly. *Nucl. Instrum. and Meth. in Phys. Res.* B **65**, 217 (1992).
169. F. Kherbouche, B. Poumellec, F. Charpentier, M. Fialin, P. Niay, presented at 4th EMAS regional workshop on electron probe microanalysis Prague, (2000).
170. V. Grubsky, D. Staradubov, W.W. Morey. *OSA 2003 Technical Digest (Bragg Gratings, Photosensitivity and Poling in Glass Waveguides*, Monterey, CA, USA), 33 (2003).
171. M.A. Fokine, B.E. Sahlgren, and R. Stubbe. *OSA 1997 Technical Digest (Bragg Gratings, Photosensitivity, and Poling in Glass Waveguides: Applications and Fundamentals*), Williamsburg, Virginia, USA, 58 (1997).
172. M.G. Sceats, P.A. Krug. *Proceedings of the SPIE meeting on Photosensitivity and Self-Organization in optical Fibers and Waveguides* **2044**, Quebec, Canada, 113 (1993).
173. N.H. Ky, H.G. Limberger, R.P. Salathé, F. Cochet, L. Dong. *Opt. Lett.* **23**, 1402 (1998).
174. H.G. Limberger, P.Y. Fonjallaz, R.P. Salathé, F. Cochet. *Appl. Phys. Lett.* **68**, 3069 (1996).
175. F. Kherbouche, B. Poumellec. *J. of Optics A: Pure and Applied Optics* **3**, 429 (2001).
176. B. Poumellec, P. Niay, M. Douay, J.F. Bayon. *J. Phys. D: Appl. Phys.* **29**, 1842 (1996).
177. B. Poumellec, P. Niay, M. Douay, J.F. Bayon. *1996 OSA Technical Digest* **22** (*Photosensitivity and Quadratic Non-Linearity in Glass Waveguides: Fundamentals and Applications*, Portland, Oregon, USA), 112, (1995).
178. B. Poumellec, P. Guénot, I. Riant, P. Sansonetti, P. Niay, P. Bernage, J.F. Bayon. *Opt. Mat.* **4**, 441 (1995).
179. M. Douay, W.X. Xie, T. Taunay, P. Bernage, P. Niay, P. Cordier, B. Poumellec, L. Dong, J.F. Bayon, H. Poignant, E. Delevaque. *J. Light. Techn.* **15**, 1329 (1997).
180. M. Fujimaki, T. Watanabe, T. Katoh, T. Kasahara, N. Miyazaki, Y. Ohki, H. Nishikawa. *Phys. Rev. Condensed Matter* B **57**, 3920 (1998).
181. B. Poumellec, P. Niay. Presented at the workshop "Photorefractive effects, materials and devices," La Colle sur Loup, France, 2003.
182. M. Fujimaki, K. Yagi, Y. Ohki, H. Nishikawa, K. Awazu. *Phys. Rev. B—Condensed Matter* **53**, 9859, (1996).
183. T.-E. Tsai and J.E. Friebele. *OSA 1997 Technical Digest (Bragg Gratings, Photosensitivity, and Poling in Glass Waveguides: Applications and Fundamentals, Williamsburg, Virginia, USA*), 101 (1997).
184. T.E. Tsai, G.M. Williams, and E.J. Friebele. *Opt. Lett.* **22**, 224 (1997).

185. M. Lancry, P. Niay, S. Bailleux, M. Douay, C. Depecker, P. Cordier, I. Riant. *Appl. Opt.* **41**, 7197 (2002).
186. T. Erdogan, V. Mizrahi, P.J. Lemaire, D. Monroe. *J. Appl. Phys.* **76**, 73 (1994).
187. S. Kannan, J.Z.Y. Guo, P.J. Lemaire. *J. Light. Technol.* **15**, 1478 (1997).
188. B. Poumellec, Tutorial lecture at summer school on Photosensitivity in Optical Waveguides and Glasses (POWAG 2000), 14L7 (2000). Available on request at Bertrand.Poumellec@lpces.u-psud.fr.
189. E. Salik, D.S. Starodubov, V. Grubsky, J. Feinberg. *Techn. Dig of Optical Fiber Communication Conference* (OFC 99), 56 (1999).
190. Q. Wang, A. Hidayat, P. Niay, M. Douay. *J. Light. Technol.* **18**, 1078 (2000).
191. K. Oh, P.S. Westbrook, R.M. Atkins, P. Reyes, R.S. Windeler, W.A. Reed, T.E. Stockert, D. Brownlow, D. DiGiovanni. *Opt. Lett.* **27**, 488 (2002).
192. E.M. Dianov, V.I. Karpov, M.V. Grekov, K.M. Golant, S.A. Vasiliev, O.I. Medvedkov, R.R. Khrapko, ECOC 97, Conf. Pub. N° **448**, 53 (1997).
193. M. Fokine. *Opt. Lett.* **27**, 1016 (2002).
194. G. Brambilla, H. Rutt. *Appl. Phys. Lett.* **80**, 3259 (2002).
195. D. Razafimahatratra, P. Niay, M. Douay, B. Poumellec. I. Riant, *Appl. Opt.* **39**, 1924 (2000).
196. B. Poumellec. *J. Non-Cryst. Solids* **239**, 108 (1998).
197. A. Hidayat, Q. Wang, P. Niay, M. Douay, B. Poumellec, F. Kherbouche. I. Riant, *Appl. Opt.* **40**, 2632 (2001).
198. M. Lancry, P. Niay, M. Douay, P. Cordier, C. Depecker, I. Riant. Proceedings of BGPP 2003, Bragg Gratings, Photosensitivity and Poling in Glass Waveguides, paper MC2, 40 (2003).
199. I. Riant. Proceedings of the SPIE, Volume 5063, pp. 463–473 (2003).
200. P. Ferdinand, "Capteurs à fibres optiques à réseaux de Bragg," Techniques de l'Ingénieur, Traité Mesures et Contrôle, R6735 (1999).
201. P. Niay, P. Bernage, M. Douay, F. Lahoreau, J.F. Bayon, T. Georges, M. Monerie, P. Ferdinand, S. Rougeault, P. Cetier, *IEEE Photon. Technol. Lett.* **6**, 1350 (1994).
202. F. Benabid, F. Couny, J.C. Knight, T.A. Birks, and P. St J. Russell. *Nature* 434, 488-49 (2005).
203. N. Groothoff, J. Canning, K. Lyytikainen, J. Zagari. *Opt. Lett.* 28, 233–235, (2003).
204. G. Brambilla, V. Pruneri, L. Reekie and D.N. Payne. *Opt. Lett.* 24, 1023–1025 (1999).
205. A. Martinez, I.Y. Khrushchev, and I. Bennion. *Elect. Lett.* 41 (2005).

16

Photorefractive Effects in Liquid Crystals

F. Simoni and L. Lucchetti

Dipartimento di Fisica e Ingegneria dei materiali e del Territorio and CNISM, Università Politecnica delle Marche, Ancona, Italy

In this chapter the main features of photorefractivity in liquid crystals are reviewed. Both experimental results and theoretical models are discussed, pointing out the complexity of the phenomenon when it occurs in these materials due to the collective character of the molecular reorientation responsible for light-induced refractive index modulation.

16.1 Introduction

Photorefractivity was reported for the first time in 1966 in inorganic crystals of $LiNbO_3$ and $LiTaO_3$ [1], and it was soon clear that this effect could be exploited for image storage and optical signal processing [2]. Driven by these potential applications, the investigation of new photorefractive materials has continued for several decades, focused on finding faster media, with higher electrooptic response. As clearly reported in this book, photorefractivity has been observed in several different materials including semiconductors, organic molecular crystals, and organic soft materials. In particular, the early studies performed on photorefractive polymers [3] gave rise to a new field of research because of the cost-effectiveness, processability, and versatility of these materials. In polymers the effect is induced by doping the polymeric matrix with a large concentration of chromophores, namely by molecules easily reorientable by external electric fields, to induce a bulk electrooptic effect. One of the most important steps in this field has been the synthesis of polymers with a low glass-transition temperature, which allows orientational alignment of nonlinear optical chromophores by the internal space-charge field within the viscous matrix. This nonlinear electrooptic effect is referred to as the quadratic electrooptic effect; it produces an additional large contribution to the total refractive index change [4]. The observation of such a quadratic electrooptic effect immediately led to the consideration of liquid crystals (LCs) as promising materials to induce photorefractivity.

The usual understanding of "liquid crystals" refers to "mesomorphic materials," namely materials showing intermediate aggregation states between the solid and the liquid phase. They are actually anisotropic fluid. Among them, interest, as regards photorefractivity, has been focused on calamitic LCs, which consist of long rod-shaped molecules, potentially able to produce a bulk birefringence larger than that of a doped polymer.

Liquid-crystalline phases are present when the chemical composition gives rise to an orientational order of the molecules, due to an interaction potential much stronger than in isotropic liquids. The rodlike molecules of calamitic LCs have a rigid or quasirigid core and two terminal groups allowing the existence of many classes of mesomorphic molecules. The chemical structure also determines the number of different mesophases that appear by increasing the temperature from the solid state to the isotropic liquid state. These phases can be distinguished by the degree of molecular order that they show. The *nematic phase* presents orientational order, but no positional order: molecules tend to align their axes along a preferred direction, which defines the unit vector **n** called the "director," producing a locally uniaxial phase. The *smectic phase* is characterized by a layered structure, and molecules may also have positional order, giving rise to different phases, which are classified according to the symmetry appearing in each layer. The *cholesteric phase* is generally described as a chiral nematic phase, where a helical arrangement is superimposed on the long-range orientational order.

In many cases, significant variations of the director **n** occur over distances much larger than the molecular scale; therefore it is useful to describe the LC has a continuum medium, taking account of the interaction that produces the long-range orientation through macroscopic elastic constants associated with the basic deformation of the structure. For a complete description of the continuum theory of LC, see the basic book of P.G. De Gennes and J. Prost [5].

The long-range directional order allows a strong collective reorientation for a given space-charge field, and in contrast to photorefractive polymers, a weak electric field is able to induce directional charge transport and to enhance the quadratic electrooptic effect.

Beginning with this initial motivation, the investigation of photorefractivity in liquid crystals pointed out several peculiar features of this phenomenon, in such a way that often the effect in liquid crystalline materials is reported as a "photorefractive-like effect" to underline similarities and differences with respect to the usual photorefractive effect. In fact, besides the "conventional" space-charge generation and migration due to diffusion anisotropy, other phenomena can play a not negligible role in LCs such as the anisotropy of the dielectric permittivity and of the conductivity (Carr–Helfrich effect, [6]), thermal diffusivity of ions, and light-induced interfacial phenomena. We will review all these effects in the following sections, limiting our discussions to low-molecular-weight nematic liquid crystals and polymer-dispersed liquid crystals (PDLCs). These latter have been proposed as suitable media for photorefractivity in order to combine the advantages of photorefractive polymers with the high refractive index change capability of LCs.

16.2 First Observations of Photorefractivity in Nematic Liquid Crystals

The first observation of photorefractivity in liquid crystals was reported by Rudenko and Sukhov in 1994 [7–9]. The effect was ascribed to a space-charge field arising from the photoinduced conductivity and diffusion anisotropies.

The authors pointed out that in a medium exhibiting photoinduced conductivity and with a local dependence on intensity of the concentration of the photoinduced carriers, optical radiation can generate a spatially uniform space-charge field stronger than that of the optical wave.

As a consequence, the question was whether a field of this kind could be excited in a nematic liquid crystal and whether such a field could produce orientational refractive index gratings, as in photorefractive materials. To answer this question, they used the nematic $4'$-(n-pentyl)-4-cyanobiphenyl (5CB) doped with a small amount of the dye rhodamine 6G (R6G) acting as the charge generator. The cell was homeotropically aligned (average orientation of the liquid crystal molecules perpendicular to the boundary surfaces). A sketch of the experimental geometry is shown in Figure 16.1. The Ar^+ ion laser beam is split into two beams that overlap at a small angle θ in the sample, which is tilted at an angle β to the bisector of θ.

In this way an optical interference pattern is created of the form $I = I_0[1 + m\cos(qx)]$, where $m = 2(I_1 I_2)^{1/2}/(I_1 + I_2)$ is the light modulation factor. This spatially modulated light photogenerates charge carriers and allows charge migration along the grating wave vector $\mathbf{q}$ and localization in the dark region of

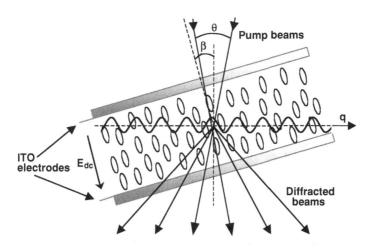

FIGURE 16.1. Schematic representation of the experimental geometry used to observe the photorefractive effect in LCs. A coherent laser beam is split into two beams that overlap at a small angle θ in the sample. The sample is tilted at an angle β to the bisector of the two beams. E_{dc} is the dc external electric field. The direction of q defines the $\mathbf{x}$-axis.

the interference pattern. The result is a sinusoidally modulated space-charge field given by

$$E_{sc} = -\frac{mk_B Tq}{2e}\left(\frac{D^+ - D^-}{D^+ + D^-}\right)\frac{\sigma - \sigma_d}{\sigma}\sin(qx) = E_{sc}^0 \sin(qx), \qquad (16.1)$$

where σ is the illuminated state conductivity (i.e., the photoconductivity), σ_d is the dark-state conductivity, k_B is the Boltzmann constant, e is the proton charge, and D^+ and D^- are the diffusion constants for the cations and anions, respectively [8, 9]. The difference between the photoconductivity and dark conductivity is linked to the ionic mobility μ^+ and μ^- and to the recombination factor γ by means of the following relation:

$$\sigma - \sigma_d = \frac{e(\mu^+ + \mu^-)(\alpha I)^{1/2}}{\gamma}. \qquad (16.2)$$

Notice that E_{sc} (which varies as $\sin(qx)$) is $\pi/2$ phase-shifted from the optical grating function ($\sim\cos(qx)$). Two factors determine the magnitude of the space-charge field in (16.1): the difference between σ and σ_d, and the difference in the diffusion coefficients of the cations and anions (in the absence of one or both of these differences, E_{sc} would be zero). The externally applied field E_{dc}, which is bigger in magnitude than the internal space-charge field, is required to keep the modulation of the internal electric field greater than zero. The coupling between the space-charge field and the uniform dc bias field E_{dc} causes reorientation of the LC director and the correspondent modulation of the refractive index. In the experiment, the application of an electric field of only 100–200 V/cm in combination with a total incident intensity on the order of 300 mW/cm^2 was sufficient to induce directional charge transport and to induce a grating spacing for the orientational grating equal to the interference pattern [7–9]. Under the approximation of small induced birefringence, the self-diffraction efficiency η of this Raman–Nath orientational grating (defined as the ratio $\eta = I_{+1}/I_0$ where I_{+1} is the intensity of the first-order diffracted beam and I_0 that of the incident beam) can be written as [6, 7]

$$\eta = \left[\frac{dmk_B T}{\lambda n_e K q e}\left(\frac{E_{dc}\varepsilon_s\varepsilon_\infty \sin\beta\cos\beta}{1 + \varepsilon_s E_{dc}^2/(2\pi K q^2)}\right)\frac{D^+ - D^-}{D^+ + D^-}\frac{\sigma - \sigma_d}{\sigma}\right]^2. \qquad (16.3)$$

In this equation d is the sample thickness, n_e is the refractive index along the extraordinary axis, K is the Frank elastic constant in the single-constant approximation [5], λ is the laser wavelength, ε_s is the static dielectric permittivity, and ε_∞ is the high-frequency dielectric permittivity. The validity of this equation was experimentally verified by measuring the diffraction efficiency versus the total intensity incident on the sample, the grating spacing, and the externally applied field intensity. The experimental curves are reported in Figures 16.2a, 16.2b and 16.2c. All the observed functional dependencies are in agreement with (16.3).

Rudenko and Sukhov also showed that the conductivity term saturated at high light intensities, and this allowed them to determine experimentally the ratio

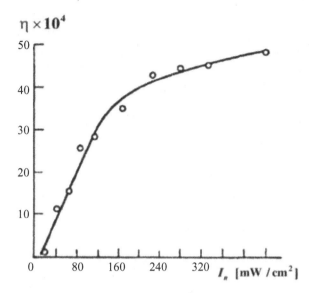

FIGURE 16.2a. Diffraction efficiency versus the total intensity of the waves. $E_{dc} = 186$ V/cm, $\beta = 1.4 \times 10^{-2}$ (after E.V. Rudenko et al. [7]).

$(D^+ - D^-)/(D^+ + D^-)$. They found a value of 0.02, indicating a very small difference in the diffusion coefficients of the positive and negative mobile ions. This fact, combined with the low solubility of the dye and the inefficient charge generation and charge transport, strongly limited the performances of the material. Significant improvements of the photorefractive performances of LCs were

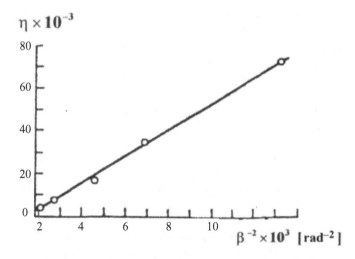

FIGURE 16.2b. η versus β^{-2} at $E_{dc} = 140$ V/cm and $I_0 = 205$ mW/cm^2 (after E.V. Rudenko et al. [7]).

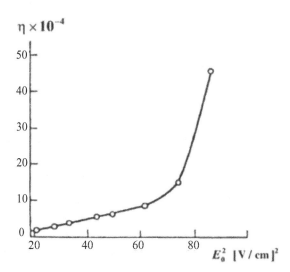

FIGURE 16.2c. η versus E^2_{dc} at $I_0 = 95$ mW/cm^2 and $\beta = 1.4 \times 10^{-2}$ (after E.V. Rudenko et al. [7]).

obtained by other authors by doping the nematics with both electron donors and acceptors having appropriate redox characteristics that produce more efficient intermolecular charge separation and bulk charge migration [10].

16.3 Effects of Dielectric and Conductivity Anisotropies

In 1994, a photorefractive effect that was derived entirely from orientational ordering due to a space-charge field was reported by Khoo and coworkers in pure nematic LCs [11]. The geometry of the experiment is similar to that used by Rudenko and Sukhov reported in Figure 16.1. The molecular reorientation effects were observed in undoped or lightly doped [12] films of several nematic LCs, including single-constituent nematics and nematic mixtures. The induced nonlocal refractive-index change responsible for these effects was attributed to LC director axis reorientation due to a laser-induced dc space-charge field arising from both conductivity and dielectric anisotropies, in the presence of an applied dc electric field.

In fact, besides the effect of photoconductivity and diffusion anisotropy already described, other mechanisms can contribute to the creation of an internal field able to reorient the liquid-crystal molecules. A theoretical discussion of the fundamental mechanisms regulating the dc field-assisted optically induced fields and the optical molecular reorientation in nematic LC films has been reported by Khoo [13].

Three processes are considered to contribute to the director-axis reorientation and the consequent refractive-index change. The first one (involving photocharge production, ion drift/diffusion, charge separation, and space-charge-field formation) is analogous to those occurring in photorefractive crystals and

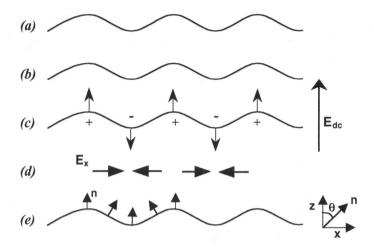

FIGURE 16.3. Schematic representation of the quantities involved in the nonlinear photorefractive effect: (a) input optical intensity; (b) photoinduced conductivity modulation; (c) flow velocity and space charge; (d) dc space-charge field; (e) director axis reorientation profile. **n** is normal to the curve and θ is the reorientation angle (after I.C. Khoo [11]).

leads to the photorefractive-like space-charge field $\mathbf{E}_{sc}$ along the grating vector $(E_{sc} = E_{sc}\mathbf{u}_x)$ given by (16.1) and already discussed. The second process involves charge generation, ionic conduction accompanied by director-axis reorientation, space-charge-field formation through dielectric and conductivity anisotropy and further director-axis reorientation. Finally, the steps involved in the third process are charge generation, material flows, formation of a velocity gradient and shear stress, and director-axis reorientation. These latter processes are unique to nematic LCs and are a consequence of the Carr–Helfrich effect occurring in these materials. Figure 16.3 shows a schematic depiction of various parameters involved in the nonlinear photorefractive effect in LCs.

Assuming an incident optical intensity distribution of the form $I = I_0(1 + \cos \mathbf{q}\cdot\mathbf{r}) = I_0(1 + \cos qx)$ and an applied dc field E_{dc} directed along the z-axis, the behavior of the consequent director axis reorientation profile $\theta(x)$, the currents, and the liquid flow velocity are schematically represented in the figure. Following the theory of Helfrich [6] for the geometry of Figure 16.3, the space-charge field arising from dc conductivity anisotropy $\Delta\sigma$ has the form $\mathbf{E}_{\Delta\sigma} = E_{\Delta\sigma}\mathbf{u}_x$, [13], with

$$E_{\Delta\sigma} = -\left(\frac{\Delta\sigma \sin\theta \cos\theta}{\sigma_{\text{II}} \sin^2\theta + \sigma_{\perp} \cos^2\theta}\right) E_{dc}, \qquad (16.4)$$

whereas the space-charge field due to the dielectric anisotropy $\Delta\varepsilon$ is $\mathbf{E}_{\Delta\varepsilon} = E_{\Delta\varepsilon}\mathbf{u}_x$, with

$$E_{\Delta\varepsilon} = -\left(\frac{\Delta\varepsilon \sin\theta \cos\theta}{\varepsilon_{\text{II}} \sin^2\theta + \varepsilon_{\perp} \cos^2\theta}\right) E_{dc}, \qquad (16.5)$$

where ε and $\varepsilon_\perp$ are the dielectric permittivities and σ and $\sigma_\perp$ the conductivities, respectively parallel and perpendicular to the director.

These space-charge fields are both increasing functions of the induced reorientation angle θ, for small θ. A small initial reorientation angle and space-charge field will thus reinforce each other; (16.4) and (16.5) show that once the reorientation is established, the space-charge field can be maintained by simply applying the dc field, without participation of the incident optical field at all. Such effect was actually observed by Khoo in his study of the holographic grating formation dynamics [11].

The fields and space charges, in conjunction with the applied dc field, create director-axis reorientation through several mechanisms. The interaction of the space charges ($\sim \cos(qx)$) with the applied field causes nematic flows in opposite directions, with positive and negative charges flowing with velocities v_z and $-v_z$, respectively. The flow gradient dv_z/dx gives rise to a flow-shear stress, which, in turn, exerts a torque on the LC molecules. This shear-induced torque τ_s is given by [13]

$$\tau_s = -\frac{dv_z}{dx}(\alpha_3 \cos^2 \theta - \alpha_2 \sin^2 \theta), \tag{16.6}$$

where α_2 and α_3 are the Leslie coefficients.

The shear torque τ_s tends to change the direction of the preferred axis, and its effect is a reorientation of the director that follows the modulation of the flow gradient ($\sim \sin(qx)$).

The space charges also influence the director-axis reorientation through the electric fields that they produce. These fields, in conjunction with the applied dc field, create a *dielectric* torque $\tau_{\Delta\varepsilon}$ given by [13]

$$\tau_{\Delta\varepsilon} = \frac{\Delta\varepsilon}{4\pi}(\hat{n} \cdot \vec{E}_{dc})(\hat{n} \times \vec{E}_{dc}) = \frac{\Delta\varepsilon}{4\pi}\left[\sin\theta\cos\theta\left(E_x^2 - E_z^2\right) + \cos 2\theta E_x E_z\right], \tag{16.7}$$

where E_x and E_z are the x and z components of the total dc electric field $E_{dc(tot)}$ (E_z coincides with the externally applied dc field E_{dc}),

$$E_{dc(tot)} = (E_{sc} + E_{\Delta\sigma} + E_{\Delta\varepsilon})\hat{u}_x + E_{dc}\hat{u}_z. \tag{16.8}$$

The dielectric torque also gives rise to a spatially phase-shifted director-axis reorientation similar to the flow-induced effect. The total conduction-induced torque is therefore partly shear-induced and partly dielectric.

The dynamics of the director axis reorientation process is governed by an interplay among the various torques produced by the fields and the elasticity of the liquid crystal and is described by the equation

$$\tau_{\Delta\varepsilon} + \tau_s + \tau_K + \tau_{opt} + \xi\frac{d\theta}{dt} = 0, \tag{16.9}$$

where τ_K is the elastic restoring torque, τ_{op} is the optical torque given by an expression that has the same form of (16.7), just by substituting the dc field with the optical field, considering the dielectric anisotropy at optical frequencies, and dividing by a factor of 2 due to time averaging [14]. Here $\xi d\theta/dt$ is a viscous damping term that accounts for the dissipation of mechanical energy caused by friction during the reorientation process (ξ is an appropriate interaction-geometry-dependent viscosity coefficient). In the experimental geometry used, the elastic torque can be written as [5, 14]

$$\tau_K = (K_1 \sin^2\theta + K_3 \cos^2\theta) \left(\frac{\partial^2\theta}{\partial x^2} + \frac{\partial^2\theta}{\partial z^2}\right)$$
$$+ [(K_1 - K_3)\sin\theta\cos\theta] \left[\left(\frac{\partial\theta}{\partial x}\right)^2 + \left(\frac{\partial\theta}{\partial z}\right)^2\right], \qquad (16.10)$$

where K_1 and K_3 are the elastic constants for splay and bend, respectively. The steady-state ($d\theta/dt = 0$) solution of the torque balance (16.9) can be easily found if one limits the analysis to small reorientation angles in the one-elastic-constant approximation ($K_1 = K_3 = K$). Within these assumptions and rewriting the shear-induced torque as (by applying the Poisson's equation, the equation of continuity, and the viscous force equation)

$$\tau_s = \frac{\alpha_3}{\eta_2} \frac{\Delta\sigma}{\sigma_p} \frac{\varepsilon_\perp}{4\pi} E_z^2\theta, \qquad (16.11)$$

(16.9) reduces to [13]

$$K\left(\frac{\partial^2\theta}{\partial x^2} + \frac{\partial^2\theta}{\partial z^2}\right) + \frac{\Delta\varepsilon}{4\pi} E_z E_{sc} \cos\beta \sin(qx)$$
$$+ \frac{\Delta\varepsilon}{4\pi} E_z^2 \left[1 + \frac{\alpha_3}{\eta_2}\frac{\Delta\sigma}{\sigma_p}\frac{\varepsilon_\perp}{\Delta\varepsilon} + \left(\frac{\Delta\varepsilon}{\varepsilon_\perp} + \frac{\Delta\sigma}{\sigma_\perp}\right)\cos\beta\right]\theta = 0, \quad (16.12)$$

where η_2 is the viscosity of the nematic fluid, and the other terms have the usual meaning. The boundary conditions over the glass plates limiting the homeotropically aligned LC cell impose $\theta(0) = \theta(d) = 0$. Then, writing the reorientation angle θ as

$$\theta = \theta_0 \sin\left(\frac{\pi z}{d}\right) \sin(qx), \qquad (16.13)$$

(16.12) can be solved to give

$$\theta_0 = \frac{\Delta\varepsilon E_z E_{sc} \cos\beta}{K\left(\frac{\pi^2}{d^2} + q^2\right) + \left(\frac{\Delta\varepsilon}{4\pi}\right) E_z^2 \left[1 + \left(\frac{\alpha_3}{\eta_2}\right)\left(\frac{\Delta\sigma}{\sigma_p}\right)\frac{\varepsilon_\perp}{\Delta\varepsilon} + \left(\frac{\Delta\varepsilon}{\varepsilon_\perp} + \frac{\Delta\sigma}{\sigma_\perp}\right)\cos\beta\right]}. \qquad (16.14)$$

The quadratic dependence on the field-induced dielectric torque $\tau_{\Delta\varepsilon}$ reflects into the appearance of the field product in $E_z E_{sc}$. Finally, under the small-θ_0 approximation, one obtains the following relation for the refractive index change Δn:

$$\Delta n \approx (n_\mathrm{II} - n_\perp)\frac{n_\mathrm{II}}{n_\perp} \sin(2\beta)\theta_0. \qquad (16.15)$$

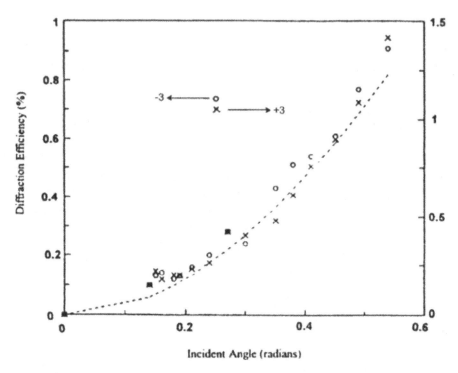

FIGURE 16.4. Dependence of the diffraction efficiency of the diffracted beams on the applied dc voltage at $\beta = 30.8$, $P_1 = 11.2$ mW, $P_{+1} = 8.8$ mW (after I.C. Khoo et al. [11]).

Since the diffraction efficiency η is proportional to the square of Δn, according to (16.14) and (16.15) the β-dependence of η is of the form $\sin^2\beta\cos^4\beta$. This result is in good agreement with the experimental observations reported by Khoo [11], as shown in Figure 16.4, where the diffraction efficiency is plotted against the incidence angle. The time evolution of the above-described mechanisms has also been examined by Khoo and coworkers in connection with the investigation of holographic grating formation in dye- and fullerene-C60-doped nematics [13].

As an example, in Figure 16.5 the dynamics of transient and permanent grating formation in a C60- doped LC is depicted. Under low optical power (few milliwatts) and low applied voltage (≤ 1.5 V), the time evolution and nature of the induced grating (as monitored by the diffraction of a probe He-Ne laser) exhibited three distinct stages depending on the illumination time. For short illumination time (few seconds) the grating had a rise and a decay time on the order of 1 s, typical of a reorientational process of liquid-crystal films with thickness a few tenths of a micron [14]. With the writing beams on, after the initial quasistationary value, the probe diffraction slowly increased in magnitude before settling down to a final steady-state value after 15–20 min. In switching off the beams before the final steady state was reached while maintaining the dc field, the probe diffraction

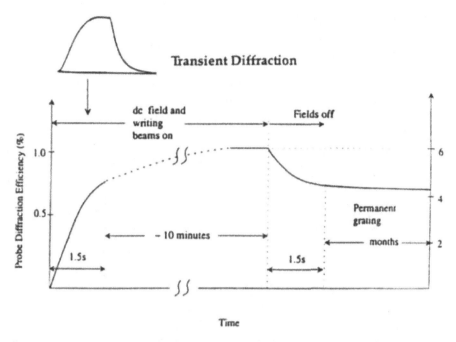

FIGURE 16.5. Schematic depiction of the dynamics of transient and permanent grating formation in a C_{60}-doped nematic liquid-crystal film. Inset shows an oscilloscope trace of the transient grating diffraction from a 25-μm-thick C_{60}-doped nematic film illuminated by 2-second-long square pump laser pulses (after I.C. Khoo [13]).

was observed to drop to a lower value. The diffraction was observed as long as the dc field was kept, going slowly (in a few min) to a lower finite value. When the dc field was turned off abruptly during this decay, the grating relaxed in about 1 s, which is again the typical orientational decay time of nematic liquid crystals. This intermediate stage was interpreted as corresponding to the case in which the reorientation angle acquires a sufficiently large magnitude so that the external-field-dependent space-charge fields $E_{\Delta\sigma}$ and $E_{\Delta\varepsilon}$ become substantial. In fact, in contrast to the usual photorefractive space-charge field E_{sc} described by (16.1), these fields depend only on the reorientation angle and on the external dc field, but not on the incident optical field: this explains why part of the induced grating can be sustained by the dc field even in the absence of the writing beams. The final stage in the grating formation was reached for an illumination time larger than 15 min. Then, if all the external fields (dc and optical) were turned off, a persistent grating was observed to last hours or several days. However, this effect should not be strictly related to photorefractivity but to photoinduced molecular adsorption on the irradiated surface.

More recently, Khoo and coworkers reported the observation of an extraordinarily large optical nonlinearity in nematic LC films doped by the azo dye methyl red [15], by detecting the diffraction of nonlinear gratings with no external field

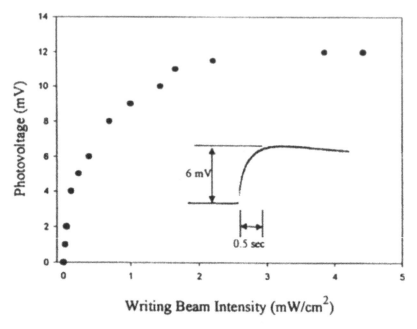

FIGURE 16.6. Observed photovoltage dependence on the optical illumination intensity in a 25-μm-thick 5CB nematic film doped by methyl red. Inset, the typical oscilloscope trace of the detected photovoltage in a 6-μm-thick film under an optical intensity of 1 mW/cm² (after I.C. Khoo et al. [15]).

applied. Refractive index gratings were generated with a total incident intensity of few mW/cm², about two orders of magnitude smaller than that used in the previously described experiments, and a nonlinear coefficient $n_2 \approx 6$ cm²/W was measured.

Tentatively, such a "supranonlinear" response was explained as a photorefractive-like effect in which the role of the external field is played by a photoinduced dc field generated by the monotonic variation of the optical intensity along the z direction, due to light absorption.

The presence of a photovoltaic effect was actually demonstrated: the photovoltage dependence on the incident optical intensity is reported in Figure 16.6, together with the oscilloscope trace of the dc voltage detected across a 6-μm-thick cell. The photovoltage reaches a saturated value of about 12 mV for intensities above 2 mW/cm². However, this qualitative model could not find a quantitative demonstration, since the photoinduced longitudinal field does not appear sufficient to sustain the photorefractivity as described above. Moreover, no significant two-beam coupling effect was observed (as usual for photorefractive materials). As a matter of fact, in the conclusions of this chapter we will recall that a different phenomenon should be involved, connected to light-induced modifications of the boundary conditions and not really related to a photorefractive effect.

16.4 Photorefractivity Based on Thermodiffusivity

Besides the usual diffusion anisotropy of charge carriers, another mechanism may be important in creating a space-charge field for the onset of a photorefractive effect. This mechanism is the thermodiffusivity, and its peculiar feature is that it does not require photoconductivity, since the light-induced charge separation takes place due to thermal diffusion [16]. This effect is proportional to the power of absorbed light and, for light intensities on the order of tenths of W/cm^2, can become stronger than the conventional photorefractive one. Because of light absorption, the thermodiffusive space-charge field may have a component along the light-propagation direction, which, combined with the field component along **q**, can lead to orientational gratings spatially resonant with the incident interference pattern, as also predicted by Rudenko and Sukhov [7, 9].

As is well known, light absorption in doped samples may induce high temperature gradients, and thermal diffusion can modulate the concentration of ions present in the liquid crystal, independently of their origin.

In order to take into account this effect, the diffusion equation must include an additional term of thermal diffusivity:

$$\vec{J}^{\pm} = -D^{\pm}\vec{\nabla}n^{\pm} - n^{\pm}D_T^{\pm}\vec{\nabla}T \pm n^{\pm}\mu^{\pm}\vec{E}_{sc}. \tag{16.16}$$

In this equation D_T is the thermal diffusion coefficient, $\vec{\nabla}T$ the temperature gradient, D the coefficient of mass diffusion, n the density of ions, and μ the mobility.

In steady state, $\vec{J}^{\pm} = 0$, and one has to solve the following system of equations:

$$-\vec{\nabla}n^{\pm} - n^{\pm}\frac{D_T^{\pm}}{D^{\pm}}\vec{\nabla}T \pm \frac{\mu^{\pm}n^{\pm}}{D^{\pm}}\vec{E}_{sc} = 0,$$
$$\vec{\nabla}\cdot\varepsilon\vec{E}_{sc} = 4\pi e(n^+ - n^-), \tag{16.17}$$

to get the space-charge field as

$$\vec{E}_{sc} = \frac{1}{n^+\mu^+ + n^-\mu^-}\left[(D^+\vec{\nabla}n^+ - D^-\vec{\nabla}n^-) + (D_T^+\vec{\nabla}n^+ - D_T^-\vec{\nabla}n^-)\vec{\nabla}T\right], \tag{16.18}$$

connecting it directly to the ion concentration and temperature modulations.

By assuming $n^+ = n^- = n$, the field can be rewritten, to a first approximation, as

$$\vec{E}_{sc} = \vec{E}_D + \vec{E}_{TD}, \tag{16.19}$$

where

$$\vec{E}_D = v\frac{k_B T}{2e}\frac{\vec{\nabla}n}{n} \quad \text{and} \quad \vec{E}_{TD} = s\frac{k_B T}{2e}\vec{\nabla}T. \tag{16.20}$$

The first term is the conventional diffusive space-charge field indicated as E_{sc} in the previous sections, proportional to the diffusion anisotropy $v \approx (D^+ - D^-)/$

$(D^+ + D^-)$, while the second term is originated by thermal diffusion and is proportional to $s \approx (D_T^+ - D_T^-)/D$.

$\vec{E}_{TD}$ can be evaluated by means of the dependence of $\vec{\nabla}T$ on the material parameters and on the light intensity. The steady-state heat-conductivity equation gives

$$\nabla^2 T = -\frac{\alpha}{\rho c \chi} I,$$ (16.21)

where, to avoid misleading formalism, we have indicated χ as the temperature-diffusion coefficient (same dimensions as D and D_T, cm^2/s). As usual, α is the absorption coefficient, ρ the density, and c the specific heat.

For an interference pattern of the usual form $I = I_0(1 + m \cos(\mathbf{q \cdot r}))$, the induced temperature spatial modulation is

$$\partial T = m \frac{\alpha I_0}{kq^2} \cos(\vec{q} \cdot \vec{r}),$$ (16.22)

and the magnitude of the thermal diffusion-originated term becomes

$$E_{TD} = s \frac{k_B T}{2e} m \frac{\alpha I_0}{kq} \sin(\vec{q} \cdot \vec{r}),$$ (16.23)

k being the thermal conductivity coefficient.

This value can be compared to the diffusive space-charge field term

$$E_D = -vm \frac{k_B T}{4e} q \sin(\vec{q} \cdot \vec{r}).$$ (16.24)

From (16.23) and (16.24) we see that the thermal diffusivity effect overcomes the conventional one if

$$\alpha I_0 \geq \frac{2\pi^2 v \chi \rho c}{s \Lambda^2},$$ (16.25)

with $\pi^2/\Lambda^2 = q^2/4$. Using typical material parameters, this condition is fulfilled by $I_0 > 30$ W/cm^2 if $\alpha = 100$ cm^{-1}, which is a reasonable absorption coefficient for liquid-crystalline samples doped with absorbing dyes.

As a matter of fact, the assumption that light modulation occurs only in the transversal plane is not fulfilled in the presence of strong light absorption. In fact, in this case the light intensity is strongly attenuated along the cell thickness. As a consequence, thermal diffusion of charges is possible also along this direction with the onset of a space-charge field normal to the LC layer.

Even if no direct experimental demonstration of photorefractivity induced by thermal diffusion has been given so far, some of the observations reported for PDLCs may find an explanation in this model. For instance, the photorefractive-like effect reported in [17] was obtained at intensities in the range 10^2–10^3 W/cm^2 without applying external fields.

16.5 The Photoelectric Effect

MacDonald and coworkers have reported a photoinduced reorientation in undoped nematics of discotic and calamitic molecules. The effect was driven by static electric field, but did not require a spatial modulated illumination as in conventional photorefraction [18–20]. Nonlinear coefficients in the range 1–10 cm^2/W were measured, comparable to the supranonlinear effect discussed in Section 16.3.

The effect was studied in 5-μm-thick nematic LC films. The discotic samples were homeotropically aligned, whereas the calamitic ones were planar (average molecular orientation parallel to the boundaries). Under application of an external field E_{dc} above the Freedericks threshold, a director reorientation occurred; since the initial reorientation and the sign of the dielectric anisotropy were different, the electric torque reoriented discotics away from E_{dc} and calamitics toward E_{dc}. In both cases, the nematic director rotated in the plane of the rubbing direction (direction of initial orientation of the molecules). The resulting changes in birefringence were detected by means of a weak He-Ne laser linearly polarized at 45° with respect to the reorientation plane. The transmitted intensity components parallel and perpendicular to the incidence plane were monitored as functions of the applied external field, and the birefringence was evaluated from the optical phase shift δ between the ordinary and extraordinary waves transmitted by the samples [21]. Under these conditions, irradiation by an Ar ion laser beam having wavelength in the range 458–514 nm gave rise to an additional reorientation. The experimental setup is reported in Figure 16.7, whereas Figure 16.8 shows the photoinduced changes in birefringence for the two kinds of cells analyzed.

The dynamics of the process was investigated as a function of the external voltage and of the laser intensity. For purely dc-field-induced reorientation, the response times were about 1 s for discotics and 0.1 s for rodlike molecules. The time constants of the photoinduced reorientation were approximately the same in the first case, while they were substantially larger for calamitics with a rise time on the order of 1 s and a decay time of 15 s. Based on a model of photoinduced charge

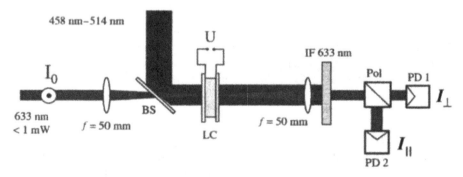

FIGURE 16.7. Experimental setup to detect the "photoelectric effect" (after R. Macdonald et al. [20]).

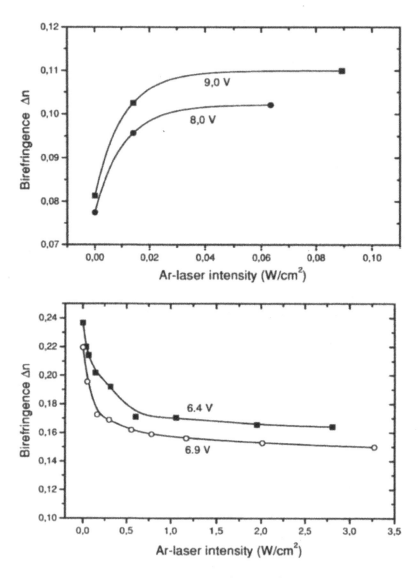

FIGURE 16.8. Photoinduced change in the birefringence for the discotic (upper) and the calamitic (lower) mesogens (after R. Macdonald et al. [20]).

generation described below, this difference was supposed to be due to the lower conductivity of calamitic molecules compared with discotics, leading to a slower space-charge recombination and, as a consequence, to slower response times. The effect was strongly dependent on the incident wavelength, this dependence being connected to light absorption and LC molecule excitation.

The phenomenon was initially described as a kind of photoelectric effect giving rise to a space-charge field resulting from optical excitation of the molecules and creation of charge carriers of different mobilities. Since holes, i.e., ions in the nematic phase, have negligible mobility compared to electrons, these latter can be assumed to migrate in a conduction-band-like electronic state. As a result, the photoexcited electrons are removed by the external electric field while holes basically remain in their positions. The observation of a strong photoconductivity confirmed the light-induced charge-carrier generation. The additional space-charge field $\mathbf{E}_{sc}$ generated in this way enhances the electric field across the cells, providing an additional electric torque acting on the LC molecules.

The electric displacement vector $\mathbf{D}_{sc}(z)$ associated with $\mathbf{E}_{sc}$ was calculated in the case of weak intensity and neglecting diffusion (i.e., considering a drift-dominated charge mobility). In these conditions the space-charge density $\rho(z)$ is proportional to the local light intensity:

$$\rho(z) = \rho_0 \exp(-\alpha z), \quad 0 \leq z \leq d, \tag{16.26}$$

where as usual, α is the absorption coefficient and the charge density at the LC surface is given by [8]

$$\rho_0 = \frac{e\sigma N_D}{h\nu\gamma N} I(0), \tag{16.27}$$

where σ is the cross section for photoexcitation, e the proton charge, N_D the LC molecule density, N the electron density, $h\nu$ the photon energy, and γ the recombination factor.

From the Maxwell equation $\nabla \cdot \mathbf{D}sc = \rho$ one gets

$$\vec{D}_{sc}(z) = \frac{\rho_0}{\alpha} \left[\frac{1}{2} - \exp(-\alpha z) + \frac{1}{2}\exp(-\alpha d) \right] \hat{u}_z, \tag{16.28}$$

which has only the z-component different from zero, assuming a space-charge density dependent only on z as in (15.5.1). The dependence of the electric displacement on z is shown in Figure 16.9. It can be seen that $\mathbf{E}_{dc}$ is shielded by the space-charge field near the positive electrode ($z \approx 0$) and enhanced in the center of the cell ($z \approx d/2$), where the reorientation mainly takes place. The enhancement depends on the number of photoinduced charges and thus on the laser intensity. Obviously, the effect becomes stronger by increasing the absorption, as expected by the observed wavelength dependence.

The effective voltage caused by the space-charge field was also experimentally measured. The results are reported in Figure 16.10 as a function of the incident intensity at $\lambda = 458$ nm.

After these first observations, other experiments with spatially inhomogeneous illumination were performed on calamitic LCs, in order to check the possibility of photorefractive grating formation associated with the photoelectric effect [22].

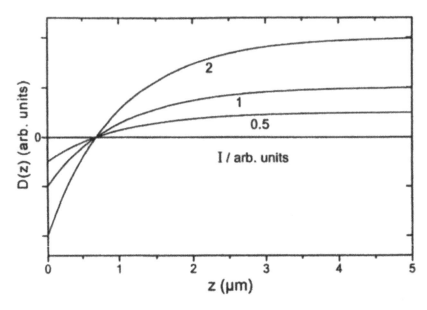

FIGURE 16.9. Change of the electric displacement D across the liquid-crystal film for different laser intensities calculated from (16.28). The absorption coefficient was assumed to be $\alpha = 1 \cdot 10^4$/cm. Intensities I are given in arbitrary units (after R. Macdonald et al. [18]).

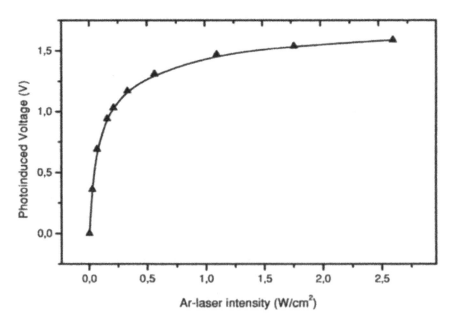

FIGURE 16.10. Photoinduced voltage versus Ar-laser intensity at $\lambda = 458$ nm (after R. Macdonald et al. [20]).

Standard wave-mixing measurements were carried out in a nontilted geometry on planar cells, with an Ar ion laser and an external dc field. When the optic axis was parallel to the grating vector, highly efficient light-induced transient gratings were obtained at pump intensities lower than 1 mW/cm^2 by simultaneous application of a dc voltage slightly above the Freedericksz threshold. The diffraction efficiency was strongly anisotropic with respect to the probe polarization direction. In particular, while η_p increased with intensity and reached values up to 19%, η_s was always lower than 0.05% and did not show any intensity dependence being η_p and η_s the diffraction efficiency for p and s polarization respectively. This anisotropy did not depend on the pump beam's polarization direction and was considered a manifestation that conventional optical field-induced reorientation processes or any isotropic mechanism (such as thermal gratings) were not responsible for the observed phenomena.

The diffraction efficiency was also studied as a function of the external dc voltage. The experimental data showed the existence of an electric field threshold for grating formation, thus underlining the importance of the biased reorientation and deformation of the director field to achieve the highly sensitive photoelectric reorientation. Above the threshold voltage, η_p increased to a peak value and then decreased to zero. This was explained by taking into account that an increase of the external voltage leads to complete or nearly complete homeotropic alignment that does not allow for any remarkable additional light-induced grating modulation. In practice, MacDonald and coworkers assumed an interplay between the external longitudinal dc field $\mathbf{E}_{dc}$ and a light-induced transversal space-charge field parallel to the grating vector. The total field creates a torque on the LC director $\mathbf{n}$, which leads to a periodic modulation when $\mathbf{n} \parallel \mathbf{q}$, but not when $\mathbf{n} \perp \mathbf{q}$, since in this case the space-charge field is perpendicular to $\mathbf{E}_{dc}$ and is not large enough to overcome the threshold for an additional twist deformation. Such a scheme was in agreement with the observed diffraction efficiency anisotropy.

The existence of the internal space-charge field was proved by two-beam coupling (TBC) experiments. Both direct measurements of the energy transfer between the two writing beams and measurements based on the grating-translation technique [23] were performed. Results of the first kind of measurements are reported in Figure 16.11.

The coupling ratio, defined as the ratio between the signal intensities with and without the pump beam i.e., $\gamma_0 = I_w/I_{w/out}$, made it possible to calculate the

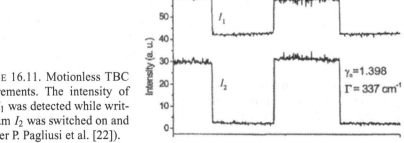

FIGURE 16.11. Motionless TBC measurements. The intensity of beam I_1 was detected while writing beam I_2 was switched on and off (after P. Pagliusi et al. [22]).

exponential gain coefficient Γ as

$$\Gamma = \left(\frac{\cos\theta}{d}\right) [\ln(\gamma_0) - \ln(2 - \gamma_0)], \tag{16.29}$$

where θ is the angle between one of the incident beams and the cell normal and d is the cell thickness. The result $\Gamma \approx 340 \text{ cm}^{-1}$ demonstrated the nonlocality of the effect under investigation, as it is clear from the relation holding between Γ and the phase shift φ between the interference pattern and the induced grating:

$$\Gamma = 4\pi \frac{\Delta n \sin\varphi}{\lambda \cos\theta}. \tag{16.30}$$

It should be noted here that it is questionable whether the use of a gain coefficient is valid in case of thin LC films, since Γ is a thick-grating concept that requires a gain increasing with thickness [24]. This is generally not true for wave-mixing experiments carried out in the Raman–Nath regime.

The second kind of experiment (grating-translation technique) made it possible to determine the modulation amplitude of the refractive index (Δn) and of the absorption coefficient ($\Delta\alpha$), and the spatial phase shifts of the phase (φ_p) and amplitude (φ_a) gratings with respect to the interference pattern. In these measurements the light-induced grating was translated along the direction of the grating vector, and the intensity of the transmitted beams was simultaneously recorded as a function of the shift. The shifting speed was fast enough to prevent decay of the grating during the experiment.

The results were analyzed taking into account that in case of low diffraction efficiency, the sum and difference of the two transmitted intensities versus the shift x are given by [23]

$$I_1 + I_2 = I_0 \exp\left(\frac{-\alpha d}{\cos\theta}\right)\left[2 - 4A\cos\left(\varphi_a + \frac{2\pi}{\Lambda}x\right)\right], \tag{16.31}$$

$$I_1 - I_2 = I_0 \exp\left(\frac{-\alpha d}{\cos\theta}\right)\left[-4P\sin\left(\varphi_p + \frac{2\pi}{\Lambda}x\right)\right], \tag{16.32}$$

where I_0 is the total incident intensity and

$$A = \frac{\Delta\alpha}{4\cos\theta}d, \qquad P = \frac{\pi\Delta n}{\lambda\cos\theta}d. \tag{16.33}$$

Figure 16.12 reports the two curves ($I_1 + I_2$) and ($I_1 - I_2$). The first curve was nearly independent of x, showing that no modulation of the absorption coefficient contributed to the grating. Moreover, from the curve ($I_1 - I_2$) a refractive index modulation $\Delta n = 1.17 \times 10^{-3}$ and a phase shift $\varphi_p = 90°$ were estimated. This phase shift is typical of pure diffusive charge transport (note that because of the nontilted geometry, the dc external field has a zero projection along the grating vector and does not give rise to any drift of the charge carriers). The obtained value of Δn corresponded to a nonlinear coefficient n_2 of about 17 cm^2/W.

It was the authors' opinion that this very high nonlinearity had the same photoelectric origin of that observed under uniform illumination. The observed TBC as well as the 90° phase shift were considered manifestations of the photorefractive

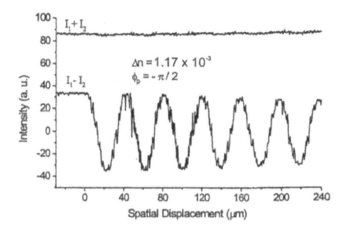

FIGURE 16.12. TBC grating-translation measurements, showing that the gratings obtained are pure phase gratings (refractive-index modulation) with a spatial phase shift of $\pi/2$ (after P. Pagliusi et al. [22]).

character of this nonlinearity. Concerning the basic mechanism, it was supposed that a longitudinal photoelectric field generated primarily from a dissociation of charge transfer states at the sample surfaces. In Section 16.6 we will see that additional experiments made it possible to propose a new model for a surface-driven photorefractive effect.

16.6 Phenomena Related to Charge Injection

The "conventional" photorefractive effect is a bulk effect based on light-induced charge generation near the irradiated surface and creation of a photo-current in the material bulk. In the last few years, several authors have reported experimental evidence of photorefractive-like effects in which the cell surfaces play a fundamental role in the process of generation of charge carriers, as also suggested at the end of [20] and [22].

In 2000, Zhang and coworkers reported the observation of a surface-mediated photorefractive effect in homeotropic LC cells both undoped and doped by a small quantity of C_{60} [25]. The effect did not involve bulk currents and was attributed to a light-induced modulation of the easy axis caused by electric charges at the interface between the LC and the alignment layer.

Using the standard geometry for grating writing and characterization, the authors performed three different kinds of experiments. When the writing beams were turned on with no applied external field, no self- or probe diffraction was observed, regardless of the incidence angle. If the writing beams were turned on *after* application of a 2-V dc field for 1 s, self-diffraction and TBC were observed. On the other hand, if the 2-V dc field was applied *after* the writing beams were turned on for 1 s, probe diffraction and TBC were also observed. The latter result

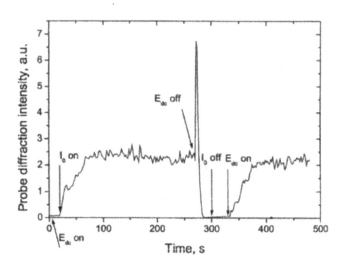

FIGURE 16.13. Surface-mediated photorefractive effect. Diffracted intensity (in arbitrary units) of a continuously incident He-Ne probe beam during a program of applying an electric field (E_{dc}) and a pair of Ar-laser writings beams (I_0) in the order indicated (after J. Zhang et al. [25]).

showed that the recorded gratings could be hidden and then revealed by turning a dc field off and on.

To gain more insight into the effect under investigation, the following experiment was performed: a dc field was applied to the cell, followed by the writing beams, and the probe diffraction was studied as a function of time. The dynamics of grating recording is shown in Figure 16.13. When the external field was switched off after about 280 s, a sharp temporary increase of probe and self-diffraction occurred. Then the signal decreased to zero before the writing beams were turned off. However, a grating remained stored in the cell, as evident from the signal increase observed when the dc field was applied again.

The model proposed by the authors was based on the light-induced generation of a charge modulation at the aligning irradiated surface, having the same spatial dependence of the incident intensity distribution:

$$\rho^{\pm}(x, z = 0) \approx (1 + \beta \cos qx), \tag{16.34}$$

where x is the direction parallel to the grating vector $\mathbf{q}$, and z the direction along the sample thickness. If charges of one sign are trapped at the surface, the external dc field leads to charge separation and produces a surface-charge grating. This grating generates in turn a modulated electric field E_ρ that has two components, one along x and one along z:

$$E_x(x, z = 0) = E_\rho \sin qx,$$
$$E_z(x, z = 0) = E_{\rho z} + E_\rho \cos qx. \tag{16.35}$$

The total electric field may in principle act on the LC director in two ways. First of all, it can create a torque on the LC bulk, as in the conventional photorefractive effect. Alternatively, it can give rise to a surface torque due to the modulation of the surface anchoring and then of the easy axis (the preferred LC orientation at the surface). This modulation may appear as a result of a locally strong x component of the surface-charge field E_ρ. In both cases a director rotation in the LC bulk would be obtained and the orientational modulation would be phase shifted with respect to the intensity distribution; the first experimental results could be explained by either of the two mechanisms.

However, an analysis of the grating decay after switching off the dc field (Figure 16.13) allows one to distinguish between them. In fact, the dc field not only creates the driving force for the grating formation, but also stabilizes the initial homeotropic alignment. Therefore, if the bulk torque were responsible for the observed effect, the grating would simply disappear in the absence of the optical and dc fields. On the other hand, if anchoring modulation were responsible for the effect, switching off the dc field would lead (as actually observed) to a temporary enhancement of the grating, because there would be no stabilizing action of the dc field, whereas there would be an easy-axis grating at the beginning of the relaxation. The authors concluded that the second mechanism dominated in their experiments.

Cipparrone and coworkers have recently proposed a new model to explain experimental observations in which the photorefractive effect appears strongly affected by the nature of the interfaces, starting from the photoelectric effect discussed in the previous section [26, 27]. In their first experiments performed in cells with different alignment agents filled with different nematic LCs, the authors found that the effect was present only in some of the analyzed samples. In particular, only certain combinations of surfactant and LC gave rise to highly efficient photorefractive gratings under the combined application of low-intensity laser beams and low dc voltage, suggesting that the observed phenomenon could be ascribed not only to the optical sensitivity of the aligning and/or the liquid-crystalline materials, but also to a chemical or physical affinity between the two components at the interface.

Figure 16.14 shows the diffraction efficiency vs. time of the obtained gratings for a p-polarized probe beam, with the time delay varying between the dc voltage application and the switching of light irradiation (referred to as preapplication time), the voltage being 4 V for 25-μm-thick cells. In all the curves, the diffraction grows in correspondence to the beginning of light irradiation with a total intensity $I = 1.8$ mW/cm^2. The solid line refers to the efficiency evolution in the case of simultaneous light illumination and voltage application. The dashed and dotted lines refer to 1-min and 2-min preapplication times, respectively. The exponential decay of the signal (with a characteristic time of about 30 s), observed by removing the pump beams, is also shown for the 2-min curve. The relaxation observed by removing the dc field is much faster, on the order of 1 s. It is evident that the rise time depends on the preapplication time. These results suggested that the grating formation was not a pure LC bulk effect, highlighting the role of

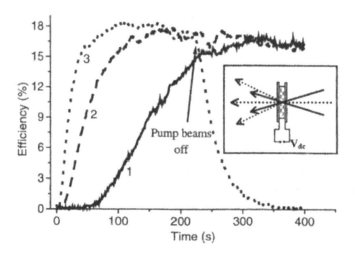

FIGURE 16.14. Surface-induced photorefractivity in a PVA-E7 cell. Time dependence of the diffraction efficiency for a p-polarized probe beam. Grating period $\Lambda = 40\,\mu m$, applied dc voltage $V_{dc} = 4.0$ V, and writing-beam intensity $I = 1.8$ mW/cm^2. In curve 1, the writing beam and the voltage are applied at the same time. In curves 2 and 3, the voltage is applied 1 and 2 min before illuminating. In the inset, the pump-probe experimental setup is shown (after P. Pagliusi et al. [26]).

the surfaces. According to the authors' interpretation, the dc field not only reoriented the LC director from the planar configuration, but also worked as a charge accumulator at the LC–substrate interface. During light irradiation the intensity pattern locally modifies the surface-charge density through recombination or diffusion mechanisms, leading to a modulated total electric field. This field may, in turn, produce a phase-shifted grating through molecular reorientation in the bulk. From this point of view, the observed photorefractive effect was interpreted as a photoelectric process activated by the interface, rather than an effect connected to the light-induced surface-anchoring modulation.

In order to give additional evidence in support of this interpretation, Cipparrone and coworkers carried out several other experiments, focusing their attention on LC cells of the nematic E7 (eutectic mixture of cyanobiphenyl compounds) aligned by PVA (poly(vinyl alcohol)) [27]. They measured the dark current in samples with and without PVA, the steady-state photocurrent, the photocurrent dynamics for different dc voltage and light-power values, and the photocurrent as a function of the incident wavelength. The first kind of measurements showed that the saturation current in dark conditions had an exponential dependence on the square root of the applied voltage, with different slopes for the low- and the high-voltage regimes. The experimental results also showed that the absence of the alignment layer did not cause a substantial difference in the dark-current curve, indicating that PVA did not prevent charge injection from the ITO electrodes (indium tin oxide coatings of the cell boundary glasses). The measurements performed on single-layer systems (only the LC film between ITO electrodes or PVA layer between ITO and Ag) did not show significant photoinduced current, thus excluding charge photogeneration in the ITO layers and in the bulk of both the LC and the polymer. Light-induced charge-injection enhancement at the ITO–PVA and ITO–LC interfaces could also be ruled out. The measured photocurrent in PVA + E7 samples pointed out the crucial role played by the PVA–LC interface. Further experiments on an asymmetric cell, in which the PVA layer was coated on one ITO glass substrate only, indicated that the ITO–LC interface was poorly photoactive and that its contribution to the photocurrent was negligible. More-over, the photosensitive interface (PVA–LC) was more efficiently activated when connected to the anode, suggesting that also in the symmetric cell, the anodic interface was quite completely responsible for the detected photocurrent signal. According to the authors' interpretation, all the experimental findings indicated a need to take into account a multilayer structure in which charge accumulation and injection at the internal interfaces are involved. In Figure 16.15 a schematic representation of the proposed light-induced processes is reported.

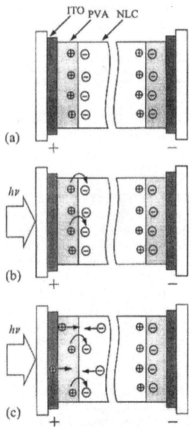

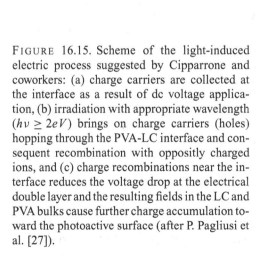

FIGURE 16.15. Scheme of the light-induced electric process suggested by Cipparrone and coworkers: (a) charge carriers are collected at the interface as a result of dc voltage applica-tion, (b) irradiation with appropriate wavelength ($h\nu \geq 2eV$) brings on charge carriers (holes) hopping through the PVA-LC interface and con-sequent recombination with oppositely charged ions, and (c) charge recombinations near the in-terface reduces the voltage drop at the electrical double layer and the resulting fields in the LC and PVA bulks cause further charge accumulation to-ward the photoactive surface (after P. Pagliusi et al. [27]).

During the first phase, in which only the dc field is applied, one can suppose the presence of an interfacial energy barrier that limits the charge carriers crossing through the surface, even if a reduced carrier flow is always present, as shown by dark-current measurements. As a result, a reservoir of one kind of charge carriers (the holes), injected by the electrode through the ITO–PVA interface, accumulates in the PVA region, while in the LC region an opposite atmosphere of ions is present due to impurities or induced molecular dissociation (Figure 16.15a). The equilibrium condition for the electrical double layer is achieved when the dark current reaches the stationary state. Then, with the onset of light irradiation of proper wavelength, the charge flow through the interface grows up due to recombination of the holes with the negative ions present on the opposite side of the interface (Figure 16.15b). This light-stimulated charge injection through the interface can occur if the irradiation induces a change in the effective interfacial potential barrier, thus allowing the jump of the holes. The depletion of surface charges reduces the voltage drop at the interface and increases the total electric field in the bulk, thus producing an accumulation of additional charges from the LC and PVA films near the surface (Figure 16.15c). A new steady state is reached when the charges hopping across the interface, the recombination process, and the charges' accumulation near the surface reach equilibrium and the photocurrent reaches a saturation value.

According to this model, the measured photocurrent is caused by the current of holes across the PVA–LC interface and by the bulk current of ions and holes in the LC and PVA layers. The flux of charges across the interface is mainly driven by the electric field inside the double layer, initially high, while the bulk current is due to the bulk electric field, initially low. This last consideration is also supported by the observation of different time constants characterizing the photocurrent dynamics.

An enhanced photorefractive effect was also reported by Bartkiewicz et al. in a hybrid structure made by a nematic LC sandwiched between two photoconductive polymeric layers [28]. The polymeric layers were coated on the ITO glass substrates and had the double function of photoconductive and alignment layers. Photorefractive gratings were obtained with a standard configuration in tilted geometry, with the aid of an externally applied dc voltage above the threshold of the Freederickzs transition. The maximum photorefractive gain obtained by TBC experiments was about 2600 cm^{-1} with an external applied field of 1 V/μm (note that even in this case the gain concept is used for evaluating the energy transfer in thin gratings). In the explanation given by the authors, the space-charge field generated along the photoconductive layers, combined with the external field, gave rise to a complex modulated pattern in the LC leading to a periodic director reorientation. The same authors have reported several other studies on this subject [29] [30]; and also dynamic effects have been investigated by them [31] and by Serak at al. [32].

A similar structure based on the simultaneous presence of a photoconductive and an insulating polymeric layer was proposed by Ono and Kawatsuki [33]. The samples used consisted of the following sequence: ITO-coated glass substrate, photoconductive layer, insulated layer, nematic LC, insulated layer, photoconductive layer, and ITO-coated glass substrate. Poly(N-vinylcarbazole) (PVK) was used as photoconductor, poly(vinyl alcohol) (PVA) was the insulator, and the nematic ZLI2061 was the LC. The total cell thickness was 50 μm. Gratings were written by means of two linearly polarized He-Ne laser beams of equal power (8 mW), which impinged on the sample at oblique incidence. An external dc voltage (0–500 V) was applied to the cells. The obtained diffraction efficiency was about 10% for a grating spacing of 20 μm. The efficiency showed also a dependence on the applied voltage, reaching its maximum for 400 V. The dynamics of grating formation was evaluated by measuring the diffraction efficiency vs. time shutting off one of the two writing beams. The rise time of the signal was 0.5 s and the decay time was 0.2 s, after 2.6 s illumination with both beams,. The decay time for usual bulk reorientation in 50-μm cells is on the order of a second [14]; then the observed slow dynamics suggested photocharge generation, transport, and trapping as dominant processes in grating generation and decay. According to the authors, the mechanism for grating formation was the following: in the absence of applied voltage the LC cell is planar, while with the voltage on, the configuration changes to homeotropic. Photocarriers are generated by the incident interference pattern in the photoconductive PVK layer and are brought to the insulated PVA region, where they are eventually trapped, producing a space-charge field. The importance of the insulating layer was confirmed by the observation of diffraction only in samples in which PVA was present. The observed gratings could not be obtained by substituting the external dc field with an ac field, or by using a normal incidence configuration. These experimental findings indicated that the dc fields actually induced charge separation via carriers drifting along the grating vector and possibly also carriers jumping from the photoconductive to the insulated layers.

16.7 Photorefractivity in Polymer-Dispersed Liquid Crystals

In polymer-dispersed liquid crystals (PDLCs), nematic liquid crystals are dispersed as micro-sized droplets in a polymeric matrix [14]. Droplets are randomly distributed in the polymer and generally, having a size close to the visible wavelength, produce a strong scattering of the incident light. A large variety of morphologies is possible, depending on the concentration, nature, and properties of the polymer and the liquid crystal. The mechanism of droplet formation is phase separation of the initial prepolymer–liquid-crystal mixture, which can be induced by polymerization (polymerization-induced phase separation: PIPS), thermal quenching (thermal-induced phase separation: TIPS) or solvent evaporation (solvent-induced phase separation: SIPS). The actual procedure used is very important in determining the overall properties of the sample. PDLCs can be switched from an opaque to a transparent state by the application of an external electric field. In fact, in the absence of an applied field, the symmetry axis of the liquid-crystal droplets (i.e., the droplets' director) is randomly oriented. In this case the refractive index mismatch between droplets and the polymer matrix produces a strong light scattering, and the sample looks opaque. When an electric field of suitable strength is applied, the droplets' directors are collectively aligned parallel to the field. If, as usual, the value of the ordinary refractive index of the droplet is close to that of the matrix, this reorientation reduces the refractive index mismatch and the sample becomes transparent.

It is challenging to use PDLCs as photorefractive media, since in these materials the required photoconductive properties can be provided by the polymer matrix, and the field-dependent refractive index change can be enhanced by reorientation of the liquid-crystal droplets. Moreover, the possibility of switching the samples from an opaque to a transparent state can be used to erase and restore the recorded gratings and thus to modulate the diffraction efficiency, opening the possibility of creating switchable holograms [34, 35]. Experimental evidence of photorefractivity in these materials was provided by Cipparrone et al., who reported the formation of permanent gratings in fully cured dye-doped PDLCs due to a photorefractive-like effect obtained without the application of external electric field [17, 36]. The necessary writing intensity ranged from 40 to 10^3 W/cm^2 depending on the exposure time and on the materials used, and the maximum diffraction efficiency was 21% for the p-polarized probe beam and for a grating period of 25 μm. Gratings could be erased by irradiating them with a single beam and rewritten in the same spot, repeating the cycle many times. The observed gratings were demonstrated to be phase gratings due to the liquid crystal reorientation inside the droplets. Morphological characterization performed by scanning electron microscope allowed the exclusion of grating-like modulations of droplet size or shape, in correspondence to the irradiated regions. Moreover, the measurements of diffraction efficiency and the optical-microscope observations showed a liquid-crystal alignment normal to the interference maxima. In particular, the observed optical behavior indicated that the LC reorientation inside the droplets was responsible for the grating structure.

These experimental results could be understood by thinking of a periodic distribution of regions in which the director was aligned perpendicular to the grating lines, and others in which a random alignment was present. Several hypotheses were considered in order to explain the physical mechanism that allowed a permanent liquid-crystal reorientation inside the droplets. Photoinduced phenomena in dye, such as photoisomerization, anisotropic molecular diffusion, enhanced optical reorientation, and orientation mechanisms of the polymeric matrix, were taken into account. However, all these hypotheses were dropped after a detailed analysis of the experimental results. Indeed, all the experimental data found a satisfactory explanation in the onset of a photorefractive-like effect causing the liquid-crystal droplets' reorientation through a space-charge field produced by light-induced charge separation in the material. In this scheme the role of the dye is to increase light absorption, producing a local temperature rise, and to act as photogenerator of charges.

Later on, several additional experiments were performed on the same dye-doped PDLCs; all of them confirmed this model and gave a deeper understanding of the overall phenomenon. TBC measurements were carried out, and the results confirmed the photorefractive nature of the observed gratings [37]. In particular, the grating-translation technique showed that gratings were essentially due to phase modulation, although weak amplitude modulations were observed in some cases. These latter were accounted for by taking into account the reorientation of the dichroic dye molecules along the director orientation, producing a weak modulation of the absorption coefficient. The phase shift between the interference pattern and the induced gratings was 90°, which is typical of charge separation induced by pure carrier diffusion. The authors also found some peculiar features different from those reported in similar systems. In fact, the memory effect had an extremely long time stability, and the spatial frequency of the obtained gratings was the same as that of the interference pattern, whereas for LC reorientation one should expect grating spatial frequency twice that of the intensity distribution due to the quadratic response to the field.

Concerning the time stability, the authors suggested that the interaction between polymer and LC at the droplet's interface plays a fundamental role. In fact, during the writing process the sample is locally heated due to the strong absorption. Under these conditions the photorefractive space-charge field and the temperature gradient can modify the droplets' interfaces and the LC orientation inside the droplets. After removing the writing beams, the new configuration is frozen by the cooling process [38]. As a matter of fact, this model allows a satisfactory explanation of the dependence of the diffraction efficiency on the incident intensity, as shown in Figure 16.16. The growth of the efficiency for intensity below 125 W/cm^2 can be associated with the rise of the local temperature necessary

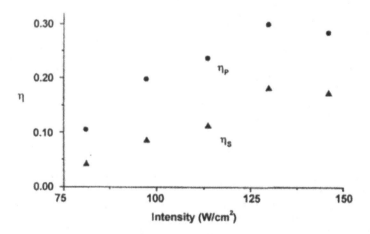

FIGURE 16.16. Diffraction efficiency versus writing intensity ($I_1 + I_2$) for PDLC made out of epoxy resin and E7 doped by D2. Writing beams p polarized, exposure time 2 s, and $\Lambda = 20\,\mu$m (after F. Simoni et al. [38]).

to obtain the memory effect. Above this value, η is independent of the incident intensity, as expected for a steady-state photorefractive space-charge field. This fact also explained the need of high-intensity values to write the gratings, due to the heating necessary to modify permanently the LC–polymer interface, changing the anchoring conditions. In this scheme the action of the photorefractive effect is combined with other mechanisms, such as local heating, to produce a permanent anchoring modification and, as a consequence, a permanent LC reorientation.

Concerning the nonconventional spatial periodicity of the gratings, the authors proposed the existence of a light-induced internal longitudinal electric field that, together with the huge number of droplet interfaces, broke the inversion symmetry of the system [34, 35]. Under these conditions, the interface modifications could be ascribed to a linear electrooptic effect that gave rise to gratings with the same spatial periodicity of the intensity pattern. The presence of the longitudinal space-charge field was tested in an experiment in which a single light beam impinged on the sample and an intensity gradient was induced in the longitudinal direction by dye absorption [38]. The transmitted intensity as a function of time was measured while an ac low-frequency voltage was simultaneously applied. The results were compared to the same curves obtained in regions irradiated by a laser beam of the same intensity used to write the gratings (see Figure 16.17).

When the ac voltage was applied, the light transmission was modulated at twice the voltage frequency, due to the LC quadratic response. In the illuminated regions,

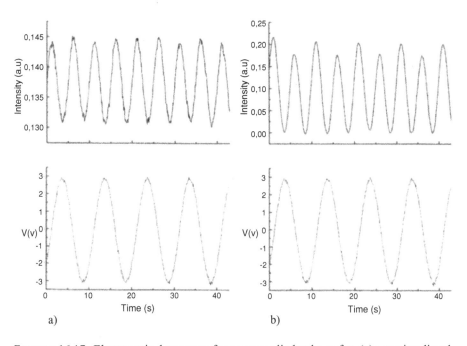

FIGURE 16.17. Electrooptical response for an ac applied voltage for: (a) a nonirradiated region; and (b) an irradiated region (after F. Simoni et al. [38]).

during the negative half-period of the applied voltage, the transmission reached values lower than those corresponding to the positive half-period, whereas in the absence of illumination the values were equal. This observation indicated the presence of an internal field alternately added or subtracted to the external voltage. After about 15 min, the effect disappeared, indicating internal field relaxation, but the overall transmission of the irradiated spot remained higher than that of the nonirradiated region, showing a permanent action induced by the internal field. The internal field was estimated by adding a bias necessary to restore the initial transmission. After few minutes from irradiation, the value 0.03 V/μm was obtained.

In a similar way, in the case of two-beam irradiation, i.e., during grating writing, the light absorption can produce two electric fields along the two directions of propagation. However, if only absorption is involved, the two intensities would remain always equal and there would be no mechanism of symmetry breaking. Actually, the energy transfer between the writing beams also plays an important role: in the direction of the amplified beam the intensity gradient is reduced as well as the light-induced longitudinal field; in the other direction the intensity gradient increases, resulting in a higher induced field. In this way the different propagation of the writing beams can be responsible for symmetry breaking during grating formation.

The space-charge fields along the two directions s_1 and s_2 were evaluated starting from the intensity distribution of each transmitted beam [39]. Taking into account the absorption and the energy transfer due to the 90° phase shift between the interference pattern and the induced grating, in case of low diffraction efficiency the following relations hold:

$$I_1(s_1) = I_0 \exp(-\alpha s_1)(1 + Cs_1),$$
$$I_2(s_2) = I_0 \exp(-\alpha s_2)(1 - Cs_2), \qquad (16.36)$$

where I_0 is the incident intensity and $C = 2\pi \Delta n/\lambda$. These intensity distributions led to the following expressions for the fields in steady state [39]:

$$E_1(s_1) = A \left(\frac{C}{Cs_1 + 1} - \alpha \right),$$
$$E_2(s_2) = A \left(\frac{C}{Cs_2 - 1} - \alpha \right), \qquad (16.37)$$

with $A = [(D^+ - D^-)/(D^+ + D^-)]2k_B T/e$. In the experiments reported, the geometry is symmetric and $s_1 = s_2$, that is, $E_2 > E_1$ for any z.

The scheme of Figure 16.18 explains how the presence of the two space-charge fields breaks the symmetry and produces a phase grating with the same spatial periodicity of the incident intensity pattern. In this figure, E_q is the photorefractive space charge field, $E_{1,2}$ is the sum of E_1 and E_2, and E_{tot} is the sum of E_q and $E_{1,2}$. As can be seen, when, for instance, δ is $\pm\pi/2$, E_{tot} changes both in amplitude and direction, inducing two different LC reorientations inside the droplets. In this way the refractive index modulation has the same periodicity as the interference

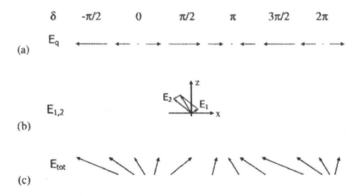

FIGURE 16.18. Space-charge electric field scheme: (a) due only to the interference light pattern contribution, (b) resultant electric field due to the components along s_1 and s_2, (c) total internal electric field resulting from the composition of all the contributions (see text) (after G. Cipparrone et al. [39]).

pattern. On the other hand, if E_q would change only the sign for $\delta = \pm\pi/2$, the grating vector would be twice that of the interference pattern.

The presence of the longitudinal internal fields was also confirmed by photocurrent measurements [36, 37]. Such measurements made it possible also to characterize the processes of photocharge generation and transport and to indicate the positive holes as the mobile carriers responsible for the phenomenon. They seemed to be generated through excitation of an already excited long-lived triplet state of dye molecules.

A more conventional approach to photorefraction in PDLCs was proposed by Golemme and coworkers, who reported the observation of photorefractive gratings in fully cured PDLCs using an applied voltage of 20 V/μm and obtaining a diffraction efficiency of 56% at 3-μm spacing [38]. The materials used were obtained from poly(methylmethacrylate) (PMMA) as polymer matrix, N-ethylcarbazole (ECZ) as hole-transport agent, a sensitizer forming a charge-transfer complex with ECZ providing photosensitivity at 633 nm, and the nematic LC TL202. TL202 is a eutectic mixture with positive dielectric anisotropy and a low intrinsic birefringence and was therefore expected to give rise to low scattering losses.

Wave mixing in tilted geometry and TBC measurements were performed, confirming the photorefractive nature of the obtained thick (Bragg) gratings. As already mentioned, the maximum value of diffraction efficiency was 56%, thanks to the reduced scattering losses of the material. The total losses for applied fields higher than 18 V/μm and for an incidence angle of 60° to the cell normal were about 40%; the decrease of scattering with applied field was due to LC reorientation inside the droplets. The diffraction efficiency and the transmission vs. the applied field are both shown in Figure 16.19. No diffraction was observed in samples without the sensitizer, thus indicating the importance of photoconduction in grating formation. Moreover, no diffraction was present for LC concentrations

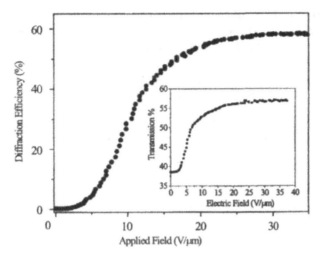

FIGURE 16.19. Diffraction efficiency vs. applied field measured in a 53-μm-thick sample of TL202:PMMA:ECZ:TNFDM at 633 nm. Grating spacing is 3 μm. The inset shows the variation of transmission vs. applied field (after A. Golemme et al. [38]).

lower than 25% in weight; scanning electron microscope observations revealed that in this case LC was dissolved in PMMA and no phase separation occurred.

16.8 Conclusion

In this chapter we have reviewed the main features of the photorefractive effect in low-molecular-weight liquid crystals and in polymer-dispersed liquid crystals, underlining the peculiarities that this phenomenon may have in these materials. In fact, their collective reorientation plays a dominant role in the light-induced modulation of the refractive index; then effects different from charge-diffusion anisotropy may be involved such as the anisotropy of the dielectric permittivity and of the conductivity, thermal diffusivity, and interfacial phenomena. We have reported the main experimental results together with the models proposed to explain them. This review certainly shows that work is in progress, that is, many answers are not the final ones and additional investigations are necessary to get a clear picture of the complex mechanisms involved. For instance, we would like to recall that after the observation of the supranonlinear response in dye-doped liquid crystals [15] (already mentioned in Section 16.3) several other experiments confirmed this behavior and a new model has been proposed to explain those amazing results [43–48]. This model does not consider photorefractivity as the basic mechanism, but gives the main role to the light-induced modifications of the anchoring conditions, which in turn produce the LC reorientation through its elasticity. This is why people speak of SINE (surface-induced nonlinear effect), which does not need a direct coupling between the field of the optical wave and the molecular director in the bulk. The major role of the interfaces in this process has been recently underlined in an experiment performed to measure the nonlinear response in thin randomly oriented nematics doped by the azo dye methyl red (MR) [49]. In this case a nonlinear refractive index exceeding 10^3 cm^2/W has been measured, over two orders of magnitude higher than the one defined as supranonlinear, and for this reason this response has been termed "colossal nonlinearity" (about 10^8 times higher than the giant optical nonlinearity [GON] of liquid crystals) [49]. Actually, TBC experiments have shown that photorefractivity may contribute to this nonlinearity, but it doesn't seem to play the predominant role.

In summary, the aim of this chapter has been mainly to show current research directions concerning photorefractivity in liquid crystals, pointing out several open problems that still need a definitive solution.

References

1. A. Ashkin, G. Boyd, J.M. Dziedzic, R.G. Smith, A.A. Ballman, J.J. Levinstein, K. Nassau. *Appl. Phys. Lett.* **9**, 72 (1966).
2. P. Gunter, J.P. Huignard, eds. *Photorefractive Materials and Their Applications* (Springer, Berlin 1988).
3. S. Ducharme, J.C. Scott, R.J. Tweig, W.E. Moerner. *Phys. Rev. Lett.* **66**, 1846 (1991).
4. W.E. Moerner, S.M. Silence. *Chem. Rev.* **94**, 127 (1994).
5. P.G. De Gennes and J. Prost. *The Physics of Liquid Crystals* (Oxford University Press, Oxford 1993).
6. W. Helfrich. *J. Chem. Phys.* **51**, 4092 (1969).
7. E.V. Rudenko and A.V. Sukhov. *JEPT Lett.* **59**, 142 (1994).
8. E.V. Rudenko and A.V. Sukhov. *Sov. Phys. JEPT* **78**, 875 (1994).
9. E.V. Rudenko and A.V. Sukhov. *Mol. Cryst. Liq. Cryst.* **282**, 125 (1996).
10. G.P. Wiederrecht, B.A. Yoon and M.R. Wasielewski. *Science* **270**, 1794 (1995).
11. I.C. Khoo, H. Li, Y. Liang. *Opt. Lett.* **19**, 1723 (1994).
12. I.C. Khoo. *Opt. Lett.* **20**, 2137 (1995).
13. I.C. Khoo. *Mol. Cryst. Liq. Cryst.* **282**, 53 (1996).
14. F. Simoni. *Nonlinear Optical Properties of Liquid Crystals and Polymer Dispersed Liquid Crystals* (World Scientific, Singapore 1997).
15. I.C. Khoo, S. Slussarenko, B.D. Guenther, M.Y. Shih, P. Chen and M.V. Wood. *Opt. Lett.* **23**, 253 (1998).
16. N.V. Tabiryan, C. Umeton. *Phys. Rev. E* **58**, 4619 (1998).
17. G. Cipparrone, A. Mazzulla, F.P. Nicoletta, L. Lucchetti, F. Simoni. *Opt. Comm.* **150**, 297 (1998).
18. R. Macdonald, P. Meindl, G. Chilaya and D. Sikharulidze. *Opt. Comm.* **150**, 195 (1998).
19. R. Macdonald, P. Meindl, G. Chilaya and D. Sikharulidze. *Mol. Cryst. Liq. Cryst.* **320**, 115 (1998).
20. R. Macdonald, P. Meindl, and S. Busch. *J. Nonlinear Opt. Phys. Mat.* **8**, 379 (1999).
21. C. Umeton, A. Sgrò and F. Simoni. *J. Opt. Soc. Am. B* **4**, 1938 (1987).

22. P. Pagliusi, R. Macdonald, S. Busch, G. Cipparrone, M. Kreuzer. *J. Opt. Soc. Am. B* **18**, 1632 (2001).
23. K. Sutter, P. Gunter. *J. Opt. Soc. Am. B* **7**, 2274 (1990).
24. W.E. Moerner, A. Grunnet-Jepsen, and C.L. Thompson. *Ann. Rev. Mater. Sci.* **27**, 585 (1997).
25. J. Zhang, V. Ostroverkhov, K.D. Singer, V. Reshetnyak, Y. Reznikov. *Opt. Lett.* **25**, 414 (2000).
26. P. Pagliusi and G. Cipparrone. *Appl. Phys. Lett.* **80**, 168 (2002).
27. P. Pagliusi and G. Cipparrone. *J. Appl. Phys.* 92, 4863 (2002).
28. S. Bartkiewicz, A. Miniewicz, F. Kajzar, M. Zagorska. *Mol. Cryst. Liq. Cryst.* **322**, 9 (1998).
29. F. Kajzar, S. Bartkiewicz, and A. Miniewicz. *Appl. Phys. Lett.* **74**, 2924 (1999).
30. S. Bartkiewicz, K. Matczyszyn, A. Miniewicz, and F. Kajzar. *Opt. Comm.* **187**, 257 (2001).
31. S. Bartkiewicz, A. Miniewicz, B.Sahraoui, and F. Kajzar. *Appl. Phys. Lett.* **81**, 3705 (2002).
32. S. Serak, A. Kovalevand, A. Agashkov. *Opt. Comm.* **181**, 391 (2000).
33. H. Ono, N. Kawatsuki. *Opt. Comm.* **147**, 237 (1998).
34. A. Golemme, B.L. Volodin, B. Kippelen, and N. Peyghambarian. *Opt. Lett*, **22**, 1226 (1997).
35. H. Ono and N. Kawatsuki. *Opt. Lett.* **22**, 1144 (1997).
36. G. Cipparrone, A. Mazzulla, F.P. Nicoletta, L. Lucchetti, F. Simoni. *Mol. Cryst. Liq. Cryst.* **320**, 249 (1998).
37. G. Cipparrone, A. Mazzulla, F. Simoni. *Opt. Lett.* **23**, 1505 (1998).
38. F. Simoni, G. Cipparrone, A. Mazzulla, P. Pagliusi. *Chem. Phys.* **245**, 429 (1999).
39. G. Cipparrone, A. Mazzulla, P. Pagliusi. *Opt. Comm.* **185**, 171 (2000).
40. G. Cipparrone, A. Mazzulla, P. Pagliusi, F. Simoni, A.V. Sukhov. *Mol. Cryst. Liq. Cryst.* **359**, 439 (2001).
41. G. Cipparrone, A. Mazzulla, P. Pagliusi, A.V. Sukhov, R.F. Ushakov. *J. Opt. Soc. Am. B* **18**, 182 (2001).
42. A. Golemme, B. Kippelen, and N. Peyghambarian. *Appl. Phys. Lett.* **73**, 2408 (1998).
43. F. Simoni, L. Lucchetti, D.E. Lucchetta, O. Francescangeli. *Opt. Express* **9**, 85 (2001).
44. L. Lucchetti, D.E. Lucchetta, O. Francescangeli, and F. Simoni. *Mol. Cryst. Liq. Cryst.* **375**, 641 (2002).
45. L. Lucchetti. D.E. Lucchetta, O. Francescangeli, and F. Simoni. *Proceedings of SPIE* **4457**, 1 (2001).
46. L. Lucchetti, M. Di Fabrizio, O. Francescangeli, and F. Simoni. *J. Nonlinear Opt. Phys. Mat.* **11**, 13 (2002).
47. A. Petrossian and S. Residori. *Europhys. Lett.* **60**, 79 (2002).
48. L. Lucchetti. *Europhys. Lett.* **61**, 573 (2003).
49. L. Lucchetti, M. Di Fabrizio, M. Gentili, and F. Simoni. *Opt. Comm.* **233**, 417 (2004).

17

Photorefractive Effects in Organic Photochromic Materials

Eunkyoung Kim

Department of Chemical Engineering, Yonsei Univ., 134, Sinchon-dong, Seodaemun-gu, Seoul, 120-749, South Korea

17.1 Introduction

Photochromism [1, 2] is reversible color change by light. This color change is usually accompanied by photoinduced transformations between two molecular states. Thus incorporating photochromic molecules into organic, inorganic, or organic-inorganic hybrid materials leads to photoresponsive systems, the properties of which can be manipulated by light. The switching of metal-ion capture [3] and energy or electron transfer [4, 5], photoinduced birefringence, dichroism, nonlinear optical (NLO) properties, phase separation [6, 7], and photostimulated conformational changes of polymer chains can occur as a consequence of a photochromic reaction.

In particular, photochromism is accompanied by a change in the index of refraction of a material in response to an optical beam [8, 9]. This change in a photochromic medium allows the reversible light-induced modulation of the refractive index and/or birefringence by the light beams interfering with a photochromic medium to produce photorefractive (PR) effects. Since the PR effects in organic photochromic materials arise from the local area (in the vicinity) of the photochromophore, the PR effects are generally *local* in nature. Therefore, the mechanism responsible for these photochromic PR effects, generated by absorption changes in photochromic materials, differs from the *non-local* nature of the PR effect mechanisms based on the physical motion of charge carriers (electric charge) in electrooptic materials. However, when the charge-carrier generation is accompanied by absorption changes, the photochromic grating (PCG) process is often mixed with typical photorefractive gratings (PRG) based on the EO effect, as observed in several photorefractive (PR) crystals [10].

Apart from such PCG, organic photochromic materials, which are generally π-conjugated systems, have potential nonlinear optical (NLO) properties that are necessary to produce the PR effect in organic PR materials. Therefore, photochromophores, such as azobenzenes, have been utilized as NLO materials in PR media to produce a reversible light-induced modulation of the refractive index via the electrooptic (Pockels) effect [11]. The photochromic control of noncentrosymmetric orders in a bulk can be achieved through the combination of photochromism and the NLO properties in the molecules to allow for new techniques such as photoassisted poling (PAP) and all-optical poling (AOP). The possibility of switching between two different NLO responses through the irradiation of a material having both NLO and photochromic properties is attractive in applications of optical signal processing [12].

Table 17.1 lists the photochromic isomerization of organic photochromophores and dipole moments, and the first hyperpolarizabilities of the colorless to the

TABLE 17.1. Dipole moments[a] and first hyperpolarizabilities[b] of photochromes [12]

Photochromophore (Most stable or colorless form)	Photochromophore (Colored form)	Method[b]
Azobenzene (e.g., DR1) trans DR1 $\mu=8.6D$ $\beta=44.6 \times 10^{-30}$ esu	cis DR1 $\mu=6.3D$ $\beta=8.4 \times 10^{-30}$ esu	AM1/FF
Spiropyran $\mu=7.5D$ $\beta=1.9 \times 10^{-30}$ esu	Merocyanine $\mu=13.6D$ $\beta=-40 \times 10^{-30}$ esu	AM1/FF
Furylfulgide (Aberchrome540®) $\mu=7.2D$ $\beta=6.6 \times 10^{-48}$ esu	Dihydrobenzofuran derivatives $\mu=6.5D$ $\beta=91.10^{-48}$ esu	EFISH
Diarylethene (open) $\beta\mu=15 \times 10^{-48}$ esu	Diarylethene (closed) $\beta\mu=55 \times 10^{-48}$ esu	EFISH at 1907 nm
Diarylethene push-pull (open) $\beta\mu=260 \times 10^{-48}$ esu	Diarylethene push-pull (closed) $\beta\mu=1100 \times 10^{-48}$ esu	EFISH at 1054 nm
N-salicylidene-4-bromoaniline (yellow, stable) $\mu=2.9D$ $\beta=2.3 \times 10^{-30}$ esu	(red, unstable) $\mu=2.7D$ $\beta=1.3 \times 10^{-30}$ esu	AM1/FF

[a] All dipole moments given in this table have been calculated using the AM1/FF method.
[b] The method of determining β is given in the third column. The values obtained by the AM1/FF method are β_0 values. All the values are moduli of the vectorial part of β projected along the direction of the ground-state dipole moment (taken from [12]).

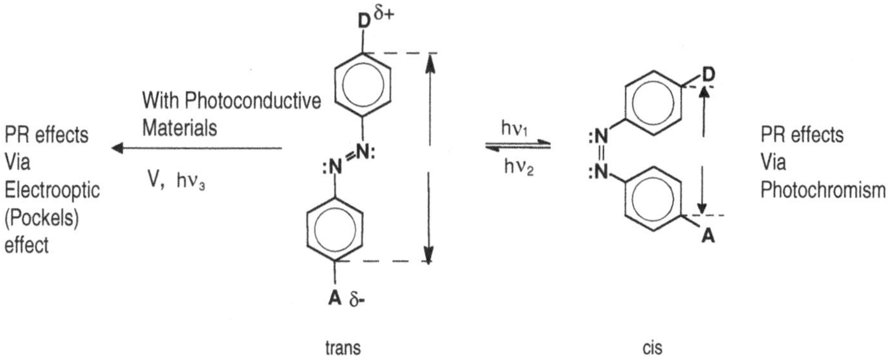

FIGURE 17.1. Isomerization (trans→cis) reaction of the azobenzene group and the process for PR effects.

colored isomers. A comprehensive review of the second-order nonlinear optical polarizabilities of photochromic molecules can be found elsewhere [12]. In this chapter, we discuss the role and effects of organic photochromic materials to generate PR effects such as optical poling, grating, and holographic recording.

17.2 Azobenzene and Azo-Containing Materials

The photoinduced phenomena of azo polymers and azo-containing hybrid materials are generally efficient due to the trans↔cis isomerization reaction to the N=N double bond of the azobenzene group (Figure 17.1). The isomerization proceeds reversibly with high quantum yields even in solid matrixes. Such photo-induced isomerization yielding to color and geometry change can lead to the PR effects based on photochromism.

Having the electron donating (D) and withdrawing groups (A) in each benzene ring, the azobenzene derivatives such as DR-1 (D=dialkyl amine, A=NO$_2$) have been studied extensively as NLO materials in PR composite based on the electrooptic effect [12, 13–17]. Therefore, the azobenzene unit in PR materials has a bifunctional role: photochromic and NLO effect, as shown in Figure 17.1.

The NLO properties of the azochromophores are originated from the π-conjugated character with electron donating and withdrawing groups, and thus the PR effects in the PR matierals containing azobenzene derivatives and photoconductive molecules are *non-local* in nature.

For example, the *non-local* nature in the PR effect of the DR1-type NLO chromophore has been reported with 4, 4′-di(N-carbazolyl)-4″-(2-N-ethyl-4-[2-(4-nitrophenyl)-1-azo]anilinoethoxy)-triphenylamin (DRDCTA, see Figure 17.2) [18–21]. Here, a disperse-red-based substituent provides the *nonlinear* optical properties necessary for the photorefractive effect. A glass doped with a bifunctional DRDCTA, a plasticizer DOP, and a sensitizer, C60 (Figure 17.2), shows the PR effect with a maximum gain coefficient of $\Gamma = 90$ cm^{-1} [18].

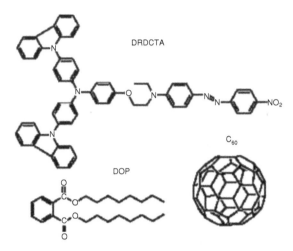

FIGURE 17.2. PR composite containing a bifunctional chromophore DRDCTA, plasticizer DOP, and sensitizer C60. The mass ratio of the three functionalities is DRD-CTA:DOP:C60 71:28:1 (see [18]).

Figure 17.3 shows a plot of the PR gain coefficients (Γ) of both the s- and p-polarized writing beams, as a function of the external electric field of the DRDCTA composite. The strong dependency of Γ on E indicates that the PR effect of this composite is *non-local* in nature. The *non-local* photorefractive nature distinguishes it from a *local* mechanism, such as photochromism, photoorientational, and dichromatic, where the grating is always in phase with the initiating light intensity distribution.

The glass-transition temperature T_g near room temperature results in the high orientational mobility of the DRDCTA molecules, which allows for the orientation of the molecules not only in the direction of the externally applied electrical field, but also along the induced space-charge field. This modulates the birefringence

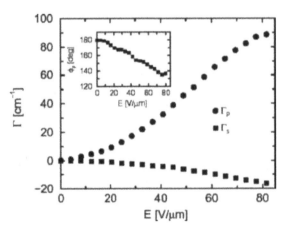

FIGURE 17.3. Photorefractive gain coefficient for s- and p-polarized writing beams versus the external electric field E. The *inset* shows the phase shift of the Δn grating with respect to the light intensity pattern for s-polarized writing (from [18]).

$(\chi^{(1)})$ and the Pockels effect $(\chi^{(2)})$ of the material, resulting in an enhancement of PR efficiency. The ratio A_{BR}/A_{EO} of both contributions to the modulation of the refractive index can be calculated from the data of Figure 17.3 by [22]

$$\frac{\Gamma_p}{\Gamma_s} = \frac{1}{2}\cos\Theta_{int}\left[\left(\frac{C}{A} - 1\right) + \left(\frac{C}{A} + 1\right)\cos\Theta_{int}\right] \tag{17.1}$$

and

$$\frac{A_{BR}}{A_{EO}} = \frac{3 - \frac{C}{A}}{\frac{C}{A} + 2}, \tag{17.2}$$

where Θ_{int} is the angle between the writing beams in the sample. For DRDCTA:DOP:C_{60} the ratio of the birefringent to the electrooptic contribution of A_{BR}/A_{EO} was calculated as -2.0 ± 0.3. It should be noted that the presence of the photoisomerization grating decreases Γ_s.

The inset of Figure 17.3 shows the phase shift ϕ_p as determined by the translation technique [23]. As in similar cases, the presence of an E-field-independent photoisomerization grating can be recognized through a phase of 180° at zero E field [24, 25]. The *trans-cis-trans* isomerization cycles cause the azo chromophores to subsequently align perpendicularly to the polarization of the writing beams in the bright regions of the interference pattern. The refractive index decreases locally, and the resulting grating consequently shifts by 180°. The phase of the superposition of both gratings is shifted toward lower values as the *non-local* photorefractive grating arises for increasing E fields.

The distinction between the *non-local* (via EO effect) and the *local* natures (via photochromic isomerization) of the PR effects has not been well documented in organic photochromic materials. However, one can roughly assume that the PR effects at zero electric field are due to its *local* nature (photochromic grating) and are generally small, as observed in the following example.

The diffraction efficiency (η) of the PR polymers containing DR1 unit in Figure 17.4 (a), as a function of the applied electric field, is presented in Figure 17.4 (b). Limited by the reflection and absorption losses, a maximum of only $\approx 12\%$ was obtained at the applied electric field of 92.4 V/μm [26]. The diffraction is $\approx 1.4\%$ instead of zero at zero electric field, due to the photoisomeric grating caused by the intensity pattern through the photoisomerization. Therefore, it can be assumed that the PR effect from the *local* nature (photochromic effect) in photochromic materials is much less than that of those of *non-local* nature. To achieve high PR effect, therefore, many azobenzene polymers having charge-transporting group have been developed [12].

To distinguish between the photorefractive effect and other mechanisms of grating formation such as photochromism in detail, a two-beam-coupling (2BC) experiment was performed using both noncentrosymmetric and centrosymmetric azo-dye-doped silica glasses [27]. The azodye 2,5-dimethyl-4-(2-hydroxyethoxy)-4′-nitroazobenzene (DMHNAB) is used as a nonlinear optical chromophore and covalently bound to a sol-gel precursor,

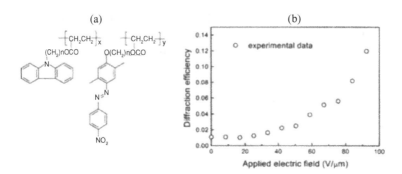

FIGURE 17.4. (a) The chemical structure of PR polymers containing DR1 unit ($n = 2$) [26]. (b) Diffraction efficiency of copolymers in Figure 17.4(a) as a function of applied electric field (taken from [26]).

3-(isocyanatopropyl)triethoxysilane (ICPTES), which was copolymerized with tetraethoxysilane (TEOS) precursors. The sol-gel precursors were mixed with other components in Figure 17.5 (a) with a ratio of pyridine(11): DMHNAB–ICPTES(1):N-ethylcarbazole (1.6):TEOS(1.1):H_2O(11):HCl(0.6):

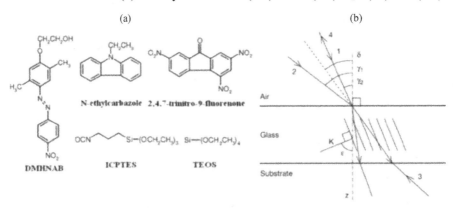

FIGURE 17.5. Chemical structure of the organic dopants and sol-gel precursors used in the synthesis of the photorefractive glass. (b) Geometry used for the two-beam-coupling and degenerate four-wave-mixing experiments in photorefractive glass. The tilt angle is $\delta = 45°$; the external angles of incidence of the writing beams 1 and 2 are $\gamma_1 = 34.5°$ and $\gamma_2 = 55.5°$, respectively. The bulk refractive index of the glass was measured as $n = 1.63$ by a prism-coupling technique. The calculated period of the grating is $\Lambda = 2.2$ μm and the angle between the grating vector $\mathbf{K}$ and the sample normal is $e = 64.6°$. The writing beams are focused to a spot of about 0.1 mm^2 and overlap in the glass. In the degenerate four-wave-mixing experiment, the readout beam 3, collinear and counterpropagating to beam 2, is focused to a spot of about 0.04 mm^2 within the grating area. The diffraction efficiency is calculated from $\eta = I_4/(I_3 - I_F)$, where I_3 and I_4 are the intensities of the readout beam 3 and diffracted beam 4, respectively, and I_F includes Fresnel losses at the glass–substrate and glass–air interfaces. The substrate is a borosilicate glass plate coated with an indium tin oxide transparent electrode (see [27]).

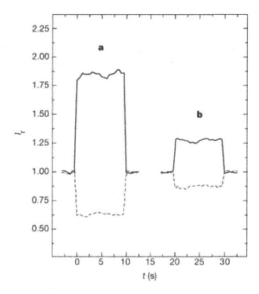

FIGURE 17.6. The two-beam-coupling signal measured in noncentrosymmetric and cen-trosymmetric glassy films. Relative intensity $I_r = I/I_0$ of one of the writing beams is monitored when the other beam is switched on and off at $t = t_0$ (in seconds) and $t = t_1$, respectively. Beam 1, signal beam (solid line); beam 2, pump beam (dashed line). I and I_0 are the intensities of a writing beam when the other writing beam is present or absent, respectively. Both beams are p-polarized, and the power density of the pump beam is 3 W cm^{-2}. The gain Γ is calculated from the experimental data using $\Gamma = \{\cos \gamma_1' \ln[r\beta/(r + 1 - \beta)]\}/L$, where $r = [I_2(I_1 = 0)]/[I_1(I_2 = 0)]$ is the ratio between the intensities of the writing beams 2 and 1 measured before they enter the sample; $\beta = (I_1(I_2 \neq 0))/(I_1(I_2 = 0))$ is the amplification factor; L is the film thickness; and γ_1' is the incidence angle of the beam 1 inside the film. **a**: Noncentrosymmetric glass, $t_0 = 0$ s, $t_1 = 10$ s, $r = 1.6$, $\beta = 1.85$, and $L = 29$ μm. **b**: Centrosymmetric glass, $t_0 = 20$ s, $t_1 = 30$ s, $r = 2$, $\beta = 1.3$, and $L = 22$ μm (taken from [27]).

2,4,7-trinitro-9-fluorenone (0.002) (bracketed numbers indicate molar equivalents) and sol-gel processed to prepare PR films.

The photorefractive nature of the grating was confirmed by asymmetric energy coupling between two writing beams in the 2BC setup shown in Figure 17.5(b). The signal beam 1 was amplified and the pump beam 2 was depleted simultaneously in the glass (Figure 17.6a), as a consequence of the phase shift between two-beam interference fringes and the refractive index grating. Using p-polarized writing beams, a gain of $\Gamma_p = 444$ cm^{-1} was measured. Given that the absorption coefficient of the film at 632.8 nm is $\alpha = 29$ cm^{-1}, a net gain of $\Gamma_p - \alpha = 415$ cm^{-1} was achieved.

The gain coefficient (Γ) was 188 cm^{-1} in an unpoled centrosymmetric sample, using p-polarized writing beams (Figure 17.6b). The centrosymmetric structure of the unpoled glass was confirmed in an independent second-harmonic-generation experiment, which eliminates the possibility of self-alignment of the chromophore, which could induce noncentrosymmetry in the glass. In the DFWM experiment on the centrosymmetric sample, the diffraction efficiency for the probe beam with polarization identical to that of the writing beams was more than 20 times higher than for the probe beam polarized perpendicularly to the writing beams.

In polarization recording according to the geometry in Figure 17.5(b), with orthogonal polarization of the writing beams 1 and 2 and s- or p-polarized readout beam 3, there is no periodic variation in light intensity over the centrosymmetric sample, and thus no periodic space-charge field can arise. Instead, there is a periodic spatial variation in the polarization state of light, which can induce orientational redistribution of dye molecules, resulting in a refractive index grating [28]. The *non-local* grating recording mechanism was not observed in the polarization recording experiment. The experiments with both polarization and intensity recording confirmed that the dye-orientational mechanism is present in the above azobenzene-type sample and that the space-charge field is required for a *non-local* response in the centrosymmetric sample.

In a 2BC experiment on a poled sample in untilted geometry (Figure 17.5(b), $\delta = 0$), a gain of $\Gamma = 193$ cm^{-1} was observed. According to photorefractive theory based on the electrooptic effect, $\Gamma = 0$ is expected for this geometry [16]. This result, together with the data from the 2BC experiment in the centrosymmetric glass, confirms the *non-local* nature of the recorded grating even in the absence of the electrooptic effect.

Reversible and efficient refractive index modulations by photochromic molecules in PR polymers can be applied to optical poling such as photoassisted poling (PAP) and all-optical poling (AOP) [12]. Figure 17.7 summarizes the photoinduced changes of photochromic molecules in polymers with different poling methods.

PAP is based on the excitation of photochromes in their absorption bands. It requires the superimposition of a dc electric field. PAP is significant in the

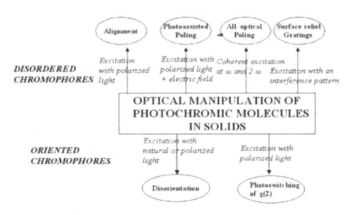

FIGURE 17.7. Photoinduced changes of photochromic molecules in polymers [12].

FIGURE 17.8. Beam configuration for AOP. During seeding, beam 3 (2ω) is combined with beam 1 (forward configuration) or beam 2 (backward configuration). After seeding, beam 3 is stopped, and SHG (beam 4) is generated due to the polar order created in the sample [29].

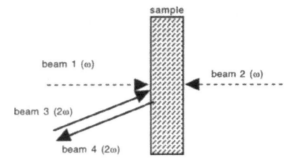

fabrication of NLO active optical waveguides or the quasi-phase-matched structures adapted for SHG. All-optical poling (AOP) is also an emerging method of poling, which can be applied to octupolar molecules. The beam configuration for AOP is shown in Figure 17.8.

The mechanism for AOP is interpreted as a light-induced excitation of chromophores, which leads to orientational hole-burning and angular redistribution. However, compared to the case of PAP, where excitation is considered at only one wavelength, each process in AOP is composed of three terms, ω-excitation, 2ω-excitation, and the interference between both [30–33]. The highest values in poly(DR1M-co-MMA) were estimated at around 70 pm^{-1} [33]. In dispersive materials such as polymers, the periodicity of the field pattern written in the sample corresponds to the coherence length, and phase matching can be achieved by AOP [31(a), 32]. SHG is thus proportional to the square of the optical path length.

Reversible and efficient refractive index modulations by azobenzene-containing polymer films can establish large surface relief gratings. They can be directly recorded on azobenzene-containing polymer films, at room temperature, using two interfering polarized Ar ion laser beams [34–35]. The irradiation of the polymer films for a few seconds, containing azo groups with intensities of 5–200 mWm^{-2}, produces reversible volume birefringence gratings with low diffraction efficiency (less than 1%). If the film is exposed to the writing beam for longer than a few seconds, an irreversible process begins, creating an overlapping and highly efficient surface grating written on the time scale of minutes. There is an initial and rapid growth (on the order of seconds) corresponding to the production of the reversible volume birefringence grating, then a slower and irreversible process (on the order of minutes), which creates surface gratings observable by AFM, with efficiencies of up to 50%. It has been established that the phase relationship between the surface and volume gratings is such that the light intensity maxima of the fringes coincide with the surface profile minima [35].

Highly efficient surface gratings have also been produced from the azo side-chain high-T_g polymers [36–38]. Figure 17.9 shows the experimental setup used by these authors and an example of a surface relief grating observed by atomic force microscopy (AFM) [39].

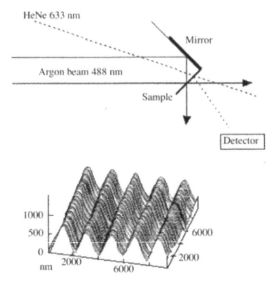

FIGURE 17.9. Optically inscribed diffraction gratings in azo polymer films. Upper part: experimental setup for grating inscription. Lower part: AFM surface profile of an optically inscribed grating in a poly(DR13A) copolymer [39].

The mechanism at the origin of the surface relief gratings is connected to the photo-isomerization process of azobenzene derivatives, which need a mass transport well below T_g. The first model was proposed by Barrett et al. [39]. It explains that the isomerization process involves the creation of a supplementary free volume when going from trans to cis, and that this creation leads to pressure gradients above the yield point of the polymer. The resulting viscoelastic flow then leads to a pressure-driven mass transport to form surface profile gratings. After several other more quantitative approaches including a 1D random walk along the excitation direction (Figure 17.10), Lagugne Labarthet et al. obtained

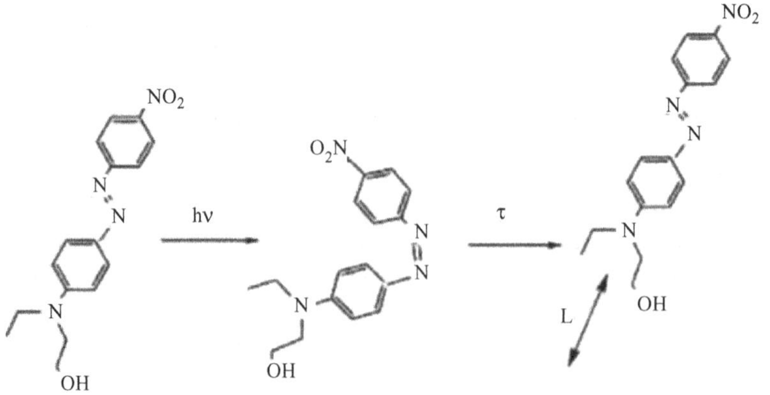

FIGURE 17.10. Photoinduced isomerization-induced caterpillar-like motion of the DR1 molecule. After one trans → cis → trans cycle, the DR1 molecule has undergone translation diffusion by an average amount L. Reprinted with permission from [42].

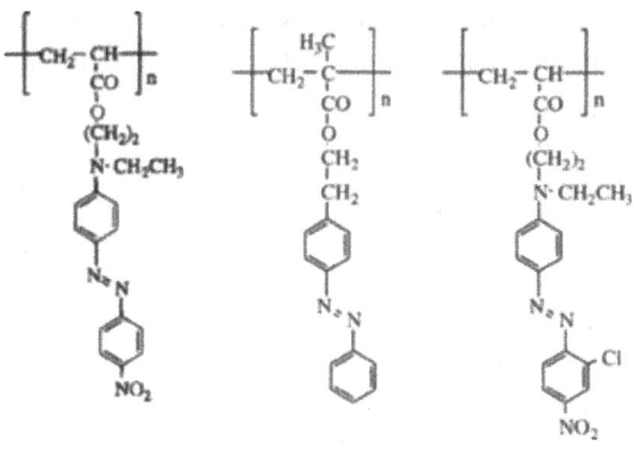

FIGURE 17.11. DRI-substituted polymethacrylates: poly(DR1A), poly(MEA), and poly(DR13A).

information about the second (T_2) and fourth (T_4) order parameters of the chromophore orientation function [40]. The information entropy theory used to build the distribution functions $f(\theta)$ from T_2 and T_4 order parameters confirms that it is important for the photoinduced orientation to be in a direction perpendicular to the polarization of the incoming light, but with very broad distribution functions even in the bottom of the gratings. They concluded that the primary orientational order generated by the angle-dependent absorption of light acts as an initializing force. This initial order is then strongly perturbed by the presence of strong pressure gradients, as suggested by Barrett et al. [41]. Different distribution profiles at the top and at the bottom of the surface relief are then created.

The structural bulkiness in the azobenzene unit affects the grating efficiency of azo polymers. For the three azobenzene-substituted polyacrylates or polymethacrylates in Figure 17.11, the diffraction efficiency varies in the order poly(DR13A) > poly(DR1A) > poly(MEA) in equivalent irradiation conditions, i.e., the bulkiest azo group induces the highest efficiency [41]. PMMA doped with DR1 also gives surface gratings, but with a much lower efficiency [43–44].

Bulk properties of the polymer matrix also have an influence on surface grating formation [39]. Barrett et al. compared poly(DR1A) with copolymers of DR1A with MMA and found similar results, regardless of azo content. They also found that compatible blends of poly(DR1A) and PMMA were not suitable for any surface grating formation, even for low PMMA content, due to the high molecular weight of PMMA used in blending (MW > 300,000). These authors established an upper experimental limit to the molecular weight of PMMA (MW ≈ 25,000), beyond which only birefringence gratings were obtained [39]. This observation supports models based on mass transport.

Surface relief gratings have been explored for applications such as holographic storage, optical filters, and resonant couplers [12, 45–49]. Subwavelength gratings could, for example, be made using this approach, which could be used in the fabrication of wave plates and antireflection surfaces for the visible. Selective-wavelength light couplers for slab waveguides were developed using this process [46, 48]. The major advantages of this process are (1) easy one-step processing, without any development, (2) large surface modulations, and (3) precision-controlled modulation depth [50]. Possible application potential in holographic cinematography was demonstrated [51].

17.3 Spiropyrans

Spiropyrans and the closely related spirooxazines undergo reversible photochemical cleavage of the C-O bond [1, 9, 52–53] in the spiropyran or spirooxazine rings when excited by UV light (see equation (17.3)). The photochromic transition from spiropyran (SP) to merocyanine (MC) is accompanied by a major change in the conjugation of the π electrons. Therefore, large changes occur in the NLO properties [53–54].

SP   MC   (17.3)

The ionic resonance form merocyanine dyes (MC). In particular, they have some of the highest known molecular second-order NLO (β) coefficients [54]. The large β values of the merocyanine-type molecules can be understood by considering its structure as an admixture of ionic and quinone resonance forms, following its conjugated π electron system. This is represented in the merocyanine molecule as equation (17.4)

16

(17.4)

In a typical polar solvent, the zwitterionic form will contribute substantially, resulting in a dipole moment of 26 D [55]. The actual contribution of each resonance form in a given merocyanine molecule, and thus dipole moment difference between the ground and excited states ($\Delta\mu$), can be solvent-dependent, leading to a large solvent dependence of β values [56]. Typically, the dipole moment for merocyanines is larger in the ground state, leading to large negative β values [55].

Merocyanine molecules have also been incorporated into a photorefractive polymer composite containing a photoconducting polymer, plasticizer, and sensitizer as well as the NLO chromophore [57]. Merocyanine itself was used as the NLO chromophore, and its photorefractive efficiencies were quite similar to those of composites containing azo dyes. An improvement of about a factor of 5 was achieved in an optimized mixture of the azo and merocyanine dyes. This improvement is ascribed both to the higher NLO efficiency of the merocyanine and to a decrease of merocyanine aggregation in the presence of the azo dye [57].

Second-order nonlinearity of the electrooptic effect of spiropyran-containing polymers can be improved by applying T_g poling process-photo-assisted poling (PAP) [58–60]. Irradiating spiropyrans, under the influence of an external electrostatic field, can produce colloidal globules and quasiliquid crystals (QLCs) [61]. SHG was observed from cooled QLC films, indicating that the merocyanines have a partial polar alignment along the direction of the electrostatic field used in their preparation. The subsequent application of an electric field to the QLC films amplified the signal, whereas application of a negative electric field (i.e., in the opposite direction) led to its cancellation. The aggregation of highly dipolar photomerocyanine under a dc electric field is responsible for a noncentrosymmetric ordering, causing the medium to exhibit NLO effects. Photomerocyanine derivatives in the liquid-crystalline phase showed $\chi^{(2)}$ values as high as 2×10^{-9} esu (ca. 1 pm·V^{-1}) and blends including push-pull molecules such as 4-(dimethylamino)-4'-nitrostilbene showed enhanced NLO properties.

All optical modulations to generate grating and holographic recoding were also investigated using the spiropyran and oxiazine derivatives [53, 62–63]. Due to a significant reverse thermal reaction, the grating and holographic recording are temporal, and therefore the mechanism of optical modulation for grating and holographic recoding involves reverse ring-closure reaction term.

Significantly, different holographic growth rates and exposure sensitivities were obtained from the three recording procedures on spirooxazine-doped polymer films using a He-Ne laser beam at 633 nm (Figure 17.12) [53]. With the most sensitive procedure (c), the exposure sensitivity was found to be 350 mJ/cm^2 of the spirooxazine-doped layers. The corresponding exposure sensitivity for the spiropyran-doped layers was 750 mJ/cm^2. These results indicate that the exposure sensitivities of a given photochromic material can be significantly increased (by a factor of about 2) through a variation in the optical recording procedure.

The experimental results of Figure 17.12 can be interpreted in terms of the relative population of the two merocyanine isomers B_1 and B_2. The photochromic film is initially exposed to the UV excitation beam alone in the recording procedures a and b, so that at steady state, the more stable colored isomers B_2 will be dominantly populated. A subsequent exposure to the visible recording beams causes decoloration to A, with the transformation from B_2 to B_1 (and/or B_2 to A in a parallel scheme) as the rate-determining step. In recording procedure c, however, the initial exposure to all beams yields a mixture of all the species associated with the scheme of equation (17.5), with a significant population of B_1. Therefore, a subsequent exposure to the recording beams causes decoloration at a significantly increased rate, determined mainly by the transformation from B_1

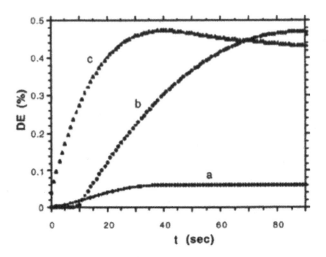

FIGURE 17.12. Real-time diffraction efficiencies (DE) versus exposure time (t) of holographic gratings recorded at 514 nm in spirooxazine–PMMA films for three recording procedures (a, b, and c). Readout was at 633 nm. (a) Initial exposure to the UV (364 nm) excitation beam of 5 mW/cm^2 and then simultaneous exposure to the UV beam and to the two recording beams of 9 mW/cm^2. (b) Initial exposure to the UV beam and then separately to the two recording beams. (c) Initial simultaneous exposure to all the beams, followed by exposure to the two recording beams only [53].

to **A**:

$$A \underset{R_{B1}(\lambda_B),\, R_{T1}}{\overset{R_A(\lambda_A)}{\rightleftharpoons}} B_1 \underset{R_{B2}(\lambda_B),\, R_{T2}}{\overset{R_{T12}}{\rightleftharpoons}} B_2 \qquad (17.5)$$

When the layers are once again simultaneously exposed to the UV and recording beams, the DE is significantly reduced, since the UV beam partially erases the gratings. These results indicate that it is possible to reversibly control the population of the photochromic isomers through optical addressing with different recording procedures. Therefore, by modulating the UV excitation beam, it is possible to modulate the holographically recorded gratings correspondingly, as shown in Figure 17.13 [53].

Rewritable holographic recording was attempted on the film of spiropyran-poly (vinylcarbazole) blends [64], using the UV line of a krypton-ion laser as the recording, an Ar-ion laser at 488 nm for the readout, and a He-Ne laser for erasure. Variations in the layer thickness or the recording-beam power significantly affected both the exposure sensitivities and the diffraction efficiencies. At optimum conditions, a DE of over 10% was achieved in a 12-μm-thick layer, at an exposure of about 300 mJ/cm^2. However, the DE dropped by more than 1 order of magnitude in a 50-μm-thick layer. The relatively high achievable DEs, at optimum conditions, may be attributed to the fact that at the readout wavelength,

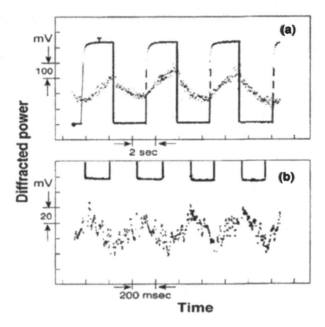

FIGURE 17.13. Modulation of holographic gratings (dotted traces) as a result of square-wave modulation of the UV beam (line traces). The modulated gratings were continuously exposed to the recording beams, each of 20 mW/cm^2: (a) with a modulation frequency of 0.2 Hz; (b) with a modulation frequency of 2.5 Hz [53].

there is negligible absorption and a refractive-index-modulated phase grating is mainly probed [65].

17.4 Diarylethenes

Diarylethens undergo reversible structural change from an open (colorless) to a closed form (colored) through irradiation with ultraviolet light and in the reverse direction through irradiation with visible light [1, 6–8] equation (17.6).

$$ (17.6) $$

Diarylethene (DA) compounds, substituted with electron donors and acceptors, can switch photochemically between low- and high-level NLO responses [66]. It was demonstrated that reversible photochemical interconversion between the two photochromic states could be used to effectively switch nonlinear optical activity. DA compounds substituted with 2-benzodithiolyl, for example, display a marked increase in nonlinear optical activity on conversion from an open to a closed form. Contrary to the thermally reversible azo and spiropyran, the colored forms of the diarylethenes are thermally stable, and they can be reversed to their colorless forms only by visible light irradiation [1]. Due to the fatigue-resistant and thermal-irreversible photochromic performance, diarylethenes are the most promising candidates for refractive-index modulation.

DA molecules, such as 1,2-bis(2-methylbenzo[b]thiophen-3-yl) hexafluorocyclopentene (BTF6, equation 17.7), were used as an active material in achieving refractive index (RI) modulation through ultraviolet or visible light irradiation [67]:

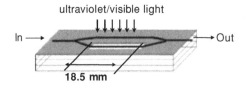

BTF6   (17.7)

The photoinduced RI changes of diarylethenes or DA-substituted polymers are in the range of 0.001–0.0001, at an 830–1550 nm wavelength [67–69]. An all-optical 1 × 2 Y-branch switch using the RI modulation in DA-doped waveguide was fabricated [70–71]. It was composed of a DA-doped polymer waveguide and a thick light-blocking metal film on one arm of the Y-branch waveguide. The switch exhibits a low crosstalk of about –14 dB, at the wavelength of 1.55 µm, using ultraviolet (UV) and visible light.

The RI modulation in DA-doped polymer films can be applied in an all-optical Mach–Zehnder modulator [72–73], shown in Figure 17.14. The photochromic dye-doped polymer film was applied as the core layer of the modulator, with a thick switchable light-blocking metal film [73]. A special feature of this device

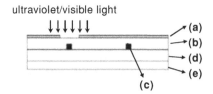

FIGURE 17.14. Configuration of Mach–Zehnder modulator: (a) Au layer (200 nm), (b) VTC-2 (25 µm), (c) Polycarbonate doped with BTF6 (2 µm, 37 wt.%), (d) Cyclotene 3022™ (18 µm), (e) Silicon wafer [72].

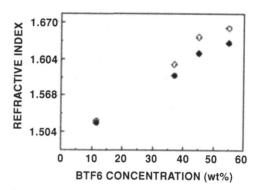

FIGURE 17.15. BTF6/polycarbonate refractive index versus BTF6 concentration measured at the wavelength of 1.55 lm in open form (●) and closed form (○) [73].

is that one arm of the Mach–Zehnder modulator was UV irradiated and the other arm was nonirradiated, allowing for a differential phase shift. The refractive index (monitored with 1.55 μm laser) of BTF6 doped (37 wt.%) polycarbonate film was $n_{TE} = 1.5671$ in dark, which changed to $n_{TE} = 1.5677$ when irradiated with a UV light of 365 nm. This resulted in a photoinduced refractive index change of 0.0006. The photoinduced refractive index changes of the film increased linearly with the diarylethene content. This indicates that the RI change originated from the photochromophore (Figure 17.15).

One arm of the Mach–Zehnder modulator was irradiated with UV light (intensity 0.4 mW/cm^2 at 365 nm) and the modulator was tested with a 1.55-μm laser. After full exposure to UV light, the following were observed: optical modulation occurred after visible light (intensity of 0.4 mW/cm^2 of 365 nm) exposure, and reversible modulation occurred after UV light exposure. The UV or visible light exposure gives a phase shift and optical modulation, as shown Figure 17.16. This is an interesting result in the sense that the modulator can work by an all-optical method. However, the response to the light pulse is too slow to apply such pulses

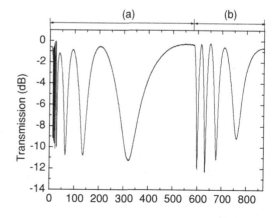

FIGURE 17.16. Characteristic of Mach–Zehnder with alternating ultraviolet and visible light. (a) Visible (514 nm) light exposure with intensity of 14.2 mW/cm^2 (b) Ultraviolet (365 nm) light exposure with intensity of 0.4 mW/cm^2 [72].

in a practical modulator. Thus it is a challenge to develop new photochromic materials having fast response in solid media under moderate light intensity.

Erasable holographic recording can be achieved using DA-doped polymer films [74]. Acetyl-substituted diarylethene (DAMBTF6, equation (17.8))-doped fluoroacrylate films were prepared using a photocuring method with a mixture of fluoroacrylate monomers and DAMBTF6, in the presence of a photoinitiator [75]. Transparent red-colored films were obtained upon exposure to a UV light but were bleached into a colorless film when exposed to visible light.

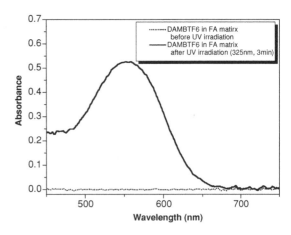

DAMBTF6

(17.8)

Figure 17.17 shows the spectral change of a DAMBTF6-doped fluoroacrylate polymer (FA) film upon UV exposure. Before irradiation with UV light, the absorbance of the film corresponded with the absorption of the ring open isomer (dotted line). The maximum absorption of the open isomer of DAMBTF6 was below 320 nm. In the visible region of the spectrum, therefore, there was no absorption band corresponding to the open isomer. The open isomer changed to a closed isomer and the absorption peak at λ_{max} (554 nm) was composed of the absorption of closed isomers upon excitation with UV light (solid line).

FIGURE 17.17. UV spectral change of diarylethene polymer films by a light 325 nm: before (dotted line) and after UV exposure (solid line) [74].

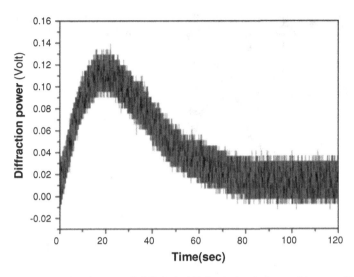

FIGURE 17.18. Intensity change of diffraction light during holographic recording [74].

To prepare for writing with a 532-nm laser, the DAMBTF6-doped FA film was initially exposed to a homogeneous and incoherent excitation beam from a UV lamp at 365 nm until maximal coloring occurred. The initial exposure was then followed by a simultaneous exposure to the two coherent 532-nm laser recording beams. The holographic interference pattern created by the two coherent beams was recorded on the photochromic film through the bleaching of the colored DAMBTF6 molecules. In the bright region of the interference pattern, a ring-opening reaction of DAMBTF6 occurred and a phase-grating was formed according to the refractive index difference.

The maximum diffraction efficiency was determined as ≈0.5 % by measuring the ratio of the intensity of the diffraction beam to the intensity of the recording beam. The diffracted light was detected within 2 seconds and the intensity of the diffracted light was maximized after 20 seconds (Figure 17.18). After prolonged exposure, the diffracted light was reduced because the recorded pattern, through the two-beam coupling, was erased by the monitoring light (633 nm). Similarly, the coupling efficiency increased and reached a maximum value of about 20 seconds. It then decreased to the initial zero point upon prolonged exposure.

Digital images were recorded onto a pixelated spatial light modulator (640 × 480 pixels) with rectangular pixel apertures and were reconstructed on the photochromic film. The readout of the stored data page involved illuminating the films and the reference beam and imaging the diffracted optical signal onto a CCD array, a MVOS CMOS camera, which converts the optical signal back into an electronic signal. Figure 17.19 shows the reconstruction of an image (the word 광변색 means photochromic in Korean) stored in the photochromic film. The original letter image with rectangular pixel apertures was completely reconstructed on the photochromic film with high resolution. The recorded image was

FIGURE 17.19. Reconstruction of an image stored by a spatial light modulator (640 × 640) in photochromic film [74].

completely erased through illumination with UV or visible light. Since the holographic recording was based on the photochromic conversion of the DAMBTF6, the volume change was much less than that in typical photopolymer media, in which the cross-linking of photoactive monomers by the writing beam results in significant volume change [76]. Erasable holographic recording based on photochromic materials is promising for real-time holography applications, but one needs to improve DE by developing new chromophores and/or by adopting a new writing system.

17.5 Conclusion

Through photoinduced reversible color change, photochromic materials change the index of refraction in response to an optical beam to produce refractive index patterns with light beams interfering in the photochromic medium. This refractive index pattern can induce the diffraction of light and thereby give rise to a spatial phase shift (displacement) between two light beams interfering in a photochromic medium. The spatial modulation of the index of refraction generated by the absorption change in the photochromic materials is rather smaller than that by the *non-local* property of the photorefractive (PR) effect mechanisms.

Organic photochromic materials, which are generally π-conjugated systems, have potential nonlinear optical (NLO) properties apart from the local photochromic refractive-index-modulation effect. Photochromophores, such as azobenzenes, have been utilized as an NLO material in PR media to produce a reversible light-induced modulation of the refractive index via the electrooptical effect. The photochromic control of a noncentrosymmetric order in a bulk can be achieved through the combination of photochromism and NLO properties in molecules. This allows for new techniques, such as photoassisted poling (PAP), all-optical polling (AOP), and surface relief grating, The ionic resonance from merocyanine dyes, generated from the spiropyran and -oxazine derivatives by UV exposure, have some of the highest known molecular second-order NLO (β) coefficients. The spiropyran and -oxazine derivatives are used for PR composites as well as in holographic recording. The thermal reverse reaction of spiro and azo derivatives limits their applications.

Due to their fatigue-resistant and thermally irreversible photochromic performance, diarylethenes are the most promising candidates for refractive index modulation. New applications toward all-optical switching, modulator, and rewritable holographic recordings have been attempted using the refractive index modulation of DA materials. These are interesting in the sense that they can work by all-optical methods. However, the response time plus sensitivity of organic materials in solid media should be improved in order to apply them in a practical information-processing device. In addition, it is also a challenge to develop new photochromophores working with new concepts to establish a larger PR effect.

References

1. (a) H. Dürr, H. Bouas-Laurent. *Photochromism: Molecules and Systems*; Elsevier: Amsterdam (1992). (b) J.C. Crano, R.J. Guglielmetti, eds. *Organic Photochromic and Thermochromic Compounds*; Plenum Publishing: New York, 1998; Vol. 1; 1999, Vol. 2.
2. G.H. Brown. *Photochromism: Techniques of Chemistry*; Wiley-Interscience: New York, (1971); Vol. III.
3. S. Shinkai, O. Manabe. *Top. Curr. Chem.* **121**, 67 (1984).
4. S.H. Kawai, S.L. Gilat, J.-M. Lehn. *J. Chem. Soc., Chem. Commun.*, **1011**. (1994).
5. T. Saika, M. Irie, T. Shimidzu. *J. Chem. Soc., Chem. Commun.*, **2123**. (1994).
6. M. Irie. *Adv. Polym. Sci.*, **94**, 27 (1990).
7. M. Irie. In *Applied Photochromic Polymer Systems*; McArdle, C.B., ed.; Blackie: Glasgow, **174** (1992).
8. E. Kim, K. Choi, S.B. Rhee. *Macromolecules*, **31**, 5726–5733 (1998).
9. E. Kim, Y.K. Choi, M.-H. Lee, S.G. Han, S.R. Keum. *Journal of the Korean Physical Society*, **35**, S615-S617 (1999) and relerences therein.
10. (a) R. Miller, D. Burland, M. Jurich, V. Lee, C. Moylan, J. Thackara, R. Twieg, T. Verbiest, W. Volksen. *Macromolecules*, **28**, 4970 (1995). (b) M.H. Davey, V.Y. Lee, T.J. Mark, R.D. Miller. *Polym. Prepr. (Am. Chem. Soc., Div. Polym. Chem.)*, **38**, 261 (1997).
11. L. Solymar, D.J. Webb, A. Grunnet-Jepsen. *The Physics and Applications of Photorefractive Materials*; Clarendon Press: Oxford, (1996).
12. J.A. Delaire, K. Nakatani. *Chem. Rev.*, **100**, 1817 (2000).
13. D.J. Williams. *Angew. Chem. Int. Ed. Engl.*, **23**, 690 (1984).
14. D.M. Burland, R.D. Miller, C. Walsh. A. *Chem. Rev.*, **94**, 31 (1994).
15. D.R. Kanis, M.A. Ratner, T. Marks. *J. Chem. Rev.*, **94**, 195 (1994).
16. H.S. Nalwa, T. Watanabe, S. Miyata. In *Nonlinear Optics of Organic Molecules and Polymers*; H.S. Nalwa, S. Miyata, eds.; CRC Press: Boca Raton, FL, 89 (1997).
17. E.G. Staring. *J. Recl. Trav. Chim. Pays-Bas*, **110**, 492 (1991).
18. S. Schloter, A. Schreiber, M. Grasruck, A. Leopold, M. Kol'chenko, J. Pan, C. Hohle, P. Strohriegl, S.J. Zilker, D. Haarer. *Appl. Phys. B*, **68**, 899 (1999).
19. C. Hohle, P. Strohriegl, S. Schloter, U. Hofmann, D. Haarer. *Proc. SPIE*, **3471**, 29 (1998).

20. U. Hofmann, M. Grasruck, A. Leopold, A. Schreiber, S. Schloter, C. Hohle, P. Strohriegl, D. Haarer, S. J. Zilker. *J. Phys. Chem. B.*, **104**, 3887–3891 (2000).
21. W.E. Moerner, S.M. Silence, F. Hache, G.C. Bjorklund. *J. Opt. Soc. Am. B*, **11**, 320 (1994).
22. A. Grunnet-Jepsen, C.L. Thompson, W.E. Moerner. *J. Opt. Soc. Am. B*, **15**, 905 (1998).
23. K. Sutter, P. Günter. *J. Opt. Soc. Am. B* **7**, 2274 (1990).
24. S. Schloter, U. Hofmann, P. Strohriegl, H.-W. Schmidt, D. Haarer. *J. Opt. Soc. Am. B*, **15**, 2473 (1998).
25. K. Anderle, R. Birenheide, M. Eich, J.H. Wendorff. *Makromol. Chem., Rapid Commun.* **10**, 477 (1989).
26. C. Yiwang, H. Yuankang, W. Feng, C. Huiying, G. Qihuang. *Polymer Volume*, **42**, 1101 (2001).
27. P. Cheben, F. Del Monte, D.J. Worsfold, D.J. Carlsson, C.P. Grover, J.D. Mackenzie. *Nature* **408**, 64–67 (2000).
28. H.J. Eichler, P. Günter, D.W. Pohl. *Laser-Induced Dynamic Gratings*, Springer, Berlin, 1986.
29. F. Charra, F. Devaux, J.-M. Nunzi, P. Raimond. *Phys. Rev. Lett.*, **68**, 2440 (1992).
30. J.-M. Nunzi, C. Fiorini, A.-C. Etilé, F. Kajzar. *Pure Appl. Opt.*, **7**, 141 (1998).
31. (a) C. Fiorini, F. Charra, J.-M. Nunzi, P. Raimond. *J. Opt. Soc. Am. B*, **14**, 1984 (1997). (b) C. Fiorini, F. Charra, J.-M. Nunzi, P. Raimond. *Nonlinear Opt.*, **9**, 339 (1995).
32. C. Fiorini, J.-M. Nunzi. *Chem. Phys. Lett.*, **286**, 415 (1998).
33. W. Chalupczak, C. Fiorini, F. Charra, J.-M., Nunzi, P. Raimond. *Opt. Commun.*, **126**, 103 (1996).
34. P. Rochon, J. Mao, A. Natansohn, E. Batalla. *Polym. Prepr.*, **35**, 154 (1994).
35. P. Rochon, E. Batalla, A. Natansohn. *Appl. Phys. Lett.*, **66**, 136 (1995).
36. D.Y. Kim, S.K. Tripathy, L. Li, J. Kumar. *Appl. Phys. Lett.*, **66**, 1166 (1995).
37. D.Y. Kim, L. Li, X. Jiang, V. Shivshankar, J. Kumar, S.K. Tripathy. *Macromolecules*, **28**, 8835 (1995).
38. S.K. Tripathy, D.Y. Kim, X.L. Jiang, L. Li, T. Lee, X. Wang, J. Kumar. *Mol. Cryst. Liq. Cryst. Sci. Technol., A*, **314**, 245 (1998).
39. C.J. Barrett, A.L. Natansohn, P.L. Rochon. *J. Phys. Chem.*, **100**, 8836 (1996).
40. F. Lagugné Labarthet, T. Buffeteau, C. Sourisseau. *J. Phys. Chem. B*, **102**, 5754 (1998).
41. C.J. Barrett, P.L. Rochon, A.L. Natansohn. *J. Chem. Phys.*, **109**, 1505 (1998).
42. P. Lefin, C. Fiorini, J.-M. Nunzi. *Pure Appl. Opt.*, **7**, 71 (1968).
43. X.L. Jiang, L. Li, J. Kumar, D.Y. Kim, S.K. Tripathy. *Appl. Phys. Lett.*, **72**, (1998).
44. F. Lagugné Labarthet, P. Rochon, A. Natansohn. *Appl. Phys. Lett.* **75**, 1377 (1999).
45. (a) M. Eich, J. Wendorff. *Makromol. Chem.* **8**, 467 (1987). (b) M. Eich, J. Wendorff, B. Reck, H. Ringsdorf. *Makromol. Chem.* **8**, 59 (1987). (c) M. Eich, J. Wendorff. *J Opt. Soc. Am. B.* **7**, 1428 (1990).
46. J. Paterson, A. Natansohn, P. Rochon, C.L. Calender, L. Robitaille. *Appl. Phys. Lett.*, **69**, 3318 (1996).
47. X.L. Xiang, L. Li, D.Y. Kim, V. Shivshankar. J. Kumar, S.K. Tripathy. *SPIE Proc.*, **2998**, 195 (1997).
48. P. Rochon, A. Natansohn, C.L. Callender, L. Robitaille. *Appl. Phys. Lett.*, **71**, 1008 (1997).
49. A. Natansohn, P. Rochon. *Chem. Rev.* **102**, 4139 (2002).
50. N.K. Viswanathan, S. Balasubramanian, L. Li, J. Kumar, S.K. Tripathy. *J. Phys. Chem. B*, **102**, 6064 (1998).
51. P.S. Ramanujam, M. Pedersen, S. Hvilsted. *Appl. Phys. Lett.*, **74**, 3227 (1999).

52. E. Fischer, Y. Hirshberg. *J. Chem. Soc.* 4522 (1952).
53. G. Berkovic, V. Krongauz, V. Weiss. *Chem. Rev.*, **100**, 1741 (2000).
54. A. Dulcic, C. Flytzanis. *Opt. Commun.*, **25**, 402 (1978).
55. B.F. Levine, C.G. Bethea, E. Wasserman, L. Leenders. *J. Chem. Phys.*, **68**, 5042 (1978).
56. R. Ortiz, S.R. Marder, L.-T. Chang, B.G. Tieman, S. Cavagnero, Z.W. Ziller. *J. Chem. Soc., Chem. Commun.*, 2263 (1994).
57. K. Meerholz, Y. De Nardin, R. Bittner, R. Wortman, F. Wurthner. *Appl. Phys. Lett.*, **73**, 4 (1998).
58. K. Nakatani, Y. Atassi, J.A. Delaire. *Nonlinear Opt.*, **15**, 351 (1996).
59. J.A. Delaire, Y. Atassi, R. Loucif-Saibi, K. Nakatani. *Nonlinear Opt.*, **9**, 317 (1995).
60. M. Dumont, G. Froc, S. Hosotte. *Nonlinear Opt.*, **9**, 327 (1995).
61. G.R. Meredith, V. Krongauz, D. Williams. *J. Chem. Phys. Lett.*, **87**, 289. 34 (1982).
62. S. Houbrechts, K. Clays, A. Persoons, Z. Prikamenou, J.-M. Lehn. *Chem. Phys. Lett.*, **258**, 485 (1996).
63. V. Weiss, A.A. Friesem, V.A. Krongauz. *Opt. Lett.*, **18**, 1089 (1993).
64. F. Ghailane, G. Manivannan, R.A. Lessard. *Opt. Eng.*, **34**, 480 (1995).
65. H. Kogelnik. *Bell. Syst. Technol. J.*, **48**, 2909 (1969).
66. S.L. Gilat, S.H. Kawai, J.-M. Lehn. *Chem. Eur. J.*, **1**, 275 (1995).
67. T. Yoshida, K. Arishima, F. Ebisawa, M. Hoshino, K. Sugakegawa, A. Ishikawa, T. Kobayashi, M. Hanazawa, Y. Horikawa. *J. Photochem. Photobiol. A: Chem*, **95**, 265 (1996).
68. E. Kim, Y.-K. Choi, M.H. Lee. *Macromolecules*, **32**, 4855 (1999).
69. M.-S. Kim, H. Maruyama, T. Kawai, M. Irie. *Chem. Mater.* **15**, 4539 (2003).
70. T. Okamoto, T. Kamiyama, I. Yamaguchi. *Opt. Lett.*, **18**, 570 (1993).
71. J.-W. Kang, J.-S. Kim, C.-M. Lee, E. Kim, J.-J. Kim. *Electron. Lett.* **36**, 1641 (2000).
72. J.-W. Kang, J.-S. Kim., C.-M. Lee, E. Kim, J.-J. Kim, *Appl. Phys. Lett.*, **80**, 1710 (2002).
73. J.-W. Kang, E. Kim, J.-J. Kim. *Optical Materials*, **21**, 543 (2002).
74. J.Y. Park, J. Kim, N. Kim, E. Kim. *ETRI Journal*, **25**, 253 (2003).
75. E. Kim, S.Y. Cho. *Mol. Cryst. Liq. Cryst.*, **377**, 385 (2002).
76. M.L. Schilling, V.L. Colvin, L. Dhar, A.L. Harris, F.C. Schilling, H.E. Katz, T. Wysocki, A. Hale, L.L. Byer, C. Boyd. *Chem. Mater.*, **11**, 247 (1997).

Index